The Mechatronics Handbook
Second Edition

Edited by
Robert H. Bishop

Mechatronic Systems, Sensors, and Actuators
Fundamentals and Modeling

Mechatronic System Control, Logic, and Data Acquisition

The Electrical Engineering Handbook Series

Series Editor
Richard C. Dorf
University of California, Davis

Titles Included in the Series

The Handbook of Ad Hoc Wireless Networks, Mohammad Ilyas
The Avionics Handbook, Second Edition, Cary R. Spitzer
The Biomedical Engineering Handbook, Third Edition, Joseph D. Bronzino
The Circuits and Filters Handbook, Second Edition, Wai-Kai Chen
The Communications Handbook, Second Edition, Jerry Gibson
The Computer Engineering Handbook, Second Edition, Vojin G. Oklobdzija
The Control Handbook, William S. Levine
The CRC Handbook of Engineering Tables, Richard C. Dorf
The Digital Avionics Handbook, Second Edition Cary R. Spitzer
The Digital Signal Processing Handbook, Vijay K. Madisetti and Douglas Williams
The Electrical Engineering Handbook, Second Edition, Richard C. Dorf
The Electric Power Engineering Handbook, Second Edition, Leonard L. Grigsby
The Electronics Handbook, Second Edition, Jerry C. Whitaker
The Engineering Handbook, Third Edition, Richard C. Dorf
The Handbook of Formulas and Tables for Signal Processing, Alexander D. Poularikas
The Handbook of Nanoscience, Engineering, and Technology, Second Edition
    William A. Goddard, III, Donald W. Brenner, Sergey E. Lyshevski, and Gerald J. Iafrate
The Handbook of Optical Communication Networks, Mohammad Ilyas and
    Hussein T. Mouftah
The Industrial Electronics Handbook, J. David Irwin
The Measurement, Instrumentation, and Sensors Handbook, John G. Webster
The Mechanical Systems Design Handbook, Osita D.I. Nwokah and Yidirim Hurmuzlu
The Mechatronics Handbook, Second Edition, Robert H. Bishop
The Mobile Communications Handbook, Second Edition, Jerry D. Gibson
The Ocean Engineering Handbook, Ferial El-Hawary
The RF and Microwave Handbook, Second Edition, Mike Golio
The Technology Management Handbook, Richard C. Dorf
The Transforms and Applications Handbook, Second Edition, Alexander D. Poularikas
The VLSI Handbook, Second Edition, Wai-Kai Chen

MECHATRONIC SYSTEMS, SENSORS, AND ACTUATORS

Fundamentals and Modeling

Edited by
Robert H. Bishop
The University of Texas at Austin
U.S.A.

CRC Press is an imprint of the
Taylor & Francis Group, an **informa** business

CRC Press
Taylor & Francis Group
6000 Broken Sound Parkway NW, Suite 300
Boca Raton, FL 33487-2742

© 2008 by Taylor & Francis Group, LLC
CRC Press is an imprint of Taylor & Francis Group, an Informa business

No claim to original U.S. Government works
Printed in the United States of America on acid-free paper
10 9 8 7 6 5 4 3 2 1

International Standard Book Number-13: 978-0-8493-9258-0 (Hardcover)

This book contains information obtained from authentic and highly regarded sources. Reprinted material is quoted with permission, and sources are indicated. A wide variety of references are listed. Reasonable efforts have been made to publish reliable data and information, but the author and the publisher cannot assume responsibility for the validity of all materials or for the consequences of their use.

No part of this book may be reprinted, reproduced, transmitted, or utilized in any form by any electronic, mechanical, or other means, now known or hereafter invented, including photocopying, microfilming, and recording, or in any information storage or retrieval system, without written permission from the publishers.

For permission to photocopy or use material electronically from this work, please access www.copyright.com (http://www.copyright.com/) or contact the Copyright Clearance Center, Inc. (CCC) 222 Rosewood Drive, Danvers, MA 01923, 978-750-8400. CCC is a not-for-profit organization that provides licenses and registration for a variety of users. For organizations that have been granted a photocopy license by the CCC, a separate system of payment has been arranged.

Trademark Notice: Product or corporate names may be trademarks or registered trademarks, and are used only for identification and explanation without intent to infringe.

Library of Congress Cataloging-in-Publication Data

Mechatronic systems, sensors, and actuators : fundamentals and modeling / editor, Robert H. Bishop.
  p. cm.
  "A CRC title."
  Includes bibliographical references and index.
  ISBN 978-0-8493-9258-0 (alk. paper)
  1. Mechatronics. 2. Detectors. 3. Actuators.

TJ163.12.M4226 2008
621--dc22   2007025435

Visit the Taylor & Francis Web site at
http://www.taylorandfrancis.com

and the CRC Press Web site at
http://www.crcpress.com

Preface

According to the original definition of mechatronics proposed by the Yasakawa Electric Company and the definitions that have appeared since, many of the engineering products designed and manufactured in the last 30 years integrating mechanical and electrical systems can be classified as *mechatronic systems*. Yet many of the engineers and researchers responsible for those products were never formally trained in mechatronics per se. The *Mechatronics Handbook, 2nd Edition* can serve as a reference resource for those very same design engineers to help connect their everyday experience in design with the vibrant field of mechatronics.

The *Handbook of Mechatronics* was originally a single-volume reference book offering a thorough coverage of the field of mechatronics. With the need to present new material covering the rapid changes in technology, especially in the area of computers and software, the single-volume reference book quickly became unwieldy. There is too much material to cover in a single book. The topical coverage in the *Mechatronics Handbook, 2nd Edition* is presented here in two books covering *Mechatronic Systems, Sensors, and Actuators: Fundamentals and Modeling* and *Mechatronic System Control, Logic, and Data Acquisition*. These two books are intended for use in research and development departments in academia, government, and industry, and as a reference source in university libraries. They can also be used as a resource for scholars interested in understanding and explaining the engineering design process.

As the historical divisions between the various branches of engineering and computer science become less clearly defined, we may well find that the mechatronics specialty provides a roadmap for nontraditional engineering students studying within the traditional structure of most engineering colleges. It is evident that there is an expansion of mechatronics laboratories and classes in the university environment worldwide. This fact is reflected in the list of contributors to these books, including an international group of academicians and engineers representing 13 countries. It is hoped that the books comprising the *Mechatronics Handbook, 2nd Edition* can serve the world community as the definitive reference source in mechatronics.

Organization

The *Mechatronics Handbook, 2nd Edition* is a collection of 56 chapters covering the key elements of mechatronics:

a. Physical Systems Modeling
b. Sensors and Actuators
c. Signals and Systems
d. Computers and Logic Systems
e. Software and Data Acquisition

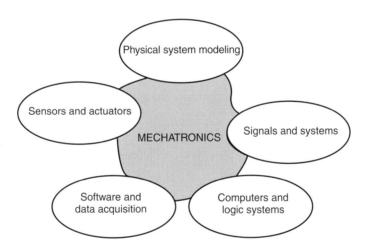

Key Elements of Mechatronics

Mechatronic Systems, Sensors, and Actuators: Fundamentals and Modeling

The book presents an overview of the field of mechatronics. It is here that the reader is first introduced to the basic definitions and the key elements of mechatronics. Also included in this book are detailed descriptions of mathematical models of the various mechanical, electrical, and fluid subsystems that comprise many mechatronic systems. Discussion of the fundamental physical relationships and mathematical models associated with commonly used sensor and actuator technologies complete the volume.

Section I—Overview of Mechatronics

In the opening section, the general subject of mechatronics is defined and organized. The chapters are overview in nature and are intended to provide an introduction to the key elements of mechatronics. For readers interested in education issues related to mechatronics, this first section concludes with a discussion on new directions in the mechatronics engineering curriculum. The chapters, listed in order of appearance, are

1. What Is Mechatronics?
2. Mechatronic Design Approach
3. System Interfacing, Instrumentation, and Control Systems
4. Microprocessor-Based Controllers and Microelectronics
5. An Introduction to Micro- and Nanotechnology
6. Mechatronics Engineering Curriculum Design

Section II—Physical System Modeling

The underlying mechanical and electrical mathematical models comprising many mechatronic systems are presented in this section. The discussion is intended to provide a detailed description of the process of physical system modeling, including topics on structures and materials, fluid systems, electrical systems, thermodynamic systems, rotational and translational systems, modeling issues associated with MEMS, and the physical basis of analogies in system models. The chapters, listed in order of appearance, are

7. Modeling Electromechanical Systems
8. Structures and Materials
9. Modeling of Mechanical Systems for Mechatronics Applications
10. Fluid Power Systems
11. Electrical Engineering
12. Engineering Thermodynamics
13. Numerical Simulation
14. Modeling and Simulation for MEMS
15. Rotational and Translational Microelectromechanical Systems: MEMS Synthesis, Microfabrication, Analysis, and Optimization
16. The Physical Basis of Analogies in Physical System Models

Section III—Mechatronic Sensors and Actuators

The basics of sensors and actuators begins with chapters on the important subject of time and frequency and on the subject of sensor and actuator characteristics. The remainder of the book is subdivided into two categories: sensors and actuators. The chapters, listed in order of appearance, are

17. Introduction to Sensors and Actuators
18. Fundamentals of Time and Frequency
19. Sensor and Actuator Characteristics
20. Sensors
    20.1 Linear and Rotational Sensors
    20.2 Acceleration Sensors
    20.3 Force Measurement
    20.4 Torque and Power Measurement
    20.5 Flow Measurement
    20.6 Temperature Measurements
    20.7 Distance Measuring and Proximity Sensors
    20.8 Light Detection, Image, and Vision Systems

    20.9   Integrated Microsensors
    20.10 Vision
21. Actuators
    21.1   Electromechanical Actuators
    21.2   Electrical Machines
    21.3   Piezoelectric Actuators
    21.4   Hydraulic and Pneumatic Actuation Systems
    21.5   MEMS: Microtransducers Analysis, Design, and Fabrication

Acknowledgments

I wish to express my heartfelt thanks to all the contributing authors. Taking time from otherwise busy and hectic schedules to author the excellent chapters appearing in this book is much appreciated.

This handbook is a result of a collaborative effort expertly managed by CRC Press. My thanks to the editorial and production staff:

Nora Konopka	Acquisitions Editor
Theresa Delforn	Project Coordinator
Joette Lynch	Project Editor

Thanks to my friend and collaborator Professor Richard C. Dorf for his continued support and guidance. And finally, a special thanks to Lynda Bishop for managing the incoming and outgoing draft manuscripts. Her organizational skills were invaluable to this project.

Editor

Robert H. Bishop is a professor of aerospace engineering and engineering mechanics at The University of Texas at Austin and holds the Joe J. King Professorship. He received his BS and MS from Texas A&M University in aerospace engineering, and his Ph.D. from Rice University in electrical and computer engineering. Prior to coming to The University of Texas at Austin, he was a member of the technical staff at the MIT Charles Stark Draper Laboratory. Dr. Bishop is a specialist in the area of planetary exploration with emphasis on spacecraft guidance, navigation and control. He is a fellow of the American Institute of Aeronautics and Astronautics. Currently, Dr. Bishop is currently working with the NASA Johnson Space Center on techniques for achieving precision landing on the moon and Mars. He is an active researcher authoring and co-authoring over 100 journal and conference papers. He was twice selected a faculty fellow at the NASA Jet Propulsion Laboratory and as a Welliver faculty fellow by The Boeing Company. Dr. Bishop co-authors *Modern Control Systems* with Professor R. C. Dorf, and he has authored two other books entitled *Learning with LabView* and *Modern Control System Design and Analysis Using Matlab and Simulink*. He received the John Leland Atwood Award by the American Society of Engineering Educators and the American Institute of Aeronautics and Astronautics that is given periodically to "a leader who has made lasting and significant contributions to aerospace engineering education." Dr. Bishop is a member of the Academy of Distinguished Teachers at The University of Texas at Austin.

List of Contributors

Raghavendra Angara
Department of Mechanical
  Engineering
University of Maryland
  Baltimore County
Baltimore, Maryland

M. Anjanappa
Department of Mechanical
  Engineering
University of Maryland
  Baltimore County
Baltimore, Maryland

Habil Ramutis Bansevicius
Department of Mechatronics
Kaunas University of Technology
Kaunas, Lithuania

Eric J. Barth
Department of Mechanical
  Engineering
Vanderbilt University
Nashville, Tennessee

Robert H. Bishop
Department of Aerospace
  Engineering and Engineering
  Mechanics
The University of Texas at
  Austin
Austin, Texas

Peter C. Breedveld
Department of Electrical
  Engineering and Control
  Engineering
University of Twente
Enschede, The Netherlands

George T.-C. Chiu
Department of Mechanical
  Engineering
Purdue University
West Lafayette, Indiana

K. Datta
Department of Mechanical
  Engineering
University of Maryland
  Baltimore County
Baltimore, Maryland

Ivan Dolezal
Technical University of Liberec
Liberec, Czech Republic

M. A. Elbestawi
Department of Mechanical
  Engineering
McMaster University
Hamilton, Ontario, Canada

Eniko T. Enikov
Aerospace and Mechanical
  Engineering
University of Arizona
Tucson, Arizona

Halit Eren
Curtin University of Technology
Perth, West Australia

H. R. (Bart) Everett
Space and Naval Warfare
  Systems Center
San Diego, California

Jeannie Sullivan Falcon
Senior Marketing Engineer
National Instruments, Inc.
Austin, Texas

Jorge Fernando Figueroa
NASA Stennis Space Center
Bay Saint Louis, Mississippi

Charles J. Fraser
University of Abertay
  Dundee
Scotland, United Kingdom

Ivan J. Garshelis
Magnova, Inc.
Pittsfield, Massachusetts

Carroll E. Goering
Department of Agricultural
  and Biological
  Engineering
University of Illinois
Urbana, Illinois

Michael Goldfarb
Department of Mechanical
  Engineering
Vanderbilt University
Nashville, Tennessee

Martin Grimheden
Department of Machine
  Design
Royal Institute of Technology
Stockholms, Sweden

Neville Hogan
Department of Mechanical Engineering
Massachusetts Institute of Technology
Cambridge, Massachusetts

Rick Homkes
Department of Computer and Information Technology
Purdue University
West Lafayette, Indiana

Bouvard Hosticka
Department of Mechanical, Aerospace, and Nuclear Engineering
School of Engineering and Applied Sciences
University of Virginia
Charlottesville, Virginia

Stanley S. Ipson
Department of Cybernetics and Virtual Systems
University of Bradford
Bradford, United Kingdom

Rolf Isermann
Laboratory for Control Engineering and Process Automation
Institute of Automatic Control
Darmstadt University of Technology
Darmstadt, Germany

S. Li
Department of Mechanical Engineering
University of Maryland Baltimore County
Baltimore, Maryland

Chang Liu
Micro and Nanotechnology Laboratory
University of Illinois
Urbana, Illinois

Michael A. Lombardi
National Institute of Standards and Technology
Boulder, Colorado

Raul G. Longoria
Department of Mechanical Engineering
The University of Texas at Austin
Austin, Texas

Kevin M. Lynch
Department Mechanical Engineering
Northwestern University
Evanston, Illinois

Sergey Edward Lyshevski
Department of Electrical Engineering
University of Rochester
Rochester, New York

Francis C. Moon
Cornell University
Ithaca, New York

Michael J. Moran
The Ohio State University
Columbus, Ohio

Dinesh Nair
The University of Texas at Austin
Austin, Texas

Pamela M. Norris
Department of Mechanical, Aerospace, and Nuclear Engineering
University of Virginia
Charlottesville, Virginia

Ondrej Novak
Technical University of Liberec
Liberec, Czech Republic

Joey Parker
Department of Mechanical Engineering
University of Alabama
Tuscaloosa, Alabama

Stefano Pastorelli
Department of Mechanics
Politecnico di Torino
Torino, Italy

Michael A. Peshkin
Department of Mechanical Engineering
Northwestern University
Evanston, Illinois

Carla Purdy
ECECS Department
University of Cincinnati
Cincinnati, Ohio

M. K. Ramasubramanian
Department of Mechanical and Aerospace Engineering
North Carolina State University
Raleigh, North Carolina

Giorgio Rizzoni
The Ohio State University
Columbus, Ohio

T. Song
Department of Mechanical Engineering
University of Maryland Baltimore County
Baltimore, Maryland

Massimo Sorli
Department of Mechanics
Politecnico di Torino
Torino, Italy

Alvin M. Strauss
Department of Mechanical Engineering
Vanderbilt University
Nashville, Tennessee

Richard Thorn
School of Engineering
University of Derby
Derby, United Kingdom

Rymantas Tadas Tolocka
Department of Engineering Mechanics
Kaunas University of Technology
Kaunas, Lithuania

Nicolas Vasquez
The University of Texas at Austin
Austin, Texas

Qin Zhang
Department of Agricultural and Biological Engineering
University of Illinois
Urbana, Illinois

Contents

SECTION I Overview of Mechatronics

1. What Is Mechatronics?
   Robert H. Bishop and M. K. Ramasubramanian 1-1

2. Mechatronic Design Approach
   Rolf Isermann 2-1

3. System Interfacing, Instrumentation, and Control Systems
   Rick Homkes 3-1

4. Microprocessor-Based Controllers and Microelectronics
   Ondrej Novak and Ivan Dolezal 4-1

5. An Introduction to Micro- and Nanotechnology
   Michael Goldfarb, Alvin M. Strauss, and Eric J. Barth 5-1

6. Mechatronics Engineering Curriculum Design
   Martin Grimheden 6-1

SECTION II Physical System Modeling

7. Modeling Electromechanical Systems
   Francis C. Moon 7-1

8 Structures and Materials
   Eniko T. Enikov 8-1

9 Modeling of Mechanical Systems for Mechatronics Applications
   Raul G. Longoria 9-1

10 Fluid Power Systems
   Qin Zhang and Carroll E. Goering 10-1

11 Electrical Engineering
   Giorgio Rizzoni 11-1

12 Engineering Thermodynamics
   Michael J. Moran 12-1

13 Numerical Simulation
   Jeannie Sullivan Falcon 13-1

14 Modeling and Simulation for MEMS
   Carla Purdy 14-1

15 Rotational and Translational Microelectromechanical Systems: MEMS Synthesis, Microfabrication, Analysis, and Optimization
   Sergey Edward Lyshevski 15-1

16 The Physical Basis of Analogies in Physical System Models
   Neville Hogan and Peter C. Breedveld 16-1

SECTION III Mechatronic Sensors and Actuators

17 Introduction to Sensors and Actuators
   M. Anjanappa, K. Datta, T. Song, Raghavendra Angara, and S. Li 17-1

18 Fundamentals of Time and Frequency
   Michael A. Lombardi 18-1

19 Sensor and Actuator Characteristics
   Joey Parker 19-1

20 Sensors

- 20.1 Linear and Rotational Sensors *Kevin M. Lynch and Michael A. Peshkin* **20**-2
- 20.2 Acceleration Sensors *Halit Eren* **20**-12
- 20.3 Force Measurement *M. A. Elbestawi* **20**-34
- 20.4 Torque and Power Measurement *Ivan J. Garshelis* **20**-48
- 20.5 Flow Measurement *Richard Thorn* **20**-62
- 20.6 Temperature Measurements *Pamela M. Norris and Bouvard Hosticka* **20**-73
- 20.7 Distance Measuring and Proximity Sensors *Jorge Fernando Figueroa and H. R. (Bart) Everett* **20**-88
- 20.8 Light Detection, Image, and Vision Systems *Stanley S. Ipson* **20**-119
- 20.9 Integrated Microsensors *Chang Liu* **20**-136
- 20.10 Vision *Nicolas Vazquez and Dinesh Nair* **20**-153

21 Actuators

- 21.1 Electromechanical Actuators *George T.-C. Chiu* **21**-1
- 21.2 ElectricalMachines *Charles J. Fraser* **21**-33
- 21.3 Piezoelectric Actuators *Habil Ramutis Bansevicius and Rymantas Tadas Tolocka* ... **21**-51
- 21.4 Hydraulic and Pneumatic Actuation Systems *Massimo Sorli and Stefano Pastorelli*.. **21**-63
- 21.5 MEMS: Microtransducers Analysis, Design, and Fabrication *Sergey Edward Lyshevski* **21**-97

Index I-1

Overview of Mechatronics

I

1 What Is Mechatronics?
Robert H. Bishop and M. K. Ramasubramanian 1-1
Basic Definitions • Key Elements of Mechatronics • Historical Perspective • The Development of the Automobile as a Mechatronic System • What Is Mechatronics? And What Is Next?

2 Mechatronic Design Approach
Rolf Isermann 2-1
Historical Development and Definition of Mechatronic Systems • Functions of Mechatronic Systems • Ways of Integration • Information Processing Systems (Basic Architecture and HW/SW Trade-offs) • Concurrent Design Procedure for Mechatronic Systems

3 System Interfacing, Instrumentation, and Control Systems
Rick Homkes 3-1
Introduction • Input Signals of a Mechatronic System • Output Signals of a Mechatronic System • Signal Conditioning • Microprocessor Control • Microprocessor Numerical Control • Microprocessor Input–Output Control • Software Control • Testing and Instrumentation • Summary

4 Microprocessor-Based Controllers and Microelectronics
Ondrej Novak and Ivan Dolezal 4-1
Introduction to Microelectronics • Digital Logic • Overview of Control Computers • Microprocessors and Microcontrollers • Programmable Logic Controllers • Digital Communications

5 An Introduction to Micro- and Nanotechnology
Michael Goldfarb, Alvin M. Strauss, and Eric J. Barth 5-1
Introduction • Microactuators • Microsensors • Nanomachines

6 Mechatronics Engineering Curriculum Design
Martin Grimheden 6-1
Introduction • The Identity of Mechatronics • Legitimacy of Mechatronics • The Selection of Mechatronics • The Communication of Mechatronics • Fine, but So What? • Putting It All Together in a Curriculum • The Evolution of Mechatronics • Where (and What) Is Mechatronics Today?

1
What Is Mechatronics?

Robert H. Bishop
The University of Texas at Austin

M. K. Ramasubramanian
North Carolina State University

1.1 Basic Definitions 1-1
1.2 Key Elements of Mechatronics 1-2
1.3 Historical Perspective 1-3
1.4 The Development of the Automobile as a
    Mechatronic System 1-7
1.5 What Is Mechatronics? And What Is Next? 1-10
References 1-11

Mechatronics is a natural stage in the evolutionary process of modern engineering design. The development of the computer, and then the microcomputer, embedded computers, and associated information technologies and software advances made mechatronics an imperative in the latter part of the twentieth century. As we begin the twenty-first century, with advances in integrated bioelectromechanical systems, quantum computers, nano- and picosystems, and other unforeseen developments, the future of mechatronics is full of potential and bright possibilities.

1.1 Basic Definitions

The definition of mechatronics has evolved since the original definition by the Yasakawa Electric Company. In trademark application documents, Yasakawa defined mechatronics in this way [1,2]:

> The word, Mechatronics, is composed of 'mecha' from mechanism and the 'tronics' from electronics. In other words, technologies and developed products will be incorporating electronics more and more into mechanisms, intimately and organically, and making it impossible to tell where one ends and the other begins.

The definition of mechatronics continued to evolve after Yasakawa suggested the original definition. One often quoted definition of mechatronics was presented by Harashima, Tomizuka, and Fukuda in 1996 [3]. In their words, mechatronics is defined as

> the synergistic integration of mechanical engineering, with electronics and intelligent computer control in the design and manufacturing of industrial products and processes.

That same year, another definition was suggested by Auslander and Kempf [4]:

> Mechatronics is the application of complex decision making to the operation of physical systems.

Yet another definition by Shetty and Kolk appeared in 1997 [5]:

> Mechatronics is a methodology used for the optimal design of electromechanical products.

We also find the suggestion by W. Bolton that [6]

> A mechatronic system is not just a marriage of electrical and mechanical systems and is more than just a control system; it is a complete integration of all of them.

Finally, from the Internet's free encyclopedia, Wikipedia, we find the description that [7]

> Mechatronics is centered on mechanics, electronics and computing which, combined, make possible the generation of simpler, more economical, reliable and versatile systems.

These definitions and descriptions of mechatronics are accurate and informative, yet each one in and of itself fails to capture the totality of mechatronics. Despite continuing efforts to define mechatronics, to classify mechatronic products, and to develop a standard mechatronics curriculum, a consensus opinion on an all-encompassing description of "what is mechatronics" eludes us. This lack of consensus is a healthy sign. It says that the field is alive, and that it is a youthful subject. Even without an unarguably definitive description of mechatronics, engineers understand from the definitions given above and from their own personal experiences the essence of the *philosophy* of mechatronics.

For many practicing engineers on the front line of engineering design, mechatronics is nothing new. Many engineering products of the past 30 years integrated mechanical, electrical, and computer systems, yet were designed by engineers that were never formally trained in mechatronics per se. It appears that modern concurrent engineering design practices, now formally viewed as part of the mechatronics specialty, are natural design processes. What is evident is that the study of mechatronics provides a mechanism for scholars interested in understanding and explaining the engineering design process to define, classify, organize, and integrate the many aspects of product design into a coherent package. As the historical divisions between mechanical, electrical, aerospace, chemical, civil, and computer engineering become less clearly defined, we should take comfort in the existence of mechatronics as a field of study in academia. The mechatronics specialty provides an educational path, that is, a roadmap, for engineering students studying within the traditional structure of most engineering colleges. Mechatronics is recognized worldwide as a vibrant area of study. Undergraduate and graduate programs in mechatronic engineering are offered in many universities. Refereed journals are being published and dedicated conferences are being organized and are well attended.

Mechatronics is not just a convenient structure for investigative studies by academicians; it is a way of life in modern engineering practice. The introduction of the microprocessor in the early 1980s and the growing desired performance to cost ratio revolutionized the paradigm of engineering design. The number of new products developed at the intersection of traditional disciplines of engineering, computer science, and the natural sciences is rising. New developments in these traditional disciplines are being absorbed into mechatronics design at an ever-increasing pace. The ongoing information technology revolution, advances in wireless communication, smart sensors design (enabled by microelectromechanical systems [MEMS] technology), and embedded systems engineering ensures that the engineering design paradigm will continue to evolve.

1.2 Key Elements of Mechatronics

The study of mechatronic systems can be divided into five areas of specialty:

1. Physical systems modeling
2. Sensors and actuators
3. Signals and systems
4. Computers and logic systems
5. Software and data acquisition

The key elements of mechatronics are illustrated in Figure 1.1. As the field of mechatronics continues to mature, the list of relevant topics associated with the area will most certainly expand.

1.3 Historical Perspective

Attempts to construct automated mechanical systems have a fascinating history. The term "automation" was not popularized until the 1940s when it was coined by the Ford Motor Company to denote a process in which a machine transferred a subassembly item from one station to another and then positioned the item precisely for additional assembly operations. But successful development of automated mechanical systems occurred long before then. For example, early applications of automatic control systems appeared in Greece from 300 to 1 BC, with the development of float regulator mechanisms [8]. Two important examples include the water clock of Ktesibios, which employed a float regulator, and an oil lamp devised by Philon, which also used a float regulator to maintain a constant level of fuel oil. Later, in the first century, Heron of Alexandria published a book entitled *Pneumatica* that described different types of water-level mechanisms using float regulators.

In Europe and Russia in the seventeenth to nineteenth centuries, many important devices that would eventually contribute to mechatronics were invented. Cornelis Drebbel (1572–1633) of Holland devised the temperature regulator representing one of the first feedback systems of that era. Subsequently, Dennis Papin (1647–1712) invented a pressure safety regulator for steam boilers in 1681. Papin's pressure regulator is similar to a modern-day pressure-cooker valve. The first mechanical calculating machine was invented by Pascal in 1642 [9]. The first historical feedback system claimed by Russia was developed by Polzunov in 1765 [10]. Polzunov's water-level float regulator, illustrated in Figure 1.2, employs a float that rises and lowers in relation to the water level, thereby controlling the valve that covers the water inlet in the boiler.

Further evolution in automation was enabled by advancements in control theory traced back to the Watt flyball governor of 1769. The flyball governor, illustrated in Figure 1.3, was used to control the speed of a steam engine [11]. Employing a measurement of the speed of the output shaft and utilizing the motion of the flyball to control the valve, the amount of steam entering the engine is controlled. As the speed of the engine increases, the metal spheres on the governor apparatus rise and extend away from the shaft axis, thereby closing the valve. This is an example of a feedback control system where the feedback signal and the control actuation are completely coupled in the mechanical hardware.

These early successful automation developments were achieved through intuition, application of practical skills, and persistence. The next step in the evolution of automation required a *theory* of automatic control. The precursor to the numerically controlled (NC) machines for automated manufacturing (to be developed in the 1950s and 1960s at MIT) appeared in the early 1800s, with the invention of feed-forward control of weaving looms by Joseph Jacquard of France. In the late 1800s, the subject now known as control theory was initiated by J. C. Maxwell through analysis of the set of differential equations describing the flyball governor [12]. Maxwell investigated the effect various system parameters had on system performance. At about the same time, Vyshnegradskii formulated a mathematical theory of regulators [13]. In the 1830s, Michael Faraday described the law of induction that would form the basis of the electric motor and the electric dynamo. Subsequently, in the late 1880s, Nikola Tesla invented the alternating-current induction motor. The basic idea of controlling a mechanical system automatically was firmly established by the end of the 1800s. The evolution of automation would accelerate significantly in the latter part of the twentieth century, expanding in capability from automation to increasingly higher levels of autonomy in the twenty-first century.

The development of pneumatic control elements in the 1930s matured to a point of finding applications in the process industries. However, before 1940, the design of control systems remained an art generally characterized by trial-and-error methods. During the 1940s, continuing advances in mathematical and analytical methods solidified the notion of control engineering as an independent engineering discipline. In the United States, the development of the telephone system and electronic feedback amplifiers spurred the use of feedback by Bode, Nyquist, and Black at Bell Telephone Laboratories [14–18]. The operation

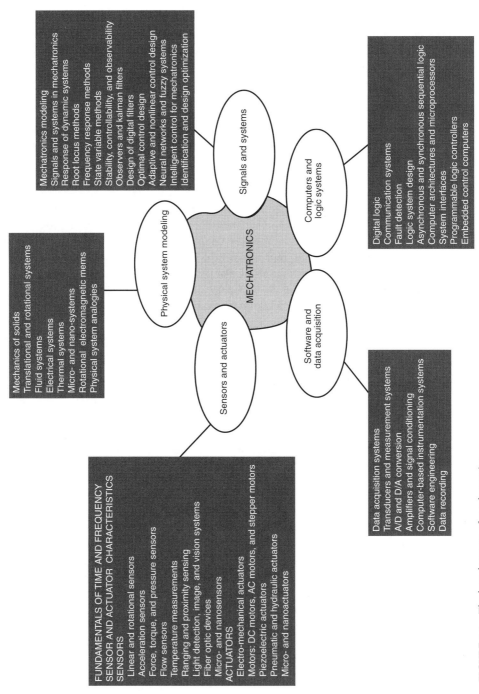

FIGURE 1.1 The key elements of mechatronics.

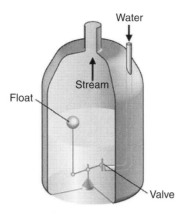

FIGURE 1.2 Water-level float regulator. (Excerted from R. C. Dorf and R. H. Bishop, *Modern Control Systems*, 11th edn., Prentice-Hall, 2008. With permission.)

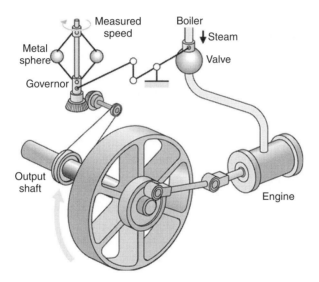

FIGURE 1.3 Watt's flyball governor. (Excerted from R. C. Dorf and R. H. Bishop, *Modern Control Systems*, 11th edn., Prentice-Hall, 2008. With permission.)

of the feedback amplifiers was described in the frequency domain, and the ensuing design and analysis practices are now generally classified as "classical control." During the same time period, control theory was also developing in Russia and Eastern Europe. Mathematicians and applied mechanicians in the former Soviet Union dominated the field of controls and concentrated on time-domain formulations and differential equation models of systems. Further developments of time-domain formulations using state variable system representations occurred in the 1960s and led to design and analysis practices now generally classified as "modern control."

The World War II war effort led to further advances in the theory and practice of automatic control in an effort to design and construct automatic airplane pilots, gun-positioning systems, radar antenna control systems, and other military systems. The complexity and expected performance of these military systems necessitated an extension of the available control techniques and fostered interest in control systems and the development of new insights and methods. Frequency-domain techniques continued to dominate the field of controls following World War II with the increased use of the Laplace transform, and the so-called s-plane methods, such as designing control systems using root locus.

FIGURE 1.4 An automated paperboard container-manufacturing machine with several transfer stations and cam-actuated indexing turrets and forming dies.

On the commercial side, driven by cost savings achieved through mass production, automation of the production process was a high priority beginning in the 1940s. During the 1950s, the invention of the cam, linkages, and chain drives became the major enabling technologies for the invention of new products and high-speed precision manufacturing and assembly. Examples include textile and printing machines, paper converting machinery, and sewing machines. High-volume precision manufacturing became a reality during this period. An example of high-volume automated manufacturing of paperboard container for packaging is shown in Figure 1.4. The automated paperboard container-manufacturing machine utilizes a sheet-fed process wherein the paperboard is cut into a fan shape to form the tapered sidewall, and wrapped around a mandrel. The seam is then heat sealed and held until cured. Another sheet-fed source of paperboard is used to cut out the plate to form the bottom of the paperboard container, formed into a shallow dish through scoring and creasing operations in a die, and assembled to the cup shell. The lower edge of the cup shell is bent inwards over the edge of the bottom plate sidewall, and heat sealed under high pressure to prevent leaks and provide a precisely level edge for standup. The brim is formed on the top to provide a ring-on-shell structure to provide the stiffness needed for its functionality. All these operations are conducted while the work piece undergoes a precision transfer from one turret to another and is then ejected. The production rate of a typical machine averages over 200 cups per minute. The automated paperboard container manufacturing did not involve any nonmechanical system except an electric motor for driving the line shaft. These machines are typical of paper converting and textile machinery, and represent automated systems significantly more complex than their predecessors.

The development of the microprocessor in the late 1960s led to early forms of computer control in process and product design. Examples include NC machines and aircraft control systems. Yet the manufacturing processes were still entirely mechanical in nature and the automation and control systems were implemented only as an afterthought. The launch of Sputnik and the advent of the space age provided yet another impetus to the continued development of controlled mechanical systems. Missiles and space probes necessitated the development of complex, highly accurate control systems. Furthermore, the need to minimize satellite mass (i.e., to minimize the amount of fuel required for the mission) while providing accurate control encouraged advancements in the important field of optimal control. Time-domain methods developed by Liapunov, Minorsky, and others, as well as the theories of optimal control developed by L. S. Pontryagin in the former Soviet Union and R. Bellman in the United States, were well matched with the increasing availability of high-speed computers and new programming languages for scientific use.

Advancements in semiconductor and integrated circuits manufacturing led to the development of a new class of products that incorporated mechanical and electronics in the system and required the two together for their functionality. The term mechatronics was introduced by Yasakawa Electric in 1969 to represent such systems. Yasakawa was granted a trademark in 1972, but after widespread usage of the term, released its trademark rights in 1982 [1–3]. Initially, mechatronics referred to systems with only mechanical systems and electrical components—no computation was involved. Examples of such systems include the automatic sliding door, vending machines, and garage door openers.

In the late 1970s, the Japan Society for the Promotion of Machine Industry (JSPMI) classified mechatronics products into four categories [1]:

6. **Class I**: Primarily mechanical products with electronics incorporated to enhance functionality. Examples include NC machine tools and variable speed drives in manufacturing machines.
7. **Class II**: Traditional mechanical systems with significantly updated internal devices incorporating electronics. The external user interfaces are unaltered. Examples include the modern sewing machine and automated manufacturing systems.
8. **Class III**: Systems that retain the functionality of the traditional mechanical system, but the internal mechanisms are replaced by electronics. An example is the digital watch.
9. **Class IV**: Products designed with mechanical and electronic technologies through synergistic integration. Examples include photocopiers, intelligent washers and dryers, rice cookers, and automatic ovens.

The enabling technologies for each mechatronic product class illustrate the progression of electromechanical products in stride with developments in control theory, computation technologies, and microprocessors. Class I products were enabled by servo technology, power electronics, and control theory. Class II products were enabled by the availability of early computational and memory devices and custom circuit design capabilities. Class III products relied heavily on the microprocessor and integrated circuits to replace mechanical systems. Class IV products marked the beginning of true mechatronic systems, through integration of mechanical systems and electronics. It was not until the 1970s with the development of the microprocessor by the Intel Corporation that integration of computational systems with mechanical systems became practical.

The division between classical control and modern control was significantly reduced in the 1980s with the advent of "robust control" theory. It is now generally accepted that control engineering must consider both the time-domain and the frequency-domain approaches simultaneously in the analysis and design of control systems. Also, during the 1980s, the utilization of digital computers as integral components of control systems became routine. There are literally hundreds of thousands of digital process control computers installed worldwide [19,20]. Whatever definition of mechatronics one chooses to adopt, it is evident that modern mechatronics involves computation as a central element. In fact, the incorporation of the microprocessor to precisely modulate mechanical power and to adapt to changes in environment is the essence of modern mechatronics and smart products.

1.4 The Development of the Automobile as a Mechatronic System

The evolution of modern mechatronics can be illustrated with the example of the automobile. Until the 1960s, the radio was the only significant electronics in an automobile. All other functions were entirely mechanical or electrical, such as the starter motor and the battery charging systems. There were no "intelligent safety systems," except augmenting the bumper and structural members to protect occupants in case of accidents. Seat belts, introduced in the early 1960s, were aimed at improving occupant safety and were completely mechanically actuated. All engine systems were controlled by the driver and/or other mechanical control systems. For instance, before the introduction of sensors and microcontrollers, a mechanical distributor was used to select the specific spark plug to fire when the fuel–air mixture is compressed. The timing of the ignition was the control variable. Themechanically controlled combustion process was not optimal in terms of fuel efficiency. Modeling of the combustion process showed that for increased fuel efficiency there existed an optimal time when the fuel should be ignited. The timing depends on load, speed, and other measurable quantities. The electronic ignition system was one of the first mechatronic systems to be introduced in the automobile in the late 1970s. The electronic ignition system consists of a crankshaft position sensor, camshaft position sensor, airflow rate, throttle position, rate of throttle position change sensors, and a dedicated microcontroller determining the timing of the spark plug firings. Early implementations involved only a Hall-effect sensor to sense the position of the rotor in the distributor accurately. Subsequent implementations eliminated the distributor completely and directly controlled the firings, utilizing a microprocessor.

The Antilock Brake System (ABS) was also introduced in the late 1970s in automobiles [21]. The ABS works by sensing lockup of any of the wheels and then modulating the hydraulic pressure as needed to minimize or eliminate sliding. The Traction Control System (TCS) was introduced in automobiles in the mid-1990s. The TCS works by sensing slippage during acceleration and then modulating the power to the slipping wheel. This process ensures that the vehicle is accelerating at the maximum possible rate under given road and vehicle conditions. The Vehicle Dynamics Control (VDC) system was introduced in automobiles in the late 1990s. The VDC works similar to the TCS with the addition of a yaw rate sensor and a lateral accelerometer. The driver intention is determined by the steering wheel position and then compared with the actual direction of motion. The TCS system is then activated to control the power to the wheels and to control the vehicle velocity, and minimize the difference between the steering wheel direction and the direction of the vehicle motion [21,22]. In some cases, the ABS is used to slow the vehicle down to achieve the desired control. Automobiles use combinations of 8-, 16-, and 32-bit processors for implementation of the control systems. The microcontroller has onboard memory (EEPROM/EPROM), digital and analog inputs, A/D converters, pulse width modulation (PWM), timer functions, such as event counting and pulse width measurement, prioritized inputs, and in some cases digital signal processing. Typically, the 32-bit processor is used for engine management, transmission control, and airbags; the 16-bit processor is used for the ABS, TCS, VDC, instrument cluster, and air conditioning systems; the 8-bit processor is used for seat, mirror control, and window lift systems. Today, there are about 30–70 microcontrollers in a car [23], with luxury vehicles often hosting over 100 microprocessors.We can expect the number of onboard microprocessors to swell with the drive toward modular automotive systems, utilizing plug-n-ply mechatronic subsystems.

Mechatronics has become a necessity for product differentiation in automobiles. Since the basics of internal combustion engine were worked out almost a century ago, differences in the engine design among the various automobiles are no longer useful as a product differentiator. In the 1970s, the Japanese automakers succeeded in establishing a foothold in the U.S. automobile market by offering unsurpassed quality and fuel-efficient small automobiles. The quality of the vehicle was the product differentiator through the 1980s. In the 1990s, consumers came to expect quality and reliability in automobiles from all manufacturers. Today, mechatronic *features* are the product differentiators in traditionally mechanical systems. This is further accelerated by higher performance price ratio in electronics, market demand for innovative productswith smart features, and the drive to reduce cost of manufacturing of existing products through redesign incorporating mechatronics elements. With the prospects of low growth, automotive makers are searching for high-tech features that will differentiate their vehicles from others [24]. The world automotive nonentertainment electronics market, now at about $37 billion, is expected to reach $52 billion by 2010. Even though the North American automotive electronics market is mature, it is predicted that there will be continued growth of about 5% up to the year 2010 in electronic braking, steering, and driver information [25]. New applications of mechatronic systems in the automotive world include semiautonomous to fully autonomous automobiles, safety enhancements, emission reduction, and other features including intelligent cruise control, and brake-by-wire systems eliminating the hydraulics [26]. Another significant growth area that benefits from amechatronics design approach is wireless networking

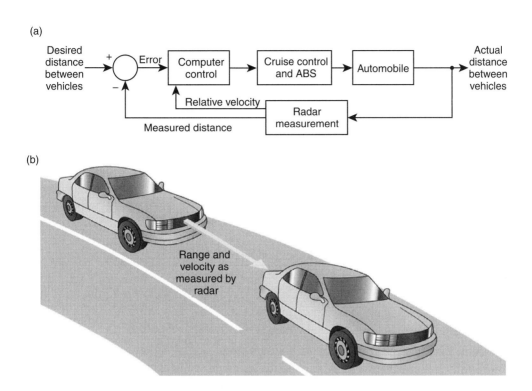

FIGURE 1.5 Using a radar to measure distance and velocity to autonomously maintain desired distance between vehicles. (Adapted from R. C. Dorf and R. H. Bishop, *Modern Control Systems*, 11th edn., Prentice-Hall, 2008. With permission.)

of automobiles to ground stations and vehicle-to-vehicle communication. Telematics combines audio, hands-free cell phone, navigation, Internet connectivity, e-mail, and voice recognition, and is perhaps the largest potential automotive growth area.

Microelectromechanical systems is an enabling technology for the cost-effective development of sensors and actuators cost for mechatronics applications. Already, several MEMS devices are in use in the automobiles, including sensors and actuators for airbag deployment and pressure sensors for manifold pressure measurement. Integrating MEMS devices with CMOS signal conditioning circuits on the same silicon is another example of development of enabling technologies that will improve mechatronic products, such as the automobile.

Millimeter-wave radar technology has found applications in automobiles. The millimeter-wave radar detects the location of objects (other vehicles) in the scenery and the distance to the obstacle and the velocity in real time. A detailed description of a working system is given by Suzuki et al. [28]. Figure 1.5 shows an illustration of the vehicle sensing capability with a millimeter-wave radar. This technology provides the capability to control the distance between the vehicle and an obstacle (or another vehicle) by integrating the sensor with the cruise control and ABS systems. The driver is able to set the speed and the desired distance between the cars ahead of him. The ABS system and the cruise control systemare coupled together to safely achieve this remarkable capability. One logical extension of the obstacle avoidance capability is slow speed semiautonomous driving, where the vehicle maintains a constant distance from the vehicle ahead in traffic jam conditions. Fully autonomous vehicles are well within the scope of mechatronics development within the next 20 years. Supporting investigations are underway in many research centers on development of semiautonomous cars with reactive path planning using GPS-based continuous traffic

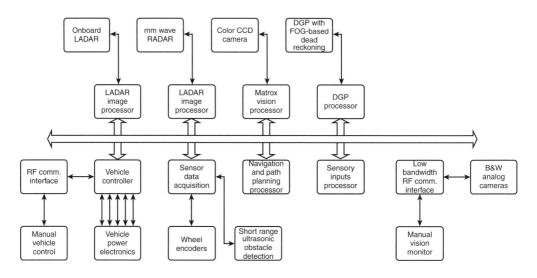

FIGURE 1.6 Autonomous vehicle system design with sensors and actuators.

model updates and stop and go automation. A proposed sensing and control system for such a vehicle, shown in Figure 1.6, involves differential global positioning systems (DGPS), real-time image processing, and dynamic path planning.

Future mechatronic systems on automobiles may include a fog-free windshield based on humidity and temperature sensing and climate control, self-parallel parking, rear parking aid, lane change assistance, fluidless electronic brake-by-wire, and replacement of hydraulic systems with electromechanical servo systems. An upcoming trend is to bring the PC into the automobile passenger compartment, including mounted tablet computers with Internet routers and storage compartments [29]. As the number of automobiles in the world increases, stricter emission standards are inevitable. Mechatronic products will in all likelihood contribute to meet the challenges in emission control and engine efficiency by providing substantial reduction in CO, NO, and HC emissions and increase in vehicle efficiency [26]. Clearly, an automobile with 30–70 microcontrollers, up to 100 electric motors, about 200 pounds of wiring, a multitude of sensors, and thousands of lines of software code can hardly be classified as a strictly mechanical system. The automobile is being transformed into a comprehensive mechatronic system.

1.5 What Is Mechatronics? And What Is Next?

Mechatronics has evolved over the past 25 years and has led to a special breed of intelligent products. What is mechatronics? It is a natural stage in the evolutionary process of modern engineering design. For some engineers, mechatronics is not new, and for others it is a philosophical approach to design, which serves as a guide for their activities. Certainly, mechatronics is an evolutionary process, not a revolutionary one. An all-encompassing definition of mechatronics does not exist, but in reality, one is not needed. It is understood that mechatronics is about the synergistic integration of mechanical, electrical, and computer systems. One can understand the extent that mechatronics reaches into various disciplines by characterizing the constituent components comprising mechatronics, which include (1) physical systems modeling, (2) sensors and actuators, (3) signals and systems, (4) computers and logic systems, and (5) software and data acquisition. Engineers and scientists from all walks of life and fields of study can contribute to mechatronics. As engineering and science boundaries become less well defined, more students will seek a multidisciplinary education with a strong design component. Academia should be moving toward a curriculum that includes coverage of mechatronic systems.

In the future, growth in mechatronic systems will be fueled by the growth in the constituent areas. Advancements in traditional disciplines fuel the growth of mechatronics systems by providing "enabling technologies." For example, the invention of the microprocessor had a profound effect on the redesign of mechanical systems and design of new mechatronics systems. We should expect continued advancements in cost-effective microprocessors and microcontrollers, sensor and actuator development enabled by advancements in applications of MEMS, adaptive control methodologies and real-time programming methods, networking and wireless technologies, mature CAE technologies for advanced system modeling, virtual prototyping, and testing. An example of a new VSLI technology permits us to construct systemson-chips (known as SoC). The SoC devices contain hundreds of millions of transistors offering especially with sophisticated functionality [30]. The continued rapid development in these areas will only accelerate pace of smart product development. The Internet is a technology that, when utilized in combination the wireless technology, may also lead to new mechatronic products. While developments in automotives provide vivid examples of mechatronics development, there are numerous examples of intelligent systems all around us, including smart home appliances such as dishwashers, vacuum cleaners, microwaves, and wireless network-enabled devices. In the area of "human-friendly machines" (a term used by H. Kobayashi [31]), we can expect advances in robot-assisted surgery, and implantable sensors and actuators. Other areas that will benefit from mechatronic advances may include robotics, manufacturing, space technology, and transportation. An area with great potential is the area of microrobotics spawned by the MEMS revolution [32]. An example of a microrobot is the so-called micromanipulation tool that can be utilized in minimal invasive surgery. The future of mechatronics is wide open.

References

1. N. Kyura and H. Oho, "Mechatronics—An Industrial Perspective," *IEEE/ASME Transactions on Mechatronics*, Vol. 1, No. 1, 1996, pp. 10–15.
2. T. Mori, "Mecha-tronics," Yasakawa Internal Trademark Application Memo 21.131.01, July 12, 1969.
3. F. Harashima, M. Tomizuka, and T. Fukuda, "Mechatronics—What is it, Why, and How?—An Editorial," *IEEE/ASME Transactions on Mechatronics*, Vol. 1, No. 1, 1996, pp. 1–4.
4. D. M. Auslander and C. J. Kempf, *Mechatronics: Mechanical System Interfacing*, Prentice Hall, Upper Saddle River, NJ, 1996.
5. D. Shetty and R. A. Kolk, *Mechatronic System Design*, PWS Publishing Company, Boston, MA, 1997.
6. W. Bolton, *Mechatronics: Electrical Control Systems in Mechanical and Electrical Engineering*, 2nd edn., Addison Wesley Longman, Harlow, England, 1999.
7. Mechatronics, *Wikipedia, The Free Encyclopedia*. Retrieved 01:00, October 10, 2006, http://en.wikipedia.org/w/index.php?title=Mechatronics&oldid=80065916.
8. I. O. Mayr, *The Origins of Feedback Control*, MIT Press, Cambridge, MA, 1970.
9. D. Tomkinson and J. Horne, *Mechatronics Engineering*, McGraw-Hill, New York, 1996.
10. E. P. Popov, *The Dynamics of Automatic Control Systems*; Gostekhizdat, Moscow, 1956; Addison-Wesley, Reading, MA, 1962.
11. R. C. Dorf and R. H. Bishop, *Modern Control Systems*, 9th edn., Prentice Hall, Upper Saddle River, NJ, 2000.
12. J. C. Maxwell, "On Governors," *Proc. of the Royal Society of London*, 16, 1868; in *Selected Papers on Mathematical Trends in Control Theory*, Dover, New York, 1964, pp. 270–283.
13. I. A. Vyshnegradskii, "On Controllers of Direct Action," *Izv. SPB Tekhnotog. Inst.*, 1877.
14. H. W. Bode, "Feedback—The History of an Idea," in *Selected Papers on Mathematical Trends in Control Theory*, Dover, New York, 1964, pp. 106–123.
15. H. S. Black, "Inventing the Negative Feedback Amplifier," *IEEE Spectrum*, December 1977, pp. 55, 60.
16. J. E. Brittain, *Turning Points in American Electrical History*, IEEE Press, New York, 1977.

17. M. D. Fagen (Ed.), A History of Engineering and Science in the Bell System, Bell Telephone Laboratories, *The Laboratories*, New York, 1975.
18. G. Newton, L. Gould, and J. Kaiser, *Analytical Design of Linear Feedback Control*, JohnWiley & Sons, New York, 1957.
19. R. C. Dorf and A. Kusiak, *Handbook of Automation and Manufacturing*, John Wiley & Sons, New York, 1994.
20. R. C. Dorf, *The Encyclopedia of Robotics*, John Wiley & Sons, New York, 1988.
21. K. Asami, Y. Nomura, and T. Naganawa, "Traction Control (TRC) System for 1987 Toyota Crown, 1989," *ABS-TCS-VDC Where Will the Technology Lead Us?* J. Mack, Ed., Society of Automotive Engineers, Inc.,Warrendale, PA, 1996, Chapter 2.
22. S. Pastor, et al., "Brake Control System," United States Patent # 5,720,533, Feb. 24, 1998, (see http://www.uspto.gov/ for more information).
23. "Chip Makers Roll Out Automotive Processors," *Design News*, Vol. 61, Issue 11, 8/14/2006, p. 46.
24. B. Jorgensen, "Shifting gears," Auto Electronics, *Electronic Business*, Feb. 2001.
25. Reed Electronics Research, *Automotive Electronics—A Profile of International Markets and Suppliers to 2010*, #IN0603375RE, See the press release at http://www.instat.com/press.asp?ID=1752&sku=IN0603375RE for more information, September 2006.
26. M. B. Barron and W. F. Powers, "The Role of Electronic Controls for Future Automotive Mechatronic Systems," *IEEE/ASME Transactions on Mechatronics*, Vol. 1, No. 1, 1996, pp. 80–88.
27. G. Kobe, "Electronics: What's Driving the Growth," *Automotive Industries*, August 2000.
28. H. Suzuki, Hiroshi, M. Shono, and O. Isaji, "Radar Apparatus for Detecting a Distance/Velocity" United States Patent #5,677,695, October 14, 1997, (see http://www.uspto.gov/ for more information).
29. J. Saranow, "In-Car Computing," Newsday, New York, March 5, 2006.
30. W. Wolf, "Embedded Systems-on-Chips,"*The Engineering Handbook*, 2nd edn., R. C. Dorf, Ed., CRC Press, Inc., Boca Raton, FL, 2005, pp. 123-1–123-11.
31. H. Kobayashi, Guest Editorial, *IEEE/ASME Transactions on Mechatronics*, Vol. 2, No. 4, 1997, p. 217.
32. T. Ebefors and G. Stemme, "Microrobotics," *The MEMS Handbook*, M. Gad-el-Hak, Ed., CRC Press, Inc., Boca Raton, FL, 2002, pp. 28-1–28-42.

2
Mechatronic Design Approach

2.1	Historical Development and Definition of Mechatronic Systems	2-1
2.2	Functions of Mechatronic Systems	2-3
	Division of Functions between Mechanics and Electronics • Improvement of Operating Properties • Addition of New Functions	
2.3	Ways of Integration	2-5
	Integration of Components (Hardware) • Integration of Information Processing (Software)	
2.4	Information Processing Systems (Basic Architecture and HW/SW Trade-Offs)	2-6
	Multilevel Control Architecture • Special Signal Processing • Model-Based and Adaptive Control Systems • Supervision and Fault Detection • Intelligent Systems (Basic Tasks)	
2.5	Concurrent Design Procedure for Mechatronic Systems	2-9
	Design Steps • Required CAD/CAE Tools • Modeling Procedure • Real-Time Simulation • Hardware-in-the-Loop Simulation • Control Prototyping	

Rolf Isermann
Darmstadt University of Technology

References 2-15

2.1 Historical Development and Definition of Mechatronic Systems

In several technical areas the integration of products or processes and electronics can be observed. This is especially true for mechanical systems which developed since about 1980. These systems changed from electro-mechanical systems with discrete electrical and mechanical parts to integrated electronic-mechanical systems with sensors, actuators, and digital microelectronics. These integrated systems, as seen in Table 2.1, are called *mechatronic systems*, with the connection of MECHAnics and elecTRONICS.

The word "mechatronics" was probably first created by a Japanese engineer in 1969 [1], with earlier definitions given by References 2 and 3. In reference, a preliminary definition is given: "Mechatronics is the synergetic integration of mechanical engineering with electronics and intelligent computer control in the design and manufacturing of industrial products and processes" [5].

All these definitions agree that mechatronics is an *interdisciplinary field*, in which the following disciplines act together (see Figure 2.1):

- *Mechanical systems* (mechanical elements, machines, precision mechanics)
- *Electronic systems* (microelectronics, power electronics, sensor and actuator technology) and
- *Information technology* (systems theory, automation, software engineering, artificial intelligence)

TABLE 2.1 Historical Development of Mechanical, Electrical, and Electronic Systems

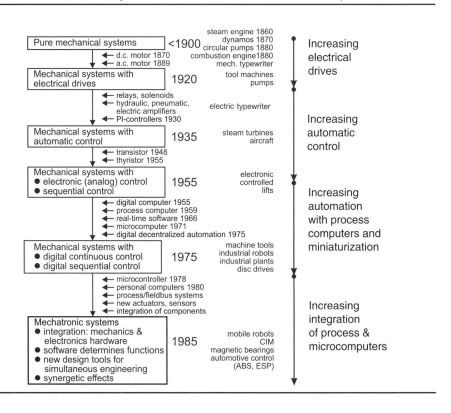

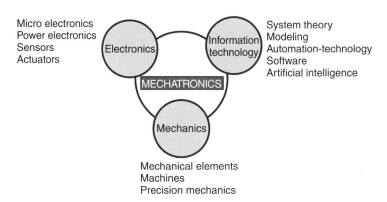

FIGURE 2.1 Mechatronics: synergetic integration of different disciplines.

Some survey contributions describe the development of mechatronics; see [5–8]. An insight into general aspects are given in the journals [4,9,10]; first conference proceedings [11–15]; and the books [16–19].

Figure 2.2 shows a general scheme of a modern mechanical process like a power producing or a power generating machine. A primary *energy flows* into the machine and is then either directly used for the energy consumer in the case of an energy transformer, or converted into another energy form in the case of an energy converter. The form of energy can be electrical, mechanical (potential or kinetic, hydraulic, pneumatic), chemical, or thermal. Machines are mostly characterized by a continuous or periodic (repetitive) energy flow. For other mechanical processes, such as mechanical elements or precision mechanical devices, piecewise or intermittent energy flows are typical.

Mechatronic Design Approach

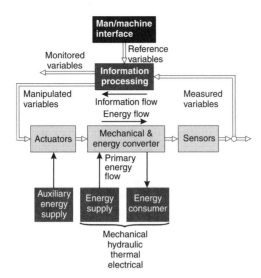

FIGURE 2.2 Mechanical process and information processing develop towards mechatronic systems.

The energy flow is generally a product of a generalized flow and a potential (effort). Information on the state of the mechanical process can be obtained by measured generalized flows (speed, volume, or mass flow) or electrical current or potentials (force, pressure, temperature, or voltage). Together with reference variables, the measured variables are the inputs for an *information flow* through the digital electronics resulting in manipulated variables for the actuators or in monitored variables on a display.

The addition and integration of feedback information flow to a feedforward energy flow in a basically mechanical system is one characteristic of many mechatronic systems. This development presently influences the design of mechanical systems. Mechatronic systems can be subdivided into:

- Mechatronic systems
- Mechatronic machines
- Mechatronic vehicles
- Precision mechatronics
- Micro mechatronics

This shows that the integration with electronics comprises many classes of technical systems. In several cases, the mechanical part of the process is coupled with an electrical, thermal, thermodynamic, chemical, or information processing part. This holds especially true for energy converters as machines where, in addition to the mechanical energy, other kinds of energy appear. Therefore, *mechatronic systems in a wider sense* comprise mechanical and also non-mechanical processes. However, the mechanical part normally dominates the system.

Because an auxiliary energy is required to change the fixed properties of formerly passive mechanical systems by feedforward or feedback control, these systems are sometimes also called *active mechanical systems*.

2.2 Functions of Mechatronic Systems

Mechatronic systems permit many improved and new functions. This will be discussed by considering some examples.

2.2.1 Division of Functions between Mechanics and Electronics

For designing mechatronic systems, the interplay for the realization of functions in the mechanical and electronic part is crucial. Compared to pure mechanical realizations, the use of amplifiers and actuators with electrical auxiliary energy led to considerable simplifications in devices, as can be seen from watches, electrical typewriters, and cameras. A further considerable *simplification in the mechanics* resulted from introducing microcomputers in connection with decentralized electrical drives, as can be seen from electronic typewriters, sewing machines, multi-axis handling systems, and automatic gears.

The design of lightweight constructions leads to elastic systems which are weakly damped through the material. An *electronic damping* through position, speed, or vibration sensors and electronic feedback can be realized with the additional advantage of an adjustable damping through the algorithms. Examples are elastic drive chains of vehicles with damping algorithms in the engine electronics, elastic robots, hydraulic systems, far reaching cranes, and space constructions (with, for example, flywheels).

The addition of closed loop control for position, speed, or force not only results in a precise tracking of reference variables, but also an approximate linear behavior, even though the mechanical systems show nonlinear behavior. By *omitting the constraint of linearization* on the mechanical side, the effort for construction and manufacturing may be reduced. Examples are simple mechanical pneumatic and electromechanical actuators and flow valves with electronic control.

With the aid of freely *programmable reference variable generation* the adaptation of nonlinear mechanical systems to the operator can be improved. This is already used for the driving pedal characteristics within the engine electronics for automobiles, telemanipulation of vehicles and aircraft, in development of hydraulic actuated excavators, and electric power steering.

With an increasing number of sensors, actuators, switches, and control units, the cable and electrical connections increase such that reliability, cost, weight, and the required space are major concerns. Therefore, the development of suitable bus systems, plug systems, and redundant and reconfigurable electronic systems are challenges for the designer.

2.2.2 Improvement of Operating Properties

By applying active feedback control, precision is obtained not only through the high mechanical precision of a passively feedforward controlled mechanical element, but by comparison of a programmed reference variable and a measured control variable. Therefore, the mechanical precision in design and manufacturing may be reduced somewhat and more simple constructions for bearings or slideways can be used. An important aspect is the compensation of a larger and time variant friction by *adaptive friction compensation* [13,20]. Also, a larger friction on cost of backlash may be intended (such as gears with pretension), because it is usually easier to compensate for friction than for backlash.

Model-based and *adaptive control* allow for a wide range of operation, compared to fixed control with unsatisfactory performance (danger of instability or sluggish behavior). A combination of robust and adaptive control allows a wide range of operation for flow-, force-, or speed-control, and for processes like engines, vehicles, or aircraft. A better control performance allows the reference variables to move closer to the constraints with an improvement in efficiencies and yields (e.g., higher temperatures, pressures for combustion engines and turbines, compressors at stalling limits, higher tensions and higher speed for paper machines and steel mills).

2.2.3 Addition of New Functions

Mechatronic systems allow functions to occur that could not be performed without digital electronics. First, *nonmeasurable quantities* can be calculated on the basis of measured signals and influenced by feedforward or feedback control. Examples are time-dependent variables such as slip for tyres, internal tensities, temperatures, slip angle and ground speed for steering control of vehicles, or parameters like damping, stiffness coefficients, and resistances. The *adaptation of parameters* such as damping and stiffness for oscillating systems (based on measurements of displacements or accelerations) is another example. Integrated *supervision and fault diagnosis* becomes more and more important with increasing automatic functions, increasing complexity, and higher demands on reliability and safety. Then, the triggering of redundant components, system reconfiguration, maintenance-on-request, and any kind of *teleservice* make the system more "intelligent." Table 2.2 summarizes some properties of mechatronic systems compared to conventional electro-mechanical systems.

TABLE 2.2 Properties of Conventional and Mechatronic Design Systems

Conventional Design	Mechatronic Design
Added components	**Integration of components (hardware)**
1 Bulky	Compact
2 Complex mechanisms	Simple mechanisms
3 Cable problems	Bus or wireless communication
4 Connected components	Autonomous units
Simple control	**Integration by information processing (software)**
5 Stiff construction	Elastic construction with damping by electronic feedback
6 Feedforward control, linear (analog) control	Programmable feedback (nonlinear) digital control
7 Precision through narrow tolerances	Precision through measurement and feedback control
8 Nonmeasurable quantities change arbitrarily	Control of nonmeasurable estimated quantities
9 Simple monitoring	Supervision with fault diagnosis
10 Fixed abilities	Learning abilities

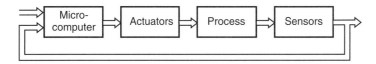

FIGURE 2.3 General scheme of a (classical) mechanical-electronic system.

2.3 Ways of Integration

Figure 2.3 shows a general scheme of a classical mechanical-electronic system. Such systems resulted from adding available sensors, actuators, and analog or digital controllers to mechanical components. The limits of this approach were given by the lack of suitable sensors and actuators, the unsatisfactory life time under rough operating conditions (acceleration, temperature, contamination), the large space requirements, the required cables, and relatively slow data processing. With increasing improvements in miniaturization, robustness, and computing power of microelectronic components, one can now put more emphasis on electronics in the design of a mechatronic system. More autonomous systems can be envisioned, such as capsuled units with touchless signal transfer or bus connections, and robust microelectronics.

The integration within a mechatronic system can be performed through the integration of components and through the integration of information processing.

2.3.1 Integration of Components (Hardware)

The integration of components (hardware integration) results from designing the mechatronic system as an overall system and imbedding the sensors, actuators, and microcomputers into the mechanical process, as seen in Figure 2.4. This spatial integration may be limited to the process and sensor, or to the process and actuator. Microcomputers can be integrated with the actuator, the process or sensor, or can be arranged at several places.

Integrated sensors and microcomputers lead to *smart sensors*, and integrated actuators and microcomputers lead to *smart actuators*. For larger systems, bus connections will replace cables. Hence, there are several possibilities to build up an integrated overall system by proper integration of the hardware.

2.3.2 Integration of Information Processing (Software)

The integration of information processing (software integration) is mostly based on advanced control functions. Besides a basic feedforward and feedback control, an additional influence may take place through the process knowledge and corresponding online information processing, as seen in Figure 2.4. This means a processing of available signals at higher levels, including the solution of tasks like supervision

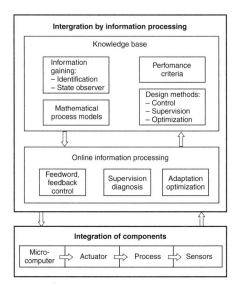

FIGURE 2.4 Ways of integration within mechatronic systems.

with fault diagnosis, optimization, and general process management. The respective problem solutions result in real-time algorithms which must be adapted to the mechanical process properties, expressed by mathematical models in the form of static characteristics, or differential equations. Therefore, a *knowledge base* is required, comprising methods for design and information gaining, process models, and performance criteria. In this way, the mechanical parts are governed in various ways through higher level information processing with intelligent properties, possibly including learning, thus forming an integration by process-adapted software.

2.4 Information Processing Systems (Basic Architecture and HW/SW Trade-Offs)

The governing of mechanical systems is usually performed through actuators for the changing of positions, speeds, flows, forces, torques, and voltages. The directly measurable output quantities are frequently positions, speeds, accelerations, forces, and currents.

2.4.1 Multilevel Control Architecture

The information processing of *direct measurable input and output signals* can be organized in several levels, as compared in Figure 2.5.

Level 1: low level control (feedforward, feedback for damping, stabilization, linearization)
Level 2: high level control (advanced feedback control strategies)
Level 3: supervision, including fault diagnosis
Level 4: optimization, coordination (of processes)
Level 5: general process management

Recent approaches to mechatronic systems use signal processing in the lower levels, such as damping, control of motions, or simple supervision. Digital information processing, however, allows for the solution of many tasks, like adaptive control, learning control, supervision with fault diagnosis, decisions

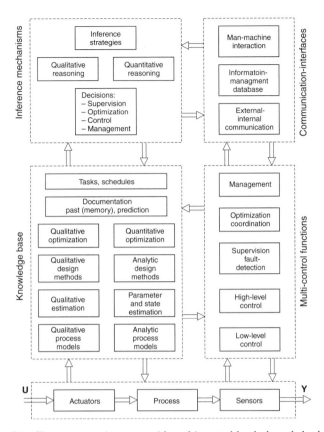

FIGURE 2.5 Advanced intelligent automatic system with multi-control levels, knowledge base, inference mechanisms, and interfaces.

for maintenance or even redundancy actions, economic optimization, and coordination. The tasks of the higher levels are sometimes summarized as "process management."

2.4.2 Special Signal Processing

The described methods are partially applicable for *nonmeasurable quantities* that are reconstructed from mathematical process models. In this way, it is possible to control damping ratios, material and heat stress, and slip, or to supervise quantities like resistances, capacitances, temperatures within components, or parameters of wear and contamination. This signal processing may require *special filters* to determine amplitudes or frequencies of vibrations, to determine derivated or integrated quantities, or *state variable observers*.

2.4.3 Model-Based and Adaptive Control Systems

The information processing is, at least in the lower levels, performed by simple algorithms or software-modules under real-time conditions. These algorithms contain free adjustable parameters, which have to be adapted to the static and dynamic behavior of the process. In contrast to manual tuning by trial and error, the use of mathematical models allows precise and fast automatic adaptation.

The mathematical models can be obtained by identification and parameter estimation, which use the measured and sampled input and output signals. These methods are not restricted to linear models, but also allow for several classes of nonlinear systems. If the parameter estimation methods are combined with appropriate control algorithm design methods, adaptive control systems result. They can be used for permanent precise controller tuning or only for commissioning [20].

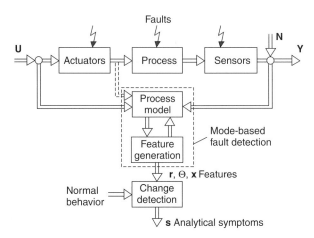

FIGURE 2.6 Scheme for a model-based fault detection.

2.4.4 Supervision and Fault Detection

With an increasing number of automatic functions (autonomy), including electronic components, sensors and actuators, increasing complexity, and increasing demands on reliability and safety, an integrated supervision with fault diagnosis becomes more and more important. This is a significant natural feature of an intelligent mechatronic system. Figure 2.6 shows a process influenced by faults. These faults indicate unpermitted deviations from normal states and can be generated either externally or internally. External faults can be caused by the power supply, contamination, or collision, internal faults by wear, missing lubrication, or actuator or sensor faults. The classical way for fault detection is the limit value checking of some few measurable variables. However, incipient and intermittant faults can not usually be detected, and an in-depth fault diagnosis is not possible by this simple approach. *Model-based fault detection* and *diagnosis methods* were developed in recent years, allowing for early detection of small faults with normally measured signals, also in closed loops [21]. Based on measured input signals, $U(t)$, and output signals, $Y(t)$, and process models, features are generated by parameter estimation, state and output observers, and parity equations, as seen in Figure 2.6.

These residuals are then compared with the residuals for normal behavior and with change detection methods analytical symptoms are obtained. Then, a fault diagnosis is performed via methods of classification or reasoning. For further details see References 22 and 23.

A considerable advantage is if the same process model can be used for both the (adaptive) *controller design and the fault detection*. In general, continuous time models are preferred if fault detection is based on parameter estimation or parity equations. For fault detection with state estimation or parity equations, discrete-time models can be used.

Advanced supervision and fault diagnosis is a basis for improving reliability and safety, state dependent maintenance, triggering of redundancies, and reconfiguration.

2.4.5 Intelligent Systems (Basic Tasks)

The information processing within mechatronic systems may range between simple control functions and intelligent control. Various definitions of intelligent control systems do exist, see References 24–30. An intelligent control system may be organized as an *online expert system*, according to Figure 2.5, and comprises

- Multi-control functions (executive functions)
- A knowledge base
- Inference mechanisms and
- Communication interfaces

The online *control functions* are usually organized in multilevels, as already described. The *knowledge base* contains quantitative and qualitative knowledge. The quantitative part operates with analytic (mathematical) process models, parameter and state estimation methods, analytic design methods (e.g., for control and fault detection), and quantitative optimization methods. Similar modules hold for the qualitative knowledge (e.g., in the form of rules for fuzzy and soft computing). Further knowledge is the past history in the memory and the possibility to predict the behavior. Finally, tasks or schedules may be included.

The *inference mechanism* draws conclusions either by quantitative reasoning (e.g., Boolean methods) or by qualitative reasoning (e.g., possibilistic methods) and takes decisions for the executive functions.

Communication between the different modules, an information management database, and the man–machine interaction has to be organized.

Based on these functions of an online expert system, an intelligent system can be built up, with the ability "to model, reason and learn the process and its automatic functions within a given frame and to govern it towards a certain goal." Hence, intelligent mechatronic systems can be developed, ranging from "low-degree intelligent" [13], such as intelligent actuators, to "fairly intelligent systems," such as self-navigating automatic guided vehicles.

An *intelligent mechatronic system* adapts the controller to the mostly nonlinear behavior (adaptation), and stores its controller parameters in dependence on the position and load (learning), supervises all relevant elements, and performs a fault diagnosis (supervision) to request maintenance or, if a failure occurs, to request a fail safe action (decisions on actions). In the case of multiple components, supervision may help to switch off the faulty component and to perform a reconfiguration of the controlled process.

2.5 Concurrent Design Procedure for Mechatronic Systems

The design of mechatronic systems requires a systematic development and use of modern design tools.

2.5.1 Design Steps

Table 2.3 shows five important development steps for mechatronic systems, starting from a purely mechanical system and resulting in a fully integrated mechatronic system. Depending on the kind of mechanical system, the intensity of the single development steps is different. For precision mechanical devices, fairly integrated mechatronic systems do exist. The influence of the electronics on *mechanical elements* may be considerable, as shown by adaptive dampers, anti-lock system brakes, and automatic gears. However, complete *machines* and *vehicles* show first a mechatronic design of their elements, and then slowly a redesign of parts of the overall structure as can be observed in the development of machine tools, robots, and vehicle bodies.

2.5.2 Required CAD/CAE Tools

The computer aided development of mechatronic systems comprises:

1. Constructive specification in the engineering development stage using CAD and CAE tools
2. Model building for obtaining static and dynamic process models
3. Transformation into computer codes for system simulation and
4. Programming and implementation of the final mechatronic software

Some software tools are described in [31]. A broad range of CAD/CAE tools is available for 2D- and 3D-mechanical design, such as Auto CAD with a direct link to CAM (computer-aided manufacturing), and PADS, for multilayer, printed-circuit board layout. However, the state of computer-aided modeling is not as advanced. Object-oriented languages such as DYMOLA and MOBILE for modeling of large combined systems are described in [31–33]. These packages are based on specified ordinary differential

TABLE 2.3 Steps in the Design of Mechatronic Systems

	Precision Mechanics	Mechanical Elements	Machines
Pure mechanical system	●	●	●
1. Addition of sensors, actuators, microelectronics, control functions	○	○	○
2. Integration of components (hardware integration)	○	○	○
3. Integration by information processing (software integration)	○	○	○
4. Redesign of mechanical system	○	○	○
5. Creation of synergetic effects	○	○	○
Fully integrated mechatronic systems	●	●	○
Examples	Sensors actuators disc-storages cameras	Suspensions dampers clutches gears brakes	Electric drives combustion engines mach. tools robots

The size of a circle indicates the present intensity of the respective mechatronic development step: ● large, ○ medium, ○ little.

equations, algebraic equations, and discontinuities. A recent description of the state of computer-aided control system design can be found in Reference 34. For system simulation (and controller design), a variety of program systems exist, like ACSL, SIMPACK, MATLAB/SIMULINK, and MATRIX-X. These simulation techniques are valuable tools for design, as they allow the designer to study the interaction of components and the variations of design parameters before manufacturing. They are, in general, not suitable for real-time simulation.

2.5.3 Modeling Procedure

Mathematical process models for static and dynamic behavior are required for various steps in the design of mechatronic systems, such as simulation, control design, and reconstruction of variables. Two ways to obtain these models are *theoretical modeling* based on first (physical) principles and *experimental modeling* (*identification*) with measured input and output variables. A basic problem of theoretical modeling of mechatronic systems is that the components originate from different domains. There exists a well-developed domain specific knowledge for the modeling of electrical circuits, multibody mechanical systems, or hydraulic systems, and corresponding software packages. However, a computer-assisted general methodology for the modeling and simulation of components from different domains is still missing [35].

The basic principles of theoretical modeling for system with energy flow are known and can be unified for components from different domains as electrical, mechanical, and thermal (see References 36–41). The modeling methodology becomes more involved if material flows are incorporated as for fluidics, thermodynamics, and chemical processes.

A general procedure for theoretical modeling of lumped parameter processes can be sketched as follows [19]:

1. Definition of flows
   - Energy flow (electrical, mechanical, thermal conductance)
   - Energy and material flow (fluidic, thermal transfer, thermodynamic, chemical)
2. Definition of process elements: flow diagrams
   - Sources, sinks (dissipative)
   - Storages, transformers, converters
3. Graphical representation of the process model
   - Multi-port diagrams (terminals, flows, and potentials, or across and through variables)
   - Block diagrams for signal flow
   - Bond graphs for energy flow
4. Statement of equations for all process elements
   - Balance equations for storage (mass, energy, momentum)
   - Constitutive equations for process elements (sources, transformers, converters)
   - Phenomenological laws for irreversible processes (dissipative systems: sinks)
5. Interconnection equations for the process elements
   - Continuity equations for parallel connections (node law)
   - Compatibility equations for serial connections (closed circuit law)
6. Overall process model calculation
   - Establishment of input and output variables
   - State space representation
   - Input/output models (differential equations, transfer functions)

An example of steps 1–3 is shown in Figure 2.7 for a drive-by-wire vehicle. A unified approach for processes with energy flow is known for electrical, mechanical, and hydraulic processes with incompressible fluids. Table 2.4 defines generalized through and across variables.

In these cases, the product of the through and across variable is power. This unification enabled the formulation of the standard *bond graph modeling* [39]. Also, for hydraulic processes with compressible fluids and thermal processes, these variables can be defined to result in powers, as seen in Table 2.4. However, using mass flows and heat flows is not engineering practice. If these variables are used, so-called pseudo bond graphs with special laws result, leaving the simplicity of standard bond graphs. Bond graphs lead to a high-level abstraction, have less flexibility, and need additional effort to generate simulation algorithms. Therefore, they are not the ideal tool for mechatronic systems [35]. Also, the tedious work needed to establish *block diagrams* with an early definition of causal input/output blocks is not suitable.

Development towards object-oriented modeling is on the way, where objects with terminals (cuts) are defined without assuming a causality in this basic state. Then, object diagrams are graphically represented, retaining an intuitive understanding of the original physical components [43,44]. Hence, theoretical modeling of mechatronic systems with a unified, transparent, and flexible procedure (from the basic components of different domains to simulation) are a challenge for further development. Many components show nonlinear behavior and nonlinearities (friction and backlash). For more complex process parts, multidimensional mappings (e.g., combustion engines, tire behavior) must be integrated.

For verification of theoretical models, several well-known identification methods can be used, such as correlation analysis and frequency response measurement, or Fourier- and spectral analysis. Since some parameters are unknown or changed with time, parameter estimation methods can be applied, both, for models with continuous time or discrete time (especially if the models are linear in the parameters) [42,45,46]. For the identification and approximation of nonlinear, multi-dimensional

TABLE 2.4 Generalized through and across Variables for Processes with Energy Flow

System	Through Variables		Across Variables	
Electrical	Electric current	I	Electric voltage	U
Magnetic	Magnetic flow	Φ	Magnetic force	Θ
Mechanical				
• Translation	Force	F	Velocity	w
• Rotation	Torque	M	Rotational speed	ω
Hydraulic	Volume flow	$\dot{V}$	Pressure	p
Thermodynamic	Entropy flow		Temperature	T

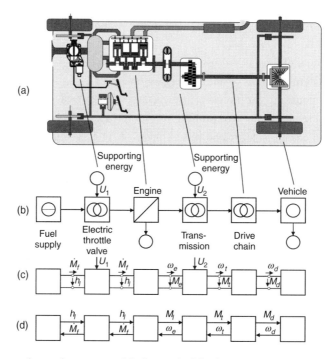

FIGURE 2.7 Different schemes for an automobile (as required for drive-by-wire-longitudinal control): (a) scheme of the components (construction map), (b) energy flow diagram (simplified), (c) multi-port diagram with flows and potentials, (d) signal flow diagram for multi-ports.

characteristics, artificial neural networks (multilayer perceptrons or radial-basis-functions) can be expanded for nonlinear dynamic processes [47].

2.5.4 Real-Time Simulation

Increasingly, real-time simulation is applied to the design of mechatronic systems. This is especially true if the process, the hardware, and the software are developed simultaneously in order to minimize iterative development cycles and to meet short time-to-market schedules. With regard to the required speed of computation *simulation methods*, it can be subdivided into

1. Simulation without (hard) time limitation
2. Real-time simulation
3. Simulation faster than real-time

Some application examples are given in Figure 2.8. Herewith, *real-time simulation* means that the simulation of a component is performed such that the input and output signals show the same time-dependent

Mechatronic Design Approach

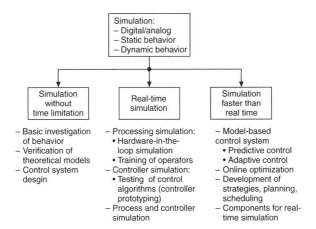

FIGURE 2.8 Classification of simulation methods with regard to speed and application examples.

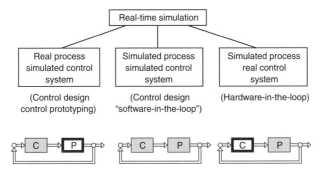

FIGURE 2.9 Classification of real-time simulation.

values as the real, dynamically operating component. This becomes a computational problem for processes which have fast dynamics compared to the required algorithms and calculation speed.

Different kinds of real-time simulation methods are shown in Figure 2.9. The reason for the real-time requirement is mostly that one part of the investigated system is not simulated but real. Three cases can be distinguished:

1. The *real process* can be operated together with the *simulated control* by using hardware other than the final hardware. This is also called "control prototyping."
2. The *simulated process* can be operated with the *real control hardware*, which is called "hardware-in-the-loop simulation."
3. The *simulated process* is run with the *simulated control* in real time. This may be required if the final hardware is not available or if a design step before the hardware-in-the-loop simulation is considered.

2.5.5 Hardware-in-the-Loop Simulation

The *hardware-in-the-loop* (HIL) simulation is characterized by operating real components in connection with real-time simulated components. Usually, the control system hardware and software is the real system, as used for series production. The controlled process (consisting of actuators, physical processes, and sensors) can either comprise simulated components or real components, as seen in Figure 2.10a. In general, mixtures of the shown cases are realized. Frequently, some actuators are real and the process

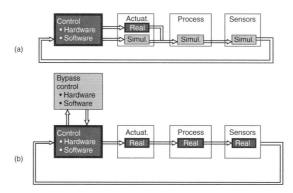

FIGURE 2.10 Real-time simulation: hybrid structures. (a) Hardware-in-the-loop simulation and (b) Control prototyping.

and the sensors are simulated. The reason is that actuators and the control hardware very often form one integrated subsystem or that actuators are difficult to model precisely and to simulate in real time. (The use of real sensors together with a simulated process may require considerable realization efforts, because the physical sensor input does not exist and must be generated artificially.) In order to change or redesign some functions of the control hardware or software, a bypass unit can be connected to the basic control hardware. Hence, hardware-in-the-loop simulators may also contain partially simulated (emulated) control functions.

The advantages of the HIL simulation are generally:

- Design and testing of the control hardware and software without operating a real process ("moving the process field into the laboratory")
- Testing of the control hardware and software under extreme environmental conditions in the laboratory (e.g., high/low temperature, high accelerations and mechanical shocks, aggressive media, electro-magnetic compatibility)
- Testing of the effects of faults and failures of actuators, sensors, and computers on the overall system
- Operating and testing of extreme and dangerous operating conditions
- Reproducible experiments, frequently repeatable
- Easy operation with different man–machine interfaces (cockpit-design and training of operators)
- Saving of cost and development time

2.5.6 Control Prototyping

For the design and testing of complex control systems and their algorithms under real-time constraints, a real-time controller simulation (emulation) with hardware (e.g., off-the-shelf signal processor) other than the final series production hardware (e.g., special ASICS) may be performed. The process, the actuators, and sensors can then be real. This is called *control prototyping* (Figure 2.10b). However, parts of the process or actuators may be simulated, resulting in a mixture of HIL-simulation and control prototyping. The advantages are mainly:

- Early development of signal processing methods, process models, and control system structure, including algorithms with high level software and high performance off-the-shelf hardware
- Testing of signal processing and control systems, together with other design of actuators, process parts, and sensor technology, in order to create synergetic effects

- Reduction of models and algorithms to meet the requirements of cheaper mass production hardware
- Defining the specifications for final hardware and software

Some of the advantages of HIL-simulation also hold for control prototyping. Some references for real-time simulation are 48 and 49.

References

1. Kyura, N. and Oho, H., Mechatronics—an industrial perspective. *IEEE/ASME Transactions on Mechatronics*, 1(1):10–15.
2. Schweitzer, G., Mechatronik-Aufgaben und Lösungen. VDI-Berichte Nr. 787. VDI-Verlag, Düsseldorf, 1989.
3. Ovaska, S. J., Electronics and information technology in high range elevator systems. *Mechatronics*, 2(1):89–99, 1992.
4. *IEEE/ASME Transactions on Mechatronics*, 1996.
5. Harashima, F., Tomizuka, M., and Fukuda, T., Mechatronics—"what is it, why and how?" An editorial. *IEEE/ASME Transactions on Mechatronics*, 1(1):1–4, 1996.
6. Schweitzer, G., Mechatronics—a concept with examples in active magnetic bearings. *Mechatronics*, 2(1):65–74, 1992.
7. Gausemeier, J., Brexel, D., Frank, Th., and Humpert, A., Integrated product development. In *Third Conf. Mechatronics and Robotics*, Paderborn, Germany, Oct. 4–6, 1995. Teubner, Stuttgart, 1995.
8. Isermann, R., Modeling and design methodology for mechatronic systems. *IEEE/ASME Transactions on Mechatronics*, 1(1):16–28, 1996.
9. *Mechatronics: An International Journal. Aims and Scope.* Pergamon Press, Oxford, 1991.
10. *Mechatronics Systems Engineering: International Journal on Design and Application of Integrated Electromechanical Systems.* Kluwer Academic Publishers, Nethol, 1993.
11. IEE, Mechatronics: Designing intelligent machines. In *Proc. IEE-Int. Conf.*, 12–13 Sept., Univ. of Cambridge, 1990.
12. Hiller, M. (ed.), *Second Conf. Mechatronics and Robotics.* Sept. 27–29, Duisburg/Moers, Germany, Moers, IMECH, 1993.
13. Isermann, R. (ed.), Integrierte mechanisch elektroni-sche Systeme. Mar. 2–3, Darmstadt, Germany, 1993. Fortschr.-Ber. VDI Reihe 12 Nr. 179. VDI-Verlag, Düsseldorf, 1993.
14. Lückel, J. (ed.), *Third Conf. Mechatronics and Robotics*, Paderborn, Germany, Oct. 4–6, Teubner, Stuttgart, 1995.
15. Kaynak, O., Özkan, M., Bekiroglu, N., and Tunay, I. (eds.), Recent advances in mechatronics. In *Proc. Int. Conf. Recent Advances in Mechatronics*, Aug. 14–16, 1995, Istanbul, Turkey.
16. Kitaura, K., *Industrial Mechatronics.* New East Business Ltd., in Japanese, 1991.
17. Bradley, D. A., Dawson, D., Burd, D., and Loader, A. J., *Mechatronics-Electronics in Products and Processes.* Chapman & Hall, London, 1991.
18. McConaill, P. A., Drews, P., and Robrock, K. H., *Mechatronics and Robotics I.* IOS-Press, Amsterdam, 1991.
19. Isermann, R., *Mechatronische Systeme.* Springer, Berlin, 1999.
20. Isermann, R., Lachmann, K. H., and Matko, D., *Adaptive Control Systems*, Prentice-Hall, London, 1992.
21. Isermann, R., Supervision, fault detection and fault diagnosis methods—advanced methods and applications. In *Proc. XIV IMEKO World Congress*, Vol. 1, pp. 1–28, Tampere, Finland, 1997.
22. Isermann, R., Supervision, fault detection and fault diagnosis methods—an introduction, special section on supervision, fault detection and diagnosis. *Control Engineering Practice*, 5(5):639–652, 1997.
23. Isermann, R. (ed.), Special section on supervision, fault detection and diagnosis. *Control Engineering Practice*, 5(5):1997.

24. Saridis, G. N., *Self Organizing Control of Stochastic Systems*. Marcel Dekker, New York, 1977.
25. Saridis, G. N. and Valavanis, K. P., Analytical design of intelligent machines. *Automatica*, 24:123–133, 1988.
26. Åström, K. J., Intelligent control. In *Proc. European Control Conf.*, Grenoble, 1991.
27. White, D. A. and Sofge, D. A. (eds.), *Handbook of Intelligent Control*. Van Norstrad, Reinhold, New York, 1992.
28. Antaklis, P., Defining intelligent control. *IEEE Control Systems*, Vol. June: 4–66, 1994.
29. Gupta, M. M. and Sinha, N. K., *Intelligent Control Systems*. IEEE-Press, New York, 1996.
30. Harris, C. J. (ed.), *Advances in Intelligent Control*. Taylor & Francis, London, 1994.
31. Otter, M. and Gruebel, G., Direct physical modeling and automatic code generation for mechatronics simulation. In *Proc. 2nd Conf. Mechatronics and Robotics*, Duisburg, Sept. 27–29, IMECH, Moers, 1993.
32. Elmquist, H., Object-oriented modeling and automatic formula manipulation in Dymola, Scandin. Simul. Society SIMS, June, Kongsberg, 1993.
33. Hiller, M., Modelling, simulation and control design for large and heavy manipulators. In *Proc. Int. Conf. Recent Advances in Mechatronics* Vol./pp. 78–85. Istanbul, Turkey, 1995.
34. James, J., Cellier, F., Pang, G., Gray, J., and Mattson, S. E., The state of computer-aided control system design (CACSD). *IEEE Transactions on Control Systems*, Special Issue, April 6–7, 1995.
35. Otter, M. and Elmqvist, H., Energy flow modeling of mechatronic systems via object diagrams. In *Proc. 2nd MATHMOD*, Vienna, pp. 705–710, 1997.
36. Paynter, H. M., *Analysis and Design of Engineering Systems*. MIT Press, Cambridge, 1961.
37. MacFarlane, A. G. J., *Engineering Systems Analysis*. G. G. Harrop, Cambridge, 1964.
38. Wellstead, P. E., *Introduction to Physical System Modelling*. Academic Press, London, 1979.
39. Karnopp, D. C., Margolis, D. L., and Rosenberg, R. C., *System Dynamics. A Unified Approach*. J. Wiley, New York, 1990.
40. Cellier, F. E., *Continuous System Modelling*. Springer, Berlin, 1991.
41. Gawtrop, F. E. and Smith, L., *Metamodelling: Bond Graphs and Dynamic Systems*. Prentice-Hall, London, 1996.
42. Eykhoff, P., *System Identification*. John Wiley & Sons, London, 1974.
43. Elmqvist, H., A structured model language for large continuous systems. Ph.D. Dissertation, Report CODEN: LUTFD2/(TFRT-1015) Dept. of Aut. Control, Lund Institute of Technology, Sweden, 1978.
44. Elmqvist, H. and Mattson, S. E., Simulator for dynamical systems using graphics and equations for modeling. *IEEE Control Systems Magazine*, 9(1):53–58, 1989.
45. Isermann, R., *Identifikation dynamischer Systeme*. 2nd Ed., Vol. 1 and 2. Springer, Berlin, 1992.
46. Ljung, L., *System Identification: Theory for the User*. Prentice-Hall, Englewood Cliffs, NJ, 1987.
47. Isermann, R., Ernst, S., and Nelles, O., Identification with dynamic neural networks—architectures, comparisons, applications—Plenary. In *Proc. IFAC Symp. System Identification (SYSID'97)*, Vol. 3, pp. 997–1022, Fukuoka, Japan, 1997.
48. Hanselmann, H., Hardware-in-the-loop simulation as a standard approach for development, customization, and production test, SAE 930207, 1993.
49. Isermann, R., Schaffnit, J., and Sinsel, S., Hardware-in-the-loop simulation for the design and testing of engine control systems. *Control Engineering Practice*, 7(7):643–653, 1999.

3
System Interfacing, Instrumentation, and Control Systems

Rick Homkes
Purdue University

3.1 Introduction 3-1
    The Mechatronic System • A Home/Office Example
    • An Automotive Example
3.2 Input Signals of a Mechatronic System 3-3
    Transducer/Sensor Input • Analog-to-Digital
    Converters
3.3 Output Signals of a Mechatronic System 3-5
    Digital-to-Analog Converters • Actuator Output
3.4 Signal Conditioning 3-6
    Sampling Rate • Filtering • Data Acquisition Boards
3.5 Microprocessor Control 3-8
    PID Control • Programmable Logic
    Controllers • Microprocessors
3.6 Microprocessor Numerical Control 3-8
    Fixed-Point Mathematics • Calibrations
3.7 Microprocessor Input–Output Control 3-9
    Polling and Interrupts • Input and Output
    Transmission • HC12 Microcontroller Input–Output
    Subsystems • Microcontroller Network Systems
3.8 Software Control 3-11
    Systems Engineering • Software Engineering
    • Software Design
3.9 Testing and Instrumentation 3-12
    Verification and Validation • Debuggers
    • Logic Analyzer
3.10 Summary 3-13

3.1 Introduction

The purpose of this chapter is to introduce a number of topics dealing with a mechatronic system. This starts with an overview of mechatronic systems and a look at the input and output signals of a mechatronic system. The special features of microprocessor input and output are next. Software, an often-neglected portion of a mechatronic system, is briefly covered with an emphasis on software engineering concepts. The chapter concludes with a short discussion of testing and instrumentation.

3.1.1 The Mechatronic System

Figure 3.1 shows a typical mechatronic system with mechanical, electrical, and computer components. The process of system data acquisition begins with the measurement of a physical value by a sensor. The sensor is able to generate some form of signal, generally an analog signal in the form of a voltage level or waveform. This analog signal is sent to an analog-to-digital converter (ADC). Commonly using a process of successive approximation, the ADC maps the analog input signal to a digital output. This digital value is composed of a set of binary values called bits (often represented by 0s and 1s). The set of bits represents a decimal or hexadecimal number that can be used by the microcontroller. The microcontroller consists of a microprocessor plus memory and other attached devices. The program in the microprocessor uses this digital value along with other inputs and preloaded values called calibrations to determine output commands. Like the input to the microprocessor, these outputs are in digital form and can be represented by a set of bits. A digital-to-analog converter (DAC) is then often used to convert the digital value into an analog signal. The analog signal is used by an actuator to control a physical device or affect the physical environment. The sensor then takes new measurements and the process repeated, thus completing a feedback control loop. Timing for this entire operation is synchronized by the use of a clock.

3.1.2 A Home/Office Example

An example of a mechatronic system is the common heating/cooling system for homes and offices. Simple systems use a bimetal thermostat with contact points controlling a mercury switch that turns on and off the furnace or air conditioner. A modern environmental control system uses these same basic components along with other components and computer program control. A temperature sensor monitors the physical environment and produces a voltage level as demonstrated in Figure 3.2 (though generally not nearly such a smooth function). After conversion by the ADC, the microcontroller uses the digitized temperature

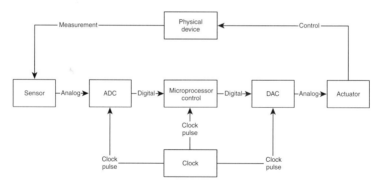

FIGURE 3.1 Microprocessor control system.

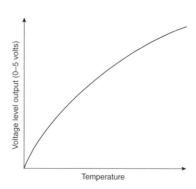

FIGURE 3.2 Voltage levels.

data along with a 24-hour clock and the user requested temperatures to produce a digital control signal. This signal directs the actuator, usually a simple electrical switch in this example. The switch, in turn, controls a motor to turn the heating or cooling unit on or off. New measurements are then taken and the cycle is repeated. While not a mechatronic product on the order of a camcorder, it is a mechatronic system because of its combination of mechanical, electrical, and computer components. This system may also incorporate some additional features. If the temperature being sensed is quite high, say 80°C, it is possible that a fire exists. It is then not a good idea to turn on the blower fan and feed the fire more oxygen. Instead the system should set off an alarm or use a data communication device to alert the fire department. Because of this type of computer control, the system is "smart," at least relative to the older mercury-switch controlled systems.

3.1.3 An Automotive Example

A second example is the antilock braking system (ABS) found in many vehicles. The entire purpose of this type of system is to prevent a wheel from locking up and thus having the driver loose directional control of the vehicle due to skidding. In this case, sensors attached to each wheel determine the rotational speed of the wheels. These data, probably in a waveform or time-varied electrical voltage, is sent to the microcontroller along with the data from sensors reporting inputs such as brake pedal position, vehicle speed, and yaw. After conversion by the ADC or input capture routine into a digital value, the program in the microprocessor then determines the necessary action. This is where the aspect of human computer interface (HCI) or human machine interface (HMI) comes into play by taking account of the "feel" of the system to the user. System calibration can adjust the response to the driver while, of course, stopping the vehicle by controlling the brakes with the actuators. There are two important things to note in this example. The first is that, in the end, the vehicle is being stopped because of hydraulic forces pressing the brake pad against a drum or rotor—a purely mechanical function. The other is that the ABS, while an "intelligent product," is not a stand-alone device. It is part of a larger system, the vehicle, with multiple microcontrollers working together through the data network of the vehicle.

3.2 Input Signals of a Mechatronic System

3.2.1 Transducer/Sensor Input

All inputs to mechatronic systems come from either some form of sensory apparatus or communications from other systems. Sensors were first introduced in the previous section and will be discussed in much more depth in Chapter 19. Transducers, devices that convert energy from one form to another, are often used synonymously with sensors. Transducers and their properties will be explained fully in Chapter 45. Sensors can be divided into two general classifications, active or passive. Active sensors emit a signal in order to estimate an attribute of the environment or device being measured. Passive sensors do not. A military example of this difference would be a strike aircraft "painting" a target using either active laser radar (LADAR) or a passive forward looking infrared (FLIR) sensor.

As stated in the Introduction section, the output of a sensor is usually an analog signal. The simplest type of analog signal is a voltage level with a direct (though not necessarily linear) correlation to the input condition. A second type is a pulse width modulated (PWM) signal, which will be explained further in a later section of this chapter when discussing microcontroller outputs. A third type is a waveform, as shown in Figure 3.3. This type of signal is modulated either in its amplitude (Figure 3.4) or its frequency (Figure 3.5) or, in some cases, both. These changes reflect the changes in the condition being monitored.

There are sensors that do not produce an analog signal. Some of these sensors produce a square wave as in Figure 3.6 that is input to the microcontroller using the EIA 232 communications standard. The square wave represents the binary values of 0 and 1. In this case the ADC is probably on-board the sensor itself, adding to the cost of the sensor. Some sensors/recorders can even create mail or TCP/IP packets as output. An example of this type of unit is the MV100 MobileCorder from Yokogawa Corporation of America.

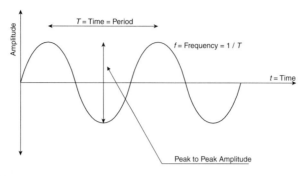

FIGURE 3.3 Sine wave.

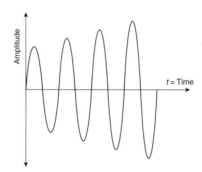

FIGURE 3.4 Amplitude modulation.

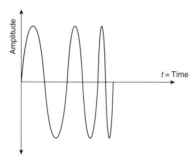

FIGURE 3.5 Frequency modulation.

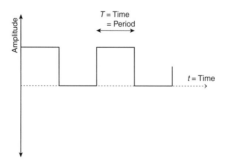

FIGURE 3.6 Square wave.

3.2.2 Analog-to-Digital Converters

The ADC can basically be typed by two parameters: the analog input range and the digital output range. As an example, consider an ADC that is converting a voltage level ranging 0–12 V into a single byte of 8 bits. In this example, each binary count increment reflects an increase in analog voltage of 1/256 of the maximum 12 V. There is an unusual twist to this conversion, however. Since a zero value represents 0 V, and a 128 value represents half of the maximum value, 6 V in this example, the maximum decimal value of 255 represents 255/256 of the maximum voltage value, or 11.953125 V. A table of the equivalent values is shown below:

Binary	Decimal	Voltage
0000 0000	0	0.0
0000 0001	1	0.00390625
1000 0000	128	6.0
1111 1111	255	11.953125

An ADC that is implemented in the Motorola HC12 microcontroller produces 10 bits. While not fitting so nicely into a single byte of data, this 10-bit ADC does give additional resolution. Using an input range from 0 to 5 V, the decimal resolution per least significant bit is 4.88 mV. If the ADC had 8 bits of output, the resolution per bit would be 19.5 mV, a fourfold difference. Larger voltages, for example, from 0 to 12 V, can be scaled with a voltage divider to fit the 0–5 V range. Smaller voltages can be amplified to span the entire range. A process known as successive approximation (using the Successive Approximation Register or SAR in the Motorola chip) is used to determine the correct digital value.

3.3 Output Signals of a Mechatronic System

3.3.1 Digital-to-Analog Converters

The output command from the microcontroller is a binary value in bit, byte (8 bits), or word (16 bits) form. This digital signal is converted to analog using a digital-to-analog converter, or DAC. Let us examine converting an 8-bit value into a voltage level between 0 and 12 V. The most significant bit in the binary value to be converted (decimal 128) creates an analog value equal to half of the maximum output, or 6 V. The next digit produces an additional one fourth, or 3 V, the next an additional one eighth, and so forth. The sum of all these weighted output values represents the appropriate analog voltage. As was mentioned in a previous section, the maximum voltage value in the range is not obtainable, as the largest value generated is 255/256 of 12 V, or 11.953125 V. The smoothness of the signal representation depends on the number of bits accepted by the DAC and the range of the output required. Figure 3.7 demonstrates a simplified step function using a one-byte binary input and 12-V analog output.

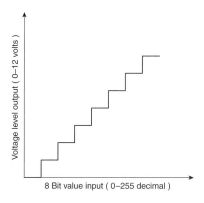

FIGURE 3.7 DAC stepped output.

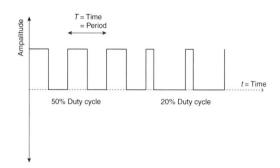

FIGURE 3.8 Pulse width modulation.

3.3.2 Actuator Output

Like sensors, actuators were first introduced in a previous section and will be described in detail in a later chapter of this handbook. The three common actuators that this section will review are switches, solenoids, and motors. Switches are simple state devices that control some activity, like turning on and off the furnace in a house. Types of switches include relays and solid-state devices. Solid-state devices include diodes, thyristors, bipolar transistors, field-effect transistors (FETs), and metal-oxide field-effect transistors (MOSFETs). A switch can also be used with a sensor, thus turning on or off the entire sensor, or a particular feature of a sensor.

Solenoids are devices containing a movable iron core that is activated by a current flow. The movement of this core can then control some form of hydraulic or pneumatic flow. Applications are many, including braking systems and industrial production of fluids. More information on solenoid actuators can be found in a later chapter. Motors are the last type of actuator that will be summarized here. There are three main types: direct current (DC), alternating current (AC), and stepper motors. DC motors may be controlled by a fixed DC voltage or by pulse width modulation (PWM). In a PWM signal, such as shown in Figure 3.8, a voltage is alternately turned on and off while changing (modulating) the width of the on-time signal, or duty cycle. AC motors are generally cheaper than DC motors, but require variable frequency drive to control the rotational speed. Stepper motors move by rotating a certain number of degrees in response to an input pulse.

3.4 Signal Conditioning

Signal conditioning is the modification of a signal to make it more useful to a system. Two important types of signal conditioning are, of course, the conversion between analog and digital, as described in the previous two sections. Other types of signal conditioning are briefly covered below, with a full coverage reserved for Chapters 46 and 47.

3.4.1 Sampling Rate

The rate at which data samples are taken obviously affects the speed at which the mechatronic system can detect a change in situation. There are several things to consider, however. For example, the response of a sensor may be limited in time or range. There is also the time required to convert the signal into a form usable by the microprocessor, the A to D conversion time. A third is the frequency of the signal being sampled. For voice digitalization, there is a very well-known sampling rate of 8000 samples per second. This is a result of the Nyquist theorem, which states that the sampling rate, to be accurate, must be at least twice the maximum frequency being measured. The 8000 samples per second rate thus works well for converting human voice over an analog telephone system where the highest frequency is approximately 3400 Hz. Lastly, the clock speed of the microprocessor must also be considered. If the ADC and DAC are

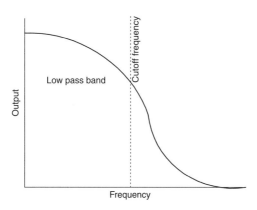

FIGURE 3.9 Low-pass filter.

on the same board as the microprocessor, they will often share a common clock. The microprocessor clock, however, may be too fast for the ADC and DAC. In this case, a prescaler is used to divide the clock frequency to a level usable by the ADC and DAC.

3.4.2 Filtering

Filtering is the attenuation (lessening) of certain frequencies from a signal. This process can remove noise from a signal and condition the line for better data transmission. Filters can be divided into analog and digital types, the analog filters being further divided into passive and active types. Analog passive filters use resistors, capacitors, and inductors. Analog active filters typically use operational amplifiers with resistors and capacitors. Digital filters may be implemented with software and/or hardware. The software component gives digital filters the feature of being easier to change. Digital filters are explained fully in Chapter 29.

Filters may also be differentiated by the type of frequencies they affect.

1. Low-pass filters allow lower set of frequencies to pass through, while high frequencies are attenuated. A simplistic example of this is shown in Figure 3.9.
2. High-pass filters, the opposite of low-pass, filter a lower frequency band while allowing higher frequencies to pass.
3. Band-pass filters allow a particular range of frequencies to pass; all others are attenuated.
4. Band-stop filters stop a particular range of frequencies while all others are allowed to pass.

There are many types and applications of filters. For example, William Ribbens in his book *Understanding Automotive Electronics* (Newnes 1998) described a software low-pass filter (sometimes also called a lag filter) that averages the last 60 fuel tank level samples taken at 1 s intervals. The filtered data are then displayed on the vehicle instrument cluster. This type of filtering reduces large and quick fluctuations in the fuel gauge due to sloshing in the tank, and thus displays a more accurate value.

3.4.3 Data Acquisition Boards

There is a special type of board that plugs into a slot in a desktop personal computer that can be used for many of the tasks above. It is called a data acquisition board, or DAQ board. This type of board can generate analog input and multiplex multiple input signals onto a single bus for transmission to the PC. It can also come with signal conditioning hardware/software and an ADC. Some units have direct memory access (DMA), where the device writes the data directly into computer memory without using the microprocessor. While desktop PCs are not usually considered as part of a mechatronic system, the DAQ board can be very useful for instrumentation.

3.5 Microprocessor Control

3.5.1 PID Control

A closed loop control system is one that determines a difference in the desired and actual condition (the error) and creates a correction control command to remove this error. PID control demonstrates three ways of looking at this error and correcting it. The first way is the P of PID, the proportional term. This term represents the control action made by the microcontroller in proportion to the error. In other words, the bigger the error, the bigger the correction. The I in PID is for the integral of the error over time. The integral term produces a correction that considers the time the error has been present. Stated in other words, the longer the error continues, the bigger the correction. Lastly, the D in PID stands for derivative. In the derivative term, the corrective action is related to the derivative or change of the error with respect to time. Stated in other words, the faster the error is changing, the bigger the correction. Control systems can use P, PI, PD, or PID in creating corrective actions. The problem generally is "tuning" the system by selecting the proper values in the terms. For more information on control design, see Chapter 31.

3.5.2 Programmable Logic Controllers

Any discussion of control systems and microprocessor control should start with the first type of "mechatronic" control, the programmable logic controller or PLC. A PLC is a simpler, more rugged microcontroller designed for environments like a factory floor. Input is usually from switches such as push buttons controlled by machine operators or position sensors. Timers can also be programmed in the PLC to run a particular process for a set amount of time. Outputs include lamps, solenoid valves, and motors, with the input–output interfacing done within the controller. A simple programming language used with a PLC is called ladder logic or ladder programming. Ladder logic is a graphical language showing logic as a combination of series (and's) and parallel (or's) blocks. Additional information can be found in Chapter 43 and in the book *Programmable Logic Controllers* by W. Bolton (Newnes 1996).

3.5.3 Microprocessors

A full explanation of a microprocessor is found in section 5.8. For this discussion of microprocessors and control, we need only know a few of the component parts of computer architecture. RAM, or random access memory, is the set of memory locations the computer uses for fast temporary storage. The radio station presets selected by the driver (or passenger) in the car radio are stored in RAM. A small electrical current maintains these stored frequencies, so disconnection of the radio from the battery will result in their loss. ROM, or read only memory, is the static memory that contains the program to run the microcontroller. Thus the radio's embedded program will not be lost when the battery is disconnected. There are several types of ROM, including erasable programmable ROM (EPROM), electrically erasable programmable ROM (EEPROM), and flash memory (a newer type of EEPROM). These types will be explained later in this handbook. There are also special memory areas in a microprocessor called registers. Registers are very fast memory locations that temporarily store the address of the program instruction being executed, intermediate values needed to complete a calculation, data needed for comparison, and data that need to be input or output. Addresses and data are moved from one point to another in RAM, ROM, and registers using a bus, a set of lines transmitting data multiple bits simultaneously.

3.6 Microprocessor Numerical Control

3.6.1 Fixed-Point Mathematics

The microprocessors in an embedded controller are generally quite small in comparison to a personal computer or computer workstation. Adding processing power in the form of a floating-point processor and additional RAM or ROM is not always an option. This means that sometimes the complex mathematical functions needed in a control system are not available. However, sometimes the values being sensed and computed, though real numbers, are of a reasonable range. Because of this situation there exists a special type of arithmetic whereby microcontrollers use integers in place of floating-point numbers to compute non-whole number (pseudo real) values.

There are several forms of fixed-point mathematics currently in use. The simplest form is based upon powers of 2, just like normal integers in binary. However, a virtual binary point is inserted into the integer to allow an approximation of real values to be stored as integers. A standard 8-bit unsigned integer is shown below along with its equivalent decimal value.

$$0001\ 0100 = (1 * 2^4) + (1 * 2^2) = (1 * 16) + (1 * 4) = 20$$

Suppose a virtual binary point is inserted between the two nibbles in the byte. There are now four bits left of the binary point with the standard positive powers of 2, and 4 bits right of the binary point with negative powers of 2. The same number now represents a real number in decimal.

$$0001\ 0100 = (1 * 2^0) + (1 * 2^{-2}) = (1 * 1) + (1 * 0.25) = 1.25$$

Obviously this method has shortcomings. The resolution of any fixed point number is limited to the power of 2 attached to the least significant bit on the right of the number, in this case 2^{-4} or 1/16 or 0.0625. Rounding is sometimes necessary. There is also a tradeoff in complexity, as the position of this virtual binary point must constantly be maintained when performing calculations. The savings in memory usage and processing time, however, often overcome these tradeoffs; so fixed-point mathematics can be very useful.

3.6.2 Calibrations

The area of calibrating a system can sometimes take on an importance not foreseen when designing a mechatronic system. The use of calibrations, numerical and logical values kept in EEPROM or ROM, allow flexibility in system tuning and implementation. For example, if different microprocessor crystal speeds may be used in a mechatronic system, but real-time values are needed, a stored calibration constant of clock cycles per microsecond will allow this calculation to be affected. Thus, calibrations are often used as a gain, the value multiplied by some input in order to produce a scaled output.

Also, as mentioned above, calibrations are often used in the testing of a mechatronic system in order to change the "feel" of the product. A transmission control unit can use a set of calibrations on engine RPM, engine load, and vehicle speed to determine when to shift gears. This is often done with hysteresis, as the shift points moving from second gear to third gear as from third gear to second gear may differ.

3.7 Microprocessor Input–Output Control

3.7.1 Polling and Interrupts

There are two basic methods for the microprocessor to control input and output. These are polling and interrupts. Polling is just that, the microprocessor periodically checking various peripheral devices to determine if input or output is waiting. If a peripheral device has some input or output that should be processed, a flag will be set. The problem is that a lot of processing time is wasted checking for inputs when they are not changing.

Servicing an interrupt is an alternative method to control inputs and outputs. In this method, a register in the microprocessor must have set an interrupt enable (IE) bit for a particular peripheral device. When an interrupt is initiated by the peripheral, a flag is set for the microprocessor. The interrupt request (IRQ) line will go active, and the microprocessor will service the interrupt. Servicing an interrupt means that the normal processing of the microprocessor is halted (i.e., interrupted) while the input/output is completed. In order to resume normal processing, the microprocessor needs to store the contents of its registers before the interrupt is serviced. This process includes saving all active register contents to a stack, a part of RAM designated for this purpose, in a process known as a push. After a push, the microprocessor can then load the address of the Interrupt Service Routine and complete the input/output. When that portion of code is complete, the contents of the stack are reloaded to the registers in an operation known as a Pop (or Pull) and normal processing resumes.

3.7.2 Input and Output Transmission

Once the input or output is ready for transmission, there are several modes that can be used. First, data can be moved in either parallel or serial mode. Parallel mode means that multiple bits (e.g., 16 bits) move in parallel down a multiple pathway or bus from source to destination. Serial mode means that the bits move one at a time, in a series, down a single pathway. Parallel mode traffic is faster in that multiple bits are moving together, but the number of pathways is a limiting factor. For this reason parallel mode is usually used for components located close to one another while serial transmission is used if any distance is involved.

Serial data transmission can also be differentiated by being asynchronous or synchronous. Asynchronous data transmission uses separate clocks between the sender and receiver of data. Since these clocks are not synchronized, additional bits called start and stop bits are required to designate the boundaries of the bytes being sent. Synchronous data transmission uses a common or synchronized timing source. Start and stop bits are thus not needed, and overall throughput is increased.

A third way of differentiating data transmission is by direction. A simplex line is a one direction only pathway. Data from a sensor to the microcontroller may use simplex mode. Half-duplex mode allows two-way traffic, but only one direction at a time. This requires a form of flow control to avoid data transmission errors. Full-duplex mode allows two-way simultaneous transmission of data.

The agreement between sending and receiving units regarding the parameters of data transmission (including transmission speed) is known as handshaking.

3.7.3 HC12 Microcontroller Input–Output Subsystems

There are four input–output subsystems on the Motorola HC12 microcontroller that can be used to exemplify the data transmission section above.

The serial communications interface (SCI) is an asynchronous serial device available on the HC12. It can be either polled or interrupt driven and is intended for communication between remote devices. Related to SCI is the serial peripheral interface (SPI). SPI is a synchronous serial interface. It is intended for communication between units that support SPI like a network of multiple microcontrollers. Because of the synchronization of timing that is required, SPI uses a system of master/slave relationships between microcontrollers.

The pulse width modulation (PWM) subsystem is often used for motor and solenoid control. Using registers that are mapped to both the PWM unit and the microprocessor, a PWM output can be commanded by setting values for the period and duty cycle in the proper registers. This will result in a particular on-time and off-time voltage command.

Last, the serial in-circuit debugger (SDI) allows the microcontroller to connect to a PC for checking and modifying embedded software.

3.7.4 Microcontroller Network Systems

There is one last topic that should be mentioned in this section on inputs and outputs. Mechatronic systems often work with other systems in a network. Data and commands are thus transmitted from one system to another. While there are many different protocols, both open and proprietary, that could be mentioned about this networking, two will serve our purposes. The first is the manufacturing automation protocol (MAP) that was developed by General Motors Corporation. This system is based on the ISO Open Systems Interconnection (OSI) model and is especially designed for computer integrated manufacturing (CIM) and multiple PLCs. The second is the controller area network (CAN). This standard for serial communications was developed by Robert Bosch GmbH for use among embedded systems in a car.

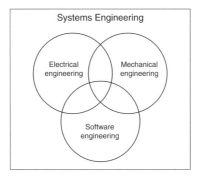

FIGURE 3.10 Mechatronics engineering disciplines.

3.8 Software Control

3.8.1 Systems Engineering

Systems engineering is the systems approach to the design and development of products and systems. As shown in Figure 3.10, a drawing that shows the relationships of the major engineering competencies with mechatronics, the systems engineering competency encompasses the mechanical, electrical, and software competencies. There are several important tasks for the systems engineers to perform, starting with requirements gathering and continuing through final product and system verification and validation. After requirements gathering and analysis, the systems engineers should partition requirements functionality between mechanical, electrical, and software components, in consultation with the three competencies involved. This is part of the implementation of concurrent engineering. As also shown by the figure, software is an equal partner in the development of a mechatronic system. It is not an add-on to the system and it is not free, the two opinions that were sometimes held in the past by engineering management. While the phrase "Hardware adds cost, software adds value" is not entirely true either, sometimes software engineers felt that their competency was not given equal weight with the traditional engineering disciplines. And one last comment—many mechatronic systems are safety related, such as an air bag system in a car. It is as important for the software to be as fault tolerant as the hardware.

3.8.2 Software Engineering

Software engineering is concerned with both the final mechatronic "product" and the mechatronic development process. Two basic approaches are used with process, with many variations upon these approaches. One is called the "waterfall" method, where the process moves (falls) from one phase to another (e.g., analysis to design) with checkpoints along the way. The other method, the "spiral" approach, is often used when the requirements are not as well fixed. In this method there is prototyping, where the customers and/or systems engineers refine requirements as more information about the system becomes known. In either approach, once the requirements for the software portion of the mechatronic system are documented, the software engineers should further partition functionality as part of software design. Metrics as to development time, development cost, memory usage, and throughput should also be projected and recorded. Here is where the Software Engineering Institute's Capability Maturity Model (SEI CMM) levels can be used for guidance. It is a truism that software is almost never developed as easily as estimated, and that a system can remain at the "90% complete" level for most of the development life cycle. The first solution attempted to solve this problem is often assigning more software engineers onto the project. This does not always work, however, because of the learning curve of the new people, as stated by Frederick Brooks in his important book *The Mythical Man Month* (Addison-Wesley 1995).

FIGURE 3.11 Mechatronic software layering.

3.8.3 Software Design

Perhaps the most important part of the software design for a mechatronic system can be seen from the hierarchy in Figure 3.11. Ranging from requirements at the top to hardware at the bottom, this layering serves several purposes. The most important is that it separates mechatronic functionality from implementation. Quite simply, an upper layer should not be concerned with how a lower layer is actually performing a task. Each layer instead is directed by the layer above and receives a service or status from a layer below it. To cross more than one layer boundary is bad technique and can cause problems later in the process. Remember that this process abstraction is quite useful, for a mechatronic system has mechanical, electrical, and software parts all in concurrent development. A change in a sensor or actuator interface should only require a change at the layer immediately above, the driver layer. There is one last reason for using a hierarchical model such as this. In the current business climate, it is unlikely that the people working at the various layers will be collocated. Instead, it is not uncommon for development to be taking place in multiple locations in multiple countries. Without a crisp division of these layers, chaos can result.

For more information on these and many other topics in software engineering such as coupling, cohesion, and software reuse, please refer to Chapter 49 of this handbook, Roger Pressman's book *Software Engineering: A Practitioner's Approach 5th Edition* (McGraw Hill 2000), and Steve McConnell's book *Code Complete* (Microsoft Press 1993).

3.9 Testing and Instrumentation

3.9.1 Verification and Validation

Verification and validation are related tasks that should be completed throughout the life cycle of the mechatronic product or system. Boehm in his book *Software Engineering Economics* (Prentice-Hall 1988) describes verification as "building the product right" while validation is "building the right product." In other words, verification is the testing of the software and product to make sure that it is built to the design. Validation, on the other hand, is to make sure the software or product is built to the requirements from the customer. As mentioned, verification and validation are life cycle tasks, not tasks completed just before the system is set for production. One of the simplest and most useful techniques is to hold hardware and software validation and verification reviews. Validation design reviews of hardware and software should include the systems engineers who have the best understanding of the customer requirements. Verification hardware design and software code reviews, or peer reviews, are an excellent means of finding errors upstream in the development process. Managers may have to decide whether to allocate resources upstream, when the errors are easier to fix, or downstream, when the ramifications can be much more drastic. Consider the difference between a code review finding a problem in code, and having the author change it and recompile, versus finding a problem after the product has been sold and in the field, where an expensive product recall may be required.

3.9.2 Debuggers

Edsgar Dijkstra, a pioneer in the development of programming as a discipline, discouraged the terms "bug" and "debug," and considered such terms harmful to the status of software engineering. They are, however, used commonly in the field. A debugger is a software program that allows a view of what is happening with the program code and data while the program is executing. Generally it runs on a PC that is connected to a special type of development microcontroller called an emulator. While debuggers can be quite useful in finding and correcting errors in code, they are not real-time, and so can actually create computer operating properly (COP) errors. However, if background debug mode (BDM) is available on the microprocessor, the debugger can be used to step through the algorithm of the program, making sure that the code is operating as expected. Intermediate and final variable values, especially those related to some analog input or output value, can be checked. Most debuggers allow multiple open windows, the setting of program execution break points in the code, and sometimes even the reflashing of the program into the microcontroller emulator. An example is the Noral debugger available for the Motorola HC12.

The software in the microcontroller can also check itself and its hardware. By programming in a checksum, or total, of designated portions of ROM and/or EEPROM, the software can check to make sure that program and data are correct. By alternately writing and reading 0x55 and 0xAA to RAM (the "checkerboard test"), the program can verify that RAM and the bus are operating properly. These startup tasks should be done with every product operation cycle.

3.9.3 Logic Analyzer

A logic analyzer is a device for nonintrusive monitoring and testing of the microcontroller. It is usually connected to both the microcontroller and a simulator. While the microcontroller is running its program and processing data, the simulator is simulating inputs and displaying outputs of the system. A "trigger word" can be entered into the logic analyzer. This is a bit pattern that will be on one of the buses monitored by the logic analyzer. With this trigger, the bus traffic around that point of interest can be captured and stored in the memory of the analyzer. An inverse assembler in the analyzer allows the machine code on the bus to be seen and analyzed in the form of the assembly level commands of the program. The analyzer can also capture the analog outputs of the microcontroller. This could be used to verify that the correct PWM duty cycle is being commanded. The simulator can introduce shorts or opens into the system, then the analyzer is used to see if the software correctly responds to the faults. The logic analyzer can also monitor the master loop of the system, making sure that the system completes all of its tasks within a designated time, for example, 15 ms. An example of a logic analyzer is the Hewlett Packard HP54620.

3.10 Summary

This chapter introduced a number of topics regarding a mechatronic system. These topics included not just mechatronic input, output, and processing, but also design, development, and testing. Future chapters will cover all of this material in much greater detail.

4
Microprocessor-Based Controllers and Microelectronics

Ondrej Novak
Ivan Dolezal
Technical University of Liberec

4.1	Introduction to Microelectronics	4-1
4.2	Digital Logic	4-2
4.3	Overview of Control Computers	4-2
4.4	Microprocessors and Microcontrollers	4-4
4.5	Programmable Logic Controllers	4-5
4.6	Digital Communications	4-6

4.1 Introduction to Microelectronics

The field of microelectronics has changed dramatically during the last two decades and digital technology has governed most of the application fields in electronics. The design of digital systems is supported by thousands of different integrated circuits supplied by many manufacturers across the world. This makes both the design and the production of electronic products much easier and cost effective. The permanent growth of integrated circuit speed, scale of integration, and reduction of costs have resulted in digital circuits being used instead of classical analog solutions of controllers, filters, and (de)modulators.

The growth in computational power can be demonstrated with the following example. One single-chip microcontroller has the computational power equal to that of one 1992 vintage computer notebook. This single-chip microcontroller has the computational power equal to four 1981 vintage IBM personal computers, or to two 1972 vintage IBM 370 mainframe computers.

Digital integrated circuits are designed to be universal and are produced in large numbers. Modern integrated circuits have many upgraded features from earlier designs, which allow for "user-friendlier" access and control. As the parameters of integrated circuits (ICs) influence not only the individually designed IC, but all the circuits that must cooperate with it, a roadmap of the future development of IC technology is updated every year. From this roadmap we can estimate future parameters of the ICs, and adapt our designs to future demands. The relative growth of the number of integrated transistors on a chip is relatively stable. In the case of memory elements, it is equal to approximately 1.5 times the current amount. In the case of other digital ICs, it is equal to approximately 1.35 times the current amount.

In digital electronics, we use quantities called logical values instead of the analog quantities of voltage and current. Logical variables usually correspond to the voltage of the signal, but they have only two values: log 1 and log 0. If a digital circuit processes a logical variable, a correct value is recognized because between the logical value voltages there is a gap (see Figure 4.1). We can arbitrarily improve the resolution of signals by simply using more bits.

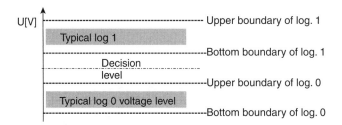

FIGURE 4.1 Voltage levels and logical values correspondence.

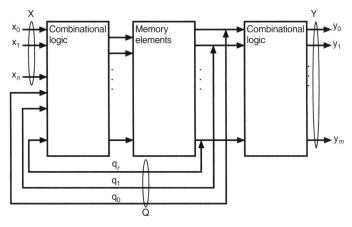

FIGURE 4.2 A finite state automaton: X—input binary vector, Y—output binary vector, Q—internal state vector.

4.2 Digital Logic

Digital circuits are composed of logic gates, such as elementary electronic circuits operating in only two states. These gates operate in such a way that the resulting logical value corresponds to the resulting value of the Boolean algebra statements. This means that with the help of gates we can realize every logical and arithmetical operation. These operations are performed in combinational circuits for which the resulting value is dependent only on the actual state of the inputs variables. Of course, logic gates are not enough for automata construction. For creating an automaton, we also need some memory elements in which we capture the responses of the arithmetical and logical blocks.

A typical scheme of a digital finite state automaton is given in Figure 4.2. The automata can be constructed from standard ICs containing logic gates, more complex combinational logic blocks and registers, counters, memories, and other standard sequential ICs assembled on a printed circuit board. Another possibility is to use application specific integrated circuits (ASIC), either programmable or full custom, for a more advanced design. This approach is suitable for designs where fast hardware solutions are preferred. Another possibility is to use microcontrollers that are designed to serve as universal automata, which function can be specified by memory programming.

4.3 Overview of Control Computers

Huge, complex, and power-consuming single-room mainframe computers and, later, single-case minicomputers were primarily used for scientific and technical computing (e.g., in FORTRAN, ALGOL) and for database applications (e.g., in COBOL). The invention in 1971 of a universal central processing unit (CPU) in a single chip microprocessor caused a revolution in the computer technology. Beginning in

FIGURE 4.3 Example of a small mechatronic system: The ALAMBETA device for measurement of thermal properties of fabrics and plastic foils (manufactured by SENSORA, Czech Republic). It employs a unique measuring method using extra thin heat flow sensors, sample thickness measurement incorporated into a head drive, microprocessor control, and connection with a PC.

1981, multi-boxes (desktop or tower case, monitor, keyboard, mouse) or single-box (notebook) microcomputers became a daily-used personal tool for word processing, spreadsheet calculation, game playing, drawing, multimedia processing, and presentations. When connected in a local area network (LAN) or over the Internet, these "personal computers (PCs)" are able to exchange data and to browse the World Wide Web (WWW).

Besides these "visible" computers, many embedded microcomputers are hidden in products such as machines, vehicles, measuring instruments, telecommunication devices, home appliances, consumer electronic products (cameras, hi-fi systems, televisions, video recorders, mobile phones, music instruments, toys, air-conditioning). They are connected with sensors, user interfaces (buttons and displays), and actuators. Programmability of such controllers brings flexibility to the devices (function program choice), some kind of intelligence (fuzzy logic), and user-friendly action. It ensures higher reliability and easier maintenance, repairs, (auto)calibration, (auto)diagnostics, and introduces the possibility of their interconnection—mutual communication or hierarchical control in a whole plant or in a smart house. A photograph of an electrically operated instrument is given in Figure 4.3.

Embedded microcomputers are based on the Harvard architecture where code and data memories are split. Firmware (program code) is cross-compiled on a development system and then resides in a nonvolatile memory. In this way, a single main program can run immediately after a supply is switched on. Relatively expensive and shock sensitive mechanical memory devices (hard disks) and vacuum tube monitors have been replaced with memory cards or solid state disks (if an archive memory is essential) and LED segment displays or LCDs. A PC-like keyboard can be replaced by a device/function specifically labeled key set and/or common keys (arrows, Enter, Escape) completed with numeric keys, if necessary. Such key sets, auxiliary switches, large buttons, the main switch, and display can be located in water and dust resistant operator panels.

Progress in circuit integration caused fast development of microcontrollers in the last two decades. Code memory, data memory, clock generator, and a diverse set of peripheral circuits are integrated with the CPU (Figure 4.4) to insert such complete single-chip microcomputers into an application specific PCB.

Digital signal processors (DSPs) are specialized embedded microprocessors with some on-chip peripherals but with external ADC/DAC, which represent the most important input/output channel. DSPs have a parallel computing architecture and a fixed point or floating point instruction set optimized for typical signal processing operations such as discrete transformations, filtering, convolution, and coding. We can find DSPs in applications like sound processing/generation, sensor (e.g., vibration) signal analysis,

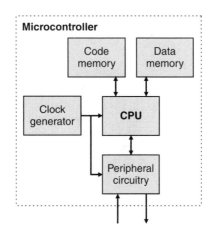

FIGURE 4.4 Block diagram of a microcontroller.

telecommunications (e.g., bandpass filter and digital modulation/demodulation in mobile phones, communication transceivers, modems), and vector control of AC motors.

Mass production (i.e., low cost), wide-spread knowledge of operation, comprehensive access to software development and debugging tools, and millions of ready-to-use code lines make PCs useful for computing-intensive measurement and control applications, although their architecture and operating systems are not well suited for this purpose.

As a result of computer expansion, there exists a broad spectrum of computing/processing means from powerful workstations, top-end PCs and VXI systems (64/32 bits, over 1000 MFLOPS/MIPS, 1000 MB of memory, input power over 100 W, cost about $10,000), downwards to PC-based computer cards/modules (32 bits, 100–300 MFLOPS/MIPS, 10–100 MB, cost less than $1000). Microprocessor cards/modules (16/8 bits, 10–30 MIPS, 1 MB, cost about $100), complex microcontroller chips (16/8 bits, 10–30 MIPS, 10–100 KB, cost about $10), and simple 8-pin microcontrollers (8 bits, 1–5 MIPS, 1 KB, 10 mW, cost about $1) are also available for very little money.

4.4 Microprocessors and Microcontrollers

There is no strict border between microprocessors and microcontrollers because certain chips can access external code and/or data memory (microprocessor mode) and are equipped with particular peripheral components.

Some microcontrollers have an internal RC oscillator and do not need an external component. However, an external quartz or ceramic resonator or RC network is frequently connected to the built-in, active element of the clock generator. Clock frequency varies from 32 kHz (extra low power) up to 75 MHz. Another auxiliary circuit generates the reset signal for an appropriate period after a supply is turned on. Watchdog circuits generate chip reset when a periodic retriggering signal does not come in time due to a program problem. There are several modes of consumption reduction activated by program instructions.

Complexity and structure of the interrupt system (total number of sources and their priority level selection), settings of level/edge sensitivity of external sources and events in internal (i.e., peripheral) sources, and handling of simultaneous interrupt events appear as some of the most important criteria of microcontroller taxonomy.

Although 16- and 32-bit microcontrollers are engaged in special, demanding applications (servo-unit control), most applications employ 8-bit chips. Some microcontrollers can internally operate with a 16-bit or even 32-bit data only in fixed-point range—microcontrollers are not provided with floating point unit (FPU). New microcontroller families are built on RISC (Reduced Instruction Set) core executing due to pipelining one instruction per few clock cycles or even per each cycle.

One can find further differences in addressing modes, number of direct accessible registers, and type of code memory (ranging from 1 to 128 KB) that are important from the view of firmware development. Flash memory enables quick and even in-system programming (ISP) using 3–5 wires, whereas classical EPROM makes chips more expensive due to windowed ceramic packaging. Some microcontrollers have built-in boot and debug capability to load code from a PC into the flash memory using UART (Universal Asynchronous Receiver/Transmitter) and RS-232C serial line. OTP (One Time Programmable) EPROM or ROM appear effective for large production series. Data EEPROM (from 64 B to 4 KB) for calibration constants, parameter tables, status storage, and passwords that can be written by firmware stand beside the standard SRAM (from 32 B to 4 KB).

The range of peripheral components is very wide. Every chip has bidirectional I/O (input/output) pins associated in 8-bit ports, but they often have an alternate function. Certain chips can set an input decision level (TTL, MOS, or Schmitt trigger) and pull-up or pull-down current sources. Output drivers vary in open collector or tri-state circuitry and maximal currents.

At least one 8-bit timer/counter (usually provided with a prescaler) counts either external events (optional pulses from an incremental position sensor) or internal clocks, to measure time intervals, and periodically generates an interrupt or variable baud rate for serial communication. General purpose 16-bit counters and appropriate registers form either capture units to store the time of input transients or compare units that generate output transients as a stepper motor drive status or PWM (pulse width modulation) signal. A real-time counter (RTC) represents a special kind of counter that runs even in sleep mode. One or two asynchronous and optionally synchronous serial interfaces (UART/USART) communicate with a master computer while other serial interfaces like SPI, CAN, and I^2C control other specific chips employed in the device or system.

Almost every microcontroller family has members that are provided with an A/D converter and a multiplexer of single-ended inputs. Input range is usually unipolar and equal to supply voltage or rarely to the on-chip voltage reference. The conversion time is given by the successive approximation principle of ADC, and the effective number of bits (ENOB) usually does not reach the nominal resolution 8, 10, or 12 bits.

There are other special interface circuits, such as field programmable gate array (FPGA), that can be configured as an arbitrary digital circuit.

Microcontroller firmware is usually programmed in an assembly language or in C language. Many software tools, including chip simulators, are available on websites of chip manufacturers or third-party companies free of charge. A professional integrated development environment and debugging hardware (in-circuit emulator) is more expensive (thousands of dollars). However, smart use of an inexpensive ROM simulator in a microprocessor system or a step-by-step development cycle using an ISP programmer of flash microcontroller can develop fairly complex applications.

4.5 Programmable Logic Controllers

A programmable logic controller (PLC) is a microprocessor-based control unit designed for an industrial installation (housing, terminals, ambient resistance, fault tolerance) in a power switchboard to control machinery or an industrial process. It consists of a CPU with memories and an I/O interface housed either in a compact box or in modules plugged in a frame and connected with proprietary buses. The compact box starts with about 16 I/O interfaces, while the module design can have thousands of I/O interfaces. Isolated inputs usually recognize industrial logic, 24 V DC or main AC voltage, while outputs are provided either with isolated solid state switches (24 V for solenoid valves and contactors) or with relays. Screw terminal boards represent connection facilities, which are preferred in PLCs to wire them to the controlled systems. I/O logical levels can be indicated with LEDs near to terminals.

Since PLCs are typically utilized to replace relays, they execute Boolean (bit, logical) operations and timer/counter functions (a finite state automaton). Analog I/O, integer or even floating point arithmetic, PWM outputs, and RTC are implemented in up-to-date PLCs. A PLC works by continually scanning a program, such as machine code, that is interpreted by an embedded microprocessor (CPU). The scan time is the time it takes to check the input status, to execute all branches (all individual rungs of a ladder

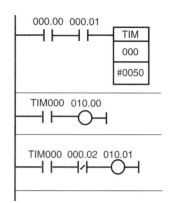

FIGURE 4.5 Example of PLC ladder diagram: 000.xx/010.xx—address group of inputs/outputs, TIM000—timer delays 5 s. 000.00—normally open input contact, 000.02—normally closed input contact.

diagram) of the program using internal (state) bit variables if any, and to update the output status. The scan time is dependent on the complexity of the program (milliseconds or tens of ms). The next scan operation either follows the previous one immediately (free running) or starts periodically.

Programming languages for PLCs are described in IEC-1131-3 nomenclature:

LD—ladder diagram (see Figure 4.5)
IL—instruction list (an assembler)
SFC—sequential function chart (usually called by the proprietary name GRAFCET)
ST—structured text (similar to a high level language)
FBD—function block diagram

PLCs are programmed using cross-compiling and debugging tools running on a PC or with programming terminals (usually using IL), both connected with a serial link. Remote operator panels can serve as a human-to-machine interface. A new alternate concept (called SoftPLC) consists of PLC-like I/O modules controlled by an industrial PC, built in a touch screen operator panel.

4.6 Digital Communications

Intercommunication among mechatronics subsystems plays a key role in their engagement of applications, both of fixed and flexible configuration (a car, a hi-fi system, a fixed manufacturing line versus a flexible plant, a wireless pico-net of computer peripheral devices). It is clear that digital communication depends on the designers demands for the amount of transferred data, the distance between the systems, and the requirements on the degree of data reliability and security.

The signal is represented by alterations of amplitude, frequency, or phase. This is accomplished by changes in voltage/current in metallic wires or by electromagnetic waves, both in radiotransmission and infrared optical transmission (either "wireless" for short distances or optical fibers over fairly long distances). Data rate or bandwidth varies from 300 b/s (teleprinter), 3.4 kHz (phone), 144 kb/s (ISDN) to tens of Mb/s (ADSL) on a metallic wire (subscriber line), up to 100 Mb/s on a twisted pair (LAN), about 30–100 MHz on a microwave channel, 1 GHz on a coaxial cable (trunk cable network, cable TV), and up to tens of Gb/s on an optical cable (backbone network).

Data transmission employs complex methods of digital modulation, data compression, and data protection against loss due to noise interference, signal distortion, and dropouts. Multilayer standard protocols (ISO/OSI 7-layer reference model or Internet 4-layer group of protocols including well-known TCP/IP), "partly hardware, partly software realized," facilitate an understanding between communication systems. They not only establish connection on a utilizable speed, check data transfer, format and compress data, but can make communication transparent for an application. For example, no difference can be seen between local and remote data sources. An example of a multilayer communication concept is depicted in Figure 4.6.

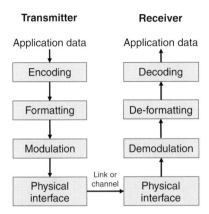

FIGURE 4.6 Example of multilayer communication.

Depending on the number of users, the communication is done either point-to-point (RS-232C from PC COM port to an instrument), point-to-multipoint (buses, networks), or even as a broadcasting (radio). Data are transferred using either switched connection (telephone network) or packet switching (computer networks, ATM). Bidirectional transmission can be full duplex (phone, RS-232C) or semi-duplex (most of digital networks). Concerning the link topology, a star connection or a tree connection employs a device ("master") mastering communication in the main node(s). A ring connection usually requires Token Passing method and a bus communication is controlled with various methods such as Master-Slave pooling, with or without Token Passing, or by using an indeterministic access (CSMA/CD in Ethernet).

An LPT PC port, SCSI for computer peripherals, and GPIB (IEEE-488) for instrumentation serve as examples of parallel (usually 8-bit) communication available for shorter distances (meters). RS-232C, RS-485, I^2C, SPI, USB, and Firewire (IEEE-1394) represent serial communication, some of which can bridge long distance (up to 1 km). Serial communication can be done either asynchronously using start and stop bits within transfer frame or synchronously using included synchronization bit patterns, if necessary. Both unipolar and bipolar voltage levels are used to drive either unbalanced lines (LPT, GPIB vs. RS-232C) or balanced twisted-pair lines (CAN vs. RS-422, RS-485).

5
An Introduction to Micro- and Nanotechnology

Michael Goldfarb
Alvin M. Strauss
Eric J. Barth
Vanderbilt University

5.1 Introduction 5-1
　The Physics of Scaling • General Mechanisms of
　Electromechanical Transduction • Sensor and Actuator
　Transduction Characteristics
5.2 Microactuators 5-3
　Electrostatic Actuation • Electromagnetic Actuation
5.3 Microsensors 5-6
　Strain • Pressure • Acceleration • Force • Angular Rate
　Sensing (Gyroscopes)
5.4 Nanomachines 5-9
References 5-11

5.1 Introduction

Originally arising from the development of processes for fabricating microelectronics, micro-scale devices are typically classified according not only to their dimensional scale, but their composition and manufacture. Nanotechnology is generally considered as ranging from the smallest of these micro-scale devices down to the assembly of individual molecules to form molecular devices. These two distinct yet overlapping fields of microelectromechanical systems (MEMS) and nanosystems or nanotechnology share a common set of engineering design considerations unique from other more typical engineering systems. Two major factors distinguish the existence, effectiveness, and development of micro-scale and nano-scale transducers from those of conventional scale. The first is the physics of scaling and the second is the suitability of manufacturing techniques and processes. The former is governed by the laws of physics and is thus a fundamental factor, while the latter is related to the development of manufacturing technology, which is a significant, though not fundamental factor. Due to the combination of these factors, effective micro-scale transducers can often not be constructed as geometrically scaled-down versions of conventional-scale transducers.

5.1.1 The Physics of Scaling

The dominant forces that influence micro-scale devices are different from those that influence their conventional-scale counterparts. This is because the size of a physical system bears a significant influence on the physical phenomena that dictate the dynamic behavior of that system. For example, larger-scale systems are influenced by inertial effects to a much greater extent than smaller-scale systems, while smaller systems are influenced more by surface effects. As an example, consider small insects that can stand on the surface of still water, supported only by surface tension. The same surface tension is present when humans come into contact with water, but on a human scale the associated forces are typically insignificant. The world in which humans live is governed by the same forces as the world in which these insects live, but the forces are present in very different proportions. This is due in general to the fact that inertial forces typically act in proportion to volume, and surface forces typically in proportion to surface area. Since volume varies with the third power of length and area with the second, geometrically similar but smaller objects have proportionally more area than larger objects.

Exact scaling relations for various types of forces can be obtained by incorporating dimensional analysis techniques [1–5]. Inertial forces, for example, can be dimensionally represented as $F_i = \rho L^3 \ddot{x}$, where F_i is a generalized inertia force, ρ is the density of an object, L is a generalized length, and x is a displacement. This relationship forms a single dimensionless group, given by

$$\Pi = \frac{F_i}{\rho L^3 \ddot{x}}$$

Scaling with geometric and kinematic similarity can be expressed as

$$\frac{L_s}{L_o} = \frac{x_s}{x_o} = N, \quad \frac{t_s}{t_o} = 1$$

where L represents the length scale, x the kinematic scale, t the time scale, the subscript o the original system, and the s represents the scaled system. Since physical similarity requires that the dimensionless group (Π) remain invariant between scales, the force relationship is given by $F_s/F_o = N^4$, assuming that the intensive property (density) remains invariant (i.e., $\rho_s = \rho_o$). An inertial force thus scales as N^4, where N is the geometric scaling factor. Alternately stated, for an inertial system that is geometrically smaller by a factor of N, the force required to produce an equivalent acceleration is smaller by a factor of N^4. A similar analysis shows that viscous forces, dimensionally represented by $F_v = \mu L \dot{x}$, scale as N^2, assuming the viscosity μ remains invariant, and elastic forces, dimensionally represented by $F_e = ELx$, scale as N^2, assuming the elastic modulus E remains invariant. Thus, for a geometrically similar but smaller system, inertial forces will become considerably less significant with respect to viscous and elastic forces.

5.1.2 General Mechanisms of Electromechanical Transduction

The fundamental mechanism for both sensing and actuation is energy transduction. The primary forms of physical electromechanical transduction can be grouped into two categories. The first is multicomponent transduction, which utilizes "action at a distance" behavior between multiple bodies, and the second is deformation-based or solid-state transduction, which utilizes mechanics-of-material phenomena such as crystalline phase changes or molecular dipole alignment. The former category includes electromagnetic transduction, which is typically based upon the Lorentz equation and Faraday's law, and electrostatic interaction, which is typically based upon Coulomb's law. The latter category includes piezoelectric effects, shape memory alloys, and magnetostrictive, electrostrictive, and photostrictive materials. Although materials exhibiting these properties are beginning to be seen in a limited number of research applications, the development of micro-scale systems is currently dominated by the exploitation of electrostatic and electromagnetic interactions. Due to their importance, electrostatic and electromagnetic transduction is treated separately in the sections that follow.

5.1.3 Sensor and Actuator Transduction Characteristics

Characteristics of concern for both microactuator and microsensor technology are repeatability, the ability to fabricate at a small scale, immunity to extraneous influences, sufficient bandwidth, and if possible, linearity. Characteristics typically of concern specifically for microactuators are achievable force, displacement, power, bandwidth (or speed of response), and efficiency. Characteristics typically of concern specifically for microsensors are high resolution and the absence of drift and hysteresis.

5.2 Microactuators

5.2.1 Electrostatic Actuation

The most widely utilized multicomponent microactuators are those based upon electrostatic transduction. These actuators can also be regarded as a variable capacitance type, since they operate in an analogous mode to variable reluctance type electromagnetic actuators (e.g., variable reluctance stepper motors). Electrostatic actuators have been developed in both linear and rotary forms. The two most common configurations of the linear type of electrostatic actuators are the normal-drive and tangential or comb-drive types, which are illustrated in Figures 5.1 and 5.2, respectively. Note that both actuators are suspended by flexures, and thus the output force is equal to the electrostatic actuation force minus the elastic force required to deflect the flexure suspension. The normal-drive type of electrostatic microactuator operates in a similar fashion to a condenser microphone. In this type of drive configuration, the actuation force is given by

$$F_x = \frac{\varepsilon A v^2}{2x^2}$$

where A is the total area of the parallel plates, ε is the permittivity of air, v is the voltage across the plates, and x is the plate separation. The actuation force of the comb-drive configuration is given by

$$F_x = \frac{\varepsilon w v^2}{2d}$$

where w is the width of the plates, ε is the permittivity of air, v is the voltage across the plates, and d is the plate separation. Dimensional examination of both relations indicates that force is independent of geometric and kinematic scaling, that is, for an electrostatic actuator that is geometrically and kinematically reduced by a factor of N, the force produced by that actuator will be the same. Since forces associated with most other physical phenomena are significantly reduced at small scales, micro-scale electrostatic forces become significant relative to other forces. Such an observation is clearly demonstrated by the fact that all intermolecular forces are electrostatic in origin, and thus the strength of all materials is a result of electrostatic forces [6].

The maximum achievable force of multicomponent electrostatic actuators is limited by the dielectric breakdown of air, which occurs in dry air at about 0.8×10^6 V/m. Fearing [7] estimates that the upper limit for force generation in electrostatic actuation is approximately 10 N/cm^2. Since electrostatic drives

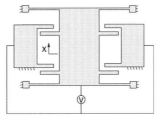

FIGURE 5.1 Schematic of a normal-drive electrostatic actuator.

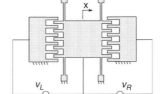

FIGURE 5.2 Comb-drive electrostatic actuator. Energizing an electrode provides motion toward that electrode.

do not have any significant actuation dynamics, and since the inertia of the moving member is usually small, the actuator bandwidth is typically quite large, on the order of a kilohertz.

The maximum achievable stroke for normal configuration actuators is limited by the elastic region of the flexure suspension and additionally by the dependence of actuation force on plate separation, as given by the above stated equations. According to Fearing, a typical stroke for a surface micromachined normal configuration actuator is on the order of a couple of microns. The achievable displacement can be increased by forming a stack of normal-configuration electrostatic actuators in series, as proposed by Bobbio et al. [8,9].

The typical stroke of a surface micromachined comb actuator is on the order of a few microns, though sometimes less. The maximum achievable stroke in a comb drive is limited primarily by the mechanics of the flexure suspension. The suspension should be compliant along the direction of actuation to enable increased displacement, but must be stiff orthogonal to this direction to avoid parallel plate contact due to misalignment. These modes of behavior are unfortunately coupled, so that increased compliance along the direction of motion entails a corresponding increase in the orthogonal direction. The net effect is that increased displacement requires increased plate separation, which results in decreased overall force.

The most common configurations of rotary electrostatic actuators are the variable capacitance motor and the wobble or harmonic drive motor, which are illustrated in Figures 5.3 and 5.4, respectively. Both motors operate in a similar manner to the comb-drive linear actuator. The variable capacitance motor is characterized by high-speed low-torque operation. Useful levels of torque for most applications therefore require some form of significant micromechanical transmission, which do not presently exist. The rotor of the wobble motor operates by rolling along the stator, which provides an inherent harmonic-drive-type transmission and thus a significant transmission ratio (on the order of several hundred times). Note that the rotor must be well insulated to roll along the stator without electrical contact. The drawback to this approach is that the rotor motion is not concentric with respect to the stator, which makes the already difficult problem of coupling a load to a micro-shaft even more difficult.

Examples of normal type linear electrostatic actuators are those by Bobbio et al. [8,9] and Yamaguchi et al. [10]. Examples of comb-drive electrostatic actuators are those by Kim et al. [11] and Matsubara et al. [12], and a larger-scale variation by Niino et al. [13]. Examples of variable capacitance rotary electrostatic motors are those by Huang et al. [14], Mehragany et al. [15], and Trimmer and Gabriel [16].

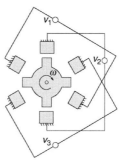

FIGURE 5.3 Variable capacitance type electrostatic motor. Opposing pairs of electrodes are energized sequentially to rotate the rotor.

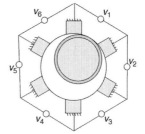

FIGURE 5.4 Harmonic drive type electrostatic motor. Adjacent electrodes are energized sequentially to roll the (insulated) rotor around the stator.

Examples of harmonic-drive motors are those by Mehragany et al. [17,18], Price et al. [19], Trimmer and Jebens [20,21], and Furuhata et al. [22]. Electrostatic microactuators remain a subject of research interest and development, and as such are not yet available on the general commercial market.

5.2.2 Electromagnetic Actuation

Electromagnetic actuation is not as omnipresent at the micro-scale as at the conventional-scale. This probably is due in part to early skepticism regarding the scaling of magnetic forces, and in part to the fabrication difficulty in replicating conventional-scale designs. Most electromagnetic transduction is based upon a current carrying conductor in a magnetic field, which is described by the Lorentz equation:

$$dF = Idl \times B$$

where F is the force on the conductor, I is the current in the conductor, l is the length of the conductor, and B is the magnetic flux density. In this relation, the magnetic flux density is an intensive variable and thus (for a given material) does not change with scale. Scaling of current, however, is not as simple. The resistance of wire is given by

$$R = \frac{\rho l}{A}$$

where ρ is the resistivity of the wire (an intensive variable), l is the length, and A the cross-sectional area. If a wire is geometrically decreased in size by a factor of N, its resistance will increase by a factor of N. Since the power dissipated in the wire is $I^2 R$, assuming the current remains constant implies that the power dissipated in the geometrically smaller wire will increase by a factor of N. Assuming the maximum power dissipation for a given wire is determined by the surface area of the wire, a wire that is smaller by a factor of N will be able to dissipate a factor of N^2 less power. Constant current is therefore a poor assumption. A better assumption is that maximum current is limited by maximum power dissipation, which is assumed to depend upon surface area of the wire. Since a wire smaller by a factor of N can dissipate a factor of N^2 less power, the current in the smaller conductor would have to be reduced by a factor of $N^{3/2}$. Incorporating this into the scaling of the Lorentz equation, an electromagnetic actuator that is geometrically smaller by a factor of N would exert a force that is smaller by a factor of $N^{5/2}$. Trimmer and Jebens have conducted a similar analysis, and demonstrated that electromagnetic forces scale as N^2 when assuming constant temperature rise in the wire, $N^{5/2}$ when assuming constant heat (power) flow (as previously described), and N^3 when assuming constant current density [23,24]. In any of these cases, the scaling of electromagnetic forces is not nearly as favorable as the scaling of electrostatic forces. Despite this, electromagnetic actuation still offers utility in microactuation, and most likely scales more favorably than does inertial or gravitational forces.

Lorentz-type approaches to microactuation utilize surface micromachined micro-coils, such as the one illustrated in Figure 5.5. One configuration of this approach is represented by the actuator of Inoue et al. [25],

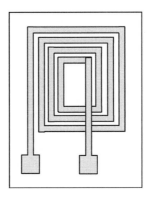

FIGURE 5.5 Schematic of surface micromachined microcoil for electromagnetic actuation.

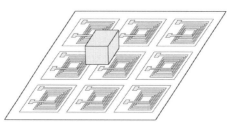

FIGURE 5.6 Microcoil array for planar positioning of a permanent micromagnet, as described by Inoue et al. [25]. Each coil produces a field, which can either attract or repel the permanent magnet, as determined by the direction of current. The magnet does not levitate, but rather slides on the insulated surface.

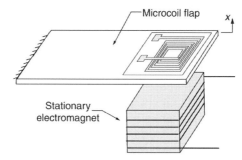

FIGURE 5.7 Cantilevered microcoil flap as described by Liu et al. [26]. The interaction between the energized coil and the stationary electromagnet deflects the flap upward or downward, depending on the direction of current through the microcoil.

which utilizes current control in an array of microcoils to position a permanent micromagnet in a plane, as illustrated in Figure 5.6. Another Lorentz-type approach is illustrated by the actuator of Liu et al. [26], which utilizes current control of a cantilevered microcoil flap in a fixed external magnetic field to effect deflection of the flap, as shown in Figure 5.7. Liu reported deflections up to 500 μm and a bandwidth of approximately 1000 Hz [26]. Other examples of Lorentz-type nonrotary actuators are those by Shinozawa et al. [27], Wagner and Benecke [28], and Yanagisawa et al. [29]. A purely magnetic approach (i.e., not fundamentally electromagnetic) is the work of Judy et al. [30], which in essence manipulates a flexure-suspended permanent micromagnet by controlling an external magnetic field.

Ahn et al. [31] and Guckel et al. [32] have both demonstrated planar rotary variable-reluctance type electromagnetic micromotors. A variable reluctance approach is advantageous because the rotor does not require commutation and need not be magnetic. The motor of Ahn et al. incorporates a 12-pole stator and 10-pole rotor, while the motor of Guckel et al. utilizes a 6-pole stator and 4-pole rotor. Both incorporate rotors of approximately 500 μm diameter. Guckel reports (no load) rotor speeds above 30,000 rev/min, and Ahn estimates maximum stall torque at 1.2 μN m. As with electrostatic microactuators, microfabricated electromagnetic actuators likewise remain a subject of research interest and development and as such are not yet available on the general commercial market.

5.3 Microsensors

Since microsensors do not transmit power, the scaling of force is not typically significant. As with conventional-scale sensing, the qualities of interest are high resolution, absence of drift and hysteresis, achieving a sufficient bandwidth, and immunity to extraneous effects not being measured.

Microsensors are typically based on either measurement of mechanical strain, measurement of mechanical displacement, or on frequency measurement of a structural resonance. The former two types are in essence analog measurements, while the latter is in essence a binary-type measurement, since the sensed quantity is typically the frequency of vibration. Since the resonant-type sensors measure frequency instead of amplitude, they are generally less susceptible to noise and thus typically provide a higher resolution measurement. According to Guckel et al. [33], resonant sensors provide as much as one hundred times the resolution of analog sensors. They are also, however, more complex and are typically more difficult to fabricate.

The primary form of strain-based measurement is piezoresistive, while the primary means of displacement measurement is capacitive. The resonant sensors require both a means of structural excitation as well as a means of resonant frequency detection. Many combinations of transduction are utilized for these purposes, including electrostatic excitation, capacitive detection, magnetic excitation and detection, thermal excitation, and optical detection.

5.3.1 Strain

Many microsensors are based upon strain measurement. The primary means of measuring strain is via piezoresistive strain gages, which is an analog form of measurement. Piezoresistive strain gages, also known as semiconductor gages, change resistance in response to a mechanical strain. Note that piezoelectric materials can also be utilized to measure strain. Recall that mechanical strain will induce an electrical charge in a piezoelectric ceramic. The primary problem with using a piezoelectric material, however, is that since measurement circuitry has limited impedance, the charge generated from a mechanical strain will gradually leak through the measurement impedance. A piezoelectric material therefore cannot provide reliable steady-state signal measurement. In constrast, the change in resistance of a piezoresistive material is stable and easily measurable for steady-state signals. One problem with piezoresistive materials, however, is that they exhibit a strong strain-temperature dependence, and so must typically be thermally compensated.

An interesting variation on the silicon piezoresistor is the resonant strain gage proposed by Ikeda et al., which provides a frequency-based form of measurement that is less susceptible to noise [34]. The resonant strain gage is a beam that is suspended slightly above the strain member and attached to it at both ends. The strain gage beam is magnetically excited with pulses, and the frequency of vibration is detected by a magnetic detection circuit. As the beam is stretched by mechanical strain, the frequency of vibration increases. These sensors provide higher resolution than typical piezoresistors and have a lower temperature coefficient. The resonant sensors, however, require a complex three-dimensional fabrication technique, unlike the typical piezoresistors which require only planar techniques.

5.3.2 Pressure

One of the most commercially successful microsensor technologies is the pressure sensor. Silicon micromachined pressure sensors are available that measure pressure ranges from around one to several thousand kPa, with resolutions as fine as one part in ten thousand. These sensors incorporate a silicon micromachined diaphragm that is subjected to fluid (i.e., liquid or gas) pressure, which causes dilation of the diaphragm. The simplest of these utilize piezoresistors mounted on the back of the diaphragm to measure deformation, which is a function of the pressure. Examples of these devices are those by Fujii et al. [35] and Mallon et al. [36]. A variation of this configuration is the device by Ikeda et al. Instead of a piezoresistor to measure strain, an electromagnetically driven and sensed resonant strain gage, as discussed in the previous section, is utilized [37]. Still another variation on the same theme is the capacitive measurement approach, which measures the capacitance between the diaphragm and an electrode that is rigidly mounted and parallel to the diaphragm. An example of this approach is by Nagata et al. [38]. A more complex approach to pressure measurement is that by Stemme and Stemme, which utilizes resonance of the diaphragm to detect pressure [39]. In this device, the diaphragm is capacitively excited and optically detected. The pressure imposes a mechanical load on the diaphragm, which increases the stiffness and, in turn, the resonant frequency.

5.3.3 Acceleration

Another commercially successful microsensor is the silicon microfabricated accelerometer, which in various forms can measure acceleration ranges from well below one to around a thousand meters per square second (i.e., sub-g to several hundred g's), with resolutions of one part in 10,000. These sensors incorporate a micromachined suspended proof mass that is subjected to an inertial force in response to an acceleration, which causes deflection of the supporting flexures. One means of measuring the deflection is by utilizing piezoresistive strain gages mounted on the flexures. The primary disadvantage to this approach is the temperature sensitivity of the piezoresistive gages. An alternative to measuring the deflection of the proof mass is via capacitive sensing. In these devices, the capacitance is measured between the proof mass and an electrode that is rigidly mounted and parallel. Examples of this approach are those by Boxenhorn and Greiff [40], Leuthold and Rudolf [41], and Seidel et al. [42]. Still another means of measuring the inertial force on the proof mass is by measuring the resonant frequency of the supporting flexures. The inertial force due to acceleration will load the flexure, which will alter its resonant frequency. The frequency of vibration is therefore a measure of the acceleration. These types of devices utilize some form of transduction to excite the structural resonance of the supporting flexures, and then utilize some other measurement technique to detect the frequency of vibration. Examples of this type of device are those by Chang et al. [43], which utilize electrostatic excitation and capacitive detection, and by Satchell and Greenwood [44], which utilize thermal excitation and piezoresistive detection. These types of accelerometers entail additional complexity, but typically offer improved measurement resolution. Still another variation of the micro-accelerometer is the force-balanced type. This type of device measures position of the proof mass (typically by capacitive means) and utilizes a feedback loop and electrostatic or electromagnetic actuation to maintain zero deflection of the mass. The acceleration is then a function of the actuation effort. These devices are characterized by a wide bandwidth and high sensitivity, but are typically more complex and more expensive than other types. Examples of force-balanced devices are those by Chau et al. [45], and Kuehnel and Sherman [46], both of which utilize capacitive sensing and electrostatic actuation.

5.3.4 Force

Silicon microfabricated force sensors incorporate measurement approaches much like the microfabricated pressure sensors and accelerometers. Various forms of these force sensors can measure forces ranging on the order of millinewtons to newtons, with resolutions of one part in 10,000. Mechanical sensing typically utilizes a beam or a flexure support which is elastically deflected by an applied force, thereby transforming force measurement into measurement of strain or displacement, which can be accomplished by piezoresistive or capacitive means. An example of this type of device is that of Despont et al. [47], which utilizes capacitive measurement. Higher resolution devices are typically of the resonating beam type, in which the applied force loads a resonating beam in tension. Increasing the applied tensile load results in an increase in resonant frequency. An example of this type of device is that of Blom et al. [48].

5.3.5 Angular Rate Sensing (Gyroscopes)

A conventional-scale gyroscope utilizes the spatial coupling of the angular momentum-based gyroscopic effect to measure angular rate. In these devices, a disk is spun at a constant high rate about its primary axis, so that when the disk is rotated about an axis not colinear with the primary (or spin) axis, a torque results in an orthogonal direction that is proportional to the angular velocity. These devices are typically mounted in gimbals with low-friction bearings, incorporate motors that maintain the spin velocity, and utilize strain gages to measure the gyroscopic torque (and thus angular velocity). Such a design would not be appropriate for a microsensor due to several factors, some of which include the diminishing effect of inertia (and thus momentum) at small scales, the lack of adequate bearings, the lack of appropriate micromotors, and the lack of an adequate three-dimensional microfabrication processes. Instead, microscale angular rate sensors are of the vibratory type, which incorporate Coriolis-type effects rather than

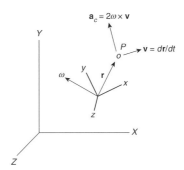

FIGURE 5.8 Illustration of Coriolis acceleration, which results from translation within a reference frame that is rotating with respect to an inertial reference frame.

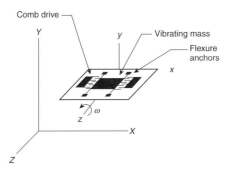

FIGURE 5.9 Schematic of a vibratory gyroscope.

the angular momentum-based gyroscopic mechanics of conventional-scale devices. A Coriolis acceleration results from linear translation within a coordinate frame that is rotating with respect to an inertial reference frame. In particular, if the particle in Figure 5.8 is moving with a velocity v within the frame xyz, and if the frame xyz is rotating with an angular velocity of ω with respect to the inertial reference frame XYZ, then a Coriolis acceleration will result equal to $\mathbf{a}_c = 2\omega \times v$. If the object has a mass m, a Coriolis inertial force will result equal to $\mathbf{F}_c = -2m\omega \times v$ (minus sign because direction is opposite $\mathbf{a}_c$). A vibratory gyroscope utilizes this effect as illustrated in Figure 5.9. A flexure-suspended inertial mass is vibrated in the x-direction, typically with an electrostatic comb drive. An angular velocity about the z-axis will generate a Coriolis acceleration, and thus force, in the y-direction. If the "external" angular velocity is constant and the velocity in the x-direction is sinusoidal, then the resulting Coriolis force will be sinusiodal, and the suspended inertial mass will vibrate in the y-direction with an amplitude proportional to the angular velocity. The motion in the y-direction, which is typically measured capacitively, is thus a measure of the angular rate. Examples of these types of devices are those by Bernstein et al. [49] and Oh et al. [50]. Note that though vibration is an essential component of these devices, they are not technically resonant sensors, since they measure amplitude of vibration rather than frequency.

5.4 Nanomachines

Nanomachines are devices that range in size from the smallest of MEMS devices down to devices assembled from individual molecules [51]. This section briefly introduces energy sources, structural hierarchy, and the projected future of the assembly of nanomachines. Built from molecular components performing individual mechanical functions, the candidates for energy sources to actuate nanomachines are limited to those that act on a molecular scale. Regarding manufacture, the assembly of nanomachines is by nature a one-molecule-at-a-time operation. Although microscopy techniques are currently used for the assembly of nanostructures, self-assembly is seen as a viable means of mass production.

In a molecular device a discrete number of molecular components are combined into a supramolecular structure where each discrete molecular component performs a single function. The combined action of these individual molecules causes the device to operate and perform its various functions. Molecular devices require an energy source to operate. This energy must ultimately be used to activate the component molecules in the device, and so the energy must be chemical in nature. The chemical energy can be obtained by adding hydrogen ions, oxidants, etc., by inducing chemical reactions by the impingement of light, or by the actions of electrical current. The latter two means of energy activation, photochemical and electrochemical energy sources, are preferred since they not only provide energy for the operation of the device, but they can also be used to locate and control the device. Additionally, such energy transduction can be used to transmit data to report on the performance and status of the device. Another reason for the preference for photochemical- and electrochemical-based molecular devices is that, as these devices are required to operate in a cyclic manner, the chemical reactions that drive the system must be reversible. Since photochemical and electrochemical processes do not lead to the accumulation of products of reaction, they readily lend themselves to application in nanodevices.

Molecular devices have recently been designed that are capable of motion and control by photochemical methods. One device is a molecular plug and socket system, and another is a piston-cylinder system [51]. The construction of such supramolecular devices belongs to the realm of the chemist who is adept at manipulating molecules.

As one proceeds upwards in size to the next level of nanomachines, one arrives at devices assembled from (or with) single-walled carbon nanotubes (SWNTs) and/or multi-walled carbon nanotubes (MWNTs) that are a few nanometers in diameter. We will restrict our discussion to carbon nanotubes (CNTs) even though there is an expanding database on nanotubes made from other materials, especially bismuth. The strength and versatility of CNTs make them superior tools for the nanomachine design engineer. They have high electrical conductivity with current carrying capacity of a billion amperes per square centimeter. They are excellent field emitters at low operating voltages. Moreover, CNTs emit light coherently and this provides for an entire new area of holographic applications. The elastic modulus of CNTs is the highest of all materials known today [52]. These electrical properties and extremely high mechanical strength make MWNTs the ultimate atomic force microscope probe tips. CNTs have the potential to be used as efficient molecular assembly devices for manufacturing nanomachines one atom at a time.

Two obvious nanotechnological applications of CNTs are nanobearings and nanosprings. Zettl and Cumings [53] have created MWNT-based linear bearings and constant force nanosprings. CNTs may potentially form the ultimate set of nanometer-sized building blocks, out of which nanomachines of all kinds can be built. These nanomachines can be used in the assembly of nanomachines, which can then be used to construct machines of all types and sizes. These machines can be competitive with, or perhaps surpass existing devices of all kinds.

SWNTs can also be used as electromechanical actuators. Baughman et al. [54] have demonstrated that sheets of SWNTs generate larger forces than natural muscle and larger strains than high-modulus ferroelectrics. They have predicted that actuators using optimized SWNT sheets may provide substantially higher work densities per cycle than any other known actuator. Kim and Lieber [55] have built SWNT and MWNT nanotweezers. These nanoscale electromechanical devices were used to manipulate and interrogate nanostructures. Electrically conducting CNTs were attached to electrodes on pulled glass micropipettes. Voltages applied to the electrodes opened and closed the free ends of the CNTs. Kim and Lieber demonstrated the capability of the nanotweezers by grabbing and manipulating submicron clusters and nanowires. This device could be used to manipulate biological cells or even manipulate organelles and clusters within human cells. Perhaps, more importantly, these tweezers can potentially be used to assemble other nanomachines.

A wide variety of nanoscale manipulators have been proposed [56] including pneumatic manipulators that can be configured to make tentacle, snake, or multi-chambered devices. Drexler has proposed telescoping nanomanipulators for precision molecular positioning and assembly work. His manipulator has a cylindrical shape with a diameter of 35 nm and an extensible length of 100 nm. A number of six degree of freedom Stewart platforms have been proposed [56], including one that allows strut lengths to be moved in 0.10 nm increments across a 100 nm work envelope. A number of other nanodevices including box-spring accelerometers, displacement accelerometers, pivoted gyroscopic accelerometers, and gimbaled nanogyroscopes have been proposed and designed [56].

Currently, much thought is being devoted to molecular assembly and self-replicating devices (self-replicating nanorobots). Self-assembly is arguably the only way for nanotechnology to advance in an engineering or technological sense. Assembling a billion or trillion atom device—one atom at a time— would be a great accomplishment. It would take a huge investment in equipment, labor, and time. Freitas [56] describes the infrastructure needed to construct a simple medical nanorobot: a 1-μm spherical respirocyte consisting of about 18 billion atoms. He estimates that a factory production line deploying a coordinated system of 100 macroscale scanning probe microscope (SPM) assemblers, where each assembler is capable of depositing one atom per second on a convergently-assembled workpiece, would result in a manufacturing throughput of two nanorobots per decade. If one conjectures about enormous increases in assembler manufacturing rates even to the extent of an output of one nanorobot per minute, it would take two million years to build the first cubic centimeter therapeutic dosage of nanorobots. Thus, it is clear that the future of medical nanotechnology and nanoengineering lies in the direction of self-assembly and self-replication.

References

1. Bridgman, P. W., *Dimensional Analysis,* 2nd Ed., Yale University Press, 1931.
2. Buckingham, E., "On physically similar systems: illustrations of the use of dimensional equations," *Physical Review,* 4(4):345–376, 1914.
3. Huntley, H. E., *Dimensional Analysis,* Dover Publications, 1967.
4. Langhaar, H. L., *Dimensional Analysis and Theory of Models,* John Wiley & Sons, 1951.
5. Taylor, E. S., *Dimensional Analysis for Engineers,* Oxford University Press, 1974.
6. Israelachvili, J. N., *Intermolecular and Surface Forces,* Academic Press, 1985, pp. 9–10.
7. Fearing, R. S., "Microactuators for microrobots: electric and magnetic," *Workshop on Micromechatronics, IEEE International Conference on Robotics and Automation,* 1997.
8. Bobbio, S. M., Keelam, M. D., Dudley, B. W., Goodwin-Hohansson, S., Jones, S. K., Jacobson, J. D., Tranjan, F. M., Dubois, T. D., "Integrated force arrays," *Proceedings of the IEEE Micro Electro Mechanical Systems,* 149–154, 1993.
9. Jacobson, J. D., Goodwin-Johansson, S. H., Bobbio, S. M., Bartlett, C. A., Yadon, L. N., "Integrated force arrays: theory and modeling of static operation," *Journal of Microelectromechanical Systems,* 4(3):139–150, 1995.
10. Yamaguchi, M., Kawamura, S., Minami, K., Esashi, M., "Distributed electrostatic micro actuators," *Proceedings of the IEEE Micro Electro Mechanical Systems,* 18–23, 1993.
11. Kim, C. J., Pisano, A. P., Muller, R. S., "Silicon-processed overhanging microgripper," *Journal of Microelectromechanical Systems,* 1(1):31–36, 1992.
12. Matsubara, T., Yamaguchi, M., Minami, K., Esashi, M., "Stepping electrostatic microactuator," *International Conference on Solid-State Sensor and Actuators,* 50–53, 1991.
13. Niino, T., Egawa, S., Kimura, H., Higuchi, T., "Electrostatic artificial muscle: compact, high-power linear actuators with multiple-layer structures," *Proceedings of the IEEE Conference on Micro Electro Mechanical Systems,* 130–135, 1994.
14. Huang, J. B., Mao, P. S., Tong, Q. Y., Zhang, R. Q., "Study on silicon electrostatic and electroquasistatic micromotors," *Sensors and Actuators,* 35:171–174, 1993.
15. Mehragany, M., Bart, S. F., Tavrow, L. S., Lang, J. H., Senturia, S. D., Schlecht, M. F., "A study of three microfabricated variable-capacitance motors," *Sensors and Actuators,* 173–179, 1990.
16. Trimmer, W., Gabriel, K., "Design considerations for a practical electrostatic micromotor," *Sensors and Actuators,* 11:189–206, 1987.

17. Mehregany, M., Nagarkar, P., Senturia, S. D., Lang, J. H., "Operation of microfabricated harmonic and ordinary side-drive motors," *Proceedings of the IEEE Conference on Micro Electro Mechanical Systems*, 1–8, 1990.
18. Dhuler, V. R., Mehregany, M., Phillips, S. M., "A comparative study of bearing designs and operational environments for harmonic side-drive micromotors," *IEEE Transactions on Electron Devices*, 40(11):1985–1989, 1993.
19. Price, R. H., Wood, J. E., Jacobsen, S. C., "Modeling considerations for electrostatic forces in electrostatic microactuators," *Sensors and Actuators*, 20:107–114, 1989.
20. Trimmer, W., Jebens, R., "An operational harmonic electrostatic motor," *Proceedings of the IEEE Conference on Micro Electro Mechanical Systems*, 13–16, 1989.
21. Trimmer, W., Jebens, R., "Harmonic electrostatic motors," *Sensors and Actuators*, 20:17–24, 1989.
22. Furuhata, T., Hirano, T., Lane, L. H., Fontanta, R. E., Fan, L. S., Fujita, H., "Outer rotor surface micromachined wobble micromotor," *Proceedings of the IEEE Conference on Micro Electro Mechanical Systems*, 161–166, 1993.
23. Trimmer, W., Jebens, R., "Actuators for microrobots," *IEEE Conference on Robotics and Automation*, 1547–1552, 1989.
24. Trimmer, W., "Microrobots and micromechanical systems," *Sensors and Actuators*, 19:267–287, 1989.
25. Inoue, T., Hamasaki, Y., Shimoyama, I., Miura, H., "Micromanipulation using a microcoil array," *Proceedings of the IEEE International Conference on Robotics and Automation*, 2208–2213, 1996.
26. Liu, C., Tsao, T., Tai, Y., Ho, C., "Surface micromachined magnetic actuators," *Proceedings of the IEEE Conference on Micro Electro Mechanical Systems*, 57–62, 1994.
27. Shinozawa, Y., Abe, T., Kondo, T., "A proportional microvalve using a bi-stable magnetic actuator," *Proceedings of the IEEE Conference on Micro Electro Mechanical Systems*, 233–237, 1997.
28. Wagner, B., Benecke, W., "Microfabricated actuator with moving permanent magnet," *Proceedings of the IEEE Conference on Micro Electro Mechanical Systems*, 27–32, 1991.
29. Yanagisawa, K., Tago, A., Ohkubo, T., Kuwano, H., "Magnetic microactuator," *Proceedings of the IEEE Conference on Micro Electro Mechanical Systems*, 120–124, 1991.
30. Judy, J., Muller, R. S., Zappe, H. H., "Magnetic microactuation of polysilicon flexure structures," *Journal of Microelectromechanical Systems*, 4(4):162–169, 1995.
31. Ahn, C. H., Kim, Y. J., Allen, M. G., "A planar variable reluctance magnetic micromotor with fully integrated stator and wrapped coils," *Proceedings of the IEEE Conference on Micro Electro Mechanical Systems*, 1–6, 1993.
32. Guckel, H., Christenson, T. R., Skrobis, K. J., Jung, T. S., Klein, J., Hartojo, K. V., Widjaja, I., "A first functional current excited planar rotational magnetic micromotor," *Proceedings of the IEEE Conference on Micro Electro Mechanical Systems*, 7–11, 1993.
33. Guckel, H., Sneigowski, J. J., Christenson, T. R., Raissi, F., "The application of fine grained, tensile polysilicon to mechanically resonant transducers," *Sensor and Actuators*, A21–A23:346–351, 1990.
34. Ikeda, K., Kuwayama, H., Kobayashi, T., Watanabe, T., Nishikawa, T., Yoshida, T., Harada, K., "Silicon pressure sensor integrates resonant strain gauge on diaphragm," *Sensors and Actuators*, A21–A23:146–150, 1990.
35. Fujii, T., Gotoh, Y., Kuroyanagi, S., "Fabrication of microdiaphragm pressure sensor utilizing micromachining," *Sensors and Actuators*, A34:217–224, 1992.
36. Mallon, J., Pourahmadi, F., Petersen, K., Barth, P., Vermeulen, T., Bryzek, J., "Low-pressure sensors employing bossed diaphragms and precision etch-stopping," *Sensors and Actuators*, A21–23:89–95, 1990.
37. Ikeda, K., Kuwayama, H., Kobayashi, T., Watanabe, T., Nishikawa, T., Yoshida, T., Harada, K., "Three-dimensional micromachining of silicon pressure sensor integrating resonant strain gauge on diaphragm," *Sensors and Actuators*, A21–A23:1007–1009, 1990.
38. Nagata, T., Terabe, H., Kuwahara, S., Sakurai, S., Tabata, O., Sugiyama, S., Esashi, M., "Digital compensated capacitive pressure sensor using cmos technology for low-pressure measurements," *Sensors and Actuators*, A34:173–177, 1992.

39. Stemme, E., Stemme, G., "A balanced resonant pressure sensor," *Sensors and Actuators*, A21–A23: 336–341, 1990.
40. Boxenhorn, B., Greiff, P., "Monolithic silicon accelerometer," *Sensors and Actuators*, A21–A23:273–277, 1990.
41. Leuthold, H., Rudolf, F., "An ASIC for high-resolution capacitive microaccelerometers," *Sensors and Actuators*, A21–A23:278–281, 1990.
42. Seidel, H., Riedel, H., Kolbeck, R., Muck, G., Kupke, W., Koniger, M., "Capacitive silicon accelerometer with highly symmetrical design," *Sensors and Actuators*, A21–A23:312–315, 1990.
43. Chang, S. C., Putty, M. W., Hicks, D. B., Li, C. H., Howe, R. T., "Resonant-bridge two-axis microaccelerometer," *Sensors and Actuators*, A21–A23:342–345, 1990.
44. Satchell, D. W., Greenwood, J. C., "A thermally-excited silicon accelerometer," *Sensors and Actuators*, A17:241–245, 1989.
45. Chau, K. H. L., Lewis, S. R., Zhao, Y., Howe, R. T., Bart, S. F., Marchesilli, R. G., "An integrated force-balanced capacitive accelerometer for low-g applications," *Sensors and Actuators*, A54:472–476, 1996.
46. Kuehnel, W., Sherman, S., "A surface micromachined silicon accelerometer with on-chip detection circuitry," *Sensors and Actuators*, A45:7–16, 1994.
47. Despont, Racine, G. A., Renaud, P., de Rooij, N. F., "New design of micromachined capacitive force sensor," *Journal of Micromechanics and Microengineering*, 3:239–242, 1993.
48. Blom, F. R., Bouwstra, S., Fluitman, J. H. J., Elwenspoek, M., "Resonating silicon beam force sensor," *Sensors and Actuators*, 17:513–519, 1989.
49. Bernstein, J., Cho, S., King, A. T., Kourepenis, A., Maciel, P., Weinberg, M., "A micromachined comb-drive tuning fork rate gyroscope," *IEEE Conference on Micro Electro Mechanical Systems*, 143–148, 1993.
50. Oh, Y., Lee, B., Baek, S., Kim, H., Kim, J., Kang, S., Song, C., "A surface-micromachined tunable vibratory gyroscope," *IEEE Conference on Micro Electro Mechanical Systems*, 272–277, 1997.
51. Venturi, M., Credi, A., Balzani, V., "Devices and machines at the molecular level," *Electronic Properties of Novel Materials, AIP Conf. Proc.*, 544:489–494, 2000.
52. Ajayan, P. M., Charlier, J. C., Rinzler, A. G., "PNAS," 96:14199–14200, 1999.
53. Zettl, A., Cumings, J., "Sharpened nanotubes, nanobearings and nanosprings," *Electronic Properties of Novel Materials, AIP Conf. Proc.*, 544:526–531, 2000.
54. Baughman, R. H., et al., "Carbon nanotube actuators," *Science*, 284:1340–1344, 1999.
55. Kim, P., Lieber, C. M., "Nanotube nanotweezers," *Science*, 286:2148–2150, 1999.
56. Freitas, R. A., "Nanomedicine," Vol. 1, *Landes Bioscience*, Austin, 1999.

6
Mechatronics Engineering Curriculum Design

6.1	Introduction	6-1
6.2	The Identity of Mechatronics	6-2
6.3	Legitimacy of Mechatronics	6-3
6.4	The Selection of Mechatronics	6-4
6.5	The Communication of Mechatronics	6-5
6.6	Fine, but So What?	6-5
6.7	Putting It All Together in a Curriculum	6-6
6.8	The Evolution of Mechatronics	6-7
	Stage 1: The Origin; No Interaction • Stage 2: Multidisciplinary Stage • Stage 3: Cross-Disciplinary Stage • Stage 4: Curriculum Stage • Stage 5 and 6: Organizational Stage	
6.9	Where (and What) Is Mechatronics Today?	6-9
References	....	6-9

Martin Grimheden
Royal Institute of Technology

6.1 Introduction

Several attempts have been made to establish a common mechatronics curriculum since the emergence of the subject. Conferences are dedicated to the task of identifying prerequisites, mandatory basic courses, and educational methods appropriate for the subject. In this chapter, we attempt to describe the subject of mechatronics according to a set of questions, or dimensions, which has been established primarily in Swedish institutes of higher education as the "Didactical Analysis." The results of this analysis can be used as guidelines for curriculum design and educational methods, which in the case of mechatronics, points toward a curriculum based on the synergistic use of previously acquired knowledge.

Mechatronics is a special subject. Most people would agree with the idea that mechatronics is built on mechanical engineering, electrical engineering, computer science, automatic control, sensors, actuators, microcontrollers, and so forth. These subjects, however, each have their own specialists, and their own educational programs and curriculum. When discussing the optimal curriculum for mechatronics, the most common mistake is to leap into discussions regarding how much attention should be given to the different subjects; how many courses should be taken in electrical engineering; and how much automatic control should the students master. These discussions are necessary, and few disagree that mechatronics students ought to study at least some of these subjects, but let us start from the beginning. We shall skip discussing the number of weeks dedicated to each subject for the time being and have a look at this didactical analysis. In this chapter, we will show that the number of weeks dedicated to each subject is not what is important. There is more to it than that.

6.2 The Identity of Mechatronics

The didactical analysis, as introduced by Dahlgren [1], can be illustrated with a set of four questions, as shown in Figure 6.1. The first of these questions relates to the identity of mechatronics: "What is mechatronics?" The question of identity is defined, or described, as "what distinguishes this particular field of knowledge?" The identity can be mapped on a scale between two extremes, from a disciplinary identity to a thematic identity. Many traditional subjects such as mathematics, chemistry, and physics are regularly viewed as disciplinary subjects, which mean that there exists a consensus in the society regarding the contents and definitions of the subjects. Commonly agreeable definitions exist, the subject is classified into sub-subjects, and universities show similarities in organizing courses, programs, and research departments.

Mechatronics shows tendencies toward the thematic identity. When participating in a mechatronics conference, lively discussions are held regarding the definition of the subject. The subject of mechatronics is defined differently at various universities and institutes; at some places mechatronics is organized as a division within mechanical engineering, and at some the "mechatronics people" consists of a research team within computer science or automatic control. Mechatronics journals have different scopes, and mechatronics educational programs have different curricula. Instead of traditional disciplines the subject is regarded as a cross-disciplinary subject, an applied subject or as a theme. The theme is used to describe applications, common denominators and aspects of the subject, and it is within these particular themes that the identity can be found.

The starting point for the analysis is based on a definition made by Comerford [2] in IEEE Spectrum: "the synergistic combination of precision mechanical engineering, electronic control and systems thinking in the design of products and manufacturing processes." Almost the same definition was published in the editorial of the first issue of the first volume of the IEEE/ASME Transactions on Mechatronics: "the synergistic integration of mechanical engineering with. . ." [3]. A similar definition, also based on synergistic use, was presented by Erkmen et al. [4] and Craig [5] in the special issue of IEEE Robotics and Automation Magazine, dedicated to mechatronics education. The keyword is synergistic: "synergistic combination of. . .." Mechatronics should not be considered as the union between mechanical and electrical engineering, automatic control and computer science, or other combinations of traditional disciplines within an engineering sphere. The identity of mechatronics is defined as synergistic combination of a number of disciplinary subjects. Mechatronics skill then relates to the ability to perform this synergistic combination.

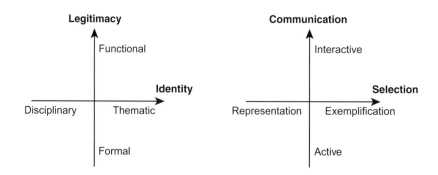

FIGURE 6.1 Illustration of the "four questions" representing the didactical analysis, together with their extremes.

A subject with a thematic identity is typically developed within disciplinary subjects where existing subject delimitations are replaced by new categories of knowledge. In some cases, cross-disciplinary activities and research fields create new subjects and, in these cases, the common denominator is the theme that joins the respective groups or areas together. Another example of a subject with a thematic identity is automatic control—a subject that has developed from the subjects of mathematics, electrical engineering, and physics.

To conclude—what is mechatronics? It is a thematic subject that clearly lacks a disciplinary tradition and academic consensus. Different universities use different themes to describe the identity. Some exemplify mechatronics with robotics, some with embedded systems, and some with the union of various subjects. Some universities use a definition based on the synergistic combination of a number of subjects. In this chapter, this definition will be used to identify mechatronics, which will affect the results of the following three questions as well.

6.3 Legitimacy of Mechatronics

The question of legitimacy is defined as "the relation between the actual outcome of the educational efforts undertaken by the university and the actual demands that are put on the students' abilities by the society and/or industry at the end of the students' education." To simplify: "Why teach mechatronics?" and "What does the industry want?" These questions will also be looked upon using the two extremes: functional and formal legitimacy.

A comparison can be made between the two extremes of legitimacy and two extremes of learning focus: theoretical knowledge and practical skills. Formal legitimacy relates to knowledge and functional legitimacy relates to skills. Students reading textbooks tend to gain knowledge, and students actively practicing in a laboratory tend to gain skills. Functional skills are rarely taught in lectures or by reading books, but rather through hands-on laboratory work, experiments, and trial and error.

In the case of mechatronics, the legitimacy can be studied by investigating the demands put on the students' abilities by the hiring industries and the actual knowledge and skills developed by the students on graduation. The legitimacy can also be studied simply by questioning the hiring industry of their level of satisfaction with the hired mechatronics engineers or by studying the employment level of mechatronics engineers and the respective positions. There is, however, a relation between the identity of the subject and the legitimacy of the subject. Since mechatronics is defined with a thematic identity, and the identity consists of the synergistic combination of various subjects, the legitimacy of mechatronics has to be viewed in the light of this identity—the ability to perform synergistic combination of various subjects. The better the results of the educational efforts, the better the ability of the students to perform synergistic combinations.

In the case of mechatronics, the subject has a functional legitimacy. The identity is thematic and basically no consensus exists about the definition of the subject. The subject is rather new in the society, at least compared to electrical engineering and mechanical engineering. All this combined could mean that industries hiring mechatronics engineers search for functional skills rather than formal knowledge.

This can also be viewed in the light of a continuously ongoing debate whether mechatronics engineers are generalists or specialists. A mechatronics engineer could be a specialist in the theme of mechatronics and a specialist in synergistic combination, particularly in the ability to synergistically combine knowledge and skills in various subjects to perform product development, to design new mechatronic products, or to analyze mechatronic systems. The mechatronics engineer will rarely be a specialist in any one subject owing to the lack of time and resources a mechatronics student can apply to a single subject compared to a student who specializes in, for example, electrical engineering.

When studying mechatronics from the didactical approach, by analyzing the identity and legitimacy, it is clear that mechatronics is a special subject that should be viewed rather differently from the subjects it is based on. Mechatronics can never be about knowing a little something about everything; mechatronics is about being able to make use of this "everything." Mechatronics is regarded by some as a philosophy [6], or as the ability to design products, which points toward legitimacy built on functional skills rather than formal knowledge.

A comparison will here be made with other subjects. Medicine for example, has been a subject undergoing major change since the 1960s. Before this, medicine on a university level was regarded with a formal legitimacy. The requirements were mainly defined by government agencies, which stipulated a list of courses and credits that students should have taken to be able to become a physician. Many of these requirements were based on formal knowledge in disciplinary subjects, such as chemistry and anatomy. In the 1960s, in Canada, the first courses and programs organized according to problem-based learning (PBL) emerged. The idea was to move from a formal legitimacy toward a functional legitimacy. The PBL programs focused on a problem, such as a disease or a patient showing certain symptoms, and thereby on the ability to diagnose a patient. The requirements changed from "knowing a certain number of indicators" to "recognizing symptoms" and "being able to diagnose."

In the case of mechatronics, one problem that arises is that most of the subjects that mechatronics rely on are regarded with a formal legitimacy and a disciplinary identity, and it is therefore difficult to apply a completely different mindset on the subject of mechatronics. This is, however, not unique to mechatronics. In medicine, the ability to recognize symptoms and diagnose diseases relies on utilization of formal knowledge in disciplinary subjects, such as chemistry.

6.4 The Selection of Mechatronics

The third question, the selection of mechatronics, can be described as "What should we teach?" or "How should we choose which material to include in our mechatronics courses?" As with the earlier questions, this can be viewed in the light of two extremes—horizontal representation or vertical exemplification.

Horizontal representation means that we choose to teach "something of everything." Each aspect of the subject should be represented in some way; all aspects should be covered in the mechatronics program. The opposite—vertical exemplification means that we choose certain aspects of the subject that we exemplify; we teach "everything of something." An example: a basic course in computer programming could teach something of many programming languages. The course could cover general programming and the major differences between some languages. The opposite, vertical exemplification would imply that the course focuses completely on one language, one platform, and so forth, with the objective that students be exceptionally qualified in one language. There are certainly advantages and disadvantages with both extremes, and the most preferable solution that most teachers adopt is to try to cover both. We usually want our students to master at least one language, but also to have knowledge about programming in general.

However, there are major differences between these two approaches. The vertical exemplification means that you study details; you become an expert in one narrow field. The horizontal representation means that the student never gets the opportunity to study details, to become an expert. The idea of vertical exemplification is that it is always easier to move from a detailed level to the whole picture than the opposite. If you master one programming language very well, it is rather easy to learn another, but no matter how much theoretical, broad knowledge you have about programming languages in general, you will not immediately be able to apply this knowledge and to transfer the knowledge into programming skills unless utilizing previous related experiences and practices in programming.

When applying the identity and legitimacy of mechatronics to the question of selection, the answer is rather obvious. Since mechatronics has a thematic identity and a functional legitimacy, the selection ought to be a vertical exemplification. The reason is that the thematic identity was based on the fact that we do not agree on what mechatronics is. We do not have a common curricula; no textbook that we consider in mechatronics covers the field, as we define it. The subject is not established in the surrounding society.

The functional legitimacy was based on the fact that the hiring industry does not consider formal curricula or require a certain number of textbooks or courses to have been taken. Basically, the industry does not have a clear view of what a mechatronics engineer should or could have studied. Instead, the hiring industry considers functional skills paramount. These skills then relate to the themes used to describe the identity. If we use robotics to describe what we mean with mechatronics (as a theme), the hiring industry should hire students with skills and abilities in robotics to design, implement, or operate robots. Therefore our selection ought to be exemplifying—a student who is really good and experienced in designing and building a robot is more desirable than a student with a broad knowledge in robotics, no matter how broad the overview is. That is, if the identity of the subject is related to this synergistic use. If the identity is disciplinary and the legitimacy is related to formal knowledge, then knowledge about robots is sufficient. Mechatronics is, however, not knowledge about mechatronics; mechatronics is knowledge and skills in mechatronics.

6.5 The Communication of Mechatronics

The fourth and final question relates to "How we should teach?" and "How the material that we select should be communicated between the faculty and the students?" Either we use a communicational model based on an active communication or an interactive communication.

An interactive communication implies that what is communicated is dependent on the recipient. What the teacher is communicating depends on the level of understanding of the student, of the existing knowledge and skill with each particular student, and of the response of the previous communication. Interactive communication is communication with feedback.

Historically, most educational communication in our universities is based on the active model. What is said in lectures, for example, is usually based on a model of student behavior and level of understanding—it is really difficult to make a lecture interactive with each student. However, there is more to education than lectures. A student performing experiments, building prototypes, and doing other hands-on activities usually adapt an interactive mode. When the student chooses how to perform the experiment and how to build the prototype, this can be regarded as an interactive communication between the student and the faculty, especially since the communication between the students and teachers tend to be driven by the students' own questions and the students' problems.

The didactical analysis therefore motivates a move from traditional lectures, textbooks, and courses toward problem-based, project-organized courses, experimental work, and hands-on approaches. This move should also reflect the selection, legitimacy, and identity of the subject. The projects, problems, and activities should then exemplify the theme of the subject. A mechatronics student could, for example, very well be designing, implementing, and operating robots in a mechatronics course—without lectures, textbooks, and exams in mechatronics theory.

6.6 Fine, but So What?

So far most people usually agree. We can accept that the subject has a thematic identity, we buy the functional legitimacy, and the vertical exemplifying and interactive communication seems fair. However, typical courses at the university level show completely opposite characteristics, as shown in Figure 6.2.

Mechatronics has a thematic identity, based on the concept of synergy. Examples of themes are robotics, embedded systems, and intelligent products. However, universities are organized in traditional disciplines. Educational programs and courses are usually categorized according to subjects and not themes. A traditional academic subject usually has higher status than a new, applied cross-disciplinary subject.

Mechatronics has a functional legitimacy. The usual requirements are related to abilities to apply knowledge and skills, to synthesis rather than analysis; the ability to design and implement a robot,

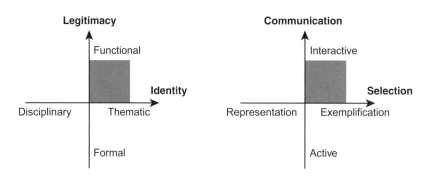

FIGURE 6.2 Illustration of the didactic analysis applied to the subject of mechatronics.

for example. However, most traditional subjects rely on a formal legitimacy. The requirements on the programs and courses are usually not specified by the industry, but rather directed by the faculty. A typical exam does not usually measure functional skills.

Mechatronics should therefore be taught with an exemplifying selection. Students should be experts in synergistic combination by practicing this. However, most traditional courses tend to give overviews of the subjects, to cover several aspects if not all. Most courses are general rather than specific.

Mechatronics should be taught in an interactive mode. This means that mechatronics courses should be problem based and project organized. The courses should enable the students to delve deep into, for example, the design of a robot. However, most courses are communicated with traditional methods such as lectures, textbooks, seminars, and written exams. This is the de facto situation at most universities today.

One conclusion from the didactical analysis is that it is not so much a matter of "What?" as of "How?" Discussions of how to teach mechatronics should be more about educational methods than content. The content is certainly important, but not as important as the educational methods. This relates also to the discussion whether the mechatronics engineer is a generalist or a specialist. If the focus is on content and not enough focus is put on the educational methods, the mechatronics engineer risks being an engineer with knowledge in some subjects, but without the possibility to compete with subject specialists, such as electrical engineers. The market for these generalist mechatronics is small, perhaps only small companies without possibilities of hiring specialists in each subject.

On the other hand, the mechatronics engineer who is a specialist in the synergistic combination of electrical engineering, mechanical engineering, automatic control, and so forth, will be able to utilize knowledge and expertise with specialists, as well as his own (though limited) competence in each area. The creation of these engineers requires appropriate educational methods and a major focus on skills such as project management, leadership skills, and communicational skills.

6.7 Putting It All Together in a Curriculum

One result of the above analysis is the conclusion that it is neither possible nor fruitful to attempt to establish a common curriculum in mechatronics until a consensus on the identity and legitimacy is reached. This is most probably also the reason why most of these attempts fail and why it is so difficult to agree on a curriculum. Each participant in the discussions has a different perception of required courses one should take.

To make matters evenmore complicated, imagine the following: A large capstone course is given during the final year of amechatronics masters' program. The students are working on a large-scale project with a corporate sponsor. The project is about designing a rather complex mechatronic product and finalizing a functional prototype. In a sense, the educational result of the course is reflected in the success of finalizing the project, since this implies motivated students and gives immediate feedback. The success of the project is however related to the diversity of the students rather than the optimal curriculum. The success of the project usually requires more skills and competencies than one curriculum can cover, but with a diverse team these skills can be distributed among several members. The project benefits greatly if at least one student has good leadership skills and if individual students have expertise in various areas. A homogenous student team does not reflect a work-like setting and therefore does not prepare students for a future career as mechatronics engineers.

Naturally, there has to be a common denominator for mechatronics engineers. To be able to create synergistic combination requires that you have something to combine. Previous studies of mechatronics educational programs, however, give the following conclusions [6–9].

Most mechatronics programs have evolved from a mechanical engineering perspective; as a specialization in a program in mechanical engineering, from an engineering design faculty team, or as a result of moving from analysis of mechanical systems to synthesis and product development. In these programs, the students usually have a high competence in mechanical engineering with some courses in electrical engineering, automatic control, computer science, and so forth. These programs are the most common, and appear to be the most in line with the thematic identity and functional legitimacy.

Some mechatronics programs have developed from an electrical engineering point of view, or from programs in computer science. These programs tend to focus more on analysis and theory, than the programs evolving from mechanical engineering, and less on project work and teamwork.

At some universities, mechatronics programs have been established and developed more detached from existing programs. It is primarily in these settings that the discussions of the optimal curriculum are the most common.

When applying the discussion of the selection of mechatronics to the argument presented above, it is more important to apply depth than width. The first two scenarios above commonly imply that the students first study for a BSc in either mechanical engineering or electrical engineering and then continue to specialize in mechatronics on a master's level. The opposite approach, to start with mechatronics from the very beginning, is much more difficult. It is difficult to make synergistic combinations, and students often will not be able to get to the detailed level required for a vertical exemplification and finish as generalists rather than the wanted specialist.

6.8 The Evolution of Mechatronics

The previous discussion regarding the didactical analysis and the "four questions" will be put into perspective by adding the dimension of time, to discuss how primarily the identity of a subject changes over time. The idea is that most traditional subjects are characterized as disciplinary and newer; cross-disciplinary subjects are more often characterized as thematic. Figure 6.3 was introduced recently to describe the evolution of the subject of mechatronics [7]. Some evidence of this model can be observed when studying how newer subjects have emerged on the basis of certain needs and on existing disciplines, which, for example, is the case for the subjects of strength of materials, solid mechanics, or automatic control. To illustrate, consider automatic control, a theme that emerged in the early-twentieth century. In this case, the originating disciplines were electrical engineering and physics. Automatic control has since developed into a well-established academic subject; thus, the original theme has been further developed and a research discipline has emerged [10].

The six stages, as shown in Figure 6.3, will here be described further.

6.8.1 Stage 1: The Origin; No Interaction

In this stage, no interaction exists between the original subjects. Usually these are subjects with a disciplinary identity, but could as well have thematic identities. In the case of mechatronics, the original subjects are represented mainly by electrical engineering, mechanical engineering, and computer science, which in this case are considered with a disciplinary identity.

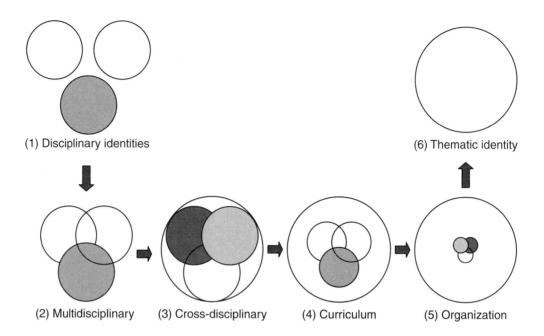

FIGURE 6.3 A framework for discussing the evolution of academic subjects. The boxed numbers are referred to in the headings below.

6.8.2 Stage 2: Multidisciplinary Stage

This is the first stage of the new subject, in this case of the evolution of mechatronics. Characteristics of this stage are that students are actively taking courses from different disciplines and areas, to either broaden their knowledge or to actively specialize in the new cross-disciplinary area. Typically, courses in electrical engineering are offered to mechanical engineering students or vice versa. The educational system is still functioning according to disciplines, which in many ways does not facilitate these alternative study patterns. In this second stage the theme has been identified, which in the case of mechatronics is about utilizing the various courses into a synergistic combination. An example would be students who choose to take a course in automatic control just to be able to design an embedded control system for a project in a design course.

6.8.3 Stage 3: Cross-Disciplinary Stage

In this stage, efforts are put in place by the university to organize activities. Courses are tailor-made for specific purposes; courses in automatic control are given specifically for mechanical engineering students with a hands-on approach and based more on applicability and implementation than theory. Courses in mechanical engineering for electrical engineering students focus more on modeling for a controller perspective. Courses in applied programming are focused on how to program a certain microcontroller, and so forth. Also, new courses are created, for example, cross-disciplinary courses such as mechatronics.

6.8.4 Stage 4: Curriculum Stage

In this stage, new programs are created to accommodate the interest and need for the new subject. Typically, MSc programs are created and offered to mechanical engineering and electrical engineering students. Also, BSc programs are created in some cases. Characteristics of this stage are the fact that the faculty is comprised of teachers with a disciplinary background. Usually, the faculty is mixed with competencies in various areas such as automatic control, mechanical, and electrical engineering.

6.8.5 Stage 5 and 6: Organizational Stage

The idea of the organizational stage is that students are graduating in the new subject, and later hired as new faculty. The idea is that the final stage is not reached until the faculty is based on people who actually specialized in mechatronics.

6.9 Where (and What) Is Mechatronics Today?

An attempt to place a number of universities on an evolutionary scale was attempted in [7], but it basically does not show much more than the authors' limited knowledge of the various universities. There are also some difficulties with the final stages of the model, which will be discussed here.

The faculty does not necessarily need to be specialized in mechatronics. The faculty should offer an education based on the didactical analysis—based on an exemplifying selection and an interactive communication. This should give students functional skills in mechatronics, meaning that the students should be able to synergistically combine knowledge and skill in various subjects. The students should be preparing for a future career in a mechatronics industry, all according to the didactical analysis.

The professors do not need to be skilled experts of synergistic combination of knowledge and skills. Compare with the coach metaphor; the athletic coach is not required to perform as well as his/her student. The coach is required to be an expert in how to guide the student to perform well. The mechatronics professors can instead be experts in the various fields and areas that mechatronics is built on: automatic control, microcontrollers, electrical engineering, and so forth. However, the professors then also need to be experts in the process of training students on how to perform this synergistic combination, and one way of gaining this expertise is to base the knowledge on own experience, just like many athletic coaches are former athletes.

This can also be seen as a clarification of the fifth and sixth stages of the evolution. Not until the thematic identity of mechatronics is fully grasped and understood, to the point that professors are transformed to coaches guiding students in synergistic combination of knowledge, will it be fruitful to discuss "new organizations." These professors then are the ones that these new organizations are built on.

References

1. Dahlgren, L. O., *Undervisningen och det meningsfulla lärandet. In Swedish only.* Linköping University, 1990.
2. Comerford, Sr. R. (Ed.), Mecha...what? *Spectrum IEEE*, 31(8): 46, 1994.
3. Harashima, F., Tomizuka, M., and Fukuda, T., Mechatronics—what is it, why and how? An editorial, *IEEE/ASME Trans Mechatronics*, 1: 1, 1996.
4. Erkmen, A. M., Tsubouchi, T., and Murphy, R., Mechatronics education, *IEEE Robot Automat Mag*, 8(2): 4, 2001.
5. Craig, K., Is anything really new in mechatronics education? *IEEE Robot Automat Mag*, 8(2): 12, 2001.
6. Grimheden, M. and Hanson, M., What is mechatronics? Proposing a didactical approach to mechatronics. In *Proc. 2nd European Workshop on Education in Mechatronics*, Kiel, Germany, 2001, p. 97.
7. Grimheden, M. and Hanson, M., Mechatronics—the evolution of an academic discipline in engineering education, *Mechatronics*, 15(2): 179, 2005.
8. Grimheden, M. and Törngren, M., What is embedded systems and how should it be taught? *ACM Trans Embedded Computing Sys*, 4: 3, 2005.
9. Wikander, J., Törngren, M., and Hanson, M., The Science and education of mechatronics engineering, *IEEE Robot Automat Mag*, 8(2): 20, 2001.
10. Abramovitch, D. Y. and Franklin, G. F., Fifty years in control—the story of the IEEE control systems society, *IEEE Control Syst Mag*, 14: 19, 2004.

Physical System Modeling

7 Modeling Electromechanical Systems
Francis C. Moon 7-1
Introduction • Models for Electromechanical Systems • Rigid Body Models • Basic Equations of Dynamics of Rigid Bodies • Simple Dynamic Models • Elastic System Modeling • Electromagnetic Forces • Dynamic Principles for Electric and Magnetic Circuits • Earnshaw's Theorem and Electromechanical Stability

8 Structures and Materials
Eniko T. Enikov 8-1
Fundamental Laws of Mechanics • Common Structures in Mechatronic Systems • Vibration and Modal Analysis • Buckling Analysis • Transducers • Future Trends

9 Modeling of Mechanical Systems for Mechatronics Applications
Raul G. Longoria 9-1
Introduction • Mechanical System Modeling in Mechatronic Systems • Descriptions of Basic Mechanical Model Components • Physical Laws for Model Formulation • Energy Methods for Mechanical System Model Formulation • Rigid Body Multidimensional Dynamics • Lagrange's Equations

10 Fluid Power Systems
Qin Zhang and Carroll E. Goering 10-1
Introduction • Hydraulic Fluids • Hydraulic Control Valves • Hydraulic Pumps • Hydraulic Cylinders • Fluid Power Systems Control • Programmable Electrohydraulic Valves

11 Electrical Engineering
Giorgio Rizzoni 11-1
Introduction • Fundamentals of Electric Circuits • Resistive Network Analysis • AC Network Analysis

12 Engineering Thermodynamics
Michael J. Moran 12-1
Fundamentals • Extensive Property Balances • Property Relations and Data • Vapor and Gas Power Cycles

13 Numerical Simulation
Jeannie Sullivan Falcon **13**-1
Introduction • Common Simulation Blocks • Textual Equations within Simulation Block Diagrams • Solvers • Simulation Timing • Visualization • Hybrid System Simulation and Control

14 Modeling and Simulation for MEMS
Carla Purdy **14**-1
Introduction • The Digital Circuit Development Process: Modeling and Simulating Systems with Micro- (or Nano-) Scale Feature Sizes • Analog and Mixed-Signal Circuit Development: Modeling and Simulating Systems with Micro- (or Nano-) Scale Feature Sizes and Mixed Digital (Discrete) and Analog (Continuous) Input, Output, and Signals • Basic Techniques and Available Tools for MEMS Modeling and Simulation • Modeling and Simulating MEMS, That Is, Systems with Micro- (or Nano-) Scale Feature Sizes, Mixed Digital (Discrete) and Analog (Continuous) Input, Output, and Signals, Two-and Three-Dimensional Phenomena, and Inclusion and Interaction of Multiple Domains and Technologies • A "Recipe" for Successful MEMS Simulation • Conclusion: Continuing Progress in MEMS Modeling and Simulation

15 Rotational and Translational Microelectromechanical Systems: MEMS Synthesis, Microfabrication, Analysis, and Optimization
Sergey Edward Lyshevski **15**-1
Introduction • MEMS Motion Microdevice Classifier and Structural Synthesis • MEMS Fabrication • MEMS Electromagnetic Fundamentals and Modeling • MEMS Mathematical Models • Control of MEMS • Conclusions

16 The Physical Basis of Analogies in Physical System Models
Neville Hogan and Peter C. Breedveld **16**-1
Introduction • History • The Force-Current Analogy: Across and Through Variables • Maxwell's Force-Voltage Analogy: Effort and Flow Variables • A Thermodynamic Basis for Analogies • Graphical Representations • Concluding Remarks

7
Modeling Electro-mechanical Systems

7.1	Introduction	7-1
7.2	Models for Electromechanical Systems	7-2
7.3	Rigid Body Models	7-2
	Kinematics of Rigid Bodies • Constraints and Generalized Coordinates • Kinematic versus Dynamic Problems	
7.4	Basic Equations of Dynamics of Rigid Bodies	7-4
	Newton–Euler Equation • Multibody Dynamics	
7.5	Simple Dynamic Models	7-6
	Compound Pendulum • Gyroscopic Motions	
7.6	Elastic System Modeling	7-8
	Piezoelastic Beam	
7.7	Electromagnetic Forces	7-10
7.8	Dynamic Principles for Electric and Magnetic Circuits	7-14
	Lagrange's Equations of Motion for Electromechanical Systems	
7.9	Earnshaw's Theorem and Electromechanical Stability	7-18
	References	7-19

Francis C. Moon
Cornell University

7.1 Introduction

Mechatronics describes the integration of mechanical, electromagnetic, and computer elements to produce devices and systems that monitor and control machine and structural systems. Examples include familiar consumer machines such as VCRs, automatic cameras, automobile air bags, and cruise control devices. A distinguishing feature of modern mechatronic devices compared to earlier controlled machines is the miniaturization of electronic information processing equipment. Increasingly computer and electronic sensors and actuators can be embedded in the structures and machines. This has led to the need for integration of mechanical and electrical design. This is true not only for sensing and signal processing but also for actuator design. In human size devices, more powerful magnetic materials and superconductors have led to the replacement of hydraulic and pneumatic actuators with servo motors, linear motors, and other electromagnetic actuators. At the material scale and in microelectromechanical systems (MEMSs), electric charge force actuators, piezoelectric actuators, and ferroelectric actuators have made great strides.

While the materials used in electromechanical design are often new, the basic dynamic principles of Newton and Maxwell still apply. In spatially extended systems one must solve continuum problems using the theory of elasticity and the partial differential equations of electromagnetic field theory. For many applications, however, it is sufficient to use lumped parameter modeling based on (i) rigid body dynamics for inertial components, (ii) Kirchhoff circuit laws for current-charge components, and (iii) magnet circuit laws for magnetic flux devices.

In this chapter we will examine the basic modeling assumptions for inertial, electric, and magnetic circuits, which are typical of mechatronic systems, and will summarize the dynamic principles and interactions between the mechanical motion, circuit, and magnetic state variables. We will also illustrate these principles with a few examples as well as provide some bibliography to more advanced references in electromechanics.

7.2 Models for Electromechanical Systems

The fundamental equations of motion for physical continua are partial differential equations (PDEs), which describe dynamic behavior in both time and space. For example, the motions of strings, elastic beams and plates, fluid flow around and through bodies, as well as magnetic and electric fields require both spatial and temporal information. These equations include those of elasticity, elastodynamics, the Navier–Stokes equations of fluid mechanics, and the Maxwell–Faraday equations of electromagnetics. Electromagnetic field problems may be found in Jackson (1968). Coupled field problems in electric fields and fluids may be found in Melcher (1980) and problems in magnetic fields and elastic structures may be found in the monograph by Moon (1984). This short article will only treat solid systems.

Many practical electromechanical devices can be modeled by lumped physical elements such as mass or inductance. The equations of motion are then integral forms of the basic PDEs and result in coupled ordinary differential equations (ODEs). This methodology will be explored in this chapter. Where physical problems have spatial distributions, one can often separate the problem into spatial and temporal parts called *separation of variables*. The spatial description is represented by a finite number of spatial or eigenmodes each of which has its modal amplitude. This method again results in a set of ODEs. Often these coupled equations can be understood in the context of simple lumped mechanical masses and electric and magnetic circuits.

7.3 Rigid Body Models

7.3.1 Kinematics of Rigid Bodies

Kinematics is the description of motion in terms of position vectors **r**, velocities **v**, acceleration **a**, rotation rate vector ω, and generalized coordinates $\{q_k(t)\}$ such as relative angular positions of one part to another in a machine (Figure 7.1). In a rigid body one generally specifies the position vector of one point, such as the center of mass $\mathbf{r}_c$, and the velocity of that point, say $\mathbf{v}_c$. The angular position of a rigid body is specified by angle sets call Euler angles. For example, in vehicles there are pitch, roll, and yaw angles (see, e.g., Moon, 1999). The angular velocity vector of a rigid body is denoted by ω. The velocity of a point in a rigid body other than the center of mass, $\mathbf{r}_p = \mathbf{r}_c + \rho$, is given by

$$\mathbf{v}_P = \mathbf{v}_c + \omega \times \rho \tag{7.1}$$

where the second term is a vector cross product. The angular velocity vector ω is a property of the entire rigid body. In general a rigid body, such as a satellite, has six degrees of freedom. But when machine elements are modeled as a rigid body, kinematic constraints often limit the number of degrees of freedom.

7.3.2 Constraints and Generalized Coordinates

Machines are often collections of rigid body elements in which each component is constrained to have one degree of freedom relative to each of its neighbors. For example, in a multi-link robot arm shown in Figure 7.2, each rigid link has a revolute degree of freedom. The degrees of freedom of each rigid link are constrained by bearings, guides, and gearing to have one type of relative motion. Thus, it is convenient

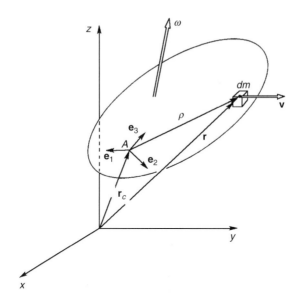

FIGURE 7.1 Sketch of a rigid body with position vector, velocity, and angular velocity vectors.

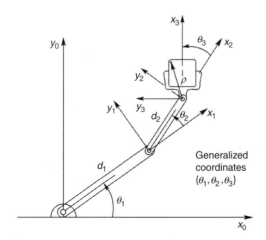

FIGURE 7.2 Multiple link robot manipulator arm.

to use these generalized motions $\{q_k: k = 1, \ldots, K\}$ to describe the dynamics. It is sometimes useful to define a vector or matrix, $J(q_k)$, called a *Jacobian*, that relates velocities of physical points in the machine to the generalized velocities $\{\dot{q}_k\}$. If the position vector to some point in the machine is $\mathbf{r}_P(q_k)$ and is determined by geometric constraints indicated by the functional dependence on the $\{q_k(t)\}$, then the velocity of that point is given by

$$\mathbf{v}_P = \sum \frac{\partial \mathbf{r}_P}{\partial q_r} \dot{q}_r = \mathbf{J} \cdot \dot{\mathbf{q}} \qquad (7.2)$$

where the sum is on the number of generalized degrees of freedom K. The 3xK matrix $\mathbf{J}$ is called a *Jacobian* and $\dot{\mathbf{q}}$ is a $K \times 1$ vector of generalized coordinates. This expression can be used to calculate the kinetic

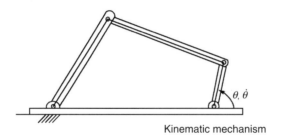

FIGURE 7.3 Example of a kinematic mechanism.

energy of the constrained machine elements, and using Lagrange's equations discussed below, derive the equations of motion (see also Moon, 1999).

7.3.3 Kinematic versus Dynamic Problems

Some machines are constructed in a closed kinematic chain so that the motion of one link determines the motion of the rest of the rigid bodies in the chain, as in the four-bar linkage shown in Figure 7.3. In these problems the designer does not have to solve differential equations of motion. Newton's laws are used to determine forces in the machine, but the motions are *kinematic*, determined through the geometric constraints.

In open link problems, such as robotic devices (Figure 7.2), the motion of one link does not determine the dynamics of the rest. The motions of these devices are inherently *dynamic*. The engineer must use both the kinematic constraints (7.2) as well as the Newton–Euler differential equation of motion or equivalent forms such as Lagrange's equation discussed below.

7.4 Basic Equations of Dynamics of Rigid Bodies

In this section we review the equations of motion for the mechanical plant in a mechatronics system. This plant could be a system of rigid bodies such as in a serial robot manipulator arm (Figure 7.2) or a magnetically levitated vehicle (Figure 7.4), or flexible structures in a MEMS accelerometer. The dynamics of flexible structural systems are described by PDEs of motion. The equation for rigid bodies involves Newton's law for the motion of the center of mass and Euler's extension of Newton's laws to the angular momentum of the rigid body. These equations can be formulated in many ways (see Moon, 1999):

1. Newton–Euler equation (vector method)
2. Lagrange's equation (scalar-energy method)
3. D'Alembert's principle (virtual work method)
4. Virtual power principle (Kane's equation, or Jourdan's principle)

7.4.1 Newton–Euler Equation

Consider the rigid body in Figure 7.1 whose center of mass is measured by the vector $\mathbf{r}_c$ in some fixed coordinate system. The velocity and acceleration of the center of mass are given by

$$\dot{\mathbf{r}}_c = \mathbf{v}_c, \qquad \dot{\mathbf{v}}_c = \mathbf{a}_c \qquad (7.3)$$

The "over dot" represents a total derivative with respect to time. We represent the total sum of vector forces on the body from both mechanical and electromagnetic sources by F. Newton's law for the motion

FIGURE 7.4 Magnetically levitated rigid body. (HSST MagLev prototype vehicle, 1998, Nagoya, Japan.)

of the center of mass of a body with mass m is given by

$$m\dot{\mathbf{v}}_c = \mathbf{F} \qquad (7.4)$$

If $\mathbf{r}$ is a vector to some point in the rigid body, we define a local position vector ρ by $\mathbf{r}_p = \mathbf{r}_c + \rho$. If a force $\mathbf{F}_i$ acts at a point $\mathbf{r}_i$ in a rigid body, then we define the moment of the force $\mathbf{M}$ about the fixed origin by

$$\mathbf{M}_i = \mathbf{r}_i \times \mathbf{F}_i \qquad (7.5)$$

The total force moment is then given by the sum over all the applied forces as the body

$$\mathbf{M} = \sum \mathbf{r}_i \times \mathbf{F}_i = \mathbf{r}_c \times \mathbf{F} + \mathbf{M}_c \quad \text{where} \quad \mathbf{M}_c = \sum \rho_i \times \mathbf{F}_i \qquad (7.6)$$

We also define the *angular momentum* of the rigid body by the product of a symmetric matrix of second moments of mass called the *inertia matrix* $\mathbf{I}_c$. The angular momentum vector about the center of mass is defined by

$$\mathbf{H}_c = \mathbf{I}_c \cdot \boldsymbol{\omega} \qquad (7.7)$$

Since $\mathbf{I}_c$ is a symmetric matrix, it can be diagonalized with principal inertias (or eigenvalues) $\{I_{ic}\}$ about principal directions (eigenvectors) $\{\mathbf{e}_1, \mathbf{e}_2, \mathbf{e}_3\}$. In these coordinates, which are attached to the body, the angular momentum about the center of mass becomes

$$\mathbf{H}_c = I_{1c}\omega_1 \mathbf{e}_1 + I_{2c}\omega_2 \mathbf{e}_2 + I_{3c}\omega_3 \mathbf{e}_3 \qquad (7.8)$$

where the angular velocity vector is written in terms of principal eigenvectors $\{\mathbf{e}_1, \mathbf{e}_2, \mathbf{e}_3\}$ attached to the rigid body.

Euler's extension of Newton's law for a rigid body is then given by

$$\dot{\mathbf{H}}_c = \mathbf{M}_c \qquad (7.9)$$

This equation says that the change in the angular momentum about the center of mass is equal to the total moment of all the forces about the center of mass. The equation can also be applied about a fixed point of rotation, which is not necessarily the center of mass, as in the example of the compound pendulum given below.

Equations 7.4 and 7.9 are known as the Newton–Euler equations of motion. Without constraints, they represent six coupled second-order differential equations for the position of the center of mass and for the angular orientation of the rigid body.

7.4.2 Multibody Dynamics

In a serial link robot arm, as shown in Figure 7.2, we have a set of connected rigid bodies. Each body is subject to both applied and constraint forces and moments. The dynamical equations of motion involve the solution of the Newton–Euler equations for each rigid link subject to the geometric or kinematics constraints between each of the bodies as in (7.2). The forces on each body will have applied terms $\mathbf{F}^a$, from actuators or external mechanical sources, and internal constraint forces $\mathbf{F}^c$. When friction is absent, the work done by these constraint forces is zero. This property can be used to write equations of motion in terms of scalar energy functions, known as Lagrange's equations (see below).

Whatever the method used to derive the equation of motions, the dynamical equations of motion for multibody systems in terms of generalized coordinates $\{q_k(t)\}$ have the form

$$\sum m_{ij}\ddot{q}_j + \sum\sum \mu_{ijk}\dot{q}_j\dot{q}_k = Q_i \qquad (7.10)$$

The first term on the left involves a generalized symmetric mass matrix $m_{ij} = m_{ji}$. The second term includes Coriolis and centripetal acceleration. The right-hand side includes all the force and control terms. This equation has a quadratic nonlinearity in the generalized velocities. These quadratic terms usually drop out for rigid body problems with a single axis of rotation. However, the nonlinear inertia terms generally appear in problems with simultaneous rotation about two or three axes as in multi-link robot arms (Figure 7.2), gyroscope problems, and slewing momentum wheels in satellites.

In modern dynamic simulation software, called multibody codes, these equations are automatically derived and integrated once the user specifies the geometry, forces, and controls. Some of these codes are called ADAMS, DADS, Working Model, and NEWEUL. However, the designer must use caution as these codes are sometimes poor at modeling friction and impacts between bodies.

7.5 Simple Dynamic Models

Two simple examples of the application of the angular momentum law are now given. The first is for rigid body rotation about a single axis and the second has two axes of rotation.

7.5.1 Compound Pendulum

When a body is constrained to a single rotary degree of freedom and is acted on by the force of gravity as in Figure 7.5, the equation of motion takes the form, where θ is the angle from the vertical,

$$I\ddot{\theta} - (m_1 L_1 - m_2 L_2)g \sin\theta = T(t) \qquad (7.11)$$

where $T(t)$ is the applied torque, $I = m_1 L_1^2 + m_2 L_2^2$ is the moment of inertia (properly called the second moment of mass). The above equation is nonlinear in the sine function of the angle. In the case of small motions about $\theta = 0$, the equation becomes a linear differential equation and one can look for solutions of the form $\theta = A\cos\omega t$, when $T(t) = 0$. For this case the pendulum exhibits sinusoidal motion with

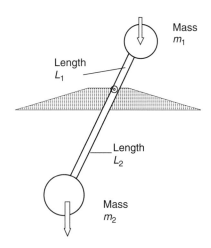

FIGURE 7.5 Sketch of a compound pendulum under gravity torques.

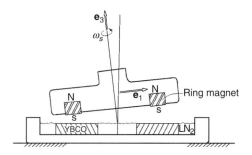

FIGURE 7.6 Sketch of a magnetically levitated flywheel on high-temperature superconducting bearings.

natural frequency

$$\omega = [g(m_2 L_2 - m_1 L_1)/I]^{1/2} \tag{7.12}$$

For the simple pendulum $m_1 = 0$, and we have the classic pendulum relation in which the natural frequency depends inversely on the square root of the length:

$$\omega = (g/L_2)^{1/2} \tag{7.13}$$

7.5.2 Gyroscopic Motions

Spinning devices such as high speed motors in robot arms or turbines in aircraft engines or magnetically levitated flywheels (Figure 7.6) carry angular momentum, devoted by the vector **H**. Euler's extension of Newton's laws says that a change in angular momentum must be accompanied by a force moment **M**,

$$\mathbf{M} = \dot{\mathbf{H}} \tag{7.14}$$

In three-dimensional problems one can often have components of angular momentum about two different axes. This leads to a Coriolis acceleration that produces a gyroscopic moment even when the two angular motions are steady. Consider the spinning motor with spin $\dot{\phi}$ about an axis with unit vector $\mathbf{e}_1$ and

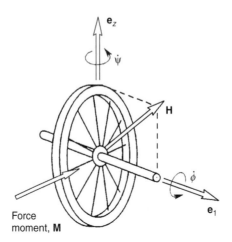

FIGURE 7.7 Gyroscopic moment on a precession, spinning rigid body.

let us imagine an angular motion of the e_1 axis, $\dot{\psi}$ about a perpendicular axis e_z called the precession axis in gyroscope parlance. Then one can show that the angular momentum is given by

$$\mathbf{H} = I_1 \dot{\phi} \mathbf{e}_1 + I_z \dot{\psi} \mathbf{e}_z \tag{7.15}$$

and the rate of change of angular momentum for constant spin and presession rates is given by

$$\dot{\mathbf{H}} = \dot{\psi} \mathbf{e}_z \times \mathbf{H} \tag{7.16}$$

There must then exist a gyroscopic moment, often produced by forces on the bearings of the axel (Figure 7.7). This moment is perpendicular to the plane formed by $\mathbf{e}_1$ and $\mathbf{e}_z$, and is proportional to the product of the rotation rates:

$$\mathbf{M} = I_1 \dot{\phi} \dot{\psi} \mathbf{e}_z \times \mathbf{e}_1 \tag{7.17}$$

This has the same form as Equation 7.10, when the generalized force Q is identified with the moment **M**, that is, the moment is the product of generalized velocities when the second derivative acceleration terms are zero.

7.6 Elastic System Modeling

Elastic structures take the form of cables, beams, plates, shells, and frames. For linear problems one can use the method of eigenmodes to represent the dynamics with a finite set of modal amplitudes for generalized degrees of freedom. These eigenmodes are found as solutions to the PDEs of the elastic structure (see, e.g., Yu, 1996).

The simplest elastic structure after the cable is a one-dimensional beam shown in Figure 7.8. For small motions we assume only transverse displacements $w(x, t)$, where x is a spatial coordinate along the beam. One usually assumes that the stresses on the beam cross section can be integrated to obtain stress vector resultants of shear V, bending moment M, and axial load T. The beam can be loaded with point or concentrated forces, end forces or moment or distributed forces as in the case of gravity, fluid forces, or electromagnetic forces. For a distributed transverse load $f(x, t)$, the equation of motion is given by

$$D\frac{\partial^4 w}{\partial x^4} - T\frac{\partial^2 w}{\partial x^2} + \rho A \frac{\partial^2 w}{\partial t^2} = f(x, t) \tag{7.18}$$

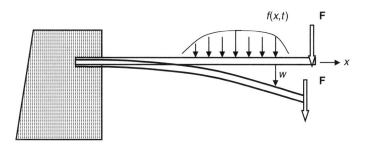

FIGURE 7.8 Sketch of an elastic cantilevered beam.

where D is the bending stiffness, A is the cross-sectional area of the beam, and ρ is the density. For a beam with Young's modulus Y, rectangular cross section of width b, and height h, $D = Ybh^3/12$. For $D = 0$, one has a cable or string under tension T, and the equation takes the form of the usual wave equation. For a beam with tension T, the natural frequencies are increased by the addition of the second term in the equation. For $T = -P$, that is, a compressive load on the end of the beam, the curvature term leads to a decrease of natural frequency with increase of the compressive force P. If the lowest natural frequency goes to zero with increasing load P, the straight configuration of the beam becomes unstable or undergoes *buckling*. The use of T or $(-P)$ to stiffen or destiffen a beam structure can be used in design of sensors to create a sensor with variable resonance. This idea has been used in a MEMS accelerometer design (see below).

Another feature of the beam structure dynamics is the fact that unlike the string or cable, the frequencies of the natural modes are not commensurate due to the presence of the fourth-order derivative term in the equation. In wave type problems this is known as *wave dispersion*. This means that waves of different wavelengths travel at different speeds so that wave pulse shapes change their form as the wave moves through the structure.

In order to solve dynamic problems in finite length beam structures, one must specify boundary conditions at the ends. Examples of boundary conditions include

$$\text{Clamped end} \quad w = 0, \quad \frac{\partial w}{\partial x} = 0$$

$$\text{Pinned end} \quad w = 0, \quad \frac{\partial^2 w}{\partial x^2} = 0 \text{ (zero moment)} \tag{7.19}$$

$$\text{Free end} \quad \frac{\partial^2 w}{\partial x^2} = 0, \quad \frac{\partial^3 w}{\partial x^3} = 0 \text{ (zero shear)}$$

7.6.1 Piezoelastic Beam

Piezoelastic materials exhibit a coupling between strain and electric polarization or voltage. Thus, these materials can be used for sensors or actuators. They have been used for active vibration suppression in elastic structures. They have also been explored for active optics space applications. Many natural materials exhibit piezoelasticity such as quartz as well as manufactured materials such as barium titanate, lead zirconate titanate (PZT), and polyvinylidene fluoride (PVDF). Unlike forces on charges and currents (see below), the electric effect takes place through a change in shape of the material. The modeling of these devices can be done by modifying the equations for elastic structures.

The following work on piezo-benders is based on the work of Lee and Moon (1989) as summarized in Miu (1993). One of the popular configurations of a piezo actuator-sensor is the piezo-bender shown in Figure 7.9. The elastic beam is of rectangular cross section as is the piezo element. The piezo element

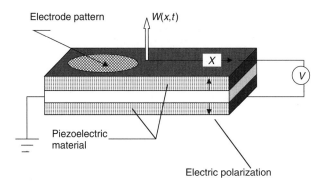

FIGURE 7.9 Elastic beam with two piezoelectric layers. (From Lee and Moon, 1989.)

can be cemented on one or both sides of the beam either partially or totally covering the surface of the non-piezo substructure.

In general the local electric dipole polarization depends on the six independent strain components produced by normal and shear stresses. However, we will assume that the transverse voltage or polarization is coupled to the axial strain in the plate-shaped piezo layers. The constitutive relations between axial stress and strain, T, S, electric field and electric displacement, E_3, D_3 (not to be confused with the bending stiffness D), are given by

$$T_1 = c_{11}S_1 - e_{31}E_3, \qquad D_3 = e_{31}S_1 + \varepsilon_3 E_3 \qquad (7.20)$$

The constants c_{11}, e_{31}, ε_3, are the elastic stiffness modulus, piezoelectric coupling constant, and the electric permittivity, respectively.

If the piezo layers are polled in the opposite directions, as shown in the Figure 7.9, an applied voltage will produce a strain extention in one layer and a strain contraction in the other layer, which has the effect of an applied moment on the beam. The electrodes applied to the top and bottom layers of the piezo layers can also be shaped so that there can be a gradient in the average voltage across the beam width. For this case the equation of motion of the composite beam can be written in the form

$$D\frac{\partial^4 w}{\partial x^4} + \rho A \frac{\partial^2 w}{\partial t^2} = -2e_{31}z_o \frac{\partial^2 V_3}{\partial x^2} \qquad (7.21)$$

where $z_o = (h_S + h_P)/2$.

The z term is the average of piezo plate and substructure thicknesses. When the voltage is uniform, then the right-hand term results in an applied moment at the end of the beam proportional to the transverse voltage.

7.7 Electromagnetic Forces

One of the keys to modeling mechatronic systems is the identification of the electric and magnetic forces. Electric forces act on charges and electric polarization (electric dipoles). Magnetic forces act on electric currents and magnetic polarization. Electric charge and current can experience a force in a uniform electric or magnetic field; however, electric and magnetic dipoles will only produce a force in an electric or magnetic field gradient.

Electric and magnetic forces can also be calculated using both direct vector methods as well as from energy principles. One of the more popular methods is *Lagrange's equation* for electromechanical systems described below.

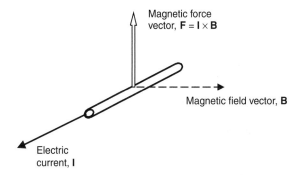

FIGURE 7.10 Electric forces on two charges (top). Magnetic force on a current carrying wire element (bottom).

Electromagnetic systems can be modeled as either distributed field quantities, such as electric field **E** or magnetic flux density **B** or as lumped element electric and magnetic circuits. The force on a point charge Q is given by the vector equation (Figure 7.10):

$$\mathbf{F} = Q\mathbf{E} \tag{7.22}$$

When **E** is generated by a single charge, the force between charges Q_1 and Q_2 is given by

$$F = \frac{Q_1 Q_2}{4\pi\varepsilon_0 r^2} \tag{7.23}$$

and is directed along the line connecting the two charges. Like charges repel and opposite charges attract one another.

The magnetic force per unit length on a current element **I** is given by the cross product

$$\mathbf{F} = \mathbf{I} \times \mathbf{B} \tag{7.24}$$

where the magnetic force is perpendicular to the plane of the current element and the magnetic field vector. The total force on a closed circuit in a uniform field can be shown to be zero. Net forces on closed circuits are produced by field gradients due to other current circuits or field sources.

Forces produced by field distributions around a volume containing electric charge or current can be calculated using the field quantities of **E**, **B** directly using the concept of magnetic and electric stresses, which was developed by Faraday and Maxwell. These electromagnetic stresses must be integrated over an area surrounding the charge or current distribution. For example, a solid containing a current distribution can experience a *magnetic pressure*, $P = B_t^2/2\mu_0$, on the surface element and a *magnetic tension*, $t_n = B_n^2/2\mu_0$, where the magnetic field components are written in terms of values tangential and normal to the surface. Thus, a one-tesla magnetic field outside of a solid will experience 40 N/cm² pressure if the field is tangential to the surface.

In general there are four principal methods to calculate electric and magnetic forces:

- Direct force vectors and moments between electric charges, currents, and dipoles
- Electric field-charge and magnetic field-current force vectors

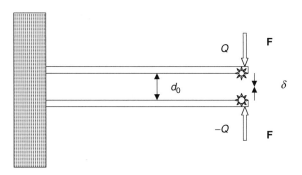

FIGURE 7.11 Two elastic beams with electric charges at the ends.

- Electromagnetic tensor, integration of electric tension, magnetic pressure over the surface of a material body
- Energy methods based on gradients of magnetic and electric energy

Examples of the direct method and stress tensor method are given below. The energy method is described in the section on Lagrange's equations.

Example 1 Charge–Charge Forces

Suppose two elastic beams in a MEMS device have electric charges Q_1, Q_2 coulombs each concentrated at their tips (Figure 7.11). The electric force between the charges is given by the vector

$$\mathbf{F} = \frac{Q_1 Q_2}{4\pi\varepsilon_0} \frac{\mathbf{r}}{r^3} \quad \text{(newtons)} \tag{7.25}$$

where $1/4\pi\varepsilon_0 = 8.99 \times 10^9 \text{ Nm}^2/\text{C}^2$.

If the initial separation between the beams is d_0, we seek the new separation under the electric force. For simplicity, we let $Q_1 = -Q_2 = Q$, where opposite charges create an attractive force between the beam tips. The deflection of the cantilevers is given by

$$\delta = \frac{FL^3}{3YI} = \frac{1}{k}F \tag{7.26}$$

where L is the length, Y the Young's modulus, I the second moment of area, and k the effective spring constant.

Under the electric force, the new separation is $d = d_0 - 2\delta$,

$$k\delta = \frac{Q^2}{4\pi\varepsilon_0} \frac{1}{(d_0 - 2\delta)^2} \tag{7.27}$$

For $\delta \ll d_0$ to first order we have

$$\delta = \frac{Q^2/4\pi\varepsilon_0 d_0^2 k}{1 - (1/d_0^3)(Q^2/k\pi\varepsilon_0)} \tag{7.28}$$

This problem shows the potential for electric field buckling because as the beam tips move closer together, the attractive force between them increases. The nondimensional expression in the denominator

$$\frac{Q^2}{\pi\varepsilon_0 d_0^3} \frac{1}{k} \tag{7.29}$$

is the ratio of the negative electric stiffness to the elastic stiffness k of the beams.

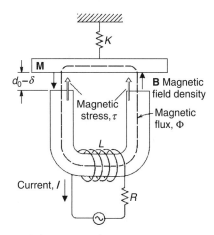

FIGURE 7.12   Force on a ferromagnetic bar near an electromagnet.

Example 2 Magnetic Force on an Electromagnet

Imagine a ferromagnetic keeper on an elastic restraint of stiffness k, as shown in Figure 7.12. Under the soft magnetic keeper, we place an electromagnet which produces N turns of current I around a soft ferromagnetic core. The current is produced by a voltage in a circuit with resistance R.

The magnetic force will be calculated using the *magnetic stress tensor* developed by Maxwell and Faraday (see, e.g., Moon, 1984, 1994). Outside a ferromagnetic body, the stress tensor is given by **t** and the stress vector on the surface defined by normal **n** is given by $\tau = \mathbf{t} \cdot \mathbf{n}$:

$$\tau = \frac{1}{\mu_0}\left(\frac{1}{2}[B_n^2 - B_t^2],\ B_n B_t\right) = (\tau_n, \tau_t) \tag{7.30}$$

For high magnetic permeability as in a ferromagnetic body, the tangential component of the magnetic field outside the surface is near zero. Thus the force is approximately normal to the surface and is found from the integral of the magnetic tension over the surface:

$$\mathbf{F} = \frac{1}{2\mu_0}\int B_n^2 \mathbf{n}\ dA \tag{7.31}$$

and $B_n^2/2\mu_0$ represents a magnetic tensile stress. Thus, if the area of the pole pieces of the electromagnet is A (neglecting fringing of the field), the force is

$$F = B_g^2 A/\mu_0 \tag{7.32}$$

where B_g is the gap field. The gap field is determined from Amperes law

$$NI = \widehat{R}\Phi,\quad \Phi = B_g A \tag{7.33}$$

where the *reluctance* is approximately given by

$$\widehat{R} = \frac{2(d_0 - \delta)}{\mu_0 A} \tag{7.34}$$

The balance of magnetic and elastic forces is then given by

$$F = \frac{1}{\mu_0 A}\Phi^2 = \frac{1}{\mu_0 A}\left(\frac{NI}{R}\right)^2 = k\delta \tag{7.35}$$

or

$$\frac{(NI)^2}{4(d_0 - \delta)^2}\mu_0 A = k\delta, \quad \frac{\mu_0 N^2 I^2 A}{4(d_0 - \delta)^2} = k\delta$$

(Note that the expression $\mu_0 N^2 I^2$ has units of force.) Again as the current is increased, the total elastic and electric stiffness goes to zero and one has the potential for buckling.

7.8 Dynamic Principles for Electric and Magnetic Circuits

The fundamental equations of electromagnetics stem from the work of nineteenth century scientists such as Faraday, Henry, and Maxwell. They take the form of partial differential equations in terms of the field quantities of electric field **E** and magnetic flux density **B**, and also involve volumetric measures of charge density q and current density **J** (see, e.g., Jackson, 1968). Most practical devices, however, can be modeled with lumped electric and magnetic circuits. The standard resistor, capacitor, inductor circuit shown in Figure 7.13 uses electric current I (amperes), charge Q (columbs), magnetic flux Φ (webers), and voltage V (volts) as dynamic variables. The voltage is the integral of the electric field along a path:

$$V_{21} = \int_1^2 \mathbf{E} \cdot d\mathbf{l} \tag{7.36}$$

The charge Q is the integral of charge density q over a volume, and electric current I is the integral of normal component of **J** across an area. The magnetic flux Φ is given as another surface integral of magnetic flux.

$$\Phi = \int \mathbf{B} \cdot d\mathbf{A} \tag{7.37}$$

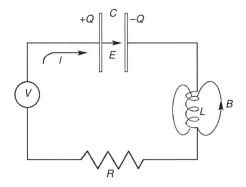

FIGURE 7.13 Electric circuit with lumped parameter capacitance, inductance, and resistance.

Modeling Electromechanical Systems

When there are no mechanical elements in the system, the dynamical equations take the form of conservation of charge and the Faraday–Henry law of flux change.

$$\frac{dQ}{dt} = I \quad \text{(Conservation of charge)} \quad (7.38)$$

$$\frac{d\phi}{dt} = V \quad \text{(Law of flux change)} \quad (7.39)$$

where $\phi = N\Phi$ is called the number of flux linkages, and N is an integer. In electromagnetic circuits the analog of mechanical constitutive properties is inductance L and capacitance C. The magnetic flux in an inductor, for example, often depends on the current I.

$$\phi = f(I) \quad (7.40)$$

For a linear inductor we have a definition of inductance L, that is, $\phi = LI$. If the system has a mechanical state variable such as displacement x, as in a magnetic solenoid actuator, then L may be a function of x.

In charge storage circuit elements, the capacitance C is defined as

$$Q = CV \quad (7.41)$$

In MEMS devices and in microphones, the capacitance may also be a function of some generalized mechanical displacement variable.

The voltages across the different circuit elements can be active or passive. A pure voltage source can maintain a given voltage, but the current depends on the passive voltages across the different circuit elements as summarized in the Kirchhoff circuit law:

$$\frac{d}{dt}L(x)I + \frac{Q}{C(x)} + RI = V(t) \quad (7.42)$$

7.8.1 Lagrange's Equations of Motion for Electromechanical Systems

It is well known that the Newton–Euler equations of motion for mechanical systems can be derived using an energy principle called Lagrange's equation. In this method one identifies generalized coordinates $\{q_k\}$, not to be confused with electric charges, and writes the kinetic energy of the system T in terms of generalized velocities and coordinates, $T(\dot{q}_k, q_k)$. Next the mechanical forces are split into so-called conservative forces, which can be derived from a potential energy function $W(q_k)$ and the rest of the forces, which are represented by a generalized force Q_k corresponding to the work done by the kth generalized coordinate. Lagrange's equations for mechanical systems then take the form:

$$\frac{d}{dt}\frac{\partial T(\dot{q}_k, q_k)}{\partial \dot{q}_k} - \frac{\partial T}{\partial q_k} + \frac{\partial W(q_k)}{\partial q_k} = Q_k \quad (7.43)$$

For example, in a linear spring–mass–damper system, with mass m, spring constant k, viscous damping constant c, and one generalized coordinate $q_1 = x$, the equation of motion can be derived using, $T = \frac{1}{2}m\dot{x}^2$, $W = \frac{1}{2}kx^2$, $Q_1 = -c\dot{x}$, in Lagrange's equation above. What is remarkable about this formulation is that it can be extended to treat both electromagnetic circuits and coupled electromechanical problems.

As an example of the application of Lagrange's equations to a coupled electromechanical problem, consider the one-dimensional mechanical device, shown in Figure 7.14, with a magnetic actuator and a capacitance actuator driven by a circuit with applied voltage $V(t)$. We can extend Lagrange's equation to

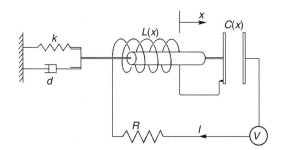

FIGURE 7.14 Coupled lumped parameter electromechanical system with single degree of freedom mechanical motion $x(t)$.

circuits by defining the charge on the capacitor, Q, as another generalized coordinate along with x, i.e., in Lagrange's formulation, $q_1 = x$, $q_2 = Q$. Then we add to the kinetic energy function a magnetic energy function $W_m(\dot{Q}, x)$, and add to the potential energy an electric field energy function $W_e(Q, x)$. The equations of both the mass and the circuit can then be derived from

$$\frac{d}{dt}\frac{\partial[T + W_m]}{\partial \dot{q}_k} - \frac{\partial[T + W_m]}{\partial q_k} + \frac{\partial[W + W_e]}{\partial q_k} = Q_k \tag{7.44}$$

The generalized force must also be modified to account for the energy dissipation in the resistor and the energy input of the applied voltage $V(t)$, that is, $Q_1 = -c\dot{x}$, $Q_2 = -R\dot{Q} + V(t)$. In this example the magnetic energy is proportional to the inductance $L(x)$, and the electric energy function is inversely proportional to the capacitance $C(x)$. Applying Lagrange's equations automatically results in expressions for the magnetic and electric forces as derivatives of the magnetic and electric energy functions, respectively, that is,

$$W_m = \frac{1}{2}L(x)\dot{Q}^2 = \frac{1}{2}LI^2, \qquad W_e = \frac{1}{2C(x)}Q^2 \tag{7.45}$$

$$F_m = \frac{\partial W_m(x, \dot{Q})}{\partial x} = \frac{1}{2}I^2\frac{dL(x)}{dx}, \qquad F_e = -\frac{\partial W_e(x, Q)}{\partial x} = -\frac{1}{2}Q^2\frac{d}{dx}\left[\frac{1}{C(x)}\right] \tag{7.46}$$

These remarkable formulii are very useful in that one can calculate the electromagnetic forces by just knowing the dependence of the inductance and capacitance on the displacement x. These functions can often be found from electrical measurements of L and C.

Example: Electric Force on a Comb-Drive MEMS Actuator

Consider the motion of an elastically constrained plate between two grounded fixed plates as in a MEMS comb-drive actuator in Figure 7.15. When the moveable plate has a voltage V applied, there is stored electric field energy in the two gaps given by

$$W_e^*(V, x) = \frac{1}{2}\varepsilon_0 V^2 A \frac{d_0}{d_0^2 - x^2} \tag{7.47}$$

In this expression the electric energy function is written in terms of the voltage V instead of the charge on the plates Q as in Equations 7.45 and 7.46. Also the initial gap is d_0, and the area of the plate is A.

Modeling Electromechanical Systems

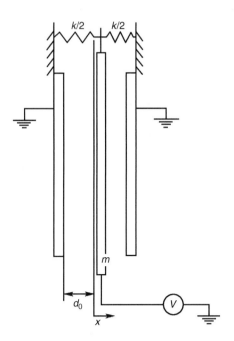

FIGURE 7.15 Example of electric force on the elements of a comb-drive actuator.

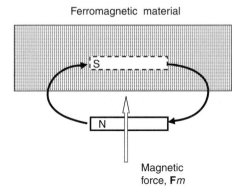

FIGURE 7.16 Decrease in natural frequency of a MEMS device with applied voltage as an example of negative electric stiffness. (From Adams, 1996.)

Using the force expressions derived from Lagrange's equations 7.44, the electric charge force on the plate is given by

$$F_e = \frac{\partial}{\partial x} W_e^*(V, x) = \frac{\varepsilon_0 V^2 A}{d_0} \frac{x}{(1 - x^2/d^2)^2} \tag{7.48}$$

This expression shows that the electric stiffness is negative for small x, which means that the voltage will decrease the natural frequency of the plate. This idea has been applied to a MEMS comb-drive actuator by Adams (1996) in which the voltage could be used to tune the natural frequency of a MEMS accelerometer, as shown in Figure 7.16.

7.9 Earnshaw's Theorem and Electromechanical Stability

It is not well known that electric and magnetic forces in mechanical systems can produce static instability, otherwise known as *elastic buckling* or *divergence*. This is a consequence of the inverse square nature of many electric and magnetic forces. It is well known that the electric and magnetic field potential Φ satisfies Laplace's equation, $\nabla^2 \Phi = 0$. There is a basic theorem in potential theory about the impossibility of a relative maximum or minimum value of a potential $\Phi(\mathbf{r})$ for solutions of Laplace's equation except at a boundary. It was stated in a theorem by Earnshaw (1829) that it is impossible for a static set of charges, magnetic and electric dipoles, and steady currents to be in a stable state of equilibrium without mechanical or other feedback or dynamic forces (see, e.g., Moon, 1984, 1994).

One example of Earnshaw's theorem is the instability of a magnetic dipole (e.g., a permanent magnet) near a ferromagnetic surface (Figure 7.17). Levitated bearings based on ferromagnetic forces, for example, require feedback control. Earnshaw's theorem also implies that if there is one degree of freedom with stable restoring forces, there must be another degree of freedom that is unstable. Thus the equilibrium positions for a pure electric or magnetic system of charges and dipoles must be saddle points. The implication for the force potentials is that the matrix of second derivatives is not positive definite. For example, suppose there are three generalized position coordinates $\{s_u\}$ for a set of electric charges. Then if the generalized forces are proportional to the gradient of the potential, $\nabla \Phi$, then the generalized electric stiffness matrix $\mathbf{K}_{ij}$, given by

$$\mathbf{K}_{ij} = \left[\frac{\partial^2 \Phi}{\partial s_i \partial s_j}\right]$$

will not be positive definite. This means that at least one of the eigenvalues will have negative stiffness.

Another example of electric buckling is a beam in an electric field with charge induced by an electric field on two nearby stationary plates as in Figure 7.15. The induced charge on the beam will be attracted to either of the two plates, but is resisted by the elastic stiffness of the beam. As the voltage is increased, the combined electric and elastic stiffnesses will decrease until the beam buckles to one or the other of the two sides. Before buckling, however, the natural frequency of the charged beam will decrease (Figure 7.16). This property has been observed experimentally in a MEMS device. A similar magneto elastic buckling is observed for a thin ferromagnetic elastic beam in a static magnetic field (see Moon, 1984). Both electroelastic and magnetoelastic buckling are derived from the same principle of Earnshaw's theorem.

There are dramatic exceptions to Earnshaw's stability theorem. One of course is the levitation of 50-ton vehicles with magnetic fields, known as MagLev, or the suspension of gas pipeline rotors using feedback controlled magnetic bearings (see Moon, 1994). Here either the device uses feedback forces, that is, the

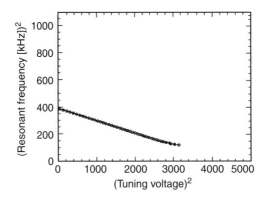

FIGURE 7.17 Magnetic force on a magnetic dipole magnet near a ferromagnetic half space with image dipole shown.

fields are not static, or the source of one of the magnetic fields is a superconductor. Diamagnetic forces are exceptions to Earnshaw's theorem, and superconducting materials have properties that behave like diamagnetic materials. Also new high-temperature superconductivity materials, such as YBaCuO, exhibit magnetic flux pinning forces that can be utilized for stable levitation in magnetic bearings without feedback (see Moon, 1994).

References

Adams, S. G. (1996), *Design of Electrostatic Actuators to Tune the Effective Stiffness of Micro-Mechanical Systems,* Ph.D. Dissertation, Cornell Unversity, Ithaca, New York.
Goldstein, H. (1980), *Classical Mechanics,* Addison-Wesley, Reading, MA.
Jackson, J. D. (1968), *Classical Electrodynamics,* J. Wiley & Sons, New York.
Lee, C. K. and Moon, F. C. (1989), "Laminated piezopolymer plates for bending sensors and actuators," *J. Acoust. Soc. Am.,* **85**(6), June 1989.
Melcher, J. R. (1981), *Continuum Electrodynamics,* MIT Press, Cambridge, MA.
Miu, D. K. (1993), *Mechatronics,* Springer-Verlag, New York.
Moon, F. C. (1984), *Magneto-Solid Mechanics,* J. Wiley & Sons, New York.
Moon, F. C. (1994), *Superconducting Levitation,* J. Wiley & Sons, New York.
Moon, F. C. (1999), *Applied Dynamics,* J. Wiley & Sons, New York.
Yu, Y.-Y. (1996), *Vibrations of Elastic Plates,* Springer-Verlag, New York.

8
Structures and Materials

8.1	Fundamental Laws of Mechanics	8-1
	Statics and Dynamics of Mechatronic Systems • Equations of Motion of Deformable Bodies • Electric Phenomena	
8.2	Common Structures in Mechatronic Systems	8-6
	Beams • Torsional Springs • Thin Plates	
8.3	Vibration and Modal Analysis	8-9
8.4	Buckling Analysis	8-10
8.5	Transducers	8-11
	Electrostatic Transducers • Electromagnetic Transducers • Thermal Actuators • Electroactive Polymer Actuators	
8.6	Future Trends	8-16
	References	8-16

Eniko T. Enikov
University of Arizona

The term mechatronics was first used by Japanese engineers to define a mechanical system with embedded electronics, capable of providing intelligence and control functions. Since then, the continued progress in integration has led to the development of microelectromechanical systems (MEMSs) in which the mechanical structures themselves are part of the electrical subsystem. The development and design of such mechatronic systems requires interdisciplinary knowledge in several disciplines—electronics, mechanics, materials, and chemistry. This section contains an overview of the main mechanical structures, the materials they are built from, and the governing laws describing the interaction between electrical and mechanical processes. It is intended for use in the initial stage of the design, when quick estimates are necessary to validate or reject a particular concept. Special attention is devoted to the newly emerging smart materials—electroactive polymer actuators. Several tables of material constants are also provided for reference.

8.1 Fundamental Laws of Mechanics

8.1.1 Statics and Dynamics of Mechatronic Systems

The fundamental laws of mechanics are the balance of linear and angular momentum. For an idealized system consisting of a point mass m moving with velocity $\mathbf{v}$, the linear momentum is defined as the product of the mass and the velocity:

$$\mathbf{L} = m\mathbf{v} \qquad (8.1)$$

The conservation of linear momentum for a single particle postulates that the rate of change of linear momentum is equal to the sum of all forces acting on the particle

$$\dot{\mathbf{L}} = m\dot{\mathbf{v}} = \sum \mathbf{F}_i \qquad (8.2)$$

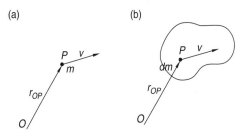

FIGURE 8.1 Difinition of velocity and position vectors for single particle (a) and rigid body (b).

where we have assumed that the mass does not change over time. The angular momentum of a particle with respect to an arbitrary reference point O is defined as

$$\mathbf{H}_O = \mathbf{r}_{OP} \times (m\mathbf{v}) \qquad (8.3)$$

where $\mathbf{r}_{OP}$ is the position vector between points O and P (see Figure 8.1a). The balance of angular momentum for a single infinitesimally small particle is automatically satisfied as a result of (8.1). In the case of multiple particles (a rigid body composed of infinite number of particles), the linear and angular momenta are defined as the sum (integral) of the momentum of individual particles (Figure 8.1b):

$$\mathbf{L} = \int_V \mathbf{v}\, dm \quad \text{and} \quad \mathbf{H}_O = \int_V \mathbf{r}_{OP} \times \mathbf{v}\, dm \qquad (8.4)$$

The second fundamental law of classical mechanics states that the rate of change of angular momentum is equal to the sum of all moments acting on the body:

$$\dot{\mathbf{H}}_O = \sum_i \mathbf{M}_i + \sum_i \mathbf{r}_i \times \mathbf{F}_i \qquad (8.5)$$

where $\mathbf{M}_i$ are the applied external force-couples in addition to the forces $\mathbf{F}_i$. If the reference point O is chosen to be the center of mass of the body G, the linear and angular momentum balance law take a simpler form:

$$m\dot{\mathbf{v}}_G = \sum_i \mathbf{F}_i \qquad (8.6)$$

$$I_G \dot{\boldsymbol{\omega}} = \sum_i \mathbf{M}_j + \sum_i \mathbf{r}_i \times \mathbf{F}_i \qquad (8.7)$$

where ω is the instantaneous vector of angular velocity and I_G is the moment of inertia about the center of mass. Equations 8.6 and 8.7 are called equations of motion and play a central role in the dynamics of rigid bodies. If there is no motion (linear and angular velocities are zero), one is faced with a *statics problem*. Conversely, when the accelerations are large, we need to solve the complete system of Equations 8.6 and 8.7 including the inertial terms. In mechatronic systems the mechanical response is generally slower than the electrical one and therefore determines the overall response. If the response time is critical to the application, one needs to consider the inertial terms in Equations 8.6 and 8.7.

8.1.2 Equations of Motion of Deformable Bodies

Rigid bodies do not change shape or size during their motion, that is, the distance between the particles they are made of is constant. In reality, all objects deform to a certain extent when subjected to external forces. Whether a body can be treated as rigid or deformable is dictated by the particular application.

In this section we will review the fundamental equations describing the motion of deformable bodies. These equations also result from the balance of linear and angular momentum applied to an infinitesimally small portion of the material volume dV. Each element dV is subjected not only to external body force $\mathbf{f}$, but also to internal forces originating from the rest of the body. These internal forces are described by a second order tensor $\mathbf{T}$, called stress tensor. The balance of linear momentum can then be stated in integral form for an *arbitrary* portion of the body occupying volume V as

$$\frac{d}{dt}\int_V \rho \mathbf{v}\, dv = \int_{\partial V} \mathbf{T} \cdot \mathbf{n}\, dA + \int_V \mathbf{f}\, dv \tag{8.8}$$

where ρ is the mass density, $\mathbf{v}$ is the velocity of the element dV, and $\mathbf{f}$ is the force per unit volume acting upon dV. The above balance law states that the rate of change of linear momentum is equal to the sum of the internal force flux (stress) acting on the boundary of V and the external body force, distributed inside V. Applying the transport theorem to (8.8) along with the mass conservation law reduces the above to

$$\int_V \rho \dot{\mathbf{v}}\, dv = \int_V \nabla \cdot \mathbf{T}\, dv + \int_V \mathbf{f}\, dv \tag{8.9}$$

Since (8.9) is valid for an arbitrary volume, it follows that the integrands are also equal. Thus the local (differential) form of linear momentum balance is

$$\rho \dot{\mathbf{v}} = \nabla \cdot \mathbf{T} + \mathbf{f} \quad \text{or with index notation} \quad \rho \dot{v}_i = T_{ij,j} + f_i \tag{8.10}$$

Using analogous procedure, the balance of angular momentum can be shown to reduce to a simple symmetry condition of the stress tensor

$$T_{ij} = T_{ji} \tag{8.11}$$

which is valid for materials without external body couples. It should be mentioned that in certain anisotropic materials, the polarization or magnetization vectors can develop body couples, for example, when $\mathbf{E} \times \mathbf{P} \neq 0$. In these cases the stress tensor is nonsymmetric and its vector invariant is equal to the body couple. Equation 8.10 are usually used in one of the three most common coordinate systems. For example, using rectangular coordinates we have

$$\begin{aligned}
\frac{\partial T_{xx}}{\partial x} + \frac{\partial T_{xy}}{\partial y} + \frac{\partial T_{xz}}{\partial z} + f_x &= \rho a_x, \quad T_{xy} = T_{yx} \\
\frac{\partial T_{yx}}{\partial x} + \frac{\partial T_{yy}}{\partial y} + \frac{\partial T_{yz}}{\partial z} + f_y &= \rho a_y, \quad T_{yz} = T_{zy} \\
\frac{\partial T_{zx}}{\partial x} + \frac{\partial T_{zy}}{\partial y} + \frac{\partial T_{zz}}{\partial z} + f_z &= \rho a_z, \quad T_{xz} = T_{zx}
\end{aligned} \tag{8.12}$$

and in cylindrical coordinates

$$\begin{aligned}
\frac{\partial T_{rr}}{\partial r} + \frac{T_{rr} - T_{\theta\theta}}{r} + \frac{1}{r}\frac{\partial T_{r\theta}}{\partial \theta} + \frac{\partial T_{rz}}{\partial z} + f_r &= \rho a_r, \quad T_{r\theta} = T_{\theta r} \\
\frac{\partial T_{r\theta}}{\partial r} + \frac{2}{r}T_{r\theta} + \frac{1}{r}\frac{\partial T_{\theta\theta}}{\partial \theta} + \frac{\partial T_{\theta z}}{\partial z} + f_\theta &= \rho a_\theta, \quad T_{\theta z} = T_{z\theta} \\
\frac{\partial T_{rz}}{\partial r} + \frac{1}{r}T_{rz} + \frac{1}{r}\frac{\partial T_{\theta z}}{\partial \theta} + \frac{\partial T_{zz}}{\partial z} + f_z &= \rho a_z, \quad T_{rz} = T_{zr}
\end{aligned} \tag{8.13}$$

where (x, y, z) and (r, θ, z) are the three coordinates, f's are the corresponding body force densities, and a's are the accelerations. In addition to Equations 8.12 or 8.13, a relation between the stress and the displacement is needed in order to determine the deformation. Since the rigid body translations and rotations do not cause deformation of the body, they do not affect the internal stress field either. In fact, the latter is a function of the gradient of the displacement, called deformation gradient. When this gradient is small, a linear relationship between the displacements and strains can be used

$$\varepsilon_x = \frac{\partial u_x}{\partial x}, \quad \varepsilon_y = \frac{\partial u_y}{\partial y}, \quad \varepsilon_z = \frac{\partial u_z}{\partial z}, \quad \varepsilon_{xy} = \frac{\partial u_x}{\partial y} + \frac{\partial u_y}{\partial x}, \quad \varepsilon_{xz} = \frac{\partial u_x}{\partial z} + \frac{\partial u_z}{\partial x}, \quad \varepsilon_{zy} = \frac{\partial u_z}{\partial y} + \frac{\partial u_y}{\partial z} \quad (8.14)$$

The conservation of momentum and kinematic relations does not contain any information about the material. Constitutive laws provide this additional information. The most common such law describes a linear elastic material and can be conveniently expressed using a symmetric matrix c_{ij}, called stiffness matrix:

$$\begin{bmatrix} T_{xx} \\ T_{yy} \\ T_{zz} \\ T_{yz} \\ T_{zx} \\ T_{xy} \end{bmatrix} = \begin{bmatrix} c_{11} & c_{12} & c_{13} & c_{14} & c_{15} & c_{16} \\ & c_{22} & c_{23} & c_{24} & c_{25} & c_{26} \\ & & c_{33} & c_{34} & c_{35} & c_{36} \\ & & & c_{44} & c_{45} & c_{46} \\ & \text{symm.} & & & c_{55} & c_{56} \\ & & & & & c_{66} \end{bmatrix} \cdot \begin{bmatrix} \varepsilon_x \\ \varepsilon_y \\ \varepsilon_z \\ \varepsilon_{yz} \\ \varepsilon_{zx} \\ \varepsilon_{xy} \end{bmatrix} \quad (8.15)$$

In the most general case, the matrix c_{ij} has 21 independent elements. When the material has a crystal symmetry, the number of independent constants is reduced. For example, single crystal Si is a common structural material in MEMS with a cubic symmetry. In this case there are only three independent constants:

$$\begin{bmatrix} T_{xx} \\ T_{yy} \\ T_{zz} \\ T_{yz} \\ T_{zx} \\ T_{xy} \end{bmatrix} = \begin{bmatrix} c_{11} & c_{12} & c_{12} & 0 & 0 & 0 \\ & c_{11} & c_{12} & 0 & 0 & 0 \\ & & c_{11} & 0 & 0 & 0 \\ & & & c_{44} & 0 & 0 \\ & \text{symm.} & & & c_{44} & 0 \\ & & & & & c_{44} \end{bmatrix} \cdot \begin{bmatrix} \varepsilon_x \\ \varepsilon_y \\ \varepsilon_z \\ \varepsilon_{yz} \\ \varepsilon_{zx} \\ \varepsilon_{xy} \end{bmatrix} \quad (8.16)$$

If the material is isotropic (amorphous or polycrystalline), the number of independent elastic constants is further reduced to two by the relation $c_{44} = (c_{11} - c_{12})/2$. The elastic constants of several most commonly used materials are listed in Table 8.1 (from Kittel 1996).

Additional information on other symmetry classes can be found in Nye (1960).

TABLE 8.1 Elastic Constants of Several Common Cubic Crystals

Crystal	Stiffness Constants at Room Temperature, 10^{11} N/m^2		
	c_{11}	c_{12}	c_{44}
W	5.233	2.045	1.607
Ta	2.609	1.574	0.818
Cu	1.684	1.214	0.754
Ag	1.249	0.937	0.461
Au	1.923	1.631	0.420
Al	1.608	0.607	0.282
K	0.0370	0.0314	0.0188
Pb	0.495	0.423	0.149
Ni	2.508	1.500	1.235
Pd	2.271	1.761	0.17
Si	1.66	0.639	0.796

8.1.3 Electric Phenomena

In the previous section the laws governing the motion of rigid and deformable bodies were reviewed. The forces entering these equations are often of electromagnetic origin; thus one has to know the distribution of electric and magnetic fields. The electromagnetic field is governed by a set of four coupled equations known as Maxwell's equations. Similarly, to the momentum equations, these can also be postulated in integral form. Here we only give the local form

$$\dot{\mathbf{B}} + \nabla \times \mathbf{E} = 0$$
$$\nabla \cdot \mathbf{D} = q^f$$
$$\nabla \times \mathbf{H} - \dot{\mathbf{D}} = \mathbf{i}$$
$$\nabla \cdot \mathbf{B} = 0$$
(8.17)

where $\mathbf{E}$ is the electric field, $\mathbf{D}$ is the electric displacement, $\mathbf{B}$ is the magnetic induction, $\mathbf{H}$ is the magnetic field strength, $\mathbf{i}$ is the electric current density, and q^f is the free charge volume density. Equations 8.17 require constitutive laws specifying the current density, electric displacement, and magnetic field in terms of electric field and magnetic induction vectors. A linear form of these laws is given by

$$\mathbf{i} = \frac{\mathbf{E}}{\rho_e}, \qquad \mathbf{D} = \varepsilon_0 \mathbf{E} + \mathbf{P}, \qquad \mathbf{B} = \mu_0 \mathbf{H} + \mu_0 \mathbf{M} = \mu_0 \mu_r \mathbf{H}$$
(8.18)

where ρ_e is the electrical resistance. The coupling between electrical and mechanical fields can be linear or nonlinear. For example, piezoelectricity is a linear phenomenon describing the generation of electric field as a result of the application of mechanical stress. Electrostriction on the other hand is a second order effect, resulting in the generation of mechanical strain proportional to the square of the electric field. Other effects include piezoresistivity, i.e., a change of the electrical resistance due to mechanical stress. In addition to these material properties, electromechanical coupling can be achieved through direct use of electromagnetic forces (Lorentz force) as is commonly done in conventional electrical machines. Lorentz force per unit volume is given by

$$\mathbf{f}^L = q^f(\mathbf{E} + \mathbf{v} \times \mathbf{B})$$
(8.19)

where q^f is the volume charge density. Equation 8.19 accounts for the forces acting on free charge only. If the fields have strong gradients, the above expression should be modified to include the polarization and magnetization terms (Maugin 1988).

$$\mathbf{f}^{EM} = q^f \mathbf{E} + \left(\mathbf{i} + \frac{\partial \mathbf{P}}{\partial t}\right) \times \mathbf{B} + \mathbf{P} \cdot \nabla \mathbf{E} + \nabla \mathbf{B} \cdot \mathbf{M}$$
(8.20)

Equation 8.19 or 8.20 can be used in the momentum equation 8.10 in place of the body force $\mathbf{f}$.

As mentioned earlier, piezoelectricity and piezoresistivity are the other commonly used effects in electromechanical systems. The piezoelectric effect occurs only in materials with certain crystal structure. Common examples include $BaTiO_3$ and lead zirconia titanate (PZT). In the quasi-electrostatic approximation (when the magnetic effects are neglected) there are four variables describing the electromechanical state of the body—electric field $\mathbf{E}$ and displacement $\mathbf{D}$, mechanical stress $\mathbf{T}$ and strain ε. The constitutive laws of piezoelectricity are given as a set of two matrix equations between the four field variables, relating one mechanical and one electrical variable to the other two in the set

$$\varepsilon_{ij} = s_{ijkl} T_{kl} + d_{ijk} E_k, \qquad D_i = d_{ikl} T_{kl} + \varepsilon_0 \Xi_{ij} E_j$$
(8.21)

where s_{ijkl} is the elastic compliance tensor, d_{ijk} is the piezoelastic tensor, Ξ_{ij} is the electric permittivity tensor. If the electric field and the polarization vectors are co-linear, the stress and strain tensors are symmetric, and the number of independent coefficients in s_{ijkl} is reduced from 81 to 21 and for the piezoelastic tensor d_{ijk} from 27 to 18. If further, the piezoelectric is poled in one direction only (e.g., index 3), the only nonzero elements are

$$d_{113}, \quad d_{223}, \quad d_{333}, \quad d_{232} = d_{322}, \quad d_{131} = d_{313}, \quad d_{123} = d_{213}$$

Numerical values for the coefficients in (8.22) for bulk $BaTiO_3$ crystals can be found in Zgonik et al. 1994.

8.2 Common Structures in Mechatronic Systems

Microelectromechanical systems traditionally use technology developed for the manufacturing of integrated circuits. As a result, the employed mechanical structures are often planar devices—springs, coils, bridges, or cantilever beams, subjected to in-plane and out-of-plane bending and torsion. Using high aspect ratio reactive ion etching combined with fusion bonding of silicon, it is possible to realize true three-dimensional structures as well. For example Figure 8.2 shows an SEM micrograph of a complex capacitive force sensor designed to accept glass fibers in an etched v-groove. In this section, we will review the fundamental relationships used in the initial designs of such electromechanical systems.

8.2.1 Beams

Microcantilevers are used in surface micromachined electrostatic switches, as "cantilever tip" for scanning probe microscopy (SPM) and in myriad of sensors, based on vibrating cantilevers. The majority of the surface micromachined beams fall into two cases—cantilever beams and bridges. Figure 8.3 illustrates a two-layer cantilever beam (Figure 8.3a) and a bridge (Figure 8.3b). The elastic force required to produce deflection d at the tip of the cantilever beam, or at the center of the bridge, is given by

$$F^{elast} = K_{eff} d \tag{8.22}$$

where

$$K_{eff} = \frac{24(EI)_{eff}}{(6l_e^3/5) + 6(l-l_e)l_e^2 + 12(l-l_e)^2 l_e + 8(l-l_e)^3} \quad \text{and} \quad K_{eff} = \frac{360(EI)_{eff}}{30l^3 - 45ll_e^2 - 5(l_e^4/l) + 3l_e^3} \tag{8.23}$$

are the effective spring constants of the composite beams for cantilever and bridge beams, respectively. The effective stiffness of the beam in both cases can be calculated from

$$(EI)_{eff} = \frac{E_1 w t_1^3}{12} + \frac{E_2 w t_2^3}{12} + \frac{E_1 E_2 t_1 t_2 w (t_1 + t_2)^2}{4(E_1 t_1 + E_2 t_2)} \tag{8.24}$$

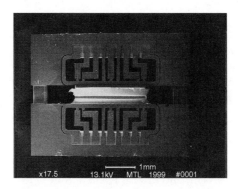

FIGURE 8.2 Capacitive force sensor using 3D micromachining.

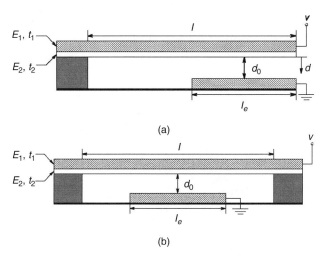

FIGURE 8.3 Surface micromachined beams: (a) Two-layer composite beam with electrostatic actuation; (b) two-layer composite bridge with electrostatic actuation.

where w is the width of the beam, t_1 the thickness of the top beam, t_2 the thickness of insulating layer (silicon oxide, silicon nitride), l the length of the beam, l_e the length of fixed electrode, E_1 the Young's modulus of the top layer, E_2 the Young's modulus of insulating layer.

8.2.2 Torsional Springs

Torsion of beams is used primarily in rotating structures such as micromirrors for optical scanning, or projection displays. The micromirror array developed by Texas Instruments for example uses polycrystalline silicon beams as hinges of the micromirror plate.

The torsion problems can be solved in a closed form for beams with elliptical or triangular cross sections (Mendleson 1968). In the case of an elliptical cross section, the moment required to produce an angular twist (angle or rotation per unit length of the beam) α [rad/m] is equal to

$$M = \frac{\pi a^3 b^3}{a^2 + b^2} G\alpha \qquad (8.25)$$

where G is the elastic shear modulus, and a and b are the lengths of the two semi-axes of the ellipse. The maximum shear stress in this case is

$$\tau^{max} = \frac{2G\alpha a^2 b}{a^2 + b^2}, \quad a > b \qquad (8.26)$$

The torsional stiffness of rectangular cross-section beams can be obtained in terms of infinite power series (Hopkins 1987). If the cross-section has dimension $a \times b$, $b < a$, the first three term of this series result in an equation similar to 8.25

$$M = 2KG\alpha, \quad \text{where} \quad K = ab^3\left[\frac{1}{3} - 0.21\frac{b}{a}\left(1 - \frac{b^4}{12a^4}\right)\right]. \qquad (8.27)$$

8.2.3 Thin Plates

Pressure sensors are one of the most popular electromechanical transducers. The basic structure used to convert mechanical pressure into electrical signal is a thin plate subjected to a pressure differential. Piezoresistive gauges are used to convert the strain in the membrane into change of resistance, which is

TABLE 8.2 Deflection and Bending Moments of Clamped Plate under Uniform Load q

b/a	$W(x=0, y=0)$	$M_x(x=a/2, y=0)$	$M_y(x=0, y=b/2)$	$M_x(x=0, y=0)$	$M_y(x=0, y=0)$
1	$0.00126 qa^4/D$	$-0.0513 qa^2$	$-0.0513 qa^2$	$0.0231 qa^2$	$0.0231 qa^2$
1.5	$0.00220 qa^4/D$	$-0.0757 qa^2$	$-0.0570 qa^2$	$0.0368 qa^2$	$0.0203 qa^2$
2	$0.00254 qa^4/D$	$-0.0829 qa^2$	$-0.0571 qa^2$	$0.0412 qa^2$	$0.0158 qa^2$
∞	$0.00260 qa^4/D$	$-0.0833 qa^2$	$-0.0571 qa^2$	$0.0417 qa^2$	$0.0125 qa^2$

Source: Evans 1939.

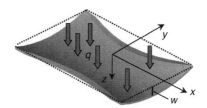

FIGURE 8.4 Thin plate subjected to positive pressure q.

read out using a conventional resistive bridge circuit. The initial pressure sensors were fabricated via anisotropic etching of silicon, which results in a rectangular diaphragm. Figure 8.4 shows a thin-plate, subjected to normal pressure q, resulting in out-of-plane displacement $w(x, y)$. The equilibrium condition for $w(x, y)$ is given by the thin plate theory (Timoshenko 1959):

$$\frac{\partial^4 w}{\partial x^4} + 2\frac{\partial^4 w}{\partial x^2 \partial y^2} + \frac{\partial^4 w}{\partial y^4} = \frac{q}{D} \tag{8.28}$$

where $D = Eh^3/12(1 - \nu^2)$ is the flexural rigidity, E is the Young's modulus, ν is the Poisson ratio, and h is the thickness of the plate. The edge-moments (moments per unit length of the edge) and the small strains are

$$\begin{aligned}
M_x(x, y) &= -D\left(\frac{\partial^2 w}{\partial x^2} - \nu\frac{\partial^2 w}{\partial y^2}\right), & \varepsilon_{xx}(x, y, z) &= -z\frac{\partial^2 w}{\partial x^2} \\
M_y(x, y) &= -D\left(\frac{\partial^2 w}{\partial y^2} - \nu\frac{\partial^2 w}{\partial x^2}\right), & \varepsilon_{yy}(x, y, z) &= -z\frac{\partial^2 w}{\partial y^2} \\
M_{xy}(x, y) &= D(1 - \nu)\frac{\partial^2 w}{\partial x \partial y}, & \varepsilon_{xy}(x, y, z) &= -z\frac{\partial^2 w}{\partial x \partial y}
\end{aligned} \tag{8.29}$$

Using (8.29), one can calculate the maximum strains occurring at the top and bottom faces of the plate in terms of the edge-moments:

$$\begin{aligned}
\varepsilon_{xx}^{max}(x, y, z) &= \frac{12z}{Eh^3}(M_x - \nu M_y)\Big|_{z=h} = \frac{12}{Eh^2}(M_x - \nu M_y) \\
\varepsilon_{yy}^{max}(x, y, z) &= \frac{12z}{Eh^3}(M_y - \nu M_x)\Big|_{z=h} = \frac{12}{Eh^2}(M_y - \nu M_x)
\end{aligned} \tag{8.30}$$

In the case of a pressure sensor with a diaphragm subjected to a uniform pressure, the boundary conditions are built-in edges: $w = 0$, $\partial w/\partial x = 0$ at $x = \pm a/2$ and $w = 0$, $\partial w/\partial y = 0$ at $y = \pm b/2$, where the diaphragm has lateral dimensions $a \times b$. The solution of this problem has been obtained by Evans (1939), showing that the maximum strains are at the center of the edges. The values of the edge-moments and the displacement of the center of plate are listed in Table 8.2.

8.3 Vibration and Modal Analysis

As mentioned earlier, the time response of a continuum structure requires the solution of Equation 8.10 with the acceleration terms present. For linear systems this solution can be represented by an infinite superposition of characteristic functions (modes). Associated with each such mode is also a characteristic number (eigenvalue) determining the time response of the mode. The analysis of these modes is called modal analysis and has a central role in the design of resonant cantilever sensors, flapping wings for micro-air-vehicles (MAVs) and micromirrors, used in laser scanners and projection systems. In the case of a cantilever beam, the flexural displacements are described by a fourth-order differential equation

$$\frac{IE}{\rho A}\frac{\partial^4 w(x,t)}{\partial x^4} + \frac{\partial^2 w(x,t)}{\partial t^2} = 0 \tag{8.31}$$

where I is the moment of inertia, E is the Young's modulus, ρ is the density, and A is the area of the cross section. When the thickness of the cantilever is much smaller than the width, E should be replaced by the reduced Young's modulus $E_1 = E/(1 - v^2)$. For a rectangular cross section, (8.31) is reduced to

$$\frac{Eh^2}{12\rho}\frac{\partial^4 w(x,t)}{\partial x^4} + \frac{\partial^2 w(x,t)}{\partial t^2} = 0 \tag{8.32}$$

where h is the thickness of the beam. The solution of (8.32) can be written in terms of an infinite series of characteristic functions representing the individual vibration modes

$$w = \sum_{i=1}^{\infty} \Phi_i(x)\sin(\omega_i t + \delta_i) \tag{8.33}$$

where the characteristic functions Φ_i are expressed with the four Rayleigh functions S, T, U, and V:

$$\begin{aligned}\Phi_i &= a_i S(\lambda_i x) + b_i T(\lambda_i x) + c_i U(\lambda_i x) + d_i V(\lambda_i x) \\ S(x) &= \frac{1}{2}(\cosh x + \cos x), \quad T(x) = \frac{1}{2}(\sinh x + \sin x) \\ U(x) &= \frac{1}{2}(\cosh x - \cos x), \quad V(x) = \frac{1}{2}(\sinh x - \sin x), \quad \lambda_i^4 = \omega_i^2 \frac{\rho A}{IE}\end{aligned} \tag{8.34}$$

The coefficients a_i, b_i, c_i, d_i, ω_i, and δ_i are determined from the boundary and initial conditions of (8.34). For a cantilever beam with a fixed end at $x = 0$ and a free end at $x = L$, the boundary conditions are

$$w(0,t) = 0, \quad \frac{\partial^2 w(L,t)}{\partial x^2} = 0$$
$$\frac{\partial w(0,t)}{\partial x} = 0, \quad \frac{\partial^3 w(L,t)}{\partial x^3} = 0 \tag{8.35}$$

Since (8.35) are to be satisfied by each of the functions Φ_i, it follows that $a_i = 0$, $b_i = 0$ and

$$\cosh(\lambda_i L)\cos(\lambda_i L) = -1 \tag{8.36}$$

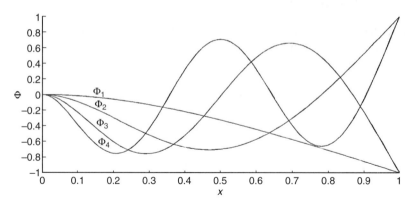

FIGURE 8.5 First four vibration modes of a cantilever beam.

From this transcendental equation the λ_i's and the circular frequencies ω_i are determined (Butt et al. 1995).

$$\lambda_i L \cong \frac{(2i-1)\pi}{2}, \quad \omega_i = \frac{(2i-1)^2 \pi^2}{4L^2} \sqrt{\frac{IE}{\rho A}} = \frac{(2i-1)^2 \pi^2}{4L^2} \sqrt{\frac{Eh^2}{12\rho}} \quad (8.37)$$

Figure 8.5 shows the first four vibrational modes of the cantilever. An important result of the modal analysis is the calculation of the amplitude of thermal vibrations of cantilevers. As the size of the cantilevers is reduced to nanometer scale, the energy of random thermal excitations becomes comparable with the energy of the individual vibration modes. This effect leads to a thermal noise in nanocantilevers. Using the equipartition theorem (Butt et al. 1995) showed that the root mean square of the amplitude of the tip of such cantilever is

$$\sqrt{\overline{z^2}} = \sqrt{\frac{kT}{K}} = \frac{0.64 \text{ Å}}{\sqrt{K}}, \quad K = \frac{Ewh^3}{4L^2} \quad (8.38)$$

Similar analysis can be performed on vibrations of thin plates such as micromirrors. The free lateral vibrations of such a plate are described by

$$\frac{\partial^4 w(x,y,t)}{\partial x^4} + 2\frac{\partial^4 w(x,y,t)}{\partial x^2 \partial y^2} + \frac{\partial^4 w(x,y,t)}{\partial y^4} = -\frac{\rho h}{D}\frac{\partial^2 w(x,y,t)}{\partial t^2} \quad (8.39)$$

The interested reader is referred to Timoshenko (1959) for further details on vibrations of plates.

8.4 Buckling Analysis

Structural instability can occur due to material failure, For example, plastic flow or fracture, or it can also occur due to large changes in the geometry of the structure (e.g., buckling, wrinkling, or collapse). The latter is the scope of this section. When short columns are subjected to a compressive load, the stress in the cross section is considered uniform. Thus for short columns, failure will occur when the plastic yield stress of the material is reached. In the case of long and slender beams under compression, due to manufacturing imperfections, the applied load or the column will have some eccentricity. As a result the force will develop a bending moment proportional to the eccentricity, resulting in additional lateral deflection. While for small loads the lateral displacement will reach equilibrium, above certain critical

TABLE 8.3 Critical Load Coefficients

	End Conditions		
	One end built-in, other free	Both ends built-in	Pin-joints at both ends
K coefficient	1/4	4	1

load, the beam will be unable to withstand the bending moment and will collapse. Consider the beam in Figure 8.5, subjected to load F with eccentricity e, resulting in lateral displacement of the tip δ. According to the beam bending equation

$$EI\frac{\partial^2 w}{\partial x^2} = M = F(\delta + e + w) \tag{8.40}$$

where the boundary conditions are $w(0) = 0$, $\partial w/\partial x \mid_{x=0} = 0$. The corresponding solution is

$$w = (e + \delta)[1 - \cos(\sqrt{IE/F}x)] \tag{8.41}$$

From $w(L) = \delta$ one has $\delta = e(1/\cos kL - 1)$, where $k = \sqrt{IE/F}$. This solution looses stability when δ grows out of bound, that is, when $\cos kL = 0$, or $kL = (2n+1)\pi/2$. From this condition the smallest critical load is

$$F^{cr} = \pi^2 IE/4L^2 \tag{8.42}$$

The above analysis and Equation 8.42 were developed by Euler. Similar conditions can be derived for other types of beam supports. A general formula for the critical load can be written as

$$F^{cr} = K\pi^2 IE/L^2 \tag{8.43}$$

where several values of the coefficient K are given in Table 8.3.

8.5 Transducers

Transducers are devices capable of converting one type of energy into another. If the output energy is mechanical work the transducer is called an actuator. The rest of the transducers are called sensors, although in most cases, a mechanical transducer can also be a sensor and vice versa. For example the capacitive transducer can be used as an actuator or position sensor. In this section the most common actuators used in micromechatronics are reviewed.

8.5.1 Electrostatic Transducers

The electrostatic transducers fall into two main categories—parallel plate electrodes and interdigitated comb electrodes. In applications where relatively large capacitance change or force is required, the parallel plate configuration is preferred. Conversely, larger displacements with linear force/displacement characteristics can be achieved with comb drives at the expense of reduced force. Parallel plate actuators are used in electrostatic micro-switches as illustrated in Figure 8.1. In this case the electrodes form a parallel plate capacitor and the force is described by

$$F_{elec} = \frac{A\varepsilon_0 \varepsilon_r^2 V^2}{2[t_2 + \varepsilon_r(d_0 - d)]^2} \tag{8.44}$$

where A is the area of overlap between the two electrodes; t_2 is the thickness of insulating layer (silicon dioxide, silicon nitride); l_e is the length of fixed electrode; ε_r is the relative permittivity of insulating layer; V is the applied voltage; d_0 is the initial separation between the capacitor plates; and d is downward

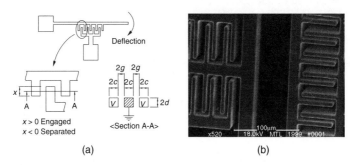

FIGURE 8.6 Lateral comb transducers: (a) Dimensions; (b) two orthogonal Si combs.

deflection of the beam. The minimum voltage required to close the gap of a cantilever actuator is known as the threshold voltage (Petersen 1978), and can be approximated as

$$V^{th} \approx \sqrt{\frac{18(IE)_{eff}\, d_0^3}{5\varepsilon_0 L^4 w}} \tag{8.45}$$

where $(IE)_{eff}$ is given by (8.24).

Comb drives also fall in two categories: symmetric and asymmetric. Symmetric comb drive is shown in Figure 8.6a. In this configuration the gaps between the individual fingers are equal. Figure 8.6(b) shows a pair of asymmetric comb capacitors, used in the force sensor shown in Figure 8.2 (Enikov 2000a). In any case, the force generated between the fingers is equal to the derivative of the total electrostatic energy with respect to the displacement

$$F^{el} = \frac{n}{2}\frac{\partial C}{\partial x} V^2 \tag{8.46}$$

where n is the number of fingers. Several authors have given approximate expressions for (8.46). One of the most accurate calculations of the force between the pair of fingers shown in Figure 8.6(a) is given by (Johnson et al. 1995) using Schwartz transforms

$$F^{el} = \begin{cases} \dfrac{\varepsilon_0 V^2}{\pi}\left\{ \ln\left[\left(\left(\dfrac{c}{g}+1\right)^2 - 1\right)\left(1+\dfrac{2g}{c}\right)^{1+c/g}\right] + \dfrac{\pi d}{g} - \dfrac{c+g}{x}\right\}, & x > \Delta_+ \text{ (engaged)} \\ -\dfrac{\varepsilon_0 V^2}{\pi}\left\{\dfrac{2(c+g)}{x}\right\}, & x < -\Delta_- \text{ (separated)} \end{cases} \tag{8.47}$$

In the transition region $x \in [-\Delta_-; \Delta_+]$, $\Delta_{+,-} \approx 2g$, the force can be approximated with a tangential line between the two branches described by (8.47).

8.5.2 Electromagnetic Transducers

Electromagnetic force has also been used extensively. It can be generated via planar coil as illustrated in Figure 8.7. The cantilever and often the coils are made of soft ferromagnetic material. Using an equivalent magnetic circuit model, the magnetic force acting on the top cantilever can be estimated as

$$F_{mag} = \frac{2n^2 I^2 (2A_2 + A_1)}{\mu_0 A_1 A_2 (2R_1 + R_2)^2} \tag{8.48}$$

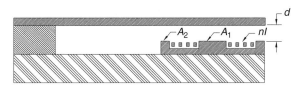

FIGURE 8.7 Electromagnetic actuation.

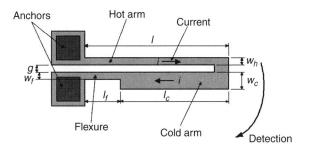

FIGURE 8.8 Lateral thermal actuator.

where

$$R_1 = \frac{d}{\mu_0 A_1} + \frac{h_1}{\mu_0 \mu_r A_1}, \qquad R_2 = \frac{d}{\mu_0 A_2} + \frac{h_1}{\mu_0 \mu_r A_2} + \frac{h_2}{\mu_0 \mu_r A_b} \qquad (8.49)$$

are the reluctances; h_1 and h_2 are the flux-path lengths inside the top and bottom permalloy layers.

8.5.3 Thermal Actuators

Thermal actuators have been investigated for positioning of micromirrors (Liew et al. 2000), and microswitch actuation (Wood et al. 1998). This actuator consists of two arms with different cross sections (see Figure 8.8). When current is passed through the two arms, the higher current density occurs in the smaller cross-section beam and thus generates more heat per unit volume. The displacement is a result of the temperature differential induced in the two arms. For the actuator shown in Figure 8.8, an approximate model for the deflection of the tip δ can be developed using the theory of thermal bimorphs (Faupel 1981)

$$\delta \approx \frac{3l^2(T^{\text{hot}} \alpha(T^{\text{hot}}) - T^{\text{cold}} \alpha(T^{\text{cold}}))}{4(w_h + w_f)} \qquad (8.50)$$

where T^{hot} and T^{cold} are the average temperatures of the hot and cold arms and $\alpha(T)$ is the temperature dependent thermal expansion coefficient. A more detailed analysis including the temperature distribution in the arms can be found in (Huang et al. 1999).

8.5.4 Electroactive Polymer Actuators

Electroactive polymer-metal composites (EAPs) are promising multi-functional materials with extremely reach physics. Recent interest towards these materials is driven by their unique ability to undergo large deformations under very low driving voltages as well as their low mass and high fracture toughness. For comparison, Table 8.4 lists several characteristic properties of EAPs and other piezoelectric ceramics.

EAPs are being tested for use in flapping-wing micro-air-vehicles (MAVs) (Rohani 1999), underwater swimming robots (Laurent 2001), and biomedical applications (Oguro 2000). An EAP actuator consists

TABLE 8.4 Comparative Properties of EAPs, Shape Memory Alloy, and Piezoceramic Actuators

Characteristic Property	EAP	Shape Memory Alloy	Piezoelectric Ceramics
Achievable strain	more than 10%	up to 8%	up to 0.3%
Young's modulus (GPa)	0.114 (wet)	75	89
Tensile strength (MPa)	34 (wet)	850	76
Response time	msec–min	sec–min	μsec–sec
Mass density (g/cm^3)	2.0	6.5	7.5
Actuation voltage	1–10 V	N/A	50–1000 V

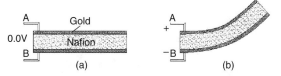

FIGURE 8.9 Polymer metal composite actuator.

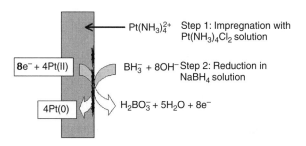

FIGURE 8.10 Two-step Pt plating process.

of an ion-exchange membrane covered with a conductive layer as illustrated in Figure 8.9a. Upon application of a potential difference at points A and B the composite bends towards the anodic side as shown in Figure 8.9b. Among the numerous ion-exchange polymers, perfluorinated sulfonic acid (Nafion Du Pont, USA) and perfluorinated carboxylic acid (Flemion, Asahi, Japan) are the most commonly used in actuator applications. The chemical formula of a unit chain of Nafion is

$$[(CF_2-CF_2)_n-CF-CF_2-]_m \\ \quad\quad\quad\quad\quad | \\ \quad\quad\quad\quad O-CF-CF_2-O-CF_2-SO_3^-M^+ \tag{8.51}$$

where M^+ is the counterion (H^+, Na^+, Li^+, ...). The ionic clusters are attached to side chains, which according to transmission electron microscopy (TEM) studies, segregate in hydrophilic nano-clusters with diameters ranging from 10 to 50 Å (Xue 1989). In 1982, Gierke proposed a structural model (Gireke 1982) according to which, the clusters are interconnected via narrow channels. The size and distribution of these channels determine the transport properties of the membrane and thus the mechanical response.

Metal–polymer composites can be produced by vapor or electrochemical deposition of metal over the surface of the membrane. The electrochemical platinization method (Fedkiw 1992), used by the author, is based on the ion-exchange properties of the Nafion. The method consists of two steps: step one—ion exchange of the protons H^+ with metal cations (e.g., Pt^{2+}); step two—chemical reduction of the Pt^{2+} ions in the membrane to metallic Pt using $NaBH_4$ solution. These steps are outlined in Figure 8.10 and an SEM microphotograph of the resulting composite is shown in Figure 8.11. The electrode surfaces are approximately 0.8 μm thick Pt deposits. Repeating the above steps several times results in dendritic growth of the electrodes into the polymer matrix (Oguro 1999) and has been shown to improve the actuation efficiency.

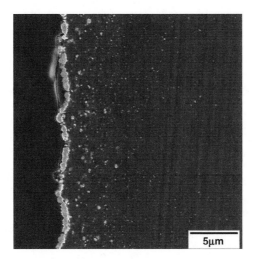

FIGURE 8.11 Nafion membrane with Pt electrode.

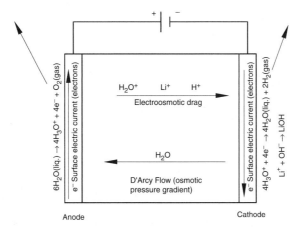

FIGURE 8.12 Ion transort in nafion.

The deformation of the polymer–metal composite can be attributed to several phenomena, the dominant one being differential swelling of the membrane due to internal osmotic pressure gradients (Eikerling 1998). A schematic representation of the ionic processes taking place inside the polymer is shown in Figure 8.12. Under the application of external electric field a flux of cations and hydroxonium ions is generated towards the cathode. At the cathode the ions pick up an electron and produce hydrogen and free water molecules. On the anodic side, the water molecules dissociate producing oxygen and hydroxonium ions. This redistribution of water within the membrane creates local expansion/contraction of the polymer matrix. Mathematically, the deformation can be described by introducing an additional strain (eigen strain) term in the expression of the total strain. Thus the total strain has two additive parts: elastic deformation of the polymer network due to external forces (mechanical, electrical) and chemical strain proportional to the compositional variables

$$\varepsilon_{ij} = \varepsilon_{ij}^{\text{elast}} + \rho_0 \sum_s \frac{\overline{V}^s}{3M^s}(c^s - c_0^s)\delta_{ij} \tag{8.52}$$

where c^s are the mass fractions, $\overline{V}^s$ are the partial molar volumes, M^s are the molar masses, and the index 0 refers to the initial value of a variable. Complete mathematical description of the polymer actuator requires the solution of mass transport (diffusion) equation, momentum balance, and Poisson equation for potential distribution, the discussion of which is beyond the scope of this book. An interesting consequence of the addition of the chemical strain in (8.46) is the explicit appearance of the pressure term in the electrochemical potential driving the diffusion. The total mass diffusion flux will have a component proportional to the negative gradient of the pressure, which for the case of water, will result in a relaxation phenomena observed experimentally. The total flux of component s is then given by

$$\mathbf{J}^s = -\frac{\rho c^s W^s}{\mathsf{M}^s} \nabla (\mu^{os}(T) + p\overline{V}^s + RT\ln(fc^s) + z^s\Phi) \tag{8.53}$$

where W^s is the mobility of component s, z^s is the valence of component s, p is the pressure, f^s is the activity coefficient, and Φ is the electric potential. We have omitted the cross-coupling terms that would appear in a fully coupled Onsager-type formulation. Interested readers are referred to (Enikov 2000b) and the references therein for further details.

8.6 Future Trends

The future MEMS are likely to be more heterogeneous in terms of materials and structures. Bio-MEMS for example, require use of nontoxic, noncorrosive materials, which is not a severe concern in standard IC components. Already departure from the traditional Si-based MEMS can be seen in the areas of optical MEMS using wide band-gap materials, nonlinear electro-optical polymers, and ceramics. As pointed earlier, the submicron size of the cantilever-based sensors brings the thermal noise issues in mechanical structures. Further reduction in size will require molecular statistic description of the interaction forces. For example, carbon nanotubes placed on highly oriented pyrolytic graphite (HOPG) experience increased adhesion force when aligned with the underlying graphite lattice (Falvo et al. 2000). The future mechatronic systems are likely to become an interface between the macro and nano domains.

References

Butt, H., Jaschke, M., "Calculation of thermal noise in atomic force microscopy," *Nanotechnology*, **6**, pp. 1–7, 1995.

Eikerling, M., Kharkats, Y.I., Kornyshev, A.A., Volfkovich, Y.M., "Phenomenological theory of electroosmotic effect and water management in polymer proton-conducting membranes," *Journal of the Electrochemical Society*, **145**(8), pp. 2684–2698, 1998.

Evans, T.H., *Journal of Applied Mechanics*, **6**, p. A-7, 1939.

Enikov, E.T., Nelson, B.J., "Three dimensional microfabrication for multi-degree of freedom capacitive force sensor using fiber chip coupling," *J. Micromech. Microeng.*, **10**, pp. 492–497, 2000.

Enikov, E.T., Nelson, B.J., "Electrotransport and deformation model of ion exhcange membrane based actuators," in *Smart Structures and Materials 2000*, Newport Beach, CA, SPIE vol. 3987, March, 2000.

Falvo, M.R., Steele, J., Taylor, R.M., Superfine, R., "Gearlike rolling motion mediated by commensurate contact: carbon nanotubes on HOPG," *Physical Review B*, **62**(6), pp. 665–667, 2000.

Faupel, J.H., Fisher, F.E., *Engineering Design: A Synthesis of Stress Analysis and Materials Engineering*, 2nd Ed., Wiley & Sons, New York, 1981.

Liu, R., Her, W.H., Fedkiw, P.S., "*In situ* electrode formation on a nafion membrane by chemical platinization," *Journal of the Electrochemical Society*, **139**(1), pp. 15–23, 1990.

Gierke, T.D., Hsu, W.S., "The cluster-network model of ion clustering in perfluorosulfonated membranes," in *Perfluorinated Ionomer Membranes*, A. Eisenberg and H.L. Yeager, Eds., vol. 180, American Chemical Society, 1982.

Johnson et al., "Electrophysics of micromechanical comb actuators," *Journal of Microelectromechanical Systems,* **4**(1), pp. 49–59, 1995.

Hopkins, *Design Analysis of Shafts and Beams,* 2nd Ed., Malabar, FL: RE Kreiger, 1987.

Huang, Q.A., Lee, N.K.S., "Analysis and design of polysilcon thermal flexture actuator," *Journal of Micromechanics and Microengineering,* **9**, pp. 64–70, 1999.

Kittel, Ch., *Introduction to Solid State Physics,* John Wiley & Sons, Inc., New York, 1996.

Laurent, G., Piat, E., "High efficiency swimming microrobot using ionic polymer metal composite actuators," to appear in 2001.

Liew, L. et al., "Modeling of thermal actuator in a bulk micromachined CMOS micromirror," *Microelectronics Journal,* **31**(9–10), pp. 791–790, 2000.

Maugin, G., *Continuum Mechanics of Electromagnetic Solids,* Elsevier, Amsterdam, The Netherlands, 1988.

Mendelson, *Plasticity: Theory and Application,* Macmillan, New York, 1968.

Nye, J.F., *Physical Properties of Crystals,* Oxford University Press, London, 1960.

Onishi, K., Sewa, Sh., Asaka, K., Fujiwara, N., Oguro, K., "Bending response of polymer electrolyte actuator," in *Smart Structures and Materials 2000,* Newport Beach, CA, SPIE vol. 3987, March, 2000.

Peterson, "Dynamic micromechanics on silicon: techniques and devices," *IEEE,* 1978.

Rohani, M.R., Hicks, G.R., "Multidisciplinary design and prototype of a micro air vehicle," *Journal of Aircraft,* **36**(1), p. 237, 1999.

Timoshenko, S., Woinowsky-Krieger, S., *Theory of Plates and Shells,* McGraw-Hill, New York, 1959.

Wood, R. et al., "MEMS microrelays," *Mechatronics,* **8**, pp. 535–547, 1998.

Xue, T., Trent, Y.S., Osseo-Asare, K., "Characterization of nafion membranes by transmision electron microscopy," *Journal of Membrane Science,* **45**, p. 261, 1989.

Zgonik et al., "Dielectric, elastic, piezoelectric, electro-optic and elasto-optic tensors of $BaTiO_3$ crystals," *Physical Review B,* **50**(9), p. 5841, 1994.

9
Modeling of Mechanical Systems for Mechatronics Applications

9.1	Introduction	9-1
9.2	Mechanical System Modeling in Mechatronic Systems	9-2
	Physical Variables and Power Bonds • Interconnection of Components • Causality	
9.3	Descriptions of Basic Mechanical Model Components	9-8
	Defining Mechanical Input and Output Model Elements • Dissipative Effects in Mechanical Systems • Potential Energy Storage Elements • Kinetic Energy Storage • Coupling Mechanisms • Impedance Relationships	
9.4	Physical Laws for Model Formulation	9-19
	Kinematic and Dynamic Laws • Identifying and Representing Motion in a Bond Graph • Assigning and Using Causality • Developing a Mathematical Model • Note on Some Difficulties in Deriving Equations	
9.5	Energy Methods for Mechanical System Model Formulation	9-28
	Multiport Models • Restrictions on Constitutive Relations • Deriving Constitutive Relations • Checking the Constitutive Relations	
9.6	Rigid Body Multidimensional Dynamics	9-31
	Kinematics of a Rigid Body • Dynamic Properties of a Rigid Body • Rigid Body Dynamics	
9.7	Lagrange's Equations	9-48
	Classical Approach • Dealing with Nonconservative Effects • Extensions for Nonholonomic Systems • Mechanical Subsystem Models Using Lagrange Methods • Methodology for Building Subsystem Model	
	References	9-53

Raul G. Longoria
The University of Texas at Austin

9.1 Introduction

Mechatronics applications are distinguished by controlled motion of mechanical systems coupled to actuators and sensors. Modeling plays a role in understanding how the properties and performance of mechanical components and systems affect the overall mechatronic system design. This chapter reviews methods for modeling systems of interconnected mechanical components, initially restricting the application to basic translational and rotational elements, which characterize a wide class of mechatronic applications. The underlying basis of mechanical motion (kinematics) is presumed known and not reviewed here, with more discussion and emphasis placed on a system dynamics perspective. More advanced applications requiring two- or three-dimensional motion is presented in Section 9.6.

Mechanical systems can be conceptualized as rigid and/or elastic bodies that may move relative to one another, depending on how they are interconnected by components such as joints, dampers, and other passive devices. This chapter focuses on those systems that can be represented using lumped-parameter descriptions, wherein bodies are treated as rigid and no dependence on spatial extent need be considered in the elastic effects. The modeling of mechanical systems in general has reached a fairly high level of maturity, being based on classical methods rooted in the Newtonian laws of motion. One benefits from the extensive and overwhelming knowledge base developed to deal with problems ranging from basic mass–spring systems to complex multibody systems. While the underlying physics are well understood, there exist many different means and ways to arrive at an end result. This can be especially true when the need arises to model a multibody system, which requires a considerable investment in methods for formulating and solving equations of motion. Those applications are not within the scope of this chapter, and the immediate focus is on modeling basic and moderately complex systems that may be of primary interest to a mechatronic system designer/analyst.

9.2 Mechanical System Modeling in Mechatronic Systems

Initial steps in modeling any physical system include defining a system boundary, and identifying how basic components can be partitioned and then put back together. In mechanical systems, these analyses can often be facilitated by identifying points in a system that have a distinct velocity. For purposes of analysis, active forces and moments are "applied" at these points, which could represent energetic interactions at a system boundary. These forces and moments are typically applied by actuators but might represent other loads applied by the environment.

A mechanical component modeled as a point mass or rigid body is readily identified by its velocity, and depending on the number of bodies and complexity of motion there is a need to introduce a coordinate system to formally describe the kinematics (e.g., see [12] or [15]). Through a kinematic analysis, additional (relative) velocities can be identified that indicate the connection with and motion of additional mechanical components such as springs, dampers, and/or actuators. The interconnection of mechanical components can generally have a dependence on geometry. Indeed, it is dependence of mechanical systems on geometry that complicates analysis in many cases and requires special consideration, especially when handling complex systems.

A preliminary description of a mechanical system should also account for any constraints on the motional states, which may be functions of time or of the states themselves. The dynamics of mechanical systems depends, in many practical cases, on the effect of constraints. Quantifying and accounting for constraints is of paramount importance, especially in multibody dynamics, and there are different schools of thought on how to develop models. Ultimately, the decision on a particular approach depends on the application needs as well as on personal preference.

It turns out that a fairly large class of systems can be understood and modeled by first understanding basic one-dimensional translation and fixed-axis rotation. These systems can be modeled using methods consistent with those used to study other systems, such as those of an electric or hydraulic type. Furthermore, building interconnected mechatronic system models is facilitated, and it is usually easier for a system analyst to conceptualize and analyze these models.

In summary, once an understanding of (a) the system components and their interconnections (including dependence on geometry), (b) applied forces/torques, and (c) the role of constraints, is developed, dynamic equations fundamentally due to Newton can be formulated. The rest of this section introduces the selection of physical variables consistent with a power flow and energy-based approach to modeling basic mechanical translational and rotational systems. In doing so, a bond graph approach [3,17,28] is introduced for developing models of mechanical systems. This provides a basis for introducing the concept of causality, which captures the input–output relationship between power-conveying variables in a system. The bond graph approach provides a way to understand and mathematically model basic as well as complex mechanical systems that is consistent with other energetic domains (electric, electromechanical, thermal, fluid, chemical, etc.).

9.2.1 Physical Variables and Power Bonds

9.2.1.1 Power and Energy Basis

One way to consistently partition and connect subsystem models is by using power and energy variables to quantify the system interaction, as illustrated for a mechanical system in Figure 9.1a. In this figure, one **port** is shown at which power flow is given by the product of force and velocity, $F \cdot V$, and another for which power is the product of torque and angular velocity, $T \cdot \omega$. These power-conjugate variables (i.e., those whose product yields power) along with those that would be used for electrical and hydraulic energy domains are summarized in Table 9.1. Similar effort (e) and flow (f) variables can be identified for other energy domains of interest (e.g., thermal, magnetic, chemical). This basis assures energetically correct models, and provides a consistent way to connect system elements together.

In modeling energetic systems, energy continuity serves as a basis to classify and to quantify systems. Paynter [28] shows how the energy continuity equation, together with a carefully defined port concept, provides a basis for a generalized modeling framework that eventually leads to a bond graph approach. Paynter's reticulated equation of energy continuity,

$$-\sum_{i=1}^{l} P_i = \sum_{j=1}^{m} \frac{dE_j}{dt} + \sum_{k=1}^{n} (P_d)_k \qquad (9.1)$$

concisely identifies the l distinct flows of power, P_i, m distinct stores of energy, E_j, and the n distinct dissipators of energy, P_d. Modeling seeks to refine the descriptions from this point. For example, in a simple mass–spring–damper system, the mass and spring store energy, a damper dissipates energy, and

TABLE 9.1 Power and Energy Variables for Mechanical Systems

Energy Domain	Effort, e	Flow, f	Power, P
General	e	f	$e \cdot f$ [W]
Translational	Force, F [N]	Velocity, V [m/s]	$F \cdot V$ [N m/s, W]
Rotational	Torque, T or τ [N m]	Angular velocity, ω [rad/s]	$T \cdot \omega$ [N m/s, W]
Electrical	Voltage, v [V]	Current, i [A]	$v \cdot i$ [W]
Hydraulic	Pressure, P [Pa]	Volumetric flowrate, Q [m³/s]	$P \cdot Q$ [W]

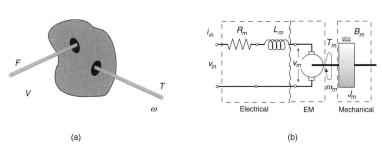

FIGURE 9.1 Basic interconnection of systems using power variables.

the interconnection of these elements would describe how power flows between them. Some of the details for accomplishing these modeling steps are presented in later sections.

One way to proceed is to define and categorize types of system elements based on the reticulated energy continuity Equation 9.1. For example, consider a system made up only of rigid bodies as energy stores (in particular of kinetic energy) for which $P_d = 0$ (we can add these later), and in general there can be l ports that could bring energy into this purely (kinetic)energy-storing system which has m distinct ways to put energy into the rigid bodies. This is a very general concept, consistent with many other ways to model physical systems. However, it is this foundation that provides for a generalized way to model and integrate different types of energetic systems.

The schematic of a permanent-magnet dc (PMDC) motor shown in Figure 9.1b illustrates how power variables would be used to identify interconnection points. This example also serves to identify the need for modeling mechanisms, such as the electromechanical (EM) interaction, that can represent the exchange of energy between two parts of a system. This model represents a simplified relationship between electrical power flow, $v \cdot i$, and mechanical power flow, $T \cdot \omega$, which forms the basis for a motor model. Further, this is an ideal power-conserving relationship that would only contain the power flows in the energy continuity equation; there are no stores or dissipators. Additional physical effects would be included later.

9.2.1.2 Power and Signal Flow

In a bond graph formulation of the PMDC motor, a **power bond** is used to identify flow of power. Power bonds quantify power flow via an effort-flow pair, which can label the bonds as shown in Figure 9.2a (convention calls for the effort to take the position above for any orientation of bond). This is a **word bond graph** model, a form used to identify the essential components in a complex system model. At this stage in a model, only the interactions of multiport systems are captured in a general fashion. Adding half-arrows on power bonds defines a power flow direction between two systems (positive in the direction of the arrow). **Signal bonds**, used in control system diagrams, have full-arrows and can be used in bond graph models to indicate interactions that convey only information (or negligible power) between multiports. For example, the word bond graph in Figure 9.2b shows a signal from the mechanical block to indicate an ideal measurement transferred to a controller as a pure signal. The controller has both signal and power flow signals, closing the loop with the electrical side of the model. These conceptual diagrams are useful for understanding and communicating the system interconnections but are not complete or adequate for quantifying system performance.

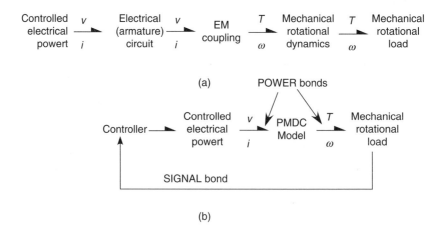

FIGURE 9.2 Power-based bond graph models: (a) PMDC motor word bond graph, (b) PMDC motor word bond graph with controller.

While it is convenient to use power and energy in formulating system models for mechanical systems, a **motional** basis is critical for identifying interconnections and when formulating quantifiable mathematical models. For many mechanical, translational, and rotational systems, it is sufficient to rely on basic one-dimensional motion and relative motion concepts to identify the interrelation between many types of practical components. Identifying network-like structure in these systems has been the basis for building electrical analogies for some time. These methods, as well as signal-flow analysis techniques, are not presented here but are the method of choice in some approaches to system dynamics [33]. Bond graph models are presented, and it will be shown in later sections how these are consistent even with more complex mechanical system formulations of three-dimensional dynamics as well as with the use of Lagrangian models.

9.2.1.3 Need for Motional Basis

In modeling mechanical translational or rotational systems, it is important to identify how the configuration changes, and a coordinate system should be defined and the effect of geometric changes identified. It is assumed that the reader is familiar with these basic concepts [12]. Usually a reference configuration is defined from which coordinates can be based. This is essential even for simple one-dimensional translation or fixed-axis rotation. The minumum number of geometrically independent coordinates required to describe the configuration of a system is traditionally defined as the **degrees of freedom**. Constraints should be identified and can be used to choose the most convenient set of coordinates for description of the system. We distinguish between degrees of freedom and the minimum number of **dynamic state variables** that might be required to describe a system. These may be related, but they are not necessarily the same variables or the same in number (e.g., a second-order system has two states but is also referred to as a single degree of freedom system).

An excellent illustration of the relevance of degrees of freedom, constraints, and the role these concepts play in modeling and realizing a practical system is shown in Figure 9.3. This illustration (adapted from Matschinsky [22]) shows four different ways to configure a wheel suspension. Case (a), which also forms the basis for a 1/4-car model clearly has only one degree of freedom. The same is true for cases (b) and (c), although there are constraints that reduce the number of coordinates to just one in each of these designs. Finally, the rigid beam axle shows how this must have two degrees of freedom in vertical and rotational motion of the beam to achieve at least one degree of freedom at each wheel.

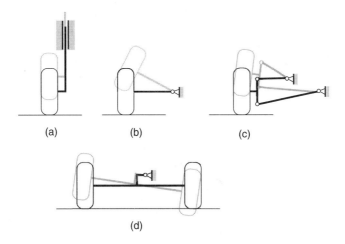

FIGURE 9.3 Wheel suspensions: (a) vertical travel only, 1 DOF; (b) swing-axle with vertical and lateral travel, 1 DOF; (c) four-bar linkage design, constrained motion, 1 DOF; (d) rigid beam axle, two wheels, vertical, and rotation travel, 2 DOF.

9.2.2 Interconnection of Components

In this chapter, we will use bond graphs to model mechanical systems. Like other graph representations used in system dynamics [33] and multibody system analysis [30,39], bond graphs require an understanding of basic model elements used to represent a system. However, once understood, graph methods provide a systematic method for representing the interconnection of multi-energetic system elements. In addition, bond graphs are unique in that they are not linear graph formulations: power bonds replace branches, multiports replace nodes [28]. In addition, they include a systematic approach for computational causality.

Recall that a single line represents power flow, and a half-arrow is used to designate positive power flow direction. Nodes in a linear graph represent across variables (e.g., velocity, voltage, flowrate); however, the multiport in a bond graph represents a system element that has a physical function defined by an energetic basis. System model elements that represent masses, springs, and other components are discussed in the next section. Two model elements that play a crucial role in describing how model elements are interconnected are the 1-junction and 0-junction. These are ideal (power-conserving) multiport elements that can represent specific physical relations in a system that are useful in interconnecting other model elements.

A point in a mechanical system that has a distinct velocity is represented by a 1-junction. When one or more model elements (e.g., a mass) have the same velocity as a given 1-junction, this is indicated by connecting them to the 1-junction with a power bond. Because the 1-junction is constrained to conserve power, it can be shown that efforts (forces, torques) on all the connected bonds must sum to zero; that is, $\sum e_i = 0$. This is illustrated in Figure 9.4a. The 1-junction enforces kinematic compatibility and introduces a way to graphically express force summation! The example in Figure 9.4b shows three systems (the blocks labeled 1, 2, and 3) connected to a point of common velocity. In the bond graph, the three systems would be connected by a 1-junction. Note that sign convention is incorporated into the sense of the power arrow.

For the purpose of analogy with electrical systems, the 1-junction can be thought of as a series electrical connection. In this way, elements connected to the 1-junction all have the same current (a flow variable) and the effort summation implied in the 1-junction conveys the Kirchhoff voltage law. In mechanical systems, 1-junctions may represent points in a system that represent the velocity of a mass, and the effort summation is a statement of Newton's law (in D'Alembert form), $\sum F - \dot{p} = 0$.

Figure 9.4 illustrates how components with common velocity are interconnected. Many physical components may be interconnected by virtue of a common effort (i.e., force or torque) or 0-junction. For example, two springs connected serially deflect and their ends have distinct rates of compression/extension; however, they have the same force across their ends (ideal, massless springs). System components that have this type of relationship are graphically represented using a 0-junction. The basic 0-junction definition is shown in Figure 9.5a. Zero junctions are especially helpful in mechanical system modeling because they can also be used to model the connection of components having relative motion. For example, the device in Figure 9.5b, like a spring, has ends that move relative to one another, but the force

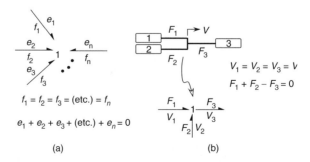

FIGURE 9.4 Mechanical 1-junction: (a) basic definition, (b) example use at a massless junction.

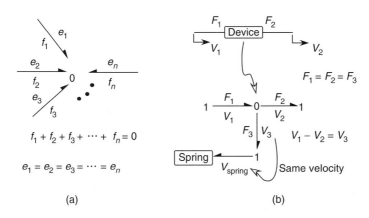

FIGURE 9.5 Mechanical 0-junction: (a) basic definition, (b) example use at a massless junction.

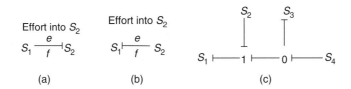

FIGURE 9.6 (a) Specifying effort from S_1 into S_2. (b) Specifying flow from S_1 into S_2. (c) A contrived example showing the constraint on causality assignment imposed by the physical definitions of 0- and 1-junctions.

on each end is the same (note this assumes there is negligible mass). The definition of the 0-junction implies that all the bonds have different velocities, so a flow difference can be formed to construct a relative velocity, V_3. All the bonds have the same force, however, and this force would be applied at the 1-junctions that identify the three distinct velocities in this example. A spring, for example, would be connected on a bond connected to the V_3 junction, as shown in Figure 9.5b, and $V_{spring} = V_3$.

The 1- and 0-junction elements graphically represent algebraic structure in a model, with distinct physical attributes from compatibility of kinematics (1-junction) and force or torque (0-junction). The graph should reflect what can be understood about the interconnection of physical devices with a bond graph. There is an advantage in forming a bond graph, since causality can then be used to form mathematical models. See the text by Karnopp, Margolis, and Rosenberg [17] for examples. There is a relation to through and across variables, which are used in linear graph methods [33].

9.2.3 Causality

Bond graph modeling was conceived with a consistent and algorithmic methodology for assignment of causality (see Paynter [28], p. 126). In the context of bond graph modeling, causality refers to the input–output relationship between variables on a power bond, and it depends on the systems connected to each end of a bond. Paynter identified the need for this concept having been extensively involved in analog computing, where solutions rely on well-defined relationships between signals. For example, if system S_1 in Figure 9.6a is a known source of effort, then when connected to a system S_2, it must specify effort into S_2, and S_2 in turn must return the flow variable, f, on the bond that connects the two systems. In a bond graph, this causal relationship is indicated by a vertical stroke drawn on the bond, as shown in Figure 9.6a. The vertical stroke at one end of a bond indicates that effort is specified into the multiport element connected at that end. In Figure 9.6b, the causality is reversed from that shown in (a).

The example in Figure 9.6c illustrates how causality "propagates" through a bond graph of interconnected bonds and systems. Note that a 1-junction with multiple ports can only have one bond specifying flow at that junction, so the other bonds specify effort into the 1-junction. A 0-junction requires one bond to specify effort, while all others specify flow. Also note that a direction for positive power flow has not been assigned on these bonds. This is intentional to emphasize the fact that power sense and causality assignment on a bond are **independent** of each other.

Causality assignment in system models will be applied in examples that follow. An extensive discussion of the successive cauality assignment procedure (sometimes referred to as SCAP) can be found in Rosenberg and Karnopp [32] or Karnopp, Margolis, and Rosenberg [17]. By using the defined bond graph elements, causality assignment is made systematically. The procedure has been programmed into several commercially available software packages that use bond graphs as formal descriptions of physical system models.

Because it reveals the input–output relationship of variables on all the bonds in a system model, causality can infer computational solvability of a bond graph model. The results are used to indicate the number of dynamic states required in a system, and the causal graph is helpful in actually deriving the mathematical model. Even if equations are not to be derived, causality can be used to derive physical insight into how a system works.

9.3 Descriptions of Basic Mechanical Model Components

Mechanical components in mechatronic systems make their presence known through motional response and by force and torque (or moment) reactions notably on support structures, actuators, and sensors. Understanding and predicting these response attributes, which arise due to combinations of frictional, elastic, and inertial effects, can be gained by identifying their inherent dissipative and energy storing nature. This emphasis on dissipation and energy storage leads to a systematic definition of constitutive relations for basic mechanical system modeling elements. These model elements form the basis for building complex nonlinear system models and for defining impedance relations useful in transfer function formulation. In the following, it is assumed that the system components can be well represented by lumped-parameter formulations.

It is presumed that a modeling decision is made so that dissipative and energy storing (kinetic and potential) elements can be identified to faithfully represent a system of interest. The reticulation is an essential part of the modeling process, but sometimes the definition and interconnection of the elements is not easy or intuitive. This section first reviews mechanical system input and output model elements, and then reviews passive dissipative elements and energy-storing elements. The section also discusses coupling elements used for modeling gears, levers, and other types of power-transforming elements. The chapter concludes by introducing impedance relationships for all of these elements.

9.3.1 Defining Mechanical Input and Output Model Elements

In dynamic system modeling, initial focus requires defining a **system boundary**, a concept borrowed from basic thermodynamics. In isolating mechanical systems, a system boundary identifies ports through which power and signal can pass. Each port is described either by a force-velocity or torque-angular velocity power conjugate pair. It is helpful, when focusing on the mechanical system modeling, to make a judgement on the causality at each port. For example, if a motor is to be attached to one port, it may be possible to define torque as the input variable and angular velocity as the output (back to the motor).

It is important to identify that these are model assumptions. We define specific elements as **sources** of effort or flow that can be attached at the boundary of a system of interest. These inputs might be known and or idealized, or they could simply be "placeholders" where we will later attach a model for an actuator or sensor. In this case, the causality specified at the port is fixed so that the (internal) system model will not change. If the causality changes, it will be necessary to reformulate a new model.

In bond graph terminology, the term **effort source** is used to define an element that specifies an effort, such as this force or torque. The symbol S_e or E can be used to represent the effort source on a bond graph.

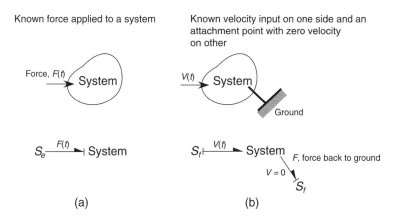

FIGURE 9.7 Two cases showing effort and flow sources on word bond graphs.

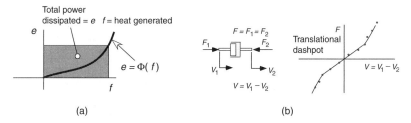

FIGURE 9.8 (a) Resistive constitutive relation. (b) Example dashpot resistive model.

A **flow source** is an element that specifies a flow on a bond, such as a translational velocity or angular or rotational velocity. The bond graph symbol is S_f or F. Two basic examples of sources are shown in Figure 9.7. Note that each bond has a defined effort or flow, depending on the source type. The causality on these model elements is always known, as shown. Further, each bond carries both pieces of information: (1) the effort or flow variable specified by the source, and (2) the *back reaction* indicated by the causality. So, for example, at the ground connection in Figure 9.7b, the source specifies the zero velocity constraint into the system, and the system, in turn, specifies an effort *back* to the ground. The symbolic representation emphasizes the causal nature of bond graph models and emphasizes which variables are available for examination. In this case, the force back into the ground might be a critical output variable.

9.3.2 Dissipative Effects in Mechanical Systems

Mechanical systems will dissipate energy due to friction in sliding contacts, dampers (passive or active), and through interaction with different energy domains (e.g., fluid loading, eddy current damping). These irreversible effects are modeled by constitutive functions between force and velocity or torque and angular velocity. In each case, the product of the effort-flow variables represents power dissipated, $P_d = e \cdot f$, and the total energy dissipated is $E_d = \int P_d \, dt = \int (e \cdot f) \, dt$. This energy can be determined given knowledge of the constitutive function, $e = \Phi(f)$, shown graphically in Figure 9.8a. We identify this as a basic *resistive* constitutive relationship that must obey the restriction imposed by the second law of thermodynamics; namely that, $e \cdot f \geq 0$. A typical mechanical dashpot that follows a resistive-type model description is summarized in Figure 9.8b.

In a bond graph model, resistive elements are symbolized by an **R** element, and a generalized, multiport R-element model is shown in Figure 9.9a. Note that the **R** element is distinguished by its ability to represent entropy production in a system. On the **R** element, a *thermal port* and bond are shown, and the power direction is always positive *away* from the **R**. In thermal systems, temperature, T, is the effort

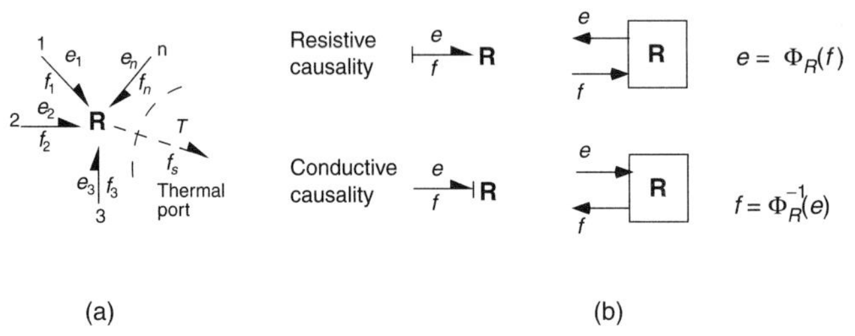

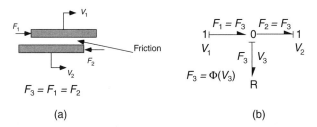

FIGURE 9.9 (a) Resistive bond graph element. (b) Resistive and conductive causality.

FIGURE 9.10 (a) Two sliding surfaces. (b) Bond graph model with causality implying velocities as known inputs.

variable and entropy flow rate, f_s is the flow variable. To compute heat generated by the **R** element, compose the calculation as Q (heat in watts) $= T \cdot f_s = \Sigma_i \, e_i \cdot f_i$ over the n ports.

The system attached to a resistive element through a power bond will generally determine the causality on that bond, since resistive elements generally have no preferred causal form.* Two possible cases on a given R-element port are shown in Figure 9.9b. A block diagram emphasizes the computational aspect of causality. For example, in a resistive case the flow (e.g., velocity) is a known input, so power dissipated is $P_d = e \cdot f = \Phi(f) \cdot f$. For the linear damper, $F = b \cdot V$, so $P_d = F \cdot V = bV^2$ (W).

In mechanical systems, many frictional effects are driven by relative motion. Hence, identifying how a dissipative effect is configured in a mechanical system requires identifying critical motion variables. Consider the example of two sliding surfaces with distinct velocities identified by 1-junctions, as shown in Figure 9.10(a). Identifying one surface with velocity V_1, and the other with V_2, the simple construction shown in Figure 9.10(b) shows how an **R** element can be connected at a *relative* velocity, V_3. Note the relevance of the causality as well. Two velocities join at the 0-junction to form a relative velocity, which is a causal input to the **R**. The causal output is a force, F_3, computed using the constitutive relation, $F = \Phi(V_3)$. The 1-junction formed to represent V_3 can be eliminated when there is only a single element attached as shown. In this case, the **R** would replace the 1-junction.

When the effort–flow relationship is linear, the proportionality constant is a **resistance**, and in mechanical systems these quantities are typically referred to as **damping constants**. Linear damping may arise in cases where two surfaces separated by a fluid slide relative to one another and induce a viscous and strictly laminar flow. In this case, it can be shown that the force and relative velocity are linearly related, and the material and geometric properties of the problem quantify the linear damping constant. Table 9.2 summarizes both translational and rotational damping elements, including the linear cases. These components are referred to as dampers, and the type of damping described here leads to the term viscous friction in mechanical applications, which is useful in many applications involving lubricated surfaces. If the relative speed is relatively high, the flow may become turbulent and this leads to nonlinear damper behavior. The constitutive relation is then a nonlinear function, but the **structure** or interconnection of

---

*This is true in most cases. Energy-storing elements, as will be shown later, have a causal form that facilitates equation formulation.

TABLE 9.2 Mechanical Dissipative Elements

Physical System	Fundamental Relations	Bond Graph
Generalized Dissipative Element $e \leftarrow$ R $\leftarrow f$ • Resistive element • **Resistance**, R	Dissipation: $\mathbf{e} \cdot \mathbf{f} = \sum_i e_i f_i = T \cdot f_s$ Resistive law: $e = \Phi_R(f)$ Conductive law: $f = \Phi_R^{-1}(e)$ Content: $P_f = \int e \cdot df$ Co-content: $P_e = \int f \cdot de$	$e_1, f_1 \searrow \quad \nearrow e_n, f_n$ $e_2, f_2 \rightarrow$ **R** $\cdots$ $\uparrow e_3, f_3$ Generalized multiport R-element
Mechanical Translation damping, b $F_1 \rightarrow \square \leftarrow F_2$ $V_1 \quad F_1 = F_2 = F \quad V_2$ • Damper $\quad V_1 - V_2 = V$ • **Damping**, b	Constitutive: $F = \Phi(V)$ Content: $P_V = \int F \cdot dV$ Co-energy: $P_F = \int V \cdot dF$ Dissipation: $P_d = P_V + P_F$	$\xrightarrow{F}$ **R**:b V Linear: $F = b \cdot V$ Dissipation: $P_d = bV^2$
Mechanical Rotation damping, B $T_1 \curvearrowright \square \curvearrowleft T_2$ $\omega_1 \quad \omega_2$ $T_1 = T_2 = T$ $\omega_1 - \omega_2 = \omega$ • Torsional damper • **Damping**, B	Constitutive: $T = \Phi(\omega)$ Content: $P_\omega = \int T \cdot d\omega$ Co-energy: $P_T = \int \omega \cdot dT$ Dissipation: $P_d = P_\omega + P_T$	$\xrightarrow{T}$ **R**:B ω Linear: $T = B \cdot \omega$ Dissipation: $P_d = B\omega^2$

TABLE 9.3 Typical Coefficient of Friction Values. Note, Actual Values Will Vary Significantly Depending on Conditions

Contacting Surfaces	Static, μ_s	Sliding or Kinetic, μ_k
Steel on steel (dry)	0.6	0.4
Steel on steel (greasy)	0.1	0.05
Teflon on steel	0.04	0.04
Teflon on teflon	0.04	—
Brass on steel (dry)	0.5	0.4
Brake lining on cast iron	0.4	0.3
Rubber on asphalt	—	0.5
Rubber on concrete	—	0.6
Rubber tires on smooth pavement (dry)	0.9	0.8
Wire rope on iron pulley (dry)	0.2	0.15
Hemp rope on metal	0.3	0.2
Metal on ice	—	0.02

the model in the system does not change. Dampers are also constructed using a piston/fluid design and are common in shock absorbers, for example. In those cases, the force–velocity characteristics are often tailored to be nonlinear.

The viscous model will not effectively model friction between dry solid bodies, which is a much more complex process and leads to performance bounds especially at lower relative velocities. One way to capture this type of friction is with the classic Coulomb model, which depends on the normal load between surfaces and on a coefficient of friction, typically denoted μ (see Table 9.3). The Coulomb model quantifies the friction force as $F = \mu N$, where N is the normal force. This function is plotted in Figure 9.11a to illustrate how it models the way the friction force always opposes motion. This model still qualifies as a resistive constitutive function relating the friction force and a relative velocity of the surfaces. In this case,

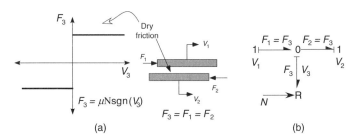

FIGURE 9.11 (a) Classic coulomb friction for sliding surfaces. (b) Bond graph showing effect of normal force as a modulation of the R-element law.

however, the velocity comes into effect only to determine the sign of the force; that is, $F = \mu N \operatorname{sgn}(V)$, where sgn is the signum function (value of 1 if $V > 0$ and −1 if $V < 0$).

This model requires a special condition when $V \to 0$. Dry friction can lead to a phenomenon referred to as stick-slip, particularly common when relative velocities between contacting surfaces approach low values. Stick-slip, or stiction, friction forces are distinguished by the way they vary as a result of other (modulating) variables, such as the normal force or other applied loads. Stick-slip is a type of system response that arises due to frictional effects. On a bond graph, a signal bond can be used to show that the normal force is determined by an external factor (e.g., weight, applied load, etc.). This is illustrated in Figure 9.11b. When the basic properties of a physical element are changed by signal bonds in this way, they are said to be **modulated**. This is a modeling technique that is very useful, but care should be taken so it is not applied in a way that violates basic energy principles.

Another difficulty with the standard dry friction model is that it has a preferred causality. In other words, if the causal input is velocity, then the constitutive relation computes a force. However, if the causal input is force then there is no unique velocity output. The function is not bi-unique. Difficulties of this sort usually indicate that additional underlying physical effects are not modeled. While the effort-flow constitutive relation is used, the form of the constitutive relation may need to be parameterized by other critical variables (temperature, humidity, etc.). More detailed models are beyond the scope of this chapter, but the reader is referred to Rabinowicz (1995) and Armstrong-Helouvry (1991) who present thorough discussions on modeling friction and its effects. Friction is usually a dominant source of uncertainty in many predictive modeling efforts (as is true in most energy domains).

9.3.3 Potential Energy Storage Elements

Part of the energy that goes into deforming any mechanical component can be associated with pure (lossless) storage of potential energy. Often the decision to model a mechanical component this way is identified through a basic constitutive relationship between an effort variable, e (force, torque), and a displacement variable, q (translational displacement, angular displacement). Such a relationship may be derived either from basic mechanics [29] or through direct measurement. An example is a translational spring in which a displacement of the ends, x, is related to an applied force, F, as $F = F(x)$.

In an energy-based lumped-parameter model, the generalized displacement variable, q, is used to define a state-determined potential energy function,

$$E = E(q) = U_q$$

This energy is related to the constitutive relationship, $e = \Phi(q)$, by

$$U(q) = U_q = \int e\, dq = \int \Phi(q)\, dq$$

It is helpful to generalize in this way, and to identify that practical devices of interest will have at least one connection (or port) in which power can flow to store potential energy. At this port the displacement

TABLE 9.4 Mechanical Potential Energy Storage Elements (Integral Form)

Physical System	Fundamental Relations	Bond Graph
Generalized Potential Energy Storage Element $e \leftarrow \boxed{C}$ $f \leftarrow$ • Capacitive element • **Capacitance**, C	State: $\mathbf{q} = $ displacement Rate: $\dot{\mathbf{q}} = \mathbf{f}$ Constitutive: $\mathbf{e} = \Phi(\mathbf{q})$ Energy: $U_q = \int \mathbf{e} \cdot d\mathbf{q}$ Co-energy: $U_e = \int \mathbf{q} \cdot d\mathbf{e}$	$\begin{array}{c} e_1 \\ f_1 = \dot{q}_1 \end{array} \searrow \quad \begin{array}{c} e_n \\ f_n = \dot{q}_n \end{array}$ $\dfrac{e_2}{f_2 = \dot{q}_2} \to C$ $\quad e_3 \uparrow f_3 = \dot{q}_3$ Generalized multiport C-element
Mechanical Translation stiffness, $k = 1/C$ $F_1 \sim\!\!\sim\!\!\sim F_2$ $V_1 \quad F_1 = F_2 = F \quad V_2$ • Spring $V_1 - V_2 = V$ • **Stiffness**, k, **compliance**, C	State: $x = $ displacement Rate: $\dot{x} = V$ Constitutive: $F = F(x)$ Energy: $U_x = \int F \cdot dx$ Co-energy: $U_F = \int x \cdot dF$	$\dfrac{F}{\dot{x}=V} \to \mathbf{C}{:}1/C{=}k$ Linear: $F = k \cdot x$ Energy: $U_x = \tfrac{1}{2} k x^2$ Co-energy: $U_F = F^2/2k$
Mechanical Rotation stiffness, $K = 1/C$ $T_1 \sim\!\!\sim\!\!\sim T_2$ $\omega_1 \quad T_1 = T_2 = T \quad \omega_2$ • Torsional spring $\omega_1 - \omega_2 = \omega$ • **Stiffness**, K, **compliance**, C	State: $\theta = $ angle Rate: $\dot{\theta} = \omega$ Constitutive: $T = T(\theta)$ Energy: $U_\theta = \int T \cdot d\theta$ Co-energy: $U_T = \int \theta \cdot dT$	$\dfrac{T}{\dot{\theta}=\omega} \to \mathbf{C}{:}1/C{=}K$ Linear: $T = K \cdot \theta$ Energy: $U_\theta = \tfrac{1}{2} K \theta^2$ Co-energy: $U_T = T^2/2K$

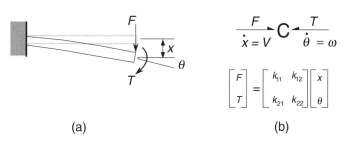

FIGURE 9.12 Example of two-port potential energy storing element: (a) cantilevered beam with translational and rotational end connections, (b) C-element, 2-port model.

variable of interest is either translational, x, or angular, θ, and the associated velocities are $V = \dot{x}$ and $\omega = \dot{\theta}$, respectively. A generalized potential energy storage element is summarized in Table 9.4, where examples are given for the translational and rotational one-port.

The linear translational spring is one in which $F = F(x) = kx = (1/C)x$, where k is the stiffness and $C \equiv 1/k$ is the *compliance* of the spring (compliance is a measure of "softness"). As shown in Table 9.4, the potential energy stored in a linear spring is $U_x = \int F\,dx = \int kx\,dx = \tfrac{1}{2} kx^2$, and the co-energy is $U_F = \int F\,dx = \int (F/k)\,dF = F^2/2k$. Since the spring is linear, you can show that $U_x = U_F$. If the spring is nonlinear due to, say, plastic deformation or work hardening, then this would not be true.

Elastic potential energy can be stored in a device through multiple ports and through different energy domains. A good example of this is the simple cantilevered beam having both tip force and moment (torque) inputs. The beam can store energy either by translational or rotational displacement of the tip. A constitutive relation for this 2-port C-element relates the force and torque to the linear and rotational displacements, as shown in Figure 9.12. A stiffness (or compliance) matrix for small deflections is derived by linear superposition.

9.3.4 Kinetic Energy Storage

All components that constitute mechanical systems have mass, but in a system analysis, where the concern is dynamic performance, it is often sufficient to focus only on those components that may store relevant amounts of kinetic energy through their motion. This presumes that an energetic basis is used for modeling, and that the tracking of kinetic energy will provide insight into the system dynamics. This is the focus of this discussion, which is concerned for the moment with one-dimensional translation and fixed-axis rotation. Later it will be shown how the formulation presented here is helpful for understanding more complex systems.

The concept of mass and its use as a model element is faciliated by Newton's relationship between the rate of change of momentum of the mass to the net forces exerted on it, $F = \dot{p}$, where p is the momentum. The energy stored in a system due to translational motion with velocity V is the kinetic energy. Using the relation from Newton's law, $dp = Fdt$, this energy is $E(p) = T(p) = T_p = \int P dt = \int FV \, dt = \int V \, dp$. If the velocity is expressed solely as a function of the momentum, p, this system is a pure translational mass, $V = \Phi(p)$. If the velocity is linearly proportional to the momentum, then $V = p/m$, where m is the mass. Similar basic definitions are made for a body in rotation about a fixed axis, and these elements are summarized in Table 9.5.

For many applications of practical interest to engineering, the velocity–momentum relation, $V = V(p)$ (the constitutive relation), is linear. Only in relativistic cases might there be a nonlinear relationship in the constitutive law for a mass. Nevertheless, this points out that for the general case of kinetic energy storage a constitutive relation is formed between the flow variable and the momentum variable, $f = f(p)$. This should help build appreciation for analogies with other energy domains, particularly in electrical systems where inductors (the mass analog) can have nonlinear relationships between current (a flow) and flux linkage (momentum).

The rotational motion of a rigid body considered here is constrained thus far to the simple case of planar and fixed-axis rotation. The mass moment of intertia of a body about an axis is defined as the sum of the products of the mass-elements and the squares of their distance from the axis. For the discrete case, $I = \sum r^2 \Delta m$, which for continuous cases becomes, $I = \int r^2 dm$ (units of kg m^2). Some common shapes

TABLE 9.5 Mechanical Kinetic Energy Storage Elements (Integral Form)

Physical System	Fundamental Relations	Bond Graph
Generalized Kinetic Energy Storage Element • Inertive element • Inertance, I	State: $\mathbf{p}$ = momentum Rate: $\dot{\mathbf{p}} = \mathbf{e}$ Constitutive: $\mathbf{f} = \Phi(\mathbf{p})$ Energy: $T_p = \int \mathbf{f} \cdot d\mathbf{p}$ Co-energy: $T_f = \int \mathbf{p} \cdot d\mathbf{f}$	Generalized multiport I-element
Mechanical Translation • Mass, M • Mass, m	State: p = momentum Rate: $\dot{p} = F$ Constitutive: $V = V(p)$ Energy: $T_p = \int f \cdot dp$ Co-energy: $T_V = \int p \cdot dV$	$\dot{p} = F$ → I:M Linear: $V = p/M$ Energy: $T_p = p^2/2M$ Co-energy: $T_V = \tfrac{1}{2} MV^2$
Mechanical Rotation • Rotational inertia • Mass moment of inertia, J	State: h = angular momentum Rate: $\dot{h} = T$ Constitutive: $\omega = \omega(h)$ Energy: $T_h = \int \omega \cdot dh$ Co-energy: $T_\omega = \int h \cdot d\omega$	$\dot{h} = T$ → I:J Linear: $\omega = h/J$ Energy: $T_h = h^2/2J$ Co-energy: $T_\omega = \tfrac{1}{2} J \omega^2$

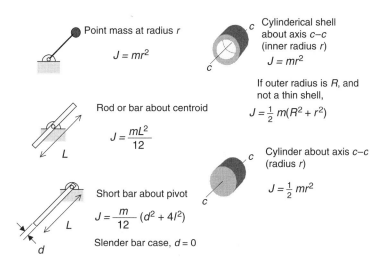

FIGURE 9.13 Mass moments of inertia for some common bodies.

and associated mass moments of inertia are given in Figure 9.13. General rigid bodies are discussed in section "Inertia Properties."

There are several useful concepts and theorems related to the properties of rigid bodies that can be helpful at this point. First, if the mass moment of inertia is known about an axis through its center of mass (I_G), then Steiner's theorem (parallel axis theorem) relates this moment of inertia to that about another axis a distance d away by $I = I_G + md^2$, where m is the mass of the body. It is also possible to build a moment of inertia for composite bodies, in those situations where the individual motion of each body is negligible. A useful concept is the radius of gyration, k, which is the radius of an imaginary cylinder of infinitely small wall thickness having the same mass, m, and the same mass moment of inertia, I, as a body in question, and given by, $k = \sqrt{I/m}$. The radius of gyration can be used to find an equivalent mass for a rolling body, say, using $m_{eq} = I/k^2$.

9.3.5 Coupling Mechanisms

Numerous types of devices serve as couplers or power transforming mechanisms, with the most common being levers, gear trains, scotch yokes, block and tackle, and chain hoists. Ideally, these devices and their analogs in other energy domains are power conserving, and it is useful to represent them using a 2-port model. In such a model element, the power in is equal to the power out, or in terms of effort-flow pairs, $e_1 f_1 = e_2 f_2$. It turns out that there are two types of basic devices that can be represented this way, based on the relationship between the power variables on the two ports. For either type, a relationship between two of the variables can usually be identified from geometry or from basic physics of the device. By imposing the restriction that there is an ideal power-conserving transformation inherent in the device, a second relationship is derived. Once one relation is established the device can usually be classified as a **transformer** or **gyrator**. It is emphasized that these model elements are used to represent the ideal power-conserving aspects of a device. Losses or dynamic effects are added to model real devices.

A device can be modeled as a **transformer** when $e_1 = me_2$ and $mf_1 = f_2$. In this relation, m is a transformer **modulus** defined by the device physics to be constant or in some cases a function of states of the system. For example, in a simple gear train the angular velocities can be ideally related by the ratio of pitch radii, and in a slider crank there can be formed a relation between the slider motion and the crank angle. Consequently, the two torques can be related, so the gear train is a transformer. A device can be modeled as a **gyrator** if $e_1 = rf_2$ and $rf_1 = e_2$, where r is the gyrator modulus. Note that this model can represent

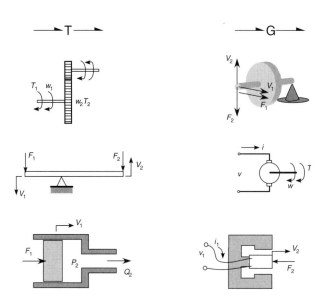

FIGURE 9.14 Common devices that can be modeled as transformers and gyrators in mechatronic systems.

the power-conserving transformation in devices for which a cross-relationship between power variables (i.e., effort related to flow) has been identified.*

Some examples of transformers and gyrators are shown in Figure 9.14. In a bond graph model, the transformer can be represented by a **TF** or **T**, while a gyrator is represented by a **GY** or **G** (note, the two letter symbol is common). The devices shown in Figure 9.14 indicate a modulus m or r, which may or may not be a constant value. Many devices may have power-conserving attributes; however, the relationship between the effort-flow variables may not be constant, so the relationship is said to be *modulated* when the modulus is a function of a dynamic variable (preferably a state of the system). On a bond graph, this can be indicated using a signal bond directed into the T or G modulus.

Examples of a modulated transformer and gyrator are given in Figure 9.15. These examples highlight useful techniques in modeling of practical devices. In the slider crank, note that the modulation is due to a change in the angular position of the crank. We can get this information from a bond that is adjacent to the transformer in question; that is, if we integrate the angular velocity found on a neighboring bond, as shown in Figure 9.15a. For the field excited dc motor shown in Figure 9.15b, the torque–current relation in the motor depends on a flux generated by the field; however, this field is excited by a circuit that is powered *independent* of the armature circuit. The signal information for modulation does not come from a neighboring bond, as in the case for the slider crank. These two examples illustrate two ways that constraints are imposed in coupling mechanisms. The modulation in the slider crank might be said to represent a holonomic constraint, and along these same lines the field excitation in the motor imposes a non-holonomic constraint. We cannot relate torque and current in the latter case without solving for the dynamics of an independent system—the field circuit. In the slider crank, the angular position required for the modulation is obtained simply by integrating the velocity, since $\dot{\theta} = \omega$. Additional discussion on constraints can be found in Section 9.7.

The system shown in Figure 9.16a is part of an all-mechanical constant-speed drive. A mechanical feedback force, F_2, will adjust the position of the middle rotor, x_2. The effect is seen in the bond graph

---

*It turns out that the gyrator model element is essential in all types of systems. The need for such an element to represent gyroscopic effects in mechanical systems was first recognized by Thomson and Tait in the late 1900s. However, it was G. D. Birkhoff (1927) and B. D. H. Tellegen (1948) who independently identified the need for this element in analysis and synthesis of systems.

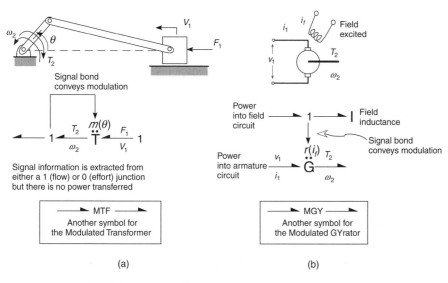

FIGURE 9.15 Concept of modulation in transformers and gyrators.

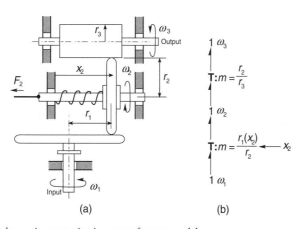

FIGURE 9.16 A nonholonomic constraint in a transformer model.

model of Figure 9.16b, which has two transformers to represent the speed ratio between the input (turntable) 1 and the mid-rotor 2, and the speed ratio between the mid-rotor and the output roller 3. The first transformer is a mechanical version of a nonholonomic transformation. Specifically, we would have to solve for the dynamics of the rotor position (x_2) in order to transform power between the input and output components of this device.

9.3.6 Impedance Relationships

The basic component descriptions presented so far are the basis for building basic models, and a very useful approach relies on impedance formulations. An impedance function, Z, is a ratio of effort to flow variables at a given system port of a physical device, and the most common application is for linear systems where $Z = Z(s)$, where s is the complex frequency variable (sometimes called the Laplace operator). An admittance is the inverse of the impedance, or $Y = 1/Z$. For each basic element defined, a linear impedance relation can be derived for use in model development. First, recall that the derivative operator can be represented by the s operator, so that dx/dt in s-domain is simply sx and $\int x\, dt$ is x/s, and so on.

TABLE 9.6 Basic Mechanical Impedance Elements

System	Resistive, Z_R	Capacitive, Z_C	Inertive, Z_I
Translation	b	k/s	$m \cdot s$
Rotation	B	K/s	$J \cdot s$

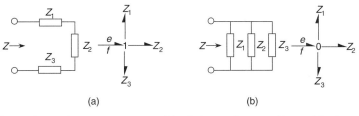

FIGURE 9.17 (a) Impedance of a series connection. (b) Admittance for a parallel combination.

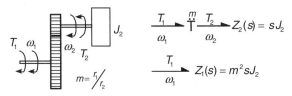

FIGURE 9.18 Rotational inertia attached to gear train, and corresponding model in impedance form. This example illustrates how a transformer can scale the gain of an impedance.

For the basic inertia element in rotation, for example, the basic rate law (see Table 9.5) is $\dot{h} = T$. In s-domain, $sh = T$. Using the linear constitutive relation, $h = J\omega$, so $sJ\omega = T$. We can observe that a rotation inertial impedance is defined by taking the ratio of effort to flow, or $T/\omega \equiv Z_I = sJ$. A similar exercise can be conducted for every basic element to construct Table 9.6.

Using the basic concept of a 0 junction and a 1 junction, which are the analogs of parallel and series circuit connections, respectively, basic impedance formulations can be derived for bond graphs in a way analogous to that done for circuits. Specifically, when impedances are connected in series, the total impedance is the sum, while admittances connected in parallel sum to give a total admittance. These basic relations are illustrated in Figure 9.17, for which

$$Z = \underbrace{Z_1 + Z_2 + \cdots + Z_n}_{n \text{ impedances in series sum to form a total impedance}}, \quad Y = \underbrace{Y_1 + Y_2 + \cdots + Y_n}_{n \text{ admittances in parallel sum to form a total admittance}} \quad (9.2)$$

Impedance relations are useful when constructing transfer functions of a system, as these can be developed directly from a circuit analog or bond graph. The transformer and gyrator elements can also be introduced in these models. A device that can be modeled with a transformer and gyrator will exhibit impedance-scaling capabilities, with the moduli serving a principal role in adjusting how an impedance attached to one "side" of the device appears when "viewed" from the other side. For example, for a device having an impedance Z_2 attached on port 2, the impedance as viewed from port 1 is derived as

$$Z_1 = \frac{e_1}{f_1} = \begin{bmatrix} e_1 \\ e_2 \end{bmatrix} \begin{bmatrix} e_2 \\ f_2 \end{bmatrix} \begin{bmatrix} f_2 \\ f_1 \end{bmatrix} = [m][Z_2(s)][m] = m^2 Z_2(s) \quad (9.3)$$

This concept is illustrated by the gear-train system in Figure 9.18. A rotational inertia is attached to the output shaft of the gear pair, which can be modeled as a transformer (losses, and other factors ignored here).

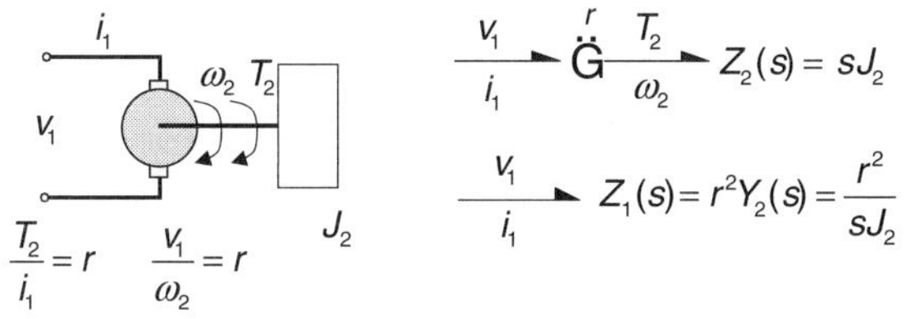

FIGURE 9.19 Rotational inertial attached to a basic rotational machine modeled as a simple gyrator. This example illustrates how a gyrator can scale the gain but also convert the impedance to an admittance form.

The impedance of the inertial is $Z_2 = sJ_2$, where J_2 is the mass moment of inertia. The gear train has an impedance-scaling capability, which can be designed through selection of the gear ratio, m.

The impedance change possible with a transformer is only in gain. The gyrator can affect gain and in addition can change the impedance into an admittance. Recall the basic gyrator relation, $e_1 = rf_2$ and $e_2 = rf_1$, then for a similar case as before,

$$\frac{e_1}{f_1} = \left[\frac{e_1}{f_2}\right]\left[\frac{f_2}{e_2}\right]\left[\frac{e_2}{f_1}\right] = [r][Y_2(s)][r] = r^2 Y_2(s) \tag{9.4}$$

This functional capability of gyrators helps identify basic motor-generator designs as integral parts of a flywheel battery system. A very simplified demonstration is shown in Figure 9.19, where a flywheel (rotational inertia) is attached to the mechanical port of a basic electromechanical gyrator. When viewed from the electrical port, you can see that the gyrator makes the inertia "look" like a potential energy storing device, since the impedance goes as $1/sC$, like a capacitive element, although here C is a mechanical inertia.

9.4 Physical Laws for Model Formulation

This section will illustrate basic equation formulation for systems ranging in complexity from mass–spring–damper models to slightly more complex models, showing how to interface with nonmechanical models.

Previous sections of this chapter provide descriptions of basic elements useful in modeling mechanical systems, with an emphasis on a dynamic system approach. The power and energy basis of a bond graph approach makes these formulations consistent with models of systems from other energy domains. An additional benefit of using a bond graph approach is that a systematic method for causality assignment is available. Together with the physical laws, causal assignment provides insight into how to develop computational models. Even without formulating equations, causality turns out to be a useful tool.

9.4.1 Kinematic and Dynamic Laws

The use of basic kinematic and dynamic equations imposes a structure on the models we build to represent mechanical translation and rotation. Dynamic equations are derived from Newton's laws, and we build free-body diagrams to understand how forces are imposed on mechanical systems. In addition, we must use geometric aspects of a system to develop kinematic equations, relying on properly defined coordinate systems. If the goal is to analyze a mechanical system alone, typically the classical application of conservation of momentum or energy methods and/or the use of kinematic analysis is required to arrive at solutions to a given problem. In a mechatronic system, it is implied that a mechanical system is coupled to other types of systems (hydraulics, electromechanical devices, etc.). Hence, we focus here on how to build models that will be easily integrated into overall system models. A detailed classical discussion of kinematics and dynamics from a fundamental perspective can be found in many introductory texts such as Meriam and Kraige [23] and Bedford and Fowler [5], or in more advanced treatments by Goldstein [11] and Greenwood [12].

When modeling simple translational systems or fixed-axis rotational systems, the basic set of laws summarized below are sufficient to build the necessary mathematical models.

Basic Dynamic and Kinematic Laws		
System	Dynamics	Kinematics
Translational	$\sum_i^N F_i = 0$	$\sum_i^N V_i = 0$
Rotational	$\sum_i^N T_i = 0$	$\sum_i^N \omega_i = 0$
Junction type	1-junction	0-junction

There is a large class of mechanical systems that can be represented using these basic equations, and in this form it is possible to see how: (a) bond graph junction elements can be used to structure these models and (b) how these equations support circuit analog equations, since they are very similar to the Kirchhoff circuit laws for voltage and current. We present here the bond graph approach, which graphically communicates these physical laws through the 0- and 1-junction elements.

9.4.2 Identifying and Representing Motion in a Bond Graph

It is helpful when studying a mechanical system to focus on identifying points in the system that have distinct velocities (V or ω). One simply can associate a 1-junction with these points. Once this is done, it becomes easier to identify connection points for other mechanical components (masses, springs, dampers, etc.) as well as points for attaching actuators or sensors. Further, it is critical to identify and to define additional velocities associated with relative motion. These may not have clear, physically identifiable points in a system, but it is necessary to localize these in order to attach components that rely on relative motion to describe their operation (e.g., suspensions).

Figure 9.20 shows how identifying velocities of interest can help identify 1-junctions at which mechanical components can be attached. For the basic mass element in part (a), the underlying premise is that a component of a system under study is idealized as a pure translational mass for which momentum and velocity are related through a constitutive relation. What this implies is that the velocity of the mass is the same throughout this element, so a 1-junction is used to identify this distinct motion. A bond attached to this 1-junction represents how any power flowing into this junction can flow into a kinetic energy storing element, **I**, which represents the mass, m. Note that the force on the bond is equal to the rate of change of momentum, $\dot{p}$, where $p = mV$.

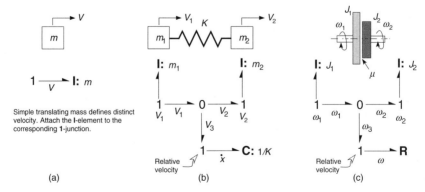

FIGURE 9.20 Identifying velocities in a mechanical system can help identify correct interconnection of components and devices: (a) basic translating mass, (b) basic two-degree of freedom system, (c) rotational frictional coupling between two rotational inertias.

The two examples in Figures 9.20b and 9.20c demonstrate how a relative velocity can be formed. Two masses each identify the two distinct velocity points in these systems. Using a 0-junction allows construction of a *velocity difference*, and in each case this forms a relative velocity. In each case the relative velocity is represented by a 1-junction, and it is critical to identify that this 1-junction is essentially an attachment point for a basic mechanical modeling element.

9.4.3 Assigning and Using Causality

Bond graphs describe how modeling decisions have been made, and how model elements (**R**, **C**, etc.) are interconnected. A power bond represents power flow, and assigning power convention using a half-arrow is an essential part of making the graph useful for modeling. A sign convention is essential for expressing the algebraic summation of effort and flow variables at 0- and 1-junctions. Power is generally assigned positive sense flowing into passive elements (resistive, capacitive, inertive), and it is usually safe to always adopt this convention. Sign convention requires consistent and careful consideration of the reference conditions, and sometimes there may be some arbitrariness, not unlike the definition of reference directions in a free-body diagram.

Causality involves an augmentation of the bond graph, but is strictly independent of power flow convention. As discussed earlier, an assignment is made on each bond that indicates the input–output relationship of the effort-flow variables. The assignment of causality follows a very consistent set of rules. A system model that has been successfully assigned causality on all bonds essentially communicates solvability of the underlying mathematical equations. To understand where this comes from, we can begin by examining the contents of Tables 9.4 and 9.5. These tables refer to the *integral* form of the energy storage elements. An energy storage element is in integral form if it has been assigned integral causality. Integral causality implies that the causal input variable (effort or flow) leads to a condition in which the state of the energy stored in that element can be determined only by *integrating* the fundamental rate law. As shown in Table 9.7, integral causality for an **I** element implies effort is the input, whereas integral causality for the **C** element implies flow is the input.

TABLE 9.7 Table Summarizing Causality for Energy Storage Elements

Integral Causality	Derivative Causality
C $\leftarrow$, $f = \dot{q}$; CONSTITUTIVE $e = \Phi_C(q)$; diagram: $\Phi_C(\bullet) \to e$, q, $\int(\bullet)dt \to f$; plot of f vs $q(t)$ with shaded area $q(t)$	**C** $\leftarrow$, $f = \dot{q}$; INVERSE CONSTITUTIVE $q = \Phi_C^{-1}(e)$; diagram: $\Phi_C^{-1}(\bullet) \to e$, q, $\frac{d}{dt} \to f$; plot with $f = dq/dt$ vs t
I $\leftarrow$, $e = \dot{p}$, f; CONSTITUTIVE $f = \Phi_I(p)$; diagram: $\int(\bullet)dt \to e$, p, $\Phi_I(\bullet) \to f$; plot with shaded area $p(t)$ vs t	**I** $\leftarrow$, $e = \dot{p}$, f; INVERSE CONSTITUTIVE $p = \Phi_I^{-1}(f)$; diagram: $\frac{d}{dt} \to e$, p, $\Phi_I^{-1}(\bullet) \to f$; plot with $e = dp/dt$ vs t

TABLE 9.8 Table of Causality Assignment Guidelines

Sources	Junctions	Ideal Coupling Elements			
$E \xrightarrow{e(t)}$	⊢ 0 ⊣ (Only one bond specifies effort.)	$\xrightarrow{e_1}$ T $\xrightarrow{e_2}$ $f_1 \quad f_2$ $e_1 = me_2$ $mf_1 = f_2$		$\xrightarrow{e_1}$ G $\xrightarrow{e_2}$ $f_1 \quad f_2$ $e_1 = rf_2$ $e_2 = rf_1$	
$F \xrightarrow{f(t)}$	⊣ 1 ⊢ (Only one bond specifies flow.)	$\xrightarrow{e_1}$ T $\xrightarrow{e_2}$ $f_1 \quad f_2$		$\xrightarrow{e_1}$ G $\xrightarrow{e_2}$ $f_1 \quad f_2$	

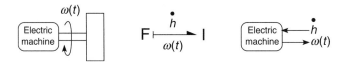

FIGURE 9.21 (a) Driving a rotational inertia with a velocity source: (b) simple bond graph with causality, (c) explanation of back effect.

As shown in this table, the alternative causality for each element leads to *derivative causality*, a condition in which the state of the energy storage element is known instantaneously and as such is said to be *dependent* on the input variable, and is in a state of dependent causality. The implication is that energy storage elements in integral causality require one differential equation (the rate law) to be solved in order to determine the value of the *state variable* (p or q). Energy storage elements in derivative causality don't require a differential equation; however, they still make their presence known through the back reaction implied. For example, if an electric machine shown in Figure 9.21a is assumed to drive a rotational inertial with a known velocity, ω, then the inertia is in derivative causality. There will also be losses, but the problem is simplified to demonstrate the causal implications. The energy is always known since, $h = J\omega$, so $T_h = h^2/2J$. However, the machine will feel an inertial back torque, $\dot{h}$, whenever a change is made to ω. This effect cannot be neglected.

Causality assignment on some of the other modeling elements is very specific, as shown in Table 9.8. For example, for sources of effort or flow, the causality is implied. On the two-port transformer and gyrator, there are two possible causality arrangements for each. Finally, for 0- and 1-junctions, the causality is also very specific since in each case only one bond can specify the effort or flow at each.

With all the guidelines established, a basic causality assignment procedure can be followed that will make sure all bonds are assigned causality (see also Rosenberg and Karnopp [32] and Karnopp, Margolis, and Rosenberg [17]):

1. For a given system, assign causality to any effort or flow sources, and for each one assign the causality as required through 0- and 1-junctions and transformer and gyrator elements. The causality should be spread through the model until a point is reached where no assignment is implied. Repeat this procedure until all sources have been assigned causality.
2. Assign causality to any **C** or **I** element, trying to assign integral causality if possible. For each assignment, propagate the causality through the system as required. Repeat this procedure until all storage elements are assigned causality.

3. Make any final assignments on **R** elements that have not had their causality assigned through steps 1 and 2, and again propagate causality as required. Any arbitrary assignment on an **R** element will indicate need for solving an algebraic equation.
4. Assign any remaining bonds arbitrarily, propagating each case as necessary.

Causality can provide information about system operation. In this sense, the bond graph provides a picture of how inputs to a system lead to certain outputs. The use of causality with a bond graph replaces ad hoc assignment of causal notions in a system. This type of information is also useful for understanding how a system can be split up into modules for simulation and/or it can confirm the actual physical boundaries of components.

Completing the assignment of causality on a bond graph will also reveal information about the solvability of the system model. The following are key results from causality assignment:

- Causality assignment will reveal the order of the system, which is equal to the number of independent energy storage elements (i.e., those with integral causality). The state variable (p or q) for any such element will be a state of the system, and one first-order differential equation will be required to describe how this state propagates through time.
- Any arbitrary assignment of causality on an **R** element indicates there is an algebraic loop. The number of arbitrary assignments can be related to the number of algebraic equations required in the model.

9.4.4 Developing a Mathematical Model

Mathematical models for lumped-parameter mechanical systems will take the form of coupled ordinary differential equations or, for a linear or linearized system, transfer functions between variables of interest and system inputs. The form of the mathematical model should match the application, and one can readily convert between the different forms. A classical approach to developing the mathematical model will involve applying Newton's second law directly to each body, taking account of the forces and torques. Commonly, the result is a second-order ordinary differential equation for each body in a system. An alternative is to use Lagrange's equations, and for multidimensional dynamics, where bodies may have combined translation and rotation, additional considerations are required as will be discussed in Section 9.6. At this point, consider those systems where a given body is either under translation or rotation.

9.4.4.1 Mass–Spring–Damper: Classical Approach

A basic mechanical system that consists of a rigid body that can translate in the z-direction is shown in Figure 9.22a. The system is modeled using a mass, a spring, and a damper, and a force, $F(t)$, is applied

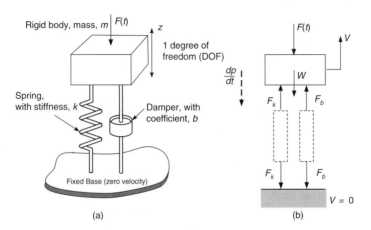

FIGURE 9.22 Basic mass–spring–damper system: (a) schematic, (b) free-body diagram.

directly to the mass. A free-body diagram in part (b) shows the forces exerted on the system. The spring and damper exert forces F_k and F_b on the mass, and these same forces are also exerted on the fixed base since the spring and damper are assumed to be massless. A component of the weight, W, resolved along the axis of motion is included. The sum of applied forces is then, $\Sigma F = F(t) + W - F_k - F_b$. The dashed arrow indicates the "inertial force" which is equal to the rate of change of the momentum in the z-direction, p_z, or, $dp_z/dt = \dot{p}_z = m\dot{V}_z$. This term is commonly used in a D'Alembert formulation, one can think of this force as opposing or resisting the effect of applied forces to accelerate the body. It is common to use the inertial force as an "applied force," especially when performing basic analysis (e.g., see Chapters 3 or 6 of [23]).

Newton's second law relates rate of change of momentum to applied forces, $\dot{p} = \Sigma F$, so, $\dot{p}_z = F(t) + W - F_k - F_b$. To derive a mathematical model, form a basic coordinate system with the z-axis positive upward. Recall the constitutive relations for each of the modeling elements, assumed here to be linear, $p_z = mV_z$, $F_k = kz_k$, and $F_b = bV_b$. In each of these elements, the associated velocity, V, or displacement, z, must be identified. The mass has a velocity, $V_z = \dot{z}$, relative to the inertial reference frame. The spring and damper have the same *relative* velocity since one end of each component is attached to the mass and the other to the base. The change in the spring length is z and the velocity is $\dot{z} - V_{base}$. However, $V_{base} = 0$ since the base is fixed, so putting this all together with Newton's second law, $m\ddot{z} = F(t) + W - kz - b\dot{z}$. A second-order ordinary differential equation (ODE) is derived for this single degree of freedom (DOF) system as

$$m\ddot{z} + b\dot{z} + kz = F(t) + W$$

In this particular example, if W is left off, z is the "oscillation" about a position established by static equilibrium, $z_{static} = W/k$.

If a transfer function is desired, a simple Laplace transform leads to (assuming zero initial conditions for motion about z_{static})

$$\frac{Z(s)}{F(s)} = \frac{1}{ms^2 + bs + k}$$

The simple mass–spring–damper example illustrates that models can be readily derived for mechanical systems with direct application of kinematics and Newton's laws. As systems become more complex either due to number of bodies and geometry, or due to interaction between many types of systems (hydraulic, electromechanical, etc.), it is helpful to employ tools that have been developed to facilitate model development. In a subsequent section, multibody problems and methods of analysis are briefly discussed. It has often been argued that the utility of bond graphs can only be seen when a very complex, multi-energetic system is analyzed. This need not be true, since a system (or mechatronics) analyst can see that a consistent formulation and efficacy of causality are very helpful in analyzing many different types of physical systems. This should be kept in mind, as these basic bond graph methods are used to re-examine the simple mass–spring–damper system.

9.4.4.2 Mass–Spring–Damper: Bond Graph Approach

Figure 9.23 illustrates the development of a bond graph model for a mass–spring–damper system. In part (a), the distinct velocity points are identified and 1-junctions are used to represent them on a bond graph. Even though the base has zero velocity, and there will be no power flow into or out of that point, it is useful to identify it at this point. A relative velocity is formed using a 0-junction, and note that all bonds have sign convention applied, so at the 0-junction, $V_{mass} - V_{relative} - V_{base} = 0$, which gives, $V_{relative} = V_{mass} - V_{base}$ as required.

The model elements needed to represent the system are connected to the 1-junctions, as shown in Figure 9.23b. Two sources are required, one to represent the applied force (effort, S_e) due to weight, and a second to represent the fixed based velocity (a flow source, S_f). The flow source is directly attached to

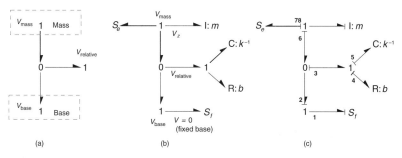

FIGURE 9.23 Basic mass–spring–damper system: (a) identifying velocity 1-junctions, (b) attaching model elements, (c) assignment of causality.

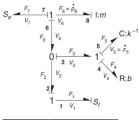

FIGURE 9.24 Equation derivation for mass–spring–damper. The '*' indicates these relations are reduced to functions of state or input. A '**' shows an intermediate variable has been reached that has elsewhere been reduced to '*'.

the 1-junction (the extra bond could be eliminated). An **I** element represents mass, a **C** represents the spring, and an **R** represents the losses in the damper. Note how the mass and the source of effort are attached to the 1-junction representing the mass velocity (the weight is always applied at that velocity). The spring and damper are attached via a power bond to the relative velocity between the mass and base.

Finally, in Figure 9.23c the eight bonds are labeled and causality is assigned. First, the fixed base source fixes the causality on bond 1, specifying the velocity at the 1-junction, and thus constraining the causality of bond 2 to have effort into the 1-junction. Since bond 2 did not specify effort into the 0-junction, causality assignment should proceed to other sources, and the effort source fixes causality on bond 7. This bond does not specify the flow at the adjoining 1-junction, so at this point we could look for other specified sources. Since there are none, we assign causality to any energy-storing elements which have a preferred integral causality. The bond 8 is assigned to give the **I** element integral causality (see Table 9.7), which then specifies the velocity at the 1-junction and thus constrains bond 6. At this point, bonds 6 and 2 both specify flow into the 0-junction, so the remaining bond 3 must specify the effort. This works out well because now bond 3 specifies flow into the remaining 1-junction (the relative velocity), which specifies velocity into the **C** and **R** elements. For the **C** element, this gives integral causality.

In summary, the causality is assigned and there are no causal conflicts (e.g., two bonds trying to specify velocity into a 1-junction). Both energy-storing elements have integral causality. This indicates that the states for the **I** (mass) and **C** (spring) will contribute to the state variables of the system. This procedure assures a minimum-size state vector, which in this case is of order 2 (a second-order system). Figure 9.24 shows a fully annotated bond graph, with force-velocity variables labeling each bond. The state for an **I** element is a momentum, in this case the translational momentum of the mass, p_8. For a **C** element,

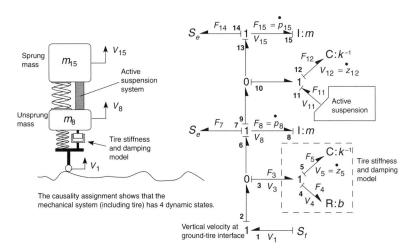

FIGURE 9.25 Example of model for vertical vibration in a quarter-car suspension model with an active suspension element. This example builds on the simple mass–spring–damper model, and shows how to integrate an actuator into a bond graph model structure.

a displacement variable is the state z_5, which here represents the change in length of the spring. The state vector is $\mathbf{x}^T = [p_8, z_5]$.

A mathematical model can be derived by referring to this bond graph, focusing on the independent energy storage elements. The **rate law** (see Tables 9.4 and 9.5) for each energy storage element in integral causality constitutes one first-order ordinary differential **state equation** for this system. In order to formulate these equations, the right-hand side of each rate law must be a function only of states or inputs to the system. The process is summarized in the table of Figure 9.24. Note that the example assumes linear constitutive relations for the elements, but it is clear in this process that this is not necessary. Of course, in some cases nonlinearity complicates the analysis as well as the modeling process in other ways.

9.4.4.3 Quarter-Car Active Suspension: Bond Graph Approach

The simple mass–spring–damper system forms a basis for building more complex models. A model for the vertical vibration of a quarter-car suspension is shown in Figure 9.25. The bond graph model illustrates the use of the mass–spring–damper model, although there are some changes required. In this case, the base is now moving with a velocity equal to the vertical velocity of the ground-tire interface (this requires knowledge of the terrain height over distance traveled as well as the longitudinal velocity of the vehicle). The power direction has changed on many of the bonds, with many now showing positive power flowing from the ground up into the suspension system.

The active suspension system is isolated to further illustrate how bond graph modeling promotes a modular approach to the study of complex systems. Most relevant is that the model identifies the required causal relation at the interface with the active suspension, specifying that the relative velocity is a causal input, and force is a causal output of the active suspension system. The active force is exerted in an equal and opposite fashion onto the sprung and unsprung mass elements.

The causality assignment identifies four states (two momentum states and two spring displacement states). Four first-order state equations can be derived using the rate laws of each of the independent energy-storing elements (C_5, I_8, C_{12}, I_{15}). At this point, depending on the goals of the analysis, either the nonlinear equations could be derived (which might include an active suspension force that depends on the velocity input), or a linearized model could be developed and impedance methods applied to derive a transfer function directly.

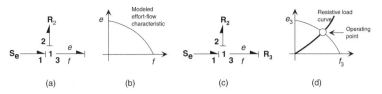

FIGURE 9.26 Algebraic loop in a simple source-load model.

9.4.5 Note on Some Difficulties in Deriving Equations

There are two common situations that can lead to difficulties in the mathematical model development. These issues will arise with any method, and is not specific to bond graphs. Both lead to a situation that may require additional algebraic manipulation in the equation derivation, and it may not be possible to accomplish this in closed form. There are also some ways to change the model in order to eliminate these problems, but this could introduce additional problems. The two issues are (1) derivative causality, and (2) algebraic loops. Both of these can be detected during causality assignment, so that a problem can be detected before too much time has been spent.

The occurence of derivative causality can be described in bond graph terms using Table 9.7. The issue is one in which the state of an energy-storing element (**I** or **C**) is dependent on the system to which it is attached. This might not seem like a problem, particularly since this implies that no differential equation need be solved to find the state. It is necessary to see that there is still a need to compute the back-effect that the system will feel in forcing the element into a given state. For example, if a mass is to be driven by a velocity, V, then it is clear that we know the energy state, $p = mV$, so all is known. However, there is an inertial force computed as $\dot{p} = m\dot{V} = ma$. Many times, it is possible to resolve this problem by performing the algebraic manipulations required to include the effect of this element (difficulty depends on complexity of the system). Sometimes, these dependent states arise because the system is not modeled in sufficient detail, and by inserting a compliance between two gears, for example, the dependence is removed. This might solve the problem, costing only the introduction of an additional state. A more serious drawback to this approach would occur if the compliance was actually very small, so that numerical stiffness problems are introduced (with modern numerical solver routines, even this problem can be tolerated). Yet another way to resolve the problem of derivative causality in mechanical systems is to employ a Lagrangian approach for mechanical system modeling. This will be discussed in Section 9.7.

Another difficulty that can arise in developing solvable systems of equations is the presence of an algebraic loop. Algebraic loops are relatively easy to generate, especially in a block diagram modeling environment. Indeed, it is often the case that algebraic loops arise because of modeling decisions, and in this way a bond graph's causality provides quick feedback regarding the system solvability. Algebraic loops imply that there is an arbitrary way to make computations in the model, and in this way they reveal themselves when an arbitrary decision must be made in assigning causality to an **R** element.*

As an example, consider the basic model of a Thevenin source in Figure 9.26a. This model uses an effort source and a resistive element to model an effort-flow (steady-state) characteristic curve, such as a motor or engine torque-speed curve or a force-velocity curve for a linear actuator. A typical characteristic is shown in Figure 9.26b. When a resistive load is attached to this source as shown in Figure 9.26c, the model is purely algebraic. When the causality is assigned, note that after applying the effort causality on bond 1, there are two resistive elements remaining. The assignment of causality is arbitrary. The solution

---

*The arbitrary assignment on an **R** element is not unlike the arbitrariness in assigning integral or derivative causality to energy-storing elements. An "arbitrary" decision to assign integral causality on an energy-storing element leads to a requirement that we solve a *differential* equation to find a state of interest. In the algebraic loop, a similar arbitary decision to assign a given causality on an **R** element implies that at least one *algebraic* equation must be solved along with any other system equations. In other words, the system is described by differential algebraic equations (DAEs).

requires analytically solving algebraic relations for the operating point, or by using a graphical approach as shown in Figure 9.26d.

This is a simple example indicating how algebraic loops are detected with a bond graph, and how the solution requires solving algebraic relations. In complex systems, this might be difficult to achieve. Sometimes it is possible to introduce or eliminate elements that are "parasitic," meaning they normally would be neglected due to their relatively small effect. However, such elements can relieve the causal bind. While this might resolve the problem, as in the case of derivative causality there are cases where such a course could introduce numerical stiffness problems. Sometimes a solution is reached by using energy methods to resolve some of these problems, as shown in the next section.

9.5 Energy Methods for Mechanical System Model Formulation

This section describes methods for using energy functions to describe basic energy-storing elements in mechanical systems, as well as a way to describe collections of energy-storing elements in multiport fields. Energy methods can be used to simplify model development, providing the means for deriving constitutive relations, and also as a basis for eliminating dependent energy storage (see last section). The introduction of these methods provides a basis for introducing the Lagrange equations in Section 9.7 as a primary approach for system equation derivation or in combination with the bond graph formulation.

9.5.1 Multiport Models

The energy-storing and resistive models introduced in Section 9.3 were summarized in Tables 9.2, 9.4, and 9.5 as multiport elements. In this section, we review how multiport elements can be used in modeling mechanical systems, and outline methods for deriving the constitutive relations. Naturally, these methods apply to the single-port elements as well.

An example of a C element with two-ports was shown in Figure 9.12 as a model for a cantilevered beam that can have both translational and rotational deflections at its tip. A 2-port is required in this model because there are two independent ways to store potential energy in the beam. A distinguishing feature in this example is that the model is based on relationships between efforts and displacement variables (for this case of a capacitive element). Multiport model elements developed in this way are categorized as explicit fields to distinguish them from implicit fields [17]. Implicit fields are formed by assembling energy-storing 1-port elements with junction structure (i.e., 1, 0, and TF elements) to form multiport models.

Explicit fields are often derived using physical laws directly, relying on an understanding of how the geometric and material properties affect the basic constitutive relation between physical variables. Geometry and material properties always govern the parametric basis of all constitutive relations, and for some cases these properties may themselves be functions of state. Indeed, these cases require the multiport description, which finds extensive use in modeling of many practical devices, especially sensors and actuators. Multiport models should follow a strict energetic basis, as described in the following.

9.5.2 Restrictions on Constitutive Relations

Energy-storing multiports must follow two basic restrictions, which are also useful in guiding the derivation of energetically-correct constitutive relations. The definition of the energy-storing descriptions summarized in Tables 9.4 and 9.5 specifies that there exists an energy state function, $E = E(\mathbf{x})$, where $\mathbf{x}$ is either a generalized displacement, $\mathbf{q}$, for capacitive (C) elements or a generalized momentum, $\mathbf{p}$, for inertive (I) elements. For the multiport energy-storing element, the specification requires the following specifications [2,3]:

1. There exists a rate law, $\dot{x}_i = u_i$, where u_i as input specifies integral causality on port i.
2. The energy stored in a multiport is determined by

$$E(\mathbf{x}) = \int \sum_{i=1}^{n} y_i dx_i \qquad (9.5)$$

3. A first restriction on a multiport constitutive relation requires that the causal output at any port is given by

$$y_i = \Phi_{si}(x) = \frac{\partial E(x)}{\partial x_i} \tag{9.6}$$

where $\Phi_{si}()$ is a single-valued function.

4. A second restriction on a multiport constitutive relation requires that the constitutive relations obey Maxwell reciprocity, or

$$\frac{\partial y_i}{\partial x_j} = \frac{\partial^2 E(x)}{\partial x_j \partial x_i} = \frac{\partial y_j}{\partial x_i} \tag{9.7}$$

9.5.3 Deriving Constitutive Relations

The first restriction on the constitutive relations, Equation (9.6), establishes how constitutive relations can be derived for a multiport if an energy function can be formulated. This restriction forms the basis for a method used in many practical applications to find constitutive relationships from energy functions (e.g., strain-energy, electromechanics, etc.). In these methods, it is assumed that at least one of the constitutive relations for an energy-storing multiport is given. Then, the energy function is formed using Equation (9.5) where, after interchanging the integral and sum,

$$E(x) = \sum_{i=1}^{n} \int y_i \, dx_i = \int y_1 \, dx_1 + \cdots + \int y_n \, dx_n \tag{9.8}$$

Presume that y_1 is a known function of the states, $y_1 = \Phi_{s1}(x)$. Since the element is conservative, any energetic state can be reached via a convenient path where $dx_i = 0$ for all i except $i = 1$. This allows the determination of $E(x)$.

To illustrate, consider the simple case of a rack and pinion system, shown in Figure 9.27. The pinion has rotational inertia, J, about its axis of rotation, and the rack has mass, m. The kinetic co-energy is easily formulated here, considering that the pinion angular velocity, ω, and the rack velocity, V, are constrained by the relationship $V = R\omega$, where R is the pinion base radius. If this basic subsystem is modeled directly, it will be found that one of the inertia elements (pinion, rack) will be in derivative causality. Say, it is desired to connect to this system through the rotational port, $T - \omega$. To form a single-port I element that includes the rack, form the kinetic co-energy as $T = T(\omega, V) = J\omega^2/2 + mV^2/2$. Use the constraint relation to write, $T = T(\omega) = (J + mR^2)\omega^2/2$. To find the constitutive relation for this 1-port rotational I element, let $h = \partial T(\omega)/\partial \omega = (J + mR^2)\omega$, where we can now define an equivalent rotational inertia as $J_{eq} = J + mR^2$.

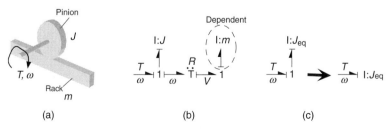

FIGURE 9.27 (a) Rack and pinion subsystem with torque input. (b) Direct model, showing dependent mass. (c) Equivalent model, derived using energy principles.

The rack and pinion example illustrates a basic method for relieving derivative causality, which can be used to build basic energy-storing element models. Some problems might arise when the kinetic co-energy depends on system configuration. In such a case, a more systematic method employing Lagrange's equations may be more suitable (see Section 9.7).

The approach described here for deriving constitutive relations is similar to Castigliano's theorem [6,9]. Castigliano's theorem relies on formulation of a strain-energy function in terms of the forces or moments, and as such employs a potential co-energy function. Specifically, the results lead to displacements (translational, rotational) as functions of efforts (forces, torques). As in the case above, these functions are found by taking partial derivatives of the co-energy with respect to force or moment. Castigliano's theorem is especially well-suited for finding force–displacment functions for curved and angled beam structures (see [6]).

Formulations using energy functions to derive constitutive relations are found in other application areas, and some references include Lyshevski [21] for electromechanics, and Karnopp, Margolis, and Rosenberg [17] for examples and applications in the context of bond graph modeling.

9.5.4 Checking the Constitutive Relations

The second restriction on the constitutive relations, Equation 9.7, provides a basis for testing or checking if the relationships are correct. This is a reciprocity condition that provides a check for energy conservation in the energy-storing element model, and a quick check for linear mechanical systems shows that either the inertia or stiffness matrix must be symmetrical.

Recall the example of the 2-port cantilevered beam, shown again in Figure 9.12. For small deflections, the total tip translational and angular deflections due to a tip force and torque can be added (using flexibility influence coefficients), which can be expressed in matrix form,

$$\begin{bmatrix} x \\ \theta \end{bmatrix} = \frac{1}{EI} \begin{bmatrix} \frac{1}{3}l^3 & \frac{1}{2}l^2 \\ \frac{1}{2}l^2 & l \end{bmatrix} \begin{bmatrix} F \\ T \end{bmatrix} = C \begin{bmatrix} F \\ T \end{bmatrix} = K^{-1} \begin{bmatrix} F \\ T \end{bmatrix}$$

where C and K are the compliance and stiffness matrices, respectively. This constitutive relation satisfies the Maxwell reciprocity since, $\partial x/\partial T = \partial \theta/\partial F$. This 2-port C element is used to model the system shown in Figure 9.28a, which consists of a bar-bell rigidly attached to the tip of the beam. Under small deflection, a bond graph shown in Figure 9.28b is assembled. Causality applied to this system reveals that each port of the 2-port C element has integral causality. On a multiport energy storing element, each port is independently assigned causality following the same rules as for 1-ports. It is possible that a multiport could have a mixed causality, where some of the ports are in derivative causality. If a multiport has mixed causality, part of the state equations will have to be inverted. This algebraic difficulty is best avoided by trying to assign integral causality to all multiport elements in a system model if possible.

In the present example, causality assignment on the I elements is also integral. In all, there are four independent energy-storing elements, so there are four state variables, $\mathbf{x} = [x, \theta, p, h]^T$. Four state equations can be derived using the rate laws indicated in Figure 9.28.

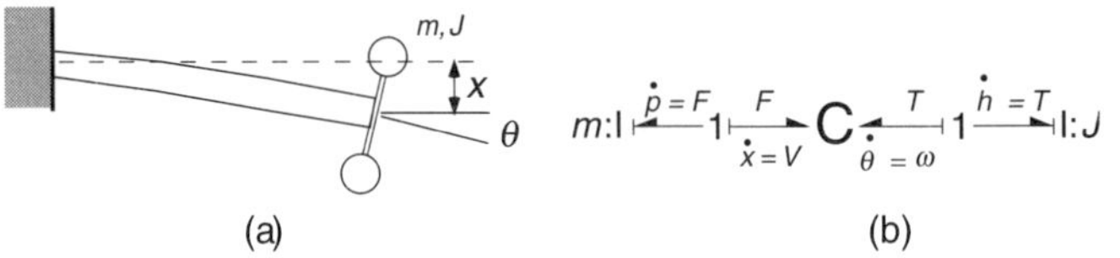

FIGURE 9.28 Model of beam rigidly supporting a bar- or dumb-bell: (a) schematic, (b) bond graph model using a 2-port C to represent beam. Dumb-bell is represented by translational mass, m, and rotational inertia, J.

9.6 Rigid Body Multidimensional Dynamics

The modeling of bodies in mechanical systems presumes adoption of a "rigid body" that can involve rotation as well as translation, and in this case the dynamic properties are more complex than those for a point mass. In earlier sections of this chapter, a simple rigid body has already been introduced, and it is especially useful for a large class of problems with rotation about a single fixed axis.

In the rigid body, the distance between any two elements of mass within a body is a constant. In some cases, it is convenient to consider a continuous distribution of mass while in others a system of discrete mass particles rigidly fixed together helps conceptualize the problem. In the latter, the rigid body properties can be found by summing over all the discrete particles, while in the continuous mass concept an integral formulation is used. Either way, basic concepts can be formulated and relations derived for use in rigid body dynamic analysis. Finally, the modeling in most engineering systems is restricted to classical Newtonian mechanics, where the linear velocity–momentum relation holds (so energy and coenergy are equal).

9.6.1 Kinematics of a Rigid Body

In this section, a brief overview is given of three-dimensional motion calculations for a rigid body. The focus here is to present methods for analyzing rotation of a rigid body about a fixed axis and methods for analyzing relative motion of a rigid body using translating and rotating axes. These concepts introduce the basis for understanding more complex formulations. While vector descriptions (denoted using an arrow over the symbol, $\vec{a}$) are useful for understanding basic problems, more complex multibody systems usually adopt a matrix formulation. The presentation here is brief and included for reference. A more extensive discussion and examples can be found in introductory dynamics textbooks (e.g., [23]), where a separate discussion is usually given on the special case of plane motion.

9.6.1.1 Rotation of a Body About a Fixed Point

Basic concepts are introduced here in relation to rotation of a rigid body about a fixed point. This basic motion specifies that any point on the body lies on the surface of a sphere with a radius centered at the fixed point. The body can be said to have spherical motion.

Euler's Theorem. Euler's theorem states that any displacement of a body in spherical motion can be expressed as a rotation about a line that passes through the center of the spherical motion. This axis can be referred to as the orientational axis of rotation [26]. For example, two rotations about different axes passing through a fixed point of rotation are equivalent to a single resultant rotation about an axis passing through that point.

Finite Rotations. If the rotations used in Euler's theorem are finite, the order of application is important because finite rotations do not obey the law of vector addition.

Infinitesimal Rotations. Infinitesimally small rotations can be added vectorially in any manner, and these are generally considered when defining rigid body motions.

Angular Velocity. A body subjected to rotation $d\vec{\theta}$ about a fixed point will have an angular velocity $\vec{\omega}$ defined by the time derivative $d\vec{\theta}/dt$, in a direction collinear with $d\vec{\theta}$. If the body is subjected to two component angular motions that define $\vec{\omega}_1$ and $\vec{\omega}_2$, then the body has a resultant angular velocity, $\vec{\omega} = \vec{\omega}_1 + \vec{\omega}_2$.

Angular Acceleration. A body's angular acceleration is found from the time derivative of the angular velocity, $\vec{\alpha} = \dot{\vec{\omega}}$, and in general the acceleration is not collinear with velocity.

Motion of Points in the Body. Given $\vec{\omega}$, the velocity of a point on the body is $\vec{v} = \vec{\omega} \times \vec{r}$, where $\vec{r}$ is a position vector to the point as measured relative to the fixed point of rotation. The acceleration of a point on the body is then, $\vec{a} = \vec{\alpha} \times \vec{r} + \vec{\omega} \times (\vec{\omega} \times \vec{r})$.

9.6.1.2 Relating Vector Time Derivatives in Coordinate Systems

It is often the case that we need to determine the time rate of change of a vector such as $\vec{A}$ in Figure 9.29 relative to different coordinate systems. Specifically, it may be easier to determine $\dot{\vec{A}}$ in x_a, y_a, z_a,

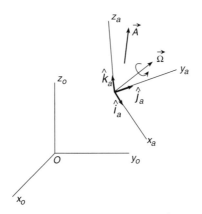

FIGURE 9.29 Often it is necessary to find the time derivative of vector $\vec{A}$ relative to a axes, x_o, y_o, z_o, given its value in the translating-rotating system x_a, y_a, z_a.

but we need to find its value in x_o, y_o, z_o. The vector $\vec{A}$ is expressed in the axes x_a, y_a, z_a using the unit vectors shown as

$$\vec{A} = A_x \hat{i}_a + A_y \hat{j}_a + A_z \hat{k}_a$$

To find the time rate of change, we identify that in the moving reference the time derivative of $\vec{A}$ is

$$\left(\frac{d\vec{A}}{dt}\right)_a = \frac{dA_x}{dt}\hat{i}_a + \frac{dA_y}{dt}\hat{j}_a + \frac{dA_z}{dt}\hat{k}_a$$

Relative to the x_o, y_o, z_o axes, the direction of the unit vectors $\hat{i}_a$, $\hat{j}_a$, and $\hat{k}_a$ change only due to rotation Ω, so,

$$\frac{d\vec{A}}{dt} = \left(\frac{d\vec{A}}{dt}\right) + A_x \frac{d\hat{i}_a}{dt} + A_y \frac{d\hat{j}_a}{dt} + A_z \frac{d\hat{k}_a}{dt}$$

$$\frac{d\hat{i}_a}{dt} = \vec{\Omega} \times \hat{i}_a, \quad \frac{d\hat{j}_a}{dt} = \vec{\Omega} \times \hat{j}_a, \quad \frac{d\hat{k}_a}{dt} = \vec{\Omega} \times \hat{k}_a$$

then,

$$\frac{d\vec{A}}{dt} = \left(\frac{d\vec{A}}{dt}\right)_a + \vec{\Omega} \times \vec{A} \tag{9.9}$$

This relationship is very useful not only for calculating derivatives, as derived here, but also for formulating basic bond graph models. This is shown in the section titled "Rigid Body Dynamics."

9.6.1.3 Motion of a Body Relative to a Coordinate System

Translating Coordinate Axes

The origin of a set of axes x_a, y_a, z_a is fixed in a rigid body at A as shown in Figure 9.30a, and translates without rotation relative to the axes x_o, y_o, z_o with known velocity and acceleration. The rigid body is subjected to angular velocity $\vec{\omega}$ and angular acceleration $\vec{\alpha}$ in three dimensions.

Motion of Point B Relative to A. The motion of point B relative to A is the same as motion about a fixed point, so $\vec{v}_{B/A} = \vec{\omega} \times \vec{v}_{B/A}$, and $\vec{a}_{B/A} = \vec{\alpha} \times \vec{r}_{B/A} + \vec{\omega} \times (\vec{\omega} \times \vec{r}_{B/A})$.

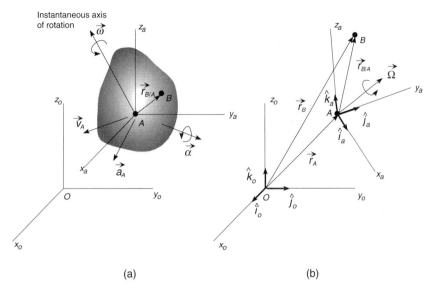

FIGURE 9.30 General rigid body motion: (a) rigid body with translating coordinate system, (b) translating and rotating coordinate system.

Motion of Point B Relative to O. For translating axes with no rotation, the velocity and acceleration of point B relative to system 0 is simply, $\vec{v}_B = \vec{v}_A + \vec{v}_{B/A}$ and $\vec{a}_B = \vec{a}_A + \vec{a}_{B/A}$ respectively, or,

$$\vec{v}_B = \vec{v}_A + \vec{\omega} \times \vec{r}_{B/A} \tag{9.10}$$

$$\vec{a}_B = \vec{a}_A + \vec{\alpha} \times \vec{r}_{B/A} + \vec{\omega} \times (\vec{\omega} \times \vec{r}_{B/A}) \tag{9.11}$$

Translating and Rotating Coordinate Axes

A general way of describing the three-dimensional motion of a rigid body uses a set of axes that can translate and rotate relative to a second set of axes, as illustrated in Figure 9.30b. Position vectors specify the locations of points A and B on the body relative to x_o, y_o, z_o, and the axes x_a, y_a, z_a have angular velocity $\vec{\Omega}$ and angular acceleration $\vec{\dot{\Omega}}$. With the position of point B given by

$$\vec{r}_B = \vec{r}_A + \vec{r}_{B/A} \tag{9.12}$$

the velocity and acceleration are found by direct differentiation as

$$\vec{v}_B = \vec{v}_A + \vec{\Omega} \times \vec{r}_{B/A} + (v_{B/A})_a \tag{9.13}$$

and

$$\vec{a}_B = \vec{a}_A + \vec{\dot{\Omega}} \times \vec{r}_{B/A} + \vec{\Omega} \times (\vec{\Omega} \times \vec{r}_{B/A}) + 2\vec{\Omega} \times (v_{B/A})_a + (\vec{a}_{B/A})_a \tag{9.14}$$

where $(v_{B/A})_a$ and $(a_{B/A})_a$ are the velocity and acceleration, respectively, of B relative to A in the x_a, y_a, z_a coordinate frame.

These equations are applicable to plane motion of the rigid body for which the analysis is simplified since $\vec{\Omega}$ and $\vec{\dot{\Omega}}$ have a constant direction. Note that for the three-dimensional case, $\vec{\dot{\Omega}}$ must be computed by using Equation 9.9.

9.6.1.4 Matrix Formulation and Coordinate Transformations

A vector in three-dimensional space characterized by the right-handed reference frame x_a, y_a, z_a, $\vec{A} = A_x \hat{i}_a + A_y \hat{j}_a + A_z \hat{k}_a$, can be represented as an ordered triplet,

$$\vec{A} = \begin{bmatrix} A_x \\ A_y \\ A_z \end{bmatrix}_a = \begin{bmatrix} A_x & A_y & A_z \end{bmatrix}_a^T$$

where the elements of the column vector represent the vector projections on the unit axes. Let $\underline{A}_a$ denote the column vector relative to the axes x_a, y_a, z_a. It can be shown that the vector $\vec{A}$ can be expressed in another right-handed reference frame x_b, y_b, z_b, by the transformation relation

$$\underline{A}_b = \overline{\underline{C}}_{ab} \underline{A}_a \tag{9.15}$$

where $\overline{\underline{C}}_{ab}$ is a 3 × 3 matrix,

$$\overline{\underline{C}}_{ab} = \begin{bmatrix} cx_a x_b & cx_a y_b & cx_a z_b \\ cy_a x_b & cy_a y_b & cy_a z_b \\ cz_a x_b & cz_a y_b & cz_a z_b \end{bmatrix} \tag{9.16}$$

The elements of this matrix are the cosines of the angles between the respective axes. For example, $cz_a y_b$ is the cosine of the angle between z_a and y_b. This is the rotational transformation matrix and it must be orthogonal, or

$$C_{ab}^{-1} = C_{ab}^T = C_{ba}$$

and for right-handed systems, let $C_{ab} = +1$.

9.6.1.5 Angle Representations of Rotation

The six degrees of freedom needed to describe general motion of a rigid body are characterized by three degrees of freedom each for translation and for rotation. The focus here is on methods for describing rotation.

Euler's theorem (9.11) confirms that only three parameters are needed to characterize rotation. Two parameters define an axis of rotation and another defines an angle about that axis. These parameters define three positional degrees of freedom for a rigid body. The three rotational parameters help construct a rotation matrix, $\overline{C}$. The following discussion describes how the rotation matrix, or direction cosine matrix, can be formulated.

General Rotation. Unit vectors for a system a, $\hat{u}_a$, are said to be carried into b, as $\hat{u}_b = \overline{\underline{C}}_{ba} \hat{u}_a$. It can be shown that a direction cosine matrix can be formulated by [30]

$$\overline{\underline{C}} = \underline{\lambda} \underline{\lambda}^T + (\overline{\underline{E}} - \underline{\lambda} \underline{\lambda}^T) \cos \psi - \overline{\underline{S}}(\underline{\lambda}) \sin \psi \tag{9.17}$$

where $\overline{\underline{E}}$ is the identity matrix, and $\underline{\lambda}$ represents a unit vector, $\underline{\lambda} = [\lambda_1, \lambda_2, \lambda_3]^T$, which is parallel to the axis of rotation, and ψ is the angle of rotation about that axis [30]. In this relation, $\overline{\underline{S}}(\underline{\lambda})$ is a **skew-symmetric matrix**, which is defined by the form

$$\overline{\underline{S}}(\underline{\lambda}) = \begin{bmatrix} 0 & -\lambda_3 & \lambda_2 \\ \lambda_3 & 0 & -\lambda_1 \\ -\lambda_2 & \lambda_1 & 0 \end{bmatrix}$$

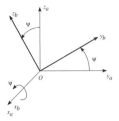

FIGURE 9.31 An elementary rotation by angle ϕ about axis x.

The matrix elements of $\overline{C}$ can be found by expanding the relation given above, using $\overline{S}(\lambda)$, to give

$$\overline{C} = \begin{bmatrix} (1-\cos\psi)\lambda_1^2 + \cos\psi & (1-\cos\psi)\lambda_1\lambda_2 + \lambda_3\sin\psi & (1-\cos\psi)\lambda_1\lambda_3 + \lambda_2\sin\psi \\ (1-\cos\psi)\lambda_2\lambda_1 + \lambda_3\sin\psi & (1-\cos\psi)\lambda_2^2 + \cos\psi & (1-\cos\psi)\lambda_2\lambda_3 + \lambda_1\sin\psi \\ (1-\cos\psi)\lambda_3\lambda_1 + \lambda_2\sin\psi & (1-\cos\psi)\lambda_3\lambda_2 + \lambda_1\sin\psi & (1-\cos\psi)\lambda_3^2 + \cos\psi \end{bmatrix} \quad (9.18)$$

The value of this formulation is in identifying that there are formally defined principle axes, characterized by the $\underline{\lambda}$, and angles of rotation, ψ, that taken together define the body orientation. These rotations describe classical angular variables formed by elementary (or principle) rotations, and it can be shown that there are two cases of particular and practical interest, formed by two different axis rotation sequences.

Elementary Rotations. Three elementary rotations are formed when the rotation axis (defined by the eigenvector) coincides with one of the base vectors of a defined coordinate system. For example, letting $\underline{\lambda} = [1, 0, 0]^T$ define an axis of rotation x, as in Figure 9.31, with an elementary rotation of ϕ gives the rotation matrix,

$$\overline{C}_{x,\phi} = \begin{bmatrix} 1 & 0 & 0 \\ 0 & \cos\phi & \sin\phi \\ 0 & -\sin\phi & \cos\phi \end{bmatrix}$$

The two elementary rotations about the other two axes, y and z, are

$$\overline{C}_{y,\theta} = \begin{bmatrix} \cos\theta & 0 & -\sin\theta \\ 0 & 1 & 0 \\ \sin\theta & 0 & \cos\theta \end{bmatrix} \quad \text{and} \quad \overline{C}_{z,\psi} = \begin{bmatrix} \cos\psi & \sin\psi & 0 \\ -\sin\psi & \cos\psi & 0 \\ 0 & 0 & 1 \end{bmatrix}$$

These three elementary rotation matrices can be used in sequence to define a direction cosine matrix, for example,

$$\overline{C} = \overline{C}_{z,\psi}\overline{C}_{y,\theta}\overline{C}_{x,\phi}$$

and the elementary rotations and the direction cosine matrix are all orthogonal; that is,

$$\overline{C}\,\overline{C}^T = \overline{C}^T\overline{C} = \overline{E}$$

where $\overline{E}$ is the identity matrix. Consequently, the inverse of the rotation or coordinate transformation matrix can be found by $\overline{C}^{-1} = \overline{C}^T$.

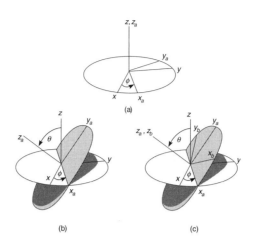

FIGURE 9.32 The rotations defining the Euler angles. (Adapted from Goldstein [11].)

It can be shown that there exist two sequences that have independent rotation sequences, and these lead to the well known Euler angle and Tait-Bryan or Cardan angle rotation descriptions [30].

Euler Angles. Euler angles are defined by a specific rotation sequence. Consider a right-handed axes system defined by the base vectors, x, y, z, as shown in Figure 9.32a. The rotation sequence of interest involves rotations about the axes in the following sequence: (1) ϕ about z, (2) θ about x_a, then (3) ψ about z_b. This set of rotation sequences is defined by the elementary rotation matrices,

$$\overline{C}_{z,\phi} = \begin{bmatrix} \cos\phi & \sin\phi & 0 \\ -\sin\phi & \cos\phi & 0 \\ 0 & 0 & 1 \end{bmatrix}, \quad \overline{C}_{x_a,\theta} = \begin{bmatrix} 1 & 0 & 0 \\ 0 & \cos\theta & \sin\theta \\ 0 & -\sin\theta & \cos\theta \end{bmatrix}, \quad \overline{C}_{z_b,\psi} = \begin{bmatrix} \cos\psi & \sin\psi & 0 \\ -\sin\psi & \cos\psi & 0 \\ 0 & 0 & 1 \end{bmatrix}$$

where the subscript on each $\overline{C}$ denotes the axis and angle of rotation. Using these transformations relates the quantity $\underline{A}$ in x, y, z to $\underline{A}_b$ in x_b, y_b, z_b, or

$$\underline{A}_b = \overline{C}_{\text{Euler}} \underline{A} = \overline{C}_{z_b,\psi} \overline{C}_{x_a,\theta} \overline{C}_{z,\phi} \underline{A}$$

where $\overline{C}_{\text{Euler}}$ is given by

$$\overline{C}_{\text{Euler}} = \begin{bmatrix} \cos\psi\cos\phi - \sin\psi\cos\theta\sin\phi & \cos\psi\sin\phi + \sin\psi\cos\theta\cos\phi & \sin\psi\sin\theta \\ -\sin\psi\cos\phi - \cos\psi\cos\theta\sin\phi & -\sin\psi\sin\phi + \cos\psi\cos\theta\cos\phi & \cos\psi\sin\theta \\ \sin\theta\sin\phi & -\sin\theta\cos\phi & \cos\theta \end{bmatrix} \quad (9.19)$$

Since $\overline{C}_{\text{Euler}}$ is orthogonal, transforming between the two coordinate systems is relatively easy since the inverse can be found simply by the transpose of Equation 9.19.

In some applications, it is desirable to derive the angles given the direction cosine matrix. So, if the (3, 3) element of $\overline{C}_{\text{Euler}}$ is given, then θ is easily found, but there can be difficulties in discerning small angles. Also, if θ goes to zero, there is a singularity in solving for ϕ and ψ, so determining body orientation becomes difficult. The problem also makes itself known when transforming angular velocities between the coordinate systems. If the problem at hand avoids this case (i.e., θ never approaches zero), then Euler angles are a viable solution. Many applications that cannot tolerate this problem adopt other representations, such as the Euler parameters to be discussed later.

In classical rigid body dynamics, ϕ is called the *precession angle*, θ is the *nutation angle*, and ψ is the *spin angle*. The relationship between the time derivative of the Euler angles, $\dot{\varphi} = [\dot{\phi}, \dot{\theta}, \dot{\psi}]^T$, and the body angular velocity, $\underline{\omega} = [\omega_x, \omega_y, \omega_z]_b^T$, is given by [11]

$$\underline{\omega}_b = \overline{T}(\underline{\varphi})\dot{\underline{\varphi}} \qquad (9.20)$$

where the transformation matrix, $\overline{T}(\underline{\varphi})$, is given by

$$\overline{T}(\underline{\varphi}) = \begin{bmatrix} \sin\theta\sin\psi & \cos\psi & 0 \\ \sin\theta\cos\psi & -\sin\psi & 0 \\ \cos\theta & 0 & 1 \end{bmatrix}$$

Note here again that $\overline{T}(\underline{\varphi})$ will become singular at $\theta = \pm\pi/2$.

Tait-Bryan or Cardan Angles. The Tait-Bryan or Cardan angles are formed when the three rotation sequences each occur about a different axis. This is the sequence preferred in flight and vehicle dynamics. Specifically, these angles are formed by the sequence: (1) ϕ about z (yaw), (2) θ about y_a (pitch), and (3) ϕ about the final x_b axis (roll), where a and b denote the second and third stage in a three-stage sequence or axes (as used in the Euler angle description). These rotations define a transformation,

$$\underline{A}_b = \overline{C}\underline{A} = \overline{C}_{x_b,\psi}\overline{C}_{y_a,\theta}\overline{C}_{z,\phi}\underline{A}$$

where

$$\overline{C}_{z,\phi} = \begin{bmatrix} \cos\phi & \sin\phi & 0 \\ -\sin\phi & \cos\phi & 0 \\ 0 & 0 & 1 \end{bmatrix}, \quad \overline{C}_{y_a,\theta} = \begin{bmatrix} \cos\theta & 0 & -\sin\theta \\ 0 & 1 & 0 \\ \sin\theta & 0 & \cos\theta \end{bmatrix}, \quad \overline{C}_{x_b,\theta} = \begin{bmatrix} 1 & 0 & 0 \\ 0 & \cos\psi & \sin\psi \\ 0 & -\sin\psi & \cos\psi \end{bmatrix}$$

and the final coordinate transformation matrix for Tait-Bryan angles is

$$\overline{C}_{\text{Tait-Bryan}} = \begin{bmatrix} \cos\theta\cos\phi & \cos\theta\sin\phi & -\sin\theta \\ \sin\psi\sin\theta\cos\phi - \cos\psi\sin\phi & \sin\psi\sin\theta\sin\phi + \cos\psi\cos\phi & \cos\theta\sin\psi \\ \cos\psi\sin\theta\cos\phi + \sin\psi\sin\phi & \cos\psi\sin\theta\sin\phi - \sin\psi\cos\phi & \cos\theta\cos\psi \end{bmatrix} \qquad (9.21)$$

A linearized form of $\overline{C}_{\text{Trait-Bryan}}$ gives a form preferred to that derived for Euler angles, making it useful in some forms of analysis and control. There remains the problem of a singularity, in this case when θ approaches $\pm\pi/2$.

For the Tait-Bryan angles, the transformation matrix relating $\dot{\underline{\varphi}}$ to $\underline{\omega}_b$ is given by

$$\overline{T}(\underline{\varphi}) = \begin{bmatrix} -\sin\theta & 0 & 1 \\ \cos\theta\sin\psi & \cos\psi & 0 \\ \cos\theta\cos\psi & -\sin\psi & 0 \end{bmatrix}$$

which becomes singular at $\theta = 0, \pi$.

9.6.1.6 Euler Parameters and Quaternions

The degenerate conditions in coordinate transformations for Euler and Tait-Bryan angles can be avoided by using more than a minimal set of parameterizing variables (beyond the three angles). The most notable set are referred to as Euler parameters, which are unit quaternions. There are many other possibilities, but this four-parameter method is used in many areas, including spacecraft/flight dynamics, robotics, and computational kinematics and dynamics. The term "quaternion" was coined by Hamilton in about 1840, but Euler himself had devised the use of Euler parameters 70 years before. Quaternions are discussed by Goldstein [11], and their use in rigid body dynamics and attitude control dates back to the late 1950s and early 1960s [13,24]. Application of quaternions is common in control applications in aerospace applications [38] as well as in ocean vehicles [10]. More recently (past 20 years or so), these methods have found their way into motion and control descriptions for robotics [34] and computational kinematics and dynamics [14,25,26]. An overview of quaternions and Euler parameters is given by Wehage [37]. Quaternions and rotational sequences and their role in a wide variety of applications areas, including sensing and graphics, are the subject of the book by Kuipers [19]. These are representative references that may guide the reader to an application area of interest where related studies can be found. In the following only a brief overview is given.

Quaternion. A quaternion is defined as the sum of a scalar, q_0, and a vector, $\vec{q}$, or,

$$q = q_0 + \vec{q} = q_0 + q_1 \hat{i} + q_2 \hat{j} + q_3 \hat{k}$$

A specific algebra and calculus exists to handle these types of mathematical objects [7,19,37]. The conjugate is defined as $\bar{q} = q_0 - \vec{q}$.

Euler Parameters. Euler parameters are normalized (unit) quaternions, and thus share the same properties, algebra and calculus. A principal eigenvector of rotation has an eigenvalue of 1 and defines the Euler axis of rotation (see Euler's theorem discussion and [11]), with angle of rotation α. Let this eigenvector be $\underline{e} = [e_1, e_2, e_3]^T$. Recall from Equation 9.17, the direction cosine matrix is now

$$\bar{\underline{C}} = \underline{e}\underline{e}^T + (I - \underline{e}\underline{e}^T)\cos\alpha - \bar{\underline{S}}(\underline{e})\sin\alpha$$

where $\bar{\underline{S}}(\underline{e})$ is a skew-symmetric matrix. The Euler parameters are defined as

$$\underline{q} = \begin{bmatrix} q_0 \\ q_1 \\ q_2 \\ q_3 \end{bmatrix} = \begin{bmatrix} \cos(\alpha/2) \\ e_1 \sin(\alpha/2) \\ e_2 \sin(\alpha/2) \\ e_3 \sin(\alpha/2) \end{bmatrix}$$

where

$$q_0^2 + q_1^2 + q_2^2 + q_3^2 = 1$$

Relating Quaternions and the Coordinate Transformation Matrix. The direction cosine matrix in terms of Euler parameters is now

$$\bar{\underline{C}}_q = (q_0^2 - \underline{q}^T\underline{q})\bar{\underline{E}} + 2\underline{q}\underline{q}^T - 2q_0\bar{\underline{S}}(\underline{q})$$

where $\underline{q} = [q_1, q_2, q_3]^T$, and $\bar{\underline{E}}$ is the identity matrix. The direction cosine matrix is now written in terms of quaternions

$$\bar{\underline{C}}_q = \begin{bmatrix} q_0^2 + q_1^2 - q_2^2 - q_3^2 & 2(q_1 q_2 + q_3 q_0) & 2(q_1 q_3 - q_2 q_4) \\ 2(q_1 q_2 - q_3 q_0) & q_0^2 - q_1^2 + q_2^2 - q_3^2 & 2(q_2 q_3 + q_1 q_4) \\ 2(q_1 q_3 + q_2 q_0) & 2(q_1 q_2 + q_3 q_0) & q_0^2 - q_1^2 - q_2^2 + q_3^2 \end{bmatrix}$$

It is possible to find the quaternions and the elements of the direction cosine matrix independently by integrating the angular rates about the principal axes of a body. Given the direction cosine matrix elements, we can find the quaternions, and vice versa. For a more extended discussion and application, the reader is referred to the listed references.

9.6.2 Dynamic Properties of a Rigid Body

9.6.2.1 Inertia Properties

The moments and products of inertia describe the distribution of mass for a body relative to a given coordinate system. This description relies on the specific orientation and reference frame. It is presumed that the reader is familiar with basic properties such as mass center, and the focus here is on those properties essential in understanding the general motion of rigid bodies, and particularly the rotational dynamics.

Moment of Inertia. For the rigid body shown in Figure 9.33a, the moment of inertia for a differential element, dm, about any of the three coordinate axes is defined as the product of the mass of the differential element and the square of the shortest distance from the axis to the element. As shown, $r_x = \sqrt{y^2 + z^2}$, so the contribution to the moment of inertia about the x-axis, I_{xx}, from dm is

$$dI_{xx} = r_x^2 = (y^2 + z^2)dm$$

The total I_{xx}, I_{yy}, and I_{zz} are found by integrating these expressions over the entire mass, m, of the body. In summary, the three moments of inertia about the x, y, and z axes are

$$I_{xx} = \int_m r_x^2 \, dm = \int_m (y^2 + z^2) \, dm$$

$$I_{yy} = \int_m r_y^2 \, dm = \int_m (x^2 + z^2) \, dm \qquad (9.22)$$

$$I_{zz} = \int_m r_z^2 \, dm = \int_m (x^2 + y^2) \, dm$$

Note that the moments of inertia, by virtue of their definition using squared distances and finite mass elements, are always positive quantities.

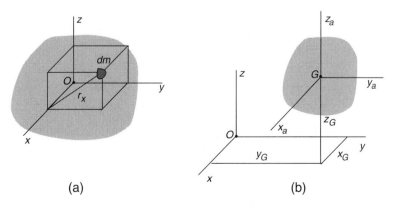

FIGURE 9.33 Rigid body properties are defined by how mass is distributed throughout the body relative to a specified coordinate system. (a) Rigid body used to describe moments and products of inertia. (b) Rigid body and axes used to describe parallel-axis and parallel-plane theorem.

Product of Inertia. The product of inertia for a differential element dm is defined with respect to a set of two orthogonal planes as the product of the mass of the element and the perpendicular (or shortest) distances from the planes to the element. So, with respect to the $y-z$ and $x-z$ planes (z common axis to these planes), the contribution from the differential element to I_{xy} is dI_{xy} and is given by $dI_{xy} = xy\,dm$.

As for the moments of inertia, by integrating over the entire mass of the body for each combination of planes, the products of inertia are

$$I_{xy} = I_{yx} = \int_m xy\,dm$$

$$I_{yz} = I_{zy} = \int_m yz\,dm \qquad (9.23)$$

$$I_{xz} = I_{zx} = \int_m xz\,dm$$

The product of inertia can be positive, negative, or zero, depending on the sign of the coordinates used to define the quantity. If either one or both of the orthogonal planes are planes of symmetry for the body, the product of inertia with respect to those planes will be zero. Basically, the mass elements would appear as pairs on each side of these planes.

Parallel-Axis and Parallel-Plane Theorems. The parallel-axis theorem can be used to transfer the moment of inertia of a body from an axis passing through its mass center to a parallel axis passing through some other point (see also the section "Kinetic Energy Storage"). Often the moments of inertia are known for axes fixed in the body, as shown in Figure 9.33b. If the center of gravity is defined by the coordinates (x_G, y_G, z_G) in the x, y, z axes, the parallel-axis theorem can be used to find moments of inertia relative to the x, y, z axes, given values based on the body-fixed axes. The relations are

$$I_{xx} = (I_{xx})_a + m(y_G^2 + z_G^2)$$
$$I_{yy} = (I_{yy})_a + m(x_G^2 + z_G^2)$$
$$I_{zz} = (I_{zz})_a + m(x_G^2 + y_G^2)$$

where, for example, $(I_{xx})_a$ is the moment of inertia relative to the x_a axis, which passes through the center of gravity. Transferring the products of inertia requires use of the parallel-plane theorem, which provides the relations

$$I_{xy} = (I_{xy})_a + mx_G y_G$$
$$I_{yz} = (I_{yz})_a + my_G z_G$$
$$I_{zx} = (I_{zx})_a + mz_G x_G$$

Inertia Tensor. The rotational dynamics of a rigid body rely on knowledge of the inertial properties, which are completely characterized by nine terms of an inertia tensor, six of which are independent. The inertia tensor is

$$\tilde{I} = \begin{bmatrix} I_{xx} & -I_{xy} & -I_{xz} \\ -I_{yx} & I_{yy} & -I_{yz} \\ -I_{zx} & -I_{zy} & I_{zz} \end{bmatrix}$$

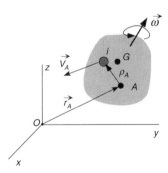

FIGURE 9.34 Rigid body in general motion relative to an inertial coordinate system, x, y, z.

and it relies on the specific location and orientation of coordinate axes in which it is defined. For a rigid body, an origin and axes orientation can be found for which the inertia tensor becomes diagonalized, or

$$\tilde{I} = \begin{bmatrix} I_x & 0 & 0 \\ 0 & I_y & 0 \\ 0 & 0 & I_z \end{bmatrix}$$

The orientation for which this is true defines the principal axes of inertia, and the principal moments of inertia are now $I_x = I_{xx}$, $I_y = I_{yy}$, and $I_z = I_{zz}$ (one should be a maximum and another a minimum of the three). Sometimes this orientation can be determined by inspection. For example, if two of the three orthogonal planes are planes of symmetry, then all of the products of inertia are zero, so this would define principal axes of inertia.

The principal axes directions can be interpreted as an eigenvalue problem, and this allows you to find the orientation that will lead to principal directions, as well as define (transform) the inertia tensor into any orientation. For details on this method, see Crandall et al. [8].

9.6.2.2 Angular Momentum

For the rigid body shown in Figure 9.34, conceptualized to be composed of particles, i, of mass, m_i, the angular momentum about the point A is defined as

$$(\vec{h}_A)_i = \vec{\rho}_A \times m_i \vec{V}_i$$

where $\vec{V}_i$ is the velocity measured relative to the inertial frame. Since $\vec{V}_i = \vec{V}_A + \vec{\omega} \times \vec{\rho}_A$, then

$$(\vec{h}_A)_i = \vec{\rho}_A \times m_i \vec{V}_i = m_i \vec{\rho}_A \times \vec{V}_A + m_i \vec{\rho}_A \times (\vec{\omega} \times \vec{\rho}_A)$$

Integrating over the mass of the body, the total angular momentum of the body is

$$\vec{h}_A = \left(\int_m \vec{\rho}_A dm \right) \times \vec{V}_A + \int_m \vec{\rho}_A \times (\vec{\omega} \times \vec{\rho}_A) \, dm \tag{9.24}$$

This equation can be used to find the angular momentum about a point of interest by setting the point A: (1) fixed, (2) at the center of mass, and (3) an arbitrary point on the mass. A general form arises in cases 1 and 2 that take the form

$$\vec{h} = \int_m \vec{\rho} \times (\vec{\omega} \times \vec{\rho}) \, dm$$

When this form is expanded for either case into x, y, z components, then

$$\vec{h} = h_x\hat{i} + h_y\hat{j} + h_z\hat{k} = \int_m (x\hat{i} + y\hat{j} + z\hat{k}) \times [(\omega_x\hat{i} + \omega_y\hat{j} + \omega_z\hat{k}) \times (x\hat{i} + y\hat{j} + z\hat{k})]\, dm$$

which can be expanded to

$$h_x\hat{i} + h_y\hat{j} + h_z\hat{k} = \left[\omega_x \int_m (y^2 + z^2)\, dm - \omega_y \int_m xy\, dm - \omega_z \int_m xz\, dm\right]\hat{i}$$

$$= \left[-\omega_x \int_m xy\, dm + \omega_y \int_m (x^2 + z^2)\, dm - \omega_z \int_m yz\, dm\right]\hat{j}$$

$$= \left[-\omega_x \int_m xy\, dm - \omega_y \int_m zy\, dm - \omega_z \int_m (x^2 + y^2)\, dm\right]\hat{k}$$

The expression for moments and products of inertia can be identified here, and then this expression leads to the three angular momentum components, written in matrix form

$$\underline{h} = \begin{bmatrix} I_{xx} & -I_{xy} & -I_{xz} \\ -I_{yx} & I_{yy} & -I_{yz} \\ -I_{zx} & -I_{zy} & I_{zz} \end{bmatrix} \begin{bmatrix} \omega_x \\ \omega_y \\ \omega_z \end{bmatrix} = \underline{I}\omega \quad (9.25)$$

Note that the case where principal axes are defined leads to the much simplified expression

$$\vec{h} = I_{xx}\omega_x\hat{i} + I_{yy}\omega_y\hat{j} + I_{zz}\omega_z\hat{k}$$

This shows that when the body rotates so that its axis of rotation is parallel to a principal axis, the angular momentum vector, $\vec{h}$, is parallel to the angular velocity vector. In general, this is not true (this is related to the discussion at the end of the section "Inertia Properties").

The angular momentum about an arbitrary point, Case 3, is the resultant of the angular momentum about the mass center (a free vector) and the moment of the *translational momentum* through the mass center,

$$\vec{p} = mV_x\hat{i} + mV_y\hat{j} + mV_z\hat{k} = m\vec{V}$$

or

$$\vec{h} = \vec{h}_G + \vec{r} \times \vec{p}$$

where $\vec{r}$ is the position vector from the arbitary point of interest to the mass center, G. This form can also be expanded into its component forms, as in Equation 9.25.

9.6.2.3 Kinetic Energy of a Rigid Body

Several forms of the kinetic energy of a rigid body are presented in this section. From the standpoint of a bond graph formulation, where kinetic energy storage is represented by an **I** element, Equation 9.25 demonstrates that the rigid body has at least three ports for rotational energy storage. Adding the three translational degrees of freedom, a rigid body can have up to six independent energy storage "ports."

A 3-port **I** element can be used to represent the rotational kinetic energy for the case of rotation about a fixed point (no translation). The constitutive relation is simply Equation 9.25. The kinetic energy is then

$$T = \frac{1}{2}\vec{\omega}\cdot\vec{h}$$

where $\vec{h}$ is the angular momentum with an inertia tensor defined about the fixed point. If the axes are aligned with principal axes, then

$$T = \frac{1}{2}I_x\omega_x^2 + \frac{1}{2}I_y\omega_y^2 + \frac{1}{2}I_z\omega_z^2$$

The total kinetic energy for a rigid body that can translate and rotate, with angular momentum defined with reference to the center of gravity, is given by

$$T = \frac{1}{2}mV_G^2 + \frac{1}{2}\vec{\omega}\cdot\vec{h}_G$$

where $V_G^2 = V_x^2 + V_y^2 + V_z^2$.

9.6.3 Rigid Body Dynamics

Given descriptions of inertial properties, translational and angular momentum, and kinetic energy of a rigid body, it is possible to describe the dynamics of a rigid body using the equations of motion using Newton's laws. The classical Euler equations are presented in this section, and these are used to show how a bond graph formulation can be used to integrate rigid body elements into a bond graph model.

9.6.3.1 Basic Equations of Motion

The translational momentum of the body in Figure 9.30 is $\underline{p} = m\underline{V}$, where m is the mass, and $\underline{V}$ is the velocity of the mass center with three components of velocity relative to the inertial reference frame x_o, y_o, z_o. In three-dimensional motion, the net force on the body is related to the rate of change of momentum by Newton's law, namely,

$$\underline{F} = \frac{d}{dt}\underline{p}$$

which can be expressed as (using Equation 9.9),

$$\underline{F} = \left.\frac{\partial \underline{p}}{\partial t}\right|_{rel} + \underline{\Omega}\times\underline{p}$$

with $\underline{p}$ now relative to the moving frame x_a, y_a, z_a, and $\underline{\Omega}$ is the absolute angular velocity of the rotating axes.

A similar expression can be written for rate of change of the angular momentum, which is related to applied torques $\underline{T}$ by

$$\underline{T} = \left.\frac{\partial \underline{h}}{\partial t}\right|_{rel} + \underline{\Omega}\times\underline{h}$$

where $\underline{h}$ is relative to the moving frame x_a, y_a, z_a.

In order to use these relations effectively, the motion of the axes x_a, y_a, z_a, must be chosen to fit the problem at hand. This choice usually comes down to three cases described by how $\underline{\Omega}$ relates to the body angular velocity $\underline{\omega}$.

1. $\underline{\Omega} = 0$. If the body has general motion and the axes are chosen to translate with the center of mass, then this case will lead to a simple set of equations with $\Omega = 0$, although it will be necessary to describe the inertia properties of the body as functions of time.
2. $\underline{\Omega} \neq 0 \neq \underline{\omega}$. In this case, axes have an angular velocity different from that of the body, a form convenient for bodies that are symmetrical about their spinning axes. The moments and products of inertia will be constant relative to the rotating axes. The equations become

$$
\begin{aligned}
F_x &= m\dot{V}_x - mV_y\Omega_z + mV_z\Omega_y \\
F_y &= m\dot{V}_y - mV_z\Omega_x + mV_x\Omega_z \\
F_z &= m\dot{V}_z - mV_x\Omega_y + mV_y\Omega_x \\
T_x &= I_x\dot{\omega}_x - I_y\omega_y\Omega_z + I_z\Omega_y\omega_z \\
T_y &= I_y\dot{\omega}_y - I_z\omega_z\Omega_x + I_x\Omega_z\omega_x \\
T_z &= I_z\dot{\omega}_z - I_x\omega_x\Omega_y + I_y\Omega_x\omega_y
\end{aligned}
\tag{9.26}
$$

3. $\underline{\Omega} = \underline{\omega}$. Here the axes are fixed and moving with the body. The moments and products of intertia relative to the moving axes will be constant. A particularly convenient case arises if the axes are chosen to be the principal axes of inertia (see the section titled "Inertia Properties"), which leads to the *Euler equations*,*

$$
\begin{aligned}
F_x &= m\dot{V}_x - mV_y\omega_z + mV_z\omega_y \\
F_y &= m\dot{V}_y - mV_z\omega_x + mV_x\omega_z \\
F_z &= m\dot{V}_z - mV_x\omega_y + mV_y\omega_x \\
T_x &= I_x\dot{\omega}_x - (I_y - I_z)\omega_y\omega_z \\
T_y &= I_y\dot{\omega}_y - (I_z - I_x)\omega_z\omega_x \\
T_z &= I_z\dot{\omega}_z - (I_x - I_y)\omega_x\omega_y
\end{aligned}
\tag{9.27}
$$

These equations of motion can be used to determine the forces and torques, given motion of the body. Textbooks on dynamics [12,23] provide extensive examples on this type of analysis. Alternatively, these can be seen as six nonlinear, coupled ordinary differential equations (ODEs). Case 3 (the Euler equations) could be solved in such a case, since these can be rewritten as six first-order ODEs. A numerical solution may need to be implemented. Modern computational software packages will readily handle these equations, and some will feature a form of these equations in a form suitable for immediate use. Case 2 requires knowledge of the axes' angular velocity, $\underline{\Omega}$.

If the rotational motion is coupled to the translational motion such that the forces and torques, say, are related, then a dynamic model is required. In some, it may be desirable to formulate the problem in a bond graph form, especially if there are actuators and sensors and other multienergetic systems to be incorporated.

---

*First developed by the Swiss mathematician L. Euler.

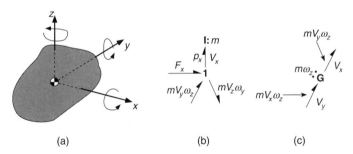

FIGURE 9.35 (a) Rigid body with angular velocity components about x, y, z axes. (b) x-direction translational dynamics in bond graph form. (c) Gyrator realization of coupling forces.

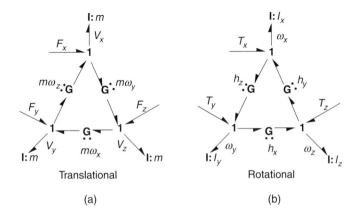

FIGURE 9.36 (a) Bond graph for rigid body translation. (b) Bond graph for rigid body rotation.

9.6.3.2 Rigid Body Bond Graph Formulation

Due to the body's rotation, there is an inherent coupling of the translational and rotational motion, which can be summarized in a bond graph form. Consider the case of Euler's equations, given in Equations 9.27. For the x-direction translational dynamics,

$$F_x = \dot{p}_x - mV_y\omega_z + mV_z\omega_y$$

where $p_x = mV_x$, and F_x is the net "external" applied forces in the x-direction. This equation, a summation of forces (efforts) is represented in bond graph form in Figure 9.35b. All of these forces are applied at a common velocity, V_x, represented by the 1-junction. The **I** element represents the storage of kinetic energy in the body associated with motion in the x-direction. The force $mV_y\omega_z$ in Figure 9.35b is induced by the y-direction velocity, V_y, and by the angular velocity component, ω_z. This physical effect is gyrational in nature, and can be captured by the gyrator, as shown in Figure 9.35c. Note that this is a modulated gyrator (could also be shown as **MGY**) with a gyrator modulus of $r = m\omega_z$ (verify that the units are force).

The six equations of motion, Equations 9.27, can be represented in bond graph form as shown in Figure 9.36. Note that these two bond graph ring formations, first shown by Karnopp and Rosenberg [18], capture the Euler equations very efficiently and provide a graphical mnemonic for rigid body motion. Indeed, Euler's equations can now be "drawn" simply in the following steps: (1) lay down three 1-junctions representing angular velocity about x, y, z (counter clockwise labeling), with **I** elements attached, (2) between each 1-junction place a gyrator, modulated by the momentum about the axis represented by the 1-junction

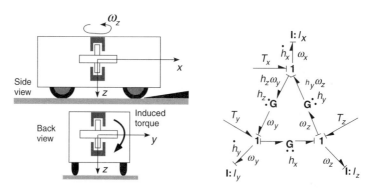

FIGURE 9.37 A cart with a rigid and internally mounted flywheel approaches a ramp.

directly opposite in the triangle, (3) draw power arrows in a counter clockwise direction. This sketch will provide the conventional Euler equations. The translational equations are also easily sketched.

These bond graph models illustrate the inherent coupling through the gyrator modulation. There are six I elements, and each can represent an independent energetic state in the form of the momenta $[p_x, p_y, p_z, h_x, h_y, h_z]$ or alternatively the analyst could focus on the associated velocities $[V_x, V_y, V_z, \omega_x, \omega_y, \omega_z]$.

If forces and torques are considered as inputs, through the indicated bonds representing $F_x, F_y, F_z, T_x, T_y, T_z$, then you can show that all the I elements are in integral causality, and the body will have six independent states described by six first-order nonlinear differential equations.

Example: Cart-Flywheel

A good example of how the rigid body bond graphs represent the basic mechanics inherent to Equations 9.27 and of how the graphical modeling can be used for "intuitive" gain is shown in Figure 9.37. The flywheel is mounted in the cart, and spins in the direction shown. The body-fixed axes are mounted in the vehicle, with the convention that z is positive into the ground (common in vehicle dynamics). The cart approaches a ramp, and the questions which arise are whether any significant loads will be applied, what their sense will be, and on which parameters or variables they are dependent.

The bond graph for rotational motion of the flywheel (assume it dominates the problem for this example) is shown in Figure 9.37. If the flywheel momentum is assumed very large, then we might just focus on its effect. At the 1-junction for ω_x, let $T_x = 0$, and since ω_z is spinning in a negative direction, you can see that the torque $h_z\omega_y$ is applied in a positive direction about the x-axis. This will tend to "roll" the vehicle to the right, and the wheels would feel an increased normal load. With the model shown, it would not be difficult to develop a full set of differential equations.

9.6.3.3 Need for Coordinate Transformations

In the cart-flywheel example, it is assumed that as the front wheels of the cart lift onto the ramp, the flywheel will react because of the direct induced motion at the bearings. Indeed, the flywheel-induced torque is also transmitted directly to the cart. The equations and basic bond graphs developed above are convenient if the forces and torques applied to the rigid body are moving with the rotating axes (assumed to be fixed to the body). The orientational changes, however, usually imply that there is a need to relate the body-fixed coordinate frames or axes to inertial coordinates. This is accomplished with a coordinate transformation, which relates the body orientation into a frame that makes it easier to interpret the motion, apply forces, understand and apply measurements, and apply feedback controls.

Example: Torquewhirl Dynamics

Figure 9.38a illustrates a cantilevered rotor that can exhibit torquewhirl. This is a good example for illustrating the need for coordinate transformations, and how Euler angles can be used in the modeling process. The whirling mode is conical and described by the angle θ. There is a drive torque, T_s, that is

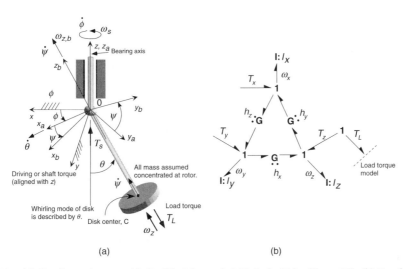

FIGURE 9.38 (a) Cantilevered rotor with flexible joint and rigid shaft. (After Vance 36). (b) Bond graph representing rigid body rotation of rotor.

aligned with the bearing axis, z, where x, y, z is the inertial coordinate frame. The bond graph in Figure 9.38b captures the rigid body motion of the rotor, represented in body-fixed axes x_b, y_b, z_b, which represent principal axes of the rotor.

The first problem seen here is that while the bond graph leads to a very convenient model formulation, the applied torque, T_s, is given relative to the inertial frame x, y, z. Also, it would be nice to know how the rotor moves relative to the inertial frame, since it is that motion that is relevant. Other issues arise, including a stiffness of the rotor that is known relative to the angle θ. These problems motivate the use of Euler angles, which will relate the motion in the body fixed to the inertial frame, and provide three additional state equations for ϕ, θ, and ψ (which are needed to quantify the motion).

In this example, the rotation sequence is (1) x, y, z (inertial) to x_a, y_a, z_a, with ϕ about the z-axis, so note, $\dot\phi = \omega_s$, (2) x_a, y_a, z_a to x_b, y_b, z_b, with θ about x_a, (3) ψ rotation about z_b. Our main interest is in the overall transformation from x, y, z (inertia) to x_b, y_b, z_b (body-fixed). In this way, we relate the body angular velocities to inertial velocities using the relation from Equation 9.20,

$$\begin{bmatrix} \omega_x \\ \omega_y \\ \omega_z \end{bmatrix}_b = \begin{bmatrix} \dot\phi \sin\theta \sin\psi + \dot\theta \cos\psi \\ \dot\phi \sin\theta \cos\psi - \dot\theta \sin\psi \\ \dot\phi \cos\theta + \dot\psi \end{bmatrix}$$

where the subscript b on the left-hand side denotes velocities relative to the x_b, y_b, z_b axes. A full and complete bond graph would include a representation of these transformations (e.g., see Karnopp, Margolis, and Rosenberg [17]). Explicit 1-junctions can be used to identify velocity junctions at which torques and forces are applied. For example, at a 1-junction for $\dot\phi = \omega_s$, the input torque T_s is properly applied. Once the bond graph is complete, causality is applied. The preferred assignment that will lead to integral causality on all the I elements is to have torques and forces applied as causal *inputs*. Note that in transforming the expression above which relates the angular velocities, a problem with Euler angles arises related to the singularity (here at $\theta = \pi/2$, e.g.).

An alternative way to proceed in the analysis is using a Lagrangian approach as in Section 9.7, as done by Vance [36], p. 292). Also, for advanced multibody systems, a multibond formulation can be more efficient and may provide insight into complex problems (see Breedveld [4] or Tiernego and Bos [35]).

9.7 Lagrange's Equations

The discussion on energy methods focuses on deriving constitutive relations for energy-storing multiports, and this can be very useful in some modeling exercises. For some cases where the constraint relationships between elements are primarily holonomic, and definitely scleronomic (not an explicit function of time), implicit multiport fields can be formulated (see Chapter 7 of [17]). The principal concern arises because of dependent energy storage, and the methods presented can be a solution in some practical cases. However, there are many mechanical systems in which geometric configuration complicates the matter. In this section, Lagrange's equations are introduced to facilitate analysis of those systems.

There are several ways to introduce, derive, and utilize the concepts and methods of Lagrange's equations. The summary presented below is provided in order to introduce fundamental concepts, and a thorough derivation can be found either in Lanczos [20] or Goldstein [11]. A derivation using energy and power flow is presented by Beaman, Paynter, and Longoria [3].

Lagrange's equations are also important because they provide a unified way to model systems from different energy domains, just like a bond graph approach. The use of scalar energy functions and minimal geometric reasoning is preferred by some analysts. It is shown in the following that the particular benefits of a Lagrange approach that make it especially useful for modeling mechanical systems enhance the bond graph approach. A combined approach exploits the benefits of both methods, and provides a methodology for treating complex mechatronic systems in a systematic fashion.

9.7.1 Classical Approach

A classical derivation of Lagrange's equations evolves from the concept of virtual displacement and virtual work developed for analyzing static systems (see Goldstein [11]). To begin with, the Lagrange equations can be derived for dynamic systems by using Hamilton's principle or D'Alembert's principle.

For example, for a system of particles, Newton's second law for the i mass, $\mathbf{F}_i = \dot{\mathbf{p}}_i$, is rewritten, $\mathbf{F}_i - \dot{\mathbf{p}}_i = 0$. The forces are classified as either applied or constraint, $\mathbf{F}_i = \mathbf{F}_i^{(a)} + \mathbf{f}_i$. The principle of virtual work is applied over the system, recognizing that constraint forces $\mathbf{f}_i$, do no work and will drop out. This leads to the D'Alembert principle [11],

$$\sum_i (\mathbf{F}_i^{(a)} - \dot{\mathbf{p}}_i) \cdot \delta \mathbf{r}_i = 0 \tag{9.28}$$

The main point in presenting this relation is to show that: (a) the constraint forces do not appear in this formulative equation and (b) the need arises for transforming relationships between, in this case, the N coordinates of the particles, $\mathbf{r}_i$, and a set of n **generalized coordinates**, $\mathbf{q}_j$, which are independent of each other (for holonomic constraints), that is,

$$\mathbf{r}_i = \mathbf{r}_i(q_1, q_2, \ldots, q_n, t) \tag{9.29}$$

By transforming to generalized coordinates, D'Alembert's principle becomes [11]

$$\sum_j \left[\left\{ \frac{d}{dt}\left(\frac{\partial T}{\partial \dot{q}_j}\right) - \frac{\partial T}{\partial q_j} \right\} - Q_j \right] \delta q_j = 0 \tag{9.30}$$

where T is the system kinetic energy, and the Q_j are components of the **generalized forces** given by

$$Q_j = \sum_i \mathbf{F}_i \cdot \frac{\partial \mathbf{r}_i}{\partial q_j}$$

Modeling of Mechanical Systems for Mechatronics Applications

If the transforming relations are restricted to be holonomic, the constraint conditions are implicit in the transforming relations, and independent coordinates are assured. Consequently, all the terms in Equation 9.30 must vanish for independent virtual displacements, δq_j, resulting in the n equations:

$$\frac{d}{dt}\left(\frac{\partial T}{\partial \dot{q}_j}\right) - \frac{\partial T}{\partial q_j} = Q_j \tag{9.31}$$

These equations become Lagrange's equations through the following development. Restrict all the applied forces, Q_j, to be derivable from a scalar function, U, where in general, $U = U(q_j, \dot{q}_j)$, and

$$Q_j = -\frac{\partial U}{\partial q_j} + \frac{d}{dt}\left(\frac{\partial U}{\partial \dot{q}_j}\right)$$

The Lagrangian is defined as $L = T - U$, and substituted into Equation 9.31 to yield the n Lagrange equations:

$$\frac{d}{dt}\left(\frac{\partial L}{\partial \dot{q}_j}\right) - \frac{\partial L}{\partial q_j} = Q_j \tag{9.32}$$

This formulation yields n second-order ODEs in the q_j.

9.7.2 Dealing with Nonconservative Effects

The derivation of Lagrange's equations assumes, to some extent, that the system is conservative, meaning that the total of kinetic and potential energy remains constant. This is not a limiting assumption because the process of reticulation provides a way to extract nonconservative effects (inputs, dissipation), and then to assemble the system later. It is necessary to recognize that the nonconservative effects can be integrated into a model based on Lagrange's equations using the Q_i's. Associating these forces with the generalized coordinates implies work is done, and this is in accord with energy conservation principles (we account for total work done on system). The generalized force associated with a coordinate, q_i, and due to external forces is then derived from $Q_i = \delta W_i/\delta q_i$, where W_i is the work done on the system by all external forces during the displacement, δq_i.

9.7.3 Extensions for Nonholonomic Systems

In the case of nonholonomic constraints, the coordinates q_j are not independent. Assume you have m nonholonomic constraints ($m \leq n$). If the equations of constraint can be put in the form

$$\sum_k \frac{\partial a_l}{\partial q_k} dq_k + \frac{\partial a_l}{\partial t} dt = \sum_k a_{lk} dq_k + a_{lt} dt = 0 \tag{9.33}$$

where l indexes up to m such constraints, then the Lagrange equations are formulated with Lagrange undetermined multipliers, λ_l. We maintain n coordinates, q_k, but the n Lagrange equations are now expressed [11] as

$$\frac{d}{dt}\left(\frac{\partial L}{\partial \dot{q}_k}\right) - \frac{\partial L}{\partial q_k} = \sum_l \lambda_l a_{lk}, \quad k = 1, 2, \ldots, n \tag{9.34}$$

However, since there are now m unknown Lagrange multipliers, λ_l, it is necessary to solve an additional m equations:

$$\sum_k a_{lk}\dot{q}_k + a_{lt} = 0 \tag{9.35}$$

The terms $\sum_l \lambda_l a_{lk}$ can be interpreted as generalized forces of constraint. These are still workless constraints. The Lagrange equations for nonholonomic constraints can be used to study holonomic systems, and this analysis would provide a solution for the constraint forces through evaluation of the Lagrange multipliers. The use of Lagrange's equations with Lagrange multipliers is one way to model complex, constrained multibody systems, as discussed in Haug [14].

9.7.4 Mechanical Subsystem Models Using Lagrange Methods

The previous sections summarize a classical formulation and application of Lagrange's equations. When formulating models of mechanical systems, these methods are well proven. Lagrange's equations are recognized as an approach useful in handling systems with complex mechanical systems, including systems with constraints. The energy-basis also makes the method attractive from the standpoint of building multi-energetic system models, and Lagrange's equations have been used extensively in electromechanics modeling, for example. For conservative systems, it is possible to arrive at solutions sometimes without worrying about forces, especially since nonconservative effects can be handled "outside" the conservative dynamics. Developing transformation equations between the coordinates, say **x**, used to describe the system and the independent coordinates, **q**, helps assure a minimal formulation. However, it is possible sometimes to lose insight into cause and effect, which is more evident in other approaches. Also, the algebraic burden can become excessive. However, it is the analytical basis of the method that makes it especially attactive. Indeed, with computer-aided symbolic processing techniques, extensive algebra becomes a non-issue.

In this section, the advantages of the Lagrange approach are merged with those of a bond graph approach. The concepts and formulations are classical in nature; however, the graphical interpretation adds to the insight provided. Further, the use of bond graphs assures a consistent formulation with causality so that some insight is provided into how the conservative dynamics described by the energy functions depend on inputs, which typically arrive from the nonconservative dynamics. The latter are very effectively dealt with using bond graph methods, and the combined approach is systematic and yields first-order differential equations, rather than the second-order ODEs in the classical approach. Also, it will be shown that in some cases the combined approach makes it relatively easy to model certain systems that would be very troublesome for a direct approach by either method independently.

A Lagrange bond graph subsystem model will capture the elements summarized with a word bond graph in Figure 9.39. The key elements are identified as follows: (a) conservative energy storage captured by kinetic and potential energy functions, (b) power-conserving transforming relations, and (c) coupling/interconnections with nonconservative and non-Lagrange system elements. Note that on the nonconservative side of the transforming relations, there are m coordinates that can be identified in the modeling, but these are not independent. The power-conserving transforming relations reduce the coordinates to a set of n independent coordinates, q_i. Associated with each independent coordinate or velocity, $\dot{q}_i$, there is an associated storage of kinetic and potential energy which can be represented by the coupled IC in Figure 9.40a [16]. An alternative is the single **C** element used to capture all the coupled energy storage [3], where the gyrator has a modulus of 1 (this is called a symplectic gyrator). In either case, this structure shows that there will be one common flow junction associated with each independent coordinate. Recall the efforts at a 1-junction sum, and at this ith junction,

$$E_{q_i} = \dot{p}_i + e_{q_i} \tag{9.36}$$

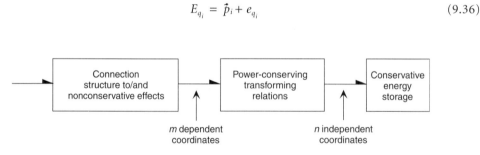

FIGURE 9.39 Block diagram illustrating the Lagrange subsystem model.

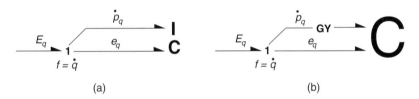

FIGURE 9.40 Elementary formulation of a flow junction in a Lagrange subsystem model. The efforts at the 1-junction for this ith independent flow variable, $\dot{q}_i$, represent Lagrange's equations.

where E_{q_i} is the net nonconservative effort at $\dot{q}_i$, e_{q_i} is a generalized conservative effort that will be determined by the Lagrange system, and the effort $\dot{p}_i$ is a rate of change of an ith generalized momentum. These terms will be defined in the next section. However, note that this effort sum is simply Newton's laws derived by virtue of a Lagrange formulation. In fact, this equation is simply a restatement of the ith Lagrange equation, as will be shown in the following. These effort sum equations give n first-order ODEs by solving for $\dot{p}_i$. The other n equations will be for the displacement variables, q_i. The following methodology is adapted from Beaman, Paynter, and Longoria [3].

9.7.5 Methodology for Building Subsystem Model

Conduct Initial Modeling. Isolate the conservative parts of the system, and make sure that any constraints are holonomic. This reticulation will identify ports to the system under study, including points in the system (typically velocities) where forces and/or torques of interest can be applied (e.g., at flow junctions). These forces and torques are either nonconservative, or they are determined by a system external to the Lagrange-type subsystem. This is a modeling decision. For example, a force due to gravity could be included in a Lagrange subsystem (being conservative) or it could be shown explicity at a velocity junction corresponding to motion modeled outside of the Lagrange subsystem. This will be illustrated in one of the examples that follow.

Define Generalized Displacement Variables. In a Lagrange approach, it is necessary to identify variables that define the configuration of a system. In mechanical system, these are translational and rotational displacements. Further, these variables are typically associated with the motion or relative motion of bodies. To facilitate a model with a minimum and independent set of coordinates, develop transforming relations between the m velocities or, more generally, flows $\dot{x}$, and n independent flows, $\dot{q}$. The form is [3],

$$\dot{x} = T(q)\dot{q} \qquad (9.37)$$

explicity showing that the matrix $T(q)$ can depend on q. This can be interpreted, in bond graph modeling terms, as a modulated transformer relationship, where q contains the modulating variables. The independent generalized displacements, q, will form possible state variables of the Lagrange subsystem.

The transforming relationships are commonly derived from (holonomic) constraints, and from considerations of geometry and basic kinematics. The matrix T is $m \times n$ and may not be invertible. The bond graph representation is shown in Figure 9.41.

Formulate the Kinetic Energy Function. Given the transforming relationships, it is now possible to express the total kinetic energy of the Lagrange subsystem using the independent flow variables, $\dot{q}$. First, the kinetic energy can be written using the $\dot{x}$ (this is usually easier), or $T = T_{\dot{x}}(\dot{x})$. Then the relations in Equation 9.37 are used to transform this kinetic energy function so it is expressed as a function of the q and $\dot{q}$ variables, $T_{\dot{x}}(\dot{x}) \rightarrow T_{\dot{q}q}(\dot{q}, q)$. For brevity, this can be indicated in the subscript, or just $T_{\dot{q}q}$. For example, a kinetic energy function that depends on x, θ, and $\dot{\theta}$ is referred to as $T_{\dot{\theta}\theta x}$ (if the number of variables is very high, certainly such a convention would not be followed).

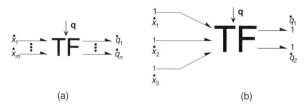

FIGURE 9.41 (a) Bond graph representation of the transforming relations. (b) Example for the case where $m = 3$ and $n = 2$.

Define Generalized Momentum Variables. With the kinetic energy function now in terms of the independent flows, $\dot{\mathbf{q}}$, generalized momenta can be defined as [3,20],

$$\tilde{\mathbf{p}} = \frac{\partial T_{\dot{q}q}}{\partial \dot{\mathbf{q}}} \tag{9.38}$$

where the "tilde" ($\tilde{\mathbf{p}}$) notation is used to distinguish these momentum variables from momentum variables defined strictly through the principles summarized in Table 9.5. In particular note that these generalized momentum variables may be functions of flow as well as of displacement (i.e., they may be configuration dependent).

Formulate the Potential Energy Function. In general, a candidate system for study by a Lagrange approach will store potential energy, in addition to kinetic energy, and the potential energy function, U, should be expressed in terms of the dependent variables, $\mathbf{x}$. Using the tranforming relations in Equation 9.37, the expression is then a function of $\mathbf{q}$, or $U = U(\mathbf{q}) = U_q$. In mechanical systems, this function is usually formed by considering energy stored in compliant members, or energy stored due to a gravitational potential. In these cases, it is usually possible to express the potential energy function in terms of the displacement variables, $\mathbf{q}$.

Derive Generalized Conservative Efforts. A conservative effort results and can be found from the expression

$$\tilde{e}_q = -\frac{\partial T_{\dot{q}q}}{\partial \mathbf{q}} + \frac{\partial U_q}{\partial \mathbf{q}} \tag{9.39}$$

where the $\mathbf{q}$ subscript is used to denote these as conservative efforts. The first term on the right-hand side represents an effect due to dependence of kinetic energy on displacement, and the second term will be recognized as the potential energy derived effort.

Identify and Express Net Power Flow into Lagrange Subsystem. At the input to the Lagrange subsystem on the "nonconservative" side, the power input can be expressed in terms of effort and flow products. Since the transforming relations are power-conserving, this power flow must equal the power flow on the "conservative" side. This fact is expressed by

$$P_x = \underbrace{\mathbf{e}_x}_{1 \times m} \underbrace{\dot{\mathbf{x}}}_{m \times 1} = \underbrace{\mathbf{e}_x}_{1 \times m} \underbrace{\mathbf{T}(\mathbf{q})}_{m \times n} \underbrace{\dot{\mathbf{q}}}_{n \times 1} = \underbrace{\mathbf{E}_q}_{1 \times n} \underbrace{\dot{\mathbf{q}}}_{n \times 1} \tag{9.40}$$

where the term $\mathbf{E}_q$ is the nonconservative effort transformed into the $\mathbf{q}$ coordinates. This term can be computed as shown by

$$\mathbf{E}_q = \mathbf{e}_x \mathbf{T}(\mathbf{q}) \tag{9.41}$$

Summary of the Method. In summary, all the terms for a Lagrange subsystem can be systematically derived. There are some difficulties that can arise. To begin with, the first step can require some geometric reasoning, and often this can be a problem in some cases, although not insurmountable. The n

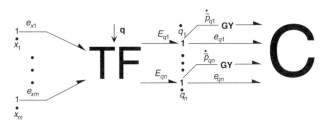

FIGURE 9.42 Lagrange subsystem model.

momentum state equations for this Lagrange subsystem are given by

$$\dot{\tilde{p}} = -e_i + E_i \tag{9.42}$$

and the state equations for the q_i must be found by inverting the generalized momentum equations, (9.38). In some cases, these n equations are coupled and must be solved simultaneously. In the end, there are $2n$ first-order state equations. In addition, the final bond graph element shown in Figure 9.42 can be coupled to other systems to build a complex system model.

Note that in order to have the $2n$ equations in integral causality, efforts (forces and torques) should be specified as causal inputs to the transforming relations. Also, this subsystem model assumes that only holonomic constraints are applied. While this might seem restrictive, it turns out that, for many practical cases, the physical effects that lead to nonholonomic constraints can be dealt with "outside" of the Lagrange model, along with dissipative effects, actuators, and so on.

References

1. Arczewski, K. and Pietrucha, J., *Mathematical Modelling of Complex Mechanical Systems*, Ellis Horwood, New York, 1993.
2. Beaman, J.J. and Rosenberg, R.C., "Constitutive and modulation structure," *Journal of Dynamic Systems, Measurement, and Control (ASME)*, Vol. 110, No. 4, pp. 395–402, 1988.
3. Beaman, J.J., Paynter, H.M., and Longoria, R.G., *Modeling of Physical Systems*, Cambridge University Press, in progress.
4. Bedford, A. and Fowler, W., *Engineering Mechanics. Dynamics*, 2nd edition, Addison Wesley Longman, Menlo Park, CA, 1999.
5. Breedveld, P.C., "Multibond graph elements in physical systems theory," *Journal of the Franklin Institute*, Vol. 319, No. 1–2, pp. 1–36, 1985.
6. Burr, A.H., *Mechanical Analysis and Design*, Elsevier Science Publishing, Co., New York, 1981.
7. Chou, J.C.K, "Quaternion kinematic and dynamic differential equations," *IEEE Transactions on Robotics and Automation*, Vol. 8, No. 1, February, 1992.
8. Crandall, S., Karnopp, D.C., Kurtz, E.F., and Pridmore-Brown, D.C., *Dynamics of Mechanical and Electromechanical Systems*, McGraw-Hill, New York, 1968 (Reprinted by Krieger Publishing Co., Malabar, FL, 1982).
9. Den Hartog, J.P., *Advanced Strength of Materials*, McGraw-Hill, New York, 1952.
10. Fjellstad, O. and Fossen, T.I., "Position and attitude tracking of AUVs: a quaternion feedback approach," *IEEE Journal of Oceanic Engineering*, Vol. 19, No. 4, pp. 512–518, 1994.
11. Goldstein, D., *Classical Mechanics*, 2nd edition, Addison-Wesley, Reading, MA, 1980.
12. Greenwood, D.T., *Principles of Dynamics*, Prentice-Hall, Englewood Cliffs, NJ, 1965.
13. Harding, C.F., "Solution to Euler's gyrodynamics-I," *Journal of Applied Mechanics*, Vol. 31, pp. 325–328, 1964.

14. Haug, E.J., *Computer Aided Kinematics and Dynamics of Mechanical Systems*, Allyn and Bacon, Needham, MA, 1989.
15. Kane, T.R. and Levinson, D.A., *Dynamics: Theory and Applications*, McGraw-Hill Publishing Co., New York, 1985.
16. Karnopp, D.C., "An approach to derivative causality in bond graph models of mechanical systems," *Journal of the Franklin Institute*, Vol. 329, No. 1, pp. 65–75, 1992.
17. Karnopp, D.C., Margolis, D., and Rosenberg, R.C., *System Dynamics: Modeling and Simulation of Mechatronic Systems*, Wiley, New York, 2000, 3rd edition, or *System Dynamics: A Unified Approach*, 1990, 2nd edition.
18. Karnopp, D.C. and Rosenberg, R.C., *Analysis and Simulation of Multiport Systems. The Bond Graph Approach to Physical System Dynamics*, MIT Press, Cambridge, MA, 1968.
19. Kuipers, J.B., *Quaternions and Rotation Sequences*, Princeton University Press, Princeton, NJ, 1998.
20. Lanczos, C., *The Variational Principles of Mechanics*, 4th edition, University of Toronto Press, Toronto, 1970. Also published by Dover, New York, 1986.
21. Lyshevski, S.E., *Electromechanical Systems, Electric Machines, and Applied Mechatronics*, CRC Press, Boca Raton, FL, 2000.
22. Matschinsky, W., *Road Vehicle Suspensions*, Professional Engineering Publishing Ltd., Suffolk, UK, 1999.
23. Meriam, J.L. and Kraige, L.G., *Engineering Mechanics. Dynamics*, 4th edition, John Wiley and Sons, New York, 1997.
24. Mortensen, R.E., "A globally stable linear regulator," *International Journal of Control*, Vol. 8, No. 3, pp. 297–302, 1968.
25. Nikravesh, P.E. and Chung, I.S., "Application of Euler parameters to the dynamic analysis of three-dimensional constrained mechanical systems," *Journal of Mechanical Design (ASME)*, Vol. 104, pp. 785–791, 1982.
26. Nikravesh, P.E., Wehage, R.A., and Kwon, O.K., "Euler parameters in computational kinematics and dynamics, Parts 1 and 2," *Journal of Mechanisms, Transmissions, and Automation in Design (ASME)*, Vol. 107, pp. 358–369, 1985.
27. Nososelov, V.S., "An example of a nonholonomic, nonlinear system not of the Chetaev type," *Vestnik Leningradskogo Universiteta*, No. 19, 1957.
28. Paynter, H., *Analysis and Design of Engineering Systems*, MIT Press, Cambridge, MA, 1961.
29. Roark, R.J. and Young, W.C., *Formulas for Stress and Strain*, McGraw-Hill, New York, 1975.
30. Roberson, R.E. and Schwertassek, *Dynamics of Multibody Systems*, Springer-Verlag, Berlin, 1988.
31. Rosenberg, R.M., *Analytical Dynamics of Discrete Systems*, Plenum Press, New York, 1977.
32. Rosenberg, R. and Karnopp, D., *Introduction to Physical System Dynamics*, McGraw-Hill, New York, 1983.
33. Rowell, D. and Wormley, D.N., *System Dynamics*, Prentice-Hall, Upper Saddle River, NJ, 1997.
34. Siciliano, B. and Villani, L., *Robot Force Control*, Kluwer Academic Publishers, Norwell, MA, 1999.
35. Tiernego, M.J.L. and Bos, A.M., "Modelling the dynamics and kinematics of mechanical systems with multibond graphs," *Journal of the Franklin Institute*, Vol. 319, No. 1–2, pp. 37–50, 1985.
36. Vance, J.M., *Rotordynamics of Turbomachinery*, John Wiley and Sons, New York, 1988.
37. Wehage, R.A., "Quaternions and Euler parameters—a brief exposition," in *Proceedings of the NATO Advanced Study Institute on Computer Aided Analysis and Optimization of Mechanical System Dynamics*, E.J. Haug (ed.), Iowa City, IA, August 1–12, 1983, pp. 147–182.
38. Wie, B. and Barba, P.M., "Quaternion feedback for spacecraft large angle maneuvers," *Journal of Guidance, Control, and Dynamics*, Vol. 8, pp. 360–365, May–June 1985.
39. Wittenburg, J., *Dynamics of Systems of Rigid Bodies*, B.G. Teubner, Studttgart, 1977.

10
Fluid Power Systems

Qin Zhang
Carroll E. Goering
University of Illinois

10.1 Introduction 10-1
   Fluid Power Systems • Electrohydraulic
   Control Systems
10.2 Hydraulic Fluids 10-2
   Density • Viscosity • Bulk Modulus
10.3 Hydraulic Control Valves 10-3
   Principle of Valve Control • Hydraulic Control Valves
10.4 Hydraulic Pumps 10-5
   Principles of Pump Operation • Pump Controls
   and Systems
10.5 Hydraulic Cylinders 10-7
   Cylinder Parameters
10.6 Fluid Power Systems Control 10-8
   System Steady-State Characteristics • System Dynamic
   Characteristics • E/H System Feedforward-Plus-PID
   Control • E/H System Generic Fuzzy Control
10.7 Programmable Electrohydraulic Valves 10-12
References 10-14

10.1 Introduction

10.1.1 Fluid Power Systems

A fluid power system uses either liquid or gas to perform desired tasks. Operation of both the liquid systems (hydraulic systems) and the gas systems (pneumatic systems) is based on the same principles. For brevity, we will focus on hydraulic systems only.

A fluid power system typically consists of a hydraulic pump, a line relief valve, a proportional direction control valve, and an actuator (Figure 10.1). Fluid power systems are widely used on aerospace, industrial, and mobile equipment because of their remarkable advantages over other control systems. The major advantages include high power-to-weight ratio, capability of being stalled, reversed, or operated intermittently, capability of fast response and acceleration, and reliable operation and long service life.

Due to differing tasks and working environments, the characteristics of fluid power systems are different for industrial and mobile applications (Lambeck, 1983). In industrial applications, low noise level is a major concern. Normally, a noise level below 70 dB is desirable and over 80 dB is excessive. Industrial systems commonly operate in the low (below 7 MPa or 1000 psi) to moderate (below 21 MPa or 3000 psi) pressure range. In mobile applications, the size is the premier concern. Therefore, mobile hydraulic systems commonly operate between 14 and 35 MPa (2000–5000 psi). Also, their allowable temperature operating range is usually higher than in industrial applications.

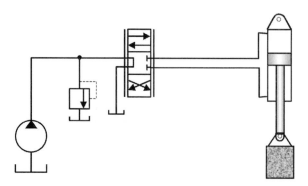

FIGURE 10.1 Schematic of a fluid power system.

10.1.2 Electrohydraulic Control Systems

The application of electronic controls to fluid power systems resulted in electrohydraulic control systems. Electrohydraulics has been widely used in aerospace, industrial, and mobile fluid power systems. Electrohydraulic controls have a few distinguishable advantages over other types of controls. First, an electrohydraulic system can be operated over a wide speed range, and its speed can be controlled continuously. More importantly, an electrohydraulic system can be stalled or operated under very large acceleration without causing its components to be damaged. A hydraulic actuator can be used in strong magnetic field without having the electromagnetic effects degrade control performance. In addition, hydraulic fluid flow can transfer heat away from system components and lubricate all moving parts continuously.

10.2 Hydraulic Fluids

Many types of fluids, for example, mineral oils, biodegradable oils, and water-based fluids, are used in fluid power systems, depending on the task and the working environment. Ideally, hydraulic fluids should be inexpensive, noncorrosive, nontoxic, noninflammable, have good lubricity, and be stable in properties. The technically important properties of hydraulic fluids include density, viscosity, and bulk modulus.

10.2.1 Density

The density, ρ, of a fluid is defined as its mass per unit volume (Welty et al., 1984):

$$\rho = \frac{m}{V} \qquad (10.1)$$

Density is approximately a linear function of pressure (P) and temperature (T) (Anderson, 1988):

$$\rho = \rho_0(1 + aP - bT) \qquad (10.2)$$

In engineering practice, the manufacturers of the hydraulic fluids often provide the relative density (i.e., the specific gravity) instead of the actual density. The specific gravity of a fluid is the ratio of its actual density to the density of water at the same temperature.

10.2.2 Viscosity

The viscosity of a fluid is a measure of its resistance to deformation rate when subjected to a shearing force (Welty et al., 1984). Manufacturers often provide two kinds of viscosity values, namely the dynamic viscosity (μ) and the kinematic viscosity (ν). The dynamic viscosity is also named the absolute viscosity and is defined by the Newtonian shear stress equation:

$$\mu = \frac{\tau}{\frac{dv}{dy}} \quad (10.3)$$

where dv is the relative velocity between two parallel layers dy apart, and τ is the shear stress.

The kinematic viscosity is the ratio of the dynamic viscosity to the density of the fluid and is defined using the following equation:

$$\nu = \frac{\mu}{\rho} \quad (10.4)$$

In the SI system, the unit of dynamic viscosity is Pascal-seconds (Pa s), and the unit of kinematic viscosity is square meter per second (m²/s). Both the dynamic and kinematic vary strongly with temperature.

10.2.3 Bulk Modulus

Bulk modulus is a measure of the compressibility or the stiffness of a fluid. The basic definition of fluid bulk modulus is the fractional reduction in fluid volume corresponding to unit increase of applied pressure, expressed using the following equation (McCloy and Martin, 1973):

$$\beta = -V\left(\frac{\partial P}{\partial V}\right) \quad (10.5)$$

The bulk modulus can either be defined as the isothermal tangent bulk modulus if the compressibility is measured under a constant temperature or as the isentropic tangent bulk modulus if the compressibility is measured under constant entropy.

In analyzing the dynamic behavior of a hydraulic system, the stiffness of the hydraulic container plays a very important role. An effective bulk modulus, β_e, is often used to consider both the fluid's compressibility, β_f, and container stiffness, β_c, at the same time (Watton, 1989).

$$\frac{1}{\beta_e} = \frac{1}{\beta_f} + \frac{1}{\beta_c} \quad (10.6)$$

10.3 Hydraulic Control Valves

10.3.1 Principle of Valve Control

In a fluid power system, hydraulic control valves are used to control the pressure, flow rate, and flow direction. There are many ways to define a hydraulic valve so that a given valve can be named differently when it is used in different applications. Commonly, hydraulic valves can be classified based on their functions, such as pressure, flow, and directional control valves, or based on their control mechanisms, such as on-off, servo, and proportional electrohydraulic valves, or based on their structures, such as spool, poppet, and needle valves.

A hydraulic valve controls a fluid power system by opening and closing the flow-passing area of the valve. Such an adjustable flow-passing area is often described using an orifice area, A_o, in engineering practice. Physically, an orifice is a controllable hydraulic resistance, R_h. Under steady-state conditions, a hydraulic resistance can be defined as a ratio of pressure drop, Δp, across the valve to the flow rate, q, through the valve.

$$R_h = \frac{d(\Delta p)}{dq} \quad (10.7)$$

Control valves make use of many configurations of orifice to realize various hydraulic resistance characteristics for different applications. Therefore, it is essential to determine the relationship between the

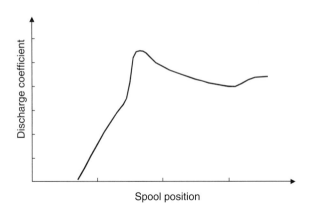

FIGURE 10.2 Discharge coefficient versus spool position in a spool valve.

pressure drop and the flow rate across the orifice. An orifice equation (McCloy and Martin, 1973) is often used to describe this relationship.

$$q = C_d A_o \sqrt{\frac{2}{\rho} \Delta P} \qquad (10.8)$$

The pressure drop across the orifice is a system pressure loss in a fluid power system. In this equation, the orifice coefficient, C_d, plays an important role, and is normally determined experimentally. It has been found that the orifice coefficient varies greatly with the spool position, but does not appear to vary much with respect to the pressure drop across the orifice in a spool valve (Figure 10.2, Viall and Zhang, 2000). Based on analytical results obtained from computational fluid dynamics simulations, the valve spool and sleeve geometries have little effect on the orifice coefficient for large spool displacements (Borghi et al., 1998).

10.3.2 Hydraulic Control Valves

There are many ways to classify hydraulic control valves. For instance, based on their structural configurations, hydraulic control valves can be grouped as cartridge valves and spool valves. This section will provide mathematical models of hydraulic control valves based on their structural configurations.

A typical cartridge valve has either a poppet or a ball to control the passing flow rate. Representing the control characteristics of a cartridge valve without loss of generality, a poppet type cartridge is analyzed (Figure 10.3).

The control characteristics of a poppet type cartridge valve can be described using an orifice equation and a force balance equation. As shown in Figure 10.3, the valve opens by lifting the poppet. Because of the cone structure of the poppet, the flow-passing area can be determined using the following equation:

$$A_x = \pi d x \sin \alpha \qquad (10.9)$$

Therefore, the passing flow can be calculated using the orifice equation. For a poppet type valve, it is recommended to use a relative higher orifice coefficient of $c_d = 0.77$–0.82 (Li et al., 2000).

$$q = c_d A_x \sqrt{\frac{2}{\rho}(P_B - P_A)} \qquad (10.10)$$

The forces acting on the poppet include the pressure, spring, and hydraulic forces. The pressure force can be determined based on the upstream, downstream, and spring chamber pressures.

$$F_P = P_A \frac{\pi d^2}{4} + P_B \frac{\pi(D^2 - d^2)}{4} - P_C \frac{\pi D^2}{4} \qquad (10.11)$$

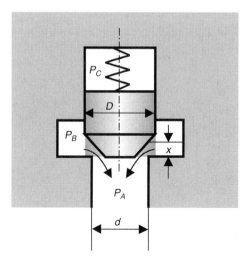

FIGURE 10.3 Operation principle of a puppet type cartridge valve.

The spring force biases the poppet towards closing. When the poppet is in the closed position, the spring force reaches its minimum value. The force increases as the poppet lifts to open the flow passage.

$$F_S = k(x_0 + x) \tag{10.12}$$

The steady-state flow force tends to open the poppet in this valve. The flow force is a function of the flow rate and fluid velocity passing through the valve orifice.

$$F_F = \rho q v \cos \alpha \tag{10.13}$$

The flow control characteristics of a spool valve are similar to those of a cartridge valve and can be described using an orifice equation. The only difference is that spool valve flow-passing area is determined by its wet perimeter, w, and spool displacement, x.

$$q = c_d w x \sqrt{\frac{2}{\rho} \Delta P} \tag{10.14}$$

If the orifice is formed by the edge of the spool and the valve body, the wet perimeter is $w = \pi d$. If the orifice is formed by n slots cut on the spool and the perimeter of each slot is n, the corresponding wet perimeter is $w = nb$. The orifice coefficient for a spool valve normally uses $c_d = 0.60$–0.65.

The forces acting on the spool also include the pressure, spring, and flow forces (Merritt, 1967). The pressure force is either balanced on the spool, because of its symmetric structure in a direct-actuator valve (actuated by a solenoid directly), or the pressure force to actuate the spool movement in a pilot actuated valve. The spring force tends to keep the spool in the central (neutral) position and can be described using Equation 10.12. The flow forces acting on the spool can be calculated using Equation 10.14. The flow velocity angle, α, is normally taken as 69°.

10.4 Hydraulic Pumps

10.4.1 Principles of Pump Operation

The pump is one of the most important components in a hydraulic system because it supplies hydraulic flow to the system. Driven by a prime mover, a hydraulic pump takes the fluid in at atmospheric pressure to fill an expanding volume of space inside the pump through an inlet port and delivers pressurized fluids to the outlet due to the reduction in internal volume near the output port. The pump capacity is determined by pump displacement (D) and operating speed (n). The displacement of a pump is defined as the theoretical volume of fluid that can be delivered in one complete revolution of the pump shaft.

$$Q = Dn \tag{10.15}$$

The pump output pressure is determined by the system load, which is the combined resistance to fluid flow in the pipeline and the resistance to move an external load. Unless the pump flow has egress either by moving a load or by passing through a relief valve back to the reservoir, excessive pressure build-up can cause serious damage to the pump and/or the connecting pipeline (Reed and Larman, 1985).

Based on their ability to change displacement, hydraulic pumps can be categorized as fixed-flow or variable-flow pumps. Based on their design, hydraulic pumps can be categorized as gear pumps, vane pumps, and piston pumps. Normally, gear pumps are fixed-flow pumps, and vane pumps and piston pumps can be either fixed-flow pumps or variable-flow pumps.

The choice of pump design varies from industry to industry. For example, the machine tool manufacturers often select vane pumps because of their low noise, and their capability to deliver a variable flow at a constant pressure. Mobile equipment manufacturers like to use piston pumps due to their high power-to-weight ratio. Some agricultural equipment manufacturers prefer gear pumps for their low cost and robustness (Reed and Larman, 1985), but piston pumps are also popular.

10.4.2 Pump Controls and Systems

Pumps are energy conversion devices that convert mechanical energy into fluid potential energy to drive various hydraulic actuators to do work. To meet the requirements of different applications, there are many types of fluid power system controls from which to choose. The design of the directional control valve must be compatible with the pump design. Normally, an open-center directional control valve is used with a fixed displacement pump and a closed-center directional control valve is used in a circuit equipped with a variable displacement pump.

A fluid power system including a fixed displacement pump and an open-center directional control valve (Figure 10.1) is an open-loop open-center system. Such a system is also called a load-sensitive system because the pump delivers only the pressure required to move the load, plus the pressure drop to overcome line losses. The open-loop open-center system is suitable for simple "on-off" controls. In such operations, the hydraulic actuator either moves the load at the maximum velocity or remains stationary with the pump unloaded. If a proportional valve is used, the open-loop open-center system can also achieve velocity control of the actuator. However, such control will increase the pressure of the extra flow for releasing it back to the tank. Such control causes significant power loss and results in low system efficiency and heat generation.

To solve this problem, an open-loop closed-center circuit is constructed using a variable displacement pump and a closed-center directional control valve. Because a variable displacement pump is commonly equipped with a pressure-limiting control or "pressure compensator," the pump displacement will be automatically increased or decreased as the system pressure decreases or increases. If the metering position of the directional control valve is used to control the actuator velocity, constant velocity can be achieved if the load is constant. However, if the load is changing, the "pressure-compensating" system will not be able to keep a constant velocity without adjusting the metering position of the control valve. To solve this problem, a "load-sensing" pump should be selected for keeping a constant velocity under changing load. The reason for a "load-sensing" pump being able to maintain a constant velocity for any valve-metering position is that it maintains a constant pressure drop across the metering orifice of the directional control valve, and automatically adjusts the pump outlet pressure to compensate for the changes in pressure caused by external load. The constant pressure drop across the valve maintains constant flow, and therefore, constant load velocity.

10.5 Hydraulic Cylinders

A hydraulic cylinder transfers the potential energy of the pressurized fluid into mechanical energy to drive the operating device performing linear motions and is the most common actuator used in hydraulic systems. A hydraulic cylinder consists of a cylinder body, a piston, a rod, and seals. Based on their structure, hydraulic cylinders can be classified as single acting (applying force in one direction only), double acting (exerts force in either direction), single rod (does not have a rod at the cap side), and double rod (has a rod at both sides of the piston) cylinders.

10.5.1 Cylinder Parameters

A hydraulic cylinder transfers energy by converting the flow rate and pressure into the force and velocity. The velocity and the force from a double-acting double-rod cylinder can be determined using the following equations:

$$v = \frac{4q}{\pi(D^2 - d^2)} \tag{10.16}$$

$$F = \frac{\pi}{4}(D^2 - d^2)(P_1 - P_2) \tag{10.17}$$

The velocity and the force from a double-acting single-rod cylinder should be determined differently for extending and retracting motions. In retraction, the velocity can be determined using Equation 10.16, and the force can be determined using the following equation:

$$F = P_1 \frac{\pi(D^2 - d^2)}{4} - P_2 \frac{\pi D^2}{4} \tag{10.18}$$

In extension, the velocity and exerting forces can be determined using the following equations:

$$v = \frac{4q}{\pi D^2} \tag{10.19}$$

$$F = (P_1 - P_2)\frac{\pi D^2}{4} + P_2 \frac{\pi d^2}{4} \tag{10.20}$$

The hydraulic stiffness, k_h, of the cylinder plays an important role in the dynamic performance of a hydraulic system. It is a function of fluid bulk modulus (β), piston areas (A_1, A_2), cylinder chamber volumes (V_1, V_2), and the volume of hydraulic hoses connected to both chambers (V_{L1}, V_{L2}). For a double-acting single-rod cylinder, the stiffness on both sides of the piston acts in parallel (Skinner and Long, 1998). The total stiffness of the cylinder is given by the following equation:

$$k_h = \beta\left(\frac{A_1^2}{V_{L1} + V_1} + \frac{A_1^2}{V_{L2} + V_2}\right) \tag{10.21}$$

The natural frequency, ω_n, of a hydraulic system is determined by the combined mass, m, of the cylinder and the load using the following equation:

$$\omega_n = \sqrt{\frac{k_h}{m}} \tag{10.22}$$

10.6 Fluid Power Systems Control

10.6.1 System Steady-State Characteristics

The steady-state characteristics of a fluid power system determine loading performance, speed control capability, and the efficiency of the system. Modeling a hydraulic system without loss of generality, a system consisting of an open-center four-way directional control valve and a single-rod double acting cylinder is used to analyze the steady-state characteristics of the system (Figure 10.1). In this system, the orifice area of the cylinder-to-tank (C-T) port in the control valve is always larger than that of the pump-to-cylinder (P-C) port. Therefore, it is reasonable to assume that the P-C orifice controls the cylinder speed during extension (Zhang, 2000).

Based on Newton's Law, the force balance on the piston is determined by the head-end chamber pressure, P_1, the head-end piston area, A_1, the rod-end chamber pressure, P_2, the rod-end piston area, A_2, and the external load, F, when the friction and leakage are neglected.

$$P_1 A_1 - P_2 A_2 = F \qquad (10.23)$$

If neglecting the line losses from actuator to reservoir, the rod-end pressure equals zero. Then, the head-end pressure is determined by the external load to the system.

$$P_1 = \frac{F}{A_1} \qquad (10.24)$$

In order to push the fluid passing the control valve and entering the head-end of the cylinder, the discharge pressure, P_P, of the hydraulic pump has to be higher than the cylinder chamber pressure. The difference between the pump discharge pressure and the cylinder chamber pressure is determined by the hydraulic resistance across the control valve. Based on the orifice equation, the flow rate entering the cylinder head-end chamber is

$$q = C_d A_o \sqrt{\frac{2}{\rho}(P_P - P_1)} \qquad (10.25)$$

Using a control coefficient, K, to represent C_d and ρ, the cylinder speed can be described using the following equation:

$$v = \frac{K A_o}{A_1} \sqrt{P_P - \frac{F}{A_1}} \qquad (10.26)$$

Equation 10.13 describes the speed-load relationship of a hydraulic cylinder under a certain fluid passing area (orifice area) of the control valve. Depicted in Figure 10.4, the cylinder speed decreases as the external load applied to the cylinder increases. When there is no external load, the cylinder speed reaches a maximum. Conversely, when the external load researches the valve of $F = P_P A_1$, then the cylinder will stall. The stall load is independent of the size of the fluid passing area in the valve. Such characteristics of a fluid power system eliminate the potential of overloading, which makes it a safer power transmission method.

In system analysis, the speed stiffness, k_v, is often used to describe the consistency of the cylinder speed under changing system load (Li et al., 2000).

$$k_v = -\frac{1}{\frac{v \partial}{\partial F}} = \frac{2(P_P A_1 - F)}{v} \qquad (10.27)$$

Fluid Power Systems

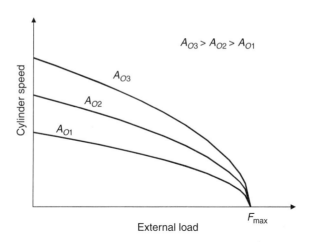

FIGURE 10.4 Hydraulic cylinder load–speed relationship under the same system pressure.

Equation 10.27 indicates that the increase in speed stiffness can be achieved either by increasing the system pressure or the cylinder size, or by decreasing the speed.

10.6.2 System Dynamic Characteristics

To analyze the dynamic characteristics of this hydraulic cylinder actuation system, one can use flow continuity and system momentum equations to model the cylinder motion. Neglecting system leakage, friction, and line loss, the following are the governing equations for the hydraulic system:

$$q = kx\sqrt{P_p - P_1} = A_1 \frac{dy}{dt} + \frac{V_1}{\beta} \frac{dP_1}{dt} \tag{10.28}$$

$$P_1 A_1 = m \frac{d^2 y}{dt^2} + F \tag{10.29}$$

To perform dynamic analysis on this hydraulic system, it is essential to derive its transfer function based on the above nonlinear equation, which can be obtained by taking the Laplace transform on the linearized form of the above equations (Watton, 1989).

$$\delta v(s) = \frac{\frac{k_1 K_i}{A_1} \delta i(s) - \left(\frac{V_1}{A_1^2 \beta} s + \frac{1}{A_1^2 k_3 R_o}\right) \delta F(s)}{\frac{V_1}{A_1^2 \beta} m s^2 + \frac{1}{A_1^2 k_2 R_o} m s + 1} \tag{10.30}$$

Making

$$\omega_n = \sqrt{\frac{A_1^2 \beta}{V_1 m}}, \quad \zeta = \frac{1}{2 k_2 R_o} \sqrt{\frac{m \beta}{V_1 A_1^2}}, \quad \text{and} \quad K_s = \frac{k_1 K_i}{A_1}$$

Equation 10.30 can be represented as

$$\delta v(s) = \frac{K_s \delta i(s)}{\frac{1}{\omega_n^2} s^2 + \frac{2\zeta}{\omega_n} s + 1} - \frac{\frac{1}{A_1^2}\left(\frac{V_1}{\beta} s + \frac{1}{k_3 R_o}\right) \delta F(s)}{\frac{1}{\omega_n^2} s^2 + \frac{2\zeta}{\omega_n} s + 1} \tag{10.31}$$

Based on the stability criterion for a second-order system, it should satisfy

$$\frac{1}{\omega_n^2}s^2 + \frac{2\zeta}{\omega_n}s + 1 = 0 \tag{10.32}$$

The speed control coefficient, K_s, is the gain between the control signal current and the cylinder speed. A higher gain can increase the system sensitivity in speed control.

10.6.3 E/H System Feedforward-Plus-PID Control

Equation 10.31 indicates that the speed control of a hydraulic cylinder is a third-order system. Its dynamic behaviors are affected by spool valve characteristics, system pressure, and cylinder size. Therefore, it is a challenging job to realize accurate and smooth speed control on a hydraulic cylinder. A feedforward plus proportional integral derivative (FPID) controller has proven capable of achieving high-speed control performance of a hydraulic cylinder (Zhang, 1999).

An FPID controller consists of a feedforward loop and a PID loop (Figure 10.5). The feedforward loop is designed to compensate for the nonlinearity of the hydraulic system, including the deadband of the system and the nonlinear flow gain of the control valve. It uses a feedforward gain to determine the basic control input based on demand speed. This feedforward gain is scheduled based on the inverse valve transform, which provides the steady-state control characteristics of the E/H valve in terms of cylinder speed and control-current to valve PWM driver.

The PID loop complements the feedforward control via the speed tracking error compensation. The PID controller is developed based on the transfer function of the linearized system for the hydraulic cylinder speed control system.

$$G(s) = \left(K_P + \frac{K_I}{s} + K_D s\right) \tag{10.33}$$

The robustness of the FPID control was evaluated based on its performance and stability. Performance robustness deals with unexpected external disturbances and stability robustness deals with internal structural or parametric changes in the system. The design of this FPID controller was based on a worst-case scenario of system operating conditions in tuning both the PID gains and the feedforward gain.

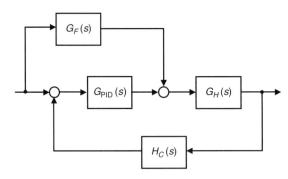

FIGURE 10.5 Schematic block diagram of the feedforward-plus-PID controller. $G_F(s)$ is the feedforward gain, $G_{PID}(s)$ is the overall gain of the feedback PID controller, $G_H(s)$ is hydraulic system gain, and $H_C(s)$ is the sensor gain.

10.6.4 E/H System Generic Fuzzy Control

Fuzzy control is an advanced control technology that can mimic a human's operating strategy in controlling complex systems and can handle systems with uncertainty and nonlinearity (Pedrycz, 1993). One common feature of fuzzy controllers is that most such controllers are designed based on natural language control laws. This feature makes it possible to design a generic controller for different plants if the control of those plants can be described using the same natural language control laws (Zhang, 2001).

The speed control on a hydraulic cylinder actually is achieved by regulating the supplied flow rate to the cylinder. In different hydraulic systems, the size of the cylinder and the capability of hydraulic system are usually different, but the control principles are very similar. Representing cylinder speed control operation, using natural language without loss of generality, the control laws are the same for all systems:

To have a fast motion, open the valve fully.
To make a slow motion, keep the valve open a little.
To hold the cylinder at its current position, return the valve to the center.
To make a reverse motion, operate the valve to the other direction.

This natural language model represents the general roles in controlling the cylinder speed via an E/H control valve on all hydraulic systems. The differences in system parameters on different systems can be handled by redefining the domain of the fuzzy variable, such as fully, a_lot, and a_little, using fuzzy membership functions (Passino and Yurkovich, 1998). This model provides the basis for designing a generic fuzzy controller for E/H systems. The adoption of the generic controller on different systems can be as easy as redefining the fuzzy membership function based on its system parameters.

Figure 10.6 shows the block diagram of a generic fuzzy controller consisting of two input variable fuzzifiers, a control rule base, and a control command defuzzifier. The two input fuzzifiers were designed to convert real-valued input variables into linguistic variables with appropriate fuzzy memberships. Each fuzzifier consists of a set of fuzzy membership functions defining the domain for each linguistic input variable. A real-valued input variable is normally converted into two linguistic values with associated memberships. The definitions of these fuzzy values play a critical role in the design of generic fuzzy controllers and are commonly defined based upon hydraulic system parameters. The fuzzy controller uses fuzzy control rules to determine control actions according to typical behaviors in the speed control of hydraulic cylinders. The control outputs are also linguistic values and associated with fuzzy memberships. For example, if the demanding speed is negative small (NS) and the error in speed was positive small (PS), the appropriate valve control action will be positive small (PS).

The appropriate control actions were determined based on predefined control rules. Since each real-valued variable commonly maps into two fuzzy values, the fuzzy inference engine fires at least two control rules containing these fuzzy values to determine the appropriate control action. Therefore, at least two appropriate fuzzy-valued control actions will be selected. However, the E/H controller can only implement one specific real-value control command at a given time. It is necessary to convert multiple fuzzy-valued control commands into one real-valued control signal in this fuzzy controller.

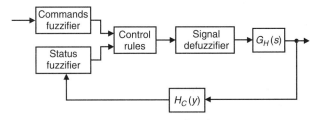

FIGURE 10.6 Block diagram of fuzzy E/H control system. The fuzzy controller consists of input variable fuzzifiers, control rules, and a signal defuzzifier.

The defuzzification process converts two or more fuzzy-valued outputs to one real-valued output. There are many defuzzification methods, such as center of gravity (COG) and center of area (COA), available for different applications (Passino and Yurkovich, 1998). By COA approach, the real-valued control signal, u, was determined by the domain and the memberships of the selected fuzzy control commands, $\mu(u_i)$, using the following equation:

$$u = \frac{\sum_{i=1}^{n} u_i \mu(u_i) du}{\sum_{i=1}^{n} \mu(u_i) du} \qquad (10.34)$$

The COA method naturally averages the domains of selected fuzzy control commands, and thus reduces the sensitivity of the system to noise. The use of a COA approach increased the robustness and accuracy of the control.

The performance of the fuzzy controller depends on the appropriation of domain definition for both input and output fuzzy variables. Properly defined fuzzy variables for a specific E/H system will improve the stability, accuracy, and nonlinearity compensation of the fuzzy controller. Normally, a triangular fuzzy membership function, μ_{FV}, was defined by domain values of a_i, a_j, and a_k, for each fuzzy value (FV) in the fuzzy controller.

$$\mu_A = \begin{bmatrix} \mu_{NL} \\ \mu_{NM} \\ \mu_{NS} \\ \mu_{ZE} \\ \mu_{PS} \\ \mu_{PM} \\ \mu_{PL} \end{bmatrix} = \begin{bmatrix} a_1 & a_1 & a_2 \\ a_1 & a_2 & a_3 \\ a_2 & a_3 & a_4 \\ a_3 & a_4 & a_5 \\ a_4 & a_5 & a_6 \\ a_5 & a_6 & a_7 \\ a_6 & a_7 & a_7 \end{bmatrix} \qquad (10.35)$$

where μ_A is a set of fuzzy membership functions for each fuzzy input or output variable; a_i, a_k are the boundaries; and a_j is the full membership point of the fuzzy value.

Equation 10.35 uses a set of seven domain values to define seven fuzzy values in the real-valued operating range. The tuning of the fuzzy controller was to determine the domain values for each of the fuzzy values. The following vector presents the domains of fuzzy membership functions for a particular variable:

$$A = \{a_1 \ a_2 \ a_3 \ a_4 \ a_5 \ a_6 \ a_7\} \qquad (10.36)$$

10.7 Programmable Electrohydraulic Valves

Proportional directional control valves are by far the most common means for motion control of hydraulic motors or cylinders in fluid power systems (McCloy, 1973). Normally, a proportional direction control valve uses a sliding spool to control the direction and the amount of fluid passing through the valve. For different applications, the spool in a proportional direction control valve is often specially designed to provide the desired control characteristics. As a result, valves are specific and cannot be interchangeable even if they are exactly of the same size. The multiplicity of such specific valves make them inconvenient and costly to manufacture, distribute, and service. To provide a solution to these problems, researchers at the University of Illinois at Urbana-Champaign (Book and Goering, 1999; Hu et al., 2001) developed a generic programmable electrohydraulic (E/H) control valve. A generic programmable valve is a set of individually

Fluid Power Systems

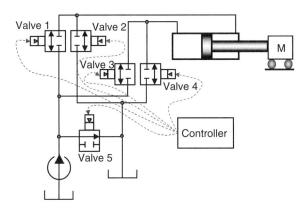

FIGURE 10.7 System schematic of a hydraulic system using generic programmable E/H valves.

controlled E/H valves capable of fulfilling flow and pressure control requirements. One set of such generic valves can replace a proportional direction control valve and other auxiliary valves, such as line release valves, in a circuit.

A generic programmable E/H valve is normally constructed using five bi-directional, proportional flow control sub-valves, three pressure sensors, and an electronic controller. Figure 10.7 shows the schematic of the generic valve circuit. Sub-valves 1 and 2 connect the pump and the head-end or the rod-end chambers of the cylinder and provide equilibrium ports of P-to-A and P-to-B as in a conventional direction control valve, while sub-valves 3 and 4 connect cylinder chambers A or B to the tank and provide equilibrium ports of A-to-T and B-to-T of a direction control valve. Sub-valve 5 connects the pump and the tank directly and provides a dual-function of line release and an equilibrium port of P-to-T of a direction control valve. By controlling the opening and closing of these sub-valves, the basic functions of the generic valve can be realized. In operation, the controller output control signals for each sub-valve are based on a predefined control logic.

With proper logic in the on-off control of all five sub-valves, the generic programmable valve was capable of realizing several basic functions, including open-center, closed-center, float-center, make-up, and pressure release functions. By applying modulation control, the generic valve can realize proportional functions such as meter-in/meter-out, load sensing, regeneration, and anti-cavitation. For example, in a conventional tandem-center or closed-center direction control valve, the ports A and B are normally closed for holding the pressure in cylinder chambers, while the ports P and T are either normally open or closed. To fulfill this function, the generic valve keeps sub-valves 1–4 closed to hold the cylinder chamber pressure, and fully opens sub-valve 5 to bleed the flow back to the tank, either at low pressure (tandem-center function) or when the system pressure exceeds a preset level (closed-center function). In conventional open-center direction control valves, all ports are normally connected. To fulfill this function, the generic valve keeps all sub-valves open. Similarly, to provide float-center function, the generic valve needs to open sub-valves 3 and 4 to release pressure in both the head-end and the rod-end chambers of the cylinder. In both cases, sub-valve 5 will be opened only when the system pressure exceeds a preset level.

It is almost impossible to achieve the regeneration function from a conventional direction control valve. In achieving this function, a generic valve needs to open sub-valves 1 and 2 to lead the returning flow of the rod-end chamber back to the head-end chamber to provide additional flow for increasing the extending speed. Make-up function in a conventional hydraulic system is provided by a separate make-up valve for supplying fluid directly from the tank in case of cavitation. The generic valve can also provide this function by actuating the corresponding cylinder-to-tank sub-valves open when the system pressure is below a certain level.

References

Anderson, W.R., *Controlling Electrohydraulic Systems*, Marcel Dekker, New York, NY, 1988.

Book, R. and Goering, C.E., Programmable electrohydraulic valve, *SAE 1999 Transactions, Journal of Commercial Vehicles* (1997), Section 2, 108:346–352.

Borghi, M.G., Cantore, G., Milani, M., and Paoluzzi, R., Analysis of hydraulic components using computational fluid dynamics models, *Proceedings of the Institution of Mechanical Engineers, Journal C* (1998), 212:619–629.

Hu, H., Zhang, Q., and Alleyne, A., Multi-function realization of a generic programmable E/H valve using flexible control logic, *Proceedings of the Fifth International Conference on Fluid Power Transmission and Control* (2001), International Academic Publishers, Beijing, China, pp. 107–110.

Lambeck, R.P., *Hydraulic Pumps and Motors: Selection and Application for Hydraulic Power Control Systems*, Marcel Dekker, New York, NY, 1983.

Li, Z., Ge, Y., and Chen, Y., *Hydraulic Components and Systems* (in Chinese), Mechanical Industry Publishing, Beijing, China, 2000.

McCloy, D. and Martin, H.R., *The Control of Fluid Power*, John Wiley & Sons, New York, NY, 1973.

Merritt, H.E., *Hydraulic Control Systems*, John Wiley & Sons, New York, NY, 1967.

Passino, K.M. and Yurkovich, S., *Fuzzy Control*, Addition-Wesley, Menlo Park, CA, 1998.

Pedrycz, W., *Fuzzy Control and Fuzzy Systems*, 2nd ed., Wiley, New York, NY, 1993.

Reed, E.W. and Larman, I.S., *Fluid Power with Microprocessor Control: An Introduction*, Prentice-Hall, New York, NY, 1985.

Skinner, S.C. and Long, R.J., *Closed Loop Electrohydraulic Systems Manual*, 2nd ed., Vickers, Rochester Hills, MI, 1998.

Viall, E.N. and Zhang, Q., Determining the discharge coefficient of a spool valve, *Proceedings of the American Control Conference* (2000), Chicago, IL, pp. 3600–3604.

Watton, J., *Fluid Power Systems, Modeling, Simulation, Analog and Microcomputer Control*, Prentice-Hall, New York, NY, 1989.

Welty, J.R., Wicks, C.E., and Wilson, R.E., *Fundamentals of Momentum, Heat, and Mass Transfer*, 3rd ed., John Wiley & Sons, New York, NY, 1984.

Zhang, Q., Hydraulic linear actuator velocity control using a feedforward-plus-PID control, *International Journal of Flexible Automation and Integrated Manufacturing* (1999), 7:275–290.

Zhang, Q., Design of a generic fuzzy controller for electrohydraulic steering, *Proceedings of the American Control Conference* (2001) (in press).

11
Electrical Engineering

11.1	Introduction	11-1
11.2	Fundamentals of Electric Circuits	11-1
	Electric Power and Sign Convention • Circuit Elements and Their i–v Characteristics • Resistance and Ohm's Law • Practical Voltage and Current Sources • Measuring Devices	
11.3	Resistive Network Analysis	11-15
	The Node Voltage Method • The Mesh Current Method • One-Port Networks and Equivalent Circuits • Nonlinear Circuit Elements	
11.4	AC Network Analysis	11-21
	Energy-Storage (Dynamic) Circuit Elements • Time-Dependent Signal Sources • Solution of Circuits Containing Dynamic Elements • Phasors and Impedance	
	References..	11-36

Giorgio Rizzoni
Ohio State University

11.1 Introduction

The role played by electrical and electronic engineering in mechanical systems has dramatically increased in importance in the past two decades, thanks to advances in integrated circuit electronics and in materials that have permitted the integration of sensing, computing, and actuation technology into industrial systems and consumer products. Examples of this integration revolution, which has been referred to as a new field called *mechatronics*, can be found in consumer electronics (auto-focus cameras, printers, microprocessor-controlled appliances), in industrial automation, and in transportation systems, most notably in passenger vehicles. The aim of this chapter is to review and summarize the foundations of electrical engineering for the purpose of providing the practicing mechanical engineer a quick and useful reference to the different fields of electrical engineering. Special emphasis has been placed on those topics that are likely to be relevant to product design.

11.2 Fundamentals of Electric Circuits

This section presents the fundamental laws of circuit analysis and serves as the foundation for the study of electrical circuits. The fundamental concepts developed in these first pages will be called on through the chapter.

The fundamental electric quantity is **charge**, and the smallest amount of charge that exists is the charge carried by an electron, equal to

$$q_e = -1.602 \times 10^{-19} \text{ coulomb} \tag{11.1}$$

As you can see, the amount of charge associated with an electron is rather small. This, of course, has to do with the size of the unit we use to measure charge, the **coulomb** (C), named after Charles Coulomb. However, the definition of the coulomb leads to an appropriate unit when we define electric current, since current consists of the flow of very large numbers of charge particles. The other charge-carrying particle in an atom, the proton, is assigned a positive sign and the same magnitude. The charge of a proton is

$$q_p = +1.602 \times 10^{-19} \text{ coulomb} \tag{11.2}$$

Electrons and protons are often referred to as **elementary charges**.

Electric current is defined as the time rate of change of charge passing through a predetermined area. If we consider the effect of the enormous number of elementary charges actually flowing, we can write this relationship in differential form:

$$i = \frac{dq}{dt} \text{ (C/s)} \tag{11.3}$$

The units of current are called **amperes** (A), where 1 A = 1 C/sec. The electrical engineering convention states that the positive direction of current flow is that of positive charges. In metallic conductors, however, current is carried by negative charges; these charges are the free electrons in the conduction band, which are only weakly attracted to the atomic structure in metallic elements and are therefore easily displaced in the presence of electric fields.

In order for current to flow there must exist a closed circuit. Figure 11.1 depicts a simple circuit, composed of a battery (e.g., a dry-cell or alkaline 1.5-V battery) and a light bulb.

Note that in the circuit of Figure 11.1, the current, i, flowing from the battery to the resistor is equal to the current flowing from the light bulb to the battery. In other words, no current (and therefore no charge) is "lost" around the closed circuit. This principle was observed by the German scientist G.R. Kirchhoff and is now known as **Kirchhoff's current law** (KCL). KCL states that because charge cannot be created but must be conserved, *the sum of the currents at a node must equal zero* (in an electrical circuit, a **node** is the junction of two or more conductors). Formally:

$$\sum_{n=1}^{N} i_n = 0 \quad \text{Kirchhoff's current law} \tag{11.4}$$

The significance of KCL is illustrated in Figure 11.2, where the simple circuit of Figure 11.2 has been augmented by the addition of two light bulbs (note how the two nodes that exist in this circuit have been emphasized by the shaded areas). In applying KCL, one usually defines currents entering a node as being negative and currents exiting the node as being positive. Thus, the resulting expression for the circuit of Figure 11.2 is

$$-i + i_1 + i_2 + i_3 = 0$$

Charge moving in an electric circuit gives rise to a current, as stated in the preceding section. Naturally, it must take some work, or energy, for the charge to move between two points in a circuit, say, from point a to point b. The total *work per unit charge* associated with the motion of charge between two points is called **voltage**. Thus, the units of voltage are those of energy per unit charge:

$$1 \text{ volt} = \frac{1 \text{ joule}}{\text{coulomb}} \tag{11.5}$$

The voltage, or **potential difference**, between two points in a circuit indicates the energy required to move charge from one point to the other. As will be presently shown, the direction, or polarity, of the voltage is closely tied to whether energy is being dissipated or generated in the process. The seemingly abstract concept of work being done in moving charges can be directly applied to the analysis of electrical circuits; consider again the simple circuit consisting of a battery and a light bulb. The circuit is drawn again for convenience in Figure 11.3, and nodes are defined by the letters a and b. A series of carefully conducted experimental observations regarding the nature of voltages in an electric circuit led Kirchhoff

Electrical Engineering

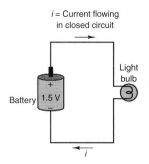

FIGURE 11.1 A simple electrical circuit.

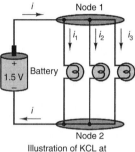

FIGURE 11.2 Illustration of Kirchhoff's current law.

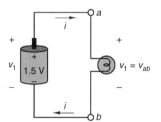

FIGURE 11.3 Voltages around a circuit.

to the formulation of the second of his laws, **Kirchhoff's voltage law**, or KVL. The principle underlying KVL is that no energy is lost or created in an electric circuit; in circuit terms, the sum of all voltages associated with sources must equal the sum of the load voltages, so that *the net voltage around a closed circuit is zero*. If this were not the case, we would need to find a physical explanation for the excess (or missing) energy not accounted for in the voltages around a circuit. KVL may be stated in a form similar to that used for KCL:

$$\sum_{n=1}^{N} v_n = 0 \quad \text{Kirchhoff's voltage law} \qquad (11.6)$$

where the v_n are the individual voltages around the closed circuit. Making reference to Figure 11.3, we can see that it must follow from KVL that the work generated by the battery is equal to the energy dissipated in the light bulb to sustain the current flow and to convert the electric energy to heat and light:

$$v_{ab} = -v_{ba}$$

or

$$v_1 = v_2$$

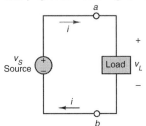

FIGURE 11.4 Sources and loads in an electrical circuit.

One may think of the work done in moving a charge from point *a* to point *b* and the work done moving it back from *b* to *a* as corresponding directly to the *voltages across individual circuit elements*. Let Q be the total charge that moves around the circuit per unit time, giving rise to the current i. Then the work done in moving Q from *b* to *a* (i.e., across the battery) is

$$W_{ba} = Q \times 1.5 \text{ V} \tag{11.7}$$

Similarly, work is done in moving Q from *a* to *b*, that is, across the light bulb. Note that the word *potential* is quite appropriate as a synonym of voltage, in that voltage represents the potential energy between two points in a circuit: if we remove the light bulb from its connections to the battery, there still exists a voltage across the (now disconnected) terminals *b* and *a*.

A moment's reflection upon the significance of voltage should suggest that it must be necessary to specify a sign for this quantity. Consider, again, the same dry-cell or alkaline battery, where, by virtue of an electrochemically induced separation of charge, a 1.5-V potential difference is generated. The potential generated by the battery may be used to move charge in a circuit. The rate at which charge is moved once a closed circuit is established (i.e., the current drawn by the circuit connected to the battery) depends now on the circuit element we choose to connect to the battery. Thus, while the voltage across the battery represents the potential for *providing energy* to a circuit, the voltage across the light bulb indicates the amount of work done in *dissipating energy*. In the first case, energy is generated; in the second, it is consumed (note that energy may also be stored, by suitable circuit elements yet to be introduced). This fundamental distinction required attention in defining the sign (or polarity) of voltages.

We shall, in general, refer to elements that provide energy as **sources**, and to elements that dissipate energy as **loads**. Standard symbols for a generalized source-and-load circuit are shown in Figure 11.4. Formal definitions will be given in a later section.

11.2.1 Electric Power and Sign Convention

The definition of voltage as work per unit charge lends itself very conveniently to the introduction of power. Recall that power is defined as the work done per unit time. Thus, the power, P, either generated or dissipated by a circuit element can be represented by the following relationship:

$$\text{Power} = \frac{\text{work}}{\text{time}} = \frac{\text{work}}{\text{unit charge}} \frac{\text{charge}}{\text{time}} = \text{voltage} \times \text{current} \tag{11.8}$$

Thus, the electrical power generated by an active element, or that dissipated or stored by a passive element, is equal to the product of the voltage across the element and the current flowing through it.

$$P = VI \tag{11.9}$$

It is easy to verify that the units of voltage (joules/coulomb) times current (coulombs/second) are indeed those of power (joules/second, or watts).

Electrical Engineering

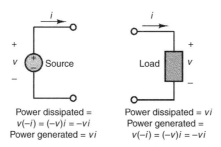

FIGURE 11.5 The passive sign convention.

It is important to realize that, just like voltage, power is a signed quantity, and that it is necessary to make a distinction between *positive* and *negative power*. This distinction can be understood with reference to Figure 11.5, in which a source and a load are shown side by side. The polarity of the voltage across the source and the direction of the current through it indicate that the voltage source *is doing work in moving charge from a lower potential to a higher potential*. On the other hand, the load is dissipating energy, because the direction of the current indicates that *charge is being displaced from a higher potential to a lower potential*. To avoid confusion with regard to the sign of power, the electrical engineering community uniformly adopts the **passive sign convention**, which simply states that *the power dissipated by a load is a positive quantity* (or, conversely, that the power generated by a source is a positive quantity). Another way of phrasing the same concept is to state that if current flows from a higher to a lower voltage (+ to −), the power dissipated will be a positive quantity.

11.2.2 Circuit Elements and Their i–v Characteristics

The relationship between current and voltage at the terminals of a circuit element defines the behavior of that element within the circuit. In this section, we shall introduce a graphical means of representing the terminal characteristics of circuit elements. Figure 11.6 depicts the representation that will be employed throughout the chapter to denote a generalized circuit element: the variable i represents the current flowing through the element, while v is the potential difference, or voltage, across the element.

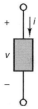

FIGURE 11.6 Generalized representation of circuit elements.

Suppose now that a known voltage were imposed across a circuit element. The current that would flow as a consequence of this voltage, and the voltage itself, form a unique pair of values. If the voltage applied to the element were varied and the resulting current measured, it would be possible to construct a functional relationship between voltage and current known as the i–v **characteristic** (or **volt–ampere characteristic**). Such a relationship defines the circuit element, in the sense that if we impose any prescribed voltage (or current), the resulting current (or voltage) is directly obtainable from the i–v characteristic. A direct consequence is that the power dissipated (or generated) by the element may also be determined from the i–v curve.

The i–v characteristics of ideal current and voltage sources can also be useful in visually representing their behavior. An ideal voltage source generates a prescribed voltage independent of the current drawn from the load; thus, its i–v characteristic is a straight vertical line with a voltage axis intercept corresponding to the source voltage. Similarly, the i–v characteristic of an ideal current source is a horizontal line with a current axis intercept corresponding to the source current. Figure 11.7 depicts this behavior.

11.2.3 Resistance and Ohm's Law

When electric current flows through a metal wire or through other circuit elements, it encounters a certain amount of **resistance**, the magnitude of which depends on the electrical properties of the material. Resistance to the flow of current may be undesired, for example, in the case of lead wires and connection

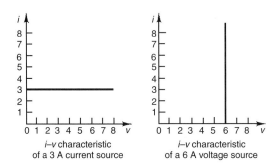

FIGURE 11.7 *i–v* characteristics of ideal sources.

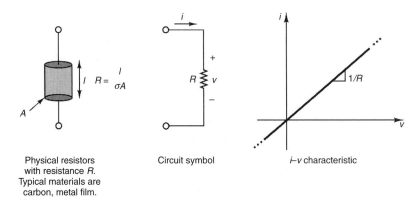

FIGURE 11.8 The resistance element.

cable—or it may be exploited in an electrical circuit in a useful way. Nevertheless, practically all circuit elements exhibit some resistance; as a consequence, current flowing through an element will cause energy to be dissipated in the form of heat. An ideal **resistor** is a device that exhibits linear resistance properties according to Ohm's law, which states that

$$V = IR \tag{11.10}$$

that is, that the voltage across an element is directly proportional to the current flow through it. R is the value of the resistance in units of ohms (Ω), where

$$1\ \Omega = 1\ \text{V/A} \tag{11.11}$$

The resistance of a material depends on a property called **resistivity**, denoted by the symbol ρ; the inverse of resistivity is called **conductivity** and is denoted by the symbol σ. For a cylindrical resistance element (shown in Figure 11.8), the resistance is proportional to the length of the sample, l, and inversely proportional to its cross-sectional area, A, and conductivity, σ.

$$v = \frac{1}{\sigma A} i \tag{11.12}$$

It is often convenient to define the **conductance** of a circuit element as the inverse of its resistance. The symbol used to denote the conductance of an element is G, where

$$G = \frac{1}{R}\ \text{siemens (S)}, \quad \text{where } 1\ \text{S} = 1\ \text{A/V} \tag{11.13}$$

TABLE 11.1 Common Resistor Values ($\frac{1}{8}$-, $\frac{1}{4}$-, $\frac{1}{2}$-, 1-, 2-W Rating)

Ω	Code	Ω	Multiplier	kΩ	Multiplier	kΩ	Multiplier	kΩ	Multiplier
10	Brn-blk-blk	100	Brown	1.0	Red	10	Orange	100	Yellow
12	Brn-red-blk	120	Brown	1.2	Red	12	Orange	120	Yellow
15	Brn-grn-blk	150	Brown	1.5	Red	15	Orange	150	Yellow
18	Brn-gry-blk	180	Brown	1.8	Red	18	Orange	180	Yellow
22	Red-red-blk	220	Brown	2.2	Red	22	Orange	220	Yellow
27	Red-vlt-blk	270	Brown	2.7	Red	27	Orange	270	Yellow
33	Org-org-blk	330	Brown	3.3	Red	33	Orange	330	Yellow
39	Org-wht-blk	390	Brown	3.9	Red	39	Orange	390	Yellow
47	Ylw-vlt-blk	470	Brown	4.7	Red	47	Orange	470	Yellow
56	Grn-blu-blk	560	Brown	5.6	Red	56	Orange	560	Yellow
68	Blu-gry-blk	680	Brown	6.8	Red	68	Orange	680	Yellow
82	Gry-red-blk	820	Brown	8.2	Red	82	Orange	820	Yellow

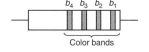

Color bands

Black	0	Blue	6
Brown	1	Violet	7
Red	2	Gray	8
Orange	3	White	9
Yellow	4	Silver	10%
Green	5	Gold	5%

Resistor value = $(b_1 b_2) \times 10^{b_3}$;
b_4 = % tolerance in actual value

FIGURE 11.9 Resistor color code.

Thus, Ohm's law can be rested in terms of conductance, as

$$I = GV \tag{11.14}$$

Ohm's law is an empirical relationship that finds widespread application in electrical engineering because of its simplicity. It is, however, only an approximation of the physics of electrically conducting materials. Typically, the linear relationship between voltage and current in electrical conductors does not apply at very high voltages and currents. Further, not all electrically conducting materials exhibit linear behavior even for small voltages and currents. It is usually true, however, that for some range of voltages and currents, most elements display a linear *i–v characteristic*.

The typical construction and the circuit symbol of the resistor are shown in Figure 11.8. Resistors made of cylindrical sections of carbon (with resistivity $\rho = 3.5 \times 10^{-5}$ Ωm) are very common and are commercially available in a wide range of values for several power ratings (as will be explained shortly). Another commonly employed construction technique for resistors employs metal film. A common power rating for resistors used in electronic circuits (e.g., in most consumer electronic appliances such as radios and television sets) is $\frac{1}{4}$ W. Table 11.1 lists the standard values for commonly used resistors and the color code associated with these values (i.e., the common combinations of the digits $b_1 b_2 b_3$ as defined in Figure 11.9. For example, if the first three color bands on a resistor show the colors red ($b_1 = 2$), violet ($b_2 = 7$), and yellow ($b_3 = 4$), the resistance value can be interpreted as follows:

$$R = 27 \times 10^4 = 270,000 \text{ Ω} = 270 \text{ kΩ}$$

In Table 11.1, the leftmost column represents the complete color code; columns to the right of it only show the third color, since this is the only one that changes. For example, a 10-Ω resistor has the code brown-black-*black*, while a 100-Ω resistor has brown-black-*brown*.

In addition to the resistance in ohms, the maximum allowable power dissipation (or **power rating**) is typically specified for commercial resistors. Exceeding this power rating leads to overheating and can cause the resistor to literally start on fire. For a resistor R, the power dissipated is given by

$$P = VI = I^2 R = \frac{V^2}{R} \tag{11.15}$$

That is, the power dissipated by a resistor is proportional to the square of the current flowing through it, as well as the square of the voltage across it. The following example illustrates a common engineering application of resistive elements: the resistance strain gauge.

Example 11.1 Resistance Strain Gauges

A common application of the resistance concept to engineering measurements is the resistance **strain gauge**. Strain gauges are devices that are bonded to the surface of an object, and whose resistance varies as a function of the surface strain experienced by the object. Strain gauges may be used to perform measurements of strain, stress, force, torque, and pressure. Recall that the resistance of a cylindrical conductor of cross-sectional area A, length L, and conductivity σ is given by the expression

$$R = \frac{L}{\sigma A}$$

If the conductor is compressed or elongated as a consequence of an external force, its dimensions will change, and with them its resistance. In particular, if the conductor is stretched, its cross-sectional area will decrease and the resistance will increase. If the conductor is compressed, its resistance decreases, since the length, L, will decrease. The relationship between change in resistance and change in length is given by the gauge factor, G, defined by

$$G = \frac{\Delta R / R}{\Delta L / L}$$

and since the strain ε is defined as the fractional change in length of an object by the formula

$$\varepsilon = \frac{\Delta L}{L}$$

the change in resistance due to an applied strain ε is given by the expression

$$\Delta R = R_0 G \varepsilon$$

where R_0 is the resistance of the strain gauge under no strain and is called the zero strain resistance. The value of G for resistance strain gauges made of metal foil is usually about 2.

Figure 11.10 depicts a typical foil strain gauge. The maximum strain that can be measured by a foil gauge is about 0.4–0.5%; that is, $\Delta L/L = 0.004$ to 0.005. For a 120-Ω gauge, this corresponds to a change in resistance of the order of 0.96–1.2 Ω. Although this change in resistance is very small, it can be detected by means of suitable circuitry. Resistance strain gauges are usually connected in a circuit called the Wheatstone bridge, which we analyze later in this section.

11.2.3.1 Open and Short Circuits

Two convenient idealizations of the resistance element are provided by the limiting cases of Ohm's law as the resistance of a circuit element approaches zero or infinity. A circuit element with resistance approaching zero is called a **short circuit**. Intuitively, one would expect a short circuit to allow for unimpeded flow of current. In fact, metallic conductors (e.g., short wires of large diameter) approximate the behavior of a short circuit. Formally, a short circuit is defined as a circuit element across which the voltage is zero, regardless of the current flowing through it. Figure 11.11 depicts the circuit symbol for an ideal short circuit.

TABLE 11.2 Resistance of Copper Wire

AWG Size	Number of Strands	Diameter per Strand	Resistance per 1000 ft (Ω)
24	Solid	0.0201	28.4
24	7	0.0080	28.4
22	Solid	0.0254	18.0
22	7	0.0100	19.0
20	Solid	0.0320	11.3
20	7	0.0126	11.9
18	Solid	0.0403	7.2
18	7	0.0159	7.5
16	Solid	0.0508	4.5
16	19	0.0113	4.7

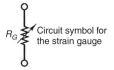

Metal-foil resistance strain gauge. The foil is formed by a photo-etching process and is less than 0.00002 in. thick. Typical resistance values are 120, 350, and 1000 Ω. The wide areas are bonding pads for electrical connections.

FIGURE 11.10 The resistance strain gauge.

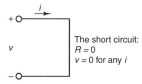

FIGURE 11.11 The short circuit.

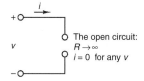

FIGURE 11.12 The open circuit.

Physically, any wire or other metallic conductor will exhibit some resistance, though small. For practical purposes, however, many elements approximate a short circuit quite accurately under certain conditions. For example, a large-diameter copper pipe is effectively a short circuit in the context of a residential electrical power supply, while in a low-power microelectronic circuit (e.g., an FM radio) a short length of 24 gauge wire (refer to Table 11.2 for the resistance of 24 gauge wire) is a more than adequate short circuit.

A circuit element whose resistance approaches infinity is called an **open circuit**. Intuitively, one would expect no current to flow through an open circuit, since it offers infinite resistance to any current. In an open circuit, we would expect to see zero current regardless of the externally applied voltage. Figure 11.12 illustrates this idea.

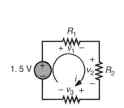
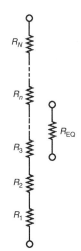

The current *i* flows through each of the four series elements. Thus, by KVL,

$$1.5 = v_1 + v_2 + v_3$$

N Series resistors are equivalent to a single resistor equal to the sum of the individual resistances.

FIGURE 11.13 Voltage divider rule.

In practice, it is not too difficult to approximate an open circuit; any break in continuity in a conducting path amounts to an open circuit. The idealization of the open circuit, as defined in Figure 11.12, does not hold, however, for very high voltages. The insulating material between two insulated terminals will break down at a sufficiently high voltage. If the insulator is air, ionized particles in the neighborhood of the two conducting elements may lead to the phenomenon of arcing; in other words, a pulse of current may be generated that momentarily jumps a gap between conductors (thanks to this principle, we are able to ignite the air–fuel mixture in a spark-ignition internal combustion engine by means of spark plugs). The ideal open and short circuits are useful concepts and find extensive use in circuit analysis.

11.2.3.2 Series Resistors and the Voltage Divider Rule

Although electrical circuits can take rather complicated forms, even the most involved circuits can be reduced to combinations of circuit elements *in parallel* and *in series*. Thus, it is important that you become acquainted with parallel and series circuits as early as possible, even before formally approaching the topic of network analysis. Parallel and series circuits have a direct relationship with Kirchhoff's laws. The objective of this section and the next is to illustrate two common circuits based on series and parallel combinations of resistors: the voltage and current dividers. These circuits form the basis of all network analysis; it is therefore important to master these topics as early as possible.

For an example of a series circuit, refer to the circuit of Figure 11.13, where a battery has been connected to resistors R_1, R_2, and R_3. The following definition applies.

Definition

Two or more circuit elements are said to be **in series** if the same current flows through each of the elements.

The three resistors could thus be replaced by a single resistor of value R_{EQ} without changing the amount of current required of the battery. From this result we may extrapolate to the more general relationship defining the equivalent resistance of N series resistors:

$$R_{EQ} = \sum_{n=1}^{N} R_n \tag{11.16}$$

which is also illustrated in Figure 11.13. A concept very closely tied to series resistors is that of the **voltage divider**.

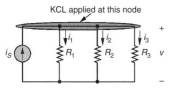
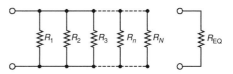

The voltage *v* appears across each parallel element; by KCL, $i_S = i_1 + i_2 + i_3$

N resistors in parallel are equivalent to a single equivalent resistor with resistance equal to the inverse of the sum of the inverse resistances.

FIGURE 11.14 Parallel circuits.

The general form of the voltage divider rule for a circuit with *N* series resistors and a voltage source is

$$v_n = \frac{R_n}{R_1 + R_2 + \cdots + R_n + \cdots + R_N} v_S \qquad (11.17)$$

11.2.3.3 Parallel Resistors and the Current Divider Rule

A concept analogous to that of the voltage may be developed by applying Kirchhoff's current law to a circuit containing only parallel resistances.

Definition

Two or more circuit elements are said to be **in parallel** if the same voltage appears across each of the elements (see Figure 11.14).

N resistors in parallel act as a single equivalent resistance, R_{EQ}, given by the expression

$$\frac{1}{R_{EQ}} = \frac{1}{R_1} + \frac{1}{R_2} + \cdots + \frac{1}{R_N} \qquad (11.18)$$

or

$$R_{EQ} = \frac{1}{1/R_1 + 1/R_2 + \cdots + 1/R_N} \qquad (11.19)$$

Very often in the remainder of this book we shall refer to the parallel combination of two or more resistors with the following notation:

$$R_1 \parallel R_2 \parallel \cdots$$

where the symbol $\parallel$ signifies "in parallel with."

The general expression for the current divider for a circuit with *N* parallel resistors is the following:

$$i_n = \frac{1/R_n}{1/R_1 + 1/R_2 + \cdots + 1/R_n + \cdots + 1/R_N} i_S \quad \text{Current divider} \qquad (11.20)$$

Example 11.2 The Wheatstone Bridge

The **Wheatstone bridge** is a resistive circuit that is frequently encountered in a variety of measurement circuits. The general form of the bridge is shown in Figure 11.15a, where R_1, R_2, and R_3 are known, while R_x is an unknown resistance, to be determined. The circuit may also be redrawn as shown in Figure 11.15b. The latter circuit will be used to demonstrate the use of the voltage divider rule in a mixed series-parallel circuit.

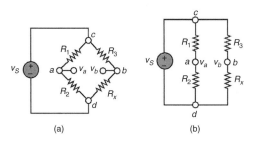

FIGURE 11.15 Wheatstone bridge circuits.

The objective is to determine the unknown resistance R_x.

1. Find the value of the voltage $v_{ad} = v_{ad} - v_{bd}$ in terms of the four resistances and the source voltage, v_S. Note that since the reference point d is the same for both voltages, we can also write $v_{ab} = v_a - v_b$.
2. If $R_1 = R_2 = R_3 = 1\ \mathrm{k}\Omega$, $v_S = 12$ V, and $v_{ab} = 12$ mV, what is the value of R_x?

Solution

1. First, we observe that the circuit consists of the parallel combination of three subcircuits: the voltage source, the series combination of R_1 and R_2, and the series combination of R_3 and R_x. Since these three subcircuits are in parallel, the same voltage will appear across each of them, namely, the source voltage, v_S.

   Thus, the source voltage divides between each resistor pair, R_1–R_2 and R_3–R_x, according to the voltage divider rule: v_a is the fraction of the source voltage appearing across R_2, while v_b is the voltage appearing across R_x:

$$v_a = v_S \frac{R_2}{R_1 + R_2} \quad \text{and} \quad v_b = v_S \frac{R_x}{R_3 + R_x}$$

   Finally, the voltage difference between points a and b is given by

$$v_{ab} = v_a - v_b = v_S \left(\frac{R_2}{R_1 + R_2} - \frac{R_x}{R_3 + R_x} \right)$$

   This result is very useful and quite general, and it finds application in numerous practical circuits.
2. In order to solve for the unknown resistance, we substitute the numerical values in the preceding equation to obtain

$$0.012 = 12 \left(\frac{1000}{2000} - \frac{R_x}{1000 + R_x} \right)$$

   which may be solved for R_x to yield

$$R_x = 996\ \Omega$$

11.2.4 Practical Voltage and Current Sources

Idealized models of voltage and current sources fail to take into consideration the finite-energy nature of practical voltage and current sources. The objective of this section is to extend the ideal models to models that are capable of describing the physical limitations of the voltage and current sources used in practice. Consider, for example, the model of an ideal voltage source. As the load resistance (R) decreases, the source is required to provide increasing amounts of current to maintain the voltage $v_S(t)$ across its terminal:

$$i(t) = \frac{v_S(t)}{R} \qquad (11.21)$$

This circuit suggests that the ideal voltage source is required to provide an infinite amount of current to the load, in the limit as the load resistance approaches zero.

Figure 11.16 depicts a model for a practical voltage source; this is composed of an ideal voltage source, v_S, in series with a resistance, r_S. The resistance r_S in effect poses a limit to the maximum current the voltage source can provide:

$$i_{S\,max} = \frac{v_S}{r_S} \qquad (11.22)$$

It should be apparent that a desirable feature of an ideal voltage source is a very small internal resistance, so that the current requirements of an arbitrary load may be satisfied.

A similar modification of the ideal current source model is useful to describe the behavior of a practical current source. The circuit illustrated in Figure 11.17 depicts a simple representation of a practical current source, consisting of an ideal source in parallel with a resistor. Note that as the load resistance approaches infinity (i.e., an open circuit), the output voltage of the current source approaches its limit,

$$v_{S\,max} = i_S r_S \qquad (11.23)$$

A good current source should be able to approximate the behavior of an ideal current source. Therefore, a desirable characteristic for the internal resistance of a current source is that it be as large as possible.

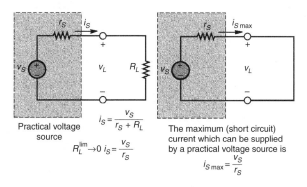

FIGURE 11.16 Practical voltage source.

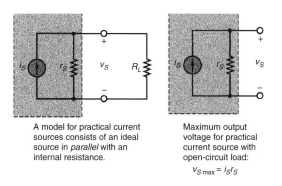

FIGURE 11.17 Practical current source.

11.2.5 Measuring Devices

11.2.5.1 The Ammeter

The **ammeter** is a device that, when connected in series with a circuit element, can measure the current flowing through the element. Figure 11.18 illustrates this idea. From Figure 11.18, two requirements are evident for obtaining a correct measurement of current:

1. The ammeter must be placed in series with the element whose current is to be measured (e.g., resistor R_2).
2. The ammeter should not resist the flow of current (i.e., cause a voltage drop), or else it will not be measuring the true current flowing the circuit. *An ideal ammeter has zero internal resistance.*

11.2.5.2 The Voltmeter

The **voltmeter** is a device that can measure the voltage across a circuit element. Since voltage is the difference in potential between two points in a circuit, the voltmeter needs to be connected across the element whose voltage we wish to measure. A voltmeter must also fulfill two requirements:

1. The voltmeter must be placed in parallel with the element whose voltage it is measuring.
2. The voltmeter should draw no current away from the element whose voltage it is measuring, or else it will not be measuring the true voltage across that element. Thus, *an ideal voltmeter has infinite internal resistance.*

Figure 11.19 illustrates these two points.

Once again, the definitions just stated for the ideal voltmeter and ammeter need to be augmented by considering the practical limitations of the devices. A practical ammeter will contribute some series resistance to the circuit in which it is measuring current; a practical voltmeter will not act as an ideal open circuit but will always draw some current from the measured circuit. Figure 11.20 depicts the circuit models for the practical ammeter and voltmeter.

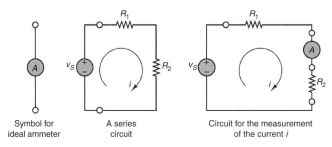

FIGURE 11.18 Measurement of current.

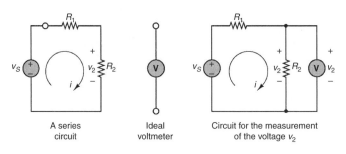

FIGURE 11.19 Measurement of voltage.

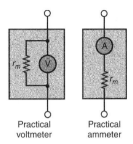

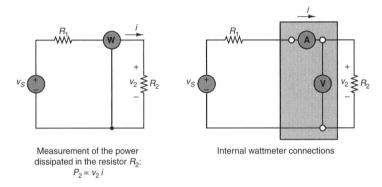

FIGURE 11.20 Models for practical ammeter and voltmeter.

FIGURE 11.21 Measurement of power.

All of the considerations that pertain to practical ammeters and voltmeters can be applied to the operation of a **wattmeter**, a measuring instrument that provides a measurement of the power dissipated by a circuit element, since the wattmeter is in effect made up of a combination of a voltmeter and an ammeter.

Figure 11.21 depicts the typical connection of a wattmeter in the same series circuit used in the preceding paragraphs. In effect, the wattmeter measures the current flowing through the load and, simultaneously, the voltage across it multiplies the two to provide a reading of the power dissipated by the load.

11.3 Resistive Network Analysis

This section will illustrate the fundamental techniques for the analysis of resistive circuits. The methods introduced are based on Kirchhoff's and Ohm's laws. The main thrust of the section is to introduce and illustrate various methods of circuit analysis that will be applied throughout the book.

11.3.1 The Node Voltage Method

Node voltage analysis is the most general method for the analysis of electrical circuits. In this section, its application to linear resistive circuits will be illustrated. The **node voltage method** is based on defining the voltage at each node as an independent variable. One of the nodes is selected as a **reference node** (usually—but not necessarily—ground), and each of the other node voltages is referenced to this node. Once each node voltage is defined, Ohm's law may be applied between any two adjacent nodes in order to determine the current flowing in each branch. In the node voltage method, *each branch current is expressed in terms of one or more node voltages*; thus, currents do not explicitly enter into the equations. Figure 11.22 illustrates how one defines branch currents in this method.

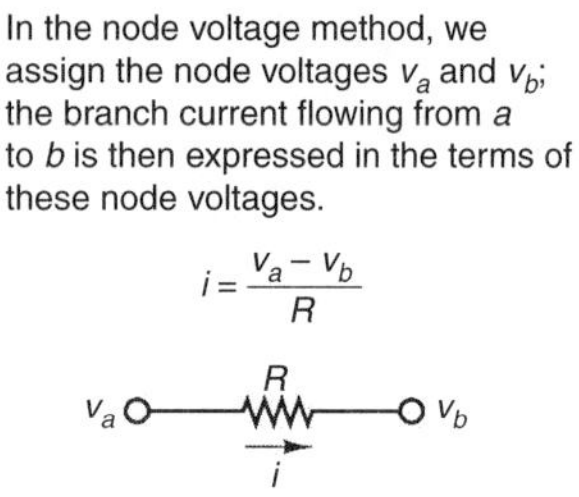

FIGURE 11.22 Branch current formulation in nodal analysis.

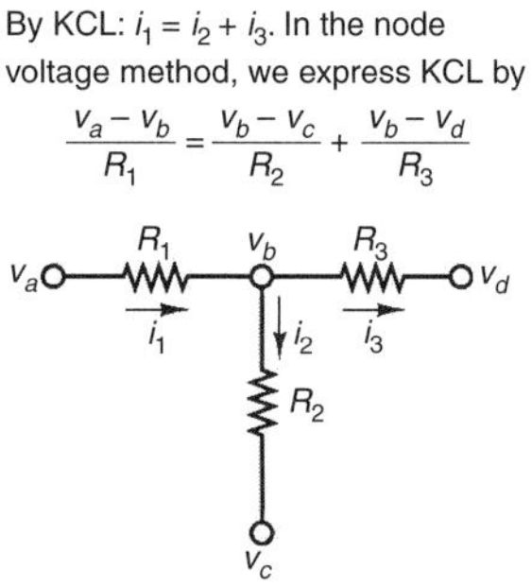

In the node voltage method, we assign the node voltages v_a and v_b; the branch current flowing from a to b is then expressed in the terms of these node voltages.

$$i = \frac{v_a - v_b}{R}$$

By KCL: $i_1 = i_2 + i_3$. In the node voltage method, we express KCL by

$$\frac{v_a - v_b}{R_1} = \frac{v_b - v_c}{R_2} + \frac{v_b - v_d}{R_3}$$

FIGURE 11.23 Use of KCL in nodal analysis.

Once each branch current is defined in terms of the node voltages, Kirchhoff's current law is applied at each node. The particular form of KCL employed in the nodal analysis equates the sum of the currents into the node to the sum of the currents leaving the node:

$$\sum i_{in} = \sum i_{out} \qquad (11.24)$$

Figure 11.23 illustrates this procedure.

The systematic application of this method to a circuit with n nodes would lead to writing n linear equations. However, one of the node voltages is the reference voltage and is therefore already known, since it is usually assumed to be zero. Thus, we can write $n - 1$ *independent linear equations* in the $n - 1$ independent variables (the node voltages). Nodal analysis provides the minimum number of equations required to solve the circuit, since any branch voltage or current may be determined from knowledge of nodal voltages.

The nodal analysis method may also be defined as a sequence of steps, as outlined below.

11.3.1.1 Node Voltage Analysis Method

1. Select a reference node (usually ground). All other node voltages will be referenced to this node.
2. Define the remaining $n - 1$ node voltages as the independent variables.
3. Apply KCL at each of the $n - 1$ nodes, expressing each current in terms of the adjacent node voltages.
4. Solve the linear system of $n - 1$ equations in $n - 1$ unknowns.

In a circuit containing n nodes we can write at most $n - 1$ independent equations.

11.3.2 The Mesh Current Method

In the mesh current method, we observe that a current flowing through a resistor in a specified direction defines the polarity of the voltage across the resistor, as illustrated in Figure 11.24, and that the sum of the voltages around a closed circuit must equal zero, by KVL. Once a convention is established regarding the direction of current flow around a mesh, simple application of KVL provides the desired equation. Figure 11.25 illustrates this point.

Electrical Engineering 11-17

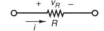

FIGURE 11.24 Basic principle of mesh analysis.

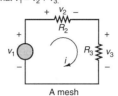

FIGURE 11.25 Use of KVL in mesh analysis.

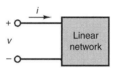

FIGURE 11.26 One-port network.

The number of equations one obtains by this technique is equal to the number of meshes in the circuit. All branch currents and voltages may subsequently be obtained from the mesh currents, as will presently be shown. Since meshes are easily identified in a circuit, this method provides a very efficient and systematic procedure for the analysis of electrical circuits. The following section outlines the procedure used in applying the mesh current method to a linear circuit.

11.3.2.1 Mesh Current Analysis Method

1. Define each mesh current consistently. We shall always define mesh currents clockwise, for convenience.
2. Apply KVL around each mesh, expressing each voltage in terms of one or more mesh currents.
3. Solve the resulting linear system of equations with mesh currents as the independent variables.

In mesh analysis, it is important to be consistent in choosing the direction of current flow. To avoid confusion in writing the circuit equations, mesh currents will be defined exclusively clockwise when we are using this method.

11.3.3 One-Port Networks and Equivalent Circuits

This general circuit representation is shown in Figure 11.26. This configuration is called a **one-port network** and is particularly useful for introducing the notion of equivalent circuits. Note that the network of Figure 11.26 is completely described by its i–v characteristic.

11.3.3.1 Thévenin and Norton Equivalent Circuits

This section discusses one of the most important topics in the analysis of electrical circuits: the concept of an **equivalent circuit**. It will be shown that it is always possible to view even a very complicated circuit in terms of much simpler *equivalent* source and load circuits, and that the transformations leading to equivalent circuits are easily managed, with a little practice. In studying node voltage and mesh current analysis, you may have observed that there is a certain correspondence (called **duality**) between current sources and voltage sources, on the one hand, and parallel and series circuits, on the other. This duality appears again very clearly in the analysis of equivalent circuits: it will shortly be shown that equivalent circuits fall into one of two classes, involving either voltage or current sources and (respectively) either

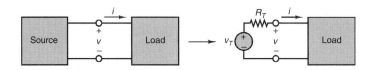

FIGURE 11.27 Illustration of Thévenin theorem.

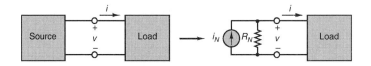

FIGURE 11.28 Illustration of Norton theorem.

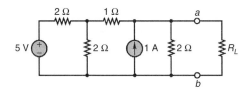

FIGURE 11.29 Computation of Thévenin resistance.

series or parallel resistors, reflecting this same principle of duality. The discussion of equivalent circuits begins with the statement of two very important theorems, summarized in Figures 11.27 and 11.28.

The Thévenin Theorem
As far as a load is concerned, any network composed of ideal voltage and current sources, and of linear resistors, may be represented by an equivalent circuit consisting of an ideal voltage source, v_T, in series with an equivalent resistance, R_T.

The Norton Theorem
As far as a load is concerned, any network composed of ideal voltage and current sources, and of linear resistors, may be represented by an equivalent circuit consisting of an ideal current source, i_N, in parallel with an equivalent resistance, R_N.

11.3.3.2 Determination of Norton or Thévenin Equivalent Resistance

The first step in computing a Thévenin or Norton equivalent circuit consists of finding the equivalent resistance presented by the circuit at its terminals. This is done by setting all sources in the circuit equal to zero and computing the effective resistance between terminals. The voltage and current sources present in the circuit are set to zero as follows: voltage sources are replaced by short circuits, current sources by open circuits. We can produce a set of simple rules as an aid in the computation of the Thévenin (or Norton) equivalent resistance for a linear resistive circuit.

Computation of Equivalent Resistance of a One-Port Network:

1. Remove the load.
2. Zero all voltage and current sources.
3. Compute the total resistance between load terminals, *with the load removed*. This resistance is equivalent to that which would be encountered by a current source connected to the circuit in place of the load.

For example, the equivalent resistance of the circuit of Figure 11.29 as seen by the load is:

$$R_{eq} = ((2\|2) + 1)\|2 = 1\ \Omega$$

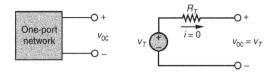

FIGURE 11.30 Equivalence of open-circuit and Thévenin voltage.

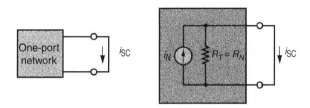

FIGURE 11.31 Illustration of Norton equivalent circuit.

11.3.3.3 Computing the Thévenin Voltage

The Thévenin equivalent voltage is defined as follows: the equivalent (Thévenin) source voltage is equal to the **open-circuit voltage** present at the load terminals with the load removed.

This states that in order to compute v_T, it is sufficient to remove the load and to compute the open-circuit voltage at the one-port terminals. Figure 11.30 illustrates that the open-circuit voltage, v_{OC}, and the Thévenin voltage, v_T, must be the same if the Thévenin theorem is to hold. This is true because in the circuit consisting of v_T and R_T, the voltage v_{OC} must equal v_T, since no current flows through R_T and therefore the voltage across R_T is zero. Kirchhoff's voltage law confirms that

$$v_T = R_T(0) + v_{OC} = v_{OC} \qquad (11.25)$$

11.3.3.4 Computing the Norton Current

The computation of the Norton equivalent current is very similar in concept to that of the Thévenin voltage. The following definition will serve as a starting point.

Definition

The Norton equivalent current is equal to the **short-circuit current** that would flow were the load replaced by a short circuit.

An explanation for the definition of the Norton current is easily found by considering, again, an arbitrary one-port network, as shown in Figure 11.31, where the one-port network is shown together with its Norton equivalent circuit.

It should be clear that the current, i_{SC}, flowing through the short circuit replacing the load is exactly the Norton current, i_N, since all of the source current in the circuit of Figure 11.31 must flow through the short circuit.

11.3.3.5 Experimental Determination of Thévenin and Norton Equivalents

Figure 11.32 illustrates the measurement of the open-circuit voltage and short-circuit current for an arbitrary network connected to any load and also illustrates that the procedure requires some special attention, because of the nonideal nature of any practical measuring instrument. The figure clearly illustrates that in the presence of finite meter resistance, r_m, one must take this quantity into account in the computation of the short-circuit current and open-circuit voltage; v_{OC} and i_{SC} appear between quotation marks in the figure specifically to illustrate that the measured "open-circuit voltage" and "short-circuit current" are, in fact, affected by the internal resistance of the measuring instrument and are not the true quantities.

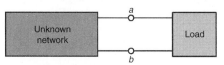
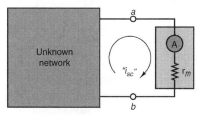
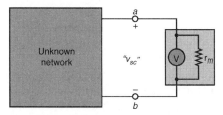

FIGURE 11.32 Measurement of open-circuit voltage and short-circuit current.

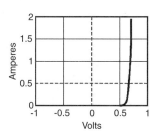

FIGURE 11.33 i–v characteristic of exponential resistor.

The following are expressions for the true short-circuit current and open-circuit voltage.

$$i_N = i_{SC}\left(1 + \frac{r_m}{R_T}\right)$$
$$v_T = v_{OC}\left(1 + \frac{R_T}{r_m}\right)$$
(11.26)

where i_N is the ideal Norton current, v_T the Thévenin voltage, and R_T the true Thévenin resistance.

11.3.4 Nonlinear Circuit Elements

11.3.4.1 Description of Nonlinear Elements

There are a number of useful cases in which a simple functional relationship exists between voltage and current in a nonlinear circuit element. For example, Figure 11.33 depicts an element with an exponential i–v characteristic, described by the following equations:

$$i = I_0 e^{\alpha v}, \quad v > 0$$
$$i = -I_0, \quad v \leq 0$$
(11.27)

Electrical Engineering

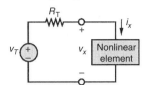

FIGURE 11.34 Representation of nonlinear element in a linear circuit.

There exists, in fact, a circuit element (the semiconductor diode) that very nearly satisfies this simple relationship. The difficulty in the i–v relationship of Equation 11.27 is that it is not possible, in general, to obtain a closed-form analytical solution, even for a very simple circuit.

One approach to analyzing a circuit containing a nonlinear element might be to treat the nonlinear element as a load, and to compute the Thévenin equivalent of the remaining circuit, as shown in Figure 11.34. Applying KVL, the following equation may then be obtained:

$$v_T = R_T i_x + v_x \tag{11.28}$$

To obtain the second equation needed to solve for both the unknown voltage, v_x, and the unknown current, i_x, it is necessary to resort to the i-v description of the nonlinear element, namely, Equation 11.27. If, for the moment, only positive voltages are considered, the circuit is completely described by the following system:

$$\begin{aligned} i_x &= I_0 e^{\alpha v_x}, \quad v > 0 \\ v_T &= R_T i_x + v_x \end{aligned} \tag{11.29}$$

The two parts of Equation 11.29 represent a system of two equations in two unknowns. Any numerical method of choice may now be applied to solve the system of Equations 11.29.

11.4 AC Network Analysis

In this section we introduce energy-storage elements, dynamic circuits, and the analysis of circuits excited by sinusoidal voltages and currents. Sinusoidal (or AC) signals constitute the most important class of signals in the analysis of electrical circuits. The simplest reason is that virtually all of the electric power used in households and industries comes in the form of sinusoidal voltages and currents.

11.4.1 Energy-Storage (Dynamic) Circuit Elements

The ideal resistor was introduced through Ohm's law in Section 11.2 as a useful idealization of many practical electrical devices. However, in addition to resistance to the flow of electric current, which is purely a dissipative (i.e., an energy-loss) phenomenon, electric devices may also exhibit energy-storage properties, much in the same way a spring or a flywheel can store mechanical energy. Two distinct mechanisms for energy storage exist in electric circuits: **capacitance** and **inductance**, both of which lead to the storage of energy in an electromagnetic field.

11.4.1.1 The Ideal Capacitor

A physical capacitor is a device that can store energy in the form of a charge separation when appropriately polarized by an electric field (i.e., a voltage). The simplest capacitor configuration consists of two parallel

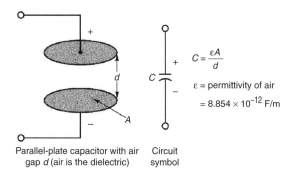

FIGURE 11.35 Structure of parallel-plate capacitor.

conducting plates of cross-sectional area A, separated by air (or another **dielectric*** material, such as mica or Teflon). Figure 11.35 depicts a typical configuration and the circuit symbol for a capacitor.

The presence of an insulating material between the conducting plates does not allow for the flow of DC current; thus, *a capacitor acts as an open circuit in the presence of DC currents*. However, if the voltage present at the capacitor terminals changes as a function of time, so will the charge that has accumulated at the two capacitor plates, since the degree of polarization is a function of the applied electric field, which is time-varying. In a capacitor, the charge separation caused by the polarization of the dielectric is proportional to the external voltage, that is, to the applied electric field:

$$Q = CV \tag{11.30}$$

where the parameter C is called the *capacitance* of the element and is a measure of the ability of the device to accumulate, or store, charge. The unit of capacitance is the coulomb/volt and is called the **farad** (F). The farad is an unpractically large unit; therefore, it is common to use microfarads (1 μF = 10^{-6} F) or picofarads (1 pF = 10^{-12} F). From Equation 11.30 it becomes apparent that if the external voltage applied to the capacitor plates changes in time, so will the charge that is internally stored by the capacitor:

$$q(t) = Cv(t) \tag{11.31}$$

Thus, although no current can flow through a capacitor if the voltage across it is constant, a time-varying voltage will cause charge to vary in time. The change with time in the stored charge is analogous to a current. The relationship between the current and voltage in a capacitor is as follows:

$$i(t) = C\frac{dv(t)}{dt} \tag{11.32}$$

If the above differential equation is integrated, one can obtain the following relationship for the voltage across a capacitor:

$$v_C(t) = \frac{1}{C}\int_{-\infty}^{t_0} i_C \, dt \tag{11.33}$$

Equation 11.33 indicates that the capacitor voltage depends on the past current through the capacitor, up until the present time, t. Of course, one does not usually have precise information regarding the flow

---

*A dielectric material contains a large number of electric dipoles, which become polarized in the presence of an electric field.

Electrical Engineering

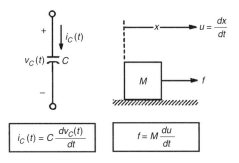

FIGURE 11.36 Defining equation for the ideal capacitor, and analogy with force–mass system.

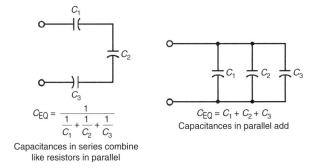

FIGURE 11.37 Combining capacitors in a circuit.

of capacitor current for all past time, and so it is useful to define the initial voltage (or *initial condition*) for the capacitor according to the following, where t_0 is an arbitrary initial time:

$$V_0 = v_C(t = t_0) = \frac{1}{C}\int_{-\infty}^{t} i_C\, dt \tag{11.34}$$

The capacitor voltage is now given by the expression

$$v_C(t) = \frac{1}{C}\int_{t_0}^{t} i_C\, dt + V_0 \quad t \geq t_0 \tag{11.35}$$

The significance of the initial voltage, V_0, is simply that at time t_0 some charge is stored in the capacitor, giving rise to a voltage, $v_C(t_0)$, according to the relationship $Q = CV$. Knowledge of this initial condition is sufficient to account for the entire past history of the capacitor current (see Figure 11.36).

From the standpoint of circuit analysis, it is important to point out that capacitors connected in series and parallel can be combined to yield a single equivalent capacitance. The rule of thumb, which is illustrated in Figure 11.37, is the following: capacitors in parallel add; capacitors in series combine according to the same rules used for resistors connected in parallel.

Physical capacitors are rarely constructed of two parallel plates separated by air, because this configuration yields very low values of capacitance, unless one is willing to tolerate very large plate areas. In order to increase the capacitance (i.e., the ability to store energy), physical capacitors are often made of tightly rolled sheets of metal film, with a dielectric (paper or Mylar) sandwiched in-between. Table 11.3 illustrates typical values, materials, maximum voltage ratings, and useful frequency ranges for various

TABLE 11.3 Capacitors

Material	Capacitance Range	Maximum Voltage (V)	Frequency Range (Hz)
Mica	1 pF–0.1 µF	100–600	10^3–10^{10}
Ceramic	10 pF–1 µF	50–1000	10^3–10^{10}
Mylar	0.001–10 µF	50–500	10^2–10^8
Paper	1000 pF–50 µF	100–105	10^2–10^8
Electrolytic	0.1 µF–0.2 F	3–600	10–10^4

types of capacitors. The voltage rating is particularly important, because any insulator will break down if a sufficiently high voltage is applied across it. The energy stored in a capacitor is given by

$$W_C(t) = \frac{1}{2} C v_C^2(t) \; (J)$$

Example 11.3 Capacitive Displacement Transducer and Microphone

As shown in Figure 11.26, the capacitance of a parallel-plate capacitor is given by the expression

$$C = \frac{\varepsilon A}{d}$$

where ε is the **permittivity** of the dielectric material, A the area of each of the plates, and d their separation. The permittivity of air is $\varepsilon_0 = 8.854 \times 10^{-12}$ F/m, so that two parallel plates of area 1 m², separated by a distance of 1 mm, would give rise to a capacitance of 8.854×10^{-3} µF, a very small value for a very large plate area. This relative inefficiency makes parallel-plate capacitors impractical for use in electronic circuits. On the other hand, parallel-plate capacitors find application as *motion transducers*, that is, as devices that can measure the motion or displacement of an object. In a capacitive motion transducer, the air gap between the plates is designed to be variable, typically by fixing one plate and connecting the other to an object in motion. Using the capacitance value just derived for a parallel-plate capacitor, one can obtain the expression

$$C = \frac{8.854 \times 10^{-3} A}{x}$$

where C is the capacitance in picofarad, A is the area of the plates in square millimeter, and x is the (variable) distance in milimeter. It is important to observe that the change in capacitance caused by the displacement of one of the plates is nonlinear, since the capacitance varies as the inverse of the displacement. For small displacements, however, the capacitance varies approximately in a linear fashion.

The *sensitivity*, S, of this motion transducer is defined as the slope of the change in capacitance per change in displacement, x, according to the relation

$$S = \frac{dC}{dx} = -\frac{8.854 \times 10^{-3} A}{2x^2} \; (pF/mm)$$

Thus, the sensitivity increases for small displacements. This behavior can be verified by plotting the capacitance as a function of x and noting that as x approaches zero, the slope of the nonlinear $C(x)$ curve becomes steeper (thus the greater sensitivity). Figure 11.38 depicts this behavior for a transducer with area equal to 10 mm².

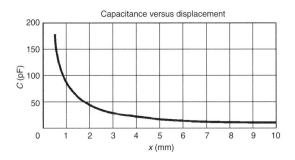

FIGURE 11.38 Response of a capacitive displacement transducer.

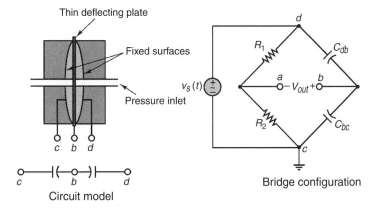

FIGURE 11.39 Capacitive pressure transducer and related bridge circuit.

This simple capacitive displacement transducer actually finds use in the popular *capacitive (or condenser) microphone*, in which the sound pressure waves act to displace one of the capacitor plates. The change in capacitance can then be converted into a change in voltage or current by means of a suitable circuit. An extension of this concept that permits measurement of differential pressures is shown in simplified form in Figure 11.39. In the figure, a three-terminal variable capacitor is shown to be made up of two fixed surfaces (typically, spherical depressions ground into glass disks and coated with a conducting material) and of a deflecting plate (typically made of steel) sandwiched between the glass disks. Pressure inlet orifices are provided, so that the deflecting plate can come into contact with the fluid whose pressure it is measuring. When the pressure on both sides of the deflecting plate is the same, the capacitance between terminals b and d, C_{bd}, will be equal to that between terminals b and c, C_{bc}. If any pressure differential exists, the two capacitances will change, with an increase on the side where the deflecting plate has come closer to the fixed surface and a corresponding decrease on the other side.

This behavior is ideally suited for the application of a bridge circuit, similar to the Wheatstone bridge circuit illustrated in Example 11.2, and also shown in Figure 11.39. In the bridge circuit, the output voltage, v_{out}, is precisely balanced when the differential pressure across the transducer is zero, but it will deviate from zero whenever the two capacitances are not identical because of a pressure differential across the transducer. We shall analyze the bridge circuit later in Example 11.4.

11.4.1.2 The Ideal Inductor

The ideal inductor is an element that has the ability to store energy in a magnetic field. Inductors are typically made by winding a coil of wire around a core, which can be an insulator or a ferromagnetic material, shown in Figure 11.40. When a current flows through the coil, a magnetic field is established, as you may recall from early physics experiments with electromagnets. In an ideal inductor, the resistance

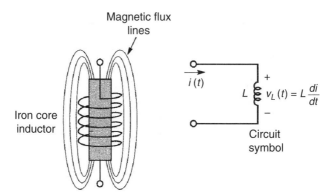

FIGURE 11.40 Iron-core inductor.

of the wire is zero, so that a constant current through the inductor will flow freely without causing a voltage drop. In other words, *the ideal inductor acts as a short circuit in the presence of DC currents*. If a time-varying voltage is established across the inductor, a corresponding current will result, according to the following relationship:

$$v_L(t) = L\frac{di_L}{dt} \tag{11.36}$$

where *L* is called the *inductance* of the coil and is measured in henry (H), where

$$1\ \text{H} = 1\ \text{V sec/A} \tag{11.37}$$

Henrys are reasonable units for practical inductors; millihenrys (mH) and microhenrys (μH) are also used.

The inductor current is found by integrating the voltage across the inductor:

$$i_L(t) = \frac{1}{L}\int_{-\infty}^{t} v_L\,dt \tag{11.38}$$

If the current flowing through the inductor at time $t = t_0$ is known to be I_0, with

$$I_0 = i_L(t = t_0) = \frac{1}{L}\int_{-\infty}^{t_0} v_L\,dt \tag{11.39}$$

then the inductor current can be found according to the equation

$$i_L(t) = \frac{1}{L}\int_{t_0}^{t} v_L\,dt + I_0 \quad t \geq t_0 \tag{11.40}$$

Inductors in series add. Inductors in parallel combine according to the same rules used for resistors connected in parallel (see Figures 11.41–11.43).

Table 11.4 and Figures 11.36, 11.41, and 11.43 illustrate a useful analogy between ideal electrical and mechanical elements.

Electrical Engineering

TABLE 11.4 Analogy between Electrical and Mechanical Variables

Mechanical System	Electrical System
Force, f (N)	Current, i (A)
Velocity, μ (m/sec)	Voltage, v (V)
Damping, B (N sec/m)	Conductance, $1/R$ (S)
Compliance, $1/k$ (m/N)	Inductance, L (H)
Mass, M (kg)	Capacitance, C (F)

The defining equation for the inductance circuit element is analogous to the equation of motion of a spring acted upon by a force.

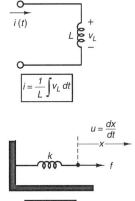

FIGURE 11.41 Defining equation for the ideal inductor and analogy with force–spring system.

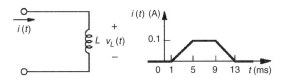

FIGURE 11.42 Combining inductors in a circuit.

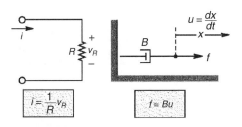

FIGURE 11.43 Analogy between electrical and mechanical elements.

11.4.2 Time-Dependent Signal Sources

Figure 11.44 illustrates the convention that will be employed to denote time-dependent signal sources.

One of the most important classes of time-dependent signals is that of **periodic signals**. These signals appear frequently in practical applications and are a useful approximation of many physical phenomena. A periodic signal $x(t)$ is a signal that satisfies the following equation:

$$x(t) = x(t + nT), \quad n = 1, 2, 3, \ldots \quad (11.41)$$

where T is the **period** of $x(t)$. Figure 11.45 illustrates a number of the periodic waveforms that are typically encountered in the study of electrical circuits. Waveforms such as the sine, triangle, square, pulse, and sawtooth waves are provided in the form of voltages (or, less frequently, currents) by commercially available **signal** (or **waveform**) **generators**. Such instruments allow for selection of the waveform peak amplitude, and of its period.

As stated in the introduction, sinusoidal waveforms constitute by far the most important class of time-dependent signals. Figure 11.46 depicts the relevant parameters of a sinusoidal waveform. A generalized sinusoid is defined as follows:

$$x(t) = A \cos(\omega t + \phi) \quad (11.42)$$

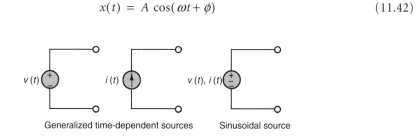

FIGURE 11.44 Time-dependent signal sources.

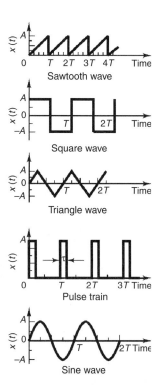

FIGURE 11.45 Periodic signal waveforms.

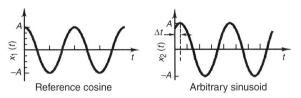

FIGURE 11.46 Sinusoidal waveforms.

where A is the **amplitude**, ω the **radian frequency**, and ϕ the **phase**. Figure 11.46 summarizes the definitions of A, ω, and ϕ for the waveforms

$$x_1(t) = A\cos(\omega t) \quad \text{and} \quad x_2(t) = A\cos(\omega t + \phi)$$

where

$$f = \text{natural frequency} = \frac{1}{T} \text{ (cycles/sec, or Hz)}$$

$$\omega = \text{radian frequency} = 2\pi f \text{ (radians/sec)} \quad (11.43)$$

$$\phi = 2\pi\frac{\Delta T}{T} \text{ (radians)} = 360\frac{\Delta T}{T} \text{ (degrees)}$$

The phase shift, ϕ, permits the representation of an arbitrary sinusoidal signal. Thus, the choice of the reference cosine function to represent sinusoidal signals—arbitrary as it may appear at first—does not restrict the ability to represent all sinusoids. For example, one can represent a sine wave in terms of a cosine wave simply by introducing a phase shift of $\pi/2$ radians:

$$A\sin(\omega t) = A\cos\left(\omega t - \frac{\pi}{2}\right) \quad (11.44)$$

It is important to note that, although one usually employs the variable ω (in units of radians per second) to denote sinusoidal frequency, it is common to refer to natural frequency, f, in units of cycles per second, or hertz (Hz). The relationship between the two is the following:

$$\omega = 2\pi f \quad (11.45)$$

11.4.2.1 Average and RMS Values

Now that a number of different signal waveforms have been defined, it is appropriate to define suitable measurements for quantifying the strength of a time-varying electrical signal. The most common types of measurements are the **average** (or **DC**) **value** of a signal waveform, which corresponds to just measuring the mean voltage or current over a period of time, and the **root-mean-square** (**rms**) **value**, which takes into account the fluctuations of the signal about its average value. Formally, the operation of computing the average value of a signal corresponds to integrating the signal waveform over some (presumably, suitably chosen) period of time. We define the time-averaged value of a signal $x(t)$ as

$$\langle x(t) \rangle = \frac{1}{T}\int_0^T x(t)\,dt \quad (11.46)$$

where T is the period of integration. Figure 11.47 illustrates how this process does, in fact, correspond to computing the average amplitude of $x(t)$ over a period of T seconds.

$$\langle A\cos(\omega t + \phi) \rangle = 0$$

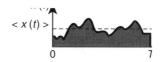

FIGURE 11.47 Averaging a signal waveform.

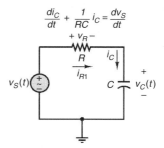

A circuit containing energy-storage elements is described by a differential equation. The differential equation describing the series RC circuit shown is

$$\frac{di_C}{dt} + \frac{1}{RC}i_C = \frac{dv_S}{dt}$$

FIGURE 11.48 Circuit containing energy-storage element.

A useful measure of the voltage of an AC waveform is the rms value of the signal, $x(t)$, defined as follows:

$$x_{rms} \sqrt{\frac{1}{T}\int_0^T x^2(t)\,dt} \tag{11.47}$$

Note immediately that if $x(t)$ is a voltage, the resulting x_{rms} will also have units of volts. If you analyze Equation 11.47, you can see that, in effect, the rms value consists of the square root of the average (or mean) of the square of the signal. Thus, the notation *rms* indicates exactly the operations performed on $x(t)$ in order to obtain its rms value.

11.4.3 Solution of Circuits Containing Dynamic Elements

The major difference between the analysis of the resistive circuits and circuits containing capacitors and inductors is now that the equations that result from applying Kirchhoff's laws are differential equations, as opposed to the algebraic equations obtained in solving resistive circuits. Consider, for example, the circuit of Figure 11.48 which consists of the series connection of a voltage source, a resistor, and a capacitor. Applying KVL around the loop, we may obtain the following equation:

$$v_S(t) = v_R(t) + v_C(t) \tag{11.48}$$

Observing that $i_R = i_C$, Equation 11.48 may be combined with the defining equation for the capacitor (Equation 4.6) to obtain

$$v_S(t) = Ri_C(t) + \frac{1}{C}\int_{-\infty}^{t} i_C\,dt \tag{11.49}$$

Equation 11.49 is an integral equation, which may be converted to the more familiar form of a differential equation by differentiating both sides of the equation, and recalling that

$$\frac{d}{dt}\left(\int_{-\infty}^{t} i_C\,dt\right) = i_C(t) \tag{11.50}$$

to obtain the following differential equation:

$$\frac{di_C}{dt} + \frac{1}{RC}i_C = \frac{1}{R}\frac{dv_S}{dt} \tag{11.51}$$

where the argument (t) has been dropped for ease of notation.

Observe that in Equation 11.51, the independent variable is the series current flowing in the circuit, and that this is not the only equation that describes the series RC circuit. If, instead of applying KVL, for example, we had applied KCL at the node connecting the resistor to the capacitor, we would have obtained the following relationship:

$$i_R = \frac{v_S - v_C}{R} = i_C = C\frac{dv_C}{dt} \tag{11.52}$$

or

$$\frac{dv_C}{dt} + \frac{1}{RC}v_C = \frac{1}{RC}v_S \tag{11.53}$$

Note the similarity between Equations 11.51 and 11.53. The left-hand side of both equations is identical, except for the dependent variable, while the right-hand side takes a slightly different form. The solution of either equation is sufficient, however, to determine all voltages and currents in the circuit.

We can generalize the results above by observing that any circuit containing a single energy-storage element can be described by a differential equation of the form

$$a_1\frac{dy(t)}{dt} + a_0(t) = F(t) \tag{11.54}$$

where $y(t)$ represents the capacitor voltage in the circuit of Figure 11.48 and where the constants a_0 and a_1 consist of combinations of circuit element parameters. Equation 11.54 is a **first-order ordinary differential equation** with constant coefficients.

Consider now a circuit that contains two energy-storage elements, such as that shown in Figure 11.49. Application of KVL results in the following equation:

$$Ri(t) + L\frac{di(t)}{dt} + \frac{1}{C}\int_{-\infty}^{t} i(t)\,dt = v_S(t) \tag{11.55}$$

Equation 11.55 is called an integro-differential equation because it contains both an integral and a derivative. This equation can be converted into a differential equation by differentiating both sides, to obtain:

$$R\frac{di(t)}{dt} + L\frac{d^2i(t)}{dt^2} + \frac{1}{C}i(t) = \frac{dv_S(t)}{dt} \tag{11.56}$$

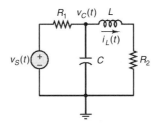

FIGURE 11.49 Second-order circuit.

or, equivalently, by observing that the current flowing in the series circuit is related to the capacitor voltage by $i(t) = C dv_C/dt$, and that Equation 11.55 can be rewritten as

$$RC\frac{dv_C}{dt} + LC\frac{d^2 v_C(t)}{dt^2} + v_C(t) = v_S(t) \tag{11.57}$$

Note that although different variables appear in the preceding differential equations, both Equations 11.55 and 11.57 can be rearranged to appear in the same general form as follows:

$$a_2 \frac{d^2 y(t)}{dt^2} + a_1 \frac{dy(t)}{dt} + a_0 y(t) = F(t) \tag{11.58}$$

where the general variable $y(t)$ represents either the series current of the circuit of Figure 11.49 or the capacitor voltage. By analogy with Equation 11.54, we call Equation 11.58 a **second-order ordinary differential equation** with constant coefficients. As the number of energy-storage elements in a circuit increases, one can therefore expect that higher-order differential equations will result.

11.4.5 Phasors and Impedance

In this section, we introduce an efficient notation to make it possible to represent sinusoidal signals as *complex numbers*, and to eliminate the need for solving differential equations.

11.4.5.1 Phasors

Let us recall that it is possible to express a generalized sinusoid as the real part of a complex vector whose **argument**, or **angle**, is given by $(\omega t + \phi)$ and whose length, or **magnitude**, is equal to the peak amplitude of the sinusoid. The **complex phasor** corresponding to the sinusoidal signal $A\cos(\omega t + \phi)$ is therefore defined to be the complex number $Ae^{j\phi}$:

$$Ae^{j\phi} = \text{complex phasor notation for } A\cos(\omega t + \phi) \tag{11.59}$$

1. Any sinusoidal signal may be mathematically represented in one of two ways: a **time-domain form**

$$v(t) = A\cos(\omega t + \phi)$$

and a **frequency-domain** (or **phasor**) **form**

$$V(j\omega) = Ae^{j\phi}$$

2. A phasor is a complex number, expressed in polar form, consisting of a *magnitude* equal to the peak amplitude of the sinusoidal signal and a *phase angle* equal to the phase shift of the sinusoidal signal *referenced to a cosine signal*.
3. When using phasor notation, it is important to make a note of the specific frequency, ω, of the sinusoidal signal, since this is not explicitly apparent in the phasor expression.

11.4.5.2 Impedance

We now analyze the *i–v* relationship of the three ideal circuit elements in light of the new phasor notation. The result will be a new formulation in which resistors, capacitors, and inductors will be described in the same notation. A direct consequence of this result will be that the circuit theorems of Section 11.3 will be extended to AC circuits. In the context of AC circuits, any one of the three ideal circuit elements

Electrical Engineering

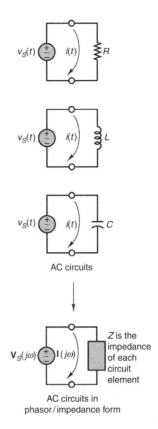

FIGURE 11.50 The impedance element.

defined so far will be described by a parameter called **impedance**, which may be viewed as a *complex resistance*. The impedance concept is equivalent to stating that capacitors and inductors act as *frequency-dependent resistors*, that is, as resistors whose resistance is a function of the frequency of the sinusoidal excitation. Figure 11.50 depicts the same circuit represented in conventional form (top) and in phasor-impedance form (bottom); the latter representation explicitly shows phasor voltages and currents and treats the circuit element as a generalized "impedance." It will presently be shown that each of the three ideal circuit elements may be represented by one such impedance element.

Let the source voltage in the circuit of Figure 11.50 be defined by

$$v_S(t) = A\cos\omega t \quad \text{or} \quad V_S(j\omega) = Ae^{j0°} \tag{11.60}$$

without loss of generality. Then the current $i(t)$ is defined by the i–v relationship for each circuit element. Let us examine the frequency-dependent properties of the resistor, inductor, and capacitor, one at a time.

The *impedance* of the resistor is defined as the ratio of the phasor voltage across the resistor to the phasor current flowing through it, and the symbol Z_R is used to denote it:

$$Z_R(j\omega) = \frac{V_S(j\omega)}{I(j\omega)} = R \tag{11.61}$$

The impedance of the inductor is defined as follows:

$$Z_L(j\omega) = \frac{V_S(j\omega)}{I(j\omega)} = \omega L e^{j90°} = j\omega L \tag{11.62}$$

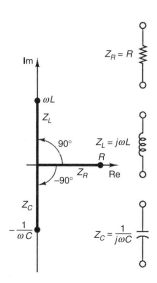

FIGURE 11.51 Impedances of R, L, and C in the complex plane.

Note that the inductor now appears to behave like a *complex frequency-dependent resistor*, and that the magnitude of this complex resistor, ωL, is proportional to the signal frequency, ω. Thus, an inductor will "impede" current flow in proportion to the sinusoidal frequency of the source signal. This means that at low signal frequencies, an inductor acts somewhat like a short circuit, while at high frequencies it tends to behave more as an open circuit. Another important point is that *the magnitude of the impedance of an inductor is always positive*, since both L and ω are positive numbers. You should verify that the units of this magnitude are also ohms.

The impedance of the ideal capacitor, $Z_C(j\omega)$, is therefore defined as follows:

$$Z_C(j\omega) = \frac{V_S(j\omega)}{I(j\omega)} = \frac{1}{\omega C} e^{-j90°} = \frac{-j}{\omega C} = \frac{1}{j\omega C} \qquad (11.63)$$

where we have used the fact that $1/j = e^{-j90°} = -j$. Thus, the impedance of a capacitor is also a frequency-dependent complex quantity, with the impedance of the capacitor varying as an inverse function of frequency, and so a capacitor acts like a short circuit at high frequencies, whereas it behaves more like an open circuit at low frequencies. Another important point is that *the impedance of a capacitor is always negative*, since both C and ω are positive numbers. You should verify that the units of impedance for a capacitor are ohms. Figure 11.51 depicts $Z_C(j\omega)$ in the complex plane, alongside $Z_R(j\omega)$ and $Z_L(j\omega)$.

The impedance parameter defined in this section is extremely useful in solving AC circuit analysis problems, because it will make it possible to take advantage of most of the network theorems developed for DC circuits by replacing resistances with complex-valued impedances. In its most general form, the impedance of a circuit element is defined as the sum of a real part and an imaginary part:

$$Z(j\omega) = R(j\omega) + jX(j\omega) \qquad (11.64)$$

where R is called the **AC resistance** and X is called the **reactance**. The frequency dependence of R and X has been indicated explicitly, since it is possible for a circuit to have a frequency-dependent resistance. The examples illustrate how a complex impedance containing both real and imaginary parts arises in a circuit.

Example 11.4 Capacitive Displacement Transducer

In Example 11.3, the idea of a capacitive displacement transducer was introduced when we considered a parallel-plate capacitor composed of a fixed plate and a movable plate. The capacitance of this variable capacitor was shown to be a *nonlinear* function of the position of the movable plate, x (see Figure 11.39).

In this example, we show that under certain conditions the impedance of the capacitor varies as a *linear* function of displacement, that is, the movable-plate capacitor can serve as a linear transducer.

Recall the expression derived in Example 11.3:

$$C = \frac{8.854 \times 10^{-3} A}{x}$$

where C is the capacitance in picofarad, A is the area of the plates in square millimeter, and x is the (variable) distance in millimeter. If the capacitor is placed in an AC circuit, its impedance will be determined by the expression

$$Z_C = \frac{1}{j\omega C}$$

so that

$$Z_C = \frac{x}{8.854 \, j\omega A}$$

Thus, at a fixed frequency ω, the impedance of the capacitor will vary linearly with displacement. This property may be exploited in the bridge circuit of Example 11.3, where a differential pressure transducer was shown as being made of two movable-plate capacitors, such that if the capacitance of one increased as a consequence of a pressure differential across the transducer, the capacitance of the other had to decrease by a corresponding amount (at least for small displacements). The circuit is shown again in Figure 11.52 where two resistors have been connected in the bridge along with the variable capacitors (denoted by $C(x)$). The bridge is excited by a sinusoidal source.

Using phasor notation, we can express the output voltage as follows:

$$\mathbf{V}_{out}(j\omega) = \mathbf{V}_S(j\omega)\left(\frac{Z_{C_{bc}(x)}}{Z_{C_{db}(x)} + Z_{C_{bc}(x)}} - \frac{R_2}{R_1 + R_2}\right)$$

If the nominal capacitance of each movable-plate capacitor with the diaphragm in the center position is given by

$$C = \frac{\varepsilon A}{d}$$

where d is the nominal (undisplaced) separation between the diaphragm and the fixed surfaces of the capacitors (in mm), the capacitors will see a change in capacitance given by

$$C_{db} = \frac{\varepsilon A}{d - x} \quad \text{and} \quad C_{bc} = \frac{\varepsilon A}{d + x}$$

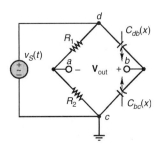

FIGURE 11.52 Bridge circuit for capacitive displacement transducer.

when a pressure differential exists across the transducer, so that the impedances of the variable capacitors change according to the displacement

$$Z_{C_{db}} = \frac{d-x}{8.854\,j\omega A} \quad \text{and} \quad Z_{C_{bc}} = \frac{d+x}{8.854\,j\omega A}$$

and we obtain the following expression for the phasor output voltage, if we choose $R_1 = R_2$.

$$\begin{aligned}
\mathbf{V}_{out}(j\omega) &= \mathbf{V}_S(j\omega)\left(\frac{\frac{d+x}{8.854\,j\omega A}}{\frac{d-x}{8.854\,j\omega A}+\frac{d+x}{8.854\,j\omega A}} - \frac{R_2}{R_1+R_2}\right) \\
&= \mathbf{V}_S(j\omega)\left(\frac{1}{2}+\frac{x}{2d}-\frac{R_2}{R_1+R_2}\right) \\
&= \mathbf{V}_S(j\omega)\frac{x}{2d}
\end{aligned}$$

Thus, the output voltage will vary as a scaled version of the input voltage in proportion to the displacement.

References

Irwin, J.D., 1989. *Basic Engineering Circuit Analysis,* 3rd ed., Macmillan, New York.
Budak, A., *Passive and Active Network Analysis and Synthesis,* Houghton Mifflin, Boston.
Nilsson, J.W., 1989. *Electric Circuits,* 3rd ed., Addison-Wesley, Reading, MA.
Rizzoni, G., 2000. *Principles and Applications of Electrical Engineering,* 3rd ed., McGraw-Hill, Burr Ridge, IL.
Smith, R.J. and Dorf, R.C., 1992. *Circuits, Devices and Systems,* 5th ed., John Wiley & Sons, New York.
1993. *The Electrical Engineering Handbook,* CRC Press, Boca Raton, FL.
Van Valkenburg, M.E., 1982, *Analog Filter Design,* Holt, Rinehart & Winston, New York.

12
Engineering Thermodynamics

12.1	Fundamentals	**12**-1
	Basic Concepts and Definitions • Laws of Thermodynamics	
12.2	Extensive Property Balances	**12**-4
	Mass Balance • Energy Balance • Entropy Balance • Control Volumes at Steady State • Exergy Balance	
12.3	Property Relations and Data	**12**-12
	P-v-T Surface • Thermodynamic Data Retrieval • Compressibility Charts • Analytical Equations of State • Ideal Gas Model	
12.4	Vapor and Gas Power Cycles	**12**-22
	Work and Heat Transfer in Internally Reversible Processes	
	References	**12**-31

Michael J. Moran
The Ohio State University

Although various aspects of what is now known as thermodynamics have been of interest since antiquity, formal study began only in the early nineteenth century through consideration of the motive power of heat: the capacity of hot bodies to produce work. Today the scope is larger, dealing generally with energy and entropy, and with relationships among the properties of matter. Moreover, in the past 25 years engineering thermodynamics has undergone a revolution, both in terms of the presentation of fundamentals and in the manner that it is applied. In particular, the second law of thermodynamics has emerged as an effective tool for engineering analysis and design.

12.1 Fundamentals

Classical thermodynamics is concerned primarily with the macrostructure of matter. It addresses the gross characteristics of large aggregations of molecules and not the behavior of individual molecules. The microstructure of matter is studied in kinetic theory and statistical mechanics (including quantum thermodynamics). In this chapter, the classical approach to thermodynamics is featured.

12.1.1 Basic Concepts and Definitions

Thermodynamics is both a branch of physics and an engineering science. The scientist is normally interested in gaining a fundamental understanding of the physical and chemical behavior of fixed, quiescent quantities of matter and uses the principles of thermodynamics to relate the properties of matter. Engineers are generally interested in studying systems and how they interact with their surroundings. To facilitate this, engineers have extended the subject of thermodynamics to the study of systems through which matter flows.

12.1.1.1 System

In a thermodynamic analysis, the *system* is the subject of the investigation. Normally the system is a specified quantity of matter and/or a region that can be separated from everything else by a well-defined surface. The defining surface is known as the *control surface* or *system boundary*. The control surface may be movable or fixed. Everything external to the system is the *surroundings*. A system of fixed mass is referred to as a *control mass* or *closed system*. When there is flow of mass through the control surface, the system is called a *control volume* or *open system*. An *isolated* system is a closed system that does not interact in any way with its surroundings.

12.1.1.2 State, Property

The condition of a system at any instant of time is called its *state*. The state at a given instant of time is described by the properties of the system. A *property* is any quantity whose numerical value depends on the state, but not the history of the system. The value of a property is determined in principle by some type of physical operation or test.

Extensive properties depend on the size or extent of the system. Volume, mass, energy, entropy, and exergy are examples of extensive properties. An extensive property is additive in the sense that its value for the whole system equals the sum of the values for its parts. *Intensive* properties are independent of the size or extent of the system. Pressure and temperature are examples of intensive properties.

12.1.1.3 Process, Cycle

Two states are identical if, and only if, the properties of the two states are identical. When any property of a system changes in value there is a change in state, and the system is said to undergo a *process*. When a system in a given initial state goes through a sequence of processes and finally returns to its initial state, it is said to have undergone a *thermodynamic cycle*.

12.1.1.4 Phase and Pure Substance

The term *phase* refers to a quantity of matter that is homogeneous throughout in both chemical composition and physical structure. Homogeneity in physical structure means that the matter is all *solid*, or all *liquid*, or all *vapor* (or equivalently all *gas*). A system can contain one or more phases. For example, a system of liquid water and water vapor (steam) contains two phases. A *pure substance* is one that is uniform and invariable in chemical composition. A pure substance can exist in more than one phase, but its chemical composition must be the same in each phase. For example, if liquid water and water vapor form a system with two phases, the system can be regarded as a pure substance because each phase has the same composition. The nature of phases that coexist in equilibrium is addressed by the *phase rule* (for discussion see Moran and Shapiro, 2000).

12.1.1.5 Equilibrium

Equilibrium means a condition of balance. In thermodynamics the concept includes not only a balance of forces, but also a balance of other influences. Each kind of influence refers to a particular aspect of thermodynamic (complete) equilibrium. *Thermal* equilibrium refers to an equality of temperature, *mechanical* equilibrium to an equality of pressure, and *phase* equilibrium to an equality of chemical potentials (for discussion see Moran and Shapiro, 2000). *Chemical* equilibrium is also established in terms of chemical potentials. For complete equilibrium the several types of equilibrium must exist individually.

12.1.1.6 Temperature

A scale of temperature independent of the thermometric substance is called a *thermodynamic temperature scale*. The Kelvin scale, a thermodynamic scale, can be elicited from the second law of thermodynamics. The definition of temperature following from the second law is valid over all temperature ranges and provides an essential connection between the several *empirical* measures of temperature. In particular, temperatures evaluated using a *constant-volume gas thermometer* are identical to those of the Kelvin scale over the range of temperatures where gas thermometry can be used. On the Kelvin scale the unit is the kelvin (K).

The Celsius temperature scale (also called the centigrade scale) uses the degree Celsius (°C), which has the same magnitude as the kelvin. Thus, temperature differences are identical on both scales. However, the zero point on the Celsius scale is shifted to 273.15 K, the *triple point* of water (Figure 12.1b), as shown by the following relationship between the Celsius temperature and the Kelvin temperature:

$$T(°C) = T(K) - 273.15 \qquad (12.1)$$

Two other temperature scales are commonly used in engineering in the United States By definition, the *Rankine scale*, the unit of which is the degree rankine (°R), is proportional to the Kelvin temperature according to

$$T(°R) = 1.8\,T(K) \qquad (12.2)$$

The Rankine scale is also an absolute thermodynamic scale with an absolute zero that coincides with the absolute zero of the Kelvin scale. In thermodynamic relationships, temperature is always in terms of the Kelvin or Rankine scale unless specifically stated otherwise.

A degree of the same size as that on the Rankine scale is used in the *Fahrenheit scale*, but the zero point is shifted according to the relation

$$T(°F) = T(°R) - 459.67 \qquad (12.3)$$

Substituting Equations 12.1 and 12.2 into Equation 12.3 gives

$$T(°F) = 1.8\,T(°C) + 32 \qquad (12.4)$$

Equation 12.4 shows that the Fahrenheit temperature of the *ice point* (0°C) is 32°F and of the *steam point* (100°C) is 212°F. The 100 Celsius or Kelvin degrees between the ice point and steam point corresponds to 180 Fahrenheit or Rankine degrees.

To provide a standard for temperature measurement taking into account both theoretical and practical considerations, the International Temperature Scale of 1990 (ITS-90) is defined in such a way that the temperature measured on it conforms with the thermodynamic temperature, the unit of which is the kelvin, to within the limits of accuracy of measurement obtainable in 1990. Further discussion of ITS-90 is provided by Preston-Thomas (1990).

12.1.1.7 Irreversibilities

A process is said to be *reversible* if it is possible for its effects to be eradicated in the sense that there is some way by which both the system and its surroundings can be exactly restored to their respective initial states. A process is *irreversible* if both the system and surroundings cannot be restored to their initial states. There are many effects whose presence during a process renders it irreversible. These include, but are not limited to, the following: heat transfer through a finite temperature difference; unrestrained expansion of a gas or liquid to a lower pressure; spontaneous chemical reaction; mixing of matter at different compositions or states; friction (sliding friction as well as friction in the flow of fluids); electric current flow through a resistance; magnetization or polarization with hysteresis; and inelastic deformation. The term *irreversibility* is used to identify effects such as these.

Irreversibilities can be divided into two classes, *internal* and *external*. Internal irreversibilities are those that occur within the system, while external irreversibilities are those that occur within the surroundings, normally the immediate surroundings. As this division depends on the location of the boundary there is some arbitrariness in the classification (by locating the boundary to take in the immediate surroundings, all irreversibilities are internal). Nonetheless, valuable insights can result when this distinction between irreversibilities is made. When internal irreversibilities are absent during a process, the process is said to be *internally reversible*. At every intermediate state of an internally reversible process of a closed system, all intensive properties are uniform throughout each phase present: the temperature, pressure, specific volume, and other intensive properties do not vary with position.

12.1.2 Laws of Thermodynamics

The first steps in a thermodynamic analysis are definition of the system and identification of the relevant interactions with the surroundings. Attention then turns to the pertinent physical laws and relationships that allow the behavior of the system to be described in terms of an engineering model, which is a simplified representation of system behavior that is sufficiently faithful for the purpose of the analysis, even if features exhibited by the actual system are ignored.

Thermodynamic analyses of control volumes and closed systems typically use, directly or indirectly, one or more of three basic laws. The laws, which are independent of the particular substance or substances under consideration, are

- Conservation of mass principle
- Conservation of energy principle
- Second law of thermodynamics

The second law may be expressed in terms of entropy or exergy.

The laws of thermodynamics must be supplemented by appropriate thermodynamic property data. For some applications a momentum equation expressing Newton's second law of motion also is required. Data for transport properties, heat transfer coefficients, and friction factors often are needed for a comprehensive engineering analysis. Principles of engineering economics and pertinent economic data also can play prominent roles.

12.2 Extensive Property Balances

The laws of thermodynamics can be expressed in terms of *extensive property balances* for mass, energy, entropy, and exergy. Engineering applications are generally analyzed on a control volume basis. Accordingly, the control volume formulations of the mass energy, entropy, and exergy balances are featured here. They are provided in the form of overall balances assuming one-dimensional flow. Equations of change for mass, energy, and entropy in the form of differential equations are also available in the literature (Bird et al., 1960).

12.2.1 Mass Balance

For applications in which inward and outward flows occur, each through one or more ports, the extensive property balance expressing the conservation of mass principle takes the form

$$\frac{dm}{dt} = \sum_i m_i - \sum_e m_e \tag{12.5}$$

where dm/dt represents the time rate of change of mass contained within the control volume, m_i denotes the mass flow rate at an inlet port, and m_e denotes the mass flow rate at an exit port.

The volumetric flow rate through a portion of the control surface with area dA is the product of the velocity component normal to the area, v_n, times the area: $v_n dA$. The mass flow rate through dA is $\rho(v_n dA)$, where ρ denotes density. The mass rate of flow through a port of area A is then found by integration over the area

$$m = \int_A \rho v_n \, dA$$

For one-dimensional flow the intensive properties are uniform with position over area A, and the last equation becomes

$$m = \rho v A = \frac{vA}{v} \tag{12.6}$$

where v denotes the specific volume (the reciprocal of density) and the subscript n has been dropped from velocity for simplicity.

12.2.2 Energy Balance

Energy is a fundamental concept of thermodynamics and one of the most significant aspects of engineering analysis. Energy can be stored within systems in various macroscopic forms: kinetic energy, gravitational potential energy, and internal energy. Energy also can be transformed from one form to another and transferred between systems. Energy can be transferred by work, by heat transfer, and by flowing matter. The total amount of energy is conserved in all transformations and transfers. The extensive property balance expressing the conservation of energy principle takes the form

$$\frac{d(U + KE + PE)}{dt} = \dot{Q} - \dot{W} + \sum_i \dot{m}_i \left(h_i + \frac{v_i^2}{2} + gz_i \right) - \sum_e \dot{m}_e \left(h_e + \frac{v_e^2}{2} + gz_e \right) \quad (12.7a)$$

where U, KE, and PE denote, respectively, the internal energy, kinetic energy, and gravitational potential energy of the overall control volume.

The right-hand side of Equation 12.7a accounts for transfers of energy across the boundary of the control volume. Energy can enter and exit control volumes by work. Because work is done on or by a control volume when matter flows across the boundary, it is convenient to separate the work rate (or power) into two contributions. One contribution is the work rate associated with the force of the fluid pressure as mass is introduced at the inlet and removed at the exit. Commonly referred to as *flow work*, this contribution is accounted for by $\dot{m}_i(p_i v_i)$ and $\dot{m}_e(p_e v_e)$, respectively, where p denotes pressure and v denotes specific volume. The other contribution, denoted by $\dot{W}$ in Equation 12.7a, includes all other work effects, such as those associated with rotating shafts, displacement of the boundary, and electrical effects. $\dot{W}$ is considered *positive* for energy transfer *from* the control volume.

Energy also can enter and exit control volumes with flowing streams of matter. On a one-dimensional flow basis, the rate at which energy enters with matter at inlet i is $\dot{m}_i(u_i + v_i^2/2 + gz_i)$, where the three terms in parentheses account, respectively, for the specific internal energy, specific kinetic energy, and specific gravitational potential energy of the substance flowing through port i. In writing Equation 12.7a the sum of the specific internal energy and specific flow work at each inlet and exit is expressed in terms of the specific enthalpy $h(=u + pv)$. Finally, $\dot{Q}$ accounts for the rate of energy transfer by heat and is considered *positive* for energy transfer *to* the control volume.

By dropping the terms of Equation 12.7a involving mass flow rates an energy rate balance for closed systems is obtained. In principle the closed system energy rate balance can be integrated for a process between two states to give the closed system energy balance:

$$(U_2 - U_1) + (KE_2 - KE_1) + (PE_2 - PE_1) = Q - W \quad (12.7b)$$
(closed systems)

where 1 and 2 denote the end states. Q and W denote the *amounts* of energy transferred by heat and work during the process, respectively.

12.2.3 Entropy Balance

Contemporary applications of engineering thermodynamics express the second law, alternatively, as an entropy balance or an exergy balance. The entropy balance is considered here.

Like mass and energy, entropy can be stored within systems and transferred across system boundaries. However, unlike mass and energy, entropy is not conserved, but generated (or produced) by *irreversibilities* within systems. A control volume form of the extensive property balance for entropy is

$$\frac{dS}{dt} = \underbrace{\sum_i \frac{\dot{Q}_j}{T_j} + \sum_i m_i s_i - \sum_e m_e s_e}_{\text{rates of entropy transfer}} + \underbrace{\dot{S}_{\text{gen}}}_{\text{rate of entropy generation}} \quad (12.8)$$

where dS/dt represents the time rate of change of entropy within the control volume. The terms $m_i s_i$ and $m_e s_e$ account, respectively, for rates of entropy transfer into and out of the control volume accompanying mass flow. $\dot{Q}_j$ represents the time rate of heat transfer at the location on the boundary where the instantaneous temperature is T_j, and $\dot{Q}_j/T_j$ accounts for the accompanying rate of entropy transfer. $\dot{S}_{\text{gen}}$ denotes the time rate of entropy generation due to irreversibilities within the control volume. An entropy rate balance for closed systems is obtained by dropping the terms of Equation 12.8 involving mass flow rates.

When applying the entropy balance in any of its forms, the objective is often to evaluate the entropy generation term. However, the value of the entropy generation for a given process of a system usually does not have much significance by itself. The significance normally is determined through comparison: the entropy generation within a given component would be compared with the entropy generation values of the other components included in an overall system formed by these components. This allows the principal contributors to the irreversibility of the overall system to be pinpointed.

12.2.4 Control Volumes at Steady State

Engineering systems are often idealized as being at *steady state*, meaning that all properties are unchanging in time. For a control volume at steady state, the identity of the matter within the control volume changes continuously, but the total amount of mass remains constant. At steady state, the mass rate balance Equation 12.5 reduces to

$$\sum_i m_i = \sum_e m_e \quad (12.9a)$$

At steady state, the energy rate balance Equation 12.7a becomes

$$0 = \dot{Q} - \dot{W} + \sum_i m_i \left(h_i + \frac{v_i^2}{2} + gz_i \right) - \sum_e m_e \left(h_e + \frac{v_e^2}{2} + gz_e \right) \quad (12.9b)$$

At steady state, the entropy rate balance Equation 12.8 reads

$$0 = \sum_j \frac{\dot{Q}_j}{T_j} + \sum_i m_i s_i - \sum_e m_e s_e + \dot{S}_{\text{gen}} \quad (12.9c)$$

Mass and energy are conserved quantities, but entropy is not generally conserved. Equation 12.9a indicates that the total rate of mass flow into the control volume equals the total rate of mass flow out of the control volume. Similarly, Equation 12.9b states that the total rate of energy transfer into the control volume equals the total rate of energy transfer out of the control volume. However, Equation 12.9c shows that the rate at which entropy is transferred out exceeds the rate at which entropy enters, the difference being the rate of entropy generation within the control volume owing to irreversibilities.

Many applications involve control volumes having a single inlet and a single exit. For such cases the mass rate balance, Equation 12.9a, reduces to $m_i = m_e$. Denoting the common mass flow rate by m,

Engineering Thermodynamics

Equations 12.9b and 12.9c give, respectively,

$$0 = \dot{Q} - \dot{W} + \dot{m}\left[(h_i - h_e) + \left(\frac{v_i^2 - v_e^2}{2}\right) + g(z_i - z_e)\right] \tag{12.10a}$$

$$0 = \frac{\dot{Q}}{T_b} + \dot{m}(s_i - s_e) + \dot{S}_{gen} \tag{12.11a}$$

where for simplicity T_b denotes the temperature, or a suitable average temperature, on the boundary where heat transfer occurs.

When energy and entropy rate balances are applied to particular cases of interest, additional simplifications are usually made. The heat transfer term $\dot{Q}$ is dropped when it is insignificant relative to other energy transfers across the boundary. This may be the result of one or more of the following: (1) the outer surface of the control volume is insulated; (2) the outer surface area is too small for there to be effective heat transfer; (3) the temperature difference between the control volume and its surroundings is small enough that the heat transfer can be ignored; (4) the gas or liquid passes through the control volume so quickly that there is not enough time for significant heat transfer to occur. The work term $\dot{W}$ drops out of the energy rate balance when there are no rotating shafts, displacements of the boundary, electrical effects, or other work mechanisms associated with the control volume being considered. The effects of kinetic and potential energy are frequently negligible relative to other terms of the energy rate balance.

The special forms of Equations 12.10a and 12.11a listed in Table 12.1 are obtained as follows: When there is no heat transfer, Equation 12.11a gives

$$s_e - s_i = \frac{\dot{S}_{gen}}{\dot{m}} \geq 0 \tag{12.11b}$$

(no heat transfer)

Accordingly, when irreversibilities are present within the control volume, the specific entropy increases as mass flows from inlet to outlet. In the *ideal* case in which no internal irreversibilities are present, mass passes through the control volume with no change in its entropy—that is, *isentropically*.

For no heat transfer, Equation 12.10a gives

$$\dot{W} = \dot{m}\left[(h_i - h_e) + \left(\frac{v_i^2 - v_e^2}{2}\right) + g(z_i - z_e)\right] \tag{12.10b}$$

(no heat transfer)

A special form that is applicable, at least approximately, to compressors, pumps, and turbines results from dropping the kinetic and potential energy terms of Equation 12.10b, leaving

$$\dot{W} = \dot{m}(h_i - h_e) \tag{12.10c}$$

(compressors, pumps, and turbines)

In *throttling devices* a significant reduction in pressure is achieved by introducing a restriction into a line through which a gas or liquid flows. For such devices $\dot{W} = 0$ and Equation 12.10c reduces further to read

$$h_i \cong h_e \tag{12.10d}$$

(throttling process)

That is, upstream and downstream of the throttling device, the specific enthalpies are equal.

TABLE 12.1 Energy and Entropy Balances for One-Inlet, One-Outlet Control Volumes at Steady State and No Heat Transfer

Energy balance

$$\dot{W} = m\left[(h_i - h_e) + \left(\frac{v_i^2 - v_e^2}{2}\right) + g(z_i - z_e)\right] \quad (12.10b)$$

Compressors, pumps, and turbines[a]

$$\dot{W} = m(h_i - h_e) \quad (12.10c)$$

Throttling

$$h_e \cong h_i \quad (12.10d)$$

Nozzles, diffusers[b]

$$v_e = \sqrt{v_i^2 + 2(h_i - h_e)} \quad (12.10e)$$

Entropy balance

$$s_e - s_i = \frac{\dot{S}_{gen}}{m} \geq 0 \quad (12.11b)$$

---

[a] For an ideal gas with constant c_p, Equation 1' of Table 12.4 allows Equation 12.10c to be written as

$$\dot{W} = mc_p(T_i - T_e) \quad (12.10c')$$

The power developed in an *isentropic process* is obtained with Equation 5' of Table 12.4 as

$$\dot{W} = mc_p T_i [1 - (p_e/p_i)^{(k-1)/k}] \quad (s = c) \quad (12.10c'')$$

where $c_p = kR/(k-1)$.

[b] For an ideal gas with constant c_p, Equation 1' of Table 12.4 allows Equation 12.10e to be written as

$$v_e = \sqrt{v_i^2 + 2c_p(T_i - T_e)} \quad (12.10e')$$

The exit velocity for an *isentropic process* is obtained with Equation 5' of Table 12.4 as

$$v_e = \sqrt{v_i^2 + 2c_p T_i [1 - (p_e/p_i)^{(k-1)/k}]} \quad (s = c) \quad (12.10e'')$$

where $c_p = kR/(k-1)$.

A *nozzle* is a flow passage of varying cross-sectional area in which the velocity of a gas or liquid increases in the direction of flow. In a *diffuser*, the gas or liquid decelerates in the direction of flow. For such devices, $\dot{W} = 0$. The heat transfer and potential energy change are generally negligible. Then Equation 12.10b reduces to

$$0 = h_i - h_e + \frac{v_i^2 - v_e^2}{2}$$

Solving for the exit velocity

$$v_e = \sqrt{v_i^2 + 2(h_i - h_e)} \quad (12.10e)$$

(nozzle, diffuser)

The steady-state forms of the mass, energy, and entropy rate balances can be applied to control volumes with multiple inlets and/or exits, for example, cases involving heat-recovery steam generators, feedwater heaters, and counterflow and crossflow heat exchangers. Transient (or unsteady) analyses can be conducted with Equations 12.5, 12.7a, and 12.8. Illustrations of all such applications are provided by Moran and Shapiro (2000).

12.2.5 Exergy Balance

Exergy provides an alternative to entropy for applying the second law. When exergy concepts are combined with principles of engineering economy, the result is known as *thermoeconomics*. Thermoeconomics allows the real cost sources to be identified: capital investment costs, operating and maintenance costs, and the costs associated with the destruction and loss of exergy. Optimization of systems can be achieved by a careful consideration of such cost sources. From this perspective thermoeconomics is *exergy-aided cost minimization*. Discussions of exergy analysis and thermoeconomics are provided by Moran (1989), Bejan et al. (1996), Moran and Tsatsaronis (2000), and Moran and Shapiro (2000). In this section salient aspects are presented.

12.2.5.1 Defining Exergy

An opportunity for doing work exists whenever two systems at different states are placed in communication because, in principle, work can be developed as the two are allowed to come into equilibrium. When one of the two systems is a suitably idealized system called an *environment* and the other is some system of interest, *exergy* is the maximum theoretical useful work (shaft work or electrical work) obtainable as the system of interest and environment interact to equilibrium, heat transfer occurring with the environment only. (Alternatively, exergy is the minimum theoretical useful work required to form a quantity of matter from substances present in the environment and bring the matter to a specified state.) Exergy is a measure of the *departure* of the state of the system from that of the environment, and is therefore an attribute of the system and environment together. Once the environment is specified, however, a value can be assigned to exergy in terms of property values for the system only, so exergy can be regarded as an extensive property of the system. Exergy can be destroyed and, like entropy, generally is not conserved.

Models with various levels of specificity are employed for describing the environment used to evaluate exergy. Models of the environment typically refer to some portion of a system's surroundings, the intensive properties of each phase of which are uniform and do not change significantly as a result of any process under consideration. The environment is regarded as composed of common substances existing in abundance within the Earth's atmosphere, oceans, and crust. The substances are in their stable forms as they exist naturally, and there is no possibility of developing work from interactions—physical or chemical—between parts of the environment. Although the intensive properties of the environment are assumed to be unchanging, the extensive properties can change as a result of interactions with other systems. Kinetic and potential energies are evaluated relative to coordinates in the environment, all parts of which are considered to be at rest with respect to one another. For computational ease, the temperature T_0 and pressure p_0 of the environment are often taken as typical ambient values, such as 1 atm and 25°C (77°F). However, these properties may be specified differently depending on the application.

When a system is in equilibrium with the environment, the state of the system is called the *dead state*. At the dead state, the conditions of mechanical, thermal, and chemical equilibrium between the system and the environment are satisfied: the pressure, temperature, and chemical potentials of the system equal those of the environment, respectively. In addition, the system has no motion or elevation relative to coordinates in the environment. Under these conditions, there is no possibility of a spontaneous change within the system or the environment, nor can there be an interaction between them. The value of exergy is zero. Another type of equilibrium between the system and environment can be identified. This is a restricted form of equilibrium where only the conditions of mechanical and thermal equilibrium must be satisfied. This state of the system is called the *restricted dead state*. At the restricted dead state, the fixed quantity of matter under consideration is imagined to be sealed in an envelope impervious to mass flow, at zero velocity and elevation relative to coordinates in the environment, and at the temperature T_0 and pressure p_0.

12.2.5.2 Exergy Transfer and Exergy Destruction

Exergy can be transferred by three means: exergy transfer associated with work, exergy transfer associated with heat transfer, and exergy transfer associated with the matter entering and exiting a control volume. All such exergy transfers are evaluated relative to the environment used to define exergy. Exergy also is destroyed by irreversibilities within the system or control volume. Exergy balances can be written in various forms, depending on whether a closed system or control volume is under consideration and whether steady-state or transient operation is of interest. Owing to its importance for a wide range of applications, an exergy rate balance for control volumes at steady state is presented alternatively as Equations 12.12a and 12.12b.

$$0 = \underbrace{\sum_i \dot{E}_{q,j} - \dot{W} + \sum_i \dot{E}_i - \sum_e \dot{E}_e}_{\text{rates of exergy transfer}} - \underbrace{\dot{E}_D}_{\text{rate of exergy destruction}} \quad (12.12a)$$

$$0 = \sum_i \left(1 - \frac{T_0}{T_j}\right)\dot{Q}_j - \dot{W} + \sum_i m_i e_i - \sum_e m_e e_e - \dot{E}_D \quad (12.12b)$$

$\dot{W}$ has the same significance as in Equation 12.7a: the work rate excluding the flow work. $\dot{Q}_j$ is the time rate of heat transfer at the location on the boundary of the control volume where the instantaneous temperature is T_j. The associated rate of exergy transfer is

$$\dot{E}_{q,j} = \left(1 - \frac{T_0}{T_j}\right)\dot{Q}_j \quad (12.13)$$

As for other control volume rate balances, the subscripts i and e denote inlets and exits, respectively. The exergy transfer rates at control volume inlets and exits are denoted, respectively, as $\dot{E}_i = m_i e_i$ and $\dot{E}_e = m_e e_e$. Finally, $\dot{E}_D$ accounts for the time rate of exergy destruction due to irreversibilities within the control volume. The exergy destruction rate is related to the entropy generation rate by

$$\dot{E}_D = T_0 \dot{S}_{\text{gen}} \quad (12.14)$$

The specific exergy transfer terms e_i and e_e are expressible in terms of four components: physical exergy e^{PH}, kinetic exergy e^{KN}, potential exergy e^{PT}, and chemical exergy e^{CH}:

$$e = e^{PH} + e^{KN} + e^{PT} + e^{CH} \quad (12.15a)$$

The first three components are evaluated as follows:

$$e^{PH} = (h - h_0) - T_0(s - s_0) \quad (12.15b)$$

$$e^{KN} = \frac{1}{2}v^2 \quad (12.15c)$$

$$e^{PT} = gz \quad (12.15d)$$

In Equation 12.15b, h_0 and s_0 denote, respectively, the specific enthalpy and specific entropy at the restricted dead state. In Equations 12.15c and 12.15d, v and z denote velocity and elevation relative to coordinates in the environment, respectively.

To evaluate the chemical exergy (the exergy component associated with the departure of the chemical composition of a system from that of the environment), alternative models of the environment can be employed depending on the application; see for example Moran (1989) and Kotas (1995). Exergy analysis is facilitated, however, by employing a *standard environment* and a corresponding table of *standard chemical exergies*. Standard chemical exergies are based on standard values of the environmental temperature T_0 and pressure p_0—for example, 298.15 K (25°C) and 1 atm, respectively. Standard environments also include a set of reference substances with standard concentrations reflecting as closely as possible the chemical makeup of the natural environment. Standard chemical exergy data is provided by Szargut et al. (1988), Bejan et al. (1996), and Moran and Shapiro (2000).

12.2.5.3 Guidelines for Improving Thermodynamic Effectiveness

To improve thermodynamic effectiveness it is necessary to deal directly with inefficiencies related to exergy destruction and exergy loss. The primary contributors to exergy destruction are chemical reaction, heat transfer, mixing, and friction, including unrestrained expansions of gases and liquids. To deal with them effectively, the principal sources of inefficiency not only should be understood qualitatively, but also determined quantitatively, at least approximately. Design changes to improve effectiveness must be done judiciously, however, for the cost associated with different sources of inefficiency can be different. For example, the unit cost of the electrical or mechanical power required to provide for the exergy destroyed owing to a pressure drop is generally higher than the unit cost of the fuel required for the exergy destruction caused by combustion or heat transfer.

Chemical reaction is a significant source of thermodynamic inefficiency. Accordingly, it is generally good practice to minimize the use of combustion. In many applications the use of combustion equipment such as boilers is unavoidable, however. In these cases a significant reduction in the combustion irreversibility by conventional means simply cannot be expected, for the major part of the exergy destruction introduced by combustion is an inevitable consequence of incorporating such equipment. Still, the exergy destruction in practical combustion systems can be reduced by minimizing the use of excess air and by preheating the reactants. In most cases only a small part of the exergy destruction in a combustion chamber can be avoided by these means. Consequently, after considering such options for reducing the exergy destruction related to combustion, efforts to improve thermodynamic performance should focus on components of the overall system that are more amenable to betterment by cost-effective measures. In other words, some exergy destructions and energy losses can be avoided, others cannot. Efforts should be centered on those that can be avoided.

Nonidealities associated with heat transfer also typically contribute heavily to inefficiency. Accordingly, unnecessary or cost-ineffective heat transfer must be avoided. Additional guidelines follow:

- The higher the temperature T at which a heat transfer occurs in cases where $T > T_0$, where T_0 denotes the temperature of the environment, the more valuable the heat transfer and, consequently, the greater the need to avoid heat transfer to the ambient, to cooling water, or to a refrigerated stream. Heat transfer across T_0 should be avoided.
- The lower the temperature T at which a heat transfer occurs in cases where $T < T_0$, the more valuable the heat transfer and, consequently, the greater the need to avoid direct heat transfer with the ambient or a heated stream.
- Since exergy destruction associated with heat transfer between streams varies inversely with the temperature level, the lower the temperature level, the greater the need to minimize the stream-to-stream temperature difference.

Although irreversibilities related to friction, unrestrained expansion, and mixing are often less significant than combustion and heat transfer, they should not be overlooked, and the following guidelines apply:

- Relatively more attention should be paid to the design of the lower temperature stages of turbines and compressors (the last stages of turbines and the first stages of compressors) than to the remaining stages of these devices. For turbines, compressors, and motors, consider the most thermodynamically efficient options.
- Minimize the use of throttling; check whether power recovery expanders are a cost-effective alternative for pressure reduction.

TABLE 12.2 Symbols and Definitions for Selected Properties

Property	Symbol	Definition	Property	Symbol	Definition
Pressure	p		Specific heat, constant volume	c_v	$(\partial u/\partial T)_v$
Temperature	T		Specific heat, constant pressure	c_p	$(\partial h/\partial T)_p$
Specific volume	v		Volume expansivity	β	$\frac{1}{v}(\partial v/\partial T)_p$
Specific internal energy	u		Isothermal compressivity	κ	$-\frac{1}{v}(\partial v/\partial p)_T$
Specific entropy	s		Isentropic compressibility	α	$-\frac{1}{v}(\partial v/\partial p)_s$
Specific enthalpy	h	$u + pv$	Isothermal bulk modulus	B	$-v(\partial p/\partial v)_T$
Specific Helmholtz function	ψ	$u - Ts$	Isentropic bulk modulus	B_s	$-v(\partial p/\partial v)_s$
Specific Gibbs function	g	$h - Ts$	Joule–Thomson coefficient	μ_J	$(\partial T/\partial p)_h$
Compressibility factor	Z	pv/RT	Joule coefficient	η	$(\partial T/\partial v)_u$
Specific heat ratio	k	c_p/c_v	Velocity of sound	c	$\sqrt{-v^2(\partial p/\partial v)_s}$

- Avoid processes using excessively large thermodynamic driving forces (differences in temperature, pressure, and chemical composition). In particular, minimize the mixing of streams differing significantly in temperature, pressure, or chemical composition.
- The greater the mass flow rate the greater the need to use the exergy of the stream effectively.

Discussion of means for improving thermodynamic effectiveness also is provided by Bejan et al. (1996) and Moran and Tsatsaronis (2000).

12.3 Property Relations and Data

Engineering thermodynamics uses a wide assortment of thermodynamic properties and relations among these properties. Table 12.2 lists several commonly encountered properties. Pressure, temperature, and specific volume can be found experimentally. Specific internal energy, entropy, and enthalpy are among those properties that are not so readily obtained in the laboratory. Values for such properties are calculated using experimental data of properties that are more amenable to measurement, together with appropriate property relations derived using the principles of thermodynamics.

Property data are provided in the publications of the National Institute of Standards and Technology (formerly the U.S. Bureau of Standards), of professional groups such as the American Society of Mechanical Engineers (ASME), the American Society of Heating, Refrigerating, and Air Conditioning Engineers (ASHRAE), and the American Chemical Society, and of corporate entities such as Dupont and Dow Chemical. Handbooks and property reference volumes such as included in the list of references for this chapter are readily accessed sources of data. Property data also are retrievable from various commercial online data bases. Computer software increasingly is available for this purpose as well.

12.3.1 *P-v-T* Surface

Considerable pressure, specific volume, and temperature data have been accumulated for industrially important gases and liquids. These data can be represented in the form $p = f(v, T)$, called an *equation of state*. Equations of state can be expressed in graphical, tabular, and analytical forms. Figure 12.1a shows the *p–v–T* relationship for water. Figure 12.1b shows the projection of the *p–v–T surface* onto the pressure–temperature plane, called the *phase diagram*. The projection onto the *p–v* plane is shown in Figure 12.1c.

Figure 12.1a has three regions labeled solid, liquid, and vapor where the substance exists only in a single phase. Between the single phase regions lie *two-phase* regions, where two phases coexist in equilibrium. The lines separating the single-phase regions from the two-phase regions are *saturation lines*. Any state represented by a point on a saturation line is a *saturation state*. The line separating the liquid

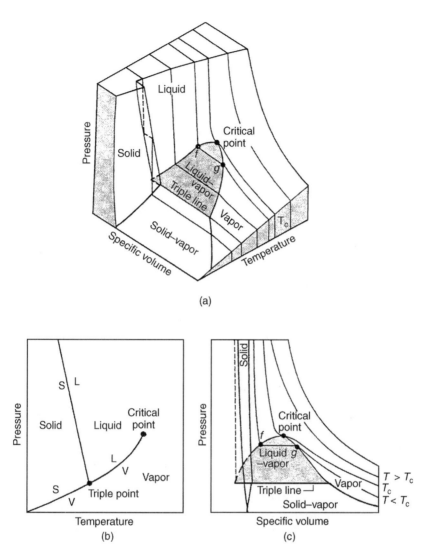

FIGURE 12.1 Pressure-specific volume–temperature surface and projections for water (not to scale).

phase and the two-phase liquid–vapor region is the saturated liquid line. The state denoted by f is a saturated liquid state. The saturated vapor line separates the vapor region and the two-phase liquid–vapor region. The state denoted by g is a saturated vapor state. The saturated liquid line and the saturated vapor line meet at the *critical point*. At the critical point, the pressure is the *critical pressure* p_c, and the temperature is the *critical temperature* T_c. Three phases can coexist in equilibrium along the line labeled *triple line*. The triple line projects onto a point on the phase diagram: the triple point.

When a phase change occurs during constant pressure heating or cooling, the temperature remains constant as long as both phases are present. Accordingly, in the two-phase liquid–vapor region, a line of constant pressure is also a line of constant temperature. For a specified pressure, the corresponding temperature is called the *saturation temperature*. For a specified temperature, the corresponding pressure is called the *saturation pressure*. The region to the right of the saturated vapor line is known as the *superheated vapor region* because the vapor exists at a temperature greater than the saturation temperature for its pressure. The region to the left of the saturated liquid line is known as the *compressed liquid region* because the liquid is at a pressure higher than the saturation pressure for its temperature.

When a mixture of liquid and vapor coexists in equilibrium, the liquid phase is a saturated liquid and the vapor phase is a saturated vapor. The total volume of any such mixture is $V = V_f + V_g$; or, alternatively, $mv = m_f v_f + m_g v_g$, where m and v denote mass and specific volume, respectively. Dividing by the total mass of the mixture m and letting the *mass fraction* of the vapor in the mixture, m_g/m, be symbolized by x, called the *quality*, the apparent specific volume v of the mixture is

$$v = (1-x)v_f + xv_g = v_f + xv_{fg} \tag{12.16a}$$

where $v_{fg} = v_g - v_f$. Expressions similar in form can be written for internal energy, enthalpy, and entropy:

$$u = (1-x)u_f + xu_g = u_f + xu_{fg} \tag{12.16b}$$

$$h = (1-x)h_f + xh_g = h_f + xh_{fg} \tag{12.16c}$$

$$s = (1-x)s_f + xs_g = s_f + xs_{fg} \tag{12.16d}$$

12.3.2 Thermodynamic Data Retrieval

Tabular presentations of pressure, specific volume, and temperature are available for practically important gases and liquids. The tables normally include other properties useful for thermodynamic analyses, such as internal energy, enthalpy, and entropy. The various *steam tables* included in the references of this chapter provide examples. Computer software for retrieving the properties of a wide range of substances is also available, as, for example, the ASME Steam Tables (1993) and Bornakke and Sonntag (1996). Increasingly, textbooks come with computer disks providing thermodynamic property data for water, certain refrigerants, and several gases modeled as ideal gases (see, e.g., Moran and Shapiro 2000).

The sample *steam table data* presented in Table 12.3 are representative of data available for substances commonly encountered in engineering practice. The form of the tables and how they are used are assumed to be familiar. In particular, the use of *linear interpolation* with such tables is assumed known.

Specific internal energy, enthalpy, and entropy data are determined relative to arbitrary datums and such datums vary from substance to substance. Referring to Table 12.3a, the datum state for the specific internal energy and specific entropy of water is seen to correspond to saturated liquid water at 0.01°C (32.02°F), the triple point temperature. The value of each of these properties is set to zero at this state. If calculations are performed involving only differences in a particular specific property, the datum cancels. When there are changes in chemical composition during the process, special care must be exercised. The approach followed when composition changes due to chemical reaction is considered in Moran and Shapiro (2000).

Liquid water data (see Table 12.3d) suggests that at fixed temperature the variation of specific volume, internal energy, and entropy with pressure is slight. The variation of specific enthalpy with pressure at fixed temperature is somewhat greater because pressure is explicit in the definition of enthalpy. This behavior for v, u, s, and h is exhibited generally by liquid data and provides the basis for the following set of equations for estimating property data at liquid states from saturated liquid data:

$$v(T, p) \approx v_f(T) \tag{12.17a}$$

$$u(T, p) \approx u_f(T) \tag{12.17b}$$

$$h(T, p) \approx h_f(T) + v_f[p - p_{\text{sat}}(T)] \tag{12.17c}$$

$$s(T, p) \approx s_f(T) \tag{12.17d}$$

The subscript f denotes the saturated liquid state at the temperature T, and p_{sat} is the corresponding saturation pressure. The underlined term of Equation 12.17c is usually negligible, giving $h(T, p) \approx h_f(T)$.

Graphical representations of property data also are commonly used. These include the p–T and p–v diagrams of Figure 12.1, the T-s diagram of Figure 12.2, the h–s (Mollier) diagram of Figure 12.3, and the p–h diagram of Figure 12.4. The compressibility charts considered next use the compressibility factor as one of the coordinates.

12.3.3 Compressibility Charts

The p–v–T relation for a wide range of common gases is illustrated by the generalized compressibility chart of Figure 12.5. In this chart, the compressibility factor, Z, is plotted versus the *reduced* pressure, p_R, *reduced* temperature, T_R, and *pseudoreduced* specific volume, v'_R where

$$Z = \frac{p\bar{v}}{\bar{R}T} \qquad (12.18)$$

In this expression $\bar{v}$ is the specific volume on a molar basis (e.g., m³/kmol) and $\bar{R}$ is the *universal gas constant* (8314 N · m/kmol · K, for example). The reduced properties are

$$p_R = \frac{p}{p_c}, \qquad T_R = \frac{T}{T_c}, \qquad v'_R = \frac{\bar{v}}{(\bar{R}T_c/p_c)} \qquad (12.19)$$

where p_c and T_c denote the critical pressure and temperature, respectively. Values of p_c and T_c are obtainable from the literature (see, eg., Moran and Shapiro 2000). The reduced isotherms of Figure 12.5 represent the best curves fitted to the data of several gases. For the 30 gases used in developing the chart, the deviation of observed values from those of the chart is at most on the order of 5% and for most ranges is much less.

12.3.4 Analytical Equations of State

Considering the isotherms of Figure 12.5, it is plausible that the variation of the compressibility factor might be expressed as an equation, at least for certain intervals of p and T. Two expressions can be written that enjoy a theoretical basis. One gives the compressibility factor as an infinite series expansion in pressure,

$$Z = 1 + \hat{B}(T)p + \hat{C}(T)p^2 + \hat{D}(T)p^3 + \cdots \qquad (12.20a)$$

and the other is a series in $1/\bar{v}$,

$$Z = 1 + \frac{B(T)}{\bar{v}} + \frac{C(T)}{\bar{v}^2} + \frac{D(T)}{\bar{v}^3} + \cdots \qquad (12.20b)$$

Such equations of state are known as *virial expansions*, and the coefficients $\hat{B}$, $\hat{C}$, $\hat{D}$,... and B, C, D,... are called *virial coefficients*. In principle, the virial coefficients can be calculated using expressions from statistical mechanics derived from consideration of the force fields around the molecules. Thus far the first few coefficients have been calculated for gases consisting of relatively simple molecules. The coefficients also can be found, in principle, by fitting p–v–T data in particular realms of interest. Only the first few coefficients can be found accurately this way, however, and the result is a *truncated* equation valid only at certain states.

Over 100 equations of state have been developed in an attempt to portray accurately the p–v–T behavior of substances and yet avoid the complexities inherent in a full virial series. In general, these equations exhibit little in the way of fundamental physical significance and are mainly empirical in character. Most are developed for gases, but some describe the p–v–T behavior of the liquid phase, at least qualitatively.

TABLE 12.3 Sample Steam Table Data

(a) Properties of Saturated Water (Liquid–Vapor): Temperature Table

Temp (°C)	Pressure (bar)	Specific Volume (m³/kg)		Internal Energy (kJ/kg)		Enthalpy (kJ/kg)			Entropy (kJ/kg · K)	
		Saturated Liquid ($v_f \times 10^3$)	Saturated Vapor (v_g)	Saturated Liquid (u_f)	Saturated Vapor (u_g)	Saturated Liquid (h_f)	Evap. (h_{fg})	Saturated Vapor (h_g)	Saturated Liquid (s_f)	Saturated Vapor (s_g)
.01	0.00611	1.0002	206.136	0.00	2375.3	0.01	2501.3	2501.4	0.0000	9.1562
4	0.00813	1.0001	157.232	16.77	2380.9	16.78	2491.9	2508.7	0.0610	9.0514
5	0.00872	1.0001	147.120	20.97	2382.3	20.98	2489.6	2510.6	0.0761	9.0257
6	0.00935	1.0001	137.734	25.19	2383.6	25.20	2487.2	2512.4	0.0912	9.0003
8	0.01072	1.0002	120.917	33.59	2386.4	33.60	2482.5	2516.1	0.1212	8.9501

(b) Properties of Saturated Water (Liquid–Vapor): Pressure Table

Pressure (bar)	Temp (°C)	Specific Volume (m³/kg)		Internal Energy (kJ/kg)		Enthalpy (kJ/kg)			Entropy (kJ/kg · K)	
		Saturated Liquid ($v_f \times 10^3$)	Saturated Vapor (v_g)	Saturated Liquid (u_f)	Saturated Vapor (u_g)	Saturated Liquid (h_f)	Evap. (h_{fg})	Saturated Vapor (h_g)	Saturated Liquid (s_f)	Saturated Vapor (s_g)
0.04	28.96	1.0040	34.800	121.45	2415.2	121.46	2432.9	2554.4	0.4226	8.4746
0.06	36.16	1.0064	23.739	151.53	2425.0	151.53	2415.9	2567.4	0.5210	8.3304
0.08	41.51	1.0084	18.103	173.87	2432.2	173.88	2403.1	2577.0	0.5926	8.2287
0.10	45.81	1.0102	14.674	191.82	2437.9	191.83	2392.8	2584.7	0.6493	8.1502
0.20	60.06	1.0172	7.649	251.38	2456.7	251.40	2358.3	2609.7	0.8320	7.9085

(c) Properties of Superheated Water Vapor

T(°C)	v(m³/kg)	u(kJ/kg)	h(kJ/kg)	s(kJ/kg·K)	v(m³/kg)	u(kJ/kg)	h(kJ/kg)	s(kJ/kg·K)
	p = 0.06 bar = 0.006 MPa (T_{sat} = 36.16°C)				p = 0.35 bar = 0.035 MPa (T_{sat} = 72.69°C)			
Sat.	23.739	2425.0	2567.4	8.3304	4.526	2473.0	2631.4	7.7158
80	27.132	2487.3	2650.1	8.5804	4.625	2483.7	2645.6	7.7564
120	30.219	2544.7	2726.0	8.7840	5.163	2542.4	2723.1	7.9644
160	33.302	2602.7	2802.5	8.9693	5.696	2601.2	2800.6	8.1519
200	36.383	2661.4	2879.7	9.1398	6.228	2660.4	2878.4	8.3237

(d) Properties of Compressed Liquid Water

T(°C)	v × 10³ (m³/kg)	u(kJ/kg)	h(kJ/kg)	s(kJ/kg·K)	v × 10³ (m³/kg)	u(kJ/kg)	h(kJ/kg)	s(kJ/kg·K)
	p = 25 bar = 2.5 MPa (T_{sat} = 223.99°C)				p = 50 bar = 5.0 MPa (T_{sat} = 263.99°C)			
20	1.0006	83.80	86.30	0.2961	0.9995	83.65	88.65	0.2956
80	1.0280	334.29	336.86	1.0737	1.0268	333.72	338.85	1.0720
140	1.0784	587.82	590.52	1.7369	1.0768	586.76	592.15	1.7343
200	1.1555	849.9	852.8	2.3294	1.1530	848.1	853.9	2.3255
Sat.	1.1973	959.1	962.1	2.5546	1.2859	1147.8	1154.2	2.9202

Source: Moran, M.J. and Shapiro, H.N. 2000. *Fundamentals of Engineering Thermodynamics*, 4th ed. Wiley, New York, as extracted from Keenan, J.H., Keyes, F.G., Hill, P.G., and Moore, J.G. 1969. *Steam Tables*. Wiley, New York.

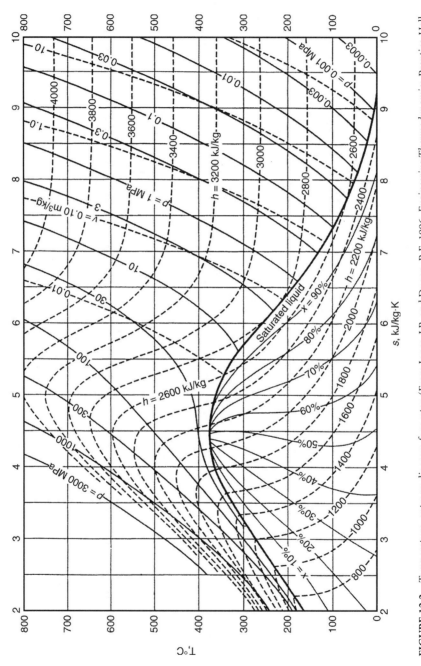

FIGURE 12.2 Temperature–entropy diagram for water. (From Jones, J.B. and Dugan, R.E. 1996. *Engineering Thermodynamics*, Prentice-Hall, Englewood Cliffs, NJ, based on data and formulations from Haar, L., Gallagher, J.S., and Kell, G.S. 1984. *NBS/NRC Steam Tables*. Hemisphere, Washington, D.C.)

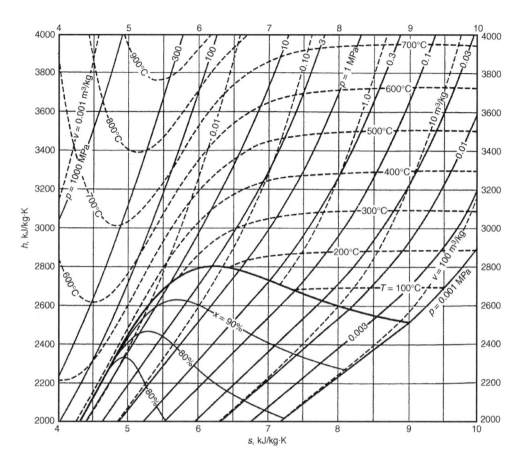

FIGURE 12.3 Enthalpy–entropy (Mollier) diagram for water. (From Jones, J.B. and Dugan, R.E. 1996. *Engineering Thermodynamics*. Prentice-Hall, Englewood Cliffs, NJ, based on data and formulations from Haar, L., Gallagher, J.S., and Kell, G.S. 1984. *NBS/NRC Steam Tables*. Hemisphere, Washington, D.C.)

Every equation of state is restricted to particular states. The realm of applicability is often indicated by giving an interval of pressure, or density, where the equation can be expected to represent the p–v–T behavior faithfully. For further discussion of equations of state see Reid and Sherwood (1966) and Reid et al. (1987).

12.3.5 Ideal Gas Model

Inspection of the generalized compressibility chart, Figure 12.5, shows that when p_R is small, and for many states when T_R is large, the value of the compressibility factor Z is close to 1. In other words, for pressures that are low relative to p_c, and for many states with temperatures high relative to T_c, the compressibility factor approaches a value of 1. Within the indicated limits, it may be assumed with reasonable accuracy that $Z = 1$ that is,

$$p\bar{v} = \bar{R}T \quad \text{or} \quad pv = RT \qquad (12.21a)$$

Other forms of this expression in common use are

$$pV = n\bar{R}T, \quad pV = mRT \qquad (12.21b)$$

In these equations, $n = m/M$, $\bar{v} = Mv$, and the *specific gas constant* is $R = \bar{R}/M$, where M denotes the molecular weight.

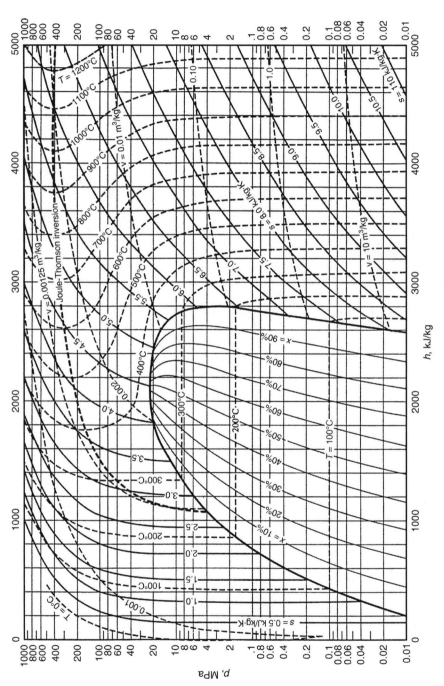

FIGURE 12.4 Pressure–enthalpy diagram for water. (From Jones, J.B. and Dugan, R.E. 1996. *Engineering Thermodynamics*. Prentice-Hall, Englewood Cliffs, NJ, based on data and formulations from Haar, L., Gallagher, J.S., and Kell, G.S. 1984. *NBS/NRC Steam Tables*. Hemisphere, Washington, D.C.)

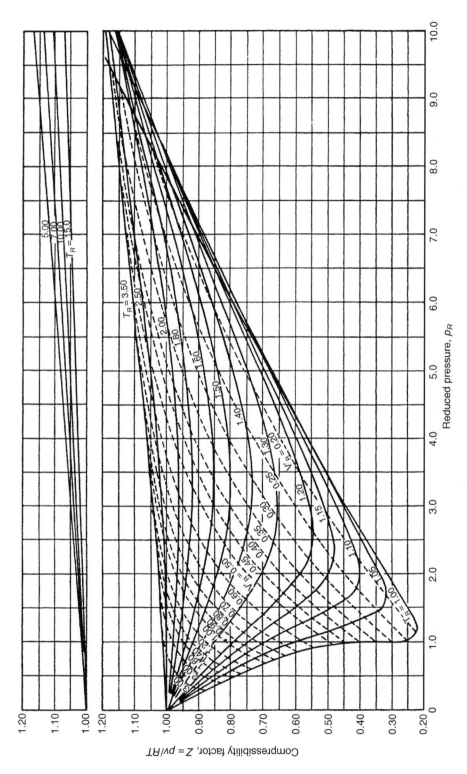

FIGURE 12.5 Generalized compressibility chart ($T_R = T/T_c$, $p_R = p/p_c$, $v_R' = v p_c / \bar{R} T_c$) for $p_R \leq 10$. (From Obert, E.F. 1960 *Concepts of Thermodynamics*. McGraw-Hill, New York.)

TABLE 12.4 Ideal Gas Expressions for Δh, Δu, and Δs

Variable Specific Heats		Constant Specific Heats[b]	
$h(T_2) - h(T_1) = \int_{T_1}^{T_2} c_p(T)\, dT$	(1)	$h(T_2) - h(T_1) = c_p(T_2 - T_1)$	(1')
$s(T_2, p_2) - s(T_1, p_1) = \int_{T_1}^{T_2} \frac{c_p(T)}{T} dT - R \ln \frac{p_2}{p_1}$	(2)[a]	$s(T_2, p_2) - s(T_1, p_1) = c_p \ln \frac{T_2}{T_1} - R \ln \frac{p_2}{p_1}$	(2')
$u(T_2) - u(T_1) = \int_{T_1}^{T_2} c_v(T)\, dT$	(3)	$u(T_2) - u(T_1) = c_v(T_2 - T_1)$	(3')
$s(T_2, v_2) - s(T_1, v_1) = \int_{T_1}^{T_2} \frac{c_v(T)}{T} dT + R \ln \frac{v_2}{v_1}$	(4)	$s(T_2, v_2) - s(T_1, v_1) = c_v \ln \frac{T_2}{T_1} + R \ln \frac{v_2}{v_1}$	(4')
$s_2 = s_1$		$s_2 = s_1$	
$\frac{p_r(T_2)}{p_r(T_1)} = \frac{p_2}{p_1}$	(5)	$\frac{T_2}{T_1} = \left(\frac{p_2}{p_1}\right)^{(k-1)/k}$	(5')
$\frac{v_r(T_2)}{v_r(T_1)} = \frac{v_2}{v_1}$	(6)	$\frac{T_2}{T_1} = \left(\frac{v_2}{v_1}\right)^{k-1}$	(6')

[a] Alternatively, $s(T_2, p_2) - s(T_1, p_1) = s°(T_2) - s°(T_1) - R \ln \frac{p_2}{p_1}$.
[b] c_p and c_v are average values over the temperature interval from T_1 to T_2.

It can be shown that $(\partial u/\partial v)_T$ vanishes identically for a gas whose equation of state is exactly given by Equation 12.21, and thus the specific internal energy depends only on temperature. This conclusion is supported by experimental observations beginning with the work of Joule, who showed that the internal energy of air at low density depends primarily on temperature.

The above considerations allow for an *ideal gas model* of each real gas: (1) the equation of state is given by Equations 12.21 and 12.22 the internal energy, enthalpy, and specific heats (Table 12.2) are functions of temperature alone. The real gas approaches the model in the limit of low reduced pressure. At other states the actual behavior may depart substantially from the predictions of the model. Accordingly, caution should be exercised when invoking the ideal gas model lest error is introduced.

Specific heat data for gases can be obtained by direct measurement. When extrapolated to zero pressure, ideal gas-specific heats result. Ideal gas-specific heats also can be calculated using molecular models of matter together with data from spectroscopic measurements. The following ideal gas-specific heat relations are frequently useful:

$$c_p(T) = c_v(T) + R \quad (12.22a)$$

$$c_p = \frac{kR}{k-1}, \quad c_v = \frac{R}{k-1} \quad (12.22b)$$

where $k = c_p/c_v$.

For processes of an ideal gas between states 1 and 2, Table 12.4 gives expressions for evaluating the changes in specific enthalpy, Δh, specific entropy, Δs, and specific internal energy, Δu. Relations also are provided for processes of an ideal gas between states having the same specific entropy: $s_2 = s_1$. Property relations and data required by the expressions of Table 12.4: h, u, c_p, c_v, p_r, v_r, and $s°$ are obtainable from the literature (see, e.g., Moran and Shapiro 2000).

12.4 Vapor and Gas Power Cycles

Vapor and gas power systems develop electrical or mechanical power from sources of chemical, solar, or nuclear origin. In *vapor* power systems the *working fluid*, normally water, undergoes a phase change from liquid to vapor, and conversely. In *gas* power systems, the working fluid remains a gas throughout, although the composition normally varies owing to the introduction of a fuel and subsequent combustion.

Engineering Thermodynamics

The processes taking place in power systems are sufficiently complicated that idealizations are typically employed to develop tractable thermodynamic models. The *air standard analysis* of gas power systems considered in the present section is a noteworthy example. Depending on the degree of idealization, such models may provide only qualitative information about the performance of the corresponding real-world systems. Yet such information frequently is useful in gauging how changes in major operating parameters might affect actual performance. Elementary thermodynamic models also can provide simple settings to assess, at least approximately, the advantages and disadvantages of features proposed to improve thermodynamic performance.

12.4.1 Work and Heat Transfer in Internally Reversible Processes

Expressions giving work and heat transfer in internally reversible processes are useful in describing the themodynamic performance of vapor and gas cycles. Important special cases are presented in the discussion to follow. For a gas as the system, the work of expansion arises from the force exerted by the system to move the boundary against the resistance offered by the surroundings:

$$W = \int_1^2 F\, dx = \int_1^2 pA\, dx$$

where the force is the product of the moving area and the pressure exerted by the system there. Noting that $A dx$ is the change in total volume of the system,

$$W = \int_1^2 p\, dV$$

This expression for work applies to both actual and internal expansion processes. However, for an internally reversible process p is not only the pressure at the moving boundary but also the pressure throughout the system. Furthermore, for an internally reversible process the volume equals mv, where the specific volume v has a single value throughout the system at a given instant. Accordingly, the work of an internally reversible expansion (or compression) process per unit of system mass is

$$\left(\frac{W}{m}\right)_{\substack{int \\ rev}} = \int_1^2 p\, dv \tag{12.23}$$

When such a process of a closed system is represented by a continuous curve on a plot of pressure vs. specific volume, the area under the curve is the magnitude of the work per unit of system mass: area a–b–c′–d′ of Figure 12.6.

For one-inlet, one-exit control volumes in the absence of internal irreversibilities, the following expression gives the work developed per unit of mass flowing:

$$\left(\frac{\dot{W}}{\dot{m}}\right)_{\substack{int \\ rev}} = -\int_i^e v\, dp + \frac{v_i^2 - v_e^2}{2} + g(z_i - z_e) \tag{12.24a}$$

where the integral is performed from inlet to exit (see Moran and Shapiro (2000) for discussion). If there is no significant change in kinetic or potential energy from inlet to exit, Equation 12.24a reads

$$\left(\frac{\dot{W}}{\dot{m}}\right)_{\substack{int \\ rev}} = -\int_i^e v\, dp \quad (\Delta ke = \Delta pe = 0) \tag{12.24b}$$

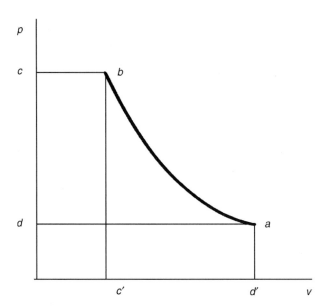

FIGURE 12.6 Internally reversible process on p–v coordinates.

The specific volume remains approximately constant in many applications with liquids. Then Equation 12.24b becomes

$$\left(\frac{\dot{W}}{m}\right)_{\substack{\text{int}\\\text{rev}}} = -v(p_e - p_i) \quad (v = \text{constant}) \tag{12.24c}$$

When the states visited by a unit of mass flowing without irreversibilities from inlet to outlet are described by a continuous curve on a plot pressure vs. specific volume, as shown in Figure 12.6, the magnitude of the integral $\int v\,dp$ of Equations 12.24a and b is represented by the area a–b–c–d *behind* the curve.

For an internally reversible process of a closed system between state 1 and state 2, the heat transfer per unit of system mass is

$$\left(\frac{Q}{m}\right)_{\substack{\text{int}\\\text{rev}}} = \int_1^2 T\,ds \tag{12.25}$$

For a one-inlet, one-exit control volume in the absence of internal irreversibilities, the following expression gives the heat transfer per unit of mass flowing from inlet i to exit e:

$$\left(\frac{Q}{m}\right)_{\substack{\text{int}\\\text{rev}}} = \int_i^e T\,ds \tag{12.26}$$

When any such process is represented by a continuous curve on a plot of temperature vs. specific entropy, the area *under* the curve is the magnitude of the heat transfer per unit of mass.

12.4.1.1 Polytropic Processes

An internally reversible process described by the expression pv^n = constant is called a *polytropic process* and n is the *polytropic exponent*. In certain applications n can be obtained by fitting pressure-specific volume data. Although this expression can be applied when real gases are considered, it most generally appears in practice together with the use of the ideal gas model. Table 12.5 provides several expressions applicable to polytropic processes and the special forms they take when the ideal gas model is assumed. The expressions for $\int p\,dv$ and $\int v\,dp$ have application to work evaluations with Equations 12.23 and 12.24, respectively.

TABLE 12.5 Polytropic Processes: pv^n = Constant[a]

General	Ideal Gas[b]
$\dfrac{p_2}{p_1} = \left(\dfrac{v_1}{v_2}\right)^n$ (1)	$\dfrac{p_2}{p_1} = \left(\dfrac{v_1}{v_2}\right)^n = \left(\dfrac{T_2}{T_1}\right)^{n/(n-1)}$ (1')
$n = 0$: constant pressure $n = \pm\infty$: constant specific volume	$n = 0$: constant pressure $n = \pm\infty$: constant specific volume $n = 1$: constant temperature $n = k$: constant specific entropy when k is constant
$n = 1$	$n = 1$
$\displaystyle\int_1^2 p\,dv = p_1 v_1 \ln\dfrac{v_2}{v_1}$ (2)	$\displaystyle\int_1^2 p\,dv = RT\ln\dfrac{v_2}{v_1}$ (2')
$-\displaystyle\int_1^2 v\,dp = -p_1 v_1 \ln\dfrac{p_2}{p_1}$ (3)	$-\displaystyle\int_1^2 v\,dp = -RT\ln\dfrac{p_2}{p_1}$ (3')
$n \neq 1$	$n \neq 1$
$\displaystyle\int_1^2 p\,dv = \dfrac{p_2 v_2 - p_1 v_1}{1-n}$ $= \dfrac{p_1 v_1}{n-1}\left[1 - \left(\dfrac{p_2}{p_1}\right)^{(n-1)/n}\right]$ (4)	$\displaystyle\int_1^2 p\,dv = \dfrac{R(T_2 - T_1)}{1-n}$ $= \dfrac{RT_1}{n-1}\left[1 - \left(\dfrac{p_2}{p_1}\right)^{(n-1)/n}\right]$ (4')
$-\displaystyle\int_1^2 v\,dp = \dfrac{n}{1-n}(p_2 v_2 - p_1 v_1)$ $= \dfrac{n p_1 v_1}{n-1}\left[1 - \left(\dfrac{p_2}{p_1}\right)^{(n-1)/n}\right]$ (5)	$-\displaystyle\int_1^2 v\,dp = \dfrac{nR}{1-n}(T_2 - T_1)$ $= \dfrac{nRT_1}{n-1}\left[1 - \left(\dfrac{p_2}{p_1}\right)^{(n-1)/n}\right]$ (5')

[a] For polytropic processes of closed systems where volume change is the only work mode, Equations 2, 4, and 2', 4' are applicable with Equation 12.23 to evaluate the work. When each unit of mass passing through a one-inlet, one-exit control volume at steady state undergoes a polytropic process, Equations 3, 5, and 3', 5' are applicable with Equations 12.24a and 12.24b to evaluate the power. Also note that generally, $-\int_1^2 v\,dp = n\int_1^2 p\,dv$.

[b]

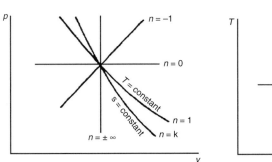
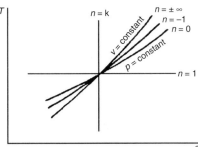

12.4.1.2 Rankine and Brayton Cycles

In their simplest embodiments vapor power and gas turbine power plants are represented conventionally in terms of four components in series, forming, respectively, the *Rankine cycle* and the *Brayton cycle* shown schematically in Table 12.6. The thermodynamically ideal counterparts of these cycles are composed of four internally reversible processes in series: two isentropic processes alternated with two constant pressure processes. Table 12.6 provides property diagrams of the actual and corresponding ideal cycles. Each actual cycle is denoted 1-2-3-4-1; the ideal cycle is 1-2s-3-4s-1. For simplicity, pressure drops through the boiler, condenser, and heat exchangers are not shown. Invoking Equation 12.26 for the ideal cycles, the heat added per unit of mass flowing is represented by the area *under* the isobar from state 2s to state 3: area a-2s-3-b-a. The heat rejected is the area *under* the isobar from state 4s to state 1: area

TABLE 12.6 Rankine and Brayton Cycles

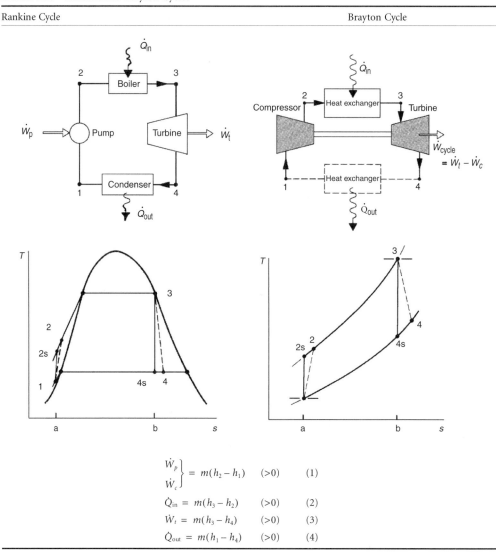

a-1-4s-b-a. Enclosed area 1-2s-3-4s-1 represents the net heat added per unit of mass flowing. For any power cycle, the net heat added equals the net work done.

Expressions for the principal energy transfers shown on the schematics of Table 12.6 are provided by Equations 1 to 4 of the table. They are obtained by reducing Equation 12.10a with the assumptions of negligible heat loss and negligible changes in kinetic and potential energy from the inlet to the exit of each component. All quantities are positive in the directions of the arrows on the figure.

The thermal efficiency of a power cycle is defined as the ratio of the *net* work developed to the total energy added by heat transfer. Using expressions (1)–(3) of Table 12.6, the thermal efficiency is

$$\eta = \frac{(h_3 - h_4) - (h_2 - h_1)}{h_3 - h_2}$$

$$= 1 - \frac{h_4 - h_1}{h_3 - h_2} \quad (12.27)$$

To obtain the thermal efficiency of the ideal cycle, h_{2s} replaces h_2 and h_{4s} replaces h_4 in Equation 12.27.

Engineering Thermodynamics

Decisions concerning cycle operating conditions normally recognize that the thermal efficiency tends to increase as the average temperature of heat addition increases and/or the temperature of heat rejection decreases. In the Rankine cycle, a high average temperature of heat addition can be achieved by superheating the vapor prior to entering the turbine and/or by operating at an elevated steam-generator pressure. In the Brayton cycle an increase in the compressor pressure ratio p_2/p_1 tends to increase the average temperature of heat addition. Owing to materials limitations at elevated temperatures and pressures, the state of the working fluid at the turbine inlet must observe practical limits, however. The turbine inlet temperature of the Brayton cycle, for example, is controlled by providing air far in excess of what is required for combustion. In a Rankine cycle using water as the working fluid, a low temperature of heat rejection is typically achieved by operating the condenser at a pressure below 1 atm. To reduce erosion and wear by liquid droplets on the blades of the Rankine cycle steam turbine, at least 90% steam quality should be maintained at the turbine exit: $x_4 > 0.9$.

The back work ratio, bwr, is the ratio of the work required by the pump or compressor to the work developed by the turbine:

$$\text{bwr} = \frac{h_2 - h_1}{h_3 - h_4} \quad (12.28)$$

As a relatively high specific volume vapor expands through the turbine of the Rankine cycle and a much lower specific volume liquid is pumped, the back work ratio is characteristically quite low in vapor power plants—in many cases on the order of 1–2%. In the Brayton cycle, however, both the turbine and compressor handle a relatively high specific volume gas, and the back ratio is much larger, typically 40% or more.

The effect of friction and other irreversibilities for flow through turbines, compressors, and pumps is commonly accounted for by an appropriate *isentropic efficiency*. Referring to Table 12.6 for the states, the isentropic turbine efficiency is

$$\eta_t = \frac{h_3 - h_4}{h_3 - h_{4s}} \quad (12.29a)$$

The isentropic compressor efficiency is

$$\eta_c = \frac{h_{2s} - h_1}{h_2 - h_1} \quad (12.29b)$$

In the isentropic pump efficiency, η_p, which takes the same form as Equation 12.29b, the numerator is frequently approximated via Equation 12.24c as $h_{2s} - h_1 \approx v_1 \Delta p$, where Δp is the pressure rise across the pump.

Simple gas turbine power plants differ from the Brayton cycle model in significant respects. In actual operation, excess air is continuously drawn into the compressor, where it is compressed to a higher pressure; then fuel is introduced and combustion occurs; finally the mixture of combustion products and air expands through the turbine and is subsequently discharged to the surroundings. Accordingly, the low-temperature heat exchanger shown by a dashed line in the Brayton cycle schematic of Table 12.6 is not an actual component, but included only to account formally for the cooling in the surroundings of the hot gas discharged from the turbine.

Another frequently employed idealization used with gas turbine power plants is that of an *air-standard analysis*. An air-standard analysis involves two major assumptions: (1) As shown by the Brayton cycle schematic of Table 12.6, the temperature rise that would be brought about by combustion is effected instead by a heat transfer from an external source. (2) The working fluid throughout the cycle is air, which behaves as an ideal gas. In a cold air-standard analysis the specific heat ratio k for air is taken as constant. Equations 1 to 6 of Table 12.4 apply generally to air-standard analyses. Equations 1' to 6' of Table 12.4 apply to *cold* air-standard analyses, as does the following expression for the turbine power obtained from Table 12.1 (Equation 10c″):

$$\dot{W}_t = m\frac{kRT_3}{k-1}[1 - (p_4/p_3)^{(k-1)/k}] \tag{12.30}$$

An expression similar in form can be written for the power required by the compressor.

12.4.1.3 Otto, Diesel, and Dual Cycles

Although most gas turbines are also internal combustion engines, the name is usually reserved to *reciprocating* internal combustion engines of the type commonly used in automobiles, trucks, and buses. Two principal types of reciprocating internal combustion engines are the spark-ignition engine and the compression-ignition engine. In a *spark-ignition* engine a mixture of fuel and air is ignited by a spark plug. In a *compression ignition* engine air is compressed to a high-enough pressure and temperature that combustion occurs spontaneously when fuel is injected.

In a *four-stroke* internal combustion engine, a piston executes four distinct strokes within a cylinder for every two revolutions of the crankshaft. Figure 12.7 gives a pressure–displacement diagram as it might be displayed electronically. With the intake valve open, the piston makes an *intake stroke* to draw a fresh charge into the cylinder. Next, with both valves closed, the piston undergoes a *compression stroke* raising the temperature and pressure of the charge. A combustion process is then initiated, resulting in a high-pressure, high-temperature gas mixture. A *power* stroke follows the compression stroke, during which the gas mixture expands and work is done on the piston. The piston then executes an *exhaust stroke* in which the burned gases are purged from the cylinder through the open exhaust valve. Smaller engines operate on *two-stroke* cycles. In two-stroke engines, the intake, compression, expansion, and exhaust operations are accomplished in one revolution of the crankshaft. Although internal combustion engines undergo *mechanical* cycles, the cylinder contents do not execute a *thermodynamic* cycle, since matter is introduced with one composition and is later discharged at a different composition.

A parameter used to describe the performance of reciprocating piston engines is the *mean effective pressure*, or mep. The mean effective pressure is the theoretical constant pressure that, if it acted on the

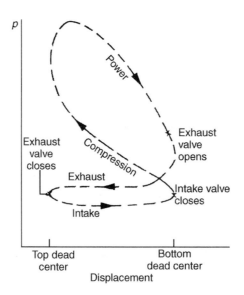

FIGURE 12.7 Pressure–displacement diagram for a reciprocating internal combustion engine.

piston during the power stroke, would produce the same net work as actually developed in one cycle. That is,

$$\text{mep} = \frac{\text{net work for one cycle}}{\text{displacement volume}} \quad (12.31)$$

where the displacement volume is the volume swept out by the piston as it moves from the top dead center to the bottom dead center. For two engines of equal displacement volume, the one with a higher mean effective pressure would produce the greater net work and, if the engines run at the same speed, greater power.

Detailed studies of the performance of reciprocating internal combustion engines may take into account many features, including the combustion process occurring within the cylinder and the effects of irreversibilities associated with friction and with pressure and temperature gradients. Heat transfer between the gases in the cylinder and the cylinder walls and the work required to charge the cylinder and exhaust the products of combustion also might be considered. Owing to these complexities, accurate modeling of reciprocating internal combustion engines normally involves computer simulation.

To conduct *elementary* thermodynamic analyses of internal combustion engines, considerable simplification is required. A procedure that allows engines to be studied *qualitatively* is to employ an *air-standard analysis* having the following elements: (1) a fixed amount of air modeled as an ideal gas is the system; (2) the combustion process is replaced by a heat transfer from an external source and represented in terms of elementary thermodynamic processes; (3) there are no exhaust and intake processes as in an actual engine: the cycle is completed by a constant-volume heat rejection process; (4) all processes are internally reversible.

The processes employed in air-standard analyses of internal combustion engines are selected to represent the events taking place within the engine simply and mimic the appearance of observed pressure–displacement diagrams. In addition to the constant volume heat rejection noted previously, the compression stroke and at least a portion of the power stroke are conventionally taken as isentropic. The heat addition is normally considered to occur at constant volume, at constant pressure, or at constant volume followed by a constant pressure process, yielding, respectively, the Otto, Diesel, and Dual cycles shown in Table 12.7.

Reducing the closed system energy balance, Equation 12.7b, gives the following expressions for work and heat applicable in each case shown in Table 12.7:

$$\frac{W_{12}}{m} = u_1 - u_2, \quad \frac{W_{34}}{m} = u_3 - u_4, \quad \frac{Q_{41}}{m} = u_1 - u_4 \quad (12.32)$$

Table 12.7 provides additional expressions for work, heat transfer, and thermal efficiency identified with each case individually. All expressions for work and heat adhere to the respective sign conventions of Equation 12.7b. Equations 1 to 6 of Table 12.4 apply generally to air-standard analyses. In a cold air-standard analysis the specific heat ratio k for air is taken as constant. Equations (1') to (6') of Table 12.4 apply to cold air-standard analyses, as does Equation 4' of Table 12.5, with $n = k$ for the isentropic processes of these cycles.

Referring to Table 12.7, the ratio of specific volumes v_1/v_2 is the *compression ratio*, r. For the Diesel cycle, the ratio v_3/v_2 is the cutoff ratio, r_c. Figure 12.8 shows the variation of the thermal efficiency with compression ratio for an Otto cycle and Diesel cycles having cutoff ratios of 2 and 3. The curves are determined on a cold air-standard basis with $k = 1.4$ using the following expression:

$$\eta = 1 - \frac{1}{r^{k-1}} \left[\frac{r_c^k - 1}{k(r_c - 1)} \right] \quad \text{(constant } k\text{)} \quad (12.33)$$

where the Otto cycle corresponds to $r_c = 1$.

TABLE 12.7 Otto, Diesel, and Dual Cycles

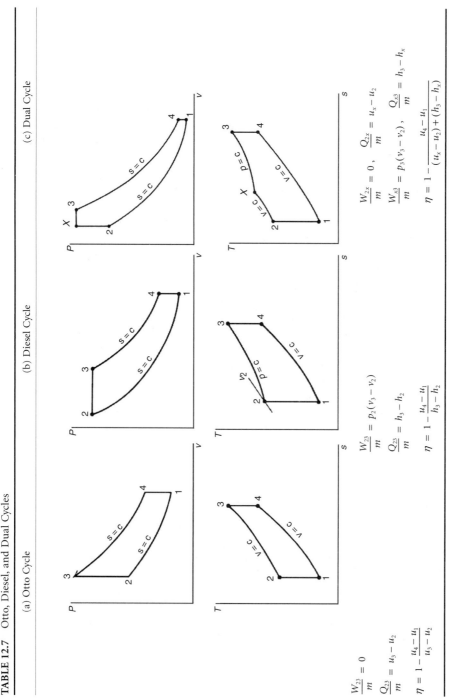

(a) Otto Cycle

$$\frac{W_{23}}{m} = 0$$
$$\frac{Q_{23}}{m} = u_3 - u_2$$
$$\eta = 1 - \frac{u_4 - u_1}{u_3 - u_2}$$

(b) Diesel Cycle

$$\frac{W_{23}}{m} = p_2(v_3 - v_2)$$
$$\frac{Q_{23}}{m} = h_3 - h_2$$
$$\eta = 1 - \frac{u_4 - u_1}{h_3 - h_2}$$

(c) Dual Cycle

$$\frac{W_{2x}}{m} = 0, \quad \frac{Q_{2x}}{m} = u_x - u_2$$
$$\frac{W_{x3}}{m} = p_3(v_3 - v_2), \quad \frac{Q_{x3}}{m} = h_3 - h_x$$
$$\eta = 1 - \frac{u_4 - u_1}{(u_x - u_2) + (h_3 - h_x)}$$

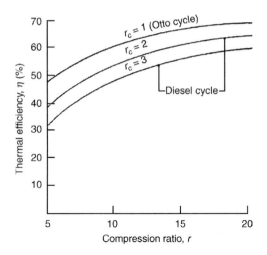

FIGURE 12.8 Thermal efficiency of the cold air-standard Otto and Diesel cycles, $k = 1.4$.

As all processes are internally reversible, areas on the p–v and T–s diagrams of Table 12.7 can be interpreted, respectively, as work and heat transfer. Invoking Equation 12.23 and referring to the p–v diagrams, the areas under process 3-4 of the Otto cycle, process 2-3-4 of the Diesel cycle, and process x-3-4 of the Dual cycle represent the work done by the gas during the power stroke, per unit of mass. For each cycle, the area under the isentropic process 1-2 represents the work done on the gas during the compression stroke, per unit of mass. The enclosed area of each cycle represents the net work done per unit mass. With Equation 12.25 and referring to the T–s diagrams, the areas under process 2-3 of the Otto and Diesel cycles and under process 2-x-3 of the Dual cycle represent the heat added per unit of mass. For each cycle, the area under the process 4-1 represents the heat rejected per unit of mass. The enclosed area of each cycle represents the net heat added, which equals the net work done, each per unit of mass.

References

ASHRAE Handbook 1993 Fundamentals. 1993. American Society of Heating, Refrigerating, and Air Conditioning Engineers, Atlanta.

ASME Steam Tables, 6th ed., 1993. ASME Press, Fairfield, NJ.

Bejan, A., Tsatsaronis, G., and Moran, M. 1996. *Thermal Design and Optimization,* John Wiley & Sons, New York.

Bird, R.B., Stewart, W.E., and Lightfoot, E.N. 1960. *Transport Phenomena,* John Wiley & Sons, New York.

Bolz, R.E. and Tuve, G.L. (Eds.). 1973. *Handbook of Tables for Applied Engineering Science,* 2nd ed., CRC Press, Boca Raton, FL.

Bornakke, C. and Sonntag, R.E. 1996. *Tables of Thermodynamic and Transport Properties,* John Wiley & Sons, New York.

Gray, D.E. (Ed.). 1972. *American Institute of Physics Handbook,* McGraw-Hill, New York.

Haar, L., Gallagher, J.S., and Kell, G.S. 1984. *NBS/NRC Steam Tables,* Hemisphere, New York.

Handbook of Chemistry and Physics, annual editions, CRC Press, Boca Raton, FL.

JANAF Thermochemical Tables, 3rd ed., 1986. American Chemical Society and the American Institute of Physics for the National Bureau of Standards.

Jones, J.B. and Dugan, R.E. 1996. *Engineering Thermodynamics,* Prentice-Hall, Englewood Cliffs, NJ.

Keenan, J.H., Keyes, F.G., Hill, P.G., and Moore, J.G. 1969 and 1978. *Steam Tables,* John Wiley & Sons, New York (1969, English Units; 1978, SI Units).

Keenan, J.H., Chao, J., and Kaye, J. 1980 and 1983. *Gas Tables—International Version*, 2nd ed., John Wiley & Sons, New York (1980, English Units; 1983, SI Units).

Knacke, O., Kubaschewski, O., and Hesselmann, K. 1991. *Thermochemical Properties of Inorganic Substances*, 2nd ed., Springer-Verlag, Berlin.

Kotas, T.J. 1995. *The Exergy Method of Thermal Plant Analysis*, Krieger, Melbourne, FL.

Liley, P.E. 1987. Thermodynamic Properties of Substances, In *Marks' Standard Handbook for Mechanical Engineers*, E.A. Avallone and T. Baumeister (Eds.), 9th ed., McGraw-Hill, New York, Sec. 4.2.

Liley, P.E., Reid, R.C., and Buck, E. 1984. Physical and chemical data. In *Perrys' Chemical Engineers, Handbook*, R.H. Perry and D.W. Green (Eds.), 6th ed., McGraw-Hill, New York, Sec. 3.

Moran, M.J. 1989. *Availability Analysis—A Guide to Efficient Energy Use*, ASME Press, New York.

Moran, M.J. 1998. Engineering Thermodynamics. In *The CRC Handbook of Mechanical Engineering*, F. Kreith (Ed.), CRC Press, Boca Raton, FL, Chap. 2.

Moran, M.J. and Shapiro, H.N. 2000. *Fundamentals of Engineering Thermodynamics*, 4th ed., John Wiley & Sons, New York.

Moran, M.J. and Shapiro, H.N. 2000. *IT: Interactive Thermodynamics*, Computer Software to Accompany Fundamentals of Engineering Thermodynamics, 4th ed., Intellipro, John Wiley & Sons, New York.

Moran, M.J. and Tsatsaronis, G. 2000. Engineering Thermodynamics. In *The CRC Handbook of Thermal Engineering*, F. Kreith (Ed.), CRC Press, Boca Raton, FL, Chap. 1.

Obert, E.F. 1960. *Concepts of Thermodynamics*, McGraw-Hill, New York.

Preston-Thomas, H. 1990. The International Temperature Scale of 1990 (ITS-90). *Metrologia*. 27: 3–10.

Reid, R.C. and Sherwood, T.K. 1966. *The Properties of Gases and Liquids*, 2nd ed., McGraw-Hill, New York.

Reid, R.C., Prausnitz, J.M., and Poling, B.E. 1987. *The Properties of Gases and Liquids*, 4th ed., McGraw-Hill, New York.

Reynolds, W.C. 1979. *Thermodynamic Properties in SI—Graphs, Tables and Computational Equations for 40 Substances*. Department of Mechanical Engineering, Stanford University, Palo Alto, CA.

Stephan, K. 1994. Tables. In *Dubbel Handbook of Mechanical Engineering*, W. Beitz and K.-H. Kuttner (Eds.), Springer-Verlag, London, Sec. C11.

Szargut, J., Morris, D.R., and Steward, F.R. 1988. *Exergy Analysis of Thermal, Chemical and Metallurgical Processes*, Hemisphere, New York.

Van Wylen, G.J., Sonntag, R.E., and Bornakke, C. 1994. *Fundamentals of Classical Thermodynamics*, 4th ed., John Wiley & Sons, New York.

Zemansky, M.W. 1972. Thermodynamic Symbols, Definitions, and Equations. In *American Institute of Physics Handbook*, D.E. Gray (Ed.), McGraw-Hill, New York, Sec. 4b.

13
Numerical Simulation

	13.1	Introduction	13-1
	13.2	Common Simulation Blocks	13-2
		Continuous Linear System Blocks • Discrete Linear System Blocks • Nonlinear Blocks and Table Lookup • Signal Generation	
	13.3	Textual Equations within Simulation Block Diagrams	13-3
	13.4	Solvers	13-4
	13.5	Simulation Timing	13-4
	13.6	Visualization	13-4
Jeannie Sullivan Falcon	13.7	Hybrid System Simulation and Control	13-5
National Instruments, Inc.	References		13-6

13.1 Introduction

Numerical simulation of dynamic systems can be used to help design and analyze mechatronics components and systems. There are several dynamic system simulation packages available including National Instruments LabVIEW® SimulationModule [1,2], The Math Works Inc. Simulink® [3], and Dymola from Dynasim [4].

Simulations can be developed to run offline on a desktop PC or they can run in real time on a desktop or an embedded target. The simulation may involve plant dynamics only or the simulation may include a control system as well. Figure 13.1 shows a V-diagram that is often used to describe a complete embedded control design process in the automotive and aerospace industries. In this diagram, offline simulation would be used in the modeling and design phase. Real-time simulation software would be used to implement a control system in the rapid prototyping phase and to simulate the plant in the hardware-in-the-loop testing phase. For the targeting phase, a real-time simulation could be converted into code that would run on an embedded target.

Numerical simulations of dynamic systems are often implemented in block diagram form with the ability to represent both feedback and feedforward paths. Figure 13.2 shows an example of a simulation created with the LabVIEW Control Design and Simulation Module [1]. Hierarchy in the block diagram may be used to represent dynamics of different subsystems.

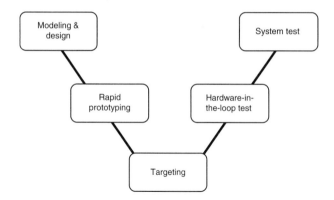

FIGURE 13.1 V-diagram for embedded control system development.

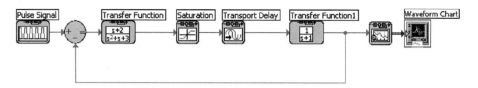

FIGURE 13.2 Example of a simulation block diagram created using the LabVIEW Control Design and Simulation Module.

13.2 Common Simulation Blocks

13.2.1 Continuous Linear System Blocks

In a simulation, continuous linear systems are typically represented in either transfer function, state space, or zero-pole-gain form. In addition, the user may also choose to add in integrator, derivative, and/or transport delay blocks.

13.2.2 Discrete Linear System Blocks

Discrete linear system blocks are used to represent sampled data systems. Discrete transfer function blocks are used to represent systems in the z-domain. Discrete state space blocks are used to represent systems as difference equations. Additional blocks in this category include discrete integrator and unit delay.

Both zero- and first-order hold blocks are used to discretize or resample an input signal. In addition, these blocks can be used to set up nested, multirate systems within a simulation. Discrete signal processing may be implemented via the use of a discrete filter block to represent an infinite impulse response (IIR) or a finite impulse response (FIR) filter.

13.2.3 Nonlinear Blocks and Table Lookup

Nonlinear system blocks include backlash, dead zone, friction, quantizer, saturation, rate limiter, relay, and switches. Custom nonlinear systems may be added through 1D, 2D, or 3D table lookup blocks or through the addition of textual equations in the block diagram.

13.2.4 Signal Generation

An input signal must always be chosen for the simulation unless an initial condition response is chosen. Input signal types include sine wave, chirp, pulse, ramp, or step signals. Random noise may be added to the input signal or may be used as the primary input signal. In addition, custom inputs can be added to the simulation through the use of an array for single-input/single-output (SISO) systems or a matrix for multi-input/multi-output (MIMO) systems, which is indexed for each time step in the simulation.

13.3 Textual Equations within Simulation Block Diagrams

Textual equations may be added to the simulation to represent linear or nonlinear equations such as kinematics calculations or nonlinear control algorithms. Figure 13.3 shows an example of a linear second-order system simulation for phase plane analysis created using the LabVIEW Control Design and Simulation Module. Both the block diagram (Figure 13.3a) and front panel (Figure 13.3b) are shown.

The block diagram shows the differential equation for the linear second-order system inside a textual input node that is called the Formula Node in LabVIEW [5]. The inputs to this node are the position (x), velocity (xdot), damping (b), and stiffness (k). The output is the acceleration (xdotdot). Integrator blocks

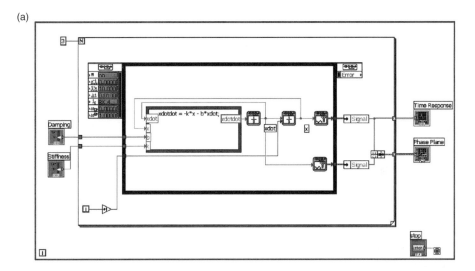

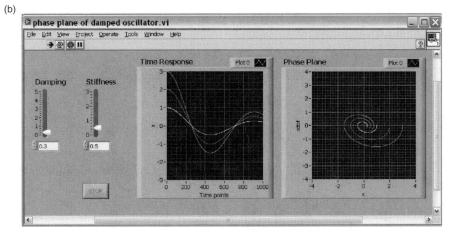

FIGURE 13.3 (a) Block diagram of simulation of a second-order system with textual node for differential equation, a "for loop" for initial condition variation, and a "while loop" for system parameter variation. This simulation was developed using the LabVIEW Control Design and Simulation Module. (b) Front panel of second-order system simulation showing system parameter inputs, time response, and phase plane analysis.

are used to calculate position and velocity and these are then fed back into the node as inputs. The system states are saved in collector blocks. The simulation is run for 10 s at each of three initial conditions. The initial condition values are given by the For Loop in LabVIEW surrounding the simulation loop. An outer While Loop surrounds the For Loop, and this allows the user to continuously vary the damping and stiffness parameters while observing the effects on the phase plane on the front panel.

13.4 Solvers

Ordinary differential equation (ODE) solvers are employed when continuous time systems or subsystems are used in a simulation. These solvers use numerical methods to approximate the solution to a differential equation over time. Both fixed-step and variable-step solvers are available. Fixed-step solvers maintain a constant time step over the duration of the simulation. Variable-step solvers take small time steps when the states vary rapidly and can take larger time steps when the states vary slowly. This ability to change step sizes can increase computational efficiency.

Variable time-step solvers are appropriate for offline simulations if computational efficiency is a concern. For real-time applications such as control prototyping or hardware-in-the-loop simulation, fixed time-step solvers should be used.

Fixed time-step solvers include Runge–Kutta 1 (Euler method), Runge–Kutta 2, Runge–Kutta 3, and Runge–Kutta 4. Variable time-step solvers can include Runge–Kutta 23, Runge–Kutta 45, BDF, Adams–Moulton, and Rosenbrock.

In order to improve computational efficiency for real-time applications, the user should consider a lower-order fixed-step solver if it meets the desired accuracy. Also, increased efficiency can be obtained by converting the continuous dynamics in the simulation to discrete time. This conversion avoids the use of a solver altogether but, again, can affect the accuracy of the result.

13.5 Simulation Timing

There are typically many timing options available with simulation packages. The user can slow down time in the simulation to study fast dynamics in detail, such as a collision between two cars. In addition, the user can speed up time to study slow dynamics more efficiently. This would be useful in studying the effects of temperature control for a large water tank. It is also possible to tie the simulation time to the actual time if the simulation is running on a real-time target.

The user may also change the start time to be negative time and may change the final time to be finite or infinite. An infinite time for final time allows the simulation to run until the user halts the simulation. With a user interface wrapped around the simulation, the user may adjust plant parameters and control gains as the simulation is running to fully explore variations in the design space of the systems.

Parallel, multirate simulations may be developed with varying timing, priorities, and phasing on simulation loops in the LabVIEW Control Design and Simulation Module [2]. Parallel, multirate simulations are useful for separating out subsystems with critical control tasks from those with less critical monitoring and data storage tasks.

13.6 Visualization

The single-body or multibody dynamics in a numerical simulation may be tied to a two- or three dimensional picture display. This is useful for both offline simulations and for providing a display for a real-time simulation or control system. Figure 13.4 shows the front panel from a LabVIEW Control Design and Simulation Module example for the F14 Tomcat fighter. The three-dimensional display was created through the importation of a VRML (virtual reality markup language). Other three-dimensional models may be created through the importation of different file formats or through the creation of OpenGL (open graphics language) objects.

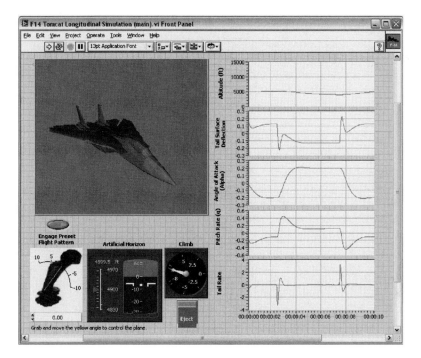

FIGURE 13.4 F14 Tomcat longitudinal simulation example in the LabVIEW Control Design and Simulation Module.

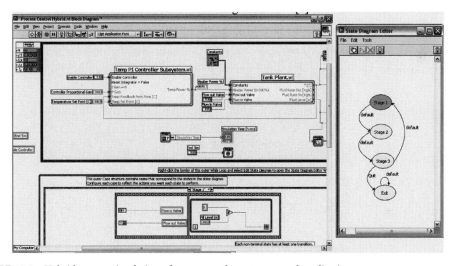

FIGURE 13.5 Hybrid system simulation of a sequenced process control application.

13.7 Hybrid System Simulation and Control

A continuous or discrete time simulation may be combined with discrete event programming to create a hybrid system or hybrid control approach. This approach is useful in applications such as sequenced process control problems or in the study of fault tolerant control systems.

Figure 13.5 shows a sequenced process control example developed using the LabVIEW Control Design and Simulation Module and the LabVIEW State Diagram Toolkit [5,6].

References

1. http://www.ni.com/embeddedcontrol: "Embedded Control Engineering—Integrated Software and Hardware for the Entire Design Process," National Instruments LabVIEW Embedded Control portal page, National Instruments, Austin, TX.
2. http://zone.ni.com/reference/en-XX/help/371894A-01/: LabVIEW Control Design and Simulation Module 8.2 Help html document, National Instruments, Austin, TX.
3. http://www.mathworks.com/products/simulink/: "Simulink—Simulation and Model-Based Design," The MathWorks Inc. Simulink product page, The MathWorks, Inc., Natick, MA.
4. http://www.dynasim.com: "Dymola Multi-Engineering Modeling and Simulation," Home page for Dynasim, Lund, Sweden.
5. http://www.ni.com/toolkits/lv_state_diagram.htm: NI LabVIEW State Diagram Toolkit information page, National Instruments, Austin, TX.
6. http://techteach.no/publications/state_diagrams/index.htm: Introduction to State Diagrams by Finn Haugen, TechTeach, Norway.
7. http://sine.ni.com/manuals/main/p/sn/n23:4.28.9533 LabVIEW Control Design and Simulation Module Help html document, National Instruments, Austin, TX.

14
Modeling and Simulation for MEMS

Carla Purdy
University of Cincinnati

14.1	Introduction	14-1
14.2	The Digital Circuit Development Process: Modeling and Simulating Systems with Micro- (or Nano-) Scale Feature Sizes	14-2
14.3	Analog and Mixed-Signal Circuit Development: Modeling and Simulating Systems with Micro- (or Nano-) Scale Feature Sizes and Mixed Digital (Discrete) and Analog (Continuous) Input, Output, and Signals	14-7
14.4	Basic Techniques and Available Tools for MEMS Modeling and Simulation	14-8
	Basic Modeling and Simulation Techniques • A Catalog of Resources for MEMS Modeling and Simulation	
14.5	Modeling and Simulating MEMS, That Is, Systems with Micro- (or Nano-) Scale Feature Sizes, Mixed Digital (Discrete) and Analog (Continuous) Input, Output, and Signals, Two- and Three-Dimensional Phenomena, and Inclusion and Interaction of Multiple Domains and Technologies	14-13
14.6	A "Recipe" for Successful MEMS Simulation	14-15
14.7	Conclusion: Continuing Progress in MEMS Modeling and Simulation	14-16
	References	14-16

14.1 Introduction

Accurate modeling and efficient simulation, in support of greatly reduced development cycle time and cost, are well established techniques in the miniaturized world of integrated circuits (ICs). Simulation accuracies of 5% or less for parameters of interest are achieved fairly regularly [1], although even much less accurate simulations (e.g., 25–30%) can still be used to obtain valuable information [2]. In the IC world, simulation can be used to predict the performance of a design, to analyze an already existing component, or to support automated synthesis of a design. Eventually, MEMS simulation environments should also be capable of these three modes of operation. The MEMS developer is, of course, most interested in quick access to particular techniques and tools to support the system currently under development. In the long run, however, consistently achieving acceptably accurate MEMS simulations will depend both on the ability of the CAD (computer-aided design) community to develop robust, efficient, user-friendly tools which will be widely available both to cutting-edge researchers and to production engineers and on the existence of readily accessible standardized processes. In this chapter we focus on fundamental approaches which will eventually lead to successful MEMS simulations becoming routine. We also survey available tools which a MEMS developer can use to achieve good simulation results. Many of these tools build MEMS development systems on platforms already in existence for other technologies, thus leveraging the extensive resources which have gone into previous development and avoiding "reinventing the wheel."

For our discussion of modeling and simulation, the salient characteristics of MEMS are:

1. Inclusion and interaction of multiple domains and technologies
2. Both two- and three-dimensional behaviors
3. Mixed digital (discrete) and analog (continuous) input, output, and signals
4. Micro- (or nano-) scale feature sizes

Techniques for the manufacture of reliable (two-dimensional) systems with micro- or nano-scale feature sizes (Characteristic 4) are very mature in the field of microelectronics, and it is logical to attempt to extend these techniques to MEMS, while incorporating necessary changes to deal with Characteristics 1–3. Here we survey some of the major principles which have made microelectronics such a rapidly evolving field, and we look at microelectronics tools which can be used or adapted to allow us to apply these principles to MEMS. We also discuss why applying such strategies to MEMS may not always be possible.

14.2 The Digital Circuit Development Process: Modeling and Simulating Systems with Micro- (or Nano-) Scale Feature Sizes

A typical VLSI digital circuit or system process flow is shown in Figure 14.1, where the dotted lines show the most optimistic point to which the developer must return if errors are discovered. Option A, for a "mature" technology, is supported by efficient and accurate simulators, so that even the first actual implementation ("first silicon") may have acceptable performance. As a process matures, the goal is to have better and better simulations, with a correspondingly smaller chance of discovering major performance flaws after implementation. However, development of models and simulators to support this goal is in itself a major task. Option B (immature technology), at its extreme, would represent an experimental technology for which not enough data are available to support even moderately robust simulations. In modern software and hardware development systems, the emphasis is on tools which provide increasingly good support for the initial stages of this process. This increases the probability that conceptual or design errors will be identified and modifications made as early in the process as possible and thus decreases both development time and overall development cost.

At the microlevel, the development cycle represented by Option A is routinely achieved today for many digital circuits. In fact, the entire process can in some cases be highly automated, so that we have "silicon compilers" or "computers designing computers." Thus, not only design analysis, but even design synthesis is possible. This would be the case for well-established silicon-based CMOS technologies, for example. There are many characteristics of digital systems which make this possible. These include:

- Existence of a small set of basic digital circuit elements. All Boolean functions can be realized by combinations of the logic functions AND, OR, NOT. In fact, all Boolean functions can be realized by combinations of just one gate, a NAND (NOT-AND) gate. So if a "model library" of basic gates (and a few other useful parts, such as I/O pins, multiplexors, and flip-flops) is developed, systems can be implemented just by combining suitable library elements.

- A small set of standardized and well-understood technologies, with well-characterized fabrication processes that are widely available. For example, in the United States, the MOSIS service [3] provides access to a range of such technologies. Similar services elsewhere include CMP in France [4], Europractice in Europe [5], VDEC in Japan [6], and CMC in Canada [7].

- A well-developed educational infrastructure and prototyping facilities. These are provided by all of the services listed above. These types of organization and educational support had their origins in

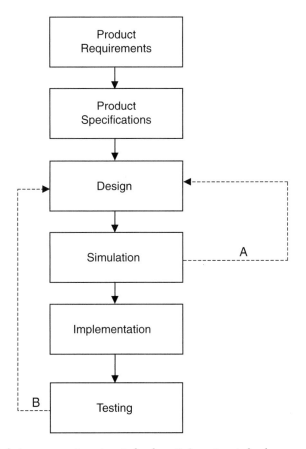

FIGURE 14.1 Product design process. A: mature technology, B: immature technology.

the work of Mead and Conway [8] and continue to produce increasingly sophisticated VLSI engineers. An important aspect of this infrastructure is that it also provides, at relatively low cost, access to example devices and systems, made with stable fabrication processes, whose behavior can be tested and compared to simulation results, thereby enabling improvements in simulation techniques.

- "Levels and views" (abstraction and encapsulation or "information hiding") (see [9]). This concept is illustrated in Figure 14.2a. For the VLSI domain, we can identify at least five useful levels of abstraction, from the lowest (layout geometry) to the highest (system specification). We can also "view" a system behaviorally, structurally, or physically. In the *behavioral domain* we describe the functionality of the circuit without specifying how this functionality will be achieved. This allows us to think clearly about what the system needs to do, what inputs are needed, and what outputs will be provided. Thus we can view the component as a "black box" that has specified responses to given inputs. The current through a MOS field effect transistor (MOSFET), given as a function of the gate voltage, is a (low-level) behavioral description, for example. In the *physical domain* we specify the actual physical parts of the circuit. At the lowest levels in this domain, we must choose what material each piece of the circuit will be made from (for example, which pieces of wire will lie in each of the metal layers usually provided in a CMOS circuit) and exactly where each piece will be placed in the actual physical layout. The physical description will be translated directly into mask layouts for the circuit. The *structural domain* is intermediate between physical and behavioral. It provides an interface between the functionality specified in the behavioral domain, which ignores geometry, and the geometry specified in the physical domain, which ignores functionality. In this intermediate domain, we can carry out logic optimization and state minimization, for example.

(a)

Levels	Views		
	Behavioral	Structural	Physical
4	Performance specifications	CPUs, memory, switches, controllers, buses	Physical partitions
3	Algorithms	Modules, data structures	Clusters
2	Register transfers	ALUs, MUXs, registers	Floorplans
1	Boolean equations, FSMs	Gates, flip-flops	Cells, modules
0	Transfer functions, timing	Transistors, wires, contacts, vias	Layout geometry

(b)

Levels	Views		
	Behavioral	Structural	Physical
4	Performance specifications	Sensors, actuators, systems	Physical partitions
3		Multiple energy domain components	Clusters
2		Domain-domain components	Floorplans
1		Single energy domain components	Cells, modules
0	Transfer functions, timing	Beams, membranes, holes, grooves, joints	Layout geometry

FIGURE 14.2 A taxonomy for component development ("levels and views"): (a) standard VLSI classifications, (b) a partial classification for MEMS components.

A schematic diagram is an example of a structural description. Of course, not all circuit characteristics can be completely encapsulated in a single one of these views. For example, if we change the physical size of a wire, we will probably affect the timing, which is a behavioral property. The principle of encapsulation leads naturally to the development of extensive IP (intellectual property), that is, libraries of increasingly sophisticated components that can be used as "black boxes" by the system developer.

- Well-developed models for basic elements that clearly delineate effects due to changes in design, fabrication process, or environment. For example, in [10], the factors in the basic first-order equations for I_{ds}, the drain-to-source current in an NMOS transistor, can clearly be divided into those under the control of the designer (W/L, the width-to-length ratio for the transistor channel), those dependent on the fabrication process (ε, the permittivity of the gate insulator, and t_{ox}, the thickness of the gate insulator), those dependent on environmental factors (V_{ds} and V_{gs}, the drain-to-source and gate-to-source voltages, respectively), and those that are a function of both the fabrication process and the environment (μ, the effective surface mobility of the carriers in the channel, and V_t, the threshold voltage). More detailed information on modeling MOSFETs can be found in [11]. Identification of fundamental parameters in one stage of the development process can be of great value in other stages. For example, the minimum feature size λ for a given technology can be used to develop a set of "design rules" that express mandatory overlaps and spacings for the different physical materials. A design tool can then be developed to "enforce" these rules, and the consequences can be used to simplify, to some extent, the modeling and simulation stages. The parameter λ can also be used to express effects due to scaling when scaling is valid.

- Mature tools for design and simulation, which have evolved over many generations and for which moderately priced versions are available from multiple sources. For example, many of today's tools incorporate versions of the design tool MAGIC [12] and the simulator SPICE (Simulation Program with Integrated Circuit Emphasis) [13], both of which were originally developed at the University of California, Berkeley. Versions of the SPICE simulator typically support several device models (currently, for example, six or more different MOS models and five different transmission line models), so that a developer can choose the level of device detail appropriate to the task at hand. Free or low-cost versions of both MAGIC and SPICE, as well as extended versions of both tools, are widely available. Many different techniques, such as model binning (optimizing models for specific ranges of model parameters) and inclusion of proprietary process information, are employed to produce better models and simulation results, especially in the HSPICE version of SPICE and in other high-end versions of these tools [11].

- Integrated development systems that are widely available and that provide support for a variety of levels and views, extensive component libraries, user-friendly interfaces and online help, as well as automatic translation between domains, along with error and constraint checking. In an integrated VLSI development system, sophisticated models, simulators, and translators keep track of circuit information for multiple levels and views, while allowing the developer to focus on one level or view at a time. Many development systems available today also support, at the higher levels of abstraction, structured "programming" languages such as VHDL (Very Large Scale Integrated Circuit Hardware Description Language) [14,15] or Verilog [16].

A digital circuit developer has many options, depending on performance constraints, number of units to be produced, desired cost, available development time, etc. At one extreme the designer may choose to develop a "custom" circuit, creating layout geometries, sizing individual transistors, modeling RC effects in individual wires, and validating design choices through extensive low-level SPICE-based simulations. At the other extreme, the developer can choose to produce a PLD (programmable logic device), with a predetermined basic layout geometry consisting of cells incorporating programmable logic and storage (Figure 14.3) that can be connected as needed to produce the desired device functionality. A high end PLD may contain as many as 100,000 (100 K) cells similar to the one in Figure 14.3 and an additional 100 K bytes of RAM (random access memory) storage. In an integrated development system, such as those

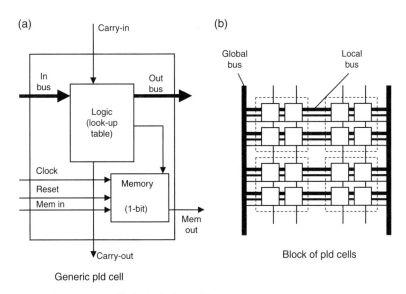

FIGURE 14.3 A generic programmable logic device architecture.

provided by References 17 and 18, the developer enters the design in either schematic form or a high level language, and then the design is automatically "compiled" and mapped to the PLD geometry, and functional and timing simulations can be run. If the simulation results are acceptable, an actual PLD can then be programmed directly, as a further step in the development process, and even tested, to some extent, with the same set of test data as was used for the simulation step. This "rapid prototyping" [19] for the production of a "chip" is not very different from the production of a working software program (and the PLD can be reprogrammed if different functionality is later desired). Such a system, of course, places many constraints on achievable designs. In addition, the automated steps, which rely on heuristics rather than exact techniques to find acceptable solutions to the many computationally complex problems that need to be solved during the development process, sacrifice performance for ease of development, so that a device designed in such a system will never achieve the ultimate performance possible for the given technology. However, the trade-offs include ease of use, much shorter development times, and the management of much larger numbers of individual circuit elements than would be possible if each individual element were tuned to its optimum performance. In addition, if a high-level language is used for input, an acceptable design can often be translated, with few changes, to a more powerful design system that will allow implementation in more flexible technologies and additional fine tuning of circuit performance. In Figure 14.4 we see some of the levels of abstraction which are present in such a development

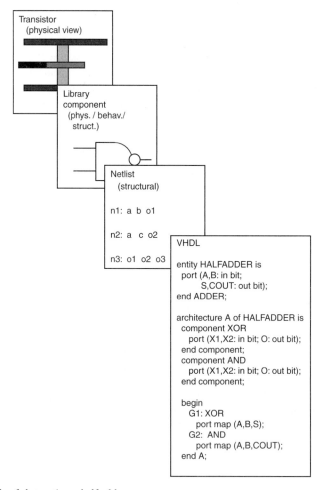

FIGURE 14.4 Levels of abstraction—half adder.

process, with the lowest level being detailed transistor models and the highest a VHDL description of a half adder.

14.3 Analog and Mixed-Signal Circuit Development: Modeling and Simulating Systems with Micro- (or Nano-) Scale Feature Sizes and Mixed Digital (Discrete) and Analog (Continuous) Input, Output, and Signals

At the lowest level, digital circuits are in fact analog devices. A CMOS inverter, for example, does not "switch" instantaneously from a voltage level representing binary 0 to a voltage level representing binary 1. However, by careful design of the inverter's physical structures, it is possible to make the switching time from the range of voltage outputs which are considered to be "0" to the range considered to be "1" (or vice versa) acceptably short. In MOSFETs, for example, the two discrete signals of interest can be identified with the transistor, modeled as a switch, being "open" or "closed," and the "switching" from one state to another can be ignored except at the very lowest levels of abstraction. In much design and simulation work, the analog aspects of the digital circuit's behavior can thus be ignored. Only at the lower levels of abstraction will the analog properties of VLSI devices or the quantum effects occurring, e.g., in a MOSFET need to be explicitly taken into account, ideally by powerful automated development tools supported by detailed models. At higher levels this behavior can be encapsulated and expressed in terms of minimum and maximum switching times with respect to a given capacitive load and given voltage levels. Even in digital systems, however, as submicron feature sizes become more common, more attention must be paid to analog effects. For example, at small feature sizes, wire delay due to RC effects and crosstalk in nearby wires become more significant factors in obtaining good simulation results [20]. It is instructive to examine how simulation support for digital systems can be extended to account for these factors.

Typically, analog circuit devices are much more likely to be "hand-crafted" than digital devices. SPICE and SPICE-like simulations are commonly used to measure performance at the level of transistors, resistors, capacitors, and inductors. For example, due to the growing importance of wireless and mobile computing, a great deal of work in analog design is currently addressing the question of how to produce circuits (digital, analog, and mixed-signal) that are "low-power," and simulations for devices to be used in these circuits are typically carried out at the SPICE level. Unless a new physical technology is to be employed, the simulations will mostly rely on the commonly available models for transistors, transmission lines, etc., thus encapsulating the lowest level behaviors.

Let us examine the factors given above for the success of digital system simulation and development to see how the analog domain compares. We assume a development cycle similar to that shown in Figure 14.1.

- Is there a small set of basic circuit elements? In the analog domain it is possible to identify sets of components, such as current mirrors, op-amps, etc. However, there is no "universal" gate or small set of gates from which all other devices can be made, as is true in the digital domain. Another complicating factor is that elementary analog circuit elements are usually defined in terms of physical performance. There is no clean notion of 0/1 behavior. Because analog signals are continuous, it is often much more difficult to untangle complex circuit behaviors and to carry out meaningful simulations where clean parameter separations give clear results. Once a preliminary analog device or circuit design has been developed, the process of using simulations to decide on exact parameter values is known as "exploring the design space." This process necessarily exhibits high computational complexity. Often heuristic methods such as simulated annealing, neural nets, or a genetic algorithm can be used to perform the necessary search efficiently [21].

- Is there a small set of well-understood technologies? In this area, the analog and mixed signal domain is similar to the digital domain. Much analog development activity focuses on a few standard and well-parameterized technologies. In general, analog devices are much more sensitive to variations in process parameters, and this must be accounted for in analog simulation.

Statistical techniques to model process variation have been included, for example, in the APLAC tool [22], which supports object-oriented design and simulation for analog circuits. Modeling and simulation methods, which incorporate probabilistic models, will become increasingly important as nanoscale devices become more common and as new technologies depending on quantum effects and biology-based computing are developed. Several current efforts, for example, are aimed at developing a "BIOSPICE" simulator, which would incorporate more stochastic system behavior [23].

- Is there a well-developed educational infrastructure and prototyping facilities? All the organizations, which support education and prototyping in the digital domain [3–7], provide similar support for analog and mixed-signal design.
- Are encapsulation and abstraction widely employed? In the past few years, a great deal of progress has been made in incorporating these concepts into analog and mixed-signal design systems. The wide availability of very powerful computers, which can perform the necessary design and simulation tasks in reasonable amounts of time, has helped to make this progress possible. In [24], for example, top-down, constraint-driven methods are described, and in [25] a rapid prototyping method for synthesizing analog and mixed signal systems, based on the tool suite VASE (VHDL-AMS Synthesis Environment), is demonstrated. These methods rely on classifications similar to those given for digital systems in Figure 14.2a.
- Are there well-developed models, mature tools, and integrated development systems which are widely available? In the analog domain, there is still much more to be done in these areas than in the digital domain, but prototypes do exist. In particular, the VHDL and Verilog languages have been extended to allow for analog and mixed-signal components. The VHDL extension, e.g., VHDL-AMS [14], will allow the inclusion of any algebraic or ordinary differential equation in a simulation. However, there does not exist a completely functional VHDL-AMS simulator, although a public domain version, incorporating many useful features, is available at [26] and many commercial versions are under development (e.g., [27]). Thus, at present, expanded versions of MAGIC and SPICE are still the most widely-used design and simulation tools. While there have been some attempts to develop design systems with configurable devices similar to the digital devices shown in Figure 14.3, these have not so far been very successful. Currently, more attention is being focused on component-based development with design reuse for SOC (systems on a chip) through initiatives such as [28].

14.4 Basic Techniques and Available Tools for MEMS Modeling and Simulation

Before trying to answer the above questions for MEMS, we need to look specifically at the tools and techniques the MEMS designer has available for the modeling and simulation tasks. As pointed out in [29,30], the bottom line is, in any simulator, all models are not created equal. The developer must be very clear about what parameters are of greatest interest and then must choose the models and simulation techniques (including implementation in a tool or tools) that are most likely to give the most accurate values for those parameters in the least amount of simulation time. For example, the model used to determine static behavior may be different from the model needed for an adequate determination of dynamic behavior. Thus, it is useful to have a range of models and techniques available.

14.4.1 Basic Modeling and Simulation Techniques

We need to make the following choices:

- What kind of behavior are we interested in? IC simulators, for example, typically support DC operating analysis, DC sweep analysis (stepping current or voltage source values) and transient sweep analysis (stepping time values), along with several other types of transient analysis [30].

- Will the computation be symbolic or numeric?
- Will use of an exact equation, nodal analysis, or finite element analysis be most appropriate? Currently, these are the techniques which are favored by most MEMS developers.

To show what these choices entail, let us look at a simple example that combines electrical and mechanical parts. The cantilever beam in Figure 14.5a, fabricated in metal, polysilicon, or a combination, may be combined with an electrically isolated plate to form a parallel plate capacitor. If a mechanical force or a varying voltage is applied to the beam (Figure 14.5b1), an accelerometer or a switch can be obtained [31]. If instead the plate can be moved back and forth, a more efficient accelerometer design results (Figure 14.5b2); this is the basic design of Analog Devices' accelerometer, probably the first truly successful commercial MEMS device [32,33]. If several beams are combined into two "combs," a comb-drive sensor or actuator results, as in Figure 14.5b3 [34]. Let us consider just the simplest case, as shown in Figure 14.5b1.

If we assume the force on the beam is concentrated at its end point, then we can use the method of [35] to calculate the "pull-in" voltage, that is, the voltage at which the plates are brought together, or to a stopper which keeps the two plates from touching. We model the beam as a dampened spring-mass system and look for the force F, which, when translated into voltage, will give the correct x value for the beam to be "pulled in."

$$F = mx'' + Bx' + kx$$

Here mass $m = \rho WTL$, where ρ is the density of the beam material, $I = WT^3/12$ is the moment of inertia, $k = 3EI/L^3$, E is the Young's modulus of the beam material, and $B = (k/EI)^{1/4}$. This second-order linear differential equation can be solved numerically to obtain the pull-down voltage. In this case, since a closed form expression can be obtained for x, symbolic computation would also be an option. In Reference 36 it is shown that for this simple problem several commonly used methods and tools will give the same result, as is to be expected.

To obtain a more accurate model of the beam we can use the method of nodal analysis, that treats the beam as a graph consisting of a set of edges or "devices," linked together at "nodes." Nodal analysis assumes that at equilibrium the sum of all values around each closed loop (the "across" quantities) will

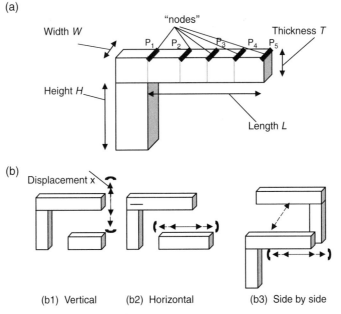

FIGURE 14.5 Cantilever beam and beam–capacitor options: (a) cantilever beam dimensions, (b) basic beam–capacitor designs.

be zero, as will the sum of all values entering or leaving a given node (the "through" quantities). Thus, for example, the sum of all forces and moments on each node must be zero, as must the sum of all currents flowing into or out of a given node. This type of modeling is sometimes referred to as "lumped parameter," since quantities such as resistance and capacitance, which are in fact distributed along a graph edge, are modeled as discrete components. In the electrical domain Kirchhoff's laws are examples of these rules. This method, which is routinely applied to electrical circuits in elementary network analysis courses (see, e.g., [37]), can easily be applied to other energy domains by using correct domain equivalents (see, e.g., [38]). A comprehensive discussion of the theory of nodal analysis can be found in [39]. In Figure 14.5a, the cantilever beam has been divided into four "devices," subbeams between node i and $i + 1$, $i =$ 1, 2, 3, 4, where the positions of nodes i and $i + 1$ are described by (x_i, y_i, θ_i) and $(x_{i+1}, y_{i+1}, \theta_{i+1})$ the coordinates and slope at P_i and P_{i+1}. The beam is assumed to have uniform width W and thickness T, and each subbeam is treated as a two-dimensional structure free to move in three-space. In [40] a modified version of nodal analysis is used to develop numerical routines to simulate several MEMS behaviors, including static and transient behavior of a beam-capacitor actuator. This modified method also adds position coordinates z_i and z_{i+1} and replaces the slope θ_i at each node with a vector of slopes, θ_{ix}, θ_{iy}, and θ_{iz}, giving each node six degrees of freedom.

Since nodal analysis is based on linear elements represented as the edges in the underlying graph, it cannot be used to model many complex structures and phenomena such as fluid flow or piezoelectricity. Even for the cantilever beam, if the beam is composed of layers of two different materials (e.g., polysilicon and metal), it cannot be adequately modeled using nodal analysis. The technique of finite element analysis (FEA) must be used instead. For example, in some follow-up work to that reported in [36], nodal analysis and symbolic computation gave essentially the same results, but the FEA results were significantly different. Finite element analysis for the beam begins with the identification of subelements, as in Figure 14.5a, but each element is treated as a true three-dimensional object. Elements need not all have the same shape, for example, tetrahedral and cubic "brick" elements could be mixed together, as appropriate. In FEA, one cubic element now has eight nodes, rather than two (Figure 14.6), so computational complexity is increased. Thus, developing efficient computer software to carry out FEA for a given structure can be a difficult task in itself. But this general method can take into account many features that cannot be adequately addressed using nodal analysis, including, for example, unaligned beam sections, and surface texture (Figure 14.7). FEA, which can incorporate static, transient, and dynamic behavior, and which can treat heat and fluid flow, as well as electrical, mechanical, and other forces, is explained in detail in [41]. The basic procedure is as follows:

- Discretize the structure or region of interest into finite elements. These need not be homogeneous, either in size or in shape. Each element, however, should be chosen so that no sharp changes in geometry or behavior occur at an interior point.
- For each element, determine the element characteristics using a "local" coordinate system. This will represent the equilibrium state (or an approximation if that state cannot be computed exactly) for the element.
- Transform the local coordinates to a global coordinate system and "assemble" the element equations into one (matrix) equation.
- Impose any constraints implied by restricted degrees of freedom (e.g., a fixed node in a mechanical problem).
- Solve (usually numerically) for the nodal unknowns.
- From the global solution, calculate the element resultants.

14.4.2 A Catalog of Resources for MEMS Modeling and Simulation

To make our discussion of the state-of-the-art of MEMS simulation less confusing, we first list some of the tools and products available. This list is by no means comprehensive, but it will provide us with a range of approaches for comparison. It should be noted that this list is accurate as of July 2001, but the MEMS development community is itself developing, with both commercial companies and university

Modeling and Simulation for MEMS   **14**-11

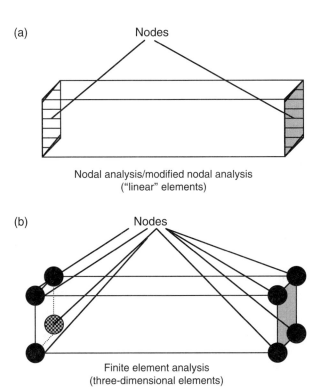

FIGURE 14.6 Nodal analysis and finite element analysis.

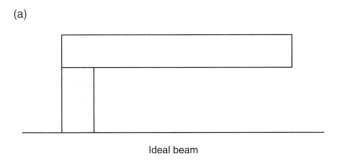

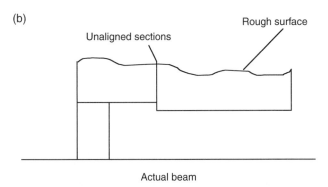

FIGURE 14.7 Ideal and actual cantilever beams (side view).

research sites frequently taking on new identities and partners and also expanding the range of services they offer.

14.4.2.1 Widely Available Tools for General Numeric and Symbolic Computation

These tools are relatively easy to learn to use. Most engineering students will have mastered at least one before obtaining a bachelor's degree. They can be used to model a device "from scratch" and to perform simple simulations. For more complex simulations, they are probably not appropriate for two reasons. First, neither is optimized to execute long computations efficiently. Second, developing the routines necessary to carry out a complex nodal or finite element analysis will in itself be a time-consuming task and will in most cases only replicate functionality already available in other tools listed here.

- Mathematica [42]. In Reference 36 Mathematica simulation results for a cantilever beam-capacitor system are compared with results from several other tools.
- Matlab (integrated with Maple) [43]. In Reference 44, for example, Matlab simulations are shown to give good approximations for a variety of parameters for microfluidic system components.

14.4.2.2 Tools Originally Developed for Specific Energy Domains

Low-cost easy to use versions of some of these tools (e.g., SPICE, ANSYS) are also readily available. Phenomena from other energy domains can be modeled using domain translation.

- SPICE (analog circuits) [13]. SPICE is the de facto standard for analog circuit simulators. It is also used to support simulation of transistors and other components for digital systems. SPICE implements numerical methods for nodal analysis. Several authors have used SPICE to simulate MEMS behavior in other energy domains. In Reference 35, for example, the equation for the motion of a damped spring, which is being used to calculate pull-in voltage, is translated into the electrical domain and reasonable simulation accuracy is obtained. In Reference 45 steady-state thermal behavior for flow-rate sensors is simulated by dividing the device to be modeled into three-dimensional "bricks," modeling each brick as a set of thermal resistors, and translating the resulting conduction and convection equations into electrical equivalents.
- APLAC [22]. This object-oriented analog and mixed-signal simulator incorporates routines, which allow statistical modeling of process variation.
- VHDL-AMS [14,26,27]. The VHDL-AMS language, designed to support digital, analog, and mixed-signal simulation, will in fact support simulation of general algebraic and ordinary differential equations. Thus mixed-energy domain simulations can be carried out. VHDL-AMS, which is typically built on a SPICE kernel, uses the technique of nodal analysis. Some VHDL-AMS MEMS models have been developed (see, e.g., [46,47]). Additional information about VHDL-AMS is available at [48].
- ANSYS [49]. Student versions of the basic ANSYS software are widely available. ANSYS is now partnering with MemsPro (see below). ANSYS models both mechanical and fluidic phenomena using FEA techniques. A survey of the ANSYS MEMS initiative can be found at [50].
- CFD software [51]. This package, which also uses FEA, was developed to model fluid flow and temperature phenomena.

14.4.2.3 Tools Developed Specifically for MEMS

The tools in this category use various simplifying techniques to provide reasonably accurate MEMS simulations without all the computational overhead of FEA.

- SUGAR [40,52]. This free package is built on a Matlab core. It uses nodal analysis and modified nodal analysis to model electrical and mechanical elements. Mechanical elements must be built from a fixed set of components including beams and gaps.

- NODAS v 1.4 [53]. This downloadable tool provides a library of parameterized components (beams, plate masses, anchors, vertical and horizontal electrostatic comb drives, and horizontal electrostatic gaps) that can be interconnected to form MEMS systems. The tool outputs parameters that can be used to perform electromechanical simulations with the Saber simulator [27]. A detailed example is available at [54], and a description of how the tool works (for v 1.3) is also available [55]. Useful information is also available in [70].

14.4.2.4 "Metatools" Which Attempt to Integrate Two or More Domain-Specific Tools into One Package

- MEMCAD, currently being supported by the firm Coventor [56]. This product was previously supported by Microcosm, Inc. It provides low-level simulation capability by integrating domain-specific FEA tools into one package to support coupled energy domain simulations. It also supports process simulation. Much of the extensive research underlying this tool is summarized in Reference 57].
- MemsPro [58], which currently incorporates links to ANSYS. MemsPro itself is an offshoot of Tanner Tools, Inc. [59], which originally produced a version of MAGIC [12] that would run on PCs. The MemsPro system provides integrated design and simulation capability. Process "design rules" can be defined by the user. SPICE simulation capability is integrated into the toolset, and a data file for use with ANSYS can also be generated. MemsPro does not do true energy domain coupling at this time. Some library components are also available.

14.4.2.5 Other Useful Resources

- The MEMS Clearinghouse website [60]. This website contains links to products, research groups, and conference information. One useful link is the Material Properties database [61], which includes results from a wide number of experiments by many different research groups. Information from this database can be used for initial "back of the envelope" calculations for component feasibility, for example.
- The Cronos website [62]. This company provides prototyping and production-level fabrication for all three process approaches (surface micromachining, bulk micromachining, and high aspect ratio manufacturing). It is also attempting to build a library of MEMS components for both surface micromachining (MUMPS, or the Multi-User MEMS Process [63]) and bulk micromachining.

14.5 Modeling and Simulating MEMS, That Is, Systems with Micro- (or Nano-) Scale Feature Sizes, Mixed Digital (Discrete) and Analog (Continuous) Input, Output, and Signals, Two- and Three-Dimensional Phenomena, and Inclusion and Interaction of Multiple Domains and Technologies

In preceding sections we briefly described the current state-of-the-art in modeling and simulation in both the digital and analog domains. While the digital tools are much more developed, in both the digital and analog domains there exist standard, well-characterized technologies, standard widely available tools, and stable educational and prototyping programs. In the much more complex realm of MEMS, this is not the case. Let us compare MEMS, point by point, with digital and analog circuits.

- Is there a small set of basic elements? The answer to this question is emphatically no. Various attempts have been made by researchers to develop a comprehensive basic set of building blocks, beginning with Petersen's identification of the fundamental component set consisting of beams, membranes, holes, grooves, and joints [64]. Most of these efforts focus on adding mechanical and electromechanical elements. In the SUGAR system, for example, the basic elements are the *beam* and the *electrostatic gap*. In the Carnegie Mellon tool MEMSYN [65], which is supported by the

NODAS simulator, basic elements include beams and gaps, as well as plate masses, anchors, and electrostatic comb drives (vertical and horizontal). For the MUMPS process there is the Consolidated Micromechanical Element Library (CaMEL), which contains both a nonparameterized cell database and a library of parameterized elements (which can be accessed through a component "generator," but not directly by the user). CaMEL supports the creation of a limited set of components, including motors and resonators, in a fixed surface-micromachined technology. But the bottom line for MEMS is that no set of basic building blocks has yet been identified which can support all the designs, in many different energy domains and in a variety of technologies, which researchers are interested in building. Moreover, there is no consensus as to how to effectively limit design options so that such a fundamental set could be identified. In addition, the continuous nature of most MEMS behavior presents the same kinds of difficulties that are faced with analog elements. Development of higher level component libraries, however, is a fairly active field, with, for example, ANSYS, CFD, MEMCAD, Carnegie Mellon, and MemsPro all providing libraries of previously designed and tested components for systems developers to use. Most of these components are in the electromechanical domain. As mentioned above, a few VHDL-AMS models are also available, but these will not be of practical value until more robust and complete VHDL-AMS simulators are developed and more experimental results can be obtained to validate these models.

- Is there a small set of well-understood technologies? Again the answer must be no. Almost all digital and analog circuits are essentially two-dimensional, but, in the case of MEMS, many designs can be developed either in the "2.5-dimensional" technology known as micromachining or in the true three-dimensional technology known as bulk micromachining. Thus, before doing any modeling or simulation, the MEMS developer must first choose not only among very different fabrication techniques but also among actual processes. Both the Carnegie Mellon and Cronos tools, for example, are based on processes that are being developed in parallel with the tools. MOSIS does provide central access to technology in which all but the final steps of surface micromachining can be done, but no other centrally maintained processing is available to the community of MEMS researchers in general. For surface micromachining, the fact that the final processing steps are performed in individual research labs is problematic for producing repeatable experimental results. For bulk micromachining examples, fabrication in small research labs rather than in a production environment is more the norm than the exception, so standardization for bulk processes is difficult to achieve. In addition, because much MEMS work is relatively low-volume, most processes are not well enough characterized for low-level modeling to be very effective. In such circumstances it is very difficult to have reliable process characterizations on which to build robust models.
- Is there a well-developed educational infrastructure and prototyping facilities? Again we must answer no. Introductory MEMS courses, especially, are much more likely to emphasize fabrication techniques than modeling and simulation. In [66] a set of teaching modules for a MEMS course emphasizing integrated design and simulation is described. However, this course requires the use of devices previously fabricated for validating design and simulation results, rather than expecting students to complete the entire design-simulate-test-fabricate sequence in one quarter or semester. In addition, well-established institutional practices make it difficult to provide the necessary support for multidisciplinary education which MEMS requires.
- Are encapsulation and abstraction widely employed? In the 1980s many researchers believed that multiple levels of abstraction were not useful for MEMS devices. Currently, however, the concept of intermediate-level "macromodels" has gained much support [57,70], and increasing emphasis is being placed on developing macromodels for MEMS components that will be a part of larger systems. In addition, there are several systems in development that are based on sets of more primitive components. But this method of development is not the norm, in large part because of the rich set of possibilities inherent in MEMS in general. In Figure 14.2b we have given a partial classification of MEMS corresponding to the classification for digital devices in Figure 14.2a. At this point it is not

Simulation tool	Levels supported
Mathematica, Matlab	all
MEMCAD	low
SPICE	low to medium
APLAC	low to medium
ANSYS, CFD	low to medium
SUGAR, NODAS	low to medium
MemsPro	low to medium
VHDL-AMS	medium to high

*Because MEMCAD incorporates process simulations, it supports both physical and behavioral views. All other tools support the behavioral view.

FIGURE 14.8 Available MEMS simulation tools, by level and view.

clear what the optimum number of levels of abstraction for MEMS would be. In Figure 14.8 we have attempted to classify some of the tools from Section 14.4 in terms of their ability to support various levels (since these are simulators, they all support the "behavioral" view. MEMCAD, which allows fabrication process simulation, also supports the "physical" view). Note that VHDL-AMS is the only tool, besides the general-purpose Mathematica and Matlab, that supports a high-level view of MEMS.

- Are there well-developed models, mature tools, and integrated development systems which are widely available? While such systems do not currently exist, it is predicted that some examples should become available within the next ten years [57].

14.6 A "Recipe" for Successful MEMS Simulation

A useful set of guidelines for analog simulation can be found in [67]. From this we can construct a set of guidelines for MEMS simulation.

1. Be sure you have access to the necessary domain-specific knowledge for all energy domains of interest before undertaking the project.
2. Never use a simulator unless you know the range of answers beforehand.
3. Never simulate more of the system than is necessary.
4. Always use the simplest model that will do the job.
5. Use the simulator exactly as you would do the experiment.
6. Use a specified procedure for exploring the design space. In most cases this means that you should change only one parameter at a time.
7. Understand the simulator you are using and all the options it makes available.
8. Use the correct multipliers for all quantities.
9. Use common sense.
10. Compare your results with experiments and make them available to the MEMS community.
11. Be sensitive to the possibility of microlevel phenomena, which may make your results invalid.

The last point is particularly important. Many phenomena, which can be ignored at larger feature sizes, will need to be taken into account at the micro level. For example, at the micro scale, fluid flow can behave in dramatically different ways [44]. Many other effects of scaling feature sizes down to the microlevel, including an analysis of why horizontal cantilever beam actuators are "better" than vertical cantilever beam actuators, are discussed in Chapter 9 of [68]. Chapters 4 and 5 of [68] also provide important information for low-level modeling and simulation.

14.7 Conclusion: Continuing Progress in MEMS Modeling and Simulation

In the past fifteen years, much progress has been made in providing MEMS designers with simulators and other tools which will give them the ability to make MEMS as useful and ubiquitous as was predicted in [64]. While there is still much to be done, the future is bright for this flexible and powerful technology. One of the main challenges remaining for modeling and simulation is to complete the design and development of a high-level MEMS description language, along with supporting models and simulators, both to speed prototyping and to provide a common user-friendly language for designers. One candidate for such a language is VHDL-AMS. In [69], the strengths and weaknesses of VHDL-AMS as a tool for MEMS development are discussed. Strengths include the ability to handle both discrete and continuous behavior, smooth transitions between levels of abstraction, the ability to handle both conservative and nonconservative systems simultaneously, and the ability to import code from other languages. Major drawbacks include the inability to do symbolic computation, the limitation to ordinary differential equations, lack of support for frequency domain simulations, and inability to do automatic unit conversions. It remains to be seen whether VHDL-AMS will eventually be extended to make it more suitable to support the MEMS domain. But it is highly likely that VHDL-AMS or some similar language will eventually come to be widely used and appreciated in the MEMS community.

References

1. Kielkowski, R.M., *SPICE: Practical Device Modeling,* McGraw-Hill, 1995.
2. Leong, S.K., Extracting MOSFET RF SPICE models, http://www.polyfet.com/MTT98.pdf (accessed July 20, 2001).
3. http://www.mosis.edu (accessed July 20, 2001).
4. http://cmp.imag.fr (accessed July 20, 2001).
5. http://www.imec.be/europractice/europractice.html (accessed July 20, 2001).
6. http://www.vdec.u-tokyo.ac.jp/English (accessed July 20, 2001).
7. http://www.cmc.ca (accessed July 20, 2001).
8. Mead, C. and Conway, L., *Introduction to VLSI Systems,* Addison-Wesley, 1980.
9. Gajski, D. and Thomas, D., Introduction to silicon compilation, in *Silicon Compilation,* D. Gajski, Ed., Addison-Wesley, 1988, 1–48.
10. Weste, N. and Esraghian, K., *Principles of CMOS VLSI Design: A Systems Perspective,* 2nd ed., Addison-Wesley, 1993.
11. Foty, D., *MOSFET Modeling with SPICE,* Prentice Hall, 1997.
12. http://www.research.compaq.com/wrl/projects/magic/magic.html (accessed July 20, 2001).
13. http://bwrc.eecs.berkeley.edu/Classes/IcBook/SPICE (accessed July 20, 2001).
14. Design Automation Standards Committee, IEEE Computer Society, *IEEE VHDL Standard Language Reference Manual (Integrated with VHDL-AMS Changes),* Standard 1076.1, IEEE, 1997.
15. Ashenden, P., *The Designer's Guide to VHDL,* 2nd ed., Morgan Kauffman, 2001.
16. Bhasker, J., *A Verilog HDL Primer,* 2nd ed., Star Galaxy Pub., 1999.
17. http://www.altera.com (accessed July 20, 2001).
18. http://www.xilinx.com (accessed July 20, 2001).
19. Hamblen, J.O. and Furman, M.D., *Rapid Prototyping of Digital Systems, A Tutorial Approach,* Kluwer, 1999.
20. Uyemura, J.P., *Introduction to VLSI Circuits and Systems,* John Wiley & Sons, Inc., 2002.
21. Sobecks, B., Performance Modeling of Analog Circuits via Neural Networks: The Design Process View, Ph.D. Dissertation, University of Cincinnati, 1998.
22. http://www.aplac.hut.fi (accessed July 20, 2001).
23. Weiss, R., Homsy, G., and Knight, T., Toward *in vivo* digital circuits, http://www.swiss.ai.mit.edu/~rweiss/bio-programming/dimacs99-evocomp-talk/ (accessed July 20, 2001).

24. Chang, H., Charbon, E., Choudhury, U., Demir, A., Liu, Felt E., Malavasi, E., Sangiovanni-Vincentelli, A., Charbon, E., and Vassiliou, I., *A Top-down, Constraint-Driven Design Methodology for Analog Integrated Circuits*, Kluwer Academic Publishers, 1996.
25. Ganesan, S., Synthesis and Rapid Prototyping of Analog and Mixed Signal Systems, Ph.D. Dissertation, University of Cincinnati, 2001.
26. SEAMS simulator project, University of Cincinnati ECECS Department, Distributed Processing Laboratory, http://www.ececs.uc.edu/~hcarter (accessed July 20, 2001).
27. http://www.analogy.com/products/Simulation/simulation.htm#Saber (accessed July 20, 2001).
28. www.design-reuse.com (accessed July 20, 2001).
29. S. M. Sandler and Analytical Engineering Inc., *The SPICE Handbook of 50 Basic Circuits*, http://dacafe.ibsystems.com/DACafe/EDATools/EDAbooks/SpiceHandBook (accessed July 20, 2001).
30. Kielkowski, R.M., *Inside Spice*, 2nd ed., McGraw Hill, 1998.
31. Gibson, D., Hare, A., Beyette, F., Jr., and Purdy, C., Design automation of MEMS systems using behavioral modeling, *Proc. Ninth Great Lakes Symposium on VLSI*, Ann Arbor Mich. (Eds. R.J. Lomax and P. Mazumder), March 1999, pp. 266–269.
32. http://www.analog.com/industry/iMEMS (accessed July 20, 2001).
33. http://www-ccrma.stanford.edu/CCRMA/Courses/252/sensors/node6.html (accessed July 20, 2001).
34. Tang, W., Electrostatic Comb Drive for Resonant Sensor and Actuator Applications, Ph.D. Dissertation, UC Berkeley, 1990.
35. Lo, N.R., Berg, E.C., Quakkelaar, S.R., Simon, J.N., Tachiki, M., Lee, H.-J., and Pister, S.J., Parameterized layout synthesis, extraction, and SPICE simulation for MEMS, *ISCAS 96*, May 1996, pp. 481–484.
36. Gibson, D., and Purdy, C.N., Extracting behavioral data from physical descriptions of MEMS for simulation, *Analog Integrated Circuits and Signal Processing 20*, 1999, pp. 227–238.
37. Hayt, W.H., Jr. and Kemmerly, J.E., *Engineering Circuit Analysis*, 5th ed., McGraw-Hill, 1993, pp. 88–95.
38. Dewey, A., Hanna, J., Hillman, B., Dussault, H., Fedder, G., Christen, E., Bakalar, K., Carter, H., and Romanowica, B., VHDL-AMS Modeling Considerations and Styles for Composite Systems, Version 2.0, http://www.ee.duke.edu/research/IMPACT/documents/model_g.pdf (accessed July 20, 2001).
39. McCalla, W.J., *Fundamentals of Computer-Aided Circuit Simulation*, Kluwer Academic, 1988.
40. Clark, J.V., Zhou, N., and Pister, K.S.J., Modified nodal analysis for MEMS with multi-energy domains, *International Conference on Modeling and Simulation of Microsystems, Semiconductors, Sensors and Actuators*, San Diego, CA, March 27–29, 2000, pp. 31–34.
41. Stasa, F.L., *Applied Finite Element Analysis for Engineers*, Holt, Rinehart and Winston, 1985.
42. http://www.wolfram.com/products/mathematica (accessed July 20, 2001).
43. http://www.mathworks.com/products/matlab (accessed July 20, 2001).
44. Mehta, A., Design and Control Oriented Approach to the Modeling of Microfluidic System Components, M.S. Thesis, University of Cincinnati, 1999.
45. Swart, N., Nathan, A., Shams, M., and Parameswaran, M., Numerical optimisation of flow-rate microsensors using circuit simulation tools, *Transducers '91*, 1991, pp. 26–29.
46. http://www.ee.duke.edu/research/IMPACT/vhdl-ams/index.html (accessed July 20, 2001).
47. Gibson, D., Carter, H., and Purdy, C., The use of hardware description languages in the development of microelectromechanical systems, *International Journal of Analog Integrated Circuits and Signal Processing*, 28(2), August 2001, pp. 173–180.
48. http://www.vhdl-ams.com/ (accessed July 20, 2001).
49. http://www.ansys.com/action/MEMSinitiative/index.htm (accessed July 20, 2001).
50. http://www.ansys.com/action/pdf/MEMS_WP.pdf (accessed July 20, 2001).
51. http://www.cfdrc.com (accessed July 20, 2001).
52. Pister, K., SUGAR V2.0, http://www-bsac.EECS.Berkeley.edu/~cfm/ mainpage.html (accessed July 20, 2001).
53. http://www.ece.cmu.edu/~mems/projects/memsyn/nodasv1_4/index.shtml (accessed July 20, 2001).
54. http://www2.ece.cmu.edu/~mems/projects/memsyn/nodasv1_4/tutorial.html (accessed July 20, 2001).

55. Jing, Q. and Fedder, G.K., NODAS 1.3-nodal design of actuators and sensors, *IEEE/VIUF International Workshop on Bahavioral Modeling and Simulation,* Orlando, FL., October 27–28, 1998.
56. http://www.coventor.com/software/coventorware/index.html (accessed July 20, 2001).
57. Senturia, S.D., Simulation and design of microsystems: a 10-year perspective, *Sensors and Actuators A,* 67, 1998, pp. 1–7.
58. www.memscap.com/index2.html (accessed July 20, 2001).
59. http://www.tanner.com/ (accessed July 20, 2001).
60. http://mems.isi.edu (accessed July 20, 2001).
61. http://mems.isi.edu/mems/materials/index.html (accessed July 20, 2001).
62. http://www.memsrus.com (accessed July 20, 2001).
63. http://www.memsrus.com/cronos/svcsmumps.html (accessed July 20, 2001).
64. Petersen, K., Silicon as a mechanical material, *IEEE Proceedings,* 70(5), May 1982, pp. 420–457.
65. http://www.ece.cmu.edu/~mems/projects/memsyn/index.shtml (accessed July 20, 2001).
66. Beyette, F., Jr. and C.N. Purdy, Teaching modules for a class in mechatronics, *European Workshop on Microelectronics Education (EWME2000),* May 2000.
67. Allen, P.E. and Holberg, D.R., *CMOS Analog Circuit Design,* Oxford University Press, 1987, pp. 142–144.
68. Madou, M., *Fundamentals of Microfabrication,* CRC Press, Boca Raton, FL, 1997.
69. Gibson, D. and Purdy, C., The strengths and weaknesses of VHDL-AMS as a tool for MEMS development, white paper, 2000, http://www.ececs.uc.edu/~cpurdy/csl.html/pub.html/weakvhdl. pdf (accessed July 20, 2001).
70. Mukherjee, T. and Fedder, G.K., Hierarchical mixed-domain circuit simulation, synthesis and extraction methodology for MEMS, *Journal of VLSI Signal Processing,* 21, 1999, pp. 233–249.

15
Rotational and Translational Microelectromechanical Systems: MEMS Synthesis, Microfabrication, Analysis, and Optimization

15.1	Introduction	15-2
15.2	MEMS Motion Microdevice Classifier and Structural Synthesis	15-3
15.3	MEMS Fabrication	15-6
	Bulk Micromachining • Surface Micromachining • LIGA and LIGA-Like Technologies	
15.4	MEMS Electromagnetic Fundamentals and Modeling	15-8
15.5	MEMS Mathematical Models	15-11
	Example 15.5.1: Mathematical Model of the Translational Microtransducer • Example 15.5.2: Mathematical Model of an Elementary Synchronous Reluctance Micromotor • Example 15.5.3: Mathematical Model of Two-Phase Permanent-Magnet Stepper Micromotors • Example 15.5.4: Mathematical Model of Two-Phase Permanent-Magnet Synchronous Micromotors	
15.6	Control of MEMS	15-22
	Proportional-Integral-Derivative Control • Tracking Control • Time-Optimal Control • Sliding Mode Control • Constrained Control of Nonlinear MEMS: Hamilton–Jacobi Method • Constrained Control of Nonlinear Uncertain MEMS: Lyapunov Method • Example 15.6.1: Control of Two-Phase Permanent-Magnet Stepper Micromotors	
15.7	Conclusions	15-35
	References	15-35

Sergey Edward Lyshevski
University of Rochester

15.1 Introduction

Electromagnetic-based MEMS are widely used in various sensing and actuation applications. For these MEMS, rotational and translational motion microdevices are needed to be devised, designed, and controlled. We introduce the classifier paradigm to perform the structural synthesis of MEMS upon electromagnetic features. As motion microdevices are devised, the following issues are emphasized: modeling, analysis, simulation, control, optimization, and validation. Innovative results are researched and studied applying the classifier, structural synthesis, design, analysis, and optimization concepts developed. The need for innovative integrated methods to perform the comprehensive analysis, high-fidelity modeling, and design of MEMS has facilitated theoretical developments within the overall spectrum of engineering and science. This chapter provides one with viable tools to perform structural synthesis, modeling, analysis, optimization, and control of MEMS.

Microelectromechanical systems integrate motion microstructures and devices as well as ICs on a single chip or on a hybrid chip. To fabricate MEMS, modified advanced microelectronics fabrication technologies, techniques, processes, and materials are used. Due to the use of complementary metal oxide semiconductor (CMOS) lithography-based technologies in fabrication microstructures, microdevices, and ICs, MEMS leverage microelectronics.

The following definition for MEMS was given in [1]:

Batch-fabricated microscale devices (ICs and motion microstructures) that convert physical parameters to electrical signals and vice versa, and in addition, microscale features of mechanical and electrical components, architectures, structures, and parameters are important elements of their operation and design.

The scope of MEMS has been further expanded towards devising novel paradigms, system-level integration high-fidelity modeling, data-intensive analysis, control, optimization, fabrication, and implementation. Therefore, we define MEMS as:

Batch-fabricated microscale systems (motion and radiating energy microdevices/microstructures—driving/sensing circuitry—controlling/processing ICs) that

1. Convert physical stimuli, events, and parameters to electrical and mechanical signals and vice versa
2. Perform actuation and sensing
3. Comprise control (intelligence, decision making, evolutionary learning, adaptation, self-organization etc.), diagnostics, signal processing, and data acquisition features

and microscale features of electromechanical, electronic, optical, and biological components (structures, devices, and subsystems), architectures, and operating principles are basics of their operation, design, analysis, and fabrication.

The integrated design, analysis, optimization, and virtual prototyping of intelligent and high-performance MEMS, system intelligence, learning, adaptation, decision making, and self-organization can be addressed, researched, and solved through the use of advanced electromechanical theory, state-of-the-art hardware, novel technologies, and leading-edge software. Many problems in MEMS can be formulated, attacked, and solved using the microelectromechanics. In particular, microelectromechanics deals with benchmarking and emerging problems in integrated electrical–mechanical–computer engineering, science, and technologies. Microelectromechanics is the integrated design, analysis, optimization, and virtual prototyping of high-performance MEMS, system intelligence, learning, adaptation, decision making, and control through the use of advanced hardware, leading-edge software, and novel fabrication technologies and processes. Integrated multidisciplinary features approach quickly, and the microelectromechanics takes place.

The computer-aided design tools are required to support MEMS analysis, simulation, design, optimization, and fabrication. Much effort has been devoted to attain the specified steady-state and dynamic performance of MEMS to meet the criteria and requirements imposed. Currently, MEMS are designed, optimized, and analyzed using available software packages based on the linear and steady-state analysis.

However, highly detailed nonlinear electromagnetic and mechanical modeling must be performed to design high-performance MEMS. Therefore, the research is concentrated on high-fidelity mathematical modeling, data intensive analysis, and nonlinear simulations, as well as control (design of control algorithms to attain the desired performance). The reported synthesis, modeling, analysis, simulation, optimization, and control concepts, tools, and paradigms ensure a cost-effective solution and can be used to guarantee rapid prototyping of high-performance state-of-the-art MEMS. It is often very difficult, and sometimes impossible, to solve a large array of nonlinear analysis and design problems for motion microdevices using conventional methods. Innovative concepts, methods, and tools that fully support the analysis, modeling, simulation, control, design, and optimization are needed. The fabrication technologies used in MEMS were developed [2,3], and micromachining technologies are discussed in this chapter. This chapter solves a number of long-standing problems for electromagnetic-based MEMS.

15.2 MEMS Motion Microdevice Classifier and Structural Synthesis

It was emphasized that the designer must design MEMS by devising novel high-performance motion microdevices, radiating energy microdevices, microscale driving/sensing circuitry, and controlling/processing ICs. A step-by-step procedure in the design of motion microdevices is:

- Define application and environmental requirements
- Specify performance specifications
- Devise motion microstructures and microdevices, radiating energy microdevices, microscale driving/sensing circuitry, and controlling/processing ICs
- Develop the fabrication process using micromachining and CMOS technologies
- Perform electromagnetic, energy conversion, mechanical, and sizing/dimension estimates
- Perform electromagnetic, mechanical, vibroacoustic, and thermodynamic design with performance analysis and outcome prediction
- Verify, modify, and refine design with ultimate goals and objectives to optimize the performance

In this section, the design and optimization of motion microdevices is reported

To illustrate the procedure, consider two-phase permanent-magnet synchronous slotless micromachines as documented in Figure 15.1.

It is evident that the electromagnetic system is *endless*, and different geometries can be utilized as shown in Figure 15.1. In contrast, in translational (linear) synchronous micromachines, the *open-ended* electromagnetic system results. The attempts to classify microelectromechanical motion devices were made in [1,4,5]; however, the qualitative and quantitative comprehensive analysis must be researched.

Motion microstructure geometry and electromagnetic systems must be integrated into the synthesis, analysis, design, and optimization. Motion microstructures can have the plate, spherical, toroidal, conical, cylindrical, and asymmetrical geometry. Using these distinct geometry and electromagnetic systems, we propose to classify MEMS. This idea is extremely useful in the study of existing MEMS as well as in the synthesis of an infinite number of innovative motion microdevices. In particular, using the possible geometry and electromagnetic systems (*endless, open-ended,* and *integrated*), novel high-performance MEMS can be synthesized.

The basic electromagnetic micromachines (microdevices) under consideration are direct- and alternating-current, induction and synchronous, rotational and translational (linear). That is, microdevices are classified using a type classifier

$$Y = \{y : y \in Y\}$$

Motion microdevices are categorized using a geometric classifier (plate *P*, spherical *S*, toroidal *T*, conical *N*, cylindrical *C*, or asymmetrical *A* geometry) and an electromagnetic system classifier (*endless E, open-ended O,* or *integrated I*). The microdevice classifier, documented in Table 15.1, is partitioned

TABLE 15.1 Classification of Electromagnetic Microdevices Using the Electromagnetic System–Geometry Classifier

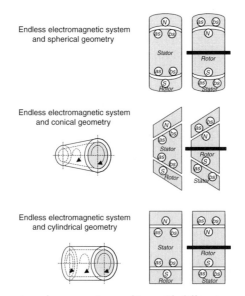

FIGURE 15.1 Permanent-magnet synchronous micromachines with different geometry.

into three horizontal and six vertical strips, and contains 18 sections, each identified by ordered pairs of characters, such as (E, P) or (O, C).

In each ordered pair, the first entry is a letter chosen from the bounded electromagnetic system set

$$M = \{E, O, I\}$$

The second entry is a letter chosen from the geometric set

$$G = \{P, S, T, N, C, A\}$$

That is, for electromagnetic microdevices, the electromagnetic system–geometric set is

$$M \times G = \{(E, F), (E, S), (E, T), \ldots, (I, N), (I, C), (I, A)\}$$

In general, we have

$$M \times G = \{(m, g) : m \in M \text{ and } g \in G\}$$

Other categorization can be applied. For example, single-, two-, three-, and multi-phase microdevices are classified using a phase classifier

$$H = \{h : h \in H\}$$

Therefore, $Y \times M \times G \times H = \{(y, m, g, h) : y \in Y, m \in M, g \in G \text{ and } h \in H\}$

Topology (radial or axial), permanent magnets shaping (strip, arc, disk, rectangular, triangular, or other shapes), permanent magnet characteristics (BH demagnetization curve, energy product, hysterisis minor loop), commutation, *emf* distribution, cooling, power, torque, size, torque-speed characteristics, as well as other distinct features of microdevices can be easily classified.

That is, the devised electromagnetic microdevices can be classified by an N-tuple as

{microdevice type, electromagnetic system, geometry, topology, phase, winding, connection, cooling}.

Using the classifier, which is given in Table 15.1 in terms of electromagnetic system–geometry, the designer can classify the existing motion microdevices as well as synthesize novel high-performance microdevices. As an example, the spherical, conical, and cylindrical geometries of a two-phase permanent-magnet synchronous microdevice are illustrated in Figure 15.2.

This section documents new results in structural synthesis which can be used to optimize the microdevice performance. The conical (existing) and spherical-conical (devised) microdevice geometries are illustrated in Figure 15.2. Using the innovative spherical-conical geometry, which is different compared to the existing conical geometry, one increases the active length L_r and average diameter D_r. For radial flux microdevices, the electromagnetic torque T_e is proportional to the squared rotor diameter and axial length. In particular, $T_e = k_T D_r^2 L_r$, where k_T is the constant. From the above relationship, it is evident

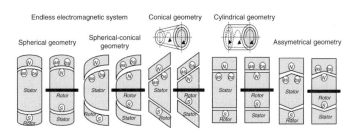

FIGURE 15.2 Two-phase permanent-magnet synchronous microdevice (micromachine) geometry.

that the spherical-conical micromotors develop higher electromagnetic torque compared with the conventional design. In addition, improved cooling, reduced undesirable torques components, as well as increased ruggedness and robustness contribute to the viability of the proposed solution. Thus, using the classifier paradigm, novel microdevices with superior performance can be devised.

15.3 MEMS Fabrication

Microelectromechanics, which integrates micromechanics and microelectronics, requires affordable, low-cost, high-yield fabrication technologies which allow one to fabricate 3-D microscale structures and devices. Micromachining is a key fabrication technology for microscale structures, devices, and MEMS. Microelectromechanical systems fabrication technologies fall into three broad categories: bulk machining, surface machining, and LIGA (LIGA-like) techniques [1–3].

15.3.1 Bulk Micromachining

Bulk and surface micromachining are based on the modified CMOS and specifically designed micromachining processes. Bulk micromachining of silicon uses wet and dry etching techniques in conjunction with etch masks and etch-stop-layers to develop microstructures from the silicon substrate. Microstructures are fabricated by etching areas of the silicon substrate to release the desired 3-D microstructures. The *anisotropic* and *isotropic* wet etching processes, as well as concentration dependent etching techniques, are widely used in bulk micromachining. The microstructures are formed by etching away the bulk of the silicon wafer to fabricate the desired 3-D structures. Bulk machining with its crystallographic and dopant-dependent etch processes, when combined with wafer-to-wafer bonding, produces complex 3-D microstructures with the desired geometry. Through bulk micromachining, one fabricates microstructures by etching deeply into the silicon wafer. There are several ways to etch the silicon wafer. The *anisotropic* etching uses etchants that etch different crystallographic directions at different rates. Through *anisotropic* etching, 3-D structures (cons, pyramids, cubes, and channels into the surface of the silicon wafer) are fabricated. In contrast, the *isotropic* etching etches all directions in the silicon wafer at same (or close) rate, and, therefore, hemisphere and cylinder structures can be made. Deep reactive ion etching uses plasma to etch straight walled structures (cubes, rectangular, triangular, etc.).

15.3.2 Surface Micromachining

Surface micromachining has become the major fabrication technology in recent years because complex 3-D microscale structures and devices can be fabricated. Surface micromachining with single-crystal silicon, polysilicon, silicon nitride, silicon oxide, and silicon dioxide (as structural and sacrificial materials which deposited and etched) is widely used to fabricate microscale structures and devices on the surface of a silicon wafer. This affordable low-cost high-yield technology is integrated with IC fabrication processes guaranteeing the needed microstructures-IC fabrication compatibility. The techniques for depositing and patterning thin films are used to produce complex microstructures and microdevices on the surface of silicon wafers (surface silicon micromachining) or on the surface of other substrates. Surface micromachining technology allows one to fabricate the structure as layers of thin films. This technology guarantees the fabrication of 3-D microdevices with high accuracy, and the surface micromachining can be called a thin film process. Each thin film is usually limited to thickness up to 5 μm, which leads to fabrication of high-performance planar-type microscale structures and devices. The advantage of surface micromachining is the use of standard CMOS fabrication processes and facilities, as well as compliance with ICs. Therefore, this technology is widely used to manufacture microscale actuators and sensors (microdevices).

Surface micromachining is based on the application of sacrificial (temporary) layers that are used to maintain subsequent layers and are removed to reveal (release) fabricated (released or suspended) microstructures. This technology was first demonstrated for ICs and applied to fabricate microstructures in the 80s. On the surface of a silicon wafer, thin layers of structural and sacrificial materials are deposited

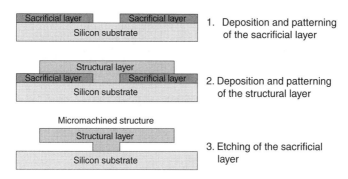

FIGURE 15.3 Surface micromachining.

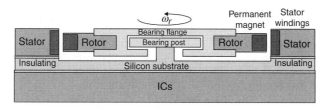

FIGURE 15.4 Cross-section schematics for slotless permanent-magnet brushless micromotor with ICs.

and patterned. Then, the sacrificial material is removed, and a micromechanical structure or device is fabricated. Figure 15.3 illustrates a typical process sequence of the surface micromachining fabrication technology.

Usually, the sacrificial layer is made of silicon dioxide (SiO_2), phosphorous-doped silicon dioxide, or silicon nitride (Si_3N_4). The structural layers are then typically formed with polysilicon, and the sacrificial layer is removed. In particular, after fabrication of the surface microstructures and microdevices (micromachines), the silicon wafer can be wet bulk etched to form cavities below the surface components, which allows a wider range of desired motion for the device. The wet etching can be done using hydrofluoric and buffered hydrofluoric acids, potassium hydroxide, ethylene-diamene-pyrocatecol, tetramethylammonium hydroxide, or sodium hydroxide. Surface micromachining technology was used to fabricate rotational micromachines [6]. For example, heavily-phosphorous-doped polysilicon can be used to fabricate rotors and stators, and silicon nitride can be applied as the structural material to attain electrical insulation. The cross-section of the slotless micromotor fabricated on the silicon substrate with polysilicon stator with deposited windings, polysilicon rotor with deposited permanent-magnets, and bearing is illustrated in Figure 15.4. The micromotor is controlled by the driving/sensing and controlling/processing ICs. To fabricate micromotor and ICs on a single- or double-sided chip (which significantly enhances the performance), similar fabrication technologies and processes are used, and the compatibility issues are addressed and resolved. The surface micromachining processes were integrated with the CMOS technology (e.g., similar materials, lithography, etching, and other techniques). To fabricate the integrated MEMS, post-, mixed-, and pre-CMOS/micromachining techniques can be applied [1–3].

15.3.3 LIGA and LIGA-Like Technologies

There is a critical need to develop the fabrication technologies allowing one to fabricate high-aspect-ratio microstructures. The LIGA process, which denotes Lithography–Galvanoforming–Molding (in German words, *Lithografie–Galvanik–Abformung*), is capable of producing 3-D microstructures of up to centimeter high with the aspect ratio (depth versus lateral dimension) more than 100 [2,7,8]. The LIGA technology is based upon X-ray lithography, which guarantees shorter wavelength (in order from few to 10 Å, which leads to negligible diffraction effects) and larger depth of focus compared with optical lithography. The ability to fabricate microstructures and microdevices in the centimeter range is particularly important in the actuators and drives applications since the specifications are imposed on the rated force and torque developed by the microdevices, and due to the limited force and torque densities, the designer faces the need to increase the actuator dimensions.

15.4 MEMS Electromagnetic Fundamentals and Modeling

The MEMS classifier, structural synthesis, and optimization were reported in Section 15.2. The classification and optimization are based on the consideration and synthesis of the electromagnetic system, analysis of the *magnetomotive* force, design of the MEMS geometry and topology, and optimization of other quantities. Different rotational (radial and axial) and translational motion microdevices are classified using *endless* (closed), *open-ended* (open), and *integrated* electromagnetic systems.

Our goal is to approach and solve a wide range of practical problems encountered in nonlinear design, modeling, analysis, control, and optimization of motion microstructures and microdevices with driving/sensing circuitry controlled by ICs for high-performance MEMS. Studying MEMS, the emphases are placed on:

- Design of high-performance MEMS through devising innovative motion microdevices with radiating energy microdevices, microscale driving/sensing circuitry, and controlling/signal processing ICs
- Optimization and analysis of rotational and translation motion microdevices
- Development of high-performance signal processing and controlling ICs for microdevices devised
- Development of mathematical models with minimum level of simplifications and assumptions in the time domain
- Design of optimal robust control algorithms
- Design of intelligent systems through self-adaptation, self-organization, evolutionary learning, decision-making, and intelligence
- Development of advanced software and hardware to attain the highest degree of intelligence, integration, efficiency, and performance

In this section, our goal is to perform nonlinear modeling, analysis, and simulation. To attain these objectives, we apply the MEMS synthesis paradigm, develop nonlinear mathematical models to model complex electromagnetic-mechanical dynamics, perform optimization, design closed-loop control systems, and perform data-intensive analysis in the time domain.

To model electromagnetic motion microdevices, using the magnetic vector and electric scalar potentials $\vec{A}$ and V, respectively, one usually solves the partial differential equations

$$-\nabla^2 \vec{A} + \mu\sigma \frac{\partial \vec{A}}{\partial t} + \mu\varepsilon \frac{\partial^2 \vec{A}}{\partial t^2} = -\mu\sigma \nabla V$$

using finite element analysis. Here, μ, σ, and ε are the permeability, conductivity, and permittivity.

However, to design electromagnetic MEMS as well as to perform electromagnetic–mechanical analysis and optimization, differential equations must be solved in the time domain. In fact, basic phenomena cannot be comprehensively modeled, analyzed, and assessed applying traditional finite element analysis, which gives the steady-state solutions and models. There is a critical need to develop the modeling tools that will allow one to augment nonlinear electromagnetics and mechanics in a single electromagnetic–mechanical modeling core to attain high-fidelity analysis with performance assessment and outcome prediction.

Operating principles of MEMS are based upon electromagnetic principles. A complete electromagnetic model is derived in terms of five electromagnetic field vectors. In particular, three electric field vectors and two magnetic field vectors are used. The electric field vectors are the electric field intensity, $\vec{E}$, the electric flux density, $\vec{D}$, and the current density, $\vec{J}$. The magnetic field vectors are the magnetic field intensity $\vec{H}$ and the magnetic field density $\vec{B}$. The differential equations for microelectromechanical motion device are found using Maxwell's equations, constitutive (auxiliary) equations, and classical mechanics.

Maxwell's partial differential equations in the $\vec{E}$- and $\vec{H}$-domain in the point form are

$$\nabla \times \vec{E}(x, y, z, t) = -\mu \frac{\partial \vec{H}(x, y, z, t)}{\partial t}$$

$$\nabla \times \vec{H}(x, y, z, t) = \varepsilon \frac{\partial \vec{E}(x, y, z, t)}{\partial t} + \vec{J}(x, y, z, t) = \varepsilon \frac{\partial \vec{E}(x, y, z, t)}{\partial t} + \sigma \vec{E}(x, y, z, t)$$

$$\nabla \cdot \vec{E}(x, y, z, t) = \frac{\rho_v(x, y, z, t)}{\varepsilon}$$

$$\nabla \cdot \vec{H}(x, y, z, t) = 0$$

where ε is the permittivity, μ is the permeability, σ is the conductivity, and ρ_v is the volume charge density.

The constitutive (auxiliary) equations are given using the permittivity ε, permeability tensor μ, and conductivity σ. In particular, one has

$$\vec{D} = \varepsilon \vec{E} \quad \text{or} \quad \vec{D} = \varepsilon \vec{E} + \vec{P}$$

$$\vec{B} = \mu \vec{H} \quad \text{or} \quad \vec{B} = \mu(\vec{H} + \vec{M})$$

$$\vec{J} = \sigma \vec{E} \quad \text{or} \quad \vec{J} = \rho_v \vec{v}$$

The Maxwell's equations can be solved using the boundary conditions on the field vectors. In two-region media, we have

$$\vec{a}_N \times (\vec{E}_2 - \vec{E}_1) = 0, \quad \vec{a}_N \times (\vec{H}_2 - \vec{H}_1) = \vec{J}_s, \quad \vec{a}_N \cdot (\vec{D}_2 - \vec{D}_1) = \rho_s, \quad \vec{a}_N \cdot (\vec{B}_2 - \vec{B}_1) = 0$$

where $\vec{J}_s$ is the surface current density vector, $\vec{a}_N$ is the surface normal unit vector at the boundary from region 2 into region 1, and ρ_s is the surface charge density.

The constitutive relations that describe media can be integrated with Maxwell's equations, which relate the fields in order to find two partial differential equations. Using the electric and magnetic field intensities $\vec{E}$ and $\vec{H}$ to model electromagnetic fields in MEMS, one has

$$\nabla \times (\nabla \times \vec{E}) = \nabla(\nabla \cdot \vec{E}) - \nabla^2 \vec{E} = -\mu \frac{\partial \vec{J}}{\partial t} - \mu \frac{\partial^2 \vec{D}}{\partial t^2} = -\mu \sigma \frac{\partial \vec{E}}{\partial t} - \mu \varepsilon \frac{\partial^2 \vec{E}}{\partial t^2}$$

$$\nabla \times (\nabla \times \vec{H}) = \nabla(\nabla \cdot \vec{H}) - \nabla^2 \vec{H} = -\mu \sigma \frac{\partial \vec{H}}{\partial t} - \mu \varepsilon \frac{\partial^2 \vec{H}}{\partial t^2}$$

The following pair of homogeneous and inhomogeneous wave equations

$$\nabla^2 \vec{E} - \mu \sigma \frac{\partial \vec{E}}{\partial t} - \mu \varepsilon \frac{\partial^2 \vec{E}}{\partial t^2} = \nabla \left(\frac{\rho_v}{\varepsilon} \right)$$

$$\nabla^2 \vec{H} - \mu \sigma \frac{\partial \vec{H}}{\partial t} - \mu \varepsilon \frac{\partial^2 \vec{H}}{\partial t^2} = 0$$

is equivalent to four Maxwell's equations and constitutive relations. For some cases, these two equations can be solved independently. It must be emphasized that it is not always possible to use the boundary conditions using only $\vec{E}$ and $\vec{H}$, and thus, the problem not always can be simplified to two electromagnetic field vectors. Therefore, the electric scalar and magnetic vector potentials are used. Denoting the magnetic vector potential as $\vec{A}$ and the electric scalar potential as V, we have

$$\nabla \times \vec{A} = \vec{B} = \mu \vec{H} \quad \text{and} \quad \vec{E} = -\frac{\partial \vec{A}}{\partial t} - \nabla V$$

The electromagnetic field is derivative from the potentials. Using the Lorentz equation

$$\nabla \cdot \vec{A} = -\frac{\partial V}{\partial t}$$

the inhomogeneous vector potential wave equation to be solved is

$$-\nabla^2 \vec{A} + \mu\sigma \frac{\partial \vec{A}}{\partial t} + \mu\varepsilon \frac{\partial^2 \vec{A}}{\partial t^2} = -\mu\sigma \nabla V$$

To model motion microdevices, the mechanical equations must be used, and Newton's second law is usually applied to derive the equations of motion.

Using the volume charge density ρ_v, the Lorenz force, which relates the electromagnetic and mechanical phenomena, is found as

$$\vec{F} = \rho_v(\vec{E} + \vec{v} \times \vec{B}) = \rho_v\vec{E} + \vec{J} \times \vec{B}$$

The electromagnetic force can be found by applying the Maxwell stress tensor method. This concept employs a volume integral to obtain the stored energy, and stress at all points of a bounding surface can be determined. The sum of local stresses gives the net force. In particular, the electromagnetic stress is

$$\vec{F} = \int_v (\rho_v\vec{E} + \vec{J} \times \vec{B})\,dv = \frac{1}{\mu}\oint_s \overset{\leftrightarrow}{T}_{\alpha\beta} \cdot d\vec{s}$$

The electromagnetic stress energy tensor (the second Maxwell stress tensor) is

$$\overset{\leftrightarrow}{T}_{\alpha\beta} = \begin{bmatrix} 0 & \vec{E}_x & \vec{E}_y & \vec{E}_z \\ -\vec{E}_x & 0 & \vec{B}_z & -\vec{B}_y \\ -\vec{E}_y & -\vec{B}_z & 0 & \vec{B}_x \\ -\vec{E}_z & \vec{B}_y & -\vec{B}_x & 0 \end{bmatrix}$$

In general, the electromagnetic torque developed by motion microstructures is found using the electromagnetic field. In particular, the electromagnetic stress tensor is given as

$$T_s = T_s^E + T_s^M$$

$$= \begin{bmatrix} E_1D_1 - \tfrac{1}{2}E_jD_j & E_1D_2 & E_1D_3 \\ E_2D_1 & E_2D_2 - \tfrac{1}{2}E_jD_j & E_2D_3 \\ E_3D_1 & E_3D_2 & E_3D_3 - \tfrac{1}{2}E_jD_j \end{bmatrix} + \begin{bmatrix} B_1H_1 - \tfrac{1}{2}B_jH_j & B_1H_2 & B_1H_3 \\ B_2H_1 & B_2H_2 - \tfrac{1}{2}B_jH_j & B_2H_3 \\ B_3H_1 & B_3H_2 & B_3H_3 - \tfrac{1}{2}B_jH_j \end{bmatrix}$$

For the Cartesian, cylindrical, and spherical coordinate systems, which can be used to develop the mathematical model, we have

$$E_x = E_1, \ E_y = E_2, \ E_z = E_3, \quad D_x = D_1, \ D_y = D_2, \ D_z = D_3,$$
$$H_x = H_1, \ H_y = H_2, \ H_z = H_3, \quad B_x = B_1, \ B_y = B_2, \ B_z = B_3$$
$$E_r = E_1, \ E_\theta = E_2, \ E_z = E_3, \quad D_r = D_1, \ D_\theta = D_2, \ D_z = D_3,$$
$$H_r = H_1, \ H_\theta = H_2, \ H_z = H_3, \quad B_r = B_1, \ B_\theta = B_2, \ B_z = B_3$$
$$E_\rho = E_1, \ E_\theta = E_2, \ E_\phi = E_3, \quad D_\rho = D_1, \ D_\theta = D_2, \ D_\phi = D_3,$$
$$H_\rho = H_1, \ H_\theta = H_2, \ H_\phi = H_3, \quad B_\rho = B_1, \ B_\theta = B_2, \ B_\phi = B_3$$

Maxwell's equations can be solved using the MATLAB environment.

In motion microdevices, the designer analyzes the torque or force production mechanisms. Newton's second law for rotational and translational motions is

$$\frac{d\omega_r}{dt} = \frac{1}{J}\sum \vec{T}_\Sigma, \quad \frac{d\theta_r}{dt} = \omega_r$$
$$\frac{dv}{dt} = \frac{1}{m}\sum \vec{F}_\Sigma, \quad \frac{dx}{dt} = v$$

where ω_r and θ_r are the angular velocity and displacement, v and x are the linear velocity and displacement, $\sum \vec{T}_\Sigma$ is the net torque, $\sum \vec{F}_\Sigma$ is the net force, J is the equivalent moment of inertia, and m is the mass.

15.5 MEMS Mathematical Models

The problems of modeling and control of MEMS are very important in many applications. A mathematical model is a mathematical description (in the form of functions or equations) of MEMS, which integrate motion microdevices (microscale actuators and sensors), radiating energy microdevices, microscale driving/sensing circuitry, and controlling/signal processing ICs. The purpose of the model development is to understand and comprehend the phenomena, as well as to analyze the end-to-end behavior.

To model MEMS, advanced analysis methods are required to accurately cope with the involved highly complex physical phenomena, effects, and processes. The need for high-fidelity analysis, computationally-efficient algorithms, and simulation time reduction increases significantly for complex microdevices, restricting the application of Maxwell's equations to problems possible to solve. As was illustrated in the previous section, nonlinear electromagnetic and energy conversion phenomena are described by the partial differential equations. The application of Maxwell's equations fulfills the need for data-intensive analysis capabilities with outcome prediction within overall modeling domains as particularly necessary for simulation and analysis of high-performance MEMS. In addition, other modeling and analysis methods are applied. The lumped mathematical models, described by ordinary differential equations, can be used. The process of mathematical modeling and model development is given below.

The first step is to formulate the modeling problem:

- Examine and analyze MEMS using a multilevel hierarchy concept, develop multivariable input-output subsystem pairs, for example, motion microstructures (microscale actuators and sensors), radiating energy microdevices, microscale circuitry, ICs, controller, input/output devices.
- Understand and comprehend the MEMS structure and system configuration.
- Gather the data and information.
- Develop input-output variable pairs, identify the independent and dependent control, disturbance, output, reference (command), state and performance variables, as well as events.

- Making accurate assumptions, simplify the problem to make the studied MEMS mathematically tractable (mathematical models, which are the idealization of physical phenomena, are never absolutely accurate, and comprehensive mathematical models simplify the reality to allow the designer to perform a thorough analysis and make accurate predictions of the system performance).

The second step is to derive equations that relate the variables and events:

- Define and specify the basic laws (Kirchhoff, Lagrange, Maxwell, Newton, and others) to be used to obtain the equations of motion. Mathematical models of electromagnetic, electronic, and mechanical microscale subsystems can be found and augmented to derive mathematical models of MEMS using defined variables and events.
- Derive mathematical models.

The third step is the simulation, analysis, and validation:

- Identify the numerical and analytic methods to be used in analysis and simulations.
- Analytically and/or numerically solve the mathematical equations (e.g., differential or difference equations, nonlinear equations, etc.).
- Using information variables (measured or observed) and events, synthesize the fitting and mismatch functionals.
- Verify the results through the comprehensive comparison of the solution (model input-state-output-event mapping sets) with the experimental data (experimental input-state-output-event mapping sets).
- Calculate the fitting and mismatch functionals.
- Examine the analytical and numerical data against new experimental data and evidence.

If the matching with the desired accuracy is not guaranteed, the mathematical model of MEMS must be refined, and the designer must start the cycle again.

Electromagnetic theory and classical mechanics form the basis for the development of mathematical models of MEMS. It was illustrated that MEMS can be modeled using Maxwell's equations and *torsional-mechanical* equations of motion. However, from modeling, analysis, design, control, and simulation perspectives, the mathematical models as given by ordinary differential equations can be derived and used.

Consider the rotational microstructure (bar magnet, current loop, and microsolenoid) in a uniform magnetic field, see Figure 15.5. The microstructure rotates if the electromagnetic torque is developed. The electromagnetic field must be studied to find the electromagnetic torque.

The torque tends to align the magnetic moment $\vec{m}$ with $\vec{B}$, and

$$\vec{T} = \vec{m} \times \vec{B}$$

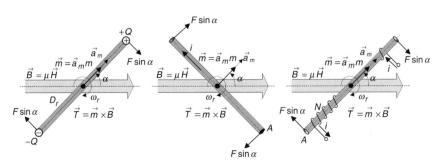

FIGURE 15.5 Clockwise rotation of the motion microstructure.

For a microstructure with outside diameter D_r, the magnet strength is Q. Hence, the magnetic moment is $m = QD_r$, and the force is found as $F = QB$.

The electromagnetic torque is

$$T = 2F\frac{1}{2}D_r \sin\alpha = QD_r B\sin\alpha = mB\sin\alpha$$

Using the unit vector in the magnetic moment direction $\vec{a}_m$, one obtains

$$\vec{T} = \vec{m} \times \vec{B} = \vec{a}_m m \times \vec{B} = QD_r \vec{a}_m \times \vec{B}$$

For a current loop with the area A, the torque is found as

$$\vec{T} = \vec{m} \times \vec{B} = \vec{a}_m m \times \vec{B} = iA\vec{a}_m \times \vec{B}$$

For a solenoid with N turns, one obtains

$$\vec{T} = \vec{m} \times \vec{B} = \vec{a}_m m \times \vec{B} = iAN\vec{a}_m \times \vec{B}$$

As the electromagnetic torque is found, using Newton's second law, one has

$$\frac{d\omega_r}{dt} = \frac{1}{J}\sum \vec{T}_\Sigma = \frac{1}{J}(\vec{T} - \vec{T}_L), \qquad \frac{d\theta_r}{dt} = \omega_r$$

where $\vec{T}_L$ is the load torque.

The *electromotive* (*emf*) and *magnetomotive* (*mmf*) forces can be used in the model development. We have

$$\text{emf} = \oint_l \vec{E} \cdot \vec{dl} = \underbrace{\oint_l (\vec{v} \times \vec{B}) \cdot \vec{dl}}_{\text{motional induction generation}} - \underbrace{\int_s \frac{\partial \vec{B}}{\partial t} \vec{ds}}_{\text{transformer induction}}$$

and

$$\text{mmf} = \int_l \vec{H} \cdot \vec{dl} = \oint_s \vec{J} \cdot \vec{ds} + \oint_s \frac{\partial \vec{D}}{\partial t} \vec{ds}$$

For preliminary design, it is sufficiently accurate to apply Faraday's or Lenz's laws, which give the electromotive force in term of the time-varying magnetic field changes. In particular,

$$\text{emf} = -\frac{d\psi}{dt} = -\frac{\partial \psi}{\partial t} - \frac{\partial \psi}{\partial \theta_r}\frac{d\theta_r}{dt} = -\frac{\partial \psi}{\partial t} - \frac{\partial \psi}{\partial \theta_r}\omega_r$$

where $\frac{\partial \psi}{\partial t}$ is the transformer term.

The total flux linkages are

$$\psi = \frac{1}{4}\pi N_S \Phi_p$$

where N_S is the number of turns and Φ_p is the flux per pole.

For radial topology micromachines, we have

$$\Phi_p = \frac{\mu i N_S}{P^2 g_e} R_{\text{in st}} L$$

where i is the current in the phase microwinding (supplied by the IC), $R_{in\,st}$ is the inner stator radius, L is the inductance, P is the number of poles, and g_e is the equivalent gap, which includes the airgap and radial thickness of the permanent magnet.

Denoting the number of turns per phase as N_S, the magnetomotive force is

$$\text{mmf} = \frac{iN_S}{P}\cos P\theta_r$$

The simplified expression for the electromagnetic torque for radial topology brushless micromachines is

$$T = \frac{1}{2}PB_{ag}i_sN_SL_rD_r$$

where B_{ag} is the air gap flux density, $B_{ag} = (\mu iN_S/2Pg_e)\cos P\theta_r$, i_s is the total current, L_r is the active length (rotor axial length), and D_r is the outside rotor diameter.

The axial topology brushless micromachines can be designed and fabricated. The electromagnetic torque is given as

$$T = k_{ax}B_{ag}i_sN_SD_a^2$$

where k_{ax} is the nonlinear coefficient, which is found in terms of active conductors and thin-film permanent magnet length; and D_a is the equivalent diameter, which is a function of windings and permanent-magnet topography.

Example 15.5.1: Mathematical Model of the Translational Microtransducer

Figure 15.6 illustrates a simple translational microstructure with a stationary member and movable translational microstructure (plunger), which can be fabricated using continuous batch-fabrication process [2]. The winding can be "printed" using the micromachining/CMOS technology.

We apply Newton's second law of motion to study the dynamics. Newton's law states that the acceleration of an object is proportional to the net force. The vector sum of all forces is found as

$$F(t) = m\frac{d^2x}{dt^2} + B_v\frac{dx}{dt} + (k_{s1}x + k_{s2}x^2) + F_e(t)$$

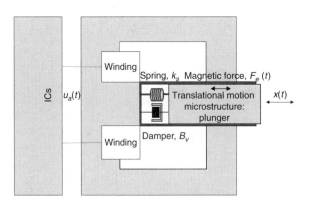

FIGURE 15.6 Microtransducer schematics with translational motion microstructure.

where x is the displacement of a translational microstructure (plunger), m is the mass of a movable plunger, B_v is the viscous friction coefficient, k_{s1} and k_{s2} are the spring constants (the spring can be made from polysilicon), and $F_e(t)$ is the magnetic force which is found using the coenergy W_c, $F_e(i,x) = \dfrac{\partial W_c(i,x)}{\partial x}$.

The stretch and restoring forces are not directly proportional to the displacement, and these forces are different on either side of the equilibrium position. The restoring/stretching force exerted by the polysilicon spring is expressed as $(k_{s1}x + k_{s2}x^2)$.

Assuming that the magnetic system is linear, the coenergy is expressed as

$$W_c(i,x) = \frac{1}{2}L(x)i^2$$

Then

$$F_e(i,x) = \frac{1}{2}i^2\frac{dL(x)}{dx}$$

The inductance is found as

$$L(x) = \frac{N^2}{\mathfrak{R}_f + \mathfrak{R}_g} = \frac{N^2 \mu_f \mu_0 A_f A_g}{A_g l_f + 2 A_f \mu_f (x + 2d)}$$

where $\mathfrak{R}_f$ and $\mathfrak{R}_g$ are the reluctances of the ferromagnetic material and air gap, A_f and A_g are the associated cross section areas, and l_f and $(x + 2d)$ are the lengths of the magnetic material and the air gap. Hence

$$\frac{dL}{dx} = -\frac{2N^2 \mu_f^2 \mu_0 A_f^2 A_g}{[A_g l_f + 2 A_f \mu_f (x + 2d)]^2}$$

Using Kirchhoff's law, the voltage equation for the phase microcircuitry is

$$u_a = ri + \frac{d\psi}{dt}$$

where the flux linkage ψ is expressed as $\psi = L(x)i$.

One obtains

$$u_a = ri + L(x)\frac{di}{dt} + i\frac{dL(x)}{dx}\frac{dx}{dt}$$

and thus

$$\frac{di}{dt} = -\frac{r}{L(x)}i + \frac{2N^2 \mu_f^2 \mu_0 A_f^2 A_g}{L(x)[A_g l_f + 2 A_f \mu_f (x + 2d)]^2}iv + \frac{1}{L(x)}u_a$$

Augmenting this equation with differential equation

$$F(t) = m\frac{d^2x}{dt^2} + B_v\frac{dx}{dt} + (k_{s1}x + k_{s2}x^2) + F_e(t)$$

three nonlinear differential equations for the studied translation microdevise are found as

$$\frac{di}{dt} = -\frac{r[A_g l_f + 2A_f \mu_f (x+2d)]}{N^2 \mu_f \mu_0 A_f A_g} i + \frac{2\mu_f A_f}{A_g l_f + 2A_f \mu_f (x+2d)} iv + \frac{A_g l_f + 2A_f \mu_f (x+2d)}{N^2 \mu_f \mu_0 A_f A_g} u_a$$

$$\frac{dx}{dt} = v$$

$$\frac{dv}{dt} = \frac{N^2 \mu_f^2 \mu_0 A_f^2 A_g}{m[A_g l_f + 2A_f \mu_f (x+2d)]^2} i^2 - \frac{1}{m}(k_{s1} x + k_{s2} x^2) - \frac{B_v}{m} v$$

Example 15.5.2: Mathematical Model of an Elementary Synchronous Reluctance Micromotor

Consider a single-phase reluctance micromotor, which can be straightforwardly fabricated using conventional CMOS, LIGA, and LIGA-like technologies. Ferromagnetic materials are used to fabricate microscale stator and rotor, and windings can be deposited on the stator, see Figure 15.7.

The *quadrature* and *direct* magnetic axes are fixed with the microrotor, which rotates with angular velocity ω_r. These magnetic axes rotate with the angular velocity ω. Assume that the initial conditions are zero. Hence, the angular displacements of the rotor θ_r and the angular displacement of the *quadrature* magnetic axis θ are equal, and

$$\theta_r = \theta = \int_{t_0}^{t} \omega_r(\tau) d\tau = \int_{t_0}^{t} \omega(\tau) d\tau.$$

The magnetizing reluctance $\Re_m$ is a function of the rotor angular displacement θ_r. Using the number of turns N_s, the magnetizing inductance is

$$L_m(\theta_r) = \frac{N_s^2}{\Re_m(\theta_r)}.$$

This magnetizing inductance varies twice per one revolution of the rotor and has minimum and maximum values, and

$$L_{m\,min} = \frac{N_s^2}{\Re_{m\,max}(\theta_r)}\bigg|_{\theta_r = 0, \pi, 2\pi, \ldots}, \quad L_{m\,max} = \frac{N_s^2}{\Re_{m\,min}(\theta_r)}\bigg|_{\theta_r = \frac{1}{2}\pi, \frac{3}{2}\pi, \frac{5}{2}\pi, \ldots}$$

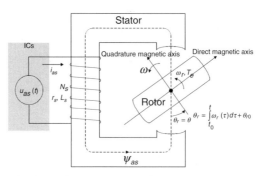

FIGURE 15.7 Microscale single-phase reluctance motor with rotational motion microstructure (microrotor).

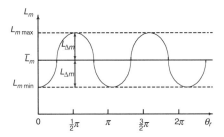

FIGURE 15.8 Magnetizing inductance $L_m(\theta_r)$.

Assume that this variation is a sinusoidal function of the rotor angular displacement. Then,

$$L_m(\theta_r) = \bar{L}_m - L_{\Delta m}\cos 2\theta_r$$

where $\bar{L}_m$ is the average value of the magnetizing inductance and $L_{\Delta m}$ is half of the amplitude of the sinusoidal variation of the magnetizing inductance.

The plot for $L_m(\theta_r)$ is documented in Figure 15.8.

The electromagnetic torque, developed by single-phase reluctance motors is found using the expression for the coenergy $W_c(i_{as}, \theta_r)$. From $W_c(i_{as}, \theta_r) = \frac{1}{2}(L_{ls} + \bar{L}_m - L_{\Delta m}\cos 2\theta_r)i_{as}^2$, one finds

$$T_e = \frac{\partial W_c(i_{as}, \theta_r)}{\partial \theta_r} = \frac{\partial [\frac{1}{2}i_{as}^2(L_{ls} + \bar{L}_m - L_{\Delta m}\cos 2\theta_r)]}{\partial \theta_r} = L_{\Delta m}i_{as}^2 \sin 2\theta_r$$

The electromagnetic torque is not developed by synchronous reluctance motors if IC feeds the dc current or voltage to the motor winding because $T_e = L_{\Delta m}i_{as}^2 \sin 2\theta_r$. Hence, conventional control algorithms cannot be applied, and new methods, which are based upon electromagnetic features must be researched. The average value of T_e is not equal to zero if the current is a function of θ_r. As an illustration, let us assume that the following current is fed to the motor winding:

$$i_{as} = i_M \operatorname{Re}(\sqrt{\sin 2\theta_r})$$

Then, the electromagnetic torque is

$$T_e = L_{\Delta m}i_{as}^2 \sin 2\theta_r = L_{\Delta m}i_M^2 (\operatorname{Re}\sqrt{\sin 2\theta_r})^2 \sin 2\theta_r \neq 0$$

and

$$T_{e\,av} = \frac{1}{\pi}\int_0^\pi L_{\Delta m}i_{as}^2 \sin 2\theta_r \, d\theta_r = \frac{1}{4}L_{\Delta m}i_M^2$$

The mathematical model of the microscale single-phase reluctance motor is found by using Kirchhoff's and Newton's second laws

$$u_{as} = r_s i_{as} + \frac{d\psi_{as}}{dt} \quad \text{(circuitry equation)}$$

$$T_e - B_m \omega_r - T_L = J\frac{d^2\theta_r}{dt^2} \quad \text{(torsional-mechanical equation)}$$

From $\psi_{as} = (L_{ls} + \bar{L}_m - L_{\Delta m} \cos 2\theta_r) i_{as}$, one obtains a set of three first-order nonlinear differential equations. In particular, we have

$$\frac{di_{as}}{dt} = \frac{r_s}{L_{ls} + \bar{L}_m - L_{\Delta m} \cos 2\theta_r} i_{as} - \frac{2L_{\Delta m}}{L_{ls} + \bar{L}_m - L_{\Delta m} \cos 2\theta_r} i_{as} \omega_r \sin 2\theta_r + \frac{1}{L_{ls} + \bar{L}_m - L_{\Delta m} \cos 2\theta_r} u_{as}$$

$$\frac{d\omega_r}{dt} = \frac{1}{J}(L_{\Delta m} i_{as}^2 \sin 2\theta_r - B_m \omega_r - T_L)$$

$$\frac{d\theta_r}{dt} = \omega_r$$

Example 15.5.3: Mathematical Model of Two-Phase Permanent-Magnet Stepper Micromotors

For two-phase permanent-magnet stepper micromotors, we have

$$u_{as} = r_s i_{as} + \frac{d\psi_{as}}{dt}$$

$$u_{bs} = r_s i_{bs} + \frac{d\psi_{bs}}{dt}$$

where the flux linkages are $\psi_{as} = L_{asas} i_{as} + L_{asbs} i_{bs} + \psi_{asm}$ and $\psi_{bs} = L_{bsas} i_{as} + L_{bsbs} i_{bs} + \psi_{bsm}$.

Here, u_{as} and u_{bs} are the phase voltages in the stator microwindings as and bs; i_{as} and i_{bs} are the phase currents in the stator microwindings; ψ_{as} and ψ_{bs} are the stator flux linkages; r_s are the resistances of the stator microwindings; L_{asas}, L_{asbs}, L_{bsas}, and L_{bsbs} are the mutual inductances.

The electrical angular velocity and displacement are found using the number of rotor tooth RT,

$$\omega_r = RT \omega_{rm}$$
$$\theta_r = RT \theta_{rm}$$

where ω_r and ω_{rm} are the electrical and rotor angular velocities, and θ_r and θ_{rm} are the electrical and rotor angular displacements.

The flux linkages are functions of the number of the rotor tooth RT, and the magnitude of the flux linkages produced by the permanent magnets ψ_m. In particular,

$$\psi_{asm} = \psi_m \cos(RT\theta_{rm}) \quad \text{and} \quad \psi_{bsm} = \psi_m \sin(RT\theta_{rm})$$

The self-inductance of the stator windings is

$$L_{ss} = L_{asas} = L_{bsbs} = L_{ls} + \bar{L}_m$$

The stator microwindings are displaced by 90 electrical degrees. Hence, the mutual inductances between the stator microwindings are zero, $L_{asbs} = L_{bsas} = 0$.
Then, we have

$$\psi_{as} = L_{ss} i_{as} + \psi_m \cos(RT\theta_{rm}) \quad \text{and} \quad \psi_{bs} = L_{ss} i_{bs} + \psi_m \sin(RT\theta_{rm})$$

Taking note of the circuitry equations, one has

$$u_{as} = r_s i_{as} + \frac{d[L_{ss} i_{as} + \psi_m \cos(RT\theta_{rm})]}{dt} = r_s i_{as} + L_{ss}\frac{di_{as}}{dt} - RT\psi_m \omega_{rm}\sin(RT\theta_{rm})$$

$$u_{bs} = r_s i_{bs} + \frac{d[L_{ss} i_{bs} + \psi_m \sin(RT\theta_{rm})]}{dt} = r_s i_{bs} + L_{ss}\frac{di_{bs}}{dt} + RT\psi_m \omega_{rm}\cos(RT\theta_{rm})$$

Therefore, we obtain

$$\frac{di_{as}}{dt} = -\frac{r_s}{L_{ss}}i_{as} + \frac{RT\psi_m}{L_{ss}}\omega_{rm}\sin(RT\theta_{rm}) + \frac{1}{L_{ss}}u_{as}$$

$$\frac{di_{bs}}{dt} = -\frac{r_s}{L_{ss}}i_{bs} - \frac{RT\psi_m}{L_{ss}}\omega_{rm}\cos(RT\theta_{rm}) + \frac{1}{L_{ss}}u_{bs}$$

Using Newton's second law, we have

$$\frac{d\omega_{rm}}{dt} = \frac{1}{J}(T_e - B_m\omega_{rm} - T_L)$$

$$\frac{d\theta_{rm}}{dt} = \omega_{rm}$$

The expression for the electromagnetic torque developed by permanent-magnet stepper micromotors must be found. Taking note of the relationship for the coenergy

$$W_c = \frac{1}{2}(L_{ss}i_{as}^2 + L_{ss}i_{bs}^2) + \psi_m i_{as}\cos(RT\theta_{rm}) + \psi_m i_{bs}\sin(RT\theta_{rm}) + W_{PM}$$

one finds the electromagnetic torque:

$$T_e = \frac{\partial W_c}{\partial \theta_{rm}} = -RT\psi_m[i_{as}\sin(RT\theta_{rm}) - i_{bs}\cos(RT\theta_{rm})]$$

Hence, the transient evolution of the phase currents i_{as} and i_{bs}, rotor angular velocity ω_{rm}, and displacement θ_{rm}, is modeled by the following differential equations:

$$\frac{di_{as}}{dt} = -\frac{r_s}{L_{ss}}i_{as} + \frac{RT\psi_m}{L_{ss}}\omega_{rm}\sin(RT\theta_{rm}) + \frac{1}{L_{ss}}u_{as}$$

$$\frac{di_{bs}}{dt} = -\frac{r_s}{L_{ss}}i_{bs} - \frac{RT\psi_m}{L_{ss}}\omega_{rm}\cos(RT\theta_{rm}) + \frac{1}{L_{ss}}u_{bs}$$

$$\frac{d\omega_{rm}}{dt} = -\frac{RT\psi_m}{J}[i_{as}\sin(RT\theta_{rm}) - i_{bs}\cos(RT\theta_{rm})] - \frac{B_m}{J}\omega_{rm} - \frac{1}{J}T_L$$

$$\frac{d\theta_{rm}}{dt} = \omega_{rm}$$

These four nonlinear differential equations are rewritten in the state-space form as

$$\begin{bmatrix} \dfrac{di_{as}}{dt} \\ \dfrac{di_{bs}}{dt} \\ \dfrac{d\omega_{rm}}{dt} \\ \dfrac{d\theta_{rm}}{dt} \end{bmatrix} = \begin{bmatrix} -\dfrac{r_s}{L_{ss}} & 0 & 0 & 0 \\ 0 & -\dfrac{r_s}{L_{ss}} & 0 & 0 \\ 0 & 0 & -\dfrac{B_m}{J} & 0 \\ 0 & 0 & 1 & 0 \end{bmatrix} \begin{bmatrix} i_{as} \\ i_{bs} \\ \omega_{rm} \\ \theta_{rm} \end{bmatrix} + \begin{bmatrix} \dfrac{RT\psi_m}{L_{ss}}\omega_{rm}\sin(RT\theta_{rm}) \\ -\dfrac{RT\psi_m}{L_{ss}}\omega_{rm}\cos(RT\theta_{rm}) \\ -\dfrac{RT\psi_m}{J}[i_{as}\sin(RT\theta_{rm}) - i_{bs}\cos(RT\theta_{rm})] \\ 0 \end{bmatrix}$$

$$+ \begin{bmatrix} \dfrac{1}{L_{ss}} & 0 \\ 0 & \dfrac{1}{L_{ss}} \\ 0 & 0 \\ 0 & 0 \end{bmatrix} \begin{bmatrix} u_{as} \\ u_{bs} \end{bmatrix} - \begin{bmatrix} 0 \\ 0 \\ \dfrac{1}{J} \\ 0 \end{bmatrix} T_L$$

The analysis of the torque equation

$$T_e = -RT\psi_m[i_{as}\sin(RT\theta_{rm}) - i_{bs}\cos(RT\theta_{rm})]$$

guides one to the conclusion that the expressions for a balanced two-phase current sinusoidal set is

$$i_{as} = -\sqrt{2}i_M\sin(RT\theta_{rm}) \quad \text{and} \quad i_{bs} = \sqrt{2}i_M\cos(RT\theta_{rm})$$

If these phase currents are fed, the electromagnetic torque is a function of the current magnitude i_M, and

$$T_e = \sqrt{2}RT\psi_m i_M$$

The phase currents needed to be fed are the functions of the rotor angular displacement. Assuming that the inductances are negligibly small, we have the following phase voltages needed to be supplied:

$$u_{as} = -\sqrt{2}u_M\sin(RT\theta_{rm}) \quad \text{and} \quad u_{bs} = \sqrt{2}u_M\cos(RT\theta_{rm})$$

Example 15.5.4: Mathematical Model of Two-Phase Permanent-Magnet Synchronous Micromotors

Consider two-phase permanent-magnet synchronous micromotors. Using Kirchhoff's voltage law, we have

$$u_{as} = r_s i_{as} + \frac{d\psi_{as}}{dt}$$

$$u_{bs} = r_s i_{bs} + \frac{d\psi_{bs}}{dt}$$

where the flux linkages are expressed as $\psi_{as} = L_{asas}i_{as} + L_{asbs}i_{bs} + \psi_{asm}$ and $\psi_{bs} = L_{bsas}i_{as} + L_{bsbs}i_{bs} + \psi_{bsm}$.
The flux linkages are periodic functions of the angular displacement (rotor position), and let

$$\psi_{asm} = \psi_m\sin\theta_{rm} \quad \text{and} \quad \psi_{bsm} = -\psi_m\cos\theta_{rm}$$

The self-inductances of the stator windings are found to be

$$L_{ss} = L_{asas} = L_{bsbs} = L_{ls} + \bar{L}_m$$

The stator windings are displaced by 90 electrical degrees, and hence, the mutual inductances between the stator windings are $L_{asbs} = L_{bsas} = 0$. Thus, we have

$$\psi_{as} = L_{ss}i_{as} + \psi_m \sin\theta_{rm} \quad \text{and} \quad \psi_{bs} = L_{ss}i_{bs} - \psi_m \cos\theta_{rm}$$

Therefore, one finds

$$u_{as} = r_s i_{as} + \frac{d(L_{ss}i_{as} + \psi_m \sin\theta_{rm})}{dt} = r_s i_{as} + L_{ss}\frac{di_{as}}{dt} + \psi_m \omega_{rm} \cos\theta_{rm}$$

$$u_{bs} = r_s i_{bs} + \frac{d(L_{ss}i_{bs} - \psi_m \cos\theta_{rm})}{dt} = r_s i_{bs} + L_{ss}\frac{di_{bs}}{dt} - \psi_m \omega_{rm} \sin\theta_{rm}$$

Using Newton's second law

$$T_e - B_m \omega_{rm} - T_L = J\frac{d^2\theta_{rm}}{dt^2}$$

we have

$$\frac{d\omega_{rm}}{dt} = \frac{1}{J}(T_e - B_m \omega_{rm} - T_L)$$

$$\frac{d\theta_{rm}}{dt} = \omega_{rm}$$

The expression for the electromagnetic torque developed by permanent-magnet motors can be obtained by using the coenergy

$$W_c = \frac{1}{2}(L_{ss}i_{as}^2 + L_{ss}i_{bs}^2) + \psi_m i_{as} \sin\theta_{rm} - \psi_m i_{bs} \cos\theta_{rm} + W_{PM}$$

Then, one has

$$T_e = \frac{\partial W_c}{\partial \theta_{rm}} = \frac{P\psi_m}{2}(i_{as}\cos\theta_{rm} + i_{bs}\sin\theta_{rm})$$

Augmenting the circuitry transients with the *torsional-mechanical* dynamics, one finds the mathematical model of two-phase permanent-magnet micromotors in the following form:

$$\frac{di_{as}}{dt} = -\frac{r_s}{L_{ss}}i_{as} - \frac{\psi_m}{L_{ss}}\omega_{rm}\cos\theta_{rm} + \frac{1}{L_{ss}}u_{as}$$

$$\frac{di_{bs}}{dt} = -\frac{r_s}{L_{ss}}i_{bs} + \frac{\psi_m}{L_{ss}}\omega_{rm}\sin\theta_{rm} + \frac{1}{L_{ss}}u_{bs}$$

$$\frac{d\omega_{rm}}{dt} = \frac{P\psi_m}{2J}(i_{as}\cos\theta_{rm} + i_{bs}\sin\theta_{rm}) - \frac{B_m}{J}\omega_{rm} - \frac{1}{J}T_L$$

$$\frac{d\theta_{rm}}{dt} = \omega_{rm}$$

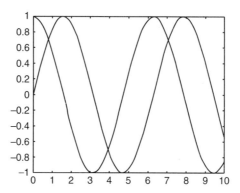

FIGURE 15.9 Air-gap mmf and the phase current waveforms.

For two-phase motors (assuming the sinusoidal winding distributions and the sinusoidal mmf waveforms), the electromagnetic torque is expressed as

$$T_e = \frac{P\psi_m}{2}(i_{as}\cos\theta_{rm} + i_{bs}\sin\theta_{rm})$$

Hence, to guarantee the balanced operation, one feeds

$$i_{as} = \sqrt{2}i_M\cos\theta_{rm} \quad \text{and} \quad i_{bs} = \sqrt{2}i_M\sin\theta_{rm}$$

to maximize the electromagnetic torque. In fact, one obtains

$$T_e = \frac{P\psi_m}{2}(i_{as}\cos\theta_{rm} + i_{bs}\sin\theta_{rm}) = \frac{P\psi_m}{2}\sqrt{2}i_M(\cos^2\theta_{rm} + \sin^2\theta_{rm}) = \frac{P\psi_m}{\sqrt{2}}i_M$$

The air-gap mmf and the phase current waveforms are plotted in Figure 15.9.

15.6 Control of MEMS

Mathematical models of MEMS can be developed with different degrees of complexity. It must be emphasized that in addition to the models of microscale motion devices, the fast dynamics of ICs should be examined. Due to the complexity of complete mathematical models of ICs, impracticality of the developed equations, and very fast dynamics, the IC dynamics can be modeled using reduced-order differential equation or as unmodeled dynamics. For MEMS, modeled using linear and nonlinear differential equations

$$\dot{x}(t) = Ax + Bu, \quad u_{min} \le u \le u_{max}, \quad y = Hx$$
$$\dot{x}(t) = F_z(t, x, r, z) + B_p(t, x, p)u, \quad u_{min} \le u \le u_{max}, \quad y = H(x)$$

different control algorithms can be designed.

Here, the state, control, output, and reference (command) vectors are denoted as x, u, y, and r; parameter uncertainties (e.g., time-varying coefficients, unmodeled dynamics, unpredicted changes, etc.) are modeled using z and p vectors.

The matrices of coefficients are A, B, and H. The smooth mapping fields of the nonlinear model are denoted as $F_z(\cdot)$, $B_p(\cdot)$, and $H(\cdot)$.

It should be emphasized that the control is bounded. For example, using the IC duty ratio d_D as the control signal, we have $0 \leq d_D \leq 1$ or $-1 \leq d_D \leq +1$. Four-quadrant ICs are used due to superior performance, and $-1 \leq d_D \leq +1$. Hence, we have $-1 \leq u \leq +1$. However, in general, $u_{min} \leq u \leq u_{max}$.

15.6.1 Proportional-Integral-Derivative Control

Many MEMS can be controlled by the proportional-integral-derivative (PID) controllers, which, taking note of control bounds, are given as [9]

$$u(t) = \text{sat}_{u_{min}}^{u_{max}}\left(e, \int e\, dt, \frac{de}{dt}\right)$$

$$= \text{sat}_{u_{min}}^{u_{max}}\left(\underbrace{\sum_{j=0}^{\varsigma} k_{pj} e^{\frac{2j+1}{2\beta+1}}}_{\text{proportional}} + \underbrace{\sum_{j=0}^{\sigma} k_{ij}\int e^{\frac{2j+1}{2\mu+1}} dt}_{\text{integral}} + \underbrace{\sum_{j=0}^{\alpha} k_{dj} \dot{e}^{\frac{2j+1}{2\gamma+1}}}_{\text{derivative}}\right), \quad u_{min} \leq u \leq u_{max}$$

where k_{pj}, k_{ij}, and k_{dj} are the matrices of the proportional, integral, and derivative feedback gains; ς, β, σ, μ, α, and γ are the nonnegative integers.

In the nonlinear PID controllers, the tracking error is used. In particular,

$$e(t) = \underset{\text{reference/command}}{r(t)} - \underset{\text{output}}{y(t)}$$

Linear bounded controllers can be straightforwardly designed. For example, letting $\varsigma = \beta = \sigma = \mu = 0$, we have the following linear PI control law:

$$u(t) = \text{sat}_{u_{min}}^{u_{max}}\left(k_{p0} e(t) + k_{i0}\int et\, dt\right)$$

The PID controllers with the state feedback extension can be synthesized as

$$u(t) = \text{sat}_{u_{min}}^{u_{max}}(e, x)$$

$$= \text{sat}_{u_{min}}^{u_{max}}\left(\underbrace{\sum_{j=0}^{\varsigma} k_{pj} e^{\frac{2j+1}{2\beta+1}}}_{\text{proportional}} + \underbrace{\sum_{j=0}^{\sigma} k_{ij}\int e^{\frac{2j+1}{2\mu+1}} dt}_{\text{integral}} + \underbrace{\sum_{j=0}^{\alpha} k_{dj} \dot{e}^{\frac{2j+1}{2\gamma+1}}}_{\text{derivative}} + G(t) B \frac{\partial V(e, x)}{\partial \begin{bmatrix} e \\ x \end{bmatrix}}\right), \quad u_{min} \leq u \leq u_{max}$$

where $V(e, x)$ is the function that satisfies the general requirements imposed on the Lyapunov pair [9], e.g., the sufficient conditions for stability are used.

It is evident that nonlinear feedback mappings result, and the nonquadratic function $V(e, x)$ can be synthesized and used to obtain the control algorithm and feedback gains.

15.6.2 Tracking Control

Tracking control is designed for the augmented systems, which are modeled using the state variables and the reference dynamics. In particular, from

$$\dot{x}(t) = Ax + Bu, \quad \dot{x}^{ref}(t) = r(t) - y(t) = r(t) - Hx(t)$$

one finds

$$\dot{x}_\Sigma(t) = A_\Sigma x_\Sigma + B_\Sigma u + N_\Sigma r, \quad y = Hx, \quad x_\Sigma = \begin{bmatrix} x \\ x_{ref} \end{bmatrix}, \quad A_\Sigma = \begin{bmatrix} A & 0 \\ -H & 0 \end{bmatrix}, \quad B_\Sigma = \begin{bmatrix} B \\ 0 \end{bmatrix}, \quad N_\Sigma = \begin{bmatrix} 0 \\ I \end{bmatrix}$$

Minimizing the quadratic performance functional

$$J = \frac{1}{2}\int_{t_0}^{t_f} (x_\Sigma^T Q x_\Sigma + u^T G u) \, dt$$

one finds the control law using the first-order necessary condition for optimality. In particular, we have

$$u = -G^{-1} B_\Sigma^T \frac{\partial V}{\partial x_\Sigma} = -G^{-1} \begin{bmatrix} B \\ 0 \end{bmatrix}^T \frac{\partial V}{\partial x_\Sigma}$$

Here, Q is the positive semi-definite constant-coefficient matrix, and G is the positive weighting constant-coefficient matrix.

The solution of the Hamilton–Jacobi equation

$$-\frac{\partial V}{\partial t} = \frac{1}{2} x_\Sigma^T Q x_\Sigma + \left(\frac{\partial V}{\partial x_\Sigma}\right)^T A x_\Sigma - \frac{1}{2}\left(\frac{\partial V}{\partial x_\Sigma}\right)^T B_\Sigma G^{-1} B_\Sigma^T \frac{\partial V}{\partial x_\Sigma}$$

is satisfied by the quadratic return function $V = \frac{1}{2} x_\Sigma^T K x_\Sigma$. Here, K is the symmetric matrix, which must be found by solving the nonlinear differential equation

$$-\dot{K} = Q + A_\Sigma^T K + K^T A_\Sigma - K^T B_\Sigma G^{-1} B_\Sigma^T K, \quad K(t_f) = K_f$$

The controller is given as

$$u = -G^{-1} B_\Sigma^T K x_\Sigma = -G^{-1} \begin{bmatrix} B \\ 0 \end{bmatrix}^T K x_\Sigma$$

From $\dot{x}_{ref}(t) = e(t)$, one has

$$x_{ref}(t) = \int e(t) \, dt$$

Therefore, we obtain the integral control law

$$u(t) = -G^{-1} \begin{bmatrix} B \\ 0 \end{bmatrix}^T K \begin{bmatrix} x(t) \\ \int e(t) \, dt \end{bmatrix}$$

In this control algorithm, the error vector is used in addition to the state feedback.

As was illustrated, the bounds are imposed on the control, and $u_{min} \leq u \leq u_{max}$. Therefore, the bounded controllers must be designed. Using the nonquadratic performance functional [9]

$$J = \int_{t_0}^{t_f} \left(x_\Sigma^T Q x_\Sigma + G \int \tan^{-1} u \, du\right) dt$$

with positive semi-definite constant-coefficient matrix Q and positive-definite matrix G, one finds

$$u(t) = -\tanh\left(G^{-1}\begin{bmatrix}B\\0\end{bmatrix}^T K \begin{bmatrix}x(t)\\ \int e(t)\,dt\end{bmatrix}\right) \approx -\mathrm{sat}_{-1}^{+1}\left(G^{-1}\begin{bmatrix}B\\0\end{bmatrix}^T K \begin{bmatrix}x(t)\\ \int e(t)\,dt\end{bmatrix}\right), \quad -1 \le u \le 1$$

This controller is obtained assuming that the solution of the functional partial differential equation can be approximated by the quadratic return function

$$V = \frac{1}{2} x_\Sigma^T K x_\Sigma$$

where K is the symmetric matrix.

15.6.3 Time-Optimal Control

A time-optimal controller can be designed using the functional

$$J = \frac{1}{2}\int_{t_0}^{t_f} (x_\Sigma^T Q x_\Sigma)\,dt$$

Taking note of the Hamilton–Jacobi equation

$$-\frac{\partial V}{\partial t} = \min_{-1 \le u \le 1}\left[\frac{1}{2} x_\Sigma^T Q x_\Sigma + \left(\frac{\partial V}{\partial x_\Sigma}\right)^T (A x_\Sigma + B_\Sigma u)\right]$$

the relay-type controller is found to be

$$u = -\mathrm{sgn}\left(B_\Sigma^T \frac{\partial V}{\partial x_\Sigma}\right), \quad -1 \le u \le 1$$

This "optimal" control algorithm cannot be implemented in practice due to the chattering phenomenon. Therefore, relay-type control laws with dead zone

$$u = -\mathrm{sgn}\left(B_\Sigma^T \frac{\partial V}{\partial x_\Sigma}\right)\Bigg|_{\text{dead zone}}, \quad -1 \le u \le 1$$

are commonly used.

15.6.4 Sliding Mode Control

Soft-switching sliding mode control laws are synthesized in [9]. Sliding mode soft-switching algorithms provide superior performance, and the chattering effect is eliminated.

To design controllers, we model the states and errors dynamics as

$$\dot{x}(t) = Ax + Bu, \quad -1 \le u \le 1$$
$$\dot{e}(t) = Nr(t) - HAx - HBu$$

The smooth sliding manifold is

$$M = \{(t, x, e) \in R_{\ge 0} \times X \times E \mid v(t, x, e) = 0\}$$
$$= \bigcap_{j=1}^{m}\{(t, x, e) \in R_{\ge 0} \times X \times E \mid v_j(t, x, e) = 0\}$$

The time-varying nonlinear switching surface is $v(t, x, e) = K_{vxe}(t, x, e) = 0$. The soft-switching control law is given as

$$u(t, x, e) = -G\phi(v), \quad -1 \leq u \leq 1, \quad G > 0$$

where $\phi(\cdot)$ is the continuous real-analytic function of class $C^\partial (\partial \geq 1)$, for example, `tanh` and `erf`.

15.6.5 Constrained Control of Nonlinear MEMS: Hamilton–Jacobi Method

Constrained optimization of MEMS is a topic of great practical interest. Using the Hamilton–Jacobi theory, the bounded controllers can be synthesized for continuous-time systems modeled as

$$\dot{x}^{MEMS}(t) = F_s(x^{MEMS}) + B_s(x^{MEMS})u^{2w+1}, \quad y = Hx^{MEMS}$$

$$u_{min} \leq u \leq u_{max}, \quad x^{MEMS}(t_0) = x_0^{MEMS}$$

Here, $x^{MEMS} \in X_s$ is the state vector; $u \in U$ is the vector of control inputs: $y \in Y$ is the measured output; $F_s(\cdot)$, $B_s(\cdot)$ and $H(\cdot)$ are the smooth mappings; $F_s(0) = 0$, $B_s(0) = 0$, and $H(0) = 0$; and w is the nonnegative integer.

To design the tracking controller, we augment the MEMS dynamics

$$\dot{x}^{MEMS}(t) = F_s(x^{MEMS}) + B_s(x^{MEMS})u^{2w+1} \quad y = H(x^{MEMS})$$

$$u_{min} \leq u \leq u_{max}, \quad x^{MEMS}(t_0) = x_0^{MEMS}$$

with the *exogenous* dynamics $\dot{x}^{ref}(t) = Nr - y = Nr - H(x^{MEMS})$.

Using the augmented state vector

$$x = \begin{bmatrix} x^{MEMS} \\ x^{ref} \end{bmatrix} \in X$$

one obtains

$$\dot{x}(t) = F(x, r) + B(x)u^{2w+1}, \quad u_{min} \leq u \leq u_{max}, \quad x(t_0) = x_0, \quad x = \begin{bmatrix} x^{MEMS} \\ x^{ref} \end{bmatrix}$$

$$F(x, r) = \begin{bmatrix} F_s(x^{MEMS}) \\ -H(x^{MEMS}) \end{bmatrix} + \begin{bmatrix} 0 \\ N \end{bmatrix} r, \quad B(x) = \begin{bmatrix} B_s(x^{MEMS}) \\ 0 \end{bmatrix}$$

The set of admissible control U consists of the Lebesgue measurable function $u(\cdot)$, and a bounded controller should be designed within the constrained control set

$$U = \{u \in \mathbb{R}^m \mid u_{imin} \leq u_i \leq u_{imax}, \quad i = 1, \ldots, m\}.$$

We map the control bounds imposed by a bounded, integrable, one-to-one, globally Lipschitz, vector-valued continuous function $\Phi \in C^\partial (\partial \geq 1)$. Our goal is to analytically design the bounded admissible state-feedback controller in the closed form as $u = \Phi(x)$. The most common Φ are the algebraic and transcendental (exponential, hyperbolic, logarithmic, trigonometric) continuously differentiable, integrable, one-to-one functions. For example, the odd one-to-one integrable function tanh with domain $(-\infty, +\infty)$ maps the control bounds. This function has the corresponding inverse function $\tanh^{-1}$ with range $(-\infty, +\infty)$.

The performance cost to be minimized is given as

$$J = \int_{t_0}^{\infty} [W_x(x) + W_u(u)]dt = \int_{t_0}^{\infty} \left[W_x(x) + (2w+1)\int (\Phi^{-1}(u))^T G^{-1} \mathrm{diag}(u^{2w}) du \right] dt$$

where $G^{-1} \in \mathbb{R}^{m \times m}$ is the positive-definite diagonal matrix.

Performance integrands $W_x(\cdot)$ and $W_u(\cdot)$ are real-valued, positive-definite, and continuously differentiable integrand functions. Using the properties of Φ one concludes that inverse function Φ^{-1} is integrable. Hence, integral

$$\int (\Phi^{-1}(u))^T G^{-1} \mathrm{diag}(u^{2w}) du$$

exists.

Example

Consider a nonlinear dynamic system

$$\frac{dx}{dt} = ax + bu^3, \quad u_{\min} \leq u \leq u_{\max}$$

Taking note of

$$W_u(u) = (2w+1)\int (\Phi^{-1}(u))^T G^{-1} \mathrm{diag}(u^{2w}) du$$

one has the positive-definite integrand

$$W_u(u) = 3\int \tanh^{-1} u \, G^{-1} u^2 du = \frac{1}{3} u^3 \tanh^{-1} u + \frac{1}{6} u^2 + \frac{1}{6} \ln(1-u^2), \quad G^{-1} = \frac{1}{3}$$

In general, if the hyperbolic tangent is used to map the saturation effect, for the single-input case, one has

$$W_u(u) = (2w+1)\int u^{2w} \tanh^{-1}\frac{u}{k} du = u^{2w+1} \tanh^{-1}\frac{u}{k} - k\int \frac{u^{2w+1}}{k^2 - u^2} du$$

Necessary conditions that the control function $u(\cdot)$ guarantees a minimum to the Hamiltonian

$$H = W_x(x) + (2w+1)\int (\Phi^{-1}(u))^T G^{-1} \mathrm{diag}(u^{2w}) du + \frac{\partial V(x)}{\partial x}^T [F(x,r) + B(x) u^{2w+1}]$$

are: first-order necessary condition $n1$,

$$\frac{\partial H}{\partial u} = 0$$

and second-order necessary condition $n2$,

$$\frac{\partial^2 H}{\partial u \times \partial u^T} > 0$$

The positive-definite return function $V(\cdot)$, $V \in C^\kappa$, $\kappa \geq 1$, is

$$V(x_0) = \inf_{u \in U} J(x_0, u) = \inf J(x_0, \Phi(\cdot)) \geq 0$$

The Hamilton–Jacobi–Bellman equation is given as

$$-\frac{\partial V}{\partial t} = \min_{u \in U} \left\{ W_x(x) + (2w+1)\int (\Phi^{-1}(u))^T G^{-1} \mathrm{diag}(u^{2w}) du + \frac{\partial V(x)}{\partial x}^T [F(x,r) + B(x) u^{2w+1}] \right\}$$

The controller should be derived by finding the control value that attains the minimum to nonquadratic functional. The first-order necessary condition (n1) leads us to an admissible bounded control law. In particular,

$$u = -\Phi\left(GB(x)^T \frac{\partial V(x)}{\partial x}\right), \quad u \in U$$

The second-order necessary condition for optimality (n2) is met because the matrix G^{-1} is positive-definite. Hence, a unique, bounded, real-analytic, and continuous control candidate is designed.

If there exists a proper function $V(x)$ which satisfies the Hamilton–Jacobi equation, the resulting closed-loop system is robustly stable in the specified state X and control U sets, and robust tracking is ensured in the convex and compact set $XY(X_0, U, R, E_0)$. That is, there exists an invariant domain of stability

$$S = \{x \in R^c, e \in R^b : \|x(t)\| \leq \varrho_x(\|x_0\|, t) + \varrho_u(\|u\|), \|e(t)\| \leq \varrho_e(\|e_0\|, t) + \varrho_r(\|r\|) + \varrho_y(\|y\|),$$
$$\forall x \in X(X_0, U), \forall t \in [t_0, \infty), \forall e \in E(E_0, R, Y)\} \subset R^c \times R^b,$$

and control $u(\cdot)$, $u \in U$ steers the tracking error to the set

$$S_E(\delta) = \{e \in R^b : e_0 \in E_0, x \in X(X_0, U), r \in R, y \in Y, t \in [t_0, \infty)$$
$$\|e(t)\| \leq \varrho_e(\|e_0\|, t) + \delta, \delta \geq 0, \forall e \in E(E_0, R, Y), \forall t \in [t_0, \infty)\} \subset R^b$$

Here ϱ_x and ϱ_e are the KL-functions; and ϱ_u, ϱ_r, and ϱ_y are the K-functions.

The solution of the functional equation should be found using nonquadratic return functions. To obtain $V(\cdot)$, the performance cost must be evaluated at the allowed values of the states and control. Linear and nonlinear functionals admit the final values, and the minimum value of the nonquadratic cost is given by power-series forms [9]. That is,

$$J_{\min} = \sum_{i=0}^{\eta} v(x_0)^{\frac{2(i+\gamma+1)}{2\gamma+1}}, \quad \eta = 0, 1, 2, \ldots, \quad \gamma = 0, 1, 2, \ldots$$

The solution of the partial differential equation is satisfied by a continuously differentiable positive-definite return function

$$V(x) = \sum_{i=0}^{\eta} \frac{2\gamma+1}{2(i+\gamma+1)} \left(x^{\frac{i+\gamma+1}{2\gamma+1}}\right)^T K_i x^{\frac{i+\gamma+1}{2\gamma+1}}$$

where matrices K_i are found by solving the Hamilton–Jacobi equation.

The quadratic return function in $V(x) = \frac{1}{2} x^T K_0 x$ is found by letting $\eta = \gamma = 0$. This quadratic candidate may be employed only if the designer enables to neglect the high-order terms in Taylor's series expansion. Using $\eta = 1$ and $\gamma = 0$, one obtains

$$V(x) = \frac{1}{2} x^T K_0 x + \frac{1}{4} (x^2)^T K_1 x^2$$

while for $\eta = 4$ and $\gamma = 1$, we have the following function:

$$V(x) = \frac{3}{4}(x^{2/3})^T K_0 x^{2/3} + \frac{1}{2} x^T K_1 x + \frac{3}{8}(x^{4/3})^T K_2 x^{4/3} + \frac{3}{10}(x^{5/3})^T K_3 x^{5/3} + \frac{1}{4}(x^2)^T K_4 x^2$$

The nonlinear bounded controller is given as

$$u = -\Phi\left(GB(x)^T \sum_{i=0}^{\eta} \text{diag}\left[x(t)^{\frac{i-\gamma}{2\gamma+1}}\right] K_i(t) x(t)^{\frac{i+\gamma+1}{2\gamma+1}}\right),$$

$$\text{diag}\left[x(t)^{\frac{i-\gamma}{2\gamma+1}}\right] = \begin{bmatrix} x_1^{\frac{i-\gamma}{2\gamma+1}} & 0 & \cdots & 0 & 0 \\ 0 & x_2^{\frac{i-\gamma}{2\gamma+1}} & \cdots & 0 & 0 \\ \vdots & \vdots & \ddots & \vdots & \vdots \\ 0 & 0 & \cdots & x_{c-1}^{\frac{i-\gamma}{2\gamma+1}} & 0 \\ 0 & 0 & \vdots & 0 & x_c^{\frac{i-\gamma}{2\gamma+1}} \end{bmatrix}$$

If matrices K_i are diagonal, we have the following control algorithm:

$$u = -\Phi\left(GB(x)^T \sum_{i=0}^{\eta} K_i x^{\frac{2i+1}{2\gamma+1}}\right)$$

15.6.6 Constrained Control of Nonlinear Uncertain MEMS: Lyapunov Method

Over the horizon $[t_0, \infty)$ we consider the dynamics of MEMS modeled as

$$x(t) = F_z(t, x, r, z) + B_p(t, x, p)u, \quad y = H(x), \quad u_{\min} \leq u \leq u_{\max}, \quad x(t_0) = x_0$$

where $t \in \mathbb{R}_{\geq 0}$ is the time; $x \in X$ is the state-space vector; $u \in U$ is the vector of bounded control inputs; $r \in R$ and $y \in Y$ are the measured reference and output vectors; $z \in Z$ and $p \in P$ are the parameter uncertainties, functions $z(\cdot)$ and $p(\cdot)$ are Lebesgue measurable and known within bounds; Z and P are the known nonempty compact sets; and $F_z(\cdot)$, $B_p(\cdot)$, and $H(\cdot)$ are the smooth mapping fields.

Let us formulate and solve the motion control problem by synthesizing robust controllers that guarantee stability and robust tracking. Our goal is to design control laws that robustly stabilize nonlinear systems with uncertain parameters and drive the tracking error $e(t) = r(t) - y(t)$, $e \in E$ robustly to the compact set. For MEMS modeled by nonlinear differential equations with parameter variations, the robust tracking of the measured output vector $y \in Y$ must be accomplished with respect to the measured uniformly bounded reference input vector $r \in R$.

The *nominal* and uncertain dynamics are mapped by $F(\cdot)$, $B(\cdot)$, and $\Xi(\cdot)$. Hence, the system evolution is described as

$$x(t) = F(t, x, r) + B(t, x)u + \Xi(t, x, u, z, p), \quad y = H(x), \quad u_{\min} \leq u \leq u_{\max}, \quad x(t_0) = x_0$$

There exists a norm of $\Xi(t, x, u, z, p)$, and $\|\Xi(t, x, u, z, p)\| \leq \rho(t, x)$, where $\rho(\cdot)$ is the continuous Lebesgue measurable function. Our goal is to solve the motion control problem, and tracking controllers must be synthesized using the tracking error vector and the state variables. Furthermore, to guarantee robustness and to expand stability margins, to improve dynamic performance, and to meet other requirements, nonqudratic Lyapunov functions $V(t, e, x)$ will be used in stability analysis and design of robust tracking control laws.

Suppose that a set of admissible control U consists of the Lebesgue measurable function $u(\cdot)$. It was demonstrated that the Hamilton–Jacobi theory can be used to find control laws, and the minimization of nonquadratic performance functionals leads one to the bounded controllers.

Letting $u = \Phi(t, e, x)$, one obtains a set of admissible controllers. Applying the error and state feedback we define a family of tracking controllers as

$$u = \Omega(x)\Phi(t, e, x) = -\Omega(x)\Phi\left(G_E(t)B_E(t,x)^T \frac{1}{s}\frac{\partial V(t,e,x)}{\partial e} + G_X(t)B(t,x)^T \frac{\partial V(t,e,x)}{\partial x}\right), \quad s = \frac{d}{dt}$$

where $\Omega(\cdot)$ is the nonlinear function; $G_E(\cdot)$ and $G_X(\cdot)$ are the diagonal matrix-functions defined on $[t_0, \infty)$; $B_E(\cdot)$ is the matrix-function; and $V(\cdot)$ is the continuous, differentiable, and real-analytic function.

Let us design the Lyapunov function. This problem is a critical one and involves well-known difficulties. The quadratic Lyapunov candidates can be used. However, for uncertain nonlinear systems, nonquadratic functions $V(t, e, x)$ allow one to realize the full potential of the Lyapunov-based theory and lead us to the nonlinear feedback maps which are needed to achieve conflicting design objectives. We introduce the following family of Lyapunov candidates:

$$V(t,e,x) = \sum_{i=0}^{\zeta} \frac{2\beta+1}{2(i+\beta+1)}\left(e^{\frac{i+\beta+1}{2\beta+1}}\right)^T K_{Ei}(t) e^{\frac{i+\beta+1}{2\beta+1}} + \sum_{i=0}^{\eta} \frac{2\gamma+1}{2(i+\gamma+1)}\left(x^{\frac{i+\gamma+1}{2\gamma+1}}\right)^T K_{Xi}(t) x^{\frac{i+\gamma+1}{2\gamma+1}}$$

where $K_{Ei}(\cdot)$ and $K_{Xi}(\cdot)$ are the symmetric matrices; ζ, β, η, and γ are the nonnegative integers; $\zeta = 0, 1, 2, \ldots$; $\beta = 0, 1, 2, \ldots$; $\eta = 0, 1, 2, \ldots$; and $\gamma = 0, 1, 2, \ldots$

The well-known quadratic form of $V(t, e, x)$ is found by letting $\zeta = \beta = \eta = \gamma = 0$, and we have

$$V(t,e,x) = \frac{1}{2} e^T K_{E0}(t) e + \frac{1}{2} x^T K_{X0}(t) x$$

By using $\zeta = 1$, $\beta = 0$, $\eta = 1$, and $\gamma = 0$, one obtains a nonquadratic candidate:

$$V(t,e,x) = \frac{1}{2} e^T K_{E0}(t) e + \frac{1}{4} e^{2^T} K_{E1}(t) e^2 + \frac{1}{2} x^T K_{X0}(t) x + \frac{1}{4} x^{2^T} K_{X1}(t) x^2$$

One obtains the following tracking control law:

$$u = -\Omega(x)\Phi\left(G_E(t)B_E(t,x)^T \sum_{i=0}^{\zeta} \mathrm{diag}\left[e(t)^{\frac{i-\beta}{2\beta+1}}\right] K_{Ei}(t) \frac{1}{s} e(t)^{\frac{i+\beta+1}{2\beta+1}}\right.$$

$$\left. + G_X(t)B(t,x)^T \sum_{i=0}^{\eta} \mathrm{diag}\left[x(t)^{\frac{i-\gamma}{2\gamma+1}}\right] K_{Xi}(t) x(t)^{\frac{i+\gamma+1}{2\gamma+1}}\right)$$

$$\mathrm{diag}\left[e(t)^{\frac{i-\beta}{2\beta+1}}\right] = \begin{bmatrix} e_1^{\frac{i-\beta}{2\beta+1}} & 0 & \cdots & 0 & 0 \\ 0 & e_2^{\frac{i-\beta}{2\beta+1}} & \cdots & 0 & 0 \\ \vdots & \vdots & \ddots & \vdots & \vdots \\ 0 & 0 & \cdots & e_{b-1}^{\frac{i-\beta}{2\beta+1}} & 0 \\ 0 & 0 & \cdots & 0 & e_b^{\frac{i-\beta}{2\beta+1}} \end{bmatrix}$$

and

$$\text{diag}\left[x(t)^{\frac{i-\gamma}{2\gamma+1}}\right] = \begin{bmatrix} x_1^{\frac{i-\gamma}{2\gamma+1}} & 0 & \cdots & 0 & 0 \\ 0 & x_2^{\frac{i-\gamma}{2\gamma+1}} & \cdots & 0 & 0 \\ \vdots & \vdots & \ddots & \vdots & \vdots \\ 0 & 0 & \cdots & x_{n-1}^{\frac{i-\gamma}{2\gamma+1}} & 0 \\ 0 & 0 & \vdots & 0 & x_n^{\frac{i-\gamma}{2\gamma+1}} \end{bmatrix}$$

If matrices K_{Ei} and K_{Xi} are diagonal, we have

$$u = -\Omega(x)\Phi\left(G_E(t)B_E(t,x)^T \sum_{i=0}^{\varsigma} K_{Ei}(t)\frac{1}{s}e(t)^{\frac{2i+1}{2\beta+1}} + G_X(t)B(t,x)^T \sum_{i=0}^{\eta} K_{Xi}(t)x(t)^{\frac{2i+1}{2\gamma+1}}\right)$$

A closed-loop uncertain system is robustly stable in $X(X_0, U, Z, P)$ and robust tracking is guaranteed in the convex and compact set $E(E_0, Y, R)$ if for reference inputs $r \in R$ and uncertainties in Z and P there exists a $C^\kappa(\kappa \geq 1)$ function $V(\cdot)$, as well as K_∞-functions $\rho_{X1}(\cdot), \rho_{X2}(\cdot), \rho_{E1}(\cdot), \rho_{E2}(\cdot)$ and K-functions $\rho_{X3}(\cdot), \rho_{E3}(\cdot)$, such that the following sufficient conditions:

$$\rho_{X1}(\|x\|) + \rho_{E1}(\|e\|) \leq V(t, e, x) \leq \rho_{X2}(\|x\|) + \rho_{E2}(\|e\|)$$

$$\frac{dV(t, e, x)}{dt} \leq -\rho_{X3}(\|x\|) - \rho_{E3}(\|e\|)$$

are guaranteed in an invariant domain of stability S, and $XE(X_0, E_0, U, R, Z, P) \subseteq S$.

The sufficient conditions under which the robust control problem is solvable were given. Computing the derivative of the $V(t, e, x)$, the unknown coefficients of $V(t, e, x)$ can be found. That is, matrices $K_{Ei}(\cdot)$ and $K_{Xi}(\cdot)$ are obtained. This problem is solved using the nonlinear inequality concept [9].

Example 15.6.1: Control of Two-Phase Permanent-Magnet Stepper Micromotors

High-performance MEMS with permanent-magnet stepper micromotors have been designed and manufactured. Controllers are needed to be designed to control permanent-magnet stepper micromotors, and the angular velocity and position are regulated by changing the magnitude of the voltages applied or currents fed to the stator windings (see Example 15.5.3). The rotor displacement is measured or observed in order to properly apply the voltages to the phase windings. To solve the motion control problem, the controller must be designed. It is illustrated that novel control algorithms are needed to be deployed to maximize the torque developed. In fact, conventional controllers

$$u = -G^{-1}B^T\frac{\partial V}{\partial x} \quad \text{and} \quad u = -\Phi\left(G^{-1}B^T\frac{\partial V}{\partial x}\right)$$

cannot be used.

Using the coenergy concept, one finds the expression for the electromagnetic torque as given by

$$T_e = -RT\psi_m[i_{as}\sin(RT\theta_{rm}) - i_{bs}\cos(RT\theta_{rm})]$$

and thus, one must fed the phase currents as sinusoidal and cosinusoidal functions of the rotor displacement.

The mathematical model of permanent-magnet stepper micromotor was found in Example 15.5.3 as

$$\frac{di_{as}}{dt} = -\frac{r_s}{L_{ss}}i_{as} + \frac{RT\psi_m}{L_{ss}}\omega_{rm}\sin(RT\theta_{rm}) + \frac{1}{L_{ss}}u_{as}$$

$$\frac{di_{bs}}{dt} = -\frac{r_s}{L_{ss}}i_{bs} - \frac{RT\psi_m}{L_{ss}}\omega_{rm}\cos(RT\theta_{rm}) + \frac{1}{L_{ss}}u_{bs}$$

$$\frac{d\omega_{rm}}{dt} = -\frac{RT\psi_m}{J}[i_{as}\sin(RT\theta_{rm}) - i_{bs}\cos(RT\theta_{rm})] - \frac{B_m}{J}\omega_{rm} - \frac{1}{J}T_L$$

$$\frac{d\theta_{rm}}{dt} = \omega_{rm}$$

The rotor resistance is a function of temperature because the resistivity is $\rho_T = \rho_0[(1 + \alpha_\rho(T° - 20°)]$. Hence, $r_s(\cdot) \in [r_{s\min}\ r_{s\max}]$. The susceptibility of the permanent magnets (thin films) decreases with increasing temperature. Other servo-system parameters also vary; in particular, $L_{ss}(\cdot) \in [L_{ss\min}\ L_{ss\max}]$ and $B_m(\cdot) \in [B_{m\min}\ B_{m\max}]$.

The following equation of motion in vector form results:

$$x(t) = F_z(t, x, r, d, z) + B_p(p)u, \quad u_{\min} \le u \le u_{\max}$$

$$x(t_0) = x_0, \quad x = \begin{bmatrix} i_{as} \\ i_{bs} \\ \omega_{rm} \\ \theta_{rm} \end{bmatrix}, \quad u = \begin{bmatrix} u_{as} \\ u_{bs} \end{bmatrix}, \quad y = \theta_{rm}$$

Here, $x \in X$ and $u \in U$ are the state and control vectors, $r \in R$ and $y \in Y$ are the measured reference and output, $d \in D$ is the disturbance, $d = T_L$, and $z \in Z$ and $p \in P$ are the unknown and bounded parameter uncertainties.

Our goal is to design the bounded control $u(\cdot)$ within the constrained set

$$U = \{u \in \mathbb{R}^2 : u_{\min} \le u \le u_{\max},\ u_{\min} < 0,\ u_{\max} > 0\} \subset \mathbb{R}^{2\check{z}}$$

An admissible control law, which guarantees a balanced two-phase voltage applied to the *ab* windings and ensures the maximal electromagnetic torque production, is synthesized as

$$u = \begin{bmatrix} u_{as} \\ u_{bs} \end{bmatrix} = \begin{bmatrix} -\sin(RT\theta_{rm}) & 0 \\ 0 & \cos(RT\theta_{rm}) \end{bmatrix}$$

$$\times \Phi\left(G_x(t)B^T\frac{\partial V(t, x, e)}{\partial x} + G_e(t)B_e^T\frac{\partial V(t, x, e)}{\partial e} + G_i(t)B_e^T\frac{1}{s}\frac{\partial V(t, x, e)}{\partial e}\right)$$

where $e \in E$ is the measured tracking error, $e(t) = r(t) - y(t)$; $\Phi(\cdot)$ is the bounded function (erf, sat, tanh), and $\Phi \in U$, $|\Phi(\cdot)| \le V_{\max}$, $V_{\max}$ is the rated voltage; $G_x(\cdot)$, $G_e(\cdot)$, and $G_i(\cdot)$ are bounded and symmetric, $G_x > 0$, $G_e > 0$, $G_i > 0$; and $V(\cdot)$ is the $C^\kappa (\kappa \ge 1)$ continuously differentiable, real-analytic function.

For $X_0 \subseteq X$, $u \in U$, $r \in R$, $d \in D$, $z \in Z$, and $p \in P$, we obtain the state evolution set X. The state-output set is

$$XY(X_0, U, R, D, Z, P) = \{(x, y) \in X \times Y : x_0 \in X_0,\ u \in U, r \in R, d \in D, z \in Z, p \in P, t \in [t_0, \infty)\}$$

and a *reference-output* map can be found. Our goal is to find the bounded controller such that the tracking error $e(\cdot):[t_0,\infty) \to E$ with $E_0 \subseteq E$ evolves in the specified closed set

$$S_e(\delta) = \{e \in \mathbf{R}^1 : e_0 \in E_0, x \in X(X_0, U, R, D, Z, P), t \in [t_0, \infty) |$$
$$\|e(t)\| \le \rho_e(t, \|e_0\|) + \rho_r(\|r\|) + \rho_d(\|d\|) + \rho_y(\|y\|) + \delta, \delta \ge 0, \forall e \in E(E_0, R, D, Y), \forall t \in [t_0, \infty) \}$$

Here, $\rho_e(\cdot)$ is the KL-function; $\rho_r(\cdot)$, $\rho_d(\cdot)$ and $\rho_y(\cdot)$ are the K-functions.

A positive-invariant domain of stability is found for the closed-loop system with $x_0 \in X_0$, $e_0 \in E_0$, $u \in U$, $r \in R$, $d \in D$, $z \in Z$ and $p \in P$. In particular,

$$S_s = \{x \in \mathbf{R}^4, e \in \mathbf{R}^1 : \|x(t)\| \le \rho_x(t, \|x_0\|) + \rho_r(\|r\|) + \rho_d(\|d\|) + \delta,$$
$$\forall x \in X(X_0, U, R, D, Z, P), \forall t \in [t_0, \infty), \|e(t)\| \le \rho_e(t, \|e_0\|) + \rho_r(\|r\|)$$
$$+ \rho_d(\|d\|) + \rho_y(\|y\|) + \delta, \forall e \in E(E_0, R, D, Y), \forall t \in [t_0, \infty) \},$$

where $\rho_x(\cdot)$ is the KL-function.

To study the robustness, tracking, and disturbance rejection, we consider a state-error set

$$XE(X_0, E_0, U, R, D, Z, P) = \{(x, e) \in X \times E : x_0 \in X_0, e_0 \in E_0, u \in U,$$
$$r \in R, d \in D, z \in Z, p \in P, t \in [t_0, \infty)\}$$

The robust tracking, stability, and disturbance rejection are guaranteed if $XE \subseteq S_s$. The *admissible* set S_s is found by using the Lyapunov stability theory [9], and

$$S_s = \left\{ x \in \mathbf{R}^4, e \in \mathbf{R}^1 : x_0 \in X_0, e_0 \in E_0, u \in U, r \in R, d \in D, z \in Z, p \in P \right.$$
$$\rho_1 \|x\| + \rho_2 \|e\| \le V(t, x, e) \le \rho_3 \|x\| + \rho_4 \|e\|, \frac{dV(t, x, e)}{dt} \le -\rho_5 \|x\| - \rho_6 \|e\|,$$
$$\left. \forall x \in X(X_0, U, R, P, Z, P), \forall e \in E(E_0, R, D, Y), \forall t \in [t_0, \infty) \right\}$$

where $\rho_1(\cdot)$, $\rho_2(\cdot)$, $\rho_3(\cdot)$ and $\rho_4(\cdot)$ are the K_∞-functions; and $\rho_5(\cdot)$ and $\rho_6(\cdot)$ are the K-functions.

If in XE there exists a C^κ Lyapunov function $V(t, x, e)$ such that for all $x_0 \in X_0$, $e_0 \in E_0$, $u \in U$, $r \in R$, $d \in D$, $z \in Z$, and $p \in P$ on $[t_0, \infty)$ sufficient condition for stability (s1)

$$\rho_1 \|x\| + \rho_2 \|e\| \le V(t, x, e) \le \rho_3 \|x\| + \rho_4 \|e\|$$

and inequality

$$\frac{dV(t, x, e)}{dt} \le -\rho_5 \|x\| - \rho_6 \|e\|$$

which is the sufficient condition for stability s2, hold, then

1. Solution $x(\cdot):[t_0, \infty) \to X$ for closed-loop system is robustly bounded and stable.
2. Convergence of the error vector $e(\cdot):[t_0, \infty) \to E$ to S_e is ensured in XE.
3. XE is convex and compact, and $XE \subseteq S_s$.

That is, if criteria (s1) and (s2) are guaranteed, we have $XE \subseteq S_s$.

Using the nonquadratic Lyapunov candidate

$$V(t, x, e) = \sum_{j=0}^{\eta} \frac{2\gamma+1}{2(j+\gamma+1)} \left(x^{\frac{j+\gamma+1}{2\gamma+1}}\right)^T K_{xj}(t) x^{\frac{j+\gamma+1}{2\gamma+1}} + \sum_{j=0}^{\varsigma} \frac{2\beta+1}{2(j+\beta+1)} \left(e^{\frac{j+\beta+1}{2\beta+1}}\right)^T K_{ej}(t) e^{\frac{j+\beta+1}{2\beta+1}}$$

$$+ \sum_{i=0}^{\sigma} \frac{2\mu+1}{2(j+\mu+1)} \left(e^{\frac{j+\mu+1}{2\mu+1}}\right)^T K_{ij}(t) e^{\frac{j+\mu+1}{2\mu+1}}$$

one obtains the bounded controller as

$$u = \begin{bmatrix} u_{as} \\ u_{bs} \end{bmatrix} = \begin{bmatrix} -\sin(RT\theta_{rm}) & 0 \\ 0 & \cos(RT\theta_{rm}) \end{bmatrix} \Phi \left(G_x(t) B^T \sum_{j=0}^{\eta} \mathrm{diag}\left[x^{\frac{j-\gamma}{2\gamma+1}}\right] K_{xj}(t) x^{\frac{j+\gamma+1}{2\gamma+1}} \right.$$

$$\left. + G_e(t) B_e^T \sum_{j=0}^{\varsigma} K_{ej}(t) e^{\frac{2j+1}{2\beta+1}} + G_i(t) B_e^T \sum_{j=0}^{\sigma} K_{ij}(t) \frac{1}{s} e^{\frac{2j+1}{2\mu+1}} \right)$$

Here, $K_{xj}(\cdot)$ are the unknown matrix-functions, and $K_{ej}(\cdot)$ and $K_{ij}(\cdot)$ are the unknown coefficients; $\eta = 0, 1, 2,\ldots$; $\gamma = 0, 1, 2,\ldots$; $\varsigma = 0, 1, 2,\ldots$; $\beta = 0, 1, 2,\ldots$; $\sigma = 0, 1, 2,\ldots$; and $\mu = 0, 1, 2,\ldots$.

Under the assumption that X_0, E_0, R, D, Z, and P are admissible, the robust tracking problem is solvable in XE. That is, the bounded real-analytic control $u(\cdot)$ guarantees the robust stability and steers the tracking error to S_e. Furthermore, stability is guaranteed, disturbance rejection is ensured, and specified input-output tracking performance can be achieved.

Applying the controller designed, one maximizes the electromagnetic torque developed by permanent-magnet stepper micromotors. This can be easily shown by using the expression for the electromagnetic torque, the balanced two-phase sinusoidal voltage set (applied phase voltages u_{as} and u_{bs}), as well as the trigonometric identity $\sin^2 a + \cos^2 a = 1$.

The tracking controller can be designed using the tracking error. In particular, we have

$$u = \begin{bmatrix} u_{as} \\ u_{bs} \end{bmatrix} = \begin{bmatrix} -\sin(RT\theta_{rm}) & 0 \\ 0 & \cos(RT\theta_{rm}) \end{bmatrix} \Phi \left(G_e(t) B_e^T \sum_{j=0}^{\varsigma} K_{ei}(t) e^{\frac{2j+1}{2\beta+1}} + G_i(t) B_e^T \sum_{i=0}^{\sigma} K_{ij}(t) \frac{1}{s} e^{\frac{2j+1}{2\mu+1}} \right)$$

The controller design, implementation, and experimental verification are reported in [9].

15.7 Conclusions

This chapter reports the current status, documents innovative results, and researches novel paradigms in synthesis, modeling, analysis, simulation, control, and optimization of high-performance MEMS. These results are obtained applying reported nonlinear modeling, analysis, synthesis, control, and optimization methods which allow one to attain performance assessment and predict outcomes. Novel MEMS were devised. The application of the plate, spherical, torroidal, conical, cylindrical, and asymmetrical motor geometry, as well as *endless, open-ended,* and *integrated* electromagnetic systems, allows one to classify MEMS. This idea is extremely useful in the studying of existing MEMS as well as in the synthesis of innovative high-performance MEMS. For example, asymmetrical (unconventional) geometry and *integrated* electromagnetic system can be applied. Optimization can be performed, and the classifier paradigm serves as a starting point from which advanced configurations can be synthesized and straightforwardly interpreted. Microscale motion devices geometry and electromagnetic systems, which play a central role, are related. Structural synthesis and optimization of MEMS are formalized and interpreted using innovative ideas. The MEMS classifier paradigm, in addition to being qualitative, leads one to quantitative analysis. In fact, using the cornerstone laws of electromagnetics and mechanics (e.g., Maxwell's, Kirchhoff and Newton equations), the differential equations to model electromagnetic and mechanical phenomena and effects can be derived and applied to attain the performance analysis with outcome prediction. Mathematical models for MEMS are found. Making use of these mathematical models, analysis and optimization were performed, and nonlinear control algorithms were designed. The electromagnetics features and phenomena were integrated into the analysis, modeling, synthesis, and optimization. It is shown that to meet the specified level of performance, novel high-performance MEMS should be synthesized, high-fidelity modeling must be performed, advanced controllers have to be synthesized, and highly detailed dynamic nonlinear simulations must be carried out. The results reported have direct application to the analysis and design of high-performance MEMS. Different MEMS can be devised, synthesized, defined, and designed, and a number of long-standing issues related to geometrical variability and electromagnetics are studied. These benchmarking results allow one to reformulate and refine extremely important problems in MEMS theory, and solve a number of very complex issues in design and optimization with the ultimate goal to synthesize innovative high-performance, high torque, and power densities MEMS.

References

1. Lyshevski, S. E., *Nano- and Micro-Electromechanical Systems: Fundamentals of Nano- and Micro-Engineering*, CRC Press, Boca Raton, FL, 2000.
2. Madou, M., *Fundamentals of Microfabrication*, CRC Press, Boca Raton, FL, 1997.
3. Campbell, S. A., *The Science and Engineering of Microelectronic Fabrication*, Oxford University Press, New York, 2001.
4. Lyshevski, S. E., *Electromechanical Systems, Electric Machines, and Applied Mechatronics*, CRC Press, Boca Raton, FL, 1999.
5. Lyshevski, S. E. and Lyshevski, M. A., "Analysis, dynamics, and control of micro-electromechanical systems," *Proc. American Control Conference*, Chicago, IL, pp. 3091–3095, 2000.
6. Mehregany, M. and Tai, Y. C., "Surface micromachined mechanisms and micro-motors," *J. Micromechanics and Microengineering*, vol. 1, pp. 73–85, 1992.
7. Becker, E. W., Ehrfeld, W., Hagmann, P., Maner, A., and Mynchmeyer, D., "Fabrication of microstructures with high aspect ratios and great structural heights by synchrotron radiation lithography, galvanoformung, and plastic molding (LIGA process)," *Microelectronic Engineering*, vol. 4, pp. 35–56, 1986.
8. Guckel, H., Christenson, T. R., Skrobis, K. J., Klein, J., and Karnowsky, M., "Design and testing of planar magnetic micromotors fabricated by deep X-ray lithography and electroplating," *Technical Digest of International Conference on Solid-State Sensors and Actuators, Transducers 93*, Yokohama, Japan, pp. 60–64, 1993.
9. Lyshevski, S. E., *Control Systems Theory with Engineering Applications*, Birkhäuser, Boston, MA, 2001.

16
The Physical Basis of Analogies in Physical System Models

Neville Hogan
Massachusetts Institute of Technology

Peter C. Breedveld
University of Twente

16.1 Introduction **16**-1
16.2 History **16**-2
16.3 The Force–Current Analogy: Across and Through Variables **16**-2
   Drawbacks of the Across-Through Classification • Measurement as a Basis for Analogies • Beyond One-Dimensional Mechanical Systems • Physical Intuition
16.4 Maxwell's Force-Voltage Analogy: Effort and Flow Variables **16**-4
   Systems of Particles • Physical Intuition • Dependence on Reference Frames
16.5 A Thermodynamic Basis for Analogies **16**-5
   Extensive and Intensive Variables • Equilibrium and Steady State • Analogies, Not Identities • Nodicity
16.6 Graphical Representations **16**-8
16.7 Concluding Remarks **16**-9
Acknowledgments **16**-10
References **16**-10

16.1 Introduction

One of the fascinating aspects of mechatronic systems is that their function depends on interactions between electrical and mechanical behavior and often magnetic, fluid, thermal, chemical, or other effects as well. At the same time, this can present a challenge as these phenomena are normally associated with different disciplines of engineering and physics. One useful approach to this multidisciplinary or "multi-physics" problem is to establish analogies between behavior in different domains—for example, resonance due to interaction between inertia and elasticity in a mechanical system is analogous to resonance due to interaction between capacitance and inductance in an electrical circuit. Analogies can provide valuable insight about how a design works, identify equivalent ways a particular function might be achieved, and facilitate detailed quantitative analysis. They are especially useful in studying dynamic behavior, which often arises from interactions between domains; for example, even in the absence of elastic effects, a mass moving in a magnetic field may exhibit resonant oscillation. However, there are many ways that analogies may be established and, unfortunately, the most appropriate analogy between electrical circuits, mechanical and fluid systems remains unresolved: is force like current, or is force more like voltage? In this contribution we examine the physical basis of the analogies in common use and how they may be extended beyond mechanical and electrical systems.

16.2 History

It is curious that one of the earliest applications of analogies between electrical and mechanical systems was to enable the demonstration and study of transients in electrical networks that were otherwise too fast to be observed by the instrumentation of the day by identifying mechanical systems with equivalent dynamic behavior; that was the topic of a series of articles on "Models and analogies for demonstrating electrical principles" (*The Engineer,* 1926). Improved methods capable of observing fast electrical transients directly (especially the cathode ray oscilloscope, still in use today) rendered this approach obsolete but enabled quantitative study of nonelectrical systems via analogous electrical circuits (Nickle, 1925). Although that method had considerably more practical importance at the time than it has today (we now have the luxury of vastly more powerful tools for numerical computation of electromechanical system responses), in the late '20s and early '30s a series of papers (Darrieus, 1929; Hähnle, 1932; Firestone, 1933) formulated a rational method to use electrical networks as a framework for establishing analogies between physical systems.

16.3 The Force–Current Analogy: Across and Through Variables

Firestone identified two types of variable in each physical domain—"across" and "through" variables—which could be distinguished based on how they were measured. An "across" variable may be measured as a difference between values at two points in space (conceptually, *across* two points); a "through" variable may be measured by a sensor in the path of power transmission between two points in space (conceptually, it is transmitted *through* the sensor). By this classification, electrical voltage is analogous to mechanical velocity and electrical current is analogous to mechanical force. Of course, this classification of variables implies a classification of network elements: a mass is analogous to a capacitor, a spring is analogous to an inductor and so forth.

The "force-is-like-current" or "mass-capacitor" analogy has a sound mathematical foundation. Kirchhoff's node law or current law, introduced in 1847 (the sum of currents into a circuit node is identically zero) can be seen as formally analogous to D'Alembert's principle, introduced in 1742 (the sum of forces on a body is identically zero, provided the sum includes the so-called "inertia force," the negative of the body mass times its acceleration). It is the analogy used in linear-graph representations of lumped-parameter systems, proposed by Trent in 1955. Linear graphs bring powerful results from mathematical graph theory to bear on the analysis of lumped-parameter systems. For example, there is a systematic procedure based on partitioning a graph into its *tree* and *links* for selecting sets of independent variables to describe a system. Graph-theoretic approaches are closely related to matrix methods that in turn facilitate computer-aided methods. Linear graphs provide a unified representation of lumped-parameter dynamic behavior in several domains that has been expounded in a number of successful textbooks (e.g., Shearer et al., 1967; Rowell & Wormley, 1997).

The mass-capacitor analogy also appears to afford some practical convenience. It is generally easier to identify points of common velocity in a mechanical system than to identify which elements experience the same force; and it is correspondingly easier to identify the nodes in an electrical circuit than all of its loops. Hence with this analogy it is straightforward to identify an electrical network equivalent to a mechanical system, at least in the one-dimensional case.

16.3.1 Drawbacks of the Across-Through Classification

Despite the obvious appeal of establishing analogies based on practical measurement procedures, the force-current analogy has some drawbacks that will be reviewed below: (i) on closer examination, measurement-based classification is ambiguous; (ii) its extension to more than one-dimensional mechanical systems is problematical; and (iii) perhaps most important, it leads to analogies (especially between mechanical and fluid systems) that defy common physical insight.

16.3.2 Measurement as a Basis for Analogies

Even a cursory review of state-of-the-art measurement technologies shows that the across-through classification may be an anachronism or, at best, an over-simplification. Velocity (an "across" variable) may be measured by an integrating accelerometer that is attached only to the point where velocity is measured—that's how the human inner ear measures head velocity. While the velocity is measured with respect to an inertial reference frame (as it should be), there is no tangible connection to that frame. As a further example, current in a conductor (a "through" variable) may be measured without inserting an ammeter in the current path; sensors that measure current by responding to the magnetic field next to the conductor are commercially available (and preferred in some applications). Moreover, in some cases similar methods can be applied to measure both "across" and "through" variables. For example, fluid flow rate is classified as a through variable, presumably by reference to its measurement by, for example, a positive-displacement meter in the flow conduit; that's the kind of fluid measurement commonly used in a household water meter. However, optical methods that are used to measure the velocity of a rigid body (classified as an across variable) are often adapted to measure the volumetric flow rate of a fluid (laser doppler velocimetry is a notable example). Apparently the same fundamental measurement technology can be associated with an across variable in one domain and a through variable in another. Thus, on closer inspection, the definition of across and through variables based on measurement procedures is, at best, ambiguous.

16.3.3 Beyond One-Dimensional Mechanical Systems

The apparent convenience of equating velocities in a mechanical system with voltages at circuit nodes diminishes rapidly as we go beyond translation in one dimension or rotation about a fixed axis. A translating body may have two or three independent velocities (in planar and spatial motion, respectively). Each independent velocity would appear to require a separate independent circuit node, but the kinetic energy associated with translation can be redistributed at will among these two or three degrees of freedom (e.g., during motion in a circle at constant speed the total kinetic energy remains constant while that associated with each degree of freedom varies). This requires some form of connection between the corresponding circuit nodes in an equivalent electrical network, but what that connection should be is not obvious.

The problem is further exacerbated when we consider rotation. Even the simple case of planar motion (i.e., a body that may rotate while translating) requires three independent velocities, hence three independent nodes in an equivalent electrical network. Reasoning as above we see that these three nodes must be connected but in a different manner from the connection between three nodes equivalent to spatial translation. Again, this connection is hardly obvious, yet translating while rotating is ubiquitous in mechanical systems—that's what a wheel usually does.

Full spatial rotation is still more daunting. In this case interaction between the independent degrees of freedom is especially important as it gives rise to gyroscopic effects, including oscillatory precession and nutation. These phenomena are important practical considerations in modern mechatronics, not arcane subtleties of classical mechanics; for example, they are the fundamental physics underlying several designs for a microelectromechanical (MEMS) vibratory rate gyroscope (Yazdi et al., 1998).

16.3.4 Physical Intuition

In our view the most important drawback of the across-through classification is that it identifies force as analogous to fluid flow rate as well as electrical current (with velocity analogous to fluid pressure as well as voltage). This is highly counter-intuitive and quite confusing. By this analogy, fluid pressure is not analogous to force despite the fact that pressure is commonly defined as force per unit area. Furthermore, stored kinetic energy due to fluid motion is not analogous to stored kinetic energy due to motion of a rigid body. Given the remarkable similarity of the physical processes underlying these two forms of energy storage, it is hard to understand why they should not be analogous.

Insight is the ultimate goal of modeling. It is a crucial factor in producing innovative and effective designs and depends on developing and maintaining a "physical intuition" about the way devices behave. It is important that analogies between physical effects in different domains can be reconciled with the physical intuition and any method that requires a counter-intuitive analogy is questionable; at a minimum it warrants careful consideration.

16.4 Maxwell's Force-Voltage Analogy: Effort and Flow Variables

An alternative analogy classifies variables in each physical domain that (loosely speaking) describe motion or cause it. Thus fluid flow rate, electrical current, and velocity are considered analogous (sometimes generically described as "flow" variables). Conversely, fluid pressure, electrical voltage, and force are considered analogous (sometimes generically described as "effort" variables).

The "force-is-like-voltage" analogy is the oldest drawn between mechanical and electrical systems. It was first proposed by Maxwell (1873) in his treatise on electricity and magnetism, where he observed the similarity between the Lagrangian equations of classical mechanics and electromechanics. That was why Firestone (1933) presented his perspective that force is like current as "A *new* analogy between mechanical and electrical systems" (emphasis added). Probably because of its age, the force-voltage analogy is deeply embedded in our language. In fact, voltage is still referred to as "electromotive force" in some contexts. Words like "resist" or "impede" also have this connotation: a large resistance or impedance implies a large force for a given motion or a large voltage for a given current.

In fact, Maxwell's classification of velocity as analogous to electrical current (with force analogous to voltage) has a deeper justification than the similarity of one mathematical form of the equations of mechanics and electromechanics; it can be traced to a similarity of the underlying physical processes.

16.4.1 Systems of Particles

Our models of the physical world are commonly introduced by describing systems of particles distributed in space. The particles may have properties such as mass, charge, etc., though in a given context we will deliberately choose to neglect most of those properties so that we may concentrate on a single physical phenomenon of interest. Thus, to describe electrical capacitance, we consider only charge, while to describe translational inertia, we consider only mass and so forth.

Given that this common conceptual model is used in different domains, it may be used to draw analogies between the variables of different physical domains. From this perspective, quantities associated with the motion of particles may be considered analogous to one another; thus mechanical velocity, electrical current, and fluid flow rate are analogous. Accordingly, mechanical displacement, displaced fluid volume, and displaced charge are analogous; and thus force, fluid pressure, and voltage are analogous. This classification of variables obviously implies a classification of network elements: a spring relates mechanical displacement and force; a capacitor relates displaced charge and voltage. Thus a spring is analogous to a capacitor, a mass to an inductor, and for this reason, this analogy is sometimes termed the "mass-inductor" analogy.

16.4.2 Physical Intuition

The "system-of-particles" models naturally lead to the "intuitive" analogy between pressure, force, and voltage. But, is such a vague and ill-defined concept as "physical intuition" an appropriate consideration in drawing analogies between physical systems? After all, physical intuition might largely be a matter of usage and familiarity, rooted in early educational and cultural background.

We think not; instead we speculate that physical intuition may be related to conformity with a mental model of the physical world. That mental model is important for thinking about physical systems and, if shared, for communicating about them. Because the "system-of-particles" model is widely assumed (sometimes explicitly, sometimes implicitly) in the textbooks and handbooks of basic science and engineering we speculate that it may account for the physical intuition shared by most engineers. If so, then conforming with that common "system-of-particles" mental model is important to facilitate designing, thinking, and communicating about mechatronic systems. The force–voltage analogy does so; the force–current analogy does not.

16.4.3 Dependence on Reference Frames

The "system-of-particles" model also leads to another important physical consideration in the choice of analogies between variables: the way they depend on reference frames. The mechanical displacement that determines the elastic potential energy stored in a spring and the displaced charge that determines the electrostatic potential energy stored in a capacitor may be defined with respect to *any* reference frame (whether time-varying or stationary). In contrast, the motion required for kinetic energy storage in a rigid body or a fluid must be defined with respect to an *inertial* frame. Though it may often be overlooked, the motion of charges required for magnetic field storage must also be defined with respect to an inertial frame (Feynman et al., 1963).

To be more precise, the constitutive equations of energy storage based on motion (e.g., in a mass or an inductor) require an inertial reference frame (or must be modified in a non-inertial reference frame). In contrast, the constitutive equations of energy storage based on displacement (e.g., in a spring or a capacitor) do not. Therefore, the mass–inductor (force–voltage) analogy is more consistent with fundamental physics than the mass–capacitor (force–current) analogy.

The modification of the constitutive equations for magnetic energy storage in a non-inertial reference frame is related to the transmission of electromagnetic radiation. However, Kirchhoff's laws (more aptly termed "Kirchhoff's approximations"), which are the foundations of electric network theory, are equivalent to assuming that electromagnetic radiation is absent or negligible. It might, therefore, be argued that the dependence of magnetic energy storage on an inertial reference frame is negligible for electrical circuits, and hence is irrelevant for any discussion of the physical basis of analogies between electrical circuits and other lumped-parameter dynamic-system models. That is undeniably true and could be used to justify the force–current analogy. Nevertheless, because of the confusion that can ensue, the value of an analogy that is fundamentally inconsistent with the underlying physics of lumped-parameter models is questionable.

16.5 A Thermodynamic Basis for Analogies

Often in the design and analysis of mechatronic systems it is necessary to consider a broader suite of phenomena than those of mechanics and electromechanics. For instance, it may be important to consider thermal conduction, convection, or even chemical reactions and more. To draw analogies between the variables of these domains it is helpful to examine the underlying physics. The analogous dynamic behavior observed in different physical domains (resonant oscillation, relaxation to equilibrium, etc.) is not merely a similarity of *mathematical* forms, it has a common *physical* basis which lies in the storage, transmission, and irreversible dissipation of energy. Consideration of energy leads us to thermodynamics; we show next that thermodynamics provides a broader basis for drawing analogies and yields some additional insight.

All of the displacements considered to be analogous above (i.e., mechanical displacement, displaced fluid volume, and displaced charge) may be associated with an energy storage function that requires equilibrium for its definition, the displacement being the argument of that energy function. Generically, these may be termed potential energy functions. To elaborate, elastic energy storage requires sustained but recoverable deformation of a material (e.g., as in a spring); the force required to sustain that deformation is determined at equilibrium, defined when the time rate of change of relative displacement of the material particles is uniformly zero (i.e., all the particles are at rest relative to each other). Electrostatic energy storage requires sustained separation of mobile charges of opposite sign (e.g., as in a capacitor); the required voltage is determined at equilibrium, defined when the time rate of change of charge motion is zero (i.e., all the charges are at rest relative to each other).

16.5.1 Extensive and Intensive Variables

In the formalism of thermodynamics, the amount of stored energy and the displacement that determines it are *extensive* variables. That is, they vary with the spatial extent (i.e., size or volume) of the object storing the energy. The total elastic energy stored in a uniform rod of constant cross-sectional area in an idealized uniform state of stress is proportional to the length (and hence volume) of the rod; so is the total relative displacement of its ends; both are extensive variables. The total electrostatic energy stored in an idealized parallel-plate capacitor (i.e., one with no fringe fields) is proportional to the area of the plates (and hence, for constant gap, the volume they enclose); so is the total separated charge on the plates; both are extensive variables (see Breedveld, 1984).

Equilibrium of these storage elements is established by an *intensive* variable that does not change with the size of the object. This variable is the gradient (partial derivative) of the stored energy with respect to the corresponding displacement. Thus, at equilibrium, the force on each cross-section of the rod is the same regardless of the length or volume of the rod; force is an intensive variable. If the total charge separated in the capacitor is proportional to area, the voltage across the plates is independent of area; voltage is an intensive variable.

Dynamics is not solely due to the storage of energy but arises from the transmission and deployment of power. The instantaneous power into an equilibrium storage element is the product of the (intensive) gradient variable (force, voltage) with the time rate of change of the (extensive) displacement variable (velocity, current). Using this thermodynamics-based approach, all intensive variables are considered analogous, as are all extensive variables and their time rates of change, and so on.

Drawing analogies from a thermodynamic classification into extensive and intensive variables may readily be applied to fluid systems. Consider the potential energy stored in an open container of incompressible fluid: The pressure at any specified depth is independent of the area at that depth and the volume of fluid above it; pressure is an intensive variable analogous to force and voltage, as our common physical intuition suggests it should be. Conversely, the energy stored in the fluid above that depth is determined by the volume of fluid; energy and volume are extensive variables, volume playing the role of displacement analogous to electrical charge and mechanical displacement. Pressure is the partial derivative of stored energy with respect to volume and the instantaneous power into storage is the product of pressure with volumetric flow rate, the time rate of change of volume flowing past the specified depth.

An important advantage of drawing analogies from a classification into extensive and intensive variables is that it may readily be generalized to domains to which the "system-of-particles" image may be less applicable. For example, most mechatronic designs require careful consideration of heating and cooling but there is no obvious flow of particles associated with heat flux. Nevertheless, extensive and intensive variables associated with equilibrium thermal energy storage can readily be identified. Drawing on classical thermodynamics, it can be seen that (total) entropy is an extensive variable and plays the role of a displacement. The gradient of energy with respect to energy is temperature, an intensive variable, which should be considered analogous to force, voltage, and pressure. Equality of temperature establishes thermal equilibrium between two bodies that may store heat (energy) and communicate it to one another.

A word of caution is appropriate here as a classification into extensive and intensive variables properly applies only to scalar quantities such as pressure, volume, etc. As outlined below, the classification can be generalized in a rigorous way to nonscalar quantities, but care is required (see Breedveld, 1984).

16.5.2 Equilibrium and Steady State

In some (though not all) domains energy storage may also be based on motion. Kinetic energy storage may be associated with rigid body motion or fluid motion; magnetic energy storage requires motion of charges. The thermodynamics-based classification properly groups these different kinds of energy storage as analogous to one another and generically they may be termed *kinetic* energy storage elements.

All of the motion variables considered to be analogous (i.e., velocity, fluid flow rate, current) may be associated with an energy storage function that is defined by *steady state* (rather than by equilibrium). For a rigid body, steady motion requires zero net force, and hence constant momentum and kinetic energy. For the magnetic field that stores energy in an inductor, steady current requires zero voltage, and hence constant magnetic flux and magnetic energy.

It might reasonably be argued that any distinction between equilibrium and steady state is purely a matter of perspective and common usage, rather than a fundamental feature of the physical world. For example, with an alternative choice of reference frames, "sustained motion" could be redefined as "rest" or "equilibrium." From this perspective, a zero-relative-velocity "equilibrium" between two rigid bodies (or between a rigid body and a reference frame) could be defined by zero force. Following this line of reasoning any distinction between the mass-inductor and mass-capacitor analogies would appear to be purely a matter of personal choice. However, while the apparent equivalence of "equilibrium" and "steady state" may be justifiable in the formal mathematical sense of zero rate of change of a variable, in a mechanical system, displacement (or position) and velocity (or momentum) are fundamentally different. For example, whereas velocity, force, and momentum may be transformed between reference frames as rank-one tensors, position (or displacement) may not be transformed as a tensor of any kind. Thus, a distinction between equilibrium and steady state reflects an important aspect of the structure of physical system models.

16.5.3 Analogies, Not Identities

It is important to remember that any classification to establish analogies is an abstraction. At most, dynamic behavior in different domains may be similar; it is not identical. We have pointed out above that if velocity or current is used as the argument of an energy storage function, care must be taken to identify an appropriate inertial reference frame and/or to understand the consequences of using a non-inertial frame. However, another important feature of these variables is that they are fundamentally vectors (i.e., they have a definable spatial orientation). One consequence is that the thermodynamic definition of extensive and intensive variables must be generalized before it may be used to classify these variables (cf., Breedveld, 1984). In contrast, a quantity such as temperature or pressure is fundamentally a scalar. Furthermore, both of these quantities are intrinsically "positive" scalars insofar as they have well-defined, unique and physically meaningful zero values (absolute zero temperature, the pressure of a perfect vacuum). Quite aside from any dependence on inertial reference frames, the across-through analogy between velocity (a vector with no unique zero value) and pressure (a scalar with a physically important zero) will cause error and confusion if used without due care.

This consideration becomes especially important when similar elements of a model are combined (for example, a number of bodies moving with identical velocity may be treated as a single rigid body) to simplify the expression of dynamic equations or improve their computability. The engineering variables used to describe energy storage can be categorized into two groups: (i) positive-valued scalar variables and (ii) nonscalar variables. Positive-valued scalar variables have a physically meaningful zero or absolute reference; examples include the volume of stored fluid, the number of moles of a chemical species, entropy, etc. Nonscalar* variables have a definable spatial orientation. Even in the one-dimensional case they can be positive or negative, the sign denoting direction with respect to some reference frame; examples include displacement, momentum, etc. These variables generally do not have a physically meaningful zero or absolute reference, though some of them must be defined with respect to an inertial frame.

Elements of a model that describe energy storage based on scalar variables can be combined in only one way: they must be in mutual equilibrium; their extensive variables are added, while the corresponding intensive variables are equal, independent of direction, and determine the equilibrium condition. For model elements that describe energy storage based on nonscalar variables there are usually two options.

---

*The term "vector variables" suggests itself but these variables may include three-dimensional spatial orientation, which may not be described as a vector.

Electrical capacitors, for instance, may be combined in parallel or in series and the resulting equivalent capacitor may readily be determined. In a parallel connection, equilibrium is determined by voltage (an intensive variable) and the electric charges (extensive variables) are added as before. However, a series connection is the "dual" in the sense that the roles of charge and voltage are exchanged: equality of charges determines equilibrium and the voltages are added. Mechanical springs may also be combined in two ways. However, that is not the case for translational masses and rotational inertias; they may only be combined into a single equivalent rigid body if their velocities are equal and in that case their momenta are added.

The existence of two "dual" ways to combine some, but not all, of the energy storage elements based on nonscalar quantities is somewhat confusing. It may have contributed to the lengthy debate (if we date its beginning to Maxwell, lasting for over a century!) on the best analogy between mechanical and electrical systems. Nevertheless, the important point is that series and parallel connections may not be generalized in a straightforward way to all domains.

16.5.4 Nodicity

As insight is the foremost goal of modeling, analogies should be chosen to promote insight. Because there may be fundamental differences between all of the physical domains, care should be exercised in drawing analogies to ensure that special properties of one domain should not be applied inappropriately to other domains. This brings us to what may well be the strongest argument against the across-through classification. History suggests that it originated with the use of equivalent electrical network representations of nonelectrical systems. Unfortunately, electrical networks provide an inappropriate basis for developing a general representation of physical system dynamics. This is because electrical networks enjoy a special property, *nodicity*, which is quite unusual among the physical system domains (except as an approximation).

Nodicity refers to the fact that any sub-network (cut-set) of an electrical network behaves as a node in the sense that a Kirchhoff current balance equation may be written for the entire sub-network. As a result of nodicity, electrical network elements can be assembled in arbitrary topologies and yet still describe a physically realizable electrical network. This property of "arbitrary connectability" is not a general property of lumped-parameter physical system models. Most notably, mass elements cannot be connected arbitrarily; they must always be referenced to an inertial frame. For that reason, electrical networks can be quite misleading when used as a basis for a general representation of physical system dynamics. This is not merely a mathematical nicety; some consequences of non-nodic behavior for control system analysis have recently been explored (Won and Hogan, 1998).

By extension, because each of the physical domains has its unique characteristics, any attempt to formulate analogies by taking one of the domains (electrical, mechanical, or otherwise) as a starting point is likely to have limitations. A more productive approach is to begin with those characteristics of physical variables common to all domains and that is the reason to turn to thermodynamics. In other words, the best way to identify analogies *between* domains may be to "step outside" *all* of them. By design, general characteristics of all domains such as the extensive nature of stored energy, the intensive nature of the variables that define equilibrium, and so forth, are not subject to the limitations of any one (such as nodicity). That is the main advantage of drawing analogies based on thermodynamic concepts such as the distinction between extensive and intensive variables.

16.6 Graphical Representations

Analogies are often associated with abstract graphical representations of multi-domain physical system models. The force-current analogy is usually associated with the linear graph representation of networks introduced by Trent (1955); the force-voltage analogy is usually associated with the bond graph representation introduced by Paynter (1960). Bond graphs classify variables into efforts (commonly force, voltage, pressure, and so forth) and flows (commonly velocity, current, fluid flow rate, and so forth). Bond graphs extend all the practical benefits of the force-current (across-through) analogy to the force-voltage (effort-flow) analogy: they provide a unified representation of lumped-parameter dynamic behavior in several domains that has been expounded in a number of successful textbooks (e.g., Karnopp et al., 1975, 1999), there are systematic methods for selecting sets of independent variables to describe a system, ways to take advantage of the ease of identifying velocities and voltages, and matrix methods to facilitate computer analysis. In fact, several computer-aided modeling support packages using the bond-graph language are now available. Furthermore, bond graphs have been applied successfully to describe the dynamics of spatial mechanisms (including gyroscopic effects) while, to the authors' knowledge, linear graphs have not.

Although the force–voltage analogy is most commonly used with bond graphs, the force–current analogy can be used just as readily; the underlying mathematical formalism is indifferent to the choice of which variables are chosen as analogous. In fact, pursuing this line of thought, the choice is unnecessary and may be avoided; doing so affords a way to clarify the potential confusion over the role of intensive variables and the dual types of connection available for some elements in some domains.

In the Generalized Bond Graph (GBG) approach (Breedveld, 1984) all energy storage becomes analogous and only one type of storage element, a (generalized) capacitor, is identified. Its displacement is an extensive variable; the gradient of its energy storage function with respect to that displacement is an intensive variable. In some (but not all) domains a particular kind of coupling known as a *gyrator* is found that gives rise to the *appearance* of a dual type of energy storage, a (generalized) inertia as well as the possibility of dual ways to connect elements. The GBG representation emphasizes the point that the presence of dual types of energy storage and dual types of connection is a special property (albeit an important one) of a limited number of domains. In principle, either a "mass–capacitor" analogy or a "mass–inductor" analogy can be derived from a GBG representation by choosing to associate the gyrating coupling with either the "equilibrium" or "steady-state" energy storage elements.

The important point to be taken here is that the basis of analogies between domains does not depend on the use of a particular abstract graphical representation. The practical value of establishing analogies between domains and the merits of a domain-independent approach based on intensive versus extensive variables remains regardless of which graph-theoretic tools (if any) are used for analysis.

16.7 Concluding Remarks

In the foregoing we articulated some important considerations in the choice of analogies between variables in different physical domains. From a strictly mathematical viewpoint there is little to choose; both analogies may be used as a basis for rigorous, self-consistent descriptions of physical systems. The substantive and important factors emerge from a physical viewpoint—considering the structured way physical behavior is described in the different domains. Summarizing:

- The "system-of-particles" model that is widely assumed in basic science and engineering naturally leads to the intuitive analogy between force and voltage, velocity and current, a mass and an inductor, and so on.
- The measurement procedures used to motivate the distinction between across and through variables at best yield an ambiguous classification.
- Nodicity (the property of "arbitrary connectability") is not a general property of lumped-parameter physical system models. Thus, electrical networks, which are nodic, can be quite misleading when used as a basis for a general representation of physical system dynamics.
- The intuitive analogy between velocity and current is consistent with a thermodynamic classification into extensive and intensive variables. As a result, the analogy can be generalized to dynamic behavior in domains to which the "system-of-particles" image may be less applicable.
- The force–voltage or mass–inductor analogy reflects an important distinction between equilibrium energy-storage phenomena and steady-state energy-storage phenomena: the constitutive equations of steady-state energy storage phenomena require an inertial reference frame (or must be modified in a non-inertial reference frame) while the constitutive equations of equilibrium energy storage phenomena do not.

Our reasoning is based on an assumption that models of physical system dynamics should properly reflect the way descriptions of physical phenomena depend on reference frames and should be compatible with thermodynamics. The across-through classification of variables does not meet these requirements. By contrast, the classification of variables based on the system-of-particles point of view that leads to an analogy between force, pressure, and voltage on the one hand and velocity, fluid flow, and current on the other not only satisfies these criteria, but is the least artificial from a common-sense point of view. We believe this facilitates communication and promotes insight, which are the ultimate benefits of using analogies.

Acknowledgments

Neville Hogan was supported in part by grant number AR40029 from the National Institutes of Health.

References

(1926). Models and analogies for demonstrating electrical principles, parts I-XIX. *The Engineer*, 142.

Breedveld, P.C. (1984). *Physical Systems Theory in Terms of Bond Graphs*, University of Twente, Enschede, Netherlands, ISBN 90-9000599-4 (distr. by author).

Darrieus, M. (1929). Les modeles mecaniques en electrotechnique. Leur application aux problemes de stabilite. *Bull. Soc. Franc. Electric.*, 36:729–809.

Feynman, R.P., Leighton, R.B., and Sands, M. (1963). *The Feynman Lectures on Physics, Volume II: Mainly Electromagnetism and Matter*, Addison-Wesley Publishing Company.

Firestone, F.A. (1933). A new analogy between mechanical and electrical system elements. *Journal of the Acoustic Society of America*, 3:249–267.

Hähnle, W. (1932). Die darstellung elektromechanischer gebilde durch rein elektrsiche schaltbilder. *Wissenschaftliche Veroffentl. Siemens Konzern*, 11:1–23.

Karnopp, D.C. and Rosenberg, R.C. (1975). *System Dynamics: A Unified Approach*, John Wiley.

Karnopp, D.C., Margolis, D.L., and Rosenberg, R.C. (1999). *System Dynamics: Modeling and Simulation of Mechatronic Systems*, 3rd ed., John Wiley & Sons

Maxwell, J.C. (1873). *Treatise on Electricity and Magnetism.*

Nickle, C.A. (1925). Oscillographic solutions of electro-mechanical systems. *Trans. A.I.E.E.*, 44:844–856.

Rowell, D. and Wormley, D.N. (1997). *System Dynamics: An Introduction*, Prentice-Hall, New Jersey.

Shearer, J.L., Murphy, A.T., and Richardson, H.H. (1967). *Introduction to System Dynamics*, Addison-Wesley Publishing Company.

Trent, H.M. (1955). Isomorphisms between oriented linear graphs and lumped physical systems. *Journal of the Acoustic Society of America*, 27:500–527.

Won, J. and Hogan, N. (1998). Coupled stability of non-nodic physical systems. *IFAC Symposium on Nonlinear Control Systems Design.*

Yazdi, N., Ayazi, F., and Najafi, K. (1998). Micromachined inertial sensors. *Proc. IEEE*, 86(8), 1640–1659.

Mechatronic Sensors and Actuators

17 **Introduction to Sensors and Actuators**
   M. Anjanappa, K. Datta, T. Song, Raghavendra Angara, and S. Li 17-1
   Introduction

18 **Fundamentals of Time and Frequency**
   Michael A. Lombardi 18-1
   Introduction • Time and Frequency Measurement • Time and Frequency Standards
   • Time and Frequency Transfer • Closing

19 **Sensor and Actuator Characteristics**
   Joey Parker 19-1
   Range • Resolution • Sensitivity • Error • Repeatability • Linearity and Accuracy
   Impedance • Nonlinearities • Static and Coulomb Friction • Eccentricity • Backlash
   Saturation • Deadband • System Response • First-Order System Response
   Underdamped Second-Order System Response • Frequency Response

20 **Sensors**
   *Kevin M. Lynch, Michael A. Peshkin, Halit Eren, M. A. Elbestawi, Ivan J. Garshelis,
   Richard Thorn, Pamela M. Norris, Bouvard Hosticka, Jorge Fernando Figueroa,
   H. R. (Bart)Everett, Stanley S. Ipson, Chang Liu, Nicolas Vazquez, and Dinesh Nair* 20-1
   Linear and Rotational Sensors • Acceleration Sensors • Force Measurement • Torque and
   Power Measurement • Flow Measurement • Temperature Measurements • Distance
   Measuring and Proximity Sensors • Light Detection, Image, and Vision Systems
   • Integrated Microsensors • Vision

21 **Actuators**
   *George T.-C. Chiu, Charles J. Fraser, Habil Ramutis Bansevicius, Rymantas Tadas Tolocka,
   Massimo Sorli, Stefano Pastorelli, and Sergey Edward Lyshevski* 21-1
   Electromechanical Actuators • Electrical Machines • Piezoelectric Actuators • Hydraulic
   and Pneumatic Actuation Systems • MEMS: Microtransducers Analysis, Design, and
   Fabrication

17
Introduction to Sensors and Actuators

M. Anjanappa
K. Datta
T. Song
Raghavendra Angara
S. Li

University of Maryland Baltimore County

17.1 Introduction 17-1
  Sensors • Actuators

17.1 Introduction

Sensors and actuators are two critical components of every closed loop control system. Such a system is also called a mechatronics system. A typical mechatronics system as shown in Figure 17.1 consists of a sensing unit, a controller, and an actuating unit. Sensing unit can be as simple as a single sensor or can consist of additional components such as filters, amplifiers, modulators, and other signal conditioners. The controller accepts the information from the sensing unit, makes decisions on the basis of the control algorithm, and outputs commands to the actuating unit. The actuating unit consists of an actuator and optionally a power supply and a coupling mechanism.

17.1.1 Sensors

Sensor is a device that when exposed to a physical phenomenon (temperature, displacement, force, etc.) produces a proportional output signal (electrical, mechanical, magnetic, etc.). The term transducer is often used synonymously with sensors. However, ideally, a sensor is a device that responds to a change in the physical phenomenon. On the other hand, a transducer is a device that converts one form of energy into another form of energy. Sensors are transducers when they sense one form of energy input and output in a different form of energy. For example, a thermocouple responds to a temperature change (thermal energy) and outputs a proportional change in electromotive force (electrical energy). Therefore, a thermocouple can be called as a sensor and or a transducer.

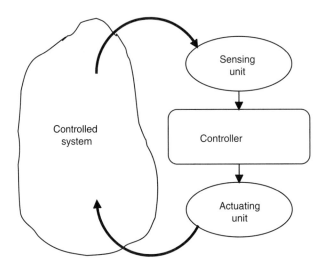

FIGURE 17.1 A typical mechatronics system.

17.1.1.1 Classification

Table 17.1 lists various types of sensors that are classified by their measurement objectives. Although this list is by no means exhaustive, it covers all the basic types including the new generation sensors such as smart material sensors, microsensors, and nanosensors.

Sensors can also be classified as *passive* or *active*. In passive sensors, the power required to produce the output is provided by the sensed physical phenomenon itself (such as a thermometer), whereas the active sensors require external power source (such as a strain gauge).

Furthermore, sensors are classified as *analog* or *digital* on the basis of the type of output signal. Analog sensors produce continuous signals that are proportional to the sensed parameter and typically require analog to digital conversion before feeding to the digital controller. Digital sensors on the other hand produce digital outputs that can be directly interfaced with the digital controller. Often, the digital outputs are produced by adding an analog to digital converter to the sensing unit. If many sensors are required, it is more economical to choose simple analog sensors and interface them to the digital controller equipped with a multichannel analog to digital converter.

17.1.1.2 Principle of Operation

Linear and Rotational Sensors: Linear and rotational position sensors are two of the most fundamental of all measurements used in a typical mechatronics system. The most common type position sensors are listed in Table 17.1. In general, the position sensors produce an electrical output that is proportional to the displacement they experience. There are contact type sensors such as strain gauge, LVDT, RVDT, tachometer, and so forth. The noncontact type includes encoders, Hall effect, capacitance, inductance, and interferometer type. They can also be classified on the basis of the range of measurement. Usually, the high-resolution type of sensors such as *Hall effect, fiber optic inductance, capacitance, and strain gauge* are suitable for only very small range (typically from 0.1 to 5 mm). The *differential transformers*, on the other hand, have a much larger range with good resolution. *Interferometer* type sensors provide both very high resolution (in terms of microns) and large range of measurements (typically up to a meter). However, interferometer type of sensors is bulky, expensive, and requires large setup time.

Among many linear displacement sensors, strain gauge provides high resolution at low noise level and is least expensive. A typical resistance strain gauge consists of resistive foil arranged as shown in Figure 17.2. A typical setup to measure the normal strain of a member loaded in tension is shown in Figure 17.3. Strain gauge #1 is bonded to the loading member, whereas strain gauge #2 is bonded to a second member made of same material, but not loaded. This arrangement compensates for any temperature effect. When

TABLE 17.1 Type of Sensors for Various Measurement Objectives

Sensors	Features
Linear/rotational sensors	
LVDT/RVDT Linear/rotational variable differential transducer	High resolution with wide range capability
	Very stable in static and quasistatic applications
Optical encoder	Simple, reliable, and low cost solution
	Good for both absolute and incremental measurements
Electrical tachometer	Resolution depends on type such as generator or magnetic pickups
Hall effect sensor	High accuracy over a small to medium range
Capacitive transducer	Very high resolution with high sensitivity
	Low power requirements
	Good for high-frequency dynamic measurements
Strain gauge elements.	Very high accuracy in small ranges
	Provides high resolution at low noise levels
Interferometer	Laser systems provide extremely high resolution in large ranges
	Very reliable and expensive
Magnetic pick up	Output is sinusoidal
Gyroscope	
Inductosyn	Very high resolution over small ranges
Acceleration sensors	
Seismic accelerometer	Good for measuring frequencies up to 40% of its natural frequency
Piezoelectric accelerometer	High sensitivity, compact, and rugged
	Very high natural frequency (100 kHz typical)
Force, torque, and pressure sensor	
Strain gauge Dynamometers/load cells	Good for both static and dynamic measurements
	They are also available as micro and nanosensors
Piezoelectric load cells	Good for high precision dynamic force measurements
Tactile sensor	Compact, has wide dynamic range
Ultrasonic stress sensor	Good for small force measurements
Flow sensors	
Pitot tube	Widely used as a flow rate sensor to determine speed in aircrafts
Orifice plate	Least expensive with limited range
Flow nozzle, venturi tubes	Accurate on wide range of flow
	More complex and expensive
Rotameter	Good for upstream flow measurements
	Used in conjunction with variable inductance sensor
Ultrasonic type	Good for very high flow rates
	Can be used for both upstream and downstream flow measurements
Turbine flow meter	Not suited for fluids containing abrasive particles
	Relationship between flow rate and angular velocity is linear
Electromagnetic flow meter	Least intrusive as it is noncontact type
	Can be used with fluids that are corrosive, contaminated, etc.
	The fluid has to be electrically conductive
Temperature sensors	
Thermocouples	This is the cheapest and the most versatile sensor
	Applicable over wide temperature ranges (-200 to $1200°C$ typical)
Thermistors	Very high sensitivity in medium ranges (up to $100°C$ typical)
	Compact but nonlinear in nature
Thermodiodes, thermo transistors	Ideally suited for chip temperature measurements
	Minimized self-heating
Resistance temperature detector (RTD)	More stable over a long period of time compared to thermocouple
	Linear over a wide range
Infrared type	Noncontact point sensor with resolution limited by wavelength.
Infrared thermography	Measures whole-field temperature distribution
Proximity sensors	
Inductance, eddy current, Hall effect, photoelectric, capacitance, etc.	Robust noncontact switching action
	The digital outputs are often directly fed to the digital controller

Continued

TABLE 17.1 Continued

Light sensors

Photoresistors, photodiodes, photo transistors, photo conductors, and so on	Measure light intensity with high sensitivity Inexpensive, reliable, and noncontact sensor
Charge-coupled diode	Captures digital image of a field of vision

Smart material sensors

Optical fiber	Alternate to strain gauges with very high accuracy and bandwidth
As strain sensor	Sensitive to the reflecting surface's orientation and status
As level sensor	Reliable and accurate
As force sensor	High resolution in wide ranges
As temperature sensor	High resolution and range (up to 2000°C)
Piezoelectric	Distributed sensing with high resolution and bandwidth
As strain sensor	Most suitable for dynamic applications
As force sensor	Least hysteresis and good setpoint accuracy
As accelerometer	
Magnetostrictive	Compact force sensor with high resolution and bandwidth
As force sensors	Good for distributed and noncontact sensing applications
As torque sensor	Accurate, high bandwidth, and noncontact sensor

Nuclear, biological, and chemical sensors

Nuclear sensors	Capable of detecting alpha, gamma, and x-rays with high sensitivity and in a wide range. Essentially counts the number of radioactive particles per minute
Biological sensors/biosensors	A specialized sensor that includes a biological entity. Has high sensitivity and selectivity in detecting glucose concentration, water contamination, airborne bacteria, pathogens, and toxic substance
Chemical sensor	Changing shape when exposed to chemicals
Chemimechanical	Changing resistance in response to volatile organic compounds, CCl_4, NH_3, NO_2, and so on
Chemiresistor	
Chemicapacitor	Changing dielectric constant in the vicinity of humidity, solvents, volatile organic compounds, toxic industrial chemicals, chemical warfare agents, and inorganic gases

Micro and nanosensors

Micro CCD image sensor	Small size, full-field image sensor
Fiberscope	Small (0.2 mm dia.) field vision scope using SMA coil actuators
Microultrasonic sensor	Detects flaws in small pipes
Microtactile sensor	Detects proximity between the end of catheter and blood vessels
MEMS accelerometers	High sensitivity, simple structure. Low power consumption, low cost, widely used in automotive industries
MEMS Gyros	Wide response range, easily mountable, low power consumption, low cost. Used in camcorders, robotics, and automobiles
MEMS pressure sensors	Fast response, high resolution, limited range. Uses piezoresistive and capacitive measuring methods
BioMEMS	Also called Lab-on-a-chip. Stand alone, onsite results, minimum sample consumption, and low cost

the member is loaded, the gauge #1 elongates thereby changing the resistance of the gauge. The change in resistance is transformed into a change in voltage by the voltage-sensitive Wheatstone bridge circuit. Assuming that the resistance of all four arms are equal initially, the change in output voltage (Δv_o) due to change in resistance (ΔR_1) of gauge #1 is

$$\frac{\Delta v_o}{v_i} = \frac{\Delta R_1/R}{4 + 2(\Delta R_1/R)}$$

Acceleration Sensors: Measurement of acceleration is important for systems subject to shock and vibration. Although acceleration can be derived from the time history data obtainable from linear or rotary sensors, the accelerometer whose output is directly proportional to the acceleration is preferred. Two

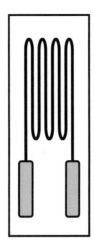

FIGURE 17.2 Bonded strain gauge.

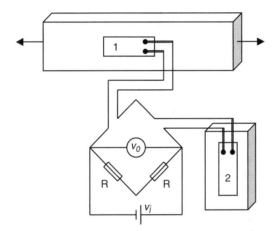

FIGURE 17.3 Experimental setup to measure normal strain using strain gauges.

common types include the *seismic mass* type and the *piezoelectric* accelerometer. The seismic mass type accelerometer is based on the relative motion between a mass and the supporting structure. The natural frequency of the seismic mass limits its use to low to medium frequency applications. The piezoelectric accelerometer, however, is compact and more suitable for high-frequency applications.

Force, Torque, and Pressure Sensors: Among many type of force/torque sensors, the *strain gauge dynamometers* and *piezoelectric type* are most common. Both are available to measure force and/or torque, either in one axis or multiple axes. The dynamometers make use of mechanical members that experience elastic deflection when loaded. These types of sensors are limited by their natural frequency. On the other hand, the piezoelectric sensors are particularly suitable for dynamic loadings in wide range of frequencies. They provide high stiffness, high resolution over a wide measurement range, and are compact.

Flow Sensors: Flow sensing is relatively a difficult task. The fluid medium can be liquid, gas, or a mixture of the two. Furthermore, the flow could be laminar or turbulent and can be a time-varying phenomenon. The *venturi meter* and *orifice plate* restrict the flow and use the pressure difference to determine the flow rate. The *pitot tube* pressure probe is another popular method of measuring flow rate. When positioned against the flow, they measure the total and static pressures. The flow velocity and in turn the flow rate can then be determined. The *rotameter* and the *turbine meters* when placed in the flow path, rotate at a

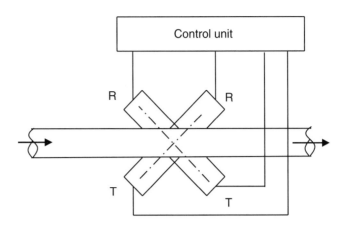

FIGURE 17.4 Ultrasonic flow sensor arrangement.

speed proportional to the flow rate. The *electromagnetic flow meters* use noncontact method. Magnetic field is applied in the transverse direction of the flow, and the fluid acts as the conductor to induce voltage proportional to the flow rate.

Ultrasonic flow meters measure fluid velocity by passing high-frequency sound waves through fluid. A schematic diagram of the ultrasonic flowmeter is as shown in Figure 17.4. The transmitters (T) provide the sound signal source. As the wave travels toward the receivers (R), its velocity is influenced by the velocity of the fluid flow due to the doppler effect. The control circuit compares the time to interpret the flow rate. This can be used for very high flow rates and also be used for both upstream and downstream flow. The other advantage is that it can be used for corrosive fluids, fluids with abrasive particles as it is like a noncontact sensor.

Temperature Sensors: A variety of devices are available to measure temperature. The most common devices are thermocouples, thermisters, resistance temperature detectors (RTDs), and infrared type.

Thermocouples are the most versatile, inexpensive, and have a wide range (up to 1200°C typical). Thermocouple simply consists of two dissimilar metal wires joined at the ends to create the sensing junction. When used in conjunction with a reference junction, the temperature difference between the reference junction and the actual temperature shows up as a voltage potential. *Thermisters* are semiconductor devices whose resistance changes as the temperature changes. They are good for very high sensitivity measurements in a limited range of up to 100°C. The relationship between the temperature and the resistance is nonlinear. The RTDs use the phenomenon that the resistance of a metal changes with temperature. They are, however, linear over a wide range and are most stable.

Infrared type sensors use the radiation heat to sense the temperature from a distance. These noncontact sensors can also be used to sense a field of vision to generate thermal map of a surface.

Proximity Sensors: They are used to sense the proximity of an object relative to another object. They usually provide a "on" or "off" signal indicating the presence or absence of an object. *Inductance, capacitance, photoelectric,* and *Hall effect* type are widely used as proximity sensors. Inductance proximity sensor consists of a coil wound around a soft iron core. The inductance of the sensor changes when a ferrous object is in its proximity. This change is converted to a voltage-triggered switch. Capacitance type is similar to inductance except the proximity of an object changes the gap and affects the capacitance. Photoelectric sensors are normally aligned with an infrared light source. The proximity of a moving object interrupts the light beam, causing the voltage level to change. Hall effect voltage is produced when a current-carrying conductor is exposed to transverse magnetic field. The voltage is proportional to transverse distance between the Hall effect sensor and an object in its proximity.

Light Sensors: Light intensity and full field vision are two important measurements used in many control applications. *Phototransistors, photoresistors,* and *photodiodes* are some of the more common type of light intensity sensors. A common photoresistor is made of cadmium sulphide, whose resistance is maximum

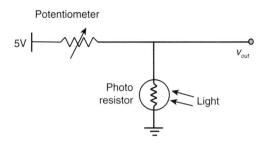

FIGURE 17.5 Light sensing with photoresistors.

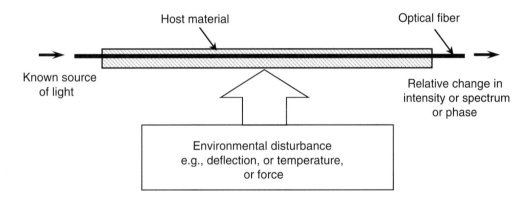

FIGURE 17.6 Principle of operation of optic fiber sensing.

when the sensor is in dark. When the photoresistor is exposed to light, its resistance drops in proportion to the intensity of light. When interfaced with a circuit as shown in Figure 17.5 and balanced, the change in light intensity will show up as change in voltage. These sensors are simple, reliable, and cheap, used widely for measuring light intensity.

Smart Material Sensors: There are many new smart materials that are gaining more applications as sensors, especially in distributed sensing circumstances. Of these, *optic fibers, piezoelectric,* and *magnetostrictive* materials have found applications. Within these, optic fibers are most used.

Optic fibers can be used to sense strain, liquid level, force, and temperature with very high resolution. Since they are economical for use as *in situ* distributed sensors on large areas, they have found numerous applications in smart structure applications such as damage sensors, vibration sensors, and cure-monitoring sensors. These sensors use the inherent material (glass and silica) property of optical fiber to sense the environment. Figure 17.6 illustrates the basic principle of operation of an embedded optic fiber used to sense displacement, force, or temperature. The relative change in the transmitted intensity or spectrum is proportional to the change in the sensed parameter.

Magnetostrictive sensor can be used to measure external load, force, pressure, vibration, temperature, and flow rates. These sensors sense the signal according to Villari effect. Figure 17.7 shows the general configuration of a C-shaped ferromagnetic sensor core wounded with the excitation and sensing coils. It can be used to detect the strain in target ferromagnetic materials. A change of strain in the target changes the magnetic circuit permeability, which shows up as a change in sensed voltage.

NBC Sensors: They have become extremely important for homeland security-related applications. The detection of nuclear, biological, and chemical agents is an active research area. The ultimate goal is to develop an all-in-one sensor that has multifunction and has a low cost.

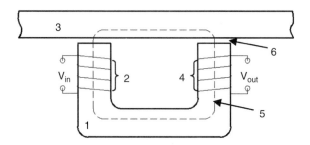

FIGURE 17.7 Magnetostrictive strain sensor. 1—Sensor core; 2—excitation coil; 3—strained ferromagnetic material; 4—sensing coil; 5—magnetic flux; 6—air gap (Teflon).

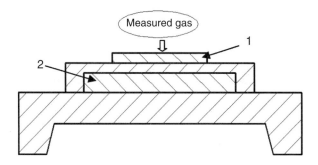

FIGURE 17.8 Metal oxide gas sensor. 1—metal oxide; 2—poly heater.

Nuclear sensors are used to detect all kinds of radiations such as alpha, gamma, beta, and x-rays. They have high sensitivity in a wide range of radiation levels. *Biological Sensors/Biosensors* that include a biological entity are used to detect the target substances such as blood glucose concentration, organophosphate, pesticides, and other health-related targets.

Chemical sensors are used to sense particular chemical components such as various gas species. It includes chemiresistor sensor, chemicapacitor sensor, chemimechanical sensor, and metal oxide gas sensor. Figure 17.8 shows a general metal oxide gas sensor diagram. Metal oxide changes electric resistance after absorbing certain gases. Catalyst deposition such as platinum (Pt) can speed up the reaction and hence increase the sensitivity of the sensor. The heater keeps the film at a constant temperature.

Micro and NanoSensors: Microsensors are the miniaturized version of the conventional macrosensors, with improved performance and reduced cost. Silicon micromachining technology has helped the development of many microsensors and continues to be one of the most active research and development topic in this area.

Vision microsensors have found applications in medical technology. A *fiberscope* of approximately 0.2 mm in diameter has been developed to inspect flaws inside tubes. Another example is a *microtactile sensor* that uses laser light to detect the contact between a catheter and the inner wall of blood vessels during insertion, which has sensitivity in the range of 1 mN.

Similarly, the progress made in the area of nanotechnology has fuelled the development of nanosensors. These are relatively new sensors that take one step further in the direction of miniaturization and are expected to open new avenues for sensing applications.

Micro and nanoeletromechanical systems, called MEMS and NEMS in short, are the state-of-the art devices in sensor technology. These devices are mainly fabricated in silicon and polymers with 3D lithographic features of various geometries by many established techniques like photolithography and surface micromachining. MEMS and NEMS have proved to be efficient, cost effective, faster, and easily mountable, with low power consumption. Their advantages over the conventional sensors enable them to rapidly widen their applications in automobile industry, drug testing laboratory, and consumer electronics.

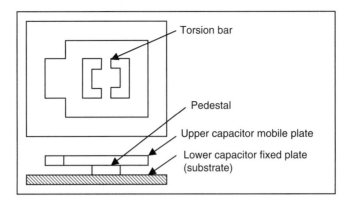

FIGURE 17.9 MEMS accelerometer.

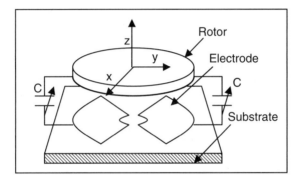

FIGURE 17.10 MEMS gyroscope.

MEMS accelerometers measure the inertial acceleration when the seismic mass of the sensor moves. Some of these MEMS devices use the principle of measuring acceleration by sensing piezoresistive stresses and others measure by sensing the change of capacitance. Figure 17.9 shows the schematic of a micromachinable accelerometer based on capacitive sensing. The device is fabricated as two layers each to act as a plate of parallel plate capacitor with top plate supported by a pedestal. Upper plate is made asymmetrical so that the center of mass falls out of the torsion bar. When the upper plate rotates along the axis of the torsion bar, it increases the distance between the plates on one side and decreases on other side. The acceleration of the body causing the motion can be measured by sensing the capacitance difference between the left and right side of the torsion bar.

There are many types of *MEMS gyros* based on the principle of their working, such as tuning fork gyro, rotating wheel gyro, wine glass resonator gyro, and Foucault pendulum gyro. Figure 17.10 shows the working principle of a rotating wheel gyro, which is similar to that of accelerometer. In this type of gyro, the top wheel is driven to vibrate about its axis of symmetry; rotation about any of the in-plane axes results in the wheel's tilting, which results in a change in capacitance that can be detected with capacitive electrodes under the wheel. With this device, it is possible to sense two axes of rotation with a single vibrating wheel.

Figure 17.11 shows a *MEMS pressure sensor* that measures pressure by sensing the capacitance. The device has a glass fixed plate and a parallel movable plate with each plate coated with a conducting material. The two layers of conducting material act as two plates of a parallel plate capacitor. When an external pressure acts on the movable plate, it changes the clearance between the plates and in turn changes the capacitance. The pressure applied can be estimated by measuring the new capacitance.

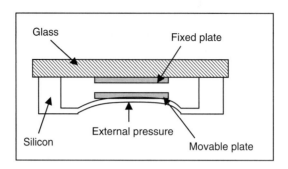

FIGURE 17.11 MEMS pressure sensor.

BioMEMS can be defined as MEMS devices used for processing, delivery, manipulation, and analysis of biological and chemical entities. Micro devices integrated with BioMEMS systems known as lab-on-a-chip and micro-total analysis systems (micro-TAS or ATAS) already have wide variety of applications. Biosensors are analytical devices that combine a biologically sensitive material with a physical or chemical transducer to selectively and quantitatively sense the presence of specific chemical compounds in a given environment. For example, amperometric biosensors detect by measuring the electric current associated with the electrons involved in redox processes, whereas potentiometric biosensors measure a change in potential at electrodes due to ions or chemical reactions at an electrode, and conductometric biosensors measure conductance changes associated with changes in the overall ionic medium between the electrodes.

17.1.1.3 Selection Criteria

A number of static and dynamic factors must be considered in selecting a suitable sensor to measure the desired physical parameter. Following is a list of typical factors:

Range	Difference between the maximum and minimum value of the sensed parameter
Resolution	The smallest change the sensor can differentiate
Accuracy	Difference between the measured value and the true value
Precision	Ability to reproduce repeatedly with a given accuracy
Sensitivity	Ratio of change in output to a unit change of the input
Zero offset	A nonzero value output for no input
Linearity	Percentage of deviation from the best-fit linear calibration curve
Zero Drift	The departure of output from zero value over a period of time for no input
Response time	The time lag between the input and output
Bandwidth	Frequency at which the output magnitude drops by 3 dB
Resonance	The frequency at which the output magnitude peak occurs
Operating temperature	The range in which the sensor performs as specified
Deadband	The range of input for which there is no output
Signal to noise ratio	Ratio between the magnitudes of the signal and the noise at the output

Choosing a sensor that satisfies all the above to the desired specification is difficult, at best. For example, finding a position sensor with micrometer resolution over a range of a meter eliminates most of the sensors. Many times, the lack of a cost-effective sensor necessitates redesigning the mechatronics system. It is therefore advisable to take a system-level approach when selecting a sensor, and avoid choosing it in isolation.

Once the above-referred functional factors are satisfied, a short list of sensors can be generated. The final selection will then depend on the size, extent of signal conditioning, reliability, robustness, maintainability, and cost.

17.1.1.4 Signal Conditioning

Normally, the output from a sensor requires postprocessing of the signals before they can be fed to the controller. The sensor output may have to be demodulated, amplified, filtered, linearized, range quantized, and isolated so that the signal can be accepted by a typical analog to digital converter of the controller. Some sensors are available with integrated signal conditioners such as the microsensors. All the electronics are integrated into one microcircuit and can be directly interfaced with the controllers.

17.1.1.5 Calibration

The sensor manufacturer usually provides the calibration curves. If the sensors are stable with no drift, there is no need to recalibrate. However, often the sensor may have to be recalibrated after integrating it with signal conditioning system. This essentially requires that a known input signal is provided to the sensor and its output recorded to establish a correct output scale. This process proves the ability to measure reliably and enhances the confidence.

If the sensor is used to measure a time-varying input, then dynamic calibration becomes necessary. Use of sinusoidal inputs is the most simple and reliable way of dynamic calibration. However, if generating sinusoidal input becomes impractical (e.g., temperature signals), then a step input can substitute for the sinusoidal signal. The transient behavior of step response should yield sufficient information about the dynamic response of the sensor.

17.1.2 Actuators

Actuators are basically the muscle behind a mechatronics system that accepts a control command (mostly in the form of an electrical signal) and produces a change in the physical system by generating force, motion, heat, flow, and so forth. Normally, the actuators are used in conjunction with the power supply and a coupling mechanism as shown in Figure 17.12. The power unit provides either ac or dc power at the rated voltage and current. The coupling mechanism acts as the interface between the actuator and the physical system. Typical mechanisms include rack and pinion, gear drive, belt drive, lead screw and nut, piston, and linkages.

17.1.2.1 Classication

Actuators can be classified on the basis of the type of energy as listed in Table 17.2. The table, although not exhaustive, lists all the basic types. They are essentially of electrical, electromechanical, electromagnetic, pneumatic, or hydraulic type. The new generations of actuators include smart material actuators, microactuators, and nanoactuators.

Actuators can also be classified as *binary* and *continuous* on the basis of the number of stable state outputs. A relay with two stable states is a good example of a binary actuator. Similarly, a stepper motor

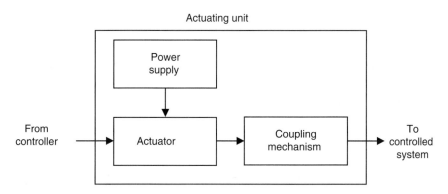

FIGURE 17.12 A typical actuating unit.

TABLE 17.2 Types of Actuators and Their Features

Actuator			Features
Electrical			
Diodes, thyristor, bipolar transistor, triacs, diacs, power MOSFET, solid state relay, etc.			Electronic type
			Very high-frequency response
			Low power consumption
Electromechanical			
Direct current Motor	Wound field	Separately excited	Speed can be controlled either by the voltage across the armature winding or by varying the field current
		Shunt	Constant-speed application
		Series	High starting torque, high acceleration torque, high speed with light load
		Compound	Low starting torque, good speed regulation
			Instability at heavy loads
	Permanent magnet	Conventional PM Motor	High efficiency, high peak power, and fast response
		Moving-coil PM motor	Higher efficiency and lower inductance than conventional dc motor
		Torque motor	Designed to run for long periods in a stalled or a low rpm condition
	Electronic commutation (Brushless motor)		Fast response
			High efficiency, often exceeding 75%
			Long life, high reliability, no maintenance needed
			Low radio frequency interference and noise production
Alternate current motor	Alternate current induction motor		The most commonly used motor in industry
			Simple, rugged, and inexpensive
	Alternate current synchronous motor		Rotor rotates at synchronous speed
			Very high efficiency over a wide range of speeds and loads
			Needs an additional system to start
	Universal motor		Can operate in dc or ac
			Very high horsepower per pound ratio
			Relatively short operating life
Stepper motor	Hybrid		Change electrical pulses into mechanical movement
			Provide accurate positioning without feedback
			Low maintenance
	Variable reluctance		
Electromagnetic			
Solenoid-type devices			Large force, short duration
Electromagnets, relay			On/Off control
Hydraulic and pneumatic			
Cylinder			Suitable for linear movement
Hydraulic motor	Gear type		Wide speed range
	Vane type		High horsepower output
	Piston type		High degree of reliability
Air motor	Rotary type		No electric shock hazard
	Reciprocating		Low maintenance
Valves	Directional Control Valves		
	Pressure Control Valves		
	Process Control Valves		
Smart material actuators			
Piezoelectric and electrostrictive			High frequency with small motion
			High voltage with low current excitation
			High resolution
Magnetostrictive			High frequency with small motion
			Low voltage with high current excitation
Shape memory alloy			Low voltage with high current excitation
			Low frequency with large motion

Continued

TABLE 17.2 Continued

Electrorheological fluids	Very high voltage excitation
	Good resistance to mechanical shock and vibration
	Low frequency with large force
Ultrasonic piezo motor	Intrinsic steady-state auto-locking capability, no servo dithering and heat generation
Micro- and Nanoactuators	
Micromotors	Suitable for micromechanical system
	Can use available silicon processing technology
MEMS thin film optical switches	Reduced size, low power requirements, high frequency
MEMS mirror deflectors	Low power consumption, high frequency
MEMS fluidic pumps and valves	Ideal for very low volume and precise manipulation of fluids. Typically, both force and stroke are small
NEMS drug dispensers	Physiological—stimuli based. Accurate, precise, and typically dispensed directly into blood stream

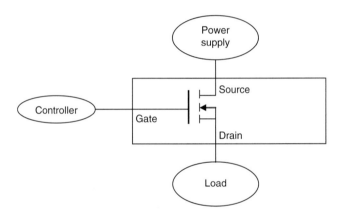

FIGURE 17.13 n-Channel power MOSFET.

is a good example of continuous actuator. When used for a position control, the stepper motor can provide stable outputs with very small incremental motion.

17.1.2.2 Principle of Operation

Electrical Actuators: Electrical switches are the choice of actuators for most of the on–off type control action. Switching devices such as *diodes, transistors, triacs, MOSFET,* and *relays* accept a low energy level command signal from the controller and switch on or off electrical devices such as motors, valves, and heating elements. For example, a MOSFET switch is shown in Figure 17.13. The gate terminal receives the low energy control signal from the controller that makes or breaks the connection between the power supply and the actuator load. When switches are used, the designer must make sure that *switch bounce* problem is eliminated either by hardware or software.

Electromechanical Actuators: The most common electromechanical actuator is a motor that converts electrical energy to mechanical motion. Motor is the principal means of converting electrical energy into mechanical energy in industry. Broadly they can be classified as *dc motors, ac motors,* and *stepper motors*. Direct current motors operate on dc voltage, and varying the voltage can easily control their speed. They are widely used in applications ranging from thousands of horsepower motors used in rolling mills to fractional horsepower motors used in automobiles (starter motors, fan motors, windshield wiper motors, etc.). They are, however, costlier, need dc power supply, and requiremore maintenance compared to ac motors.

The governing equation of motion of a dc motor can be written as:

$$T = J\frac{d\omega}{dt} + T_L + T_{\text{LOSS}}$$

where T is torque, J is the, total inertia, ω is the angular mechanical speed of the rotor, T_L is the torque applied to the motor shaft, and T_{loss} is the internal mechanical losses such as friction.

Alternate current motors are the most popular motors since they use standard ac power, do not require brushes and commutator, and are therefore less expensive. Alternate current motors can be further classified as the *induction motors, synchronous motors,* and *universal motors* according to their physical construction. The induction motor is simple, rugged, and maintenance free. They are available in many sizes and shapes on the basis of the number of phases used. For example, a three-phase induction motor is used in large horsepower applications, such as pump drives, steel mill drives, hoist drives, and vehicle drives. The two-phase servomotor is used extensively in position control system. Single-phase induction motors are widely used in many household appliances. The synchronous motor is one of the most efficient electrical motors in industry, so it is used in industry to reduce the cost of electrical power. In addition, synchronous motor rotates at synchronous speed, so it is also used in applications that require synchronous operations. The universal motors operate with either ac or dc power supply. It is normally used in fractional horsepower application. The dc universal motor has the highest horsepower-per-pound ratio, but has a relatively short operating life.

The *stepper motor* is a discrete (incremental) positioning device that moves one step at a time for each pulse command input. Since they accept direct digital commands and produce a mechanical motion, the stepper motors are used widely in industrial control applications. They are mostly used in fractional horsepower applications. With the rapid progress in low cost and high-frequency solid-state drives, they are finding increased applications.

Figure 17.14 shows a simplified unipolar stepper motor. The winding-1 is between the top and bottom stator pole, and the winding-2 is between the left and right motor poles. The rotor is a permanent magnet with six poles resulting in a single-step angle of 30°. With appropriate excitation of winding-1, the top stator pole becomes a north pole and the bottom stator pole becomes a south pole. This attracts the rotor into the position as shown. Now if the winding-1 is de-energized and winding-2 is energized, the rotor will turn 30°. With appropriate choice of current flow through winding-2, the rotor can be rotated either clockwise or counterclockwise. By exciting the two windings in sequence, the motor can be made to rotate at a desired speed continuously.

Electromagnetic Actuators: Solenoid is the most common electromagnetic actuator. Direct current solenoid actuator consists of a soft iron core enclosed within a current-carrying coil. When the coil is energized, a magnetic field is established that provides the force to push or pull the iron core. Alternate current solenoid devices are also encountered, such as ac excitation relay.

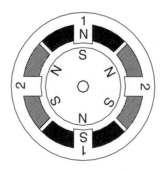

FIGURE 17.14 Unipolar stepper motor.

A solenoid-operated directional control valve is shown in Figure 17.15. Normally, due to the spring force the soft iron core is pushed to the extreme left position as shown. When the solenoid is excited, the soft iron core will move to the right extreme position, thus providing the electromagnetic actuation.

Another important type is the *electromagnet*. The electromagnets are used extensively in applications that require large forces.

Hydraulic and Pneumatic Actuators: Hydraulic and pneumatic actuators are normally either *rotary motors* or *linear piston/cylinder* or *control valves*. They are ideally suited for generating very large forces coupled with large motion. Pneumatic actuators use air under pressure that is most suitable for low to medium force, short stroke, and high-speed applications. Hydraulic actuators use pressurized oil that is incompressible. They can produce very large forces coupled with large motion in a cost-effective manner. The disadvantage with the hydraulic actuators is that they are more complex and need more maintenance.

The rotary motors are usually used in applications where low speed and high torque are required. The cylinder/piston actuators are suited for application of linear motion such as aircraft flap control. Control valves in the form of directional control valves are used in conjunction with rotary motors and cylinders to control the fluid flow direction as shown in Figure 17.15. In this solenoid-operated directional control valve, the valve position dictates the direction motion of the cylinder/piston arrangement.

Smart Material Actuators: Unlike the conventional actuators, the smart material actuators typically become part of the load-bearing structures. This is achieved by embedding the actuators in a distributed manner and integrating into the load-bearing structure that could be used to suppress vibration, cancel the noise, and change shape. Of the many smart material actuators, *shape memory alloys* (SMA), *piezoelectric* (PZT), *magnetostrictive*, *electrorheological fluids*, and *ion exchange polymers* are most common.

Shape memory alloys are alloys of nickel and titanium, which undergo phase transformation when subjected to thermal field. The SMAs are also known as NITINOL for Nickel Titanium Naval Ordnance Laboratory. When cooled below a critical temperature, their crystal structure enters martensitic phase as shown in Figure 17.16. In this state, the alloy is plastic and can easily be manipulated. When the alloy is heated above the critical temperature (in the range of 50–80°C), the phase changes to austenitic phase. Here the alloy resumes the shape that it formally had at the higher temperature. For example, a straight wire at room temperature can be made to regain its programmed semicircle shape when heated, which has found applications in orthodontics and other tensioning devices. The wires are typically heated by passing a current (up to several amperes at very low voltage, 2–10 V typical).

The *piezoelectric* actuators are essentially piezocrystals with top and bottom conducting films as shown in Figure 17.17. When an electric voltage is applied across the two conducting films, the crystal expands in the transverse direction as shown by the dotted lines. When the voltage polarity is reversed, the crystal contracts thereby providing bidirectional actuation. The interaction between the mechanical and electrical

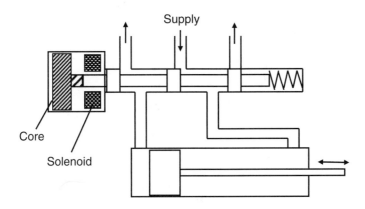

FIGURE 17.15 Solenoid-operated directional control valve.

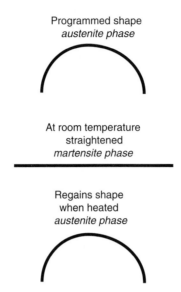

FIGURE 17.16 Phase changes of shape memory alloy.

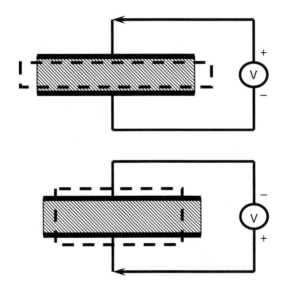

FIGURE 17.17 Piezoelectric actuator.

behavior of the piezoelectric materials can be expressed as

$$T = c^E S - eE$$

where T is the stress, c^E is the elastic coefficient at constant electric field, S is the strain, e is the dielectric permitivity, and E is the electric field.

One application of these actuators is as shown in Figure 17.18. The two piezoelectric patches are excited with opposite polarity to create transverse vibration in the cantilever beam. These actuators provide high bandwidth (0–10 kHz typical) with small displacement. Since there are no moving parts to the actuator, it is compact and ideally suited for micro and nanoactuation.

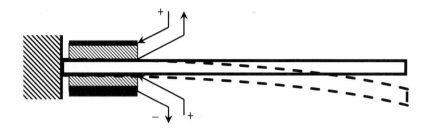

FIGURE 17.18 Vibration of beam using piezoelectric actuators.

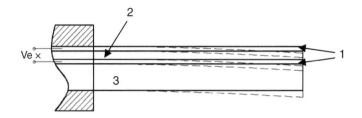

FIGURE 17.19 Piezoelectric thin film actuator. 1—Top and bottomelectrodes; 2—piezoelectric thin film; 3—Silicon structure.

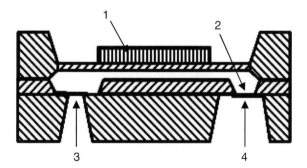

FIGURE 17.20 Silicon cantilever check valve pump. 1—Piezo disk; 2—check valve; 3—inlet; 4—outlet.

Unlike the bidirectional actuation of piezoelectric actuators, the *electrostriction* effect is a second order effect, that is, it responds to an electric field with unidirectional expansion regardless of polarity.

Figure 17.19 illustrates a piezo thin film actuator. Two conductors that served as top and bottom electrodes are attached on the piezo material. When an excitation voltage is applied across the conductors, the piezo material expands or contracts along the cantilever length direction. Thus the silicon substrate deflects downwards or upwards. Figure 17.20 shows a typical micropump that uses piezoelectric material as the microactuator. When the voltage is applied across the piezo disk, the disk deforms to either increase or decrease the volume of the chamber causing the fluid to either enter or leave the chamber through the check valves. It can operate at high frequencies in KHz range.

Magnetostrictive material is an alloy of terbium, dysprosium, and iron, which generates mechanical strains up to 2000 microstrain in response to applied magnetic fields. They are available in the form of rods, plates, washers, and powder. Figure 17.21 shows a typical magnetostrictive rod actuator that is surrounded by a magnetic coil. When the coil is excited, the rod elongates in proportion to the intensity of the magnetic field established.

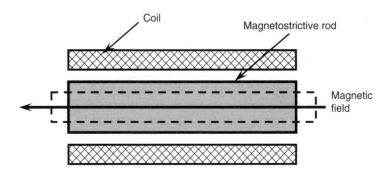

FIGURE 17.21 Magnetostrictive rod actuator.

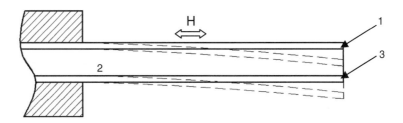

FIGURE 17.22 Magnetostrictive thin film actuator. 1—Tb–Fe; 2—substrate; 3—Sm–Fe.

The magneto–mechanical relationship is given as

$$\varepsilon = S^H \sigma + dH$$

where ε is the strain, S^H is the compliance at constant magnetic filed, σ is the stress, d is the magnetostriction constant, and H is the magnetic field intensity.

Figure 17.22 shows the schematic view of the magnetostrictive thin film actuator. On the substrate (silicon, polyimide, gallium arsenide, etc.), Tb–Fe thin film is sputtered and on the opposite side Sm–Fe thin film is sputtered. When a magnetic field is applied parallel to the cantilever length direction, Tb–Fe film expands and Sm–Fe film contracts in the length direction; thus, the cantilever deflects downwards. In the same way, when a magnetic field is applied parallel to the cantilever width direction, it deflects upwards.

Ion exchange polymers exploit the electro-osmosis phenomenon of the natural ionic polymers for purposes of actuation. When a voltage potential is applied across the cross-linked polyelectrolytic network, the ionizable groups attain a net charge generating a mechanical deformation. These types of actuators have been used to develop artificial muscles and artificial limbs. The primary advantage is their capacity to produce large deformation with a relatively low voltage excitation.

Micro and Nanoactuators: Microactuators, also called micromachines, MEMS, and microsystems, are the tiny mobile devices being developed utilizing the standard microelectronics processes with the integration of semiconductors and machined mircomechanical elements. Another definition states that any device produced by assembling extremely small functional parts of around 1–15 mm is called a mircomachine.

In *electrostatic motor* electrostatic force is dominant, unlike the conventional motors that are based on magnetic forces. For smaller micromechanical systems the electrostatic forces are well suited as an actuating force. Figure 17.23 shows one type of electrostatic motor. The rotor is an annular disk with uniform permitivity and conductivity. In operation, a voltage is applied to the two conducting parallel plates separated by an insulation layer. The rotor rotates with a constant velocity between the two coplanar concentric arrays of stator electrodes.

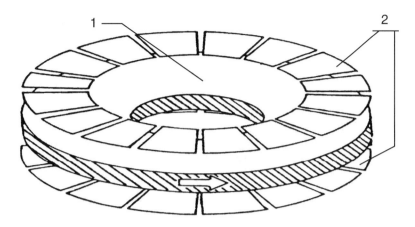

FIGURE 17.23 Electrostatic motor. 1—Rotor; 2—stator electrodes.

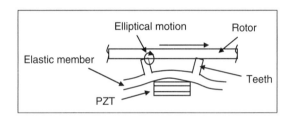

FIGURE 17.24 Ultrasonic motor.

In *ultrasonic motors*, the stator of the motor is arranged with a series of PZT crystals along the circumference at regular distances. Each of the stator PZT is made to excite in a fashion that the combined effect of all the PZTs generate a wave motion at their surface. The generated wave motion is propagated to the elastic member of the device, which is closely packed to the PZT crystals. Teeth of the elastic member magnify the wave and then transmit the motion to the rotor in contact. The detailed view of the mechanism at the tip of the teeth is given in Figure 17.24. The tip of the teeth makes an elliptical motion; friction between tip of the teeth and the rotor makes it rotate in a direction tangential to the ellipse generated. Frequency of excitation of the PZTs determines the speed of the rotor. However, the movement depends on many other factors such as dimensions of the elliptical shape generated, number of teeth, and distance between the teeth.

One of the targeted applications of NEMS is in *NEMS drug dispensing devices*. This technology allows a nanofabricated drug delivery device with physiologically directed accurate delivery mechanism. The main approach to achieve this is by fabricating nanocontainers capable of releasing drugs in the body in response to stimuli. Further, as this device is systemically injectable into the body, no surgery is required. As this is stimuli-based delivery, added advantage of this novel device is that it dispenses drug only to the cells that need treatment, avoiding any uncalled delivery to the normal cells.

17.1.2.3 Selection Criteria

The selection of proper actuator is more complicated than the sensors, primarily due to their effect on the dynamic behavior of the overall system. Furthermore, the selection of actuator dominates the power needs and the coupling mechanisms of the entire system. The coupling mechanism can sometime be completely avoided if the actuator provides the output that can be directly interfaced to the physical system. For example, choosing a linear motor in place of a rotary motor can eliminate the need of a coupling mechanism to convert rotary motion to linear motion. In general, the following performance parameters must be addressed before choosing an actuator for a specific need:

Continuous power output: The maximum force/torque attainable continuously without exceeding the temperature limits.
Range of motion: The range of linear/rotary motion.
Resolution: The minimum increment of force/torque attainable.
Accuracy: Linearity of the relationship between the input and output.
Peak force/torque: The force/torque at which the actuator stalls.
Heat dissipation: Maximum wattage of heat dissipation in continuous operation.
Speed characteristics: Force/torque versus speed relationship.
No load speed: Typical operating speed/velocity with no external load.
Frequency response: The range of frequency over which the output follows the input faithfully, applicable to linear actuators.
Power requirement: Type of power (ac or dc), number of phases, voltage level, and current capacity.

In addition to the above-referred criteria, many other factors become important depending on the type of power and the coupling mechanism required. For example, if a rack and pinion coupling mechanism is chosen, the backlash and friction will affect the resolution of the actuating unit.

18

Fundamentals of Time and Frequency

Michael A. Lombardi
National Institute of Standards and Technology

18.1 Introduction 18-1
    Coordinated Universal Time
18.2 Time and Frequency Measurement 18-2
    Accuracy • Stability
18.3 Time and Frequency Standards 18-9
    Quartz Oscillators • Rubidium Oscillators • Cesium Oscillators
18.4 Time and Frequency Transfer 18-13
    Fundamentals of Time and Frequency Transfer • Radio Time and Frequency Transfer Signals
18.5 Closing 18-17
References 18-17

18.1 Introduction

Time and frequency standards supply three basic types of information: *time-of-day*, *time interval*, and *frequency*. Time-of-day information is provided in hours, minutes, and seconds, but often also includes the *date* (month, day, and year). A device that displays or records time-of-day information is called a *clock*. If a clock is used to label when an event happened, this label is sometimes called a *time tag* or *time stamp*. Date and time-of-day can also be used to ensure that events are *synchronized*, or happen at the same time.

Time interval is the duration or elapsed time between two events. The standard unit of time interval is the second(s). However, many engineering applications require the measurement of shorter time intervals, such as milliseconds (1 ms = 10^{-3} s), microseconds (1 μs = 10^{-6} s), nanoseconds (1 ns = 10^{-9} s), and picoseconds (1 ps = 10^{-12} s). Time is one of the seven base physical quantities, and the second is one of seven base units defined in the International System of Units (SI). The definitions of many other physical quantities rely upon the definition of the second. The second was once defined based on the earth's rotational rate or as a fraction of the tropical year. That changed in 1967 when the era of atomic time keeping formally began. The current definition of the SI second is:

> The duration of 9,192,631,770 periods of the radiation corresponding to the transition between two hyperfine levels of the ground state of the cesium-133 atom.

Frequency is the rate of a repetitive event. If T is the period of a repetitive event, then the frequency f is its reciprocal, $1/T$. Conversely, the period is the reciprocal of the frequency, $T = 1/f$. Since the period is a time interval expressed in seconds (s), it is easy to see the close relationship between time interval and frequency. The standard unit for frequency is the hertz (Hz), defined as events or cycles per second. The frequency of electrical signals is often measured in multiples of hertz, including kilohertz (kHz), megahertz (MHz), or gigahertz (GHz), where 1 kHz equals one thousand (10^3) events per second, 1 MHz

TABLE 18.1 Uncertainties of Physical Realizations of the Base SI Units

SI Base Unit	Physical Quantity	Uncertainty
Candela	Luminous intensity	1×10^{-4}
Kelvin	Temperature	3×10^{-7}
Mole	Amount of substance	8×10^{-8}
Ampere	Electric current	4×10^{-8}
Kilogram	Mass	1×10^{-8}
Meter	Length	1×10^{-12}
Second	Time interval	1×10^{-15}

equals one million (10^6) events per second, and 1 GHz equals one billion (10^9) events per second. A device that produces frequency is called an *oscillator*. The process of setting multiple oscillators to the same frequency is called *syntonization*.

Of course, the three types of time and frequency information are closely related. As mentioned, the standard unit of time interval is the second. By counting seconds, we can determine the date and the time-of-day. And by counting events or cycles per second, we can measure frequency.

Time interval and frequency can now be measured with less uncertainty and more resolution than any other physical quantity. Today, the best time and frequency standards can realize the SI second with uncertainties of $\cong 1 \times 10^{-15}$. Physical realizations of the other base SI units have much larger uncertainties, as shown in Table 18.1 [1–5].

18.1.1 Coordinated Universal Time (UTC)

The world's major metrology laboratories routinely measure their time and frequency standards and send the measurement data to the Bureau International des Poids et Measures (BIPM) in Sevres, France. The BIPM averages data collected from more than 200 atomic time and frequency standards located at more than 40 laboratories, including the National Institute of Standards and Technology (NIST). As a result of this averaging, the BIPM generates two time scales, International Atomic Time (TAI), and Coordinated Universal Time (UTC). These time scales realize the SI second as closely as possible.

UTC runs at the same frequency as TAI. However, it differs from TAI by an integral number of seconds. This difference increases when *leap seconds* occur. When necessary, leap seconds are added to UTC on either June 30 or December 31. The purpose of adding leap seconds is to keep atomic time (UTC) within ±0.9 s of an older time scale called UT1, which is based on the rotational rate of the earth. Leap seconds have been added to UTC at a rate of slightly less than once per year, beginning in 1972 [3,5].

Keep in mind that the BIPM maintains TAI and UTC as "paper" time scales. The major metrology laboratories use the published data from the BIPM to steer their clocks and oscillators and generate real-time versions of UTC. Many of these laboratories distribute their versions of UTC via radio signals, which are discussed in section 18.4.

You can think of UTC as the ultimate standard for time-of-day, time interval, and frequency. Clocks synchronized to UTC display the same hour, minute, and second all over the world (and remain within one second of UT1). Oscillators syntonized to UTC generate signals that serve as reference standards for time interval and frequency.

18.2 Time and Frequency Measurement

Time and frequency measurements follow the conventions used in other areas of metrology. The frequency standard or clock being measured is called the *device under test* (DUT). A measurement compares the DUT to a *standard* or *reference*. The standard should outperform the DUT by a specified ratio, called the *test uncertainty ratio* (TUR). Ideally, the TUR should be 10:1 or higher. The higher the ratio, the less averaging is required to get valid measurement results.

Fundamentals of Time and Frequency

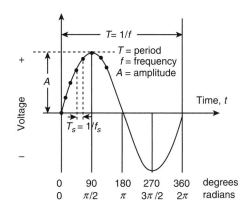

FIGURE 18.1 An oscillating sine wave.

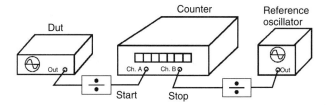

FIGURE 18.2 Measurement using a time interval counter.

The test signal for time measurements is usually a pulse that occurs once per second (1 pps). The pulse width and polarity varies from device to device, but TTL levels are commonly used. The test signal for frequency measurements is usually at a frequency of 1 MHz or higher, with 5 or 10 MHz being common. Frequency signals are usually sine waves, but can also be pulses or square waves. If the frequency signal is an oscillating sine wave, it might look like the one shown in Figure 18.1. This signal produces one cycle (360° or 2π radians of phase) in one period. The signal amplitude is expressed in volts, and must be compatible with the measuring instrument. If the amplitude is too small, it might not be able to drive the measuring instrument. If the amplitude is too large, the signal must be attenuated to prevent overdriving the measuring instrument.

This section examines the two main specifications of time and frequency measurements—*accuracy* and *stability*. It also discusses some instruments used to measure time and frequency.

18.2.1 Accuracy

Accuracy is the degree of conformity of a measured or calculated value to its definition. Accuracy is related to the offset from an ideal value. For example, *time offset* is the difference between a measured on-time pulse and an ideal on-time pulse that coincides exactly with UTC. *Frequency offset* is the difference between a measured frequency and an ideal frequency with zero uncertainty. This ideal frequency is called the *nominal frequency*.

Time offset is usually measured with a *time interval counter* (*TIC*), as shown in Figure 18.2. A TIC has inputs for two signals. One signal starts the counter and the other signal stops it. The time interval between the start and stop signals is measured by counting cycles from the time base oscillator. The resolution of a low cost TIC is limited to the period of its time base. For example, a TIC with a 10-MHz time base oscillator would have a resolution of 100 ns. More elaborate TICs use interpolation schemes to detect parts of a time base cycle and have much higher resolution—1 ns resolution is commonplace, and 20 ps resolution is available.

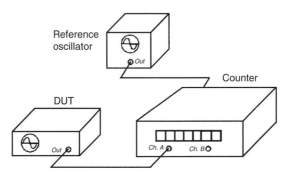

FIGURE 18.3 Measurement using a frequency counter.

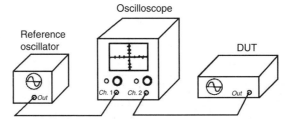

FIGURE 18.4 Phase comparison using an oscilloscope.

Frequency offset can be measured in either the *frequency domain* or *time domain*. A simple frequency domain measurement involves directly counting and displaying the frequency output of the DUT with a *frequency counter*. The reference for this measurement is either the counter's internal time base oscillator, or an external time base (Figure 18.3). The counter's resolution, or the number of digits it can display, limits its ability to measure frequency offset. For example, a 9-digit frequency counter can detect a frequency offset no smaller than 0.1 Hz at 10 MHz (1×10^{-8}). The frequency offset is determined as

$$f(\text{offset}) = \frac{f_{\text{measured}} - f_{\text{nominal}}}{f_{\text{nominal}}}$$

where f_{measured} is the reading from the frequency counter, and f_{nominal} is the frequency labeled on the oscillator's nameplate, or specified output frequency.

Frequency offset measurements in the time domain involve a *phase comparison* between the DUT and the reference. A simple phase comparison can be made with an oscilloscope (Figure 18.4). The oscilloscope will display two sine waves (Figure 18.5). The top sine wave represents a signal from the DUT, and the bottom sine wave represents a signal from the reference. If the two frequencies were exactly the same, their phase relationship would not change and both would appear to be stationary on the oscilloscope display. Since the two frequencies are not exactly the same, the reference appears to be stationary and the DUT signal moves. By measuring the rate of motion of the DUT signal we can determine its frequency offset. Vertical lines have been drawn through the points where each sine wave passes through zero. The bottom of the figure shows bars whose width represents the phase difference between the signals. In this case the phase difference is increasing, indicating that the DUT is lower in frequency than the reference.

Measuring high accuracy signals with an oscilloscope is impractical, since the phase relationship between signals changes very slowly and the resolution of the oscilloscope display is limited. More precise phase comparisons can be made with a TIC, using a setup similar to Figure 18.2. If the two input signals have the same frequency, the time interval will not change. If the two signals have different frequencies,

Fundamentals of Time and Frequency

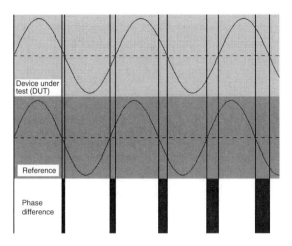

FIGURE 18.5 Two sine waves with a changing phase relationship.

the time interval will change, and the rate of change is the frequency offset. The resolution of a TIC determines the smallest frequency change that it can detect without averaging. For example, a low cost TIC with a single-shot resolution of 100 ns can detect frequency changes of 1×10^{-7} in 1 s. The current limit for TIC resolution is about 20 ps, which means that a frequency change of 2×10^{-11} can be detected in 1 s. Averaging over longer intervals can improve the resolution to <1 ps in some units [6].

Since standard frequencies like 5 or 10 MHz are not practical to measure with a TIC, *frequency dividers* (shown in Figure 18.2) or *frequency mixers* are used to convert the test frequency to a lower frequency. Divider systems are simpler and more versatile, since they can be easily built or programmed to accommodate different frequencies. Mixer systems are more expensive, require more hardware including an additional reference oscillator, and can often measure only one input frequency (e.g., 10 MHz), but they have a higher signal-to-noise ratio than divider systems.

If dividers are used, measurements are made from the TIC, but instead of using these measurements directly, we determine the rate of change from reading to reading. This rate of change is called the *phase deviation*. We can estimate frequency offset as follows:

$$f(\text{offset}) = \frac{-\Delta t}{T}$$

where Δt is the amount of phase deviation, and T is the measurement period.

To illustrate, consider a measurement of +1 μs of phase deviation over a measurement period of 24 h. The unit used for measurement period (h) must be converted to the unit used for phase deviation (μs). The equation becomes

$$f(\text{offset}) = \frac{-\Delta t}{T} = \frac{-1 \; \mu s}{86{,}400{,}000{,}000 \; \mu s} = -1.16 \times 10^{-11}$$

As shown, a device that accumulates 1 μs of phase deviation/day has a frequency offset of -1.16×10^{-11} with respect to the reference. This simple example requires only two time interval readings to be made, and Δt is simply the difference between the two readings. Often, multiple readings are taken and the frequency offset is estimated by using least squares linear regression on the data set, and obtaining Δt from the slope of the least squares line. This information is usually presented as a phase plot, as shown in Figure 18.6. The device under test is high in frequency by exactly 1×10^{-9}, as indicated by a phase deviation of 1 ns/s [2,7,8].

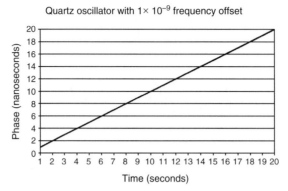

FIGURE 18.6 A sample phase plot.

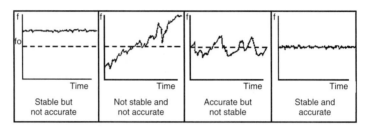

FIGURE 18.7 The relationship between accuracy and stability.

Dimensionless frequency offset values can be converted to units of frequency (Hz) if the nominal frequency is known. To illustrate this, consider an oscillator with a nominal frequency of 5 MHz and a frequency offset of $+1.16 \times 10^{-11}$. To find the frequency offset in hertz, multiply the nominal frequency by the offset:

$$(5 \times 10^6)\,(+1.16 \times 10^{-11}) = 5.80 \times 10^{-5} = +0.0000580 \text{ Hz}$$

Then, add the offset to the nominal frequency to get the actual frequency:

$$5{,}000{,}000 \text{ Hz} + 0.0000580 \text{ Hz} = 5{,}000{,}000.0000580 \text{ Hz}$$

18.2.2 Stability

Stability indicates how well an oscillator can produce the same time or frequency offset over a given time interval. It doesn't indicate whether the time or frequency is "right" or "wrong," but only whether it *stays the same*. In contrast, accuracy indicates how well an oscillator has been set on time or on frequency. To understand this difference, consider that a stable oscillator that needs adjustment might produce a frequency with a large offset. Or, an unstable oscillator that was just adjusted might temporarily produce a frequency near its nominal value. Figure 18.7 shows the relationship between accuracy and stability.

Stability is defined as the statistical estimate of the frequency or time fluctuations of a signal over a given time interval. These fluctuations are measured with respect to a mean frequency or time offset. *Short-term* stability usually refers to fluctuations over intervals less than 100 s. *Long-term* stability can refer to measurement intervals greater than 100 s, but usually refers to periods longer than 1 day.

Stability estimates can be made in either the frequency domain or time domain, and can be calculated from a set of either frequency offset or time interval measurements. In some fields of measurement, stability is estimated by taking the standard deviation of the data set. However, standard deviation only works with stationary data, where the results are time independent, and the noise is *white*, meaning that it is evenly distributed across the frequency band of the measurement. Oscillator data is usually nonstationary, since it contains time dependent noise contributed by the frequency offset. With stationary data, the mean and standard deviation will converge to particular values as more measurements are made. With nonstationary data, the mean and standard deviation never converge to any particular values. Instead, there is a moving mean that changes each time we add a measurement.

For these reasons, a non-classical statistic is often used to estimate stability in the time domain. This statistic is sometimes called the *Allan variance*, but since it is the square root of the variance, its proper name is the *Allan deviation*. The equation for the Allan deviation ($\sigma_y(\tau)$) is

$$\sigma_y(\tau) = \sqrt{\frac{1}{2(M-1)} \sum_{i=1}^{M-1} (y_{i+1} - y_i)^2}$$

where y_i is a set of frequency offset measurements containing y_1, y_2, y_3, and so on, M is the number of values in the y_i series, and the data are equally spaced in segments τ seconds long. Or

$$\sigma_y(\tau) = \sqrt{\frac{1}{2(N-2)\tau^2} \sum_{i=1}^{N-2} [x_{i+2} - 2x_{i+1} + x_i]^2}$$

where x_i is a set of phase measurements in time units containing x_1, x_2, x_3, and so on, N is the number of values in the x_i series, and the data are equally spaced in segments τ seconds long. Note that while standard deviation subtracts the mean from each measurement before squaring their summation, the Allan deviation subtracts the previous data point. This differencing of successive data points removes the time dependent noise contributed by the frequency offset.

An Allan deviation graph is shown in Figure 18.8. It shows the stability of the device improving as the averaging period (τ) gets longer, since some noise types can be removed by averaging. At some point, however, more averaging no longer improves the results. This point is called the *noise floor*, or the point where the remaining noise consists of nonstationary processes such as flicker noise or random walk. The device measured in Figure 18.8 has a noise floor of ~5 × 10^{-11} at τ = 100 s.

Practically speaking, a frequency stability graph also tells us how long we need to average to get rid of the noise contributed by the reference and the measurement system. The noise floor provides some indication of the amount of averaging required to obtain a TUR high enough to show us the true frequency

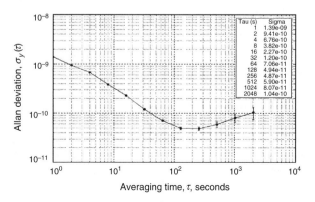

FIGURE 18.8 A frequency stability graph.

TABLE 18.2 Statistics Used to Estimate Time and Frequency Stability and Noise Types

Name	Mathematical Notation	Description
Allan deviation	$\sigma_y(\tau)$	Estimates frequency stability. Particularly suited for intermediate- to long-term measurements
Modified Allan deviation	MOD $\sigma_y(\tau)$	Estimates frequency stability. Unlike the normal Allan deviation, it can distinguish between white and flicker phase noise, which makes it more suitable for short-term stability estimates
Time deviation	$\sigma_x(\tau)$	Used to measure time stability. Clearly identifies both white and flicker phase noise, the noise types of most interest when measuring time or phase
Total deviation	$\sigma_{y,\,\mathrm{TOTAL}}(\tau)$	Estimates frequency stability. Particularly suited for long-term estimates where τ exceeds 10% of the total data sample

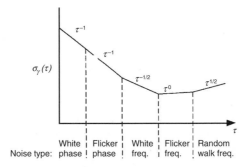

FIGURE 18.9 Using a frequency stability graph to identify noise types.

offset of the DUT. If the DUT is an atomic oscillator (section 18.4) and the reference is a radio controlled transfer standard (section 18.5) we might have to average for 24 h or longer to have confidence in the measurement result.

Five noise types are commonly discussed in the time and frequency literature: *white phase, flicker phase, white frequency, flicker frequency,* and *random walk frequency.* The slope of the Allan deviation line can help identify the amount of averaging needed to remove these noise types (Figure 18.9). The first type of noise to be removed by averaging is phase noise, or the rapid, random fluctuations in the phase of the signal. Ideally, only the device under test would contribute phase noise to the measurement, but in practice, some phase noise from the measurement system and reference needs to be removed through averaging. Note that the Allan deviation does not distinguish between white phase noise and flicker phase noise. Table 18.2 shows several other statistics used to estimate stability and identify noise types for various applications.

Identifying and eliminating sources of oscillator noise can be a complex subject, but plotting the first order differences of a set of time domain measurements can provide a basic understanding of how noise is removed by averaging. Figure 18.10 was made using a segment of the data from the stability graph in Figure 18.8. It shows phase plots dominated by white phase noise (1 s averaging), white frequency noise (64 s averages), flicker frequency noise (256 s averages), and random walk frequency (1024 s averages). Note that the white phase noise plot has a 2 ns scale, and the other plots use a 100 ps scale [8–12].

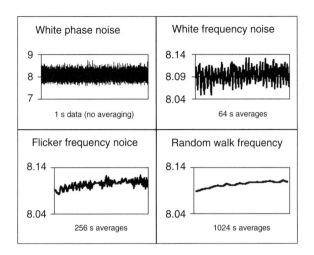

FIGURE 18.10 Phase plots of four noise types.

18.3 Time and Frequency Standards

All time and frequency standards are based on a *periodic event* that repeats at a constant rate. The device that produces this event is called a *resonator*. In the simple case of a pendulum clock, the pendulum is the resonator. Of course, a resonator needs an energy source before it can move back and forth. Taken together, the energy source and resonator form an *oscillator*. The oscillator runs at a rate called the *resonance frequency*. For example, a clock's pendulum can be set to swing back and forth at a rate of once per second. Counting one complete swing of the pendulum produces a time interval of 1 s. Counting the total number of swings creates a *time scale* that establishes longer time intervals, such as minutes, hours, and days. The device that does the counting and displays or records the results is called a *clock*. Table 18.3 shows how the frequency uncertainty of a clock's resonator corresponds to the timing uncertainty of a clock.

Throughout history, clock designers have searched for more stable resonators, and the evolution of time and frequency standards is summarized in Table 18.4. The uncertainties listed for modern standards represent current (year 2001) devices, and not the original prototypes. Note that the performance of time and frequency standards has improved by 13 orders of magnitude in the past 700 years, and by about nine orders of magnitude in the past 100 years.

The stability of time and frequency standards is closely related to their quality factor, or Q. The Q of an oscillator is its resonance frequency divided by its resonance width. The resonance frequency is the natural frequency of the oscillator. The resonance width is the range of possible frequencies where the oscillator will oscillate. A high-Q resonator will not oscillate at all unless it is near its resonance frequency. Obviously, a high resonance frequency and a narrow resonance width are both advantages when seeking a high Q. Generally speaking, the higher the Q, the more stable the oscillator, since a high Q means that an oscillator will stay close to its natural resonance frequency.

This section begins by discussing quartz oscillators, which achieve the highest Q of any mechanical-type device. It then discusses oscillators with much higher Q factors, based on the atomic resonance of rubidium and cesium. Atomic oscillators use the quantized energy levels in atoms and molecules as the source of their resonance. The laws of quantum mechanics dictate that the energies of a bound system, such as an atom, have certain discrete values. An electromagnetic field at a particular frequency can boost an atom from one energy level to a higher one. Or, an atom at a high energy level can drop to a lower level by emitting energy. The resonance frequency (f) of an atomic oscillator is the difference between

Table 18.3 Relationship of Frequency Uncertainty to Time Uncertainty

Frequency Uncertainty	Measurement Period	Time Uncertainty
$\pm 1.00 \times 10^{-3}$	1 s	± 1 ms
$\pm 1.00 \times 10^{-6}$	1 s	± 1 μs
$\pm 1.00 \times 10^{-9}$	1 s	± 1 ns
$\pm 2.78 \times 10^{-7}$	1 h	± 1 ms
$\pm 2.78 \times 10^{-10}$	1 h	± 1 μs
$\pm 2.78 \times 10^{-13}$	1 h	± 1 ns
$\pm 1.16 \times 10^{-8}$	1 day	± 1 ms
$\pm 1.16 \times 10^{-11}$	1 day	± 1 μs
$\pm 1.16 \times 10^{-14}$	1 day	± 1 ns

TABLE 18.4 Evolution of Time and Frequency Standards

Standard	Resonator	Date of Origin	Timing Uncertainty (24 h)	Frequency Uncertainty (24 h)
Sundial	Apparent motion of the sun	3500 BC	NA	NA
Verge escapement	Verge and foliet mechanism	14th century	15 min	1×10^{-2}
Pendulum	Pendulum	1656	10 s	1×10^{-4}
Harrison chronometer (H4)	Spring and balance wheel	1759	350 ms	4×10^{-6}
Shortt pendulum	Two pendulums, slave and master	1921	10 ms	1×10^{-7}
Quartz crystal	Quartz crystal	1927	10 μs	1×10^{-10}
Rubidium gas cell	^{87}Rb resonance (6,834,682,608 Hz)	1958	100 ns	1×10^{-12}
Cesium beam	^{133}Cs resonance (9,192,631,770 Hz)	1952	1 ns	1×10^{-14}
Hydrogen maser	Hydrogen resonance (1,420,405,752 Hz)	1960	1 ns	1×10^{-14}
Cesium fountain	^{133}Cs resonance (9,192,631,770 Hz)	1991	100 ps	1×10^{-15}

the two energy levels divided by Planck's constant (h):

$$f = \frac{E_2 - E_1}{h}$$

The principle underlying the atomic oscillator is that since all atoms of a specific element are identical, they should produce exactly the same frequency when they absorb or release energy. In theory, the atom is a perfect "pendulum" whose oscillations are counted to measure time interval. The discussion of atomic oscillators is limited to devices that are commercially available, and excludes the primary and experimental standards found in laboratories such as NIST. Table 18.5 provides a summary [1,4,8].

18.3.1 Quartz Oscillators

Quartz crystal oscillators are by far the most common time and frequency standards. An estimated two billion (2×10^9) quartz oscillators are manufactured annually. Most are small devices built for wristwatches, clocks, and electronic circuits. However, they are also found inside test and measurement equipment, such as counters, signal generators, and oscilloscopes; and interestingly enough, inside every atomic oscillator.

TABLE 18.5 Summary of Oscillator Types

Oscillator Type	Quartz (TCXO)	Quartz (OCXO)	Rubidium	Commercial Cesium Beam	Hydrogen Maser
Q	10^4 to 10^6	3.2×10^6 (5 MHz)	10^7	10^8	10^9
Resonance frequency	Various	Various	6.834682608 GHz	9.192631770 GHz	1.420405752 GHz
Leading cause of failure	None	None	Rubidium lamp (life expectancy >15 years)	Cesium beam tube (life expectancy of 3 to 25 years)	Hydrogen depletion (life expectancy >7 years)
Stability, $\sigma_y(\tau)$, $\tau = 1$ s	1×10^{-8} to 1×10^{-9}	1×10^{-12}	5×10^{-11} to 5×10^{-12}	5×10^{-11} to 5×10^{-12}	1×10^{-12}
Noise floor, $\sigma_y(\tau)$	1×10^{-9} ($\tau = 1$ to 10^2 s)	1×10^{-12} ($\tau = 1$ to 10^2 s)	1×10^{-12} ($\tau = 10^3$ to 10^5 s)	1×10^{-14} ($\tau = 10^5$ to 10^7 s)	1×10^{-15} ($\tau = 10^3$ to 10^5 s)
Aging/year	5×10^{-7}	5×10^{-9}	1×10^{-10}	None	$\sim 1 \times 10^{-13}$
Frequency offset after warm-up	1×10^{-6}	1×10^{-8} to 1×10^{-10}	5×10^{-10} to 5×10^{-12}	5×10^{-12} to 1×10^{-14}	1×10^{-12} to 1×10^{-13}
Warm-up period	<10 s to 1×10^{-6}	<5 min to 1×10^{-8}	<5 min to 5×10^{-10}	30 min to 5×10^{-12}	24 h to 1×10^{-12}

A quartz crystal inside the oscillator is the resonator. It can be made of either natural or synthetic quartz, but all modern devices use synthetic quartz. The crystal strains (expands or contracts) when a voltage is applied. When the voltage is reversed, the strain is reversed. This is known as the *piezoelectric effect*. Oscillation is sustained by taking a voltage signal from the resonator, amplifying it, and feeding it back to the resonator. The rate of expansion and contraction is the resonance frequency and is determined by the cut and size of the crystal. The output frequency of a quartz oscillator is either the fundamental resonance or a multiple of the resonance, called an *overtone frequency*. Most high stability units use either the third or fifth overtone to achieve a high Q. Overtones higher than fifth are rarely used because they make it harder to tune the device to the desired frequency. A typical Q for a quartz oscillator ranges from 10^4 to 10^6. The maximum Q for a high stability quartz oscillator can be estimated as $Q = 1.6 \times 10^7/f$, where f is the resonance frequency in megahertz.

Environmental changes due to temperature, humidity, pressure, and vibration can change the resonance frequency of a quartz crystal, but there are several designs that reduce these environmental effects. The *oven-controlled crystal oscillator* (OCXO) encloses the crystal in a temperature-controlled chamber called an oven. When an OCXO is turned on, it goes through a "warm-up" period while the temperatures of the crystal resonator and its oven stabilize. During this time, the performance of the oscillator continuously changes until it reaches its normal operating temperature. The temperature within the oven then remains constant, even when the outside temperature varies. An alternate solution to the temperature problem is the *temperature-compensated crystal oscillator* (TCXO). In a TCXO, the signal from a temperature sensor is used to generate a correction voltage that is applied to a voltage-variable reactance, or varactor. The varactor then produces a frequency change equal and opposite to the frequency change produced by temperature. This technique does not work as well as oven control, but is less expensive. Therefore, TCXOs are used when high stability over a wide temperature range is not required.

Quartz oscillators have excellent short-term stability. An OCXO might be stable ($\sigma_y(\tau)$, at $\tau = 1$ s) to 1×10^{-12}. The limitations in short-term stability are due mainly to noise from electronic components in the oscillator circuits. Long-term stability is limited by *aging*, or a change in frequency with time due to internal changes in the oscillator. Aging is usually a nearly linear change in the resonance frequency that can be either positive or negative, and occasionally, a reversal in direction of aging occurs. Aging has many possible causes including a build-up of foreign material on the crystal, changes in the oscillator circuitry,

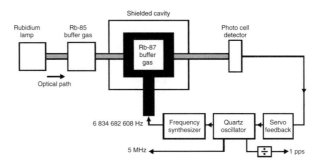

FIGURE 18.11 Rubidium oscillator.

or changes in the quartz material or crystal structure. A high quality OCXO might age at a rate of <5 × 10^{-9} per year, while a TCXO might age 100 times faster.

Due to aging and environmental factors such as temperature and vibration, it is hard to keep even the best quartz oscillators within 1×10^{-10} of their nominal frequency without constant adjustment. For this reason, atomic oscillators are used for applications that require better long-term accuracy and stability [4,13,14].

18.3.2 Rubidium Oscillators

Rubidium oscillators are the lowest priced members of the atomic oscillator family. They operate at 6,834,682,608 Hz, the resonance frequency of the rubidium atom (^{87}Rb), and use the rubidium frequency to control the frequency of a quartz oscillator. A microwave signal derived from the crystal oscillator is applied to the ^{87}Rb vapor within a cell, forcing the atoms into a particular energy state. An optical beam is then pumped into the cell and is absorbed by the atoms as it forces them into a separate energy state. A photo cell detector measures how much of the beam is absorbed, and its output is used to tune a quartz oscillator to a frequency that maximizes the amount of light absorption. The quartz oscillator is then locked to the resonance frequency of rubidium, and standard frequencies are derived from the quartz oscillator and provided as outputs (Figure 18.11).

Rubidium oscillators continue to get smaller and less expensive, and offer perhaps the best price-to-performance ratio of any oscillator. Their long-term stability is much better than that of a quartz oscillator and they are also smaller, more reliable, and less expensive than cesium oscillators.

The Q of a rubidium oscillator is about 10^7. The shifts in the resonance frequency are due mainly to collisions of the rubidium atoms with other gas molecules. These shifts limit the long-term stability. Stability ($\sigma_y(\tau)$), at $\tau = 1$ s) is typically 1×10^{-11}, and about 1×10^{-12} at 1 day. The frequency offset of a rubidium oscillator ranges from 5×10^{-10} to 5×10^{-12} after a warm-up period of a few minutes or hours, so they meet the accuracy requirements of most applications without adjustment.

18.3.3 Cesium Oscillators

Cesium oscillators are *primary frequency standards* since the SI second is defined from the resonance frequency of the cesium atom (^{133}Cs), which is 9,192,631,770 Hz. A properly working cesium oscillator should be close to its nominal frequency without adjustment, and there should be no change in frequency due to aging.

Commercially available oscillators use *cesium beam* technology. Inside a cesium oscillator, ^{133}Cs atoms are heated to a gas in an oven. Atoms from the gas leave the oven in a high-velocity beam that travels through a vacuum tube toward a pair of magnets. The magnets serve as a gate that allows only atoms of a particular magnetic energy state to pass into a microwave cavity, where they are exposed to a microwave

Fundamentals of Time and Frequency

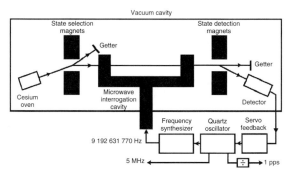

FIGURE 18.12 Cesium beam oscillator.

frequency derived from a quartz oscillator. If the microwave frequency matches the resonance frequency of cesium, the cesium atoms change their magnetic energy state.

The atomic beam then passes through another magnetic gate near the end of the tube. Those atoms that changed their energy state while passing through the microwave cavity are allowed to proceed to a detector at the end of the tube. Atoms that did not change state are deflected away from the detector. The detector produces a feedback signal that continually tunes the quartz oscillator in a way that maximizes the number of state changes so that the greatest number of atoms reaches the detector. Standard output frequencies are derived from the locked quartz oscillator (Figure 18.12).

The Q of a commercial cesium standard is a few parts in 10^8. The beam tube is typically <0.5 m in length, and the atoms travel at velocities of >100 m/s inside the tube. This limits the observation time to a few milliseconds, and the resonance width to a few hundred hertz. Stability ($\sigma_y(\tau)$, at $\tau = 1$ s) is typically 5×10^{-12} and reaches a noise floor near 1×10^{-14} at about 1 day, extending out to weeks or months. The frequency offset is typically near 1×10^{-12} after a warm-up period of 30 min.

18.4 Time and Frequency Transfer

Many applications require clocks or oscillators at different locations to be set to the same time (*synchronization*), or the same frequency (*syntonization*). *Time and frequency transfer* techniques are used to compare and adjust clocks and oscillators at different locations. Time and frequency transfer can be as simple as setting your wristwatch to an audio time signal, or as complex as controlling the frequency of oscillators in a network to parts in 10^{13}.

Time and frequency transfer can use signals broadcast through many different media, including coaxial cables, optical fiber, radio signals (at numerous places in the radio spectrum), telephone lines, and the Internet. Synchronization requires both an on-time pulse and a time code. Syntonization requires extracting a stable frequency from the broadcast. The frequency can come from the carrier itself, or from a time code or other information modulated onto the carrier.

This section discusses both the fundamentals of time and frequency transfer and the radio signals used as calibration references. Table 18.6 provides a summary.

18.4.1 Fundamentals of Time and Frequency Transfer

Signals used for time and frequency transfer are generally referenced to atomic oscillators that are steered to agree as closely as possible with UTC. Information is sent from a transmitter (A) to a receiver (B) and is delayed by τ_{ab}, commonly called the *path delay* (Figure 18.13).

To illustrate path delay, consider a radio signal broadcast over a path 1000 km long. Since radio signals travel at the speed of light (~3.3 µs/km), we can calibrate the path by applying a 3.3-ms correction to our

TABLE 18.6 Summary of Time and Frequency Transfer Signals and Methods

Signal or Link	Receiving Equipment	Time Uncertainty (24 h)	Frequency Uncertainty (24 h)
Dial-up computer time service	Computer, client software, modem, and phone line	<15 ms	Not recommended for frequency measurements
Internet time service	Computer, client software, and Internet connection	<1 s	Not recommended for frequency measurements
HF radio (3–30 MHz)	HF receiver and antenna	1–20 ms	10^{-6}–10^{-9}
LF radio (30–300 kHz)	LF receiver and antenna	1–100 μs	10^{-10}–10^{-12}
Global positioning system (GPS)	GPS receiver antenna	<20 ns	$<2 \times 10^{-13}$

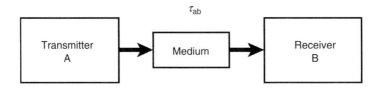

FIGURE 18.13 One-way time and frequency transfer.

measurement. Of course, for many applications the path delay is simply ignored. For example, if our goal is simply to synchronize a computer clock within 1 s of UTC, there is no need to worry about a 100-ms path delay through a network. And, of course, path delay is not important to frequency transfer systems, since on-time pulses are not required. Instead, frequency transfer requires only a stable path where the delays remain relatively constant.

More sophisticated transfer systems estimate and remove all or part of the path delay. This is usually done in one of two ways. The first way is to estimate τ_{ab} and send the time out early by this amount. For example, if τ_{ab} is at least 20 ms for all users, the time can be sent 20 ms early. This advancement of the timing signal removes at least some of the delay for all users.

A better technique is to compute τ_{ab} and to apply a correction to the received signal. A correction for τ_{ab} can be computed if the position of both the transmitter and receiver are known. If the transmitter is stationary, a constant can be used for the transmitter position. If the transmitter is moving (a satellite, for example) it must broadcast its position in addition to broadcasting time. The Global Positioning System (GPS) provides the best of both worlds—each GPS satellite broadcasts its position and the receiver can use coordinates from multiple satellites to compute its own position.

The transmitted information often includes a *time code* so that a clock can be set to the correct time-of-day. Most time codes contain the UTC hour, minute, and second; the month, day, and year; and advance warning of daylight saving time and leap seconds.

18.4.2 Radio Time and Frequency Transfer Signals

There are many types of radio receivers designed to receive time and frequency signals. Some are designed primarily to produce time-of-day information or an on-time pulse, others are designed to output standard frequencies, and some can be used for both time and frequency transfer. The following sections look at three types of time and frequency radio signals that distribute UTC—high frequency (HF), low frequency (LF), and GPS satellite signals.

Fundamentals of Time and Frequency

18.4.2.1 HF Radio Signals (Including WWV and WWVH)

High frequency radio broadcasts occupy the radio spectrum from 3 to 30 MHz. These signals are commonly used for time and frequency transfer at moderate performance levels. Some HF broadcasts provide audio time announcements and digital time codes. Other broadcasts simply provide a carrier frequency for use as a reference.

HF time and frequency stations include NIST radio stations WWV and WWVH. WWV is located near Fort Collins, Colorado, and WWVH is on the island of Kauai, Hawaii. Both stations broadcast continuous time and frequency signals on 2.5, 5, 10, and 15 MHz, and WWV also broadcasts on 20 MHz. All frequencies broadcast the same program, and at least one frequency should be usable at all times. The stations can also be heard by telephone; dial (303) 499-7111 for WWV or (808) 335-4363 for WWVH.

WWV and WWVH signals can be used in one of three modes:

- The audio portion of the broadcast includes seconds pulses or ticks, standard audio frequencies, and voice announcements of the UTC hour and minute. WWV uses a male voice, and WWVH uses a female voice.
- A binary time code is sent on a 100 Hz subcarrier at a rate of 1 bit per second. The time code contains the hour, minute, second, year, day of year, leap second and Daylight Saving Time (DST) indicators, and a UT1 correction. This code can be read and displayed by radio clocks.
- The carrier frequency can be used as a reference for the calibration of oscillators. This is done most often with the 5 and 10 MHz carrier signals, since they match the output frequencies of standard oscillators.

The time broadcast by WWV and WWVH will be late when it arrives at the user's location. The time offset depends upon the receiver's distance from the transmitter, but should be <15 ms in the continental United States. A good estimate of the time offset requires knowledge of HF radio propagation. Most users receive a signal that has traveled up to the ionosphere and was then reflected back to earth. Since the height of the ionosphere changes throughout the day, the path delay also changes. Path delay variations limit the received frequency uncertainty to parts in 10^9 when averaged for 1 day.

HF radio stations such as WWV and WWVH are useful for low level applications, such as the manual synchronization of analog and digital clocks, simple frequency calibrations, and calibrations of stop watches and timers. However, LF and GPS signals are better choices for more demanding applications [2,7,15].

18.4.2.2 LF Radio Signals (Including WWVB)

Before the advent of satellites, low frequency signals were the method of choice for time and frequency transfer. While the use of LF signals has diminished in the laboratory, they still have two major advantages—they can often be received indoors without an external antenna and several stations broadcast a time code. This makes them ideal for many consumer electronic products that display time-of-day information.

Many time and frequency stations operate in the LF band from 30 to 300 kHz (Table 18.7). The performance of the received signal is influenced by the path length and signal strength. Path length is important because the signal is divided into ground wave and sky wave. The ground wave signal is more stable. Since it travels the shortest path between the transmitter and receiver, it arrives first and its path delay is much easier to estimate. The sky wave is reflected from the ionosphere and produces results similar to those obtained with HF reception. Short paths make it possible to continuously track the ground wave. Longer paths produce a mixture of sky wave and ground wave. And over very long paths, only sky wave reception is possible.

Signal strength is also important. If the signal is weak, the receiver might search for a new cycle of the carrier to track. Each time the receiver adjusts its tracking point by one cycle, it introduces a phase step equal to the period of the carrier. For example, a cycle slip on a 60 kHz carrier introduces a 16.67 μs phase step. However, a strong ground wave signal can produce very good results. An LF receiver that

TABLE 18.7 LF Time and Frequency Broadcast Stations

Call Sign	Country	Frequency (kHz)	Always On?
DCF77	Germany	77.5	Yes
DGI	Germany	177	Yes
HBG	Switzerland	75	Yes
JG2AS	Japan	40	Yes
MSF	United Kingdom	60	Yes
RBU	Russia	66.666	No
RTZ	Russia	50	Yes
TDF	France	162	Yes
WWVB	United States	60	Yes

continuously tracks the same cycle of a ground wave signal can transfer frequency with an uncertainty of about 1×10^{-12} when averaged for 1 day.

NIST operates LF radio station WWVB from Fort Collins, Colorado at a transmission frequency of 60 kHz. The station broadcasts 24 h per day, with an effective radiated output power of 50 kW. The WWVB time code is synchronized with the 60 kHz carrier and contains the year, day of year, hour, minute, second, and flags that indicate the status of daylight saving time, leap years, and leap seconds. The time code is received and displayed by wristwatches, alarm clocks, wall clocks, and other consumer electronic products [2,7,15].

18.4.2.3 Global Positioning System

The Global Positioning System (GPS) is a navigation system developed and operated by the U.S. Department of Defense (DoD) that is usable nearly anywhere on the earth. The system consists of a constellation of at least 24 satellites that orbit the earth at a height of 20,200 km in six fixed planes inclined 55° from the equator. The orbital period is 11 h 58 m, which means that each satellite will pass over the same place on earth twice per day. By processing signals received from the satellites, a GPS receiver can determine its position with an uncertainty of <10 m.

The satellites broadcast on two carrier frequencies, L1 at 1575.42 MHz and L2 at 1227.6 MHz. Each satellite broadcasts a spread spectrum waveform, called a *pseudo random noise* (PRN) code on L1 and L2, and each satellite is identified by the PRN code it transmits. There are two types of PRN codes. The first type is a *coarse acquisition* (C/A) code with a chipping rate of 1023 chips per millisecond. The second is a *precision* (P) code with a chipping rate of 10230 chips per millisecond. The C/A code is broadcast on L1, and the P code is broadcast on both L1 and L2. GPS reception is line-of-sight, which means that the receiving antenna must have a clear view of the sky [16].

Each satellite carries either rubidium or cesium oscillators, or a combination of both. These oscillators are steered from DoD ground stations and are referenced to the United States Naval Observatory time scale, UTC (USNO), which by agreement is always maintained within 100 ns of UTC (NIST). The oscillators provide the reference for both the carrier and the code broadcasts.

GPS signals now dominate the world of high performance time and frequency transfer, since they provide reliable reception and exceptional results with minimal effort. A GPS receiver can automatically compute its latitude, longitude, and altitude from position data received from the satellites. The receiver can then calibrate the radio path and synchronize its on-time pulse. In addition to the on-time pulse, many receivers provide standard frequencies such as 5 or 10 MHz by steering an OCXO or rubidium oscillator using the satellite signals. GPS receivers also produce time-of-day and date information.

A GPS receiver calibrated for equipment delays has a timing uncertainty of <20 ns relative to UTC (NIST), and the frequency uncertainty is often $<2 \times 10^{-13}$ when averaged for 1 day. Figure 18.14 shows an Allan deviation plot of the output of a low cost GPS receiver. The stability is near 1×10^{-13} after about 1 day of averaging.

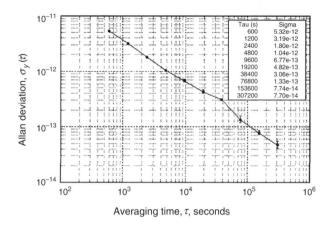

FIGURE 18.14 Frequency stability of GPS receiver.

18.5 Closing

As noted earlier, time and frequency standards and measurements have improved by about nine orders of magnitude in the past 100 years. This rapid advance has made many new products and technologies possible. While it is impossible to predict what the future holds, we can be certain that oscillator Qs will continue to increase, measurement uncertainties will continue to decrease, and new technologies will continue to emerge.

References

1. Jespersen, J., and Fitz-Randolph, J., *From Sundials to Atomic Clocks: Understanding Time and Frequency*, 2nd ed., Dover, Mineola, New York, 1999.
2. Kamas, G., and Lombardi, M. A., *Time and Frequency Users Manual*, NIST Special Publication 559, U.S. Government Printing Office, Washington, DC, 1990.
3. Levine, J., Introduction to time and frequency metrology, *Rev. Sci. Instrum.*, 70, 2567, 1999.
4. Hackman, C., and Sullivan, D. B., Eds., *Time and Frequency Measurement*, American Association of Physics Teachers, College Park, Maryland, 1996.
5. ITU Radiocommunication Study Group 7, *Selection and Use of Precise Frequency and Time Systems*, International Telecommunications Union, Geneva, Switzerland, 1997.
6. Novick, A. N., Lombardi, M. A., Zhang, V. S., and Carpentier, A., A high performance multi-channel time interval counter with an integrated GPS receiver, in *Proc. 31st Annu. Precise Time and Time Interval (PTTI) Meeting*, Dana Point, California, p. 561, 1999.
7. Lombardi, M. A., Time measurement and frequency measurement, in *The Measurement, Instrumentation, and Sensors Handbook*, Webster, J. G., Ed., CRC Press, Boca Raton, FL, 1999, chaps. 18–19.
8. Sullivan, D. B., Allan, D. W., Howe, D. A., and Walls, F. L., Eds., *Characterization of Clocks and Oscillators*, NIST Technical Note 1337, U.S. Government Printing Office, Washington, DC, 1990.
9. Jespersen, J., Introduction to the time domain characterization of frequency standards, in *Proc. 23rd Annu. Precise Time and Time Interval (PTTI) Meeting*, Pasadena, California, p. 83, 1991.
10. IEEE Standards Coordinating Committee 27, *IEEE Standard Definitions of Physical Quantities for Fundamental Frequency and Time Metrology—Random Instabilities*, Institute of Electrical and Electronics Engineers, New York, 1999.
11. Walls, F. L., and Ferre-Pikal, E. S., Measurement of frequency, phase noise, and amplitude noise, in *Wiley Encyclopedia of Electrical and Electronics Engineering*, John Wiley & Sons, New York, 1999, 12, 459.

12. Howe, D. A., An extension of the Allan variance with increased confidence at long term, *IEEE Int. Freq. Control Symp.*, 321, 1995.
13. Vig, J. R., Introduction to quartz frequency standards, *Army Research and Development Technical Report,* SLCET-TR-92-1, October 1992.
14. Hewlett-Packard Company, *Fundamentals of Quartz Oscillators,* HP Application Note 200-2, 1997.
15. Carr, J. J., *Elements of Electronic Instrumentation and Measurement,* 3rd ed., Prentice-Hall, NJ, 1996.
16. Hoffmann-Wellenhof, B., Lichtenegger, H., and Collins, J., *GPS: Theory and Practice,* 3rd ed., Springer-Verlag, New York, 1994.

19
Sensor and Actuator Characteristics

19.1	Range	**19**-1
19.2	Resolution	**19**-2
19.3	Sensitivity	**19**-2
19.4	Error	**19**-2
19.5	Repeatability	**19**-3
19.6	Linearity and Accuracy	**19**-3
19.7	Impedance	**19**-4
19.8	Nonlinearities	**19**-5
19.9	Static and Coulomb Friction	**19**-5
19.10	Eccentricity	**19**-6
19.11	Backlash	**19**-6
19.12	Saturation	**19**-7
19.13	Deadband	**19**-7
19.14	System Response	**19**-8
19.15	First-Order System Response	**19**-8
19.16	Underdamped Second-Order System Response	**19**-9
19.17	Frequency Response	**19**-12
	Reference	**19**-14

Joey Parker
University of Alabama

Mechatronic systems use a variety of sensors and actuators to measure and manipulate mechanical, electrical, and thermal systems. Sensors have many characteristics that affect their measurement capabilities and their suitability for each application. Analog sensors have an output that is continuous over a finite region of inputs. Examples of analog sensors include potentiometers, LVDTs (linear variable differential transformers), load cells, and thermistors. Digital sensors have a fixed or countable number of different output values. A common digital sensor often found in mechatronic systems is the incremental encoder. An analog sensor output conditioned by an analog-to-digital converter (ADC) has the same digital output characteristics, as seen in Figure 19.1.

19.1 Range

The range (or span) of a sensor is the difference between the minimum (or most negative) and maximum inputs that will give a valid output. Range is typically specified by the manufacturer of the sensor. For example, a common type K thermocouple has a range of 800°C (from −50°C to 750°C). A ten-turn potentiometer would have a range of 3600 degrees.

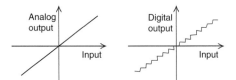

FIGURE 19.1 Analog and digital sensor outputs.

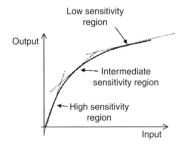

FIGURE 19.2 Sensor sensitivity.

19.2 Resolution

The resolution of a sensor is the smallest increment of input that can be reliably detected. Resolution is also frequently known as the least count of the sensor. Resolution of digital sensors is easily determined. A 1024 ppr (pulse per revolution) incremental encoder would have a resolution of

$$\frac{1 \text{ revolution}}{1024 \text{ pulses}} \times \frac{360 \text{ degrees}}{1 \text{ revolution}} = 0.3516 \frac{\text{degrees}}{\text{pulse}}$$

The resolution of analog sensors is usually limited only by low-level electrical noise and is often much better than equivalent digital sensors.

19.3 Sensitivity

Sensor sensitivity is defined as the change in output per change in input. The sensitivity of digital sensors is closely related to the resolution. The sensitivity of an analog sensor is the slope of the output versus input line. A sensor exhibiting truly linear behavior has a constant sensitivity over the entire input range. Other sensors exhibit nonlinear behavior where the sensitivity either increases or decreases as the input is changed, as shown in Figure 19.2.

19.4 Error

Error is the difference between a measured value and the true input value. Two classifications of errors are bias (or systematic) errors and precision (or random) errors. Bias errors are present in all measurements made with a given sensor, and cannot be detected or removed by statistical means. These bias errors can be further subdivided into

- Calibration errors (a zero or null point error is a common type of bias error created by a nonzero output value when the input is zero)
- Loading errors (adding the sensor to the measured system changes the system)
- Errors due to sensor sensitivity to variables other than the desired one (e.g., temperature effects on strain gages)

19.5 Repeatability

Repeatability (or reproducibility) refers to a sensor's ability to give identical outputs for the same input. Precision (or random) errors cause a lack of repeatability. Fortunately, precision errors can be accounted for by averaging several measurements or other operations such as low-pass filtering. Electrical noise and hysteresis (described later) both contribute to a loss of repeatability.

19.6 Linearity and Accuracy

The accuracy of a sensor is inversely proportional to error, that is, a highly accurate sensor produces low errors. Many manufacturers specify accuracy in terms of the sensor's linearity. A least-squares straight-line fit between all output measurements and their corresponding inputs determines the nominal output of the sensor. Linearity (or accuracy) is specified as a percentage of full scale (maximum valid input), as shown in Figure 19.3, or as a percentage of the sensor reading, as shown in Figure 19.4. Figures 19.3 and 19.4 show both of these specifications for 10% linearity, which is much larger than most actual sensors.

Accuracy and precision are two terms that are frequently confused. Figure 19.5 shows four sets of histograms for ten measurements of angular velocity of an actuator turning at a constant 100 rad/s. The first set of data shows a high degree of precision (low standard deviation) and repeatability, but the

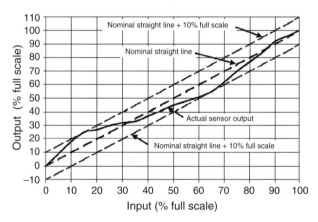

FIGURE 19.3 Linearity specified at full scale.

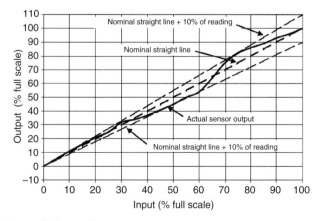

FIGURE 19.4 Linearity specified at reading.

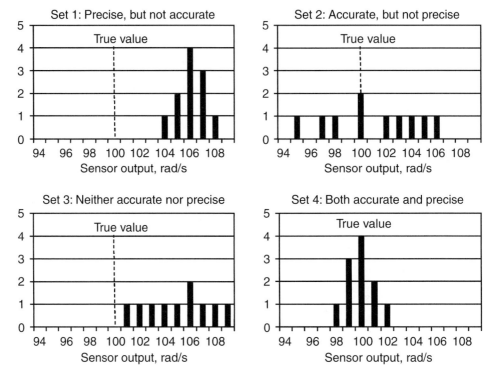

FIGURE 19.5 Examples of accuracy and precision.

average accuracy is poor. The second set of data shows a low degree of precision (high standard deviation), but the average accuracy is good. The third set of data shows both low precision and low accuracy, while the fourth set of data shows both high precision, high repeatability, and high accuracy.

19.7 Impedance

Impedance is the ratio of voltage and current flow for a sensor. For a simple resistive sensor (such as a strain gage or a thermistor), the impedance Z is the same as the resistance R, which has units of ohms (Ω),

$$Z_R = \frac{V}{I} = R$$

For more complicated sensors, impedance includes the effects of capacitance, C, and inductance, L. Inclusion of these terms makes the impedance frequency sensitive, but the units remain ohms:

$$Z_C = \frac{V}{I} = \frac{1}{jC\omega} \quad \text{and} \quad Z_L = \frac{V}{I} = jL\omega$$

where $j = \sqrt{-1}$ is the imaginary number and ω is the driving frequency. The impedance form is particularly nice for analyzing simple circuits, as parallel and series inductances can be treated just like resistances. Two types of impedance are important in sensor applications: input impedance and output impedance. Input impedance is a measure of how much current must be drawn to power a sensor (or signal conditioning circuit). Input impedance is frequently modeled as a resistor in parallel with the input terminals. High input impedance is desirable, since the device will then draw less current from the source. Oscilloscopes and data acquisition equipment frequently have input impedances of 1 MΩ or more to minimize this current draw. Output impedance is a measure of a sensor's (or signal conditioning circuit's) ability to provide current for the next stage of the system. Output impedance is frequently modeled as a resistor in series with the sensor output. Low output impedance is desirable, but is often not available directly from a sensor. Piezoelectric sensors in particular have high output impedances and cannot source much current (typically micro-amps or less). Op-amp circuits are frequently used to buffer sensor outputs for this reason. Op-amp circuits (especially voltage followers) provide nearly ideal circumstances for many sensors, since they have high input impedance but can substantially lower output impedance.

19.8 Nonlinearities

Linear systems have the property of superposition. If the response of the system to input A is output A, and the response to input B is output B, then the response to input C (= input A + input B) will be output C (= output A + output B). Many real systems will exhibit linear or nearly linear behavior over some range of operation. Therefore, linear system analysis is correct, at least over these portions of a system's operating envelope. Unfortunately, most real systems have nonlinearities that cause them to operate outside of this linear region, and many common assumptions about system behavior, such as superposition, no longer apply. Several nonlinearities commonly found in mechatronic systems include static and coulomb friction, eccentricity, backlash (or hysteresis), saturation, and deadband.

19.9 Static and Coulomb Friction

In classic linear system analysis, friction forces are assumed to be proportional to velocity, that is, viscous friction. With an actuator velocity of zero, there should be no friction. In reality, a small amount of static (no velocity) or Coulomb friction is almost always present, even in roller or ball type anti-friction bearings. A typical plot of friction force versus velocity is given in Figure 19.6. Note that the static friction force can assume any value between some upper and lower limit at zero velocity. Static friction has two primary effects on mechatronic systems:

1. Some of the actuator torque or force is wasted overcoming friction forces, which leads to inefficiency from an energy viewpoint.
2. As the actuator moves the system to its final location, the velocity approaches zero and the actuator force/torque will approach a value that exactly balances frictional and gravity loads. Since static friction can assume any value at zero velocity, the actuator will come to slightly different final resting positions each time—depending on the final value of static friction. This effect contributes to some loss of repeatability in mechatronic systems.

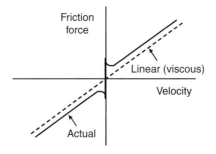

FIGURE 19.6 Static and Coulomb friction.

19.10 Eccentricity

The ideal relationships for gears, pulleys, and chain drives assume that the point of gear contact remains at a fixed distance from the center of rotation for each gear. In reality, the true center of the gears pitch circle and the center of rotation will be separated by a small amount, known as the eccentricity. Small tooth-to-tooth errors can also cause local variations in the pitch circle radius. The combination of these two effects can lead to a nonlinear geometrical relationship between two gears like that of Figure 19.7, where the nonlinear behavior is greatly exaggerated for clarity. Eccentricity impacts the accuracy of position measurements made on the input side of the gear pair, as the output gear is not exactly where the sensor measurement indicates.

19.11 Backlash

If two otherwise perfect gears are not mounted on a center-to-center distance that exactly matches the sum of the pitch radii, there will be a small clearance, or backlash, between the teeth. When the input gear reverses direction, a small rotation is required before this clearance is removed and the output gear begins to move. Gear backlash is just one of many phenomena that can be characterized as hysteresis, as shown in Figure 19.8. Clearance between shafts and bearings can cause hysteretic effects also. Backlash exhibits effects similar to those for eccentricity, that is, a loss of repeatability, particularly when approaching a measured point from different directions. The gear backlash problem is so prevalent and potentially harmful that many manufacturers go to great lengths to minimize or reduce the effect:

- Gears mounted closer together than the theoretically ideal spacing
- Split "anti-backlash" gears that are spring loaded to force teeth to maintain engagement at all times
- External spring-loaded mounts for one of the gears to force engagement
- Specially designed gears with anti-backlash features

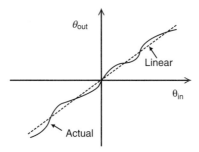

FIGURE 19.7 Gear eccentricity.

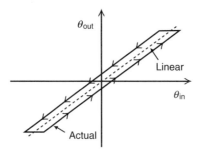

FIGURE 19.8 Gear backlash.

19.12 Saturation

All real actuators have some maximum output capability, regardless of the input. This violates the linearity assumption, since at some point the input command can be increased without significantly changing the output; see Figure 19.9. This type of nonlinearity must be considered in mechatronic control system design, since maximum velocity and force or torque limitations affect system performance. Control systems modeled with linear system theory must be carefully tested or analyzed to determine the impact of saturation on system performance.

19.13 Deadband

Another nonlinear characteristic of some actuators and sensors is known as deadband. The deadband is typically a region of input close to zero at which the output remains zero. Once the input travels outside the deadband, then the output varies with input, as shown in Figure 19.10. Analog joystick inputs frequently use a small amount of deadband to reduce the effect of noise from human inputs. A very small movement of the joystick produces no output, but the joystick acts normally with larger inputs.

Deadband is also commonly found in household thermostats and other process type controllers, as shown in Figure 19.11. When a room warms and the temperature reaches the setpoint (or desired value)

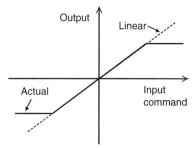

FIGURE 19.9 Saturation.

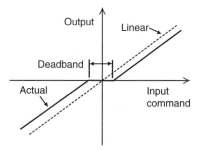

FIGURE 19.10 Deadband.

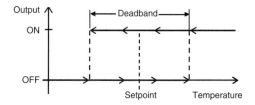

FIGURE 19.11 Thermostat deadband.

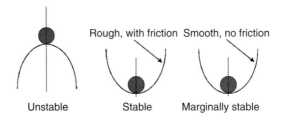

FIGURE 19.12 System stability.

on the thermostat, the output remains off. Once room temperature has increased to the setpoint plus half the deadband, then the cooling system output goes to fully on. As the room cools, the output stays fully on until the temperature reaches the setpoint minus half the deadband. At this point the cooling system output goes fully off.

19.14 System Response

Sensors and actuators respond to inputs that change with time. Any system that changes with time is considered a dynamic system. Understanding the response of dynamic systems to different types of inputs is important in mechatronic system design. The most important concept in system response is stability. The term stability has many different definitions and uses, but the most common definition is related to equilibrium. A system in equilibrium will remain in the same state in the absence of external disturbances. A stable system will return to an equilibrium state if a "small" disturbance moves the system away from the initial state. An unstable system will not return to an equilibrium position, and frequently will move "far" from the initial state.

Figure 19.12 illustrates three stability conditions with a simple ball and hill system. In each case an equilibrium position is easily identified—either the top of the hill or the bottom of the valley. In the unstable case, a small motion of the ball away from the equilibrium position will cause the ball to move "far" away, as it rolls down the hill. In the stable case, a small movement of the ball away from the equilibrium position will eventually result in the ball returning, perhaps after a few oscillations. In the third case, the absence of friction causes the ball to oscillate continuously about the equilibrium position once a small movement has occurred. This special case is often known as marginal stability, since the system never quite returns to the equilibrium position.

Most sensors and actuators are inherently stable. However, the addition of active control systems can cause a system of stable devices to exhibit overall unstable behavior. Careful analysis and testing is required to ensure that a mechatronic system acts in a stable manner. The complex response of stable dynamic systems is frequently approximated by much simpler systems. Understanding both first-order and second-order system responses to either instantaneous (or step) changes in inputs or sinusoidal inputs will suffice for most situations.

19.15 First-Order System Response

First-order systems contain two primary elements: an energy storing element and an element which dissipates (or removes) energy. Typical first-order systems include resistor–capacitor filters and resistor–inductor networks (e.g., a coil of a stepper motor). Thermocouples and thermistors also form first-order systems, due to thermal capacitance and resistance. The differential equation describing the time response of a generic first-order system is

$$\frac{dy(t)}{dt} + \frac{1}{\tau}y(t) = f(t)$$

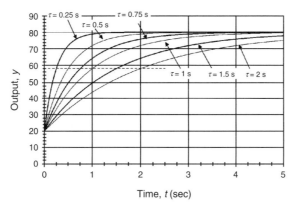

FIGURE 19.13 First-order system—step response.

where $y(t)$ is the dependent output variable (velocity, acceleration, temperature, voltage, etc.), t is the independent input variable (time), τ is the time constant (units of seconds), and $f(t)$ is the forcing function (or system input).

The solution to this equation for a step or constant input is given by

$$y(t) = y_\infty + (y_0 - y_\infty)e^{-t/\tau}$$

where y_∞ is the limiting or final (steady-state) value, y_0 is the initial value of the independent variable at $t = 0$.

A set of typical first-order system step responses is shown in Figure 19.13. The initial value is arbitrarily selected as 20 with final values of 80. Time constants ranging from 0.25 to 2 s are shown. Each of these curves directly indicates its time constant at a key point on the curve. Substituting $t = \tau$ into the first-order response equation with $y_0 = 20$ and $y_\infty = 80$ gives

$$y(\tau) = 80 + (20 - 80)e^{-1} = 57.9$$

Each curve crosses the $y(\tau) \approx 57.9$ line when its time constant τ equals the time t. This concept is frequently used to experimentally determine time constants for first-order systems.

19.16 Underdamped Second-Order System Response

Second-order systems contain three primary elements: two energy storing elements and an element which dissipates (or removes) energy. The two energy storing elements must store different types of energy. A typical mechanical second-order system is the spring–mass–damper combination shown in Figure 19.14. The spring stores potential energy (PE = $\frac{1}{2}kx^2$), while the mass stores kinetic energy (KE = $\frac{1}{2}mv^2$), where k is the spring stiffness (typical units of N/m), x is the spring deflection (typical units of m), m is the mass (typical units of kg), and v is the absolute velocity of the mass (typical units of m/s).

A common electrical second-order system is the resistor–inductor–capacitor (RLC) network, where the capacitor and inductor store electrical energy in two different forms. The generic form of the dynamic equation for an underdamped second-order system is

$$\frac{d^2y(t)}{dt^2} + 2\zeta\omega_n\frac{dy(t)}{dt} + \omega_n^2 y(t) = f(t)$$

where $y(t)$ is the dependent variable (velocity, acceleration, temperature, voltage, etc.), t is the independent variable (time), ζ is the damping ratio (a dimensionless quantity), ω_n is the natural frequency (typical units of rad/s), and $f(t)$ is the forcing function (or input).

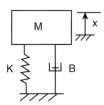

FIGURE 19.14 Spring–mass–damper system.

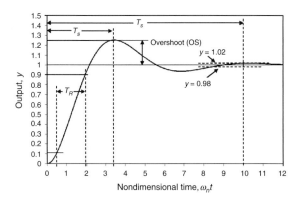

FIGURE 19.15 Second-order system—step response.

The response of an underdamped ($0 \leq \zeta < 1$) second-order system to a *unit* step input can be determined as:

$$y(t) = 1 - e^{-\zeta \omega_n t}\left(\cos \omega_n \sqrt{1-\zeta^2}\, t + \frac{\zeta}{\sqrt{1-\zeta^2}} \sin \omega_n \sqrt{1-\zeta^2}\, t\right)$$

This second-order system step response is often characterized by a set of time response parameters illustrated in Figure 19.15.

These time response parameters are functions of the damping ratio ζ and the natural frequency ω_n:

- Peak time, T_P: the time required to reach the first (or maximum) peak

$$T_P = \frac{\pi}{\omega_n \sqrt{1-\zeta^2}}$$

- Percent overshoot, %OS: amount the response exceeds or overshoots the steady-state value

$$\%OS = 100 e^{-(\zeta \pi / \sqrt{1-\zeta^2})}$$

- Settling time, T_S: the time when the system response remains within ±2% of the steady-state value

$$T_S = \frac{4}{\zeta \omega_n}$$

- Rise time, T_R: time required for the response to go from 10% to 90% of the steady-state value. Figure 19.16 shows the nondimensional rise time ($\omega_n T_R$) as a function of damping ratio, z. A frequently used approximation relating these two parameters is

$$\omega_n T_R \approx 2.16 \zeta + 0.6 \quad 0.3 \leq \zeta \leq 0.8$$

Figures 19.17 and 19.18 show the unit step response of a second-order system as a function of damping ratio ζ.

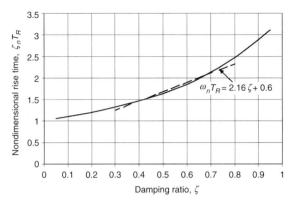

FIGURE 19.16 Rise time versus damping ratio, ζ.

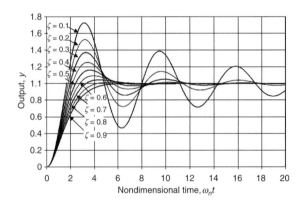

FIGURE 19.17 Second-order system step response versus damping ratio, ζ.

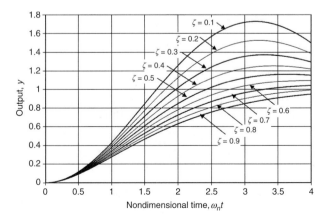

FIGURE 19.18 Initial second-order system step response versus damping ratio, ζ.

19.17 Frequency Response

The response of any dynamic system to a sinusoidal input is called the frequency response. A generic first-order system with a sinusoidal input of amplitude A would have the dynamic equation of

$$\frac{dy(t)}{dt} + \frac{1}{\tau}y(t) = f(t) = A\sin(\omega t)$$

where ω is the frequency of the sinusoidal input and τ is the first-order time constant. The steady-state solution to this equation is

$$y(t) = AM\sin(\omega t + \Phi)$$

where $M = 1/\sqrt{(\tau\omega)^2 + 1}$ is the amplitude ratio (a dimensionless quantity), and $\Phi = -\tan^{-1}(\tau\omega)$ is the phase angle.

Figure 19.19 is a plot of the magnitude ratio M_{dB} and the phase angle Φ as a function of the non-dimensional frequency, $\tau\omega$. Note that the magnitude is frequently plotted in terms of decibels, where $M_{dB} = 20\log_{10}(M)$.

The frequency at which the magnitude ratio equals 0.707 (or −3 dB) is called the bandwidth. For a first-order system, the bandwidth is inversely proportional to the time constant. So, $\omega = 1/\tau$.

A generic second-order system with a sinusoidal input of amplitude A and frequency ω would have the dynamic equation of

$$\frac{d^2y(t)}{dt^2} + 2\zeta\omega_n\frac{dy(t)}{dt} + \omega_n^2 y(t) = A\sin(\omega t)$$

The steady-state solution to this equation is

$$y(t) = \frac{AM}{\omega_n^2}\sin(\omega t + \Phi)$$

where

$$M = \frac{1}{\sqrt{[1-(\omega^2/\omega_n^2)]^2 + [2\zeta(\omega/\omega_n)]^2}}$$

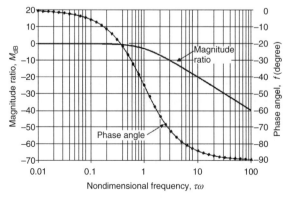

FIGURE 19.19 Frequency response for first-order system.

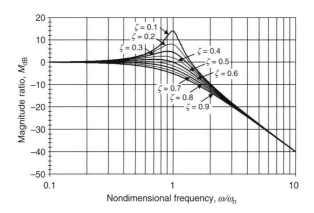

FIGURE 19.20 Frequency response magnitude for second-order system.

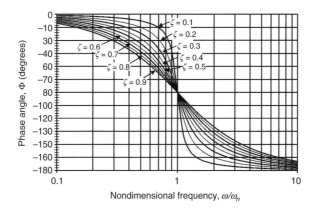

FIGURE 19.21 Frequency response phase angle for second-order system.

is the amplitude ratio (a dimensionless quantity), and

$$\Phi = -\tan^{-1}\left[\frac{2\zeta(\omega/\omega_n)}{1-(\omega^2/\omega_n^2)}\right]$$

is the phase angle.

Figures 19.20 and 19.21 are plots of the magnitude response $M_{dB} = 20\log_{10}(M)$ and the phase angle Φ for the second-order system as a function of damping ratio, ζ. The peak value in the magnitude response, M_P, can be found by taking the derivative of M with respect to ω and setting the result to zero to find (Nise, 1995)

$$M_P = \frac{1}{2\zeta\sqrt{1-\zeta^2}}$$

This peak value in M occurs at the frequency ω_P given by

$$\omega_P = \omega_n\sqrt{1-2\zeta^2}$$

The peak value in an experimentally determined frequency response can be used to estimate both the natural frequency and damping ratio for a second-order system. These parameters can then be used to estimate time domain responses such as peak time and percent overshoot.

Reference

Nise, N. S., *Control Systems Engineering*, 2nd ed., Benjamin/Cummings, 1995.

20
Sensors

20.1	Linear and Rotational Sensors		20-2
	Contact • Infrared • Resistive • Tilt (Gravity) • Capacitive • AC Inductive • DC Magnetic • Ultrasonic • Magnetostrictive Time-of-Flight • Laser Interferometry • References		
20.2	Acceleration Sensors		20-12
	Overview of Accelerometer Types • Dynamics and Characteristics of Accelerometers • Vibrations • Typical Error Sources and Error Modeling • Inertial Accelerometers • Electromechanical Accelerometers • Piezoelectric Accelerometers • Piezoresistive Accelerometers • Strain-Gauge Accelerometers • Electrostatic Accelerometers • Micro- and Nanoaccelerometers • Signal Conditioning and Biasing • References		
20.3	Force Measurement		20-34
	General Considerations • Hooke's Law • Force Sensors		
20.4	Torque and Power Measurement		20-48
	Fundamental Concepts • Arrangements of Apparatus for Torque and Power Measurement • Torque Transducer Technologies • Torque Transducer Construction, Operation, and Application • Apparatus for Power Measurement • References		
20.5	Flow Measurement		20-62
	Introduction • Terminology • Flow Characteristics • Flowmeter Classification • Differential Pressure Flowmeter • Variable Area Flowmeter • Positive Displacement Flowmeter • Turbine Flowmeter • Vortex Shedding Flowmeter • Electromagnetic Flowmeter • Ultrasonic Flowmeter • Coriolis Flowmeter • Two-Phase Flow • Flowmeter Installation • Flowmeter Selection • References		
20.6	Temperature Measurements		20-73
	Introduction • Thermometers That Rely Upon Differential Expansion Coefficients • Thermometers That Rely Upon Phase Changes • Electrical Temperature Sensors and Transducers • Noncontact Thermometers • Microscale Temperature Measurements • Closing Comments • References		
20.7	Distance Measuring and Proximity Sensors		20-88
	Distance Measuring Sensors • Proximity Sensors		
20.8	Light Detection, Image, and Vision Systems		20-119
	Introduction • Basic Radiometry • Light Sources • Light Detectors • Image Formation • Image Sensors • Vision Systems • References		
20.9	Integrated Microsensors		20-136
	Introduction • Examples of Micro- and Nanosensors • Future Development Trends • Conclusions • References		
20.10	Vision		20-153
	Digital Images • System Setup and Calibration • Machine Vision • References		

Kevin M. Lynch
Northwestern University

Michael A. Peshkin
Northwestern University

Halit Eren
Curtin University of Technology

M. A. Elbestawi
McMaster University

Ivan J. Garshelis
Magnova, Inc.

Richard Thorn
University of Derby

Pamela M. Norris
University of Virginia

Bouvard Hosticka
University of Virginia

Jorge Fernando Figueroa
NASA Stennis Space Center

H. R. (Bart) Everett
Space and Naval Warfare Systems Center

Stanley S. Ipson
University of Bradford

Chang Liu
University of Illinois

Nicolas Vazquez
Dinesh Nair
University of Texas

20.1 Linear and Rotational Sensors

Kevin M. Lynch and Michael A. Peshkin

By far the most common motions in mechanical systems are linear translation along a fixed axis and angular rotation about a fixed axis. More complex motions are usually accomplished by composing these simpler motions. In this chapter we provide a summary of some of the many technologies available for sensing linear and rotational motion along a single axis. We have arranged the sensing modalities according to the physical effect exploited to provide the measurement.

20.1.1 Contact

The simplest kind of displacement sensor is a mechanical switch which returns one bit of information: touching or not touching. A typical *microswitch* consists of a lever which, when depressed, creates a mechanical contact within the switch, which closes an electrical connection (Figure 20.1). Microswitches may be used as bump sensors for mobile robots, often by attaching a compliant material to the lever (such as a whisker) to protect the robot body from impact with a rigid obstacle. Another popular application of the microswitch in robotics is as a *limit switch*, indicating that a joint has reached the limit of its allowable travel.

Figure 20.2 shows a typical configuration for a microswitch. The pull-up resistor keeps the signal at +V until the switch closes, sending the signal to ground. As the switch closes, however, a series of micro-impacts may lead to "bounce" in the signal. A "debouncing" circuit may be necessary to clean up the output signal.

Switches may be designated NO or NC for normally open or normally closed, where "normally" indicates the unactivated or unpressed state of the switch. A switch may also have multiple *poles* (P) and one or two *throws* (T) for each pole. A pole moves as the switch is activated, and the throws are the possible contact points for the pole. Thus an SPDT (single pole double throw) switch switches a single pole from contact with one throw to the other, and a DPST (double pole single throw) switch switches two poles from open to closed circuit (Figure 20.3).

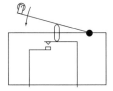

FIGURE 20.1 A typical microswitch.

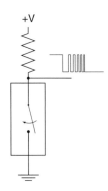

FIGURE 20.2 Signal bounce at a closing switch.

Sensors

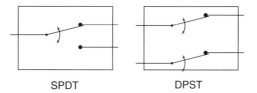

FIGURE 20.3 SPDT and DPST switch configurations.

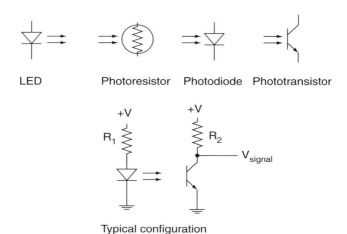

FIGURE 20.4 Optoelectronic circuit symbols and a typical emitter/detector configuration.

20.1.2 Infrared

Infrared light can be used in a variety of ways to measure linear and rotational displacement. Typically, an infrared light-emitting diode (LED), or *photoemitter*, is used as a source, and an infrared sensitive device is used to detect the emitted light. The detector could be a *photoresistor* or *photocell*, a variable resistor which changes resistance depending on the strength of the incident light (possibly infrared or visible); a *photodiode*, which allows the flow of electrical current in one direction in the presence of infrared light, and otherwise acts as an open circuit; or a *phototransistor*. In a phototransistor, the incident infrared light acts as the base current for the transistor, allowing the flow of collector current proportional to the strength of the received infrared light (up to saturation of the transistor). Circuit symbols for the various elements are shown in Figure 20.4.

If the emitter and detector are facing each other, they can be used as a beam-breaker, to detect if something passes between. This is called a *photointerrupter* (Figure 20.5). If the emitter and detector are free to move along the line connecting them, the strength of the received signal can be used to measure the distance separating them. Infrared photodetectors may be sensitive to ambient light, however. To distinguish the photoemitter light from background light, the source can be modulated (i.e., switched on and off at a high frequency), and the detector circuitry designed to respond only to the modulated infrared.

An emitter and detector facing the same direction can be used to roughly measure the distance to a nearby surface by the strength of the returned light reflecting off the surface. This is called a *photoreflector* (Figure 20.6). Alternatively, such a sensor could be used to detect light absorbing or light reflecting surfaces at a constant distance, as in mobile robot line following. Light polarizing filters can also be used on the emitter and detector so that the detector only recognizes light reflected by a special "optically active" retroreflecting surface.

FIGURE 20.5 The Fairchild semiconductor QVA11234 photointerrupter.

FIGURE 20.6 The Fairchild semiconductor QRB1114 photoreflective sensor.

FIGURE 20.7 A position sensitive detector, UDT Sensors, Inc.

Photointerrupters and photoreflectors can be bought prepackaged or constructed separately from an infrared LED and a photodiode or phototransistor, after making certain the detector is sensitive to the wavelength produced by the LED.

Photoreflector units are also available with more advanced position sensitive detectors (PSDs), which report the location of infrared light incident on the sensing surface (Figure 20.7). The fixed location of the LED relative to the PSD, as well as the location of the image of the infrared light on the PSD, allows the

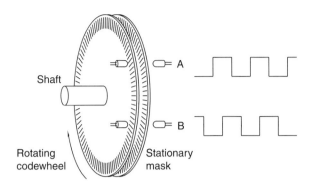

FIGURE 20.8 Schematic of an incremental encoder.

use of triangulation to determine the distance to the target. Such distance measuring sensors are manufactured by Sharp and Hamamatsu.

20.1.2.1 Optical Encoders

An optical encoder uses photointerrupters to convert motion into an electrical pulse train. These electrical pulses "encode" the motion, and the pulses are counted or "decoded" by circuitry to produce the displacement measurement. The motion may be either linear or rotational, but we focus on more common rotary optical encoders.

There are two basic configurations for rotary optical encoders, the *incremental* encoder and the *absolute* encoder. In an incremental encoder, a disk (or codewheel) attached to a rotating shaft spins between two photointerrupters (Figure 20.8). The disk has a radial pattern of lines, deposited on a clear plastic or glass disk or cut out of an opaque disk, so that as the disk spins, the radial lines alternately pass and block the infrared light to the photodetectors. (Typically there is also a stationary mask, with the same pattern as the rotating codewheel, in the light path from the emitters to the detectors.) This results in pulse trains from each of the photodetectors at a frequency proportional to the angular velocity of the disk. These signals are labeled A and B, and they are 1/4 cycle out of phase with each other. The signals may come from photointerrupters aligned with two separate tracks of lines at different radii on the disk, or they may be generated by the same track, with the photointerrupters placed relative to each other to give out of phase pulse trains.

By counting the number of pulses and knowing the number of radial lines in the disk, the rotation of the shaft can be measured. The direction of rotation is determined by the phase relationship of the A and B pulse trains, that is, which signal leads the other. For example, a rising edge of A while B = 1 may indicate counterclockwise rotation, while a rising edge of A while B = 0 indicates clockwise rotation. The two out-of-phase signals are known as quadrature signals.

Incremental encoders commonly have a third output signal called the index signal, labeled I or Z. The index signal is derived from a separate track yielding a single pulse per revolution of the disk, providing a home signal for absolute orientation. In practice, multiple photointerrupters can be replaced by a single source and a single array detecting device.

IC decoder chips are available to decode the pulse trains. The inputs to the chip are the A and B signals, and the outputs are one or more pulse trains to be fed into a counter chip. For example, the US Digital LS7083 outputs two pulse trains, one each for clockwise and counterclockwise rotation, which can be sent to the inputs of a 74193 counter chip (Figure 20.9). Standard decoding methods for the quadrature input are 1X, 2X, and 4X resolution. In 1X resolution, a single count is generated for the rising or falling edge of just one of the pulse trains, so that the total number of encoder counts for a single revolution of the disk is equal to the number of lines in the disk. In 4X resolution, a count is generated for each rising and falling edge of both pulse trains, resulting in four times the angular resolution. An encoder

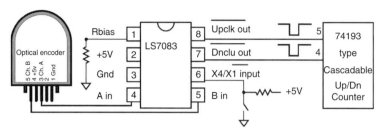

FIGURE 20.9 An optical encoder, US Digital LS7083 quadrature decoder chip, and counter. (Courtesy of US Digital, Inc.)

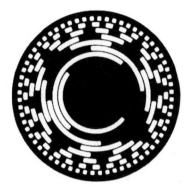

FIGURE 20.10 An 8-bit Gray code absolute encoder disk. (Courtesy of BEI Technologies Industrial Encoder Division.)

with 1000 lines on the code wheel being decoded at 4X resolution yields an angular resolution of $360°/(4 \times 1000) = 0.09°$.

While a *single-ended output* encoder provides the signals A, B, and possibly Z, a *differential output* encoder also provides the complementary outputs A′, B′, and Z′. Differential outputs, when used with a differential receiver, can increase the electrical noise immunity of the encoder.

A drawback of the incremental encoder is that there is no way to know the absolute position of the shaft at power-up without rotating it until the index pulse is received. Also, if pulses are momentarily garbled due to electrical noise, the estimate of the shaft rotation is lost until the index pulse is received. A solution to these problems is the absolute encoder. An absolute encoder uses k photointerrupters and k code tracks to produce a k-bit binary word uniquely representing 2^k different orientations of the disk, giving an angular resolution of $360°/2^k$ (Figure 20.10). Unlike an incremental encoder, an absolute encoder always reports the absolute angle of the encoder.

The radial patterns on the tracks are arranged so that as the encoder rotates in one direction, the binary word increments or decrements according to a binary code. Although natural binary code is a possibility, the Gray code is a more common solution. With natural binary code, incrementing by one may change many or all of the bits, for example, 7 to 8 in decimal is 0111 to 1000 in natural binary. With the Gray code, only one bit changes as the number increments or decrements, for example, 7 to 8 in decimal is 0100 to 1100 in Gray code. The rotational uncertainty during a Gray code transition is only a single count, or $360°/2^k$. With the natural binary code, an infinitesimal misalignment between the lines and the photointerrupters may cause the reading to briefly go from 0111 (7) to 1111 (15) during the transition to 1000 (8).

In general, incremental encoders provide higher resolution at a lower cost and are the most common choice for many industrial and robotic applications.

20.1.3 Resistive

One of the simplest and least expensive ways to measure rotational or linear motion is using a variable resistor called a *potentiometer* or *rheostat*. We focus on rotary potentiometers, or "pots" for short, but the principle of operation is the same in the linear case.

A pot consists of three terminals (Figure 20.11a,b). Two end terminals, call them terminals 1 and 3, connect to either end of a length of resistive material, such as partially conductive plastic, ceramic, or a long thin wire. (For compactness, the long wire is wound around in loops to make a coil, leading to the name *wirewound* potentiometer.)

The other terminal, terminal 2, is connected to a *wiper*, which slides over the material as the pot shaft rotates. The total resistance of the pot R_{13} is equal to the sum of the resistance R_{12} between terminal 1 and the wiper, and the resistance R_{23} between the wiper and terminal 3. Typically the wiper can rotate from one end of the resistive material ($R_{13} = R_{12}$) to the other ($R_{13} = R_{23}$). If the full motion of the wiper is caused by one revolution of the shaft or less, the pot is called a *single-turn* pot. If the full motion is caused by multiple revolutions, it is called a *multi-turn* pot.

Typically a pot is used by connecting terminal 1 to a voltage V, terminal 3 to ground, and using the voltage at the wiper as a measure of the rotation. The voltage observed at the wiper is $V(R_{23}/R_{13})$ and is a linear function of the rotation of the shaft.

A remarkably simple absolute sensor for a wide range of distances is the string pot or draw-wire sensor (Figure 20.12). It consists of a string wrapped on a spool, with a potentiometer to monitor rotations of the spool. A return spring keeps the string taut. Lengths up to many meters may be measured, using sensors incorporating multi-turn pots. The same technique is similarly useful for short distances (a few centimeters) using compact single-turn pots and a small spool. Both tolerate misalignment or arc-like motion well. String pots are susceptible to damage to the string in exposed applications, but the sensor element is small and unobtrusive. Manufacturers include RDP Electronics, SpaceAge Control, and UniMeasure.

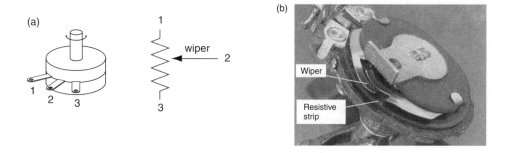

FIGURE 20.11 (a) As the shaft of the potentiometer rotates, the wiper moves from one end of the resistive material to the other. (b) The inside of a typical potentiometer, showing the wiper contacting a resistive strip.

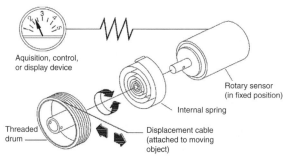

FIGURE 20.12 A string pot. (Courtesy of Space Age Control, Inc.)

Another type of resistive sensor is the flexible bend sensor. Conductive ink between two electrical contacts on a flexible material changes resistance as the material bends and stretches. Used in a voltage divider with a fixed resistor, the analog voltage may be used as a measure of the bend. Such a sensor could be used to detect contact (like a whisker) or as a rough measure of the deformation of a surface to which it is attached.

20.1.4 Tilt (Gravity)

A *mercury switch* can be used to provide one bit of information about orientation relative to the gravity vector. A small drop of mercury enclosed in a glass bulb opens or closes the electrical connection between two leads depending on the orientation of the sensor. Several mercury switches at different orientations may be used to get a rough estimate of tilt. The signal from a mercury switch may "bounce" much like the signal from a mechanical contact switch (Figure 20.2).

An *inclinometer* can be used to measure the amount of tilt. One example is the *electrolytic tilt sensor*. Manufacturers include The Fredericks Company and Spectron Glass. Two-axis models have five parallel rod-like electrodes in a sealed capsule, partially filled with a conductive liquid. Four of the electrodes are at the corners of a square, with one in the middle. Tilting the sensor changes the distribution of current injected via the center electrode in favor of the electrodes which are more deeply immersed.

Tilt sensors may be obtained with liquids of varying viscosity, to minimize sloshing. Because a DC current through the liquid would cause electrolysis and eventually destroy the sensor, AC measurements of conductivity are used. As a result, the support electronics are not trivial.

The liquid conductivity is highly temperature dependent. The support electronics for the tilt sensor must rely on a ratio of conductivity between pairs of rods. Also, although the electrolytic tilt sensor operates over a wide temperature range, it is greatly disturbed by nonuniformities of temperature across the cell.

Another kind of simple inclinometer can be constructed from a rotary potentiometer with a pendulum bob attached. A problem with this solution is that friction may stop the bob's motion when it is not vertical. A related idea is to use an absolute optical encoder with a pendulum bob. Complete sensors operating on this principle can be purchased with advanced options, such as magnetic damping to reduce overshoot and oscillation. An example is US Digital's 12-bit A2I absolute inclinometer.

Of course, gravity acting on a device is indistinguishable from acceleration. If the steady-state tilt of a device is the measurement of interest, simple signal conditioning should be used to ensure that the readings have settled.

Other more sophisticated tilt sensors include gyroscopes and microelectromechanical (MEMS) devices, which are not discussed here.

20.1.5 Capacitive

Capacitance can be used to measure proximity or linear motions on the order of millimeters. The capacitance C of a parallel plate capacitor is given by $C = \varepsilon_r \varepsilon_o A/d$, where ε_r is the relative permittivity of the dielectric between the plates, ε_o is the permittivity of free space, A is the area of overlap of the two plates, and d is the plate separation. As the plates translate in the direction normal to their planes, C is a nonlinear function of the distance d. As the plates translate relative to each other in their planes, C is a linear function of the area of overlap A. Used as proximity sensors, capacitive sensors can detect metallic or nonmetallic objects, liquids, or any object with a dielectric constant greater than air.

One common sensing configuration has one plate of the capacitor inside a probe, sealed in an insulator. The external target object forms the other plate of the capacitor, and it must be grounded to the proximity sensor ground. As the sensor approaches the target, the capacitance increases, modifying the oscillation of a detector circuit including the capacitor. This altered oscillation may be used to signal proximity or to obtain a distance measurement.

Manufacturers of capacitive sensors include Cutler-Hammer and RDP Electronics.

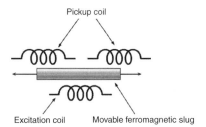

FIGURE 20.13 Operating principle of an LVDT.

20.1.6 AC Inductive

20.1.6.1 LVDT

The best known AC inductive sensor is the *linear variable differential transformer*, or LVDT. The LVDT is a tube with a plunger, the displacement of the plunger being the variable to be measured (Figure 20.13). The tube is wrapped with at least two coils, an excitation coil and a pickup coil. An AC current (typically 1 kHz) is passed through the excitation coil, and an AC signal is detected from the pickup coil and compared in magnitude and in phase (0° or 180°) to the excitation current. Support electronics are needed for the demodulation, which is called synchronous detection. The plunger carries a ferromagnetic slug, which enhances the magnetic coupling from the excitation coil to the pickup coil. Depending on the position of the slug within the pickup coil, the detected signal may be zero (when the ferrite slug is centered in the pickup coil), or increasing in amplitude in one or the other phase, depending on displacement of the slug.

LVDTs are a highly evolved technology and can be very accurate, in some cases to the micron level. They have displacement ranges of millimeters up to a meter. They do not tolerate misalignment or nonlinear motion, as a string pot does.

20.1.6.2 Resolvers

A *resolver* provides a measure of shaft angle, typically with sine and cosine analog outputs. It uses an AC magnetic technique similar to the LVDT, and similar support electronics to provide synchronous detection. Resolvers are very rugged and for this reason are often preferred over optical encoders on motor shafts, although they are not as accurate and they have greater support electronics requirements. Some resolver drives have extra outputs as if they were incremental encoders, for compatibility. Additionally, resolvers provide an absolute measure of shaft angle. The resolver, like the LVDT, is a well established and evolved technology.

20.1.7 DC Magnetic

A magnetic field acting on moving electrons (e.g., a current in a semiconductor) produces a sideways force on the electrons, and this force can be detected as a voltage perpendicular to the current. The effect is small, even in semiconductors, but has become the basis of a class of very rugged, inexpensive, and versatile sensors.

20.1.7.1 Hall Effect Switches

Hall effect switches refers to devices which produce a binary output, depending on whether the magnetic field intensity exceeds a threshold or not. In their component form, these switches may be packaged as 3-terminal devices the size of a transistor package (TO-92) or surface mount, having only a power lead (3–24 V), a ground lead, and an output lead. Typically the output is pulled to ground, or not, depending on the magnetic state. Hall effect switches are also available in environmental packages of all sorts.

The actuation threshold ranges from a few gauss (the Earth's magnetic field is 1/2 G) up to the hundreds of gauss levels typical of permanent magnets. Often there is a fair degree of unit-to-unit variability in threshold.

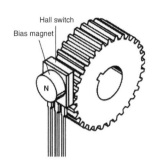

FIGURE 20.14 Detecting gear teeth in a ferrous material using a Hall switch and a bias magnet. (Courtesy of Allegro Microsystems, Inc.)

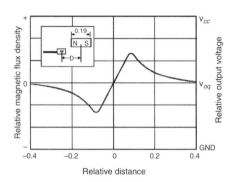

FIGURE 20.15 Output of an analog Hall sensor versus position relative to a magnet. (Courtesy of Allegro Microsystems, Inc.)

Hall effect switches are hysteretic: their "turn-on" threshold may be different than the "turn-off" threshold. Sometimes hysteresis is used to make a switch latching, so that it stays in its last state (on or off) until the applied magnetic field is reversed. Non-latching Hall switches may be unipolar (responding only to one orientation of magnetic field) or bipolar (responding to a field of either polarity). Turn-on and turn-off times are in microseconds.

Hall switches have wide operating temperature ranges and are often used in automobile engine compartments. Another advantage is that they are not susceptible to most of the fouling mechanisms of optical or mechanical switches, such as liquids or dirt. While often the moving part that is detected is a magnet, it can also be arranged that a stationary "bias" magnet is intensified in its effect on the hall switch by the approach of a ferrous part, such as a gear tooth, thus allowing nonmagnetized objects to be detected (Figure 20.14).

Typical applications are the detection of a moving part, replacing a mechanical limit switch. The Hall switch has no moving or exposed parts and is wear-free. Another common use is in indexing of rotational or translational motion. The Hall switch is installed to detect one position, and its output pulse is used as a reference for an incremental encoder which can count distance from that reference point. Hall switches are inexpensive and small, so a number of them can be spaced at intervals of millimeters, forming a low-resolution linear or rotational encoder or multi-position switch. Such an encoder or switch has the ruggedness advantages of Hall switches.

20.1.7.2 Analog Hall Sensors

In a package the same small size as Hall switches, and costing little more, one can also get Hall devices that have an analog output proportional to magnetic field strength (Figure 20.15). Typically, these have full-scale magnetic field sensitivity in the 100 G range. These are not useful as a compass in the Earth's sub-gauss magnetic field.

Hall sensors are useful as linear or rotational encoders. Two Hall sensors may be arranged at right angles to detect the sine and cosine of the angle of a rotating magnet, thus forming an absolute rotation sensor.

Commercially available devices of this nature are called "Hall potentiometers" and have a variety of outputs (e.g., sine and cosine, or a linear ramp repeating with each revolution). In contrast to potentiometers with resistive strips and sliders, Hall pots allow continuous 360° rotation and experience no wear. All Hall effect devices are susceptible to external magnetic fields, however.

Hall sensors are also excellent transducers of short linear or arc-like motions. The motion of a bar magnet past a Hall sensor exposes the sensor to a magnetic field—which can be arranged to vary linearly with displacement—over a range of several millimeters up to several centimeters. (The bar magnet travels less than its own length.) Commercial implementations are known as throttle position sensors.

20.1.7.3 Tape-Based Sensors

There are a number of linear and rotational sensors, both incremental and absolute, which are similar to optical encoders but use magnetic patterns rather than optical ones. Linear applications are likely to require an exposed strip. In exposed applications, magnetic sensors have advantages in resistance to dirt, although the magnetic stripes must be protected from damage.

20.1.8 Ultrasonic

Ultrasonic (US) sensors use the time-of-flight of a pulse of ultrasonic sound through air or liquid to measure distance. Sensors are available with ranges from a few centimeters to 10 m. A great advantage of US sensors is that all of the sensor's electronic and transducer components are in one location, out of harm's way. The corresponding disadvantages are that US sensors tend to be indiscriminate: they may detect spurious targets, even very small ones, especially if these are near the transducer. Sensors are available with carefully shaped beams (down to 7°) to minimize detection of spurious targets. Some include compensation for variation in air temperature, which affects sound velocity. US sensors can be used in surprising geometries. For instance, they can be used to detect the liquid level in a vertical pipe; back-reflection of sound pulses from the walls of a smooth pipe are minimal.

There is also an inexpensive and easily interfaced US sensor from Polaroid, derived from a ranging device for an instant camera, which is popular with experimenters.

Ultrasonic sensors typically have an analog output proportional to distance to target. Accuracies of the 1% level can be expected in a well-controlled environment.

20.1.9 Magnetostrictive Time-of-Flight

A more accurate technique for using time-of-flight to infer distance is the *magnetostrictive wire transducer* (MTS). A moving magnet forms the "target" corresponding to the acoustic target in US sensors, and need not touch the magnetostrictive wire which is the heart of the device. The magnet's field acting on the magnetostrictive wire creates an ultrasonic pulse in the wire when a current pulse is passed through the wire. The time interval from the current pulse to the detection of the ultrasonic pulse at the end of the wire is used to determine the position of the magnet along the wire (Figure 20.16).

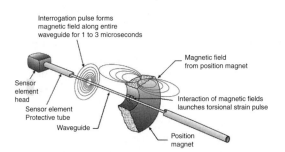

FIGURE 20.16 Principle of operation of a magnetostrictive linear position sensor. (Courtesy of Temposonics, Inc.)

The magnetostrictive transducer does not have the inherent compactness and ruggedness of ultrasonic through air, but does achieve similarly large measurement lengths, up to several meters. Accuracy and stability are excellent, far better than ultrasonic. Some misalignment or nonlinear motion is tolerated, because the target magnet does not need to be in very close proximity to the magnetostrictive wire.

20.1.10 Laser Interferometry

Laser interferometers are capable of measuring incremental linear motions with resolution on the order of nanometers. In an interferometer, collimated laser light passes through a beam-splitter, sending the light energy on two different paths. One path is directly reflected to the detector, such as an optical sensing array, giving a flight path of fixed length. The other path reflects back to the detector from a retroreflector (mirror) attached to the target to be measured. The two beams constructively or destructively interfere with each other at the detector, creating a pattern of light and dark fringes. The interference pattern can be interpreted to find the phase relationship between the two beams, which depends on the relative lengths of the two paths, and therefore the distance to the moving target. As the target moves, the pattern repeats when the length of the variable path changes by the wavelength of the laser. Thus the laser interferometer is inherently an incremental measuring device.

Laser interferometers are easily the most expensive sensors discussed in this chapter. They also have the highest resolution. Laser interferometers are very sensitive to mechanical misalignment and vibrations.

More information about sensors may be found in Sensors magazine (http://www.sensorsmag.com/).

References

1. Histand, M. B. and Alciatore, D. G., *Introduction to Mechatronics and Measurement Systems*, McGraw-Hill, Boston, MA, 1999.
2. Bolton, W., *Mechatronics*, 2nd ed., Addison Wesley Longman, New York, NY, 1999.
3. Horowitz, P. and Hill, W., *The Art of Electronics*, 2nd ed., Cambridge University Press, Cambridge, UK, 1998.
4. Auslander, D. M. and Kempf, C. J., *Mechatronics: Mechanical System Interfacing*, Prentice-Hall, Upper Saddle River, NJ, 1996.
5. Jones, J. L., Flynn, A. M., and Seiger, B. A., *Mobile Robots: Inspiration to Implementation*, 2nd ed., A. K. Peters, Boston, MA, 1999.

20.2 Acceleration Sensors

Halit Eren

Acceleration relating to motion is an important section of kinematic quantities: position, velocity, acceleration, and jerk. Each one of these quantities has a linear relationship with the neighboring ones. That is, all the kinematic quantities can be derived from a single quantity. For example, acceleration can be obtained by differentiating the corresponding velocity or by integrating the jerk. Likewise, velocity can be obtained by differentiating the position or by integrating the acceleration. In practice, only integration is widely used since it provides better noise characteristics and attenuation.

There are two classes of acceleration measurements techniques: *direct* measurements by specific accelerometers and *indirect* measurements where velocity is differentiated. The applicability of these techniques depends on the type of motion (rectilinear, angular, or curvilinear motion) or equilibrium centered vibration. For rectilinear and curvilinear motions, the direct measurement accelerometers are preferred. However, the angular acceleration is usually measured by indirect methods.

Acceleration is an important parameter for general-purpose absolute motion measurements, vibration, and shock sensing. For these measurements, accelerometers are commercially available in a wide range and many different types to meet diverse application requirements, mainly in three areas: (1) *Commercial applications*—automobiles, ships, appliances, sports and other hobbies; (2) *Industrial applications*—robotics, machine control, vibration testing and instrumentation; and (3) *High reliability applications*—military, space and aerospace, seismic monitoring, tilt, vibration and shock measurements.

Accelerometers have been in use for many years. Early accelerometers were mechanical types relying on analog electronics. Although early accelerometers still find many applications, modern accelerometers are essentially semiconductor devices within electronic chips integrated with the signal processing circuitry. Mechanical accelerometers detect the force imposed on a mass when acceleration occurs. A new type of accelerometer, the thermal type, senses the position through heat transfer.

20.2.1 Overview of Accelerometer Types

A basic accelerometer consists of a mass that is free to move along a sensitive axis within a case. The technology is largely based on this basic accelerometer and can be classified in a number of ways, such as mechanical or electrical, active or passive, deflection or null-balance accelerometers, etc. The majority of industrial accelerometers are classified as either deflection or null-balance types. Accelerometers used in vibration and shock measurements are usually the deflection types, whereas those used for the measurement of motions of vehicles, aircraft, and so on for navigation purposes may be either deflection or null-balance type.

This article will concentrate on the direct measurements of acceleration, which can be achieved by the accelerometers of the following types:

- Inertial and mechanical
- Electromechanical
- Piezoelectric
- Piezoresistive
- Strain gauges
- Capacitive and electrostatic force balance
- Micro- and nanoaccelerometers

Depending on the principles of operations, these accelerometers have their own subclasses.

20.2.2 Dynamics and Characteristics of Accelerometers

Acceleration is related to motion, a vector quantity, exhibiting a direction as well as magnitude. The direction of motion is described in terms of some arbitrary Cartesian or orthogonal coordinate systems. Typical rectilinear, angular, and curvilinear motions are illustrated in Figures 20.17a–c, respectively.

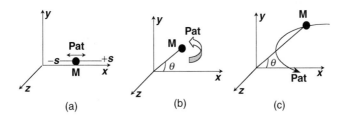

FIGURE 20.17 Types of motions to which accelerometers are commonly applied.

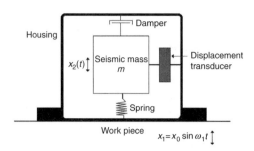

FIGURE 20.18 A typical seismic accelerometer.

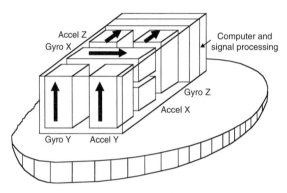

FIGURE 20.19 Arrangements of accelerometers in navigation and guidance systems.

The governing equations of these motions are as follows:

$$\text{Rectilinear acceleration} \quad a = \lim_{\Delta t \to 0} \frac{\Delta v}{\Delta t} = \frac{dv}{dt} = \frac{d(ds/dt)}{dt} = \frac{d^2 s}{dt^2} \quad (20.1)$$

$$\text{Angular acceleration} \quad \alpha = \lim_{\Delta t \to 0} \frac{\Delta \omega}{\Delta t} = \frac{d\omega}{dt} = \frac{d(d\theta/dt)}{dt} = \frac{d^2 \theta}{dt^2} \quad (20.2)$$

$$\text{Curvilinear acceleration} \quad a = \frac{dv}{dt} = \frac{d^2 x}{dt^2} i + \frac{d^2 y}{dt^2} j + \frac{d^2 z}{dt^2} k \quad (20.3)$$

where a and α are the accelerations; v and ω are the speeds; s is the distance; θ is the angle; i, j, and k are the unit vectors in x, y, and z directions, respectively.

For the correct applications of the accelerometers, a sound understanding of the characteristics of the motion under investigation is very important. The application areas may be linear and vibratory motion, angular motion, monitoring of the tilt of an object, or various forms of combinations. In each case, the correct selection and mounting of the accelerometer is necessary.

The majority of accelerometers can be viewed and analyzed as a single-degree-of-freedom seismic instrument that can be characterized by a mass, a spring, and a damper arrangement as shown in Figure 20.18. In the case of multi-degrees-of-freedom systems, the principles of curvilinear motion can be applied as in Equation 20.3 and multiple transducers must be used to create uniaxial, biaxial, or triaxial sensing points of the measurements. A typical example is the inertial navigation and guidance systems as illustrated in Figure 20.20. In such applications, acceleration sensors play an important role in orientation and direction finding. Usually, miniature triaxial sensors detect changes in roll, pitch, and azimuth in x, y, and z directions.

If a single-degree-of-freedom system behaves linearly in a time invariant manner, the basic second-order differential equation describing the motion of the forced mass–spring system can be written as

$$f(t) = m\frac{d^2 x}{dt^2} + c\frac{dx}{dt} + kx \tag{20.4}$$

where $f(t)$ is the force, m is the mass, c is the velocity constant, and k is the spring constant.

Nevertheless, the base of the accelerometer is in motion too. When the base is in motion, the force is transmitted through the spring to the suspended mass, depending on the transmissibility of the force to the mass. Equation 20.4 may be generalized by taking the effect motion of the base into account as

$$m\frac{d^2 z}{dt^2} + c\frac{dz}{dt} + kz = mg\cos\theta - m\frac{d^2 x_1}{dt^2} \tag{20.5}$$

where $z = x_2 - x_1$ is the relative motion between the mass and the base, x_1 is the displacement of the base, x_2 is the displacement of mass, and θ is the angle between the sense axis and gravity.

The complete solution to Equation 20.5 can be obtained by applying the superposition principle. The sup-erposition principle states that if there are simultaneously superimposed actions on a body, the total effect can be obtained by summing the effects of each individual action. Using superposition and using Laplace transforms gives

$$\frac{X(s)}{F(s)} = \frac{1}{ms^2} + cs + k \tag{20.6}$$

or

$$\frac{X(s)}{F(s)} = \frac{K}{s^2/\omega_n^2 + 2\zeta s/\omega_n + 1} \tag{20.7}$$

where s is the Laplace operator, $K = 1/k$ is the static sensitivity, $\omega_n = \sqrt{k/m}$ is the undamped critical frequency (rad/s), and $\zeta = (c/2)\sqrt{km}$ is the damping ratio.

As can be seen in the performance of accelerometers, the important parameters are the static sensitivity, the natural frequency, and the damping ratio, which are all functions of mass, velocity, and spring constants. Accelerometers are designed and manufactured to have different characteristics by suitable selection of these parameters. A short list of major manufacturers is given in Table 20.1.

20.2.3 Vibrations

This section is concerned with applications of accelerometers to measure physical properties such as acceleration, vibration and shock, and the motion in general. Although there may be fundamental differences in the types of motions, a sound understanding of the basic principles of the vibration will lead to the applications of accelerometers in different situations by making appropriate corrective measures.

Vibration is an oscillatory motion resulting from application of varying forces to a structure. The vibrations can be periodic, stationary random, nonstationary random, or transient.

20.2.3.1 Periodic Vibrations

In periodic vibrations, the motion of an object repeats itself in an oscillatory manner. This can be represented by a sinusoidal waveform

$$x(t) = X_p \sin(\omega t) \tag{20.8}$$

TABLE 20.1 List of Manufacturers

Analog Devices, Inc. 1 Technology Way, P.O. Box 9106 Norwood, MA 02062-9106 USA Tel: 781-329-4700 Fax: 781-326-8703	Kistler Instrument Corp. 75 John Glenn Dr. Amherst, NY 14228 2119 USA Tel: 888-KISTLER (547-8537) Fax: 716-691-5226
Aydin Telemetry 47 Friends Lane & Penns Trail Newtown, PA 18940 0328 USA Tel: 215-968-4271 Fax: 215-968-3214	Oceana Sensor Technologies, Inc. 1632-T Corporate Landing Pkwy. Virginia Beach, VA 23454 USA Tel: 757-426-3678 Fax: 757-426-3633
Bruel & Kjaer 2815-A Colonnades Court Norcross, GA 30071 USA Tel: 800-332-2040 Fax: 770-447-8440	PCB Piezotronics, Inc. 3425-T Walden Ave. Depew, NY 14043 2495 USA Tel: 716-684-0001 Fax: 716-684-0987
Dytran Instruments, Inc. 21592 Marilla St. Chatsworth, CA 91311 USA Tel: 800-899-7818 Fax: 800-899-7088	Piezo Systems, Inc. 186 Massachusetts Ave. Cambridge, MA 02139-4229 USA Tel: 617-547-1777 Fax: 617-354-2200
Endevco Corp. 30700 Rancho Viejo Rd. San Juan Capistrano, CA 92675 USA Tel: 800-982-6732 Fax: 949-661-7231	Rieker Instrument Co. P.O. Box 128 Folcroft, PA 19032 0128 USA Tel: 610-534-9000 Fax: 610-534-4670
Entran Devices, Inc. 10-T Washington Ave. Fairfield, NJ 07004 USA Tel: 888-8-ENTRAN (836-8726) Fax: 973-227-6865	Sensotec, Inc. 2080 Arlingate Lane Columbus, OH 43228 USA Tel: 800-858-6184 Fax: 614-850-1111
GS Sensors, Inc. 16 W. Chestnut St. Ephrata, PA 17522 USA Tel: 717-721-9727 Fax: 717-721-9859	Techkor Instrumentation 2001 Fulling Mill Rd. P.O. Box 70 Dept. T-1 New Cumberland, PA 17057-0070 USA Tel: 800-697-4567 Fax: 717-939-7170

where $x(t)$ is the time-dependent displacement, $\omega = 2\pi f t$ is the angular frequency, and X_p is the maximum displacement from a reference point.

The velocity of the object is the time rate of change of displacement,

$$v(t) = \frac{dx}{dt} = \omega X_p \cos(\omega t) = V_p \sin(\omega t + \pi/2) \qquad (20.9)$$

where $v(t)$ is the time-dependent velocity, and $V_p = \omega X_p$ is the maximum velocity.

The acceleration of the object is the time rate of change of velocity,

$$a(t) = \frac{dv}{dt} = \frac{d^2 x}{dt^2} = -\omega^2 X_p \sin(\omega t) = A_p \sin(\omega t + \pi) \qquad (20.10)$$

where $a(t)$ is the time-dependent acceleration, and $A_p = \omega^2 X_p = \omega V_p$ is the maximum acceleration.

Sensors

From the preceding equations it can be seen that the basic form and the period of vibration remains the same in acceleration, velocity, and displacement. But velocity leads displacement by a phase angle of 90° and acceleration leads velocity by another 90°.

In nature, vibrations can be periodic, but not necessarily sinusoidal. If they are periodic but nonsinusoidal, they can be expressed as a combination of a number of pure sinusoidal curves, determined by Fourier analysis as

$$x(t) = X_0 + X_1 \sin(\omega_1 t + \Phi_1) + X_2 \sin(\omega_2 t + \Phi_2) + \cdots + X_n \sin(\omega_n t + \Phi_n) \qquad (20.11)$$

where $\omega_1, \omega_2, \ldots, \omega_n$ are the frequencies (rad/s), $X_0, X_1, \ldots, X_n$ are the maximum amplitudes of respective frequencies, and $\phi_1, \phi_2, \ldots, \phi_n$ are the phase angles.

The number of terms may be infinite, and the higher the number of elements, the better the approximation. These elements constitute the frequency spectrum. The vibrations can be represented in the time domain or frequency domain, both of which are extremely useful in analysis.

20.2.3.2 Stationary Random Vibrations

Random vibrations are often met in nature, where they constitute irregular cycles of motion that never repeat themselves exactly. Theoretically, an infinitely long time record is necessary to obtain a complete description of these vibrations. However, statistical methods and probability theory can be used for the analysis by taking representative samples. Mathematical tools such as probability distributions, probability densities, frequency spectra, cross-correlations, auto-correlations, digital Fourier transforms (DFTs), fast Fourier transforms (FFTs), auto-spectral-analysis, root mean squared (rms) values, and digital filter analysis are some of the techniques that can be employed.

20.2.3.3 Nonstationary Random Vibrations

In this case, the statistical properties of vibrations vary in time. Methods such as time averaging and other statistical techniques can be employed.

20.2.3.4 Transients and Shocks

Often, short duration and sudden occurrence vibrations need to be measured. Shock and transient vibrations may be described in terms of force, acceleration, velocity, or displacement. As in the case of random transients and shocks, statistical methods and Fourier transforms are used in the analysis.

20.2.4 Typical Error Sources and Error Modeling

Acceleration measurement errors occur due to four primary reasons: sensors, acquisition electronics, signal processing, and application specific errors. In the direct acceleration measurements, the main error sources are the sensors and data acquisition electronics. These errors will be discussed in the biasing section and in some cases, sensor and acquisition electronic errors may be as high as 5%. Apart from these errors, sampling and A/D converters introduce the usual errors, which are inherent in them and exist in all computerized data acquisition systems. However, the errors may be minimized by the careful selection of multiplexers, sample-and-hold circuits, and A/D converters.

When direct measurements are made, ultimate care must be exercised for the selection of the correct accelerometer to meet the requirements of a particular application. In order to reduce the errors, once the characteristics of the motion are studied, the following particulars of the accelerometers need to be considered: the transient response or cross-axis sensitivity; frequency range; sensitivity, mass and dynamic range; cross-axis response; and environmental conditions, such as temperature, cable noise, stability of bias, scale factor, and misalignment, etc.

20.2.4.1 Sensitivity of Accelerometers

During measurements, the transverse motions of the system affect most accelerometers and the sensitivity to these motions plays a major role in the accuracy of the measurement. The transverse, also known as cross-axis, sensitivity of an accelerometer is its response to acceleration in a plane perpendicular to the

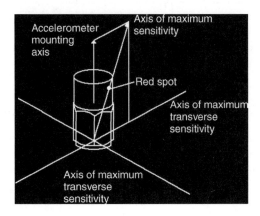

FIGURE 20.20 Illustration of cross-axis sensitivity.

main accelerometer axis as shown in Figure 20.20. The cross-axis sensitivity is normally expressed as a percentage of the main-axis sensitivity and should be as low as possible. The manufacturers usually supply the direction of minimum sensitivity.

The measurement of the maximum cross-axis sensitivity is part of the individual calibration process and should always be less than 3% or 4%. If high levels of transverse vibration are present, this may result in erroneous overall results. In this case, separate arrangements should be made to establish the level and frequency contents of the cross-axis motions. The cross-axis sensitivity of typical accelerometers mentioned in the relevant sections were 2–3% for piezoelectric types and less than 1% for most others.

20.2.4.2   The Frequency Range

Acceleration measurements are normally confined to using the linear portion of the response curve of the accelerometer. The response is limited at the low frequencies as well as at the high frequencies by the natural resonances. As a rule of thumb, the upper-frequency limit for the measurement can be set to one-third of the accelerometer's resonance frequency such that the vibrations measured will be less than 1 dB in linearity. It should be noted that an accelerometer's useful frequency range may be significantly higher; that is, one-half or two-thirds of its resonant frequency. The measurement frequencies may be set to higher values in applications in which lower linearity (say 3 dB) is acceptable. The lower measuring frequency limit is determined by two factors. The first is the low-frequency cutoff of the associated preamplifiers. The second is the effect of ambient temperature fluctuations to which the accelerometer may be sensitive.

20.2.4.3   The Mass of Accelerometer and Dynamic Range

Ideally, the higher the transducer sensitivity, the better. Compromises may have to be made for sensitivity versus frequency, range, overload capacity, size, and so on.

In some cases, high errors will be introduced due to wrong selection of the sensor that is suitable for a specific application. For example, accelerometer mass becomes important when using small and light test objects. The accelerometer should not load the structural member, since additional mass can significantly change the levels and frequency presence at measuring points and invalidate the results. As a general rule, the accelerometer mass should not be greater than one-tenth the effective mass of the part or the structure that it is mounted onto for measurements.

The dynamic range of the accelerometer should match the high or low acceleration levels of the measured objects. General-purpose accelerometers can be linear from $5000g$ to $10,000g$, which is well in the range of most mechanical shocks. Special accelerometers can measure up to $100,000g$.

20.2.4.4 The Transient Response

Shocks are characterized as sudden releases of energy in the form of short-duration pulses exhibiting various shapes and rise times. They have high magnitudes and wide frequency contents. In applications where transient and shock measurements are involved, the overall linearity of the measuring system may be limited to high and low frequencies by a phenomena known as *zero shift* and *ringing*, respectively. The zero shift is caused by both the phase nonlinearity in the preamplifiers and the accelerometer not returning to steady-state operation conditions after being subjected to high shocks. Ringing is caused by high-frequency components of the excitation near-resonance frequency, preventing the accelerometer from returning back to its steady-state operation condition. To avoid measuring errors due to these effects, the operational frequency of the measuring system should be limited to the linear range.

20.2.4.5 Full Scale Range and Overload Capability

Most accelerometers are able to measure acceleration in both positive and negative directions. They are also designed to be able to accommodate overload capacity. Manufacturers also supply information on these two characteristics.

20.2.4.6 Environmental Conditions

In selection and implementation of accelerometers, environmental conditions such as temperature ranges, temperature transients, cable noise, magnetic field effects, humidity, and acoustic noise need to be considered. Manufacturers supply information on environmental conditions.

20.2.5 Inertial Accelerometers

Inertial accelerometers are mechanical accelerometers that make use of a seismic mass that is suspended by a spring or a lever inside a rigid frame as shown in Figure 20.17. The frame carrying the seismic mass is connected firmly to the vibrating source whose characteristics are to be measured. As the system vibrates, the mass tends to remain fixed in its position so that the motion can be registered as a relative displacement between the mass and the frame. An appropriate transducer senses this displacement and the output signal is processed further. The displacement sensing element can be made from a variety of materials exhibiting resistive, capacitive, inductive, piezoelectric, piezoresistive, and optical capabilities. In practice, the seismic mass does not remain absolutely steady, but it can satisfactorily act as a reference position for selected frequencies.

By proper selection of mass, spring, and damper combinations, the seismic instrument may be used for either acceleration or displacement measurements. In general, a large mass and soft spring are suitable for vibration and displacement measurement, while a relatively small mass and a stiff spring are used in accelerometers. However, the term seismic is commonly applied to instruments, which sense very low levels of vibration in the ground or structures.

In order to describe the response of the seismic accelerometer in Figure 20.17, Newton's second law, the equation of motion, may be written as

$$m\frac{d^2 x_2}{dt^2} + c\frac{dx_2}{dt} + kx_2 = c\frac{dx_1}{dt} + kx_1 + mg\cos(\theta) \tag{20.12}$$

where x_1 is the displacement of the vibration frame, x_2 is the displacement of the seismic mass, c is the velocity constant, θ is the angle between the sense axis and gravity, and k is the spring constant.

Taking $md^2 x_1/dt^2$ from both sides of the equation and rearranging gives

$$m\frac{d^2 z}{dt^2} + c\frac{dz}{dt} + kz = mg\cos(\theta) - m\frac{d^2 x_1}{dt^2} \tag{20.13}$$

where $z = x_2 - x_1$ is the relative motion between the mass and the base.

In Equation 20.12, it is assumed that the damping force on the seismic mass is proportional to velocity only. If a harmonic vibratory motion is impressed on the instrument such that

$$x_1(t) = X_0 \sin(\omega_1 t) \qquad (20.14)$$

where ω_1 is the frequency of vibration (rad/s), writing

$$-m\frac{d^2 x_1}{dt^2} = mX_0 \sin \omega_1 t$$

modifies Equation 20.13 as

$$m\frac{d^2 z}{dt^2} + c\frac{dz}{dt} + kz = mg\cos(\theta) + ma_1 \sin \omega_1 t \qquad (20.15)$$

where $a_1 = mX_0 \omega_1^2$.

Equation 20.15 will have transient and steady-state solutions. The steady-state solution of the differential equation 20.15 may be determined as

$$z = \frac{mg\cos(\theta)}{k} + \frac{ma_1 \sin \omega_1 t}{(k - m\omega_1^2 + jc\omega_1)} \qquad (20.16)$$

Rearranging Equation 20.16 results in

$$z = \frac{mg\cos(\theta)}{\omega_n} + \frac{a_1 \sin(\omega_1 t - \phi)}{\sqrt{\omega_n^2(1 - r^2)^2 + (2\zeta r)^2}} \qquad (20.17)$$

where $\omega_n \, (=\sqrt{k/m})$ is the natural frequency of the seismic mass, $\zeta \, (=c/2\sqrt{km})$ is the damping ratio. The damping ratio can be written in terms of the critical damping ratio as $\zeta = c/c_c$, where $c_c = 2\sqrt{km}$, $\phi \, (= \tan^{-1} [c\omega_1/(k - m\omega_1^2)])$ is the phase angle, and $r \, (=\omega_1/\omega_n)$ is the frequency ratio.

A plot of Equation 20.17 $(x_1 - x_2)_0/x_0$ against frequency ratio ω_1/ω_n, is illustrated in Figure 20.21. This figure shows that the output amplitude is equal to the input amplitude when $c/c_c = 0.7$ and $\omega_1/\omega_n > 2$. The output becomes essentially a linear function of the input at high frequency ratios. For satisfactory system performance, the instrument constant c/c_c and ω_n should be carefully calculated or obtained from calibrations.

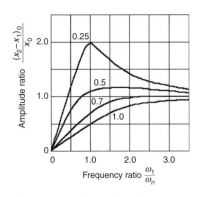

FIGURE 20.21 Frequency versus amplitude ratio of seismic accelerometers.

In this way the anticipated accuracy of measurement may be predicted for frequencies of interest. A comprehensive treatment of the analysis has been performed by McConnell [7]; interested readers should refer to this text for further details.

Seismic instruments are constructed in a variety of ways. In a potentiometric instrument, a voltage divider potentiometer is used for sensing the relative displacement between the frame and the seismic mass. In the majority of potentiometric accelerometers, the device is filled with a viscous liquid that interacts continuously with the frame and the seismic mass to provide damping. These accelerometers have a low frequency of operation (less than 100 Hz) and are mainly intended for slowly varying accelerations, and low-frequency vibrations. A typical family of such instruments offers many different models, covering the range of ±1g to ±50g full scale. The natural frequency ranges from 12 to 89 Hz, and the damping ratio ζ can be kept between 0.5 and 0.8 by using a temperature compensated liquid-damping arrangement. Potentiometer resistance may be selected in the range of 1000–10,000 Ω, with a corresponding resolution of 0.45–0.25% of full scale. The cross-axis sensitivity is less than ±1%. The overall accuracy is ±1% of full scale or less at room temperatures. The size is about 50 mm^3 with a total mass of about 1/2 kg.

Linear variable differential transformers (LVDTs) offer another convenient means of measurement of the relative displacement between the seismic mass and the accelerometer housing. These devices have higher natural frequencies than potentiometer devices, up to 300 Hz. Since the LVDT has lower resistance to motion, it offers much better resolution. A typical family of liquid-damped differential-transformer accelerometers exhibits the following characteristics. The full scale ranges from ±2g to ±700g, the natural frequency from 35 to 620 Hz, the nonlinearity 1% of full scale. The full-scale output is about 1 V with an LVDT excitation of 10 V at 2000 Hz, the damping ratio ranges from 0.6 to 0.7, the residual voltage at the null position is less than 1%, and the hysteresis is less than 1% of full scale. The size is about 50 mm^3 with a mass of about 120 g.

Electrical resistance strain gages are also used for displacement sensing of the seismic mass. In this case, the seismic mass is mounted on a cantilever beam rather than on springs. Resistance strain gauges are bonded on each side of the beam to sense the strain in the beam resulting from the vibrational displacement of the mass. A viscous liquid that entirely fills the housing provides damping of the system. The output of the strain gauges is connected to an appropriate bridge circuit. The natural frequency of such a system is about 300 Hz. The low natural frequency is due to the need for a sufficiently large cantilever beam to accommodate the mounting of the strain gauges.

One serious drawback of the seismic instruments is temperature effects requiring additional compensation circuits. The damping of the instrument may also be affected by changes in the viscosity of the fluid due to temperature. For instance, the viscosity of silicone oil, often used in these instruments, is strongly dependent on temperature.

20.2.5.1 Suspended-Mass, Cantilever, and Pendulum-Type Inertial Accelerometers

There are a number of different inertial-type accelerometers, most of which are in development stages or used under very special circumstances, such as gyropendulum, reaction-rotor, vibrating-string, and centrifugal-force-balance.

The vibrating-string instrument, Figure 20.22, makes use of a proof mass supported longitudinally by a pair of tensioned, transversely vibrating strings with uniform cross section and equal lengths and masses. The frequency of vibration of the strings is set to several thousand cycles per second. The proof mass is supported radially in such a way that the acceleration normal to the strings does not affect the string tension. In the presence of acceleration along the sensing axis, a deferential tension exists on the two strings, thus altering the frequency of vibration. From the second law of motion the frequencies may be written as

$$f_1^2 = \frac{T_1}{4m_s l} \quad \text{and} \quad f_2^2 = \frac{T_2}{4m_s l} \qquad (20.18)$$

where T is the tension, m_s are the mass, and l is the lengths of strings.

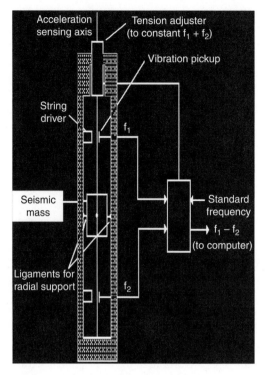

FIGURE 20.22 A typical suspended-mass vibrating-string accelerometer.

The quantity $T_1 - T_2$ is proportional to ma where a is the acceleration along the axis of the strings. An expression for the difference of the frequency-squared terms may be written as

$$f_1^2 - f_2^2 = \frac{T_1 - T_2}{4 m_s l} = \frac{ma}{4 m_s l} \tag{20.19}$$

Hence

$$f_1 - f_2 = \frac{ma}{4 m_s l (f_1 + f_2)} \tag{20.20}$$

The sum of frequencies $(f_1 + f_2)$ can be held constant by serving the tension in the strings with reference to the frequency of a standard oscillator. Then, the difference between the frequencies becomes linearly proportional to acceleration. In some versions, the beamlike property of the vibratory elements is used by gripping them at nodal points corresponding to the fundamental mode of the vibration of the beam, and at the respective centers of percussion of the common proof mass. The output frequency is proportional to acceleration and the velocity is proportional to phase, thus offering an important advantage. The improved versions of these devices lead to cantilever-type accelerometers, discussed next.

In a cantilever-type accelerometer, a small cantilever beam mounted on the block is placed against the vibrating surface, and an appropriate mechanism is provided for varying the beam length. The beam length is adjusted such that its natural frequency is equal to the frequency of the vibrating surface, and hence the resonance condition is obtained. Recently, slight variations of cantilever-beam arrangements are finding new applications in microaccelerometers.

In a different type of suspended-mass configuration, a pendulum is used that is pivoted to a shaft rotating about a vertical axis. Pickoff mechanisms are provided for the pendulum and the shaft speed.

The system is servo controlled to maintain it at null position. Gravitational acceleration is balanced by the centrifugal acceleration. The shaft speed is proportional to the square root of the local value of the acceleration.

20.2.6 Electromechanical Accelerometers

Electromechanical accelerometers, essentially servo or null-balance types, rely on the principle of feedback. In these instruments, an acceleration-sensitive mass is kept very close to a neutral position or zero displacement point by sensing the displacement and feeding back the effect of this displacement. A proportional magnetic force is generated to oppose the motion of the mass displaced from the neutral position, thus restoring this position just as a mechanical spring in a conventional accelerometer would do. The advantages of this approach are better linearity and elimination of hysteresis effects, as compared to the mechanical springs. Also, in some cases, electrical damping can be provided, which is much less sensitive to temperature variations.

One very important feature of electromechanical accelerometers is the capability of testing the static and dynamic performances of the devices by introducing electrically excited test forces into the system. This remote self-checking feature can be quite convenient in complex and expensive tests where accuracy is essential. These instruments are also useful in acceleration control systems, since the reference value of acceleration can be introduced by means of a proportional current from an external source. They are used for general-purpose motion measurements and monitoring low-frequency vibrations.

There are a number of different electromechanical accelerometers: coil-and-magnetic types, induction types, etc.

20.2.6.1 Coil-and-Magnetic Accelerometers

These accelerometers are based on Ampere's law, that is, "a current-carrying conductor disposed within a magnetic field experiences a force proportional to the current, the length of the conductor within the field, the magnetic field density, and the sine of the angle between the conductor and the field." The coils of these accelerometers are located within the cylindrical gap defined by a permanent magnet and a cylindrical soft iron flux return path. They are mounted by means of an arm situated on a minimum friction bearing or flexure so as to constitute an acceleration-sensitive seismic mass. A pickoff mechanism senses the displacement of the coil under acceleration and causes the coil to be supplied with a direct current via a suitable servo controller to restore or maintain a null condition. The electrical currents in the restoring circuit are linearly proportional to acceleration, provided (1) armature reaction affects are negligible and fully neutralized by a compensating coil in opposition to the moving coil, and (2) the gain of the servo system is large enough to prevent displacement of the coil from the region in which the magnetic field is constant.

In these accelerometers, the magnetic structure must be shielded adequately to make the system insensitive to external disturbances or the earth's magnetic field. Also, in the presence of acceleration there will be a temperature rise due to i^2R losses. The effects of these i^2R losses on the performance are determined by the thermal design and heat-transfer properties of the accelerometers.

20.2.6.2 Induction Accelerometers

The cross-product relationship of current, magnetic field, and force is the basis for induction-type electromagnetic accelerometers. These accelerometers are essentially generators rather than motors. One type of instrument, the cup-and-magnet design, includes a pendulous element with a pickoff mechanism and a servo controller driving a tachometer coupling. A permanent magnet and a flux return ring, closely spaced with respect to an electrically conductive cylinder, are attached to the pendulous element. A rate-proportional drag force is obtained by the electromagnetic induction effect between the magnet and the conductor. The pickoff mechanism senses pendulum deflection under acceleration and causes the servo controller to turn the rotor to drag the pendulous element toward the null position. Under steady-state conditions motor speed is a measure of the acceleration acting on the instrument. Stable servo operation is achieved by employing a time-lead network to compensate the inertial time lag of the motor and magnet combination. The accuracy of the servo-type accelerometers is ultimately limited by consistency and stability of scale factors of coupling and cup-and-magnet devices as a function of time and temperature.

Another accelerometer based on induction design uses the eddy-current induction torque generation. The force-generating mechanism of an induction accelerometer consists of a stable magnetic field, usually supplied by a permanent magnet, which penetrates orthogonally through a uniform conduction sheet. The movement of the conducting sheet relative to the magnetic field in response to acceleration results in a generated electromotive potential in each circuit in the conductor. This action is in accordance with Faraday's principle. In induction-type accelerometers, the induced eddy currents are confined to the conductor sheet, making the system essentially a drag coupling. Since angular rate is proportional to acceleration, angular position represents change in velocity. This is a particularly useful feature in navigation applications.

A typical commercial instrument based on the servo-accelerometer principle might have a micromachined quartz flexure suspension, differential capacitance angle pick-off, air-squeeze film plus servo-lead compensation for system damping. Of the available models, as an example, a typical 30g unit has a threshold and resolution of 1 μg, a frequency response that is flat to within 0.05% at 10 Hz and 2% at 100 Hz, a natural frequency of 1500 Hz, a damping ratio from 0.3 to 0.8, and transverse or cross-axis sensitivity of 0.1%. If, for example, the output current is about 1.3 mA/g, a 250 Ω readout resistor would give about ±10 V full scale at 30g. These accelerometers are good for precision work and used in many applications such as aircraft and missile control systems, measurement of tilt angles for borehole navigation, and axle angular bending in aircraft weight and balance systems.

20.2.7 Piezoelectric Accelerometers

Piezoelectric accelerometers are widely used for general-purpose acceleration, shock, and vibration measurements. They are basically motion transducers with large output signals and comparatively small sizes and they are self generators not requiring external power sources. They are available with very high natural frequencies and are therefore suitable for high-frequency applications and shock measurements.

These devices utilize a mass in direct contact with the piezoelectric component or crystal as shown in Figure 20.23. When a varying motion is applied to the accelerometer, the crystal experiences a varying force excitation ($F = ma$), causing a proportional electric charge q to be developed across it. So,

$$q = d_{ij}F = d_{ij}ma \qquad (20.21)$$

where q is the charge developed and d_{ij} is the piezoelectric coefficient of the material.

As this equation shows, the output from the piezoelectric material is dependent on its mechanical properties, d_{ij}. Two commonly used piezoelectric crystals are lead-zirconate titanate ceramic (PZT) and crystalline quartz. They are both self-generating materials and produce a large electric charge for their size. The piezoelectric strain constant of PZT is about 150 times that of quartz. As a result, PZTs are much more sensitive and smaller in size than quartz counterparts. These accelerometers are useful for

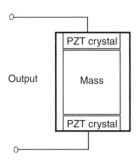

FIGURE 20.23 A compression type piezoelectric accelerometer arrangement.

high-frequency applications. The roll-off typically starts near 100 Hz. These active devices have no DC response. Since piezoelectric accelerometers have comparatively low mechanical impedances, their effect on the motion of most structures is negligible.

Mathematically, their transfer function approximates a third-order system that can be expressed as

$$\frac{e_0(s)}{a(s)} = \frac{K_q \tau s}{C\omega_n^2(\tau s + 1)(s^2/\omega_n^2 + 2\zeta s/\omega_n + 1)} \quad (20.22)$$

where K_q is the piezoelectric constant related to charge (C cm), τ is the time constant of the crystal, and s is the Laplace variable. It is worth noting that the crystal itself does not have a time constant τ, but the time constant is observed when the accelerometer is connected to an electric circuit, for example, an RC circuit.

The low-frequency response is limited by the piezoelectric characteristic $\tau s/(\tau s + 1)$, while the high-frequency response is related to mechanical response. The damping factor ζ is very small, usually less than 0.01 or near zero. Accurate low-frequency response requires large τ, which is usually achieved by use of high-impedance voltage amplifiers. At very low frequencies thermal effects can have severe influences on the operation characteristics.

In piezoelectric accelerometers, two basic design configurations are used: compression types and shear-stress types. In compression-type accelerometers, the crystal is held in compression by a preload element; therefore the vibration varies the stress in compressed mode. In a shear-stress accelerometer, vibration simply deforms the crystal in shear mode. The compression accelerometer has a relatively good mass to sensitivity ratio and hence exhibits better performance. But, since the housing acts as an integral part of the spring–mass system, it may produce spurious interfaces in the accelerometer output if excited around its natural frequency.

Piezoelectric accelerometers are available in a wide range of specifications and are offered by a large number of manufacturers. For example, the specifications of a shock accelerometer may have 0.004 pC/g in sensitivity and a natural frequency of up to 250,000 Hz, while a unit designed for low-level seismic measurements might have 1000 pC/g in sensitivity and only 7000 Hz natural frequency. They are manufactured as small as 3 × 3 mm in dimension with about 0.5 g in mass, including cables. They have excellent temperature ranges and some of them are designed to survive the intensive radiation environment of nuclear reactors. However, piezoelectric accelerometers tend to have larger cross-axis sensitivity than other types, about 2–4%. In some cases, large cross-axis sensitivity may be minimized during installations by the correct orientation of the device. These accelerometers may be mounted with threaded studs, with cement or wax adhesives, or with magnetic holders.

20.2.8 Piezoresistive Accelerometers

Piezoresistive accelerometers are essentially semiconductor strain gauges with large gauge factors. High gauge factors are obtained since the material resistivity is dependent primarily on the stress, not only on the dimensions. This effect can be greatly enhanced by appropriate doping of semiconductors such as silicon. Most piezoresistive accelerometers use two or four active gauges arranged in a Wheatstone bridge. Extra precision resistors are used, as part of the circuit, in series with the input to control the sensitivity, for balancing, and for offsetting temperature effects. The sensitivity of a piezoresistive sensor comes from the elastic response of its structure and resistivity of the material. Wire and thick or thin film resistors have low gauge factors, that is, the resistance change due to strain is small. The mechanical construction of a piezoresistive accelerometer is shown in Figure 20.24.

Piezoresistive accelerometers are useful for acquiring vibration information at low frequencies, for example, below 1 Hz. In fact, they are inherently true non-vibrational acceleration sensors. They generally have wider bandwidth, smaller nonlinearities and zero shifting, and better hysteresis characteristics compared to piezoelectric counterparts. They are suitable to measure shocks well above 100,000g. Typical characteristics

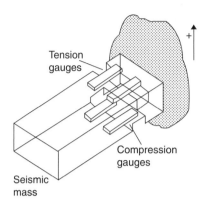

FIGURE 20.24 Bonding of piezoresistive and piezoelectric accelerometers to the inertial systems.

of piezoresistive accelerometers may be listed to be 100 mV/g as the sensitivity, 0–750 Hz as the frequency range, 2500 Hz in resonance frequency, 25g as the amplitude range, 2000g as the shock rating, and 0–95°C as the temperature range, with a total mass of about 25 g.

Most contemporary piezoresistive sensors are manufactured from a single piece of silicon. This gives better stability and less thermal mismatch between parts. In a typical monolithic sensing element a 1-mm silicon chip incorporates the spring, mass, and four-arm bridge assembly. The elements are formed by a pattern of dopant in the originally flat silicon. Subsequent etching of channels frees the gauges and simultaneously defines the masses as regions of silicon of original thickness.

20.2.9 Strain-Gauge Accelerometers

Strain-gauge accelerometers are based on resistance properties of electrical conductors. If a conductor is stretched or compressed, its resistance alters due to (a) dimensional changes, and (b) the changes in the fundamental property of material called piezoresistance. This indicates that the resistivity ρ of the conductor depends on the mechanical strain applied onto it. The dependence is expressed as the gauge factor:

$$\frac{dR/R}{dL/L} = 1 + 2v + \frac{d\rho/\rho}{dL/L} \tag{20.23}$$

where 1 indicates the resistance change due to length, $2v$ indicates resistance change due to area, and $(d\rho/\rho)/(dL/L)$ indicates the resistance change due to piezoresistivity.

There are many types of strain-gauges: unbonded metal-wire gauges, bonded metal-wire gauges, bonded metal-foil gauges, vacuum-deposited thin-metal-film gauges, bonded semiconductor gauges, and diffused semiconductor gauges. However, usually bonded and unbonded metal-wire gauges find wider applications. A section of the strain-gauge accelerometers, particularly bonded semiconductor types, known as the piezoresistive transducers, are used, but they suffer from high temperature sensitivities, nonlinearities, and some mounting difficulties. Nevertheless, with the recent developments of micromachine technology, these sensors have been improved considerably, thus finding many new applications.

Unbonded-strain-gauge accelerometers use the strain wires as the spring element and as the motion transducer, using similar arrangements as in Figure 20.25. They are useful for general-purpose motion and vibration measurements from low to medium frequencies. They are available in wide ranges and characteristics: typically ±5g to ±200g full scale, a natural frequency of 17–800 Hz, a 10-V excitation voltage AC or DC, full scale output ±20 to ±50 mV, a resolution less than 0.1%, an inaccuracy less than 1% full scale, and a cross-axis sensitivity less than 2%. The damping ratio (using silicone oil damping) is 0.6–0.8 at room temperature. These instruments are small and light, usually with a mass less than 25 g.

Bonded-strain-gauge accelerometers generally use a mass supported by a thin flexure beam. The strain-gauges are cemented onto the beam to achieve maximum sensitivity, temperature compensation, and sensitivity to both cross-axis and angular accelerations. Their characteristics are similar to the unbonded-strain-gauge accelerometers but have greater sizes and weights. Often silicone oil is used for damping. Semiconductor strain gauges are widely used as strain sensors in cantilever-beams and mass types of accelerometers. They allow high outputs (0.2–0.5 V full scale). Typically, a $\pm 25g$ acceleration unit has a flat response from 0 to 750 Hz, a damping ratio of 0.7, a mass of about 28 g, and an operational temperature of $-18°C$ to $+93°C$. A triaxial $\pm 20{,}000g$ model has a flat response from 0 to 15 kHz, a damping ratio of 0.01, and a compensation temperature range of 0–45°C, and is $13 \times 10 \times 13$ mm^3 in size and 10 g in mass.

20.2.10 Electrostatic Accelerometers

Electrostatic accelerometers are based on Coulomb's law between two charged electrodes; therefore, they are capacitive types. Depending on the operation principles and external circuits they can be broadly classified as (a) electrostatic-force-feedback accelerometers, and (b) differential-capacitance accelerometers.

20.2.10.1 Electrostatic-Force-Feedback Accelerometers

An electrostatic-force-feedback accelerometer consists of an electrode, with mass m and area S, mounted on a light pivoted arm that moves relative to some fixed electrodes. The nominal gap h between the pivoted and fixed electrodes is maintained by means of a force-balancing servo system, which is capable of varying the electrode potential in response to signals from a pickoff mechanism that senses relative changes in the gap. Mathematically, the field between the electrodes may be expressed by

$$E = \frac{Q}{\varepsilon k S} \tag{20.24}$$

where E is the intensity or potential gradient (dV/dx), Q is the charge, S is the area of the conductor, and k is the dielectric constant of the space outside the conductor.

From this expression, it can be shown that the force per unit area of the charged conductor (in N/m^2) is given by

$$\frac{F}{S} = \frac{Q^2}{2\varepsilon k S^2} = \frac{\varepsilon k E^2}{2} \tag{20.25}$$

Consider one movable and one stationary electrode and assume that the movable electrode is maintained at a bias potential V_1 and the stationary one at a potential V_2. The electrical intensity E in the gap, h, can be expressed as

$$E_1 = \frac{V_1 - V_2}{h} \tag{20.26}$$

so that the force of attraction may be found as

$$F_1 = \frac{\varepsilon k E^2 S}{2h^2} = \frac{\varepsilon k (V_1 - V_2)^2 S}{2h^2} \tag{20.27}$$

In the presence of acceleration, if V_2 is adjusted to restrain the movable electrode to the null position, the expression relating acceleration and electrical potential may be given by

$$a = \frac{F_1}{m} = \frac{\varepsilon k (V_1 - V_2)^2 S}{2h^2 m} \tag{20.28}$$

The device so far described can measure acceleration in one direction only, and the output is quadratic in character, that is,

$$(V_1 - V_2) = D\sqrt{a} \tag{20.29}$$

where D is the constant of proportionality.

The output may be linearized in a number of ways, one of which is the quarter-square method. If the servo controller applies a potential $-V_2$ to the other fixed electrode, the force of attraction between this electrode and the movable electrode becomes

$$a = \frac{F_2}{m} = \frac{\varepsilon k (V_1 + V_2)^2 S}{2 h^2 m} \tag{20.30}$$

and the force-balance equation of the movable electrode when the instrument experiences a downward acceleration a now is

$$ma = F_2 - F_1 = \frac{\varepsilon k S [(V_1 + V_2)^2 - (V_1 - V_2)^2]}{2 h^2}$$

or

$$ma = F_2 - F_1 = \frac{2 \varepsilon k S V_1 V_2}{h^2} \tag{20.31}$$

Hence, if the bias potential V_1 is held constant and the gain of the control loop is high so that variations in the gap are negligible, the acceleration becomes a linear function of the controller output voltage V_2.

The principal difficulty in mechanizing the electrostatic force accelerometer is the relatively high electric field intensity required to obtain an adequate force. Damping can be provided electrically or by viscosity of the gaseous atmosphere in the inter-electrode space if the gap h is sufficiently small. The scheme works best in micromachined instruments. Nonlinearity in the voltage breakdown phenomenon permits larger gradients in very small gaps.

A typical electrostatic accelerometer has the following characteristics: range $\pm 50g$, resolution $10^{-3}g$, sensitivity 100 mV/g, nonlinearity <1% FS, transverse sensitivity <1% FS, thermal sensitivity 6×10^{-4}/K, mechanical shock 10,000g, operating temperature $-45°C$ to $90°C$, supply voltage 5 V DC, and weight 45 g. The main advantages of electrostatic accelerometers are their extreme mechanical simplicity, low power requirements, absence of inherent sources of hysteresis errors, zero temperature coefficients, and ease of shielding from stray fields.

20.2.10.2 Differential-Capacitance Accelerometers

Differential-capacitance accelerometers are based on the principle of the change of capacitance in proportion to applied acceleration. In one type, the seismic mass of the accelerometer is made as the movable element of an electrical oscillator. The seismic mass is supported by a resilient parallel-motion beam arrangement from the base. The system is set to have a certain defined nominal frequency when undisturbed. If the instrument is accelerated, the frequency varies above and below the nominal value depending on the direction of acceleration.

The seismic mass carries an electrode located in opposition to a number of base-fixed electrodes that define variable capacitors. The base-fixed electrodes are resistances coupled in the feedback path of a wideband, phase-inverting amplifier. The gain of the amplifier is predetermined to ensure maintenance of oscillations over the range of variation of the capacitance determined by the applied acceleration. The value of the capacitance C for each of the variable capacitors is given by

$$C = \frac{\varepsilon k S}{h} \tag{20.32}$$

where k is the dielectric constant, ε is the permittivity of free space, S is the area of the electrode, and h is the variable gap.

Denoting the magnitude of the gap for zero acceleration as h_0, the value of h in the presence of acceleration a may be written as

$$h = h_0 + \frac{ma}{K} \tag{20.33}$$

where m is the value of the proof mass and K is the spring constant. Thus,

$$C = \frac{\varepsilon k S}{h_0 + (ma/K)} \tag{20.34}$$

For example, the frequency of oscillation of the resistance–capacitance type circuit is given by the expression

$$f = \frac{\sqrt{6}}{2\pi RC} \tag{20.35}$$

Substituting this value of C in Equation 20.34 gives

$$f = \frac{\sqrt{6}[h_0 + (ma/K)]}{2\pi R \varepsilon k S} \tag{20.36}$$

Denote the constant quantity $\sqrt{6}/(2\pi R \varepsilon k S)$ as B and rewrite Equation 20.36 as

$$f = Bh_0 + \frac{Bma}{K} \tag{20.37}$$

The first term on the right-hand side expresses the fixed bias frequency f_0 and the second term denotes the change in frequency resulting from acceleration, so that the expression may be written as

$$f = f_0 + f_a \tag{20.38}$$

If the output frequency is compared with an independent source of a constant frequency of f_0, then f_a can be determined easily.

A commonly used capacitive-type accelerometer is based on a thin diaphragm with spiral flexures that provide the spring, proof mass, and moving plate necessary for the differential capacitor. Plate motion between the electrodes pumps air parallel to the plate surface and through holes in the plate to provide squeeze film damping. Since air viscosity is less temperature sensitive than oil, the desired damping ratio of 0.7 hardly changes more than 15%. A family of such instruments are easily available with full-scale ranges from ±0.2g (4 Hz flat response) to ±1000g (3000 Hz), a cross-axis sensitivity less than 1%, and a full-scale output of ±1.5 V. The size of a typical device is about 25 mm^3 with a mass of 50 g.

20.2.11 Micro- and Nanoaccelerometers

By the end of the 1970s it became apparent that the essentially planar processing integrated-circuit (IC) technology could be modified to fabricate three-dimensional electromechanical structures by the micromachining process. Accelerometers and pressure sensors were among the first IC sensors. The first accelerometer was developed in 1979. Since then the technology has been progressing steadily and now an extremely diverse range of accelerometers are readily available. Most sensors use bulk micromachining rather than surface micromachining techniques. In bulk micromachining the flexures, resonant beams, and all other critical components of the accelerometer are made from bulk silicon in order to exploit the full mechanical properties of silicon crystals. With proper design and film process, bulk micromachining yields extremely stable and robust accelerometers.

The selective etching of multiple layers of deposited thin films, or surface micromachining, allows movable microstructures to be fabricated on silicon wafers. With surface micromachining, layers of structure material are disposed and patterned as shown in Figure 20.25. These structures are formed by polysilicons and sacrificial materials such as silicon dioxides. The sacrificial material acts as an intermediate spacer layer and is etched away to produce a freestanding structure. Surface machining technology also allows smaller and more complex structures to be built in multiple layers on a single substrate. A typical example of modern micromachined accelerometer is given in Figure 20.26. Multiple accelerometers can be mounted on a single chip, sensing accelerations in x, y, and z directions. The primary signal conditioning is also provided in the same chip. The output from the chip is usually read in the digital form.

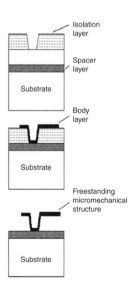

FIGURE 20.25 Steps of micromachining to manufacture micro- and nanoaccelerometers.

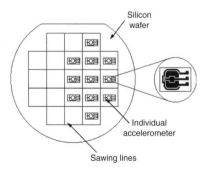

FIGURE 20.26 Multiple accelerometers in a single chip.

Most micro- and nanoaccelerometers detect acceleration by measuring the relative motion between proof mass and mounting substrate. The proof mass is suspended above the substrate by a mechanical spring suspension. When the sensor undergoes acceleration, the proof mass tends to remain stationary and therefore displaces with respect to the moving substrate. This displacement is measured capacitively or by means of piezoresistive or piezoelectric methods using CMOS technology. Chip circuits provide offset cancellation for bias stability, gain scale factor stability, zero acceleration bias stability, temperature compensation, prefiltering, noise immune digital output, and so on.

The operational principles of some of the microaccelerometers are very similar to capacitive force-balance or vibrating-beam accelerometers, discussed earlier. Manufacturing techniques may change from one manufacturer to another. However, in general, vibrating-beam accelerometers are preferred because of better air-gap properties and improved bias performance characteristics.

Vibrating-beam accelerometers, also termed resonant-beam force transducers, are made in such a way that an acceleration along a positive input axis places the vibrating beam in tension. Thus, the resonant frequency of the vibrating beam increases or decreases with the applied acceleration.

In DETF, an electronic oscillator capacitively couples energy into two vibrating beams to keep them oscillating at their resonant frequency. The beams vibrate 180° out of phase to cancel reaction forces at the ends. The dynamic cancellation effect of the DETF design prevents energy from being lost through the ends of the beam. Hence, the dynamically balanced DETF resonator has a high Q factor, which leads to a stable oscillator circuit. The acceleration signal is produced from the oscillator as a frequency-modulated square wave that can be used for a digital interface.

The frequency of resonance of the system must be much higher than any input acceleration, and this limits the measurable range. In a micromachined accelerometer, used in military applications, the following characteristics are given: a range of $\pm 1200g$, a sensitivity of 1.11 Hz/g, a bandwidth of 2500 Hz, an unloaded DETF frequency of 9952 Hz. The frequency at $+1200g$ is 11,221 Hz, the frequency at $-1200g$ is 8544 Hz, and the temperature sensitivity is 5 mg/°C. The accelerometer size is 6 mm diameter by 4.3 mm length, with a mass of about 9 g. It has a turn-on time of less than 60 s, the accelerometer is powered with +9 to +16 V DC, and the nominal output is a 9000-Hz square wave.

Surface micromachining has also been used to manufacture specific application accelerometers, such as air-bag applications in the automotive industry. In one type, a three-layer differential capacitor is created by alternate layers of polysilicon and phosphosilicate glass (PSG) on a 0.38-mm thick, 100-mm long wafer. A silicon wafer serves as the substrate for the mechanical structure. The trampoline-shaped middle layer is suspended by four supporting arms. This movable structure is the seismic mass for the accelerometer. The upper and lower polysilicon layers are fixed plates for the differential capacitors. The glass is sacrificially etched by hydrofluoric acid (HF).

20.2.12 Signal Conditioning and Biasing

Common signal conditioners are appropriate for interfacing accelerometers to computers or other instruments for further signal processing. Generally, the generated raw signals are amplified and filtered suitably by the circuits within the accelerometer casing supplied by manufacturers. Nevertheless, piezoelectric and piezoresistive transducers require special signal conditioners with certain characteristics that will be discussed next.

20.2.12.1 Piezoelectric Accelerometers

Piezoelectric accelerometers supply small energy to the signal conditioners since they have high capacitive source impedances. The equivalent circuit of a piezoelectric accelerometer can be regarded as an active capacitor that charges itself when mechanically loaded. The selection of the elements of the external signal conditioning circuit is dependent on the characteristics of the equivalent circuit. A most common approach is the charge amplifier since the system gain and low-frequency responses of these amplifiers are well defined. The performance of the circuit is also independent of cable length and capacitance of the accelerometer. In many applications, noise-treated cables are necessary to avoid the triboelectric charges occurring due to movement of cables.

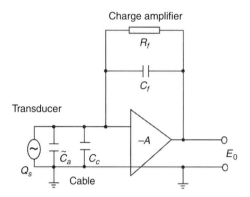

FIGURE 20.27 A typical charge amplifier.

The charge amplifier (with about 1000 MΩ input impedance) basically converts the input charge to voltage first and then amplifies this voltage. It consists of a charge converter output voltage, which occurs as a result of the charge input signal returning through the feedback capacitor to maintain the input voltage at the input level close to zero, as shown in Figure 20.27. With the help of basic operational-type feedback, the amplifier input is maintained at essentially 0 V; therefore, it looks like a short circuit to the input. Thus, the charge input is stored in the feedback capacitor, producing a voltage across it that is equal to the value of the charge input divided by the capacitance of the feedback capacitor. The complete transfer function of the circuit describing the relationship between the output voltage and the input acceleration magnitude may be determined by the following complex transform:

$$\frac{E_0}{a_0} = S_a j R_f C_f \omega \left[1 + j R_f C_f \omega \left(1 + C_f + \frac{C_a + C_c}{1 + G} \right) \right] \quad (20.39)$$

where E_0 is the charge converter output (V), a_0 is the magnitude of acceleration (m/s^2), S_a is the accelerometer sensitivity (mV/g), C_a is the accelerometer capacitance (F), C_c is the cable capacitance (F), C_f is the feedback capacitance (F), R_f is the feedback loop resistance, and G is the amplifier open-loop gain.

In most applications, since C_f is selected to be large compared to $(C_a + C_c)/(1 + G)$, the system gain becomes independent of the cable length. In this case the denominator of the equation can be simplified to give a first-order system with roll off at

$$f_{-3dB} = \frac{1}{2 \pi R_e C_f} \quad (20.40)$$

with a slope of 10 dB per decade. For practical purposes, the low-frequency response of this system is a function of well-defined electronic components and does not vary by cable length. This is an important feature when measuring low-frequency vibrations.

Many piezoelectric accelerometers are manufactured with preamplifiers and other signal-conditioning circuits enclosed in the same casing. Some accelerometer preamplifiers include integrators to convert the acceleration proportional outputs to either velocity or displacement proportional signals. To attenuate noise and vibration signals that lie outside the frequency range of interest, most preamplifiers are equipped with a range of high- and low-pass filters. This avoids interference from electrical noise or signals inside the linear portion of the accelerometer frequency range. Nevertheless, it is worth mentioning that these devices usually have two time constants, external and internal. The mixture of these two time constants can lead to problems particularly at low frequencies. Manufacturers through design and construction usually fix the internal time constants. However, care must be observed to account for the effect of external time constants through impedance matching.

Sensors

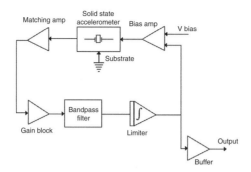

FIGURE 20.28 A typical signal conditioning arrangement for single chip microaccelerometers.

20.2.12.2 Piezoresistive Transducers

Piezoresistive transducers generally have high-amplitude outputs, low-output impedance, and low intrinsic noise. Most of these transducers are designed for constant-voltage excitations. They are usually calibrated for constant-current excitations to avoid external interference. Many piezoresistive transducers are configured as full-bridge devices. Some have four active piezoresistive arms, together with two fixed precision resistors to permit shunt calibration.

20.2.12.3 Microaccelerometers

In microaccelerometers signal-conditioning circuitry is integrated within the same chip as the sensor. A typical example of the signal-conditioning circuitry is given in Figure 20.28 in block diagram form. In this type of accelerometer, the electronic system is essentially a crystal-controlled oscillator circuit and the output signal of the oscillator is a frequency-modulated acceleration signal. Some circuits provide a buffered square-wave output that can be directly interfaced digitally. In these cases the need for analog-to-digital (A/D) conversion is eliminated, thus removing one of the major sources of errors. In other types of accelerometers, signal conditioning circuits such as A/D converters are retained within the chip.

20.2.12.4 Force Feedback Accelerometers

Signals from force feedback accelerometers often must be digitized for use in digital systems. A common solution is to use voltage to frequency or current to frequency converters to convert the analog signals to train pulses. These converters are expensive, often as much as the accelerometer, and add as much to the error budget.

Here, it is worth mentioning that GPS systems are becoming add-ons to many position sensing mechanisms. Because of antenna dynamics, shadowing, multipath effects, and to provide redundancy for critical systems such as aircraft, many of these systems require inertial aiding, tied-in with accelerometers and gyros. With the development of micromachining, small and cost-effective GPS assisted inertial systems will be available in the near future. These developments will require extensive signal processing with a high degree of accuracy. Dynamic ranges on the order of a million to one (e.g., 30–32 bits) need to be dealt with. In order to achieve accuracy requirements, a great challenge awaits the signal processing practitioner.

References

1. Bentley, J. P., *Principles of Measurement Systems*, 2nd ed., Burnt Mill, UK: Longman Scientific and Technical, 1988.
2. Doebelin, E. O., *Measurement Systems: Application and Design*, 4th ed., Singapore: McGraw-Hill, 1990.
3. Frank, R., *Understanding Smart Sensors*, Boston: Artech House, 1996.
4. Harris, C., *Shock and Vibration Handbook*, 4th ed., McGraw-Hill, 1995.
5. Holman, J. P., *Experimental Methods for Engineers*, 5th ed., Singapore: McGraw-Hill, 1989.

6. Lawrance, A., *Modern Inertial Technology-Navigation, Guidance, and Control,* Springer-Verlag: New York, 1993.
7. McConnell, K. G., *Vibration Testing: Theory and Practice,* New York: Wiley, 1995.
8. *Machine Vibration: Dynamics and Control,* London: Springler, 1992–1996.
9. *Measuring Vibration,* Bruel & Kjaer, 1982.
10. Sydenham, P. H., Hancock, N. H., and Thorn, R., *Introduction to Measurement Science and Engineering,* New York: Wiley, 1989.
11. Tompkins, W. J. and Webster, J. G., *Interfacing Sensors to the IBM PC,* Englewood Cliffs, NJ: Prentice-Hall, 1988.

20.3 Force Measurement

M. A. Elbestawi

Force, which is a vector quantity, can be defined as an action that will cause an acceleration or a certain reaction of a body. This chapter will outline the methods that can be employed to determine the magnitude of these forces.

20.3.1 General Considerations

The determination or measurement of forces must yield to the following considerations: if the forces acting on a body do not produce any acceleration, they must form a *system of forces in equilibrium.* The system is then considered to be in static equilibrium. The forces experienced by a body can be classified into two categories: internal, where the individual particles of a body act on each other, and external otherwise. If a body is supported by other bodies while subject to the action of forces, deformations and/or displacements will be produced at the points of support or contact. The internal forces will be distributed throughout the body until equilibrium is established, and then the body is said to be in a state of tension, compression, or shear. In considering a body at a definite section, it is evident that all the internal forces act in pairs, the two forces being equal and opposite, whereas the external forces act singly.

20.3.2 Hooke's Law

The basis for force measurement results from the physical behavior of a body under external forces. Therefore, it is useful to review briefly the mechanical behavior of materials. When a metal is loaded in uniaxial tension, uniaxial compression, or simple shear (Figure 20.29), it will behave elastically until a critical value of normal stress (S) or shear stress (τ) is reached, and then it will deform plastically [1]. In the

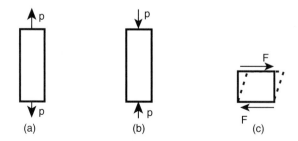

FIGURE 20.29 When a metal is loaded in uniaxial tension (a), uniaxial compression (b), or simple shear (c), it will behave elastically until a critical value of normal stress or shear stress is reached.

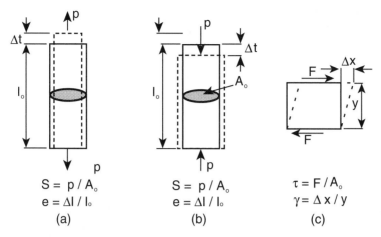

FIGURE 20.30 Elastic stress and strain for: (a) uniaxial tension; (b) uniaxial compression; (c) simple shear [1].

elastic region, the atoms are temporarily displaced but return to their equilibrium positions when the load is removed. Stress (S or τ) and strain (e or γ) in the elastic region are defined as indicated in Figure 20.30.

$$v = -\frac{e_2}{e_1} \tag{20.41}$$

Poisson's ratio (v) is the ratio of transverse (e_2) to direct (e_1) strain in tension or compression. In the elastic region, v is between 1/4 and 1/3 for metals. The relation between stress and strain in the elastic region is given by Hooke's law:

$$S = Ee \quad \text{(tension or compression)} \tag{20.42}$$

$$\tau = G\gamma \quad \text{(simple shear)} \tag{20.43}$$

where E and G are the Young's and shear modulus of elasticity, respectively. A small change in specific volume ($\Delta Vol/Vol$) can be related to the elastic deformation, which is shown to be as follows for an isotropic material (same properties in all directions):

$$\frac{\Delta Vol}{Vol} = e_1(1 - 2v) \tag{20.44}$$

The bulk modulus (K = reciprocal of compressibility) is defined as follows:

$$K = \Delta p \bigg/ \left(\frac{\Delta Vol}{Vol}\right) \tag{20.45}$$

where Δp is the pressure acting at a particular point. For an elastic solid loaded in uniaxial compression (S):

$$K = S \bigg/ \left(\frac{\Delta Vol}{Vol}\right) = \frac{S}{e_1(1 - 2v)} = \frac{E}{1 - 2v} \tag{20.46}$$

Thus, an elastic solid is compressible as long as v is less than 1/2, which is normally the case for metals. Hooke's law, Equation 20.42, for uniaxial tension can be generalized for a three-dimensional elastic condition.

The theory of elasticity is well established and is used as a basis for force measuring techniques. Note that the measurement of forces in separate engineering applications is very application specific, and care must be taken in the selection of the measuring techniques outlined below.

20.3.2.1 Basic Methods of Force Measurement

An unknown force may be measured by the following means:

1. Balancing the unknown force against a standard mass through a system of levers
2. Measuring the acceleration of a known mass
3. Equalizing it to a magnetic force generated by the interaction of a current-carrying coil and a magnet
4. Distributing the force on a specific area to generate pressure and then measuring the pressure
5. Converting the applied force into the deformation of an elastic element

The aforementioned methods used for measuring forces yield a variety of designs of measuring equipment. The challenge involved with the task of measuring force resides primarily in sensor design. The basics of sensor design can be resolved into two problems:

1. Primary geometric, or physical constraints, governed by the application of the force sensor device
2. The means by which the force can be converted into a workable signal form (such as electronic signals or graduated displacements)

The remaining sections will discuss the types of devices used for force-to-signal conversion and finally illustrate some examples of applications of these devices for measuring forces.

20.3.3 Force Sensors

Force sensors are required for a basic understanding of the response of a system. For example, cutting forces generated by a machining process can be monitored to detect a tool failure or to diagnose the causes of this failure in controlling the process parameters, and in evaluating the quality of the surface produced. Force sensors are used to monitor impact forces in the automotive industry. Robotic handling and assembly tasks are controlled by detecting the forces generated at the end effector. Direct measurement of forces is useful in controlling many mechanical systems.

Some types of force sensors are based on measuring a deflection caused by the force. Relatively high deflections (typically, several micrometers) would be necessary for this technique to be feasible. The excellent elastic properties of helical springs make it possible to apply them successfully as force sensors that transform the load to be measured into a deflection. The relation between force and deflection in the elastic region is demonstrated by Hooke's law. Force sensors that employ strain gage elements or piezoelectric (quartz) crystals with built-in microelectronics are common. Both impulsive forces and slowly varying forces can be monitored using these sensors.

Of the available force measuring techniques, a general subgroup can be defined as that of load cells. Load cells are comprised generally of a rigid outer structure, some medium that is used for measuring the applied force, and the measuring gage. Load cells are used for sensing large, static, or slowly varying forces with little deflection and are a relatively accurate means of sensing forces. Typical accuracies are of the order of 0.1% of the full-scale readings. Various strategies can be employed for measuring forces that are strongly dependent on the design of the load cell. For example, Figure 20.31 illustrates different types of load cells that can be employed in sensing large forces for relatively little cost. The hydraulic load cell employs a very stiff outer structure with an internal cavity filled with a fluid. Application of a load increases the oil pressure, which can be read off an accurate gage.

Other sensing techniques can be utilized to monitor forces, such as piezoelectric transducers for quicker response of varying loads, pneumatic methods, strain gages, etc. The proper sensing technique needs special consideration based on the conditions required for monitoring.

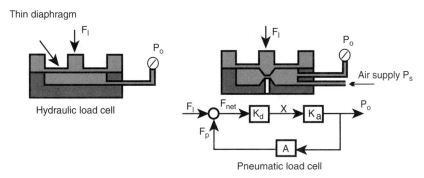

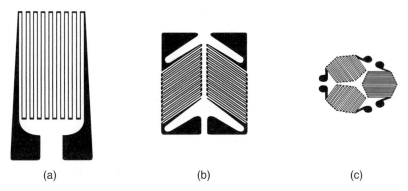

FIGURE 20.31 Different types of load cells [2].

FIGURE 20.32 Configuration of metal-foil resistance strain gages: (a) single element, (b) two element, and (c) three element.

20.3.3.1 Strain Gage Load Cell

The strain gage load cell consists of a structure that elastically deforms when subjected to a force and a strain gage network that produces an electrical signal proportional to this deformation. Examples of this are beam and ring types of load cells.

Strain Gages

Strain gages use a length of gage wire to produce the desired resistance (which is usually about 120 Ω) in the form of a flat coil. This coil is then cemented (bonded) between two thin insulating sheets of paper or plastic. Such a gage cannot be used directly to measure deflection. It has to be first fixed properly to a member to be strained. After bonding the gage to the member, they are baked at about 195°F (90°C) to remove moisture. Coating the unit with wax or resin will provide some mechanical protection. The resistance between the member under test and the gage itself must be at least 50 MΩ. The total area of all conductors must remain small so that the cement can easily transmit the force necessary to deform the wire. As the member is stressed, the resulting strain deforms the strain gage and the cross-sectional area diminishes. This causes an increase in resistivity of the gage that is easily determined. In order to measure very small strains, it is necessary to measure small changes of the resistance per unit resistance ($\Delta R/R$). The change in the resistance of a bonded strain gage is usually less than 0.5%. A wide variety of gage sizes and grid shapes are available, and typical examples are shown in Figure 20.32.

The use of strain gages to measure force requires careful consideration with respect to rigidity and environment. By virtue of their design, strain gages of shorter length generally possess higher response frequencies (examples: 660 kHz for a gage of 0.2 mm and 20 kHz for a gage of 60 mm in length). The environmental considerations focus mainly on the temperature of the gage. It is well known that resistance is a function of temperature and, thus, strain gages are susceptible to variations in temperature.

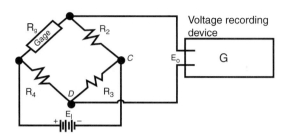

FIGURE 20.33 The Wheatstone bridge.

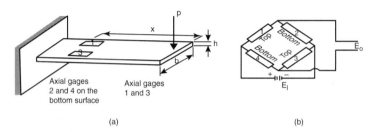

FIGURE 20.34 Beam-type load cells: (a) a selection of beam-type load cells (elastic element with strain gages), and (b) gage positions in the Wheatstone bridge [3].

Thus, if it is known that the temperature of the gage will vary due to any influence, temperature compensation is required in order to ensure that the force measurement is accurate.

A Wheatstone bridge (Figure 20.33) is usually used to measure this small order of magnitude. In Figure 20.33, no current will flow through the galvanometer (G) if the four resistances satisfy a certain condition. In order to demonstrate how a Wheatstone bridge operates [3], a voltage scale has been drawn at points C and D of Figure 20.33. Assume that R_1 is a bonded gage and that initially Equation 20.47 is satisfied. If R_1 is now stretched so that its resistance increases by one unit ($+\Delta R$), the voltage at point D will be increased from zero to plus one unit of voltage ($+\Delta V$), and there will be a voltage difference of one unit between C and D that will give rise to a current through C. If R_4 is also a bonded gage, and at the same time that R_1 changes by $+\Delta R$, R_4 changes by $-\Delta R$, the voltage at D will move to $+2\Delta V$. Also, if at the same time, R_2 changes by $-\Delta R$, and R_3 changes by $+\Delta R$, then the voltage of point C will move to $-2\Delta V$, and the voltage difference between C and D will now be $4\Delta V$. It is then apparent that although a single gage can be used, the sensitivity can be increased fourfold if two gages are used in tension while two others are used in compression.

$$\frac{R_1}{R_4} = \frac{R_2}{R_3} \tag{20.47}$$

The grid configuration of the metal-foil resistance strain gages is formed by a photo-etching process. The shortest gage available is 0.20 mm; the longest is 102 mm. Standard gage resistances are 120 Ω and 350 Ω. A strain gage exhibits a resistance change $\Delta R/R$ that is related to the strain in the direction of the grid lines by the expression in Equation 20.48 (where S_g is the gage factor or calibration constant for the gage).

$$\frac{\Delta R}{R} = S_g \varepsilon \tag{20.48}$$

Beam-Type Load Cell

Beam-type load cells are commonly employed for measuring low-level loads [3]. A simple cantilever beam (see Figure 20.34a) with four strain gages, two on the top surface and two on the bottom surface (all oriented along the axis of the beam) is used as the elastic member (sensor) for the load cell. The gages are wired into a Wheatstone bridge as shown in Figure 20.34b. The load P produces a moment $M = Px$ at the gage location (x) that results in the following strains:

$$\varepsilon_1 = -\varepsilon_2 = \varepsilon_3 = -\varepsilon_4 = \frac{6M}{Ebh^2} = \frac{6Px}{Ebh^2} \quad (20.49)$$

where b is the width of the cross-section of the beam and h is the height of the cross-section of the beam. Thus, the response of the strain gages is obtained from Equation 20.50.

$$\frac{\Delta R_1}{R_1} = -\frac{\Delta R_2}{R_2} = \frac{\Delta R_3}{R_3} = -\frac{\Delta R_4}{R_4} = \frac{6S_g Px}{Ebh^2} \quad (20.50)$$

The output voltage E_o from the Wheatstone bridge, resulting from application of the load P, is obtained from Equation 20.51. If the four strain gages on the beam are assumed to be identical, then Equation 20.51 holds.

$$E_o = \frac{6S_g PxE_1}{Ebh^2} \quad (20.51)$$

The range and sensitivity of a beam-type load cell depends on the shape of the cross-section of the beam, the location of the point of application of the load, and the fatigue strength of the material from which the beam is fabricated.

Ring-Type Load Cell

Ring-type load cells incorporate a proving ring (see Figure 20.35) as the elastic element. The ring element can be designed to cover a very wide range of loads by varying the diameter D, the thickness t, or the depth w of the ring. Either strain gages or a linear variable-differential transformer (LVDT) can be used as the sensor.

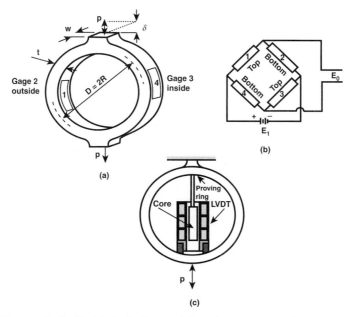

FIGURE 20.35 Ring-type load cells: (a) elastic element with strain-gage sensors; (b) gage positions in the Wheatstone bridge; and (c) elastic element with an LVDT sensor [3].

The load P is linearly proportional to the output voltage E_o. The sensitivity of the ring-type load cell with an LVDT sensor depends on the geometry of the ring (R, t, and w), the material from which the ring is fabricated (E), and the characteristics of the LVDT (S and E_i). The range of a ring-type load cell is controlled by the strength of the material used in fabricating the ring.

20.3.3.2 Piezoelectric Methods

A piezoelectric material exhibits a phenomenon known as the *piezoelectric effect*. This effect states that when asymmetrical, elastic crystals are deformed by a force, an electrical potential will be developed within the distorted crystal lattice. This effect is reversible. That is, if a potential is applied between the surfaces of the crystal, it will change its physical dimensions [4]. Elements exhibiting piezoelectric qualities are sometimes known as electrorestrictive elements.

The magnitude and polarity of the induced surface charges are proportional to the magnitude and direction of the applied force [4]:

$$Q = dF \qquad (20.52)$$

where d is the charge sensitivity (a constant for a given crystal) of the crystal in C/N. The force F causes a thickness variation Δt meters of the crystal:

$$F = \frac{aY}{t}\Delta t \qquad (20.53)$$

where a is area of crystal, t is thickness of crystal, and Y is Young's modulus.

$$Y = \frac{\text{stress}}{\text{strain}} = \frac{Ft}{a\Delta t} \qquad (20.54)$$

The charge at the electrodes gives rise to a voltage $E_o = Q/C$, where C is capacitance in farads between the electrodes and $C = \varepsilon a/t$ where ε is the absolute permittivity.

$$E_o = \frac{dF}{C} = \frac{dtF}{\varepsilon a} \qquad (20.55)$$

The voltage sensitivity $= g = d/\varepsilon$ in volt meter per newton can be obtained as:

$$E_o = g\frac{t}{a}F = gtP \qquad (20.56)$$

The piezoelectric materials used are quartz, tourmaline, Rochelle salt, ammonium dihydrogen phosphate (ADP), lithium sulfate, barium titanate, and lead zirconate titanate (PZT) [4]. Quartz and other earthly piezoelectric crystals are naturally polarized. However, synthetic piezoelectric material, such as barium titanate ceramic, are made by baking small crystallites under pressure and then placing the resultant material in a strong dc electric field [4]. After that, the crystal is polarized, along the axis on which the force will be applied, to exhibit piezoelectric properties. Artificial piezoelectric elements are free from the limitations imposed by the crystal structure and can be molded into any size and shape. The direction of polarization is designated during their production process.

The different modes of operation of a piezoelectric device for a simple plate are shown in Figure 20.36 [4]. By adhering two crystals together so that their electrical axes are perpendicular, bending moments or torque can be applied to the piezoelectric transducer and a voltage output can be produced (Figure 20.37) [4]. The range of forces that can be measured using piezoelectric transducers are from 1 to 200 kN and at a ratio of 2×10^5.

Piezoelectric crystals can also be used in measuring an instantaneous change in the force (dynamic forces). A thin plate of quartz can be used as an electronic oscillator. The frequency of these oscillations will be dominated by the natural frequency of the thin plate. Any distortion in the shape of the plate

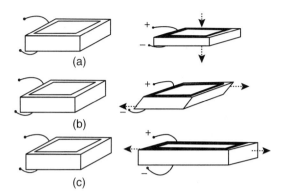

FIGURE 20.36 Modes of operation for a simple plate as a piezoelectric device [4].

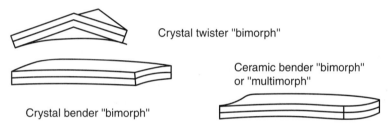

FIGURE 20.37 Curvature of "twister" and "bender" piezoelectric transducers when voltage applied [4].

caused by an external force, alters the oscillation frequency. Hence, a dynamic force can be measured by the change in frequency of the oscillator.

Resistive Method

The resistive method employs the fact that when the multiple contact area between semiconducting particles (usually carbon) and the distance between the particles are changed, the total resistance is altered. The design of such transducers yields a very small displacement when a force is applied. A transducer might consist of 2–60 thin carbon disks mounted between a fixed and a movable electrode. When a force is applied to the movable electrode and the carbon disks move together by 5–250 μm per interface, the transfer function of their resistance against the applied force is approximately hyperbolic, that is, highly nonlinear. The device is also subject to large hysteresis and drift together with a high transverse sensitivity.

In order to reduce hysteresis and drift, rings are used instead of disks. The rings are mounted on an insulated rigid core and prestressed. This almost completely eliminates any transverse sensitivity error. The core's resonant frequency is high and can occur at a frequency as high as 10 kHz. The possible measuring range of such a transducer is from 0.1 to 10 kg. The accuracy and linear sensitivity of this transducer is very poor.

Inductive Method

The inductive method utilizes the fact that a change in mechanical stress of a ferromagnetic material causes its permeability to alter. The changes in magnetic flux are converted into induced voltages in the pickup coils as the movement takes place. This phenomenon is known as the *Villari effect* or *magnetostriction*. It is known to be particularly strong in nickel–iron alloys.

Transducers utilizing the Villari effect consist of a coil wound on a core of magnetostrictive material. The force to be measured is applied on this core, stressing it and causing a change in its permeability and inductance. This change can be monitored and used for determining the force.

The applicable range for this type of transducer is a function of the cross-sectional area of the core. The accuracy of the device is determined by a calibration process. This transducer has poor linearity and is subject to hysteresis. The permeability of a magnetostrictive material increases when it is subject to pure torsion, regardless of direction. A flat frequency response is obtained over a wide range from 150 to 15,000 Hz.

Piezotransistor Method

Devices that utilize *anisotropic stress effects* are described as piezotransistors. In this effect, if the upper surface of a *p-n* diode is subjected to a localized stress, a significant reversible change occurs in the current across the junction. These transistors are usually silicon nonplanar type, with an emitter base junction. This junction is mechanically connected to a diaphragm positioned on the upper surface of a typical TO-type can [4]. When a pressure or a force is applied to the diaphragm, an electronic charge is produced. It is advisable to use these force-measuring devices at a constant temperature by virtue of the fact that semiconducting materials also change their electric properties with temperature variations. The attractive characteristic of piezotransistors is that they can withstand a 500% overload.

Multicomponent Dynamometers Using Quartz Crystals as Sensing Elements

The Piezoelectric Effects in Quartz

For force measurements, the *direct piezoelectric effect* is utilized. The direct longitudinal effect measures compressive force; the direct shear effect measures shear force in one direction. For example, if a disk of crystalline quartz (SiO_2) cut normally to the crystallographic *x*-axis is loaded by a compression force, it will yield an electric charge, nominally 2.26 pC/N. If a disk of crystalline quartz is cut normally to the crystallographic *y*-axis, it will yield an electric charge (4.52 pC/N) if loaded by a shear force in one specific direction. Forces applied in the other directions will not generate any output [5].

A charge amplifier is used to convert the charge yielded by a quartz crystal element into a proportional voltage. The range of a charge amplifier with respect to its conversion factor is determined by a feedback capacitor. Adjustment to mechanical units is obtained by additional operational amplifiers with variable gain.

The Design of Quartz Multicomponent Dynamometers

The main element for designing multicomponent dynamometers is the three-component force transducer (Figure 20.38). It contains a pair of X-cut quartz disks for the normal force component and a pair of Y-cut quartz disks (shear-sensitive) for each shear force component.

Three-component dynamometers can be used for measuring cutting forces during machining. Four three-component force transducers sandwiched between a base plate and a top plate are shown in Figure 20.38. The force transducer is subjected to a preload as shear forces are transmitted by friction. The four force transducers experience a drastic change in their load, depending on the type and position of force application. An overhanging introduction of the force develops a tensile force for some transducers, thus reducing the preload. Bending of the dynamometer top plate causes bending and shearing stresses. The measuring ranges of a dynamometer depend not only on the individual forces, but also on the individul bending stresses.

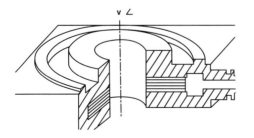
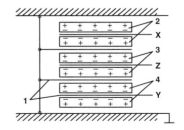

FIGURE 20.38 Three-component force transducer.

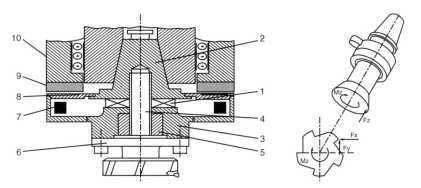

FIGURE 20.39 Force measuring system to determine the tool-related cutting forces in five-axis milling [6].

Measuring Signals Transmitted by Telemetry

Figure 20.39 shows the newly designed force measuring system RCD (rotating cutting force dynamometer). A ring-shaped sensor (1) is fitted in a steep angle taper socket (2) and a base ring (3) allowing sensing of the three force components F_x, F_y, and F_z at the cutting edge as well as the moment M_z. The physical operating principle of this measuring cell is based on the piezoelectric effect in quartz plates. The quartz plates incorporated in the sensor are aligned so that the maximum cross-sensitivity between the force components is 1%. As a result of the rigid design of the sensor, the resonant frequencies of the force measuring system range from 1200 to 3000 Hz and the measuring ranges cover a maximum of 10 kN [6].

Force-proportional charges produced at the surfaces of the quartz plates are converted into voltages by four miniature charge amplifiers (7) in hybrid construction. These signals are then filtered by specific electrical circuitry to prevent aliasing effects, and digitized with 8-bit resolution using a high sampling rate (pulse-code modulation). The digitized signals are transmitted by a telemetric unit consisting of a receiver and transmitter module, an antenna at the top of the rotating force measuring system (8), as well as a fixed antenna (9) on the splash cover of the two-axis milling head (10). The electrical components, charge amplifier, and transmitter module are mounted on the circumference of the force measuring system [6].

The cutting forces and the moment measured are digitized with the force measuring system described above. They are modulated on an FM carrier and transmitted by the rotating transmitter to the stationary receiver. The signals transmitted are fed to an external measured-variable conditioning unit.

Measuring Dynamic Forces

Any mechanical system can be considered in the first approximation as a weakly damped oscillator consisting of a spring and a mass. If a mechanical system has more than one resonant frequency, the lowest one must be taken into consideration. As long as the test frequency remains below 10% of the resonant frequency of the reference transducer (used for calibration), the difference between the dynamic sensitivity obtained from static calibration will be less than 1%. The above considerations assume a sinusoidal force signal. The static calibration of a reference transducer is also valid for dynamic calibration purposes if the test frequency is much lower (at least 10 times lower) than the resonant frequency of the system.

20.3.3.3 Capacitive Force Transducer

A transducer that uses capacitance variation can be used to measure force. The force is directed onto a membrane whose elastic deflection is detected by a capacitance variation. A highly sensitive force transducer can be constructed because the capacitive transducer senses very small deflections accurately. An electronic circuit converts the capacitance variations into DC-voltage variations [7].

A capacitance sensor consists of two metal plates separated by an air gap. The capacitance C between terminals is given by the expression:

$$C = \varepsilon_o \varepsilon_r \frac{A}{h} \qquad (20.57)$$

where

C = capacitance in farads (F)
ε_o = dielectric constant of free space
ε_r = relative dielectric constant of the insulator
A = overlapping area for the two plates
h = thickness of the gap between the two plates

The sensitivity of capacitance-type sensors is inherently low. Theoretically, decreasing the gap h should increase the sensitivity; however, there are practical electrical and mechanical conditions that preclude high sensitivities. One of the main advantages of the capacitive transducer is that moving of one of its plates relative to the other requires an extremely small force to be applied. A second advantage is stability and the sensitivity of the sensor is not influenced by pressure or temperature of the environment.

20.3.3.4 Force Sensing Resistors (Conductive Polymers)

Force sensing resistors (FSRs) utilize the fact that certain polymer thick-film devices exhibit decreasing resistance with the increase of an applied force. A force sensing resistor is made up of two parts. The first is a resistive material applied to a film. The second is a set of digitating contacts applied to another film. Figure 20.40 shows this configuration. The resistive material completes the electrical circuit between the two sets of conductors on the other film. When a force is applied to this sensor, a better connection is made between the contacts; hence, the conductivity is increased. Over a wide range of forces, it turns out that the conductivity is approximately a linear function of force. Figure 20.41 shows the resistance of the sensor as a function of force. It is important to note that there are three possible regions for the sensor to operate. The first abrupt transition occurs somewhere in the vicinity of 10 g of force. In this region, the resistance changes very rapidly. This behavior is useful when one is designing switches using force sensing resistors.

FSRs should not be used for accurate measurements of force because sensor parts may exhibit 15–25% variation in resistance between each other. However, FSRs exhibit little hysteresis and are considered far less costly than other sensing devices. Compared to piezofilm, the FSR is far less sensitive to vibration and heat.

20.3.3.5 Magnetoresistive Force Sensors

The principle of *magnetoresistive force sensors* is based on the fact that metals, when cooled to low temperatures, show a change of resistivity when subjected to an applied magnetic field. Bismuth, in particular, is quite sensitive in this respect. In practice, these devices are severely limited because of their high sensitivity to ambient temperature changes.

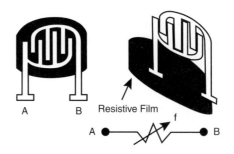

FIGURE 20.40 Diagram of a typical force sensing resistor (FSR).

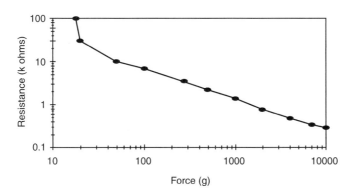

FIGURE 20.41 Resistance as a function of force for a typical force sensing resistor.

20.3.3.6 Magnetoelastic Force Sensors

Magnetoelastic transducer devices operate based on the Joule effect, that is, a ferromagnetic material is dimensionally altered when subjected to a magnetic field. The principle of operation is as follows: Initially, a current pulse is applied to the conductor within the waveguide. This sets up a magnetic field circumference-wise around the waveguide over its entire length. There is another magnetic field generated by the permanent magnet that exists only where the magnet is located. This field has a longitudinal component. These two fields join vectorally to form a helical field near the magnet which, in turn, causes the waveguide to experience a minute torsional strain or twist only at the location of the magnet. This twist effect is known as the *Wiedemann effect* [8].

Magnetoelastic force transducers have a high frequency response (on the order of 20 kHz). Some of the materials that exhibit magnetoelastic include Monel metal, Permalloy, Cekas, Alfer, and a number of nickel–iron alloys. Disadvantages of these transducers include: (1) the fact that excessive stress and aging may cause permanent changes, (2) zero drift and sensitivity changes due to temperature sensitivity, and (3) hysteresis errors.

20.3.3.7 Torsional Balances

Balancing devices that utilize the deflection of a spring may also be used to determine forces. *Torsional balances* are equal arm-scale-force measuring devices. They are comprised of horizontal steel bands instead of pivots and bearings. The principle of operation is based on force application on one of the arms that will deflect the torsional spring (within its design limits) in proportion to the applied force. This type of instrument is susceptible to hysteresis and temperature errors and, therefore, is not used for precise measurements.

Tactile Sensors

Tactile sensors are usually interpreted as a touch sensing technique. Tactile sensors cannot be considered as simple touch sensors, where very few discrete force measurements are made. In tactile sensing, a force "distribution" is measured using a closely spaced array of force sensors.

Tactile sensing is important in both grasping and object identification operations. Grasping an object must be done in a stable manner so that the object is not allowed to slip or get damaged. Object identification includes recognizing the shape, location, and orientation of a product, as well as identifying surface properties and defects. Ideally, these tasks would require two types of sensing [9]:

1. Continuous sensing of contact forces
2. Sensing of the surface deformation profile

These two types of data are generally related through stress–strain relations of the tactile sensor. As a result, almost continuous variable sensing of tactile forces (the sensing of the tactile deflection profile) is achieved.

Tactile Sensor Requirements

Significant advances in tactile sensing are taking place in the robotics area. Applications include automated inspection of surface profiles, material handling or parts transfer, parts assembly, and parts identification and gaging in manufacturing applications and fine-manipulation tasks. Some of these applications may need only simple touch (force–torque) sensing if the parts being grasped are properly oriented and if adequate information about the process is already available.

Naturally, the main design objective for tactile sensing devices has been to mimic the capabilities of human fingers [9]. Typical specifications for an industrial tactile sensor include:

1. Spatial resolution of about 2 mm
2. Force resolution (sensitivity) of about 2 g
3. Maximum touch force of about 1 kg
4. Low response time of 5 ms
5. Low hysteresis
6. Durability under extremely difficult working conditions
7. Insensitivity to change in environmental conditions (temperature, dust, humidity, vibration, etc.)
8. Ability to monitor slip

Tactile Array Sensor

Tactile array sensors (Figure 20.42) consist of a regular pattern of sensing elements to measure the distribution of pressure across the fingertip of a robot. The 8 × 8 array of elements at 2 mm spacing in each direction provides 64 force sensitive elements. Table 20.2 outlines some of the characteristics of early tactile array sensors. The sensor is composed of two crossed layers of copper strips separated by strips of thin silicone rubber. The sensor forms a thin, compliant layer that can be easily attached to a variety of fingertip shapes and sizes. The entire array is sampled by computer.

TABLE 20.2 Summary of Some of the Characteristics of Early Tactile Array Sensors

	Size of Array		
Device Parameter	(4 × 4)	(8 × 8)	(16 × 16)
Cell spacing (mm)	4.00	2.00	1.00
Zero-pressure capacitance (fF)	6.48	1.62	0.40
Rupture force (N)	18.90	1.88	0.19
Max. linear capacitance (fF)	4.80	1.20	0.30
Max. output voltage (V)	1.20	0.60	0.30
Max. resolution (bit)	9.00	8.00	8.00
Readout (access) time (μs)	—	<20	—

©IEEE 1985.

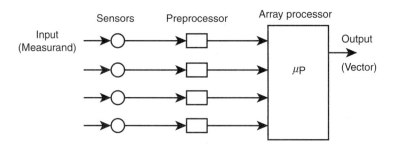

FIGURE 20.42 Tactile array sensor.

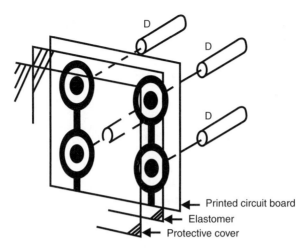

FIGURE 20.43 Typical taxel sensor array.

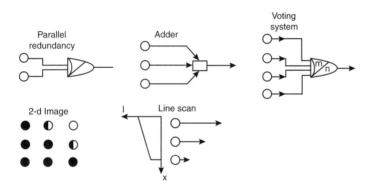

FIGURE 20.44 General arrangement of an intelligent sensor array system [9].

A typical tactile sensor array can consist of several sensing elements. Each element or taxel (Figure 20.43) is used to sense the forces present. Since tactile sensors are implemented in applications where sensitivity providing semblance to human touch is desired, an elastomer is utilized to mimic the human skin. The elastomer is generally a conductive material whose electrical conductivity changes locally when pressure is applied. The sensor itself consists of three layers: a protective covering, a sheet of conductive elastomer, and a printed circuit board. The printed circuit board consists of two rows of two "bullseyes," each with conductive inner and outer rings that compromise the taxels of the sensor. The outer rings are connected together and to a column-select transistor. The inner rings are connected to diodes (D) in Figure 20.43. Once the column in the array is selected, the current flows through the diodes, through the elastomer, and thence through a transistor to ground. As such, it is generally not possible to excite just one taxel because the pressure applied causes a local deformation in neighboring taxels. This situation is called *crosstalk* and is eliminated by the diodes [10].

Tactile array sensor signals are used to provide information about the contact kinematics. Several feature parameters, such as contact location, object shape, and the pressure distribution, can be obtained. The general layout of a sensor array system can be seen in Figure 20.44. An example of this is a contact and force sensing finger. This tactile finger has four contact sensors made of piezoelectric polymer strips on the surface of the fingertip that provide dynamic contact information. A strain gage force sensor provides static grasp force information.

References

1. Shaw, M. C., *Metal Cutting Principles*, Oxford: Oxford Science Publications, Clarendon Press, 1989.
2. Doebelin, E. O., *Measurement Systems, Application and Design*, 4th ed., New York: McGraw-Hill, 1990.
3. Dally, J. W., Riley, W. F., and McConnel, K. G., *Instrumentation for Engineering Measurements*, New York: John Wiley & Sons, 1984.
4. Mansfield, P. H., *Electrical Transducers for Industrial Measurement*, London: The Butterworth Group, 1973.
5. Martini, K. H., Multicomponent dynamometers using quartz crystals as sensing elements, *ISA Trans.*, 22(1), 1983.
6. Spur, G., Al-Badrawy, S. J., and Stirnimann, J., Measuring the Cutting Force in Five-Axis Milling, Translated paper "Zerpankraftmessung bei der funfachsigen Frasbearbeitung," Zeitschrift fur wirtschaftliche Fertigung und Automatisierung 9/93 Carl Hanser, Munchen, Kistler Piezo-Instrumentation, 20.162e 9.94.
7. Nachtigal, C. L., *Instrumentation and Control, Fundamentals and Applications*, Wiley Series in Mechanical Engineering Practice, New York: Wiley Interscience, John Wiley & Sons, 1990.
8. DeSilva, C. W., *Control Sensors and Actuators*, Englewood Cliffs, NJ: Prentice-Hall, 1989.
9. Gardner, J. W., *Microsensors Principles and Applications*, New York: John Wiley & Sons, 1995.
10. Stadler, W., *Analytical Robotics and Mechatronics*, New York: McGraw-Hill, 1995.

Further Information

Wright, C. P., *Applied Measurement Engineering, How to Design Effective Mechanical Measurement Systems*, Englewood Cliffs, NJ: Prentice Hall, 1995.

Herceg, E. E., *Handbook of Measurement and Control*, Pennsauken, NJ: Schavitz Engineering, 1972.

Considine, D. M., *Encyclopedia of Instrumentation and Control*, New York: McGraw-Hill, 1971.

Norton, H. N., *Sensor and Analyzer Handbook*, Englewood Cliffs, NJ: Prentice-Hall, 1982.

Sze, S. M., *Semiconductor Sensors*, New York: John Wiley & Sons, 1994.

Lindberg, B., and Lindstrom, B., Measurements of the segmentation frequency in the chip formation process, *Ann. CIRP*, 32(1), 1983.

Tlusty, J., and Andrews, G. C., A critical review of sensors for unmanned machinning, *Ann. CIRP*, 32(2), 1983.

20.4 Torque and Power Measurement

Ivan J. Garshelis

Torque, speed, and power are the defining mechanical variables associated with the functional performance of rotating machinery. The ability to accurately measure these quantities is essential for determining a machine's efficiency and for establishing operating regimes that are both safe and conducive to long and reliable services. Online measurements of these quantities enable real-time control, help ensure consistency in product quality, and can provide early indications of impending problems. Torque and power measurements are used in testing advanced designs of new machines and in the development of new machine components. Torque measurements also provide a well-established basis for controlling and verifying the tightness of many types of threaded fasteners. This chapter describes the basic concepts as well as the various methods and apparati in current use for the measurement of torque and power; the measurement of speed, or more precisely, angular velocity, is discussed elsewhere [1].

20.4.1 Fundamental Concepts

20.4.1.1 Angular Displacement, Velocity, and Acceleration

The concept of *rotational* motion is readily formalized: all points within a rotating rigid body move in parallel or coincident planes while remaining at fixed distances from a line called the *axis*. In a perfectly rigid body, all points also remain at fixed distances from each other. Rotation is perceived as a change in the angular position of a reference point on the body, i.e., as its *angular displacement*, $\Delta\theta$, over some time interval, Δt. The motion of that point, and therefore of the whole body, is characterized by its clockwise (CW) or counterclockwise (CCW) *direction* and by its *angular velocity*, $\omega = \Delta\theta/\Delta t$. If during a time interval Δt, the velocity changes by $\Delta\omega$, the body is undergoing an *angular acceleration*, $\alpha = \Delta\omega/\Delta t$. With angles measured in radians, and time in seconds, units of ω become radians per second (rad s^{-1}) and of α, radians per second per second (rad s^{-2}). Angular velocity is often referred to as *rotational speed* and measured in numbers of complete revolutions per minute (rpm) or per second (rps).

20.4.1.2 Force, Torque, and Equilibrium

Rotational motion, as with motion in general, is controlled by *forces* in accordance with Newton's laws. Because a force directly affects only that component of motion in its line of action, forces or components of forces acting in any plane that includes the axis produce no tendency for rotation about that axis. Rotation can be initiated, altered in velocity, or terminated only by a *tangential force* F_t acting at a finite radial distance l from the axis. The effectiveness of such forces increases with both F_t and l; hence, their product, called a *moment*, is the activating quantity for rotational motion. A moment about the rotational axis constitutes a *torque*. Figure 20.45(a) shows a force F acting at an angle β to the tangent at a point P, distant l (the moment arm) from the axis. The torque T is found from the *tangential component* of F as

$$T = F_t l = (F \cos \beta) l \tag{20.58}$$

The combined effect, known as the *resultant*, of any number of torques acting at different locations along a body is found from their *algebraic sum*, wherein torques tending to cause rotation in CW and CCW directions are assigned opposite signs. Forces, hence torques, arise from physical contact with other solid bodies, motional interaction with fluids, or via gravitational (including inertial), electric, or magnetic force fields. The *source* of each such torque is subjected to an equal, but oppositely directed, *reaction* torque. With force measured in newtons and distance in meters, Equation 20.58 shows the unit of torque to be a Newton meter (Nm).

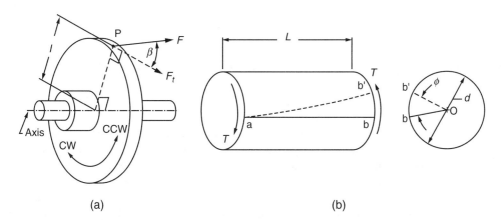

FIGURE 20.45 (a) The off-axis force F at P produces a torque $T = (F \cos \beta) l$ tending to rotate the body in the CW direction. (b) Transmitting torque T over length L twists the shaft through angle ϕ.

A nonzero resultant torque will cause the body to undergo a proportional angular acceleration, found, by application of Newton's second law, from

$$T_r = I\alpha \tag{20.59}$$

where I, having units of kilogram square meter (kg m^2), is the moment of inertia of the body around the axis (i.e., its *polar* moment of inertia). Equation 20.59 is applicable to any body regardless of its state of motion. When $\alpha = 0$, Equation 20.59 shows that T_r is also zero; the body is said to be in *equilibrium*. For a body to be in equilibrium, there must be either more than one *applied* torque, or none at all.

20.4.1.3 Stress, Rigidity, and Strain

Any portion of a rigid body in equilibrium is also in equilibrium; hence, as a condition for equilibrium of the portion, any torques applied thereto from *external* sources must be balanced by equal and directionally opposite *internal* torques from adjoining portions of the body. Internal torques are *transmitted* between adjoining portions by the collective action of *stresses* over their common cross-sections. In a solid body having a round cross-section (e.g., a typical shaft), the *shear stress* τ varies linearly from zero at the axis to a maximum value at the surface. The shear stress, τ_m, at the surface of a shaft of diameter, d, transmitting a torque, T, is found from

$$\tau_m = \frac{16\,T}{\pi d^3} \tag{20.60}$$

Real materials are not *perfectly* rigid but have instead a *modulus of rigidity*, G, which expresses the finite ratio between τ and *shear strain*, γ. The maximum strain in a solid round shaft therefore also exists at its surface and can be found from

$$\gamma_m = \frac{\tau_m}{G} = \frac{16\,T}{\pi d^3 G} \tag{20.61}$$

Figure 20.45b shows the manifestation of shear strain as an angular displacement between axially separated cross sections. Over the length L, the solid round shaft shown will be *twisted* by the torque through an angle ϕ found from

$$\phi = \frac{32\,L\,T}{\pi d^4 G} \tag{20.62}$$

20.4.1.4 Work, Energy, and Power

If during the time of application of a torque, T, the body rotates through some angle θ, mechanical work

$$W = T\theta \tag{20.63}$$

is performed. If the torque acts in the same CW or CCW sense as the displacement, the work is said to be done *on* the body, or else it is done *by* the body. Work done *on* the body causes it to accelerate, thereby appearing as an increase in *kinetic energy* (KE = $I\omega^2/2$). Work done *by* the body causes deceleration with a corresponding decrease in kinetic energy. If the body is not accelerating, any work done on it at one location must be done by it at another location. Work and energy are each measured in units called a joule (J). Equation 20.63 shows that 1 J is equivalent to 1 N m rad, which, since a radian is a dimensionless ratio, $\equiv$ 1 N m. To avoid confusion with torque, it is preferable to quantify mechanical work in units of m N, or better yet, in J.

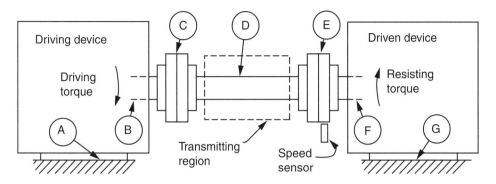

FIGURE 20.46 Schematic arrangement of devices used for the measurement of torque and power.

The *rate* at which work is performed is termed *power*, P. If a torque T acts over a small interval of time Δt, during which there is an angular displacement $\Delta \theta$, work equal to $T\Delta\theta$ is performed at the rate $T\Delta\theta/\Delta t$. Replacing $\Delta\theta/\Delta t$ by ω, power is found simply as

$$P = T\omega \qquad (20.64)$$

The unit of power follows from its definition and is given the special name watt (W). 1 W = 1 J s^{-1} = 1 m N s^{-1}. Historically, power has also been measured in horsepower (Hp), where 1 Hp = 746 W. Rotating bodies effectively transmit power between locations where torques from external sources are applied.

20.4.2 Arrangements of Apparatus for Torque and Power Measurement

Equations 20.58 through 20.64 express the physical bases for torque and power measurement. Figure 20.46 illustrates a generalized measurement arrangement. The actual apparatus used is selected to fulfill the specific measurement purposes. In general, a driving torque originating within a device at one location (B in Figure 20.46) is resisted by an opposing torque developed by a different device at another location (F). The driving torque (from, e.g., an electric motor, a gasoline engine, a steam turbine, muscular effort, etc.) is coupled through connecting members C, transmitting region D, and additional couplings, E, to the driven device (an electric generator, a pump, a machine tool, mated threaded fasteners, etc.) within which the resisting torque is met at F. The torque at B or F is the quantity to be measured. These torques may be *indirectly* determined from a correlated physical quantity, for example, an electrical current or fluid pressure associated with the operation of the driving or driven device, or more directly by measuring either the *reaction* torque at A or G, or the *transmitted* torque through D. It follows from the cause-and-effect relationship between torque and rotational motion that most interest in transmitted torque will involve rotating bodies.

To the extent that the frames of the driving and driven devices and their mountings to the "Earth" are *perfectly* rigid, the reaction at A will *at every instant* equal the torque at B, as will the reaction at G equal the torque at F. Under equilibrium conditions, these equalities are independent of the compliance of any member. Also under equilibrium conditions, and except for usually minor *parasitic* torques (due, e.g., to bearing friction and air drag over rapidly moving surfaces), the driving torque at B will equal the resisting torque at F.

Reaction torque at A or G is often determined, using Equation 20.58, from measurements of the forces acting at known distances fixed by the apparatus. Transmitted torque is determined from measurements, on a suitable member within region D, of τ_m, γ_m, or ϕ and applying Equations 20.60–20.62 (or analogous expressions for members having other than solid round cross sections [2]). *Calibration*, the measurement of the stress, strain, or twist angle resulting from the application of a *known* torque, makes it unnecessary to know any details about the member within D. When $\alpha \neq 0$, and is measurable, T may also be determined from Equation 20.59. Requiring only noninvasive, observational measurements, this method is especially useful for determining transitory torques; for example those associated with firing events in multicylinder internal combustion engines [3].

Equations 20.63 and 20.64 are applicable *only* during rotation because, in the absence of motion, no work is done and power transfer is zero. Equation 20.63 can be used to determine *average* torque from calorimetric measurements of the heat generated (equal to the mechanical work W) during a totalized number of revolutions ($\equiv \theta/2\pi$). Equation 20.64 is routinely applied in power measurement, wherein T is determined by methods based on Equations 20.58, 20.60, 20.61, or 20.62, and ω is measured by any suitable means [4].

F, T, and ϕ are sometimes measured by simple mechanical methods. For example, a "torque wrench" is often used for the controlled tightening of threaded fasteners. In these devices, torque is indicated by the position of a needle moving over a calibrated scale in response to the elastic deflection of a spring member, in the simplest case, the bending of the wrench handle [5]. More generally, instruments, variously called *sensors* or *transducers*, are used to convert the desired (torque or speed related) quantity into a linearly proportional electrical signal. (Force sensors are also known as *load cells*.) The determination of P most usually requires multiplication of the two signals from separate sensors of T and ω. A transducer, wherein the amplitude of a *single* signal proportional to the power being transmitted along a shaft, has also been described [6].

20.4.3 Torque Transducer Technologies

Various physical interactions serve to convert F, τ, γ, or ϕ into proportional electrical signals. Each requires that some axial portion of the shaft be dedicated to the torque sensing function. Figure 20.47 shows typical features of sensing regions for four sensing technologies in present use.

20.4.3.1 Surface Strain

Figure 20.47a illustrates a sensing region configured to convert surface strain (γ_m) into an electrical signal proportional to the transmitted torque. Surface strain became the key basis for measuring both force and torque following the invention of bonded wire strain gages by E. E. Simmons, Jr. and Arthur C. Ruge in 1938 [7]. A modern strain gage consists simply of an elongated electrical conductor, generally formed in a serpentine pattern in a very thin foil or film, bonded to a thin insulating carrier. The carrier is attached, usually with an adhesive, to the surface of the load carrying member. Strain is sensed as a change in gage resistance. These changes are generally too small to be accurately measured directly and so it is common to employ two to four gages arranged in a Wheatstone bridge circuit. Independence from axial and bending loads as well as from temperature variations are obtained by using a four-gage bridge comprised of two diametrically opposite pairs of matched strain gages, each aligned along a *principal strain* direction. In round shafts (and other shapes used to transmit torque), tensile and compressive principal strains occur at 45° angles to the axis. Limiting strains, as determined from Equation 20.61 (with τ_m equal to the shear proportional limit of the shaft material), rarely exceed a few parts in 10^3. Typical practice is to increase the compliance of the sensing region (e.g., by reducing its diameter or with hollow or specially shaped sections) in order to attain the limiting strain at the highest value of the torque to be measured. This maximizes the measurement sensitivity.

20.4.3.2 Twist Angle

If the shaft is *slender* enough (e.g., $L > 5\ d$), ϕ, at limiting values of τ_m for typical shaft materials, can exceed 1°, enough to be resolved with sufficient accuracy for practical torque measurements (ϕ at τ_m can be found by manipulating Equations 20.60–20.62). Figure 20.47(b) shows a common arrangement wherein torque is determined from the difference in tooth-space phasing between two identical "toothed" wheels attached at opposite ends of a compliant "torsion bar." The phase displacement of the periodic electrical signals from the two "pickups" is proportional to the peripheral displacement of salient features on the two wheels, and hence to the twist angle of the torsion bar and thus to the torque. These features are chosen to be sensible by any of a variety of noncontacting magnetic, optical, or capacitive techniques.

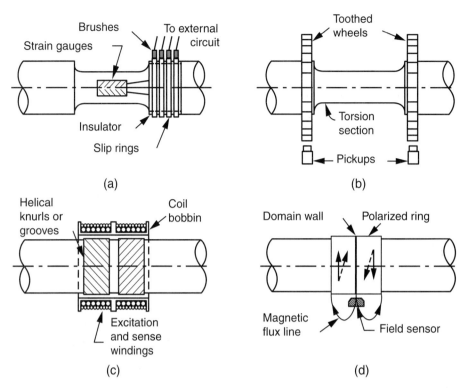

FIGURE 20.47 Four techniques in present use for measuring transmitted torque. (a) Torsional strain in the shaft alters the electrical resistance for four strain gages (two not seen) connected in a Wheatstone bridge circuit. In the embodiment shown, electrical connections are made to the bridge through slip rings and brushes. (b) Twist of the torsion section causes angular displacement of the surface features on the toothed wheels. This creates a phase difference in the signals from the two pickups. (c) The permeabilities of the two grooved regions of the shaft change oppositely with torsional stress. This is sensed as a difference in the output voltages of the two sense windings. (d) Torsional stress causes the initially circumferential magnetizations in the ring (solid arrows) to tilt (dashed arrows). These helical magnetizations cause magnetic poles to appear at the domain wall and ring ends. The resulting magnetic field is sensed by the field sensor.

With more elaborate pickups, the relative angular position of the two wheels appears as the amplitude of a *single* electrical signal, thus providing for the measurement of torque even on a stationary shaft (e.g., [13–15]). In still other constructions, a shaft-mounted variable displacement transformer or a related type of electrical device is used to provide speed independent output signals proportional to ϕ.

20.4.3.3 Stress

In addition to elastic strain, the stresses by which torque is transmitted are manifested by changes in the magnetic properties of ferromagnetic shaft materials. This "magnetoelastic interaction" [8] provides an inherently noncontacting basis for measuring torque. Two types of magnetoelastic (sometimes called magnetostrictive) torque transducers are in present use: Type 1 derive output signals from torque-induced variations in magnetic circuit permeances; Type 2 create a magnetic field in response to torque. Type 1 transducers typically employ "branch," "cross," or "solenoidal" constructions [9]. In branch and cross designs, torque is detected as an imbalance in the permeabilities along orthogonal 45° helical paths (the principal stress directions) on the shaft surface or on the surface of an *ad hoc* material attached to the shaft. In solenoidal constructions torque is detected by differences in the *axial* permeabilities of two adjacent surface regions, preendowed with symmetrical magnetic "easy" axes (typically along the 45° principal stress directions). While branch and cross-type sensors are readily miniaturized [10], local variations in magnetic properties of typical shaft surfaces limit their accuracy. Solenoidal designs, illustrated in Figure 20.47c, avoid this pitfall by effectively averaging these variations. Type 2 transducers are generally constructed with a ring of magnetoelastically active material rigidly attached to the shaft. The ring is magnetized during manufacture of the transducer, usually with each axial half polarized in an opposite circumferential direction as indicated by the solid arrows in Figure 20.47(d) [11]. When torque is applied, the magnetizations tilt into helical directions (dashed arrows), causing magnetic poles to develop at the central domain wall and (of opposite polarity) at the ring end faces. Torque is determined from the output signal of one or more magnetic field sensors (e.g., Hall effect, magnetoresistive, or flux gate devices) mounted so as to sense the intensity and polarity of the magnetic field that arises in the space near the ring.

20.4.4 Torque Transducer Construction, Operation, and Application

Although a torque sensing region can be created directly on a desired shaft, it is more usual to install a preassembled *modular* torque transducer into the driveline. Transducers of this type are available with capacities from 0.001 to 200,000 Nm. Operating principle descriptions and detailed installation and operating instructions can be found in the catalogs and literature of the various manufacturers [12–20]. Tradenames often identify a specific type of transducers; for example, *Torquemeters* [13] refers to a family of noncontact strain gage models; *Torkducer*® [18] identifies a line of Type 1 magnetoelastic transducers; *Torqstar*^{TM} [12] identifies a line of Type 2 magnetoelastic transducers; *Torquetronic* [16] is a class of transducers using wrap-around twist angle sensors; and *TorXimitor*^{TM} [20] identifies optoelectronic-based, noncontact, strain gage transducers. Many of these devices show generic similarities transcending their specific sensing technology as well as their range. Figure 20.48 illustrates many of these common features.

20.4.4.1 Mechanical Considerations

Maximum operating speeds vary widely; upper limits depend on the size, operating principle, type of bearings, lubrication, and dynamic balance of the rotating assembly. Ball bearings, lubricated by grease, oil, or oil mist, are typical. Parasitic torques associated with bearing lubricants and seals limit the accuracy of low-end torque measurements. (Minute capacity units have no bearings [15].) Forced lubrication can

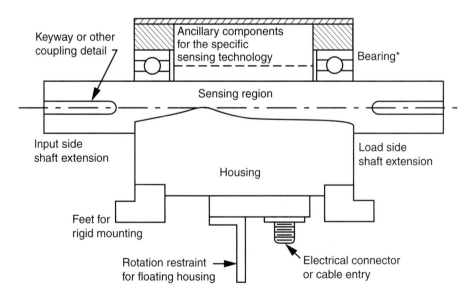

FIGURE 20.48 Modular torque transducer showing generic features and alternative arrangements for free floating or rigid mounting. Bearings* are used only on rotational models. Shaft extensions have keyways or other features to facilitate torque coupling.

allow operation up to 80,000 rpm [16]. High-speed operation requires careful consideration of the effects of centrifugal stresses on the sensed quantity as well as of critical (vibration inducing) speed ranges. Torsional oscillations associated with resonances of the shaft elasticity (characterized by its spring constant) with the rotational inertia of coupled masses can corrupt the measurement, damage the transducer by dynamic excursions above its rated overload torque, and *even be physically dangerous*.

Housings either *float* on the shaft bearings or are *rigidly mounted*. Free floating housings are restrained from rotating by such "soft" means as a cable, spring, or compliant bracket, or by an eccentric external feature simply resting against a fixed surface. In free floating installations, the axes of the driving and driven shafts must be carefully aligned. Torsionally rigid "flexible" couplings at each shaft end are used to accommodate small angular and/or radial misalignments. Alternatively, the use of dual flexible couplings at one end will allow direct coupling of the other end. Rigidly mounted housings are equipped with mounting feet or lugs similar to those found on the frame of electric motors. Free-floating models are sometimes rigidly mounted using adapter plates fastened to the housing. Rigid mountings are preferred when it is difficult or impractical to align the driving and driven shafts, as for example when driving or driven machines are changed often. Rigidly mounted housings *require* the use of dual flexible couplings at *both* shaft ends.

Modular transducers designed for zero or limited rotation applications have no need for bearings. To ensure that *all* of the torque applied at the ends is sensed, it is important in such "reaction"-type torque transducers to limit attachment of the housing to the shaft to only one side of the sensing region. Whether rotating or stationary, the external shaft ends generally include such torque coupling details as flats, keyways, splines, tapers, flanges, male/female squares drives, etc.

20.4.4.2 Electrical Considerations

By their very nature, transducers require some electrical input power or *excitation*. The "raw" output signal of the actual sensing device also generally requires "conditioning" into a level and format appropriate for display on a digital or analog meter or to meet the input requirements of data acquisition equipment. Excitation and signal conditioning are supplied by electronic circuits designed to match the characteristics of the specific sensing technology. For example, strain gage bridges are typically powered with 10–20 V (DC or AC) and have outputs in the range of 1.5–3.0 mV/V of excitation at the rated load. Raising these millivolt signals to more usable levels requires amplifiers having gains of 100 or more. With AC excitation, oscillators and demodulators (or rectifiers) are also needed. Circuit elements of these types are normal when inductive elements are used either as a necessary part of the sensor or simply to implement noncontact constructions.

Strain gages, differential transformers, and related sensing technologies require that electrical components be mounted *on* the torqued member. Bringing electrical power to and output signals from these components on rotating shafts require special methods. The most direct and common approach is to use conductive means wherein brushes (typically of silver graphite) bear against (silver) slip rings. Useful life is extended by providing means to lift the brushes off the rotating rings when measurements are not being made. Several "noncontacting" methods are also used. For example, power can be supplied via inductive coupling between stationary and rotating transformer windings [12–15], by the illumination of shaft-mounted photovoltaic cells [20], or even by batteries strapped to the shaft [21] (limited by centrifugal force to relatively low speeds). Output signals are coupled off the shaft through rotary transformers, by frequency-modulated (infrared) LEDs [19,20], or by radio-frequency (FM) telemetry [21]. Where shaft rotation is limited to no more than a few full rotations, as in steering gear, valve actuators or oscillating mechanisms, hard wiring both power and signal circuits is often suitable. Flexible cabling minimizes incidental torques and makes for a long and reliable service life. All such wiring considerations are avoided when noncontact technologies or constructions are used.

20.4.4.3 Costs and Options

Prices of torque transducers reflect the wide range of available capacities, performance ratings, types, styles, optional features, and accessories. In general, prices of any one type increase with increasing capacity. Reaction types cost about half of similarly rated rotating units. A typical foot-mounted, 565 N m capacity, strain gage transducer with either slip rings or rotary transformers and integral speed sensor, specified nonlinearity and hysteresis each within ±0.1%, costs about $4000 (1997). Compatible instrumentation prividing transducer excitation, conditioning, and analog output with digital display of torque and speed costs about $2000. A comparable magnetoelastic transducer with ±0.5% accuracy costs about $1300. High-capacity transducers for extreme speed service with appropriate lubrication options can cost more than $50,000. Type 2 magnetoelastic transducers, mass produced for automotive power steering applications, cost approximately $10.

20.4.5 Apparatus for Power Measurement

Rotating machinery exists in specific types without limit and can operate at power levels from fractions of a watt to some tens of megawatts, a range spanning more than 10^8. Apparatus for power measurement exists in a similarly wide range of types and sizes. Mechanical power flows from a *driver* to a *load*. This power can be determined *directly* by application of Equation 20.64, simply by measuring, in addition to ω, the output torque of the driver or the input torque to the load, whichever is the device under test (DUT). When the DUT is a driver, measurements are usually required over its full service range of speed and torque. The test apparatus therefore must act as a controllable load and be able to *absorb* the delivered power. Similarly, when the DUT is a pump or fan or other type of load, or one whose function is simply to alter speed and torque (e.g., a gear box), the test apparatus must include a *driver* capable of supplying power over the DUT's full rated range of torque and speed. Mechanical power can also be determined *indirectly* by conversion into (or from) another form of energy (e.g., heat or electricity) and measuring the relevant calorimetric or electrical quantities. In view of the wide range of readily available methods and apparatus for accurately measuring both torque and speed, indirect methods need only be considered when special circumstances make direct methods difficult.

Dynamometer is the special name given to the power-measuring apparatus that includes absorbing and/or driving means and wherein torque is determined by the reaction forces on a stationary part (the *stator*). An effective dynamometer is conveniently assembled by mounting the DUT in such a manner as to allow measurement of the reaction torque on its frame. Figure 20.49 shows a device designed to facilitate such measurements. Commercial models (Torque Table® [12]) rated to support DUTs weighing 222–4900 N are available with torque capacities from 1.3 to 226 Nm. "Torque tubes" [4] or other DUT mounting arrangements are also used. Other than for possible rotational/elastic resonances, these systems

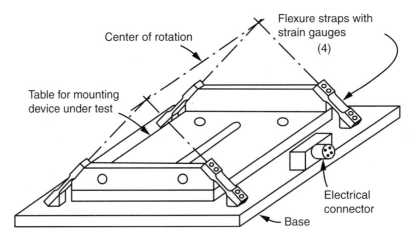

FIGURE 20.49 Support system for measuring the reaction torque of a rotating machine. The axis of the machine must be accurately set on the "center of rotation." The holes and keyway in the table facilitate machine mounting and alignment. Holes in the front upright provide for attaching a lever arm from which calibrating weights may be hung [4,11].

Sensors 20-57

have no speed limitations. More generally, and especially for large machinery, dynamometers include a specialized driving or absorbing machine. Such dynamometers are classified according to their function as *absorbing* or *driving* (sometimes *motoring*). A *universal dynamometer* can function as either a driver or an absorber.

20.4.5.1 Absorption Dynamometers

Absorption dynamometers, often called *brakes* because their operation depends on the creation of a controllable *drag* torque, convert mechanical work into heat. A drag torque, as distinguished from an active torque, can act only to restrain and not to initiate rotational motion. Temperature rise within a dynamometer is controlled by carrying away the heat energy, usually by transfer to a moving fluid, typically air or water. Drag torque is created by inherently dissipative processes such as: friction between rubbing surfaces, shear or turbulence of viscous liquids, the flow of electrical current, or magnetic hysteresis. Gaspard Riche de Prony (1755–1839), in 1821 [22], invented a highly useful form of a friction brake to meet the needs for testing the steam engines that were then becoming prevalent. Brakes of this type are often used for instructional purposes, for they embody the general principles and major operating considerations for all types of absorption dynamometers. Figure 20.50 shows the basic form and constructional features of a *prony brake*. The power that would normally be delivered by the shaft of the driving engine to the driven load is (for measurement purposes) converted instead into heat via the work done by the frictional forces between the friction blocks and the flywheel rim. Adjusting the tightness of the clamping bolts varies the frictional drag torque as required. Heat is removed from the inside surface of the rim by arrangements (not shown) utilizing either a continuous flow or evaporation of water. There is no need to know the magnitude of the frictional forces nor even the radius of the flywheel (facts recognized by Prony), because, while the drag torque tends to rotate the clamped-on apparatus, it is held stationary by the equal but opposite reaction torque Fr. F at the end of the torque arm of radius r (a fixed dimension of the apparatus) is monitored by a scale or load cell. The power is found from Equations 20.58 and 20.64 as $P = Fr\omega = Fr2\pi N/60$ where N is in rpm.

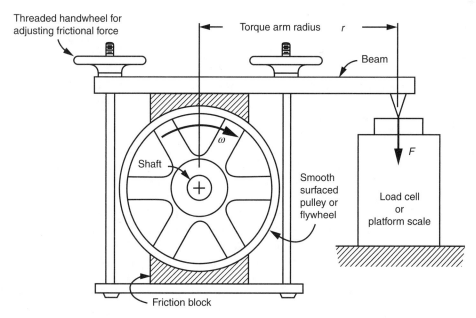

FIGURE 20.50 A classical prony brake. This brake embodies the defining features of all absorbing dynamometers: conversion of mechanical work into heat and determination of power from measured values of reaction torque and rotational velocity.

Uneven retarding forces associated with fluctuating coefficients of friction generally make rubbing friction a poor way to generate drag torque. Nevertheless, because they can be easily constructed, *ad hoc* variations of prony brakes, often using only bare ropes or wooden cleats connected by ropes or straps, find use in the laboratory or wherever undemanding or infrequent power measurements are to be made. More sophisticated prony brake constructions are used in standalone dynamometers with self-contained cooling water tanks in sizes up to 746 kW (1000 Hp) for operation up to 3600 rpm with torques to 5400 Nm [23]. Available in stationary and mobile models, they find use in testing large electric motors as well as engines and transmissions on agricultural vehicles. Prony brakes allow full drag torque to be imposed down to zero speed.

William Froude (1810–1879) [24] invented a *water brake* (1877) that does not depend on rubbing friction. Drag torque within a *Froude brake* is developed between the rotor and the stator by the momentum imparted by the rotor to water contained within the brake casing. Rotor rotation forces the water to circulate between cup-like pockets cast into facing surfaces of both rotor and stator. The rotor is supported in the stator by bearings that also fix its axial position. Labyrinth-type seals prevent water leakage while minimizing frictional drag and wear. The stator casing is supported in the dynamometer frame in cradle fashion by *trunnion* bearings. The torque that prevents rotation of the stator is measured by reaction forces in much the same manner as with the prony brake. Drag torque is adjusted by a valve, controlling either the back pressure in the water outlet piping [25] or the inlet flow rate [26] or sometimes (to allow very rapid torque changes) with two valves controlling both [27]. In any case, the absorbed energy is carried away by the continuous water flow. Other types of cradle-mounted water brakes, while externally similar, have substantially different internal constructions and depend on other principles for developing the drag torque (e.g., smooth rotors develop viscous drag by shearing and turbulence). Nevertheless, all *hydraulic dynamometers* purposefully function as *inefficient* centrifugal pumps. Regardless of internal design and valve settings, maximum drag torque is low at low speeds (zero at standstill) but can rise rapidly, typically varying with the square of rotational speed. The irreducible presence of some water, as well as windage, places a speed-dependent lower limit on the *controllable* drag torque. In any one design, wear and vibration caused by cavitation place upper limits on the speed and power level. Hydraulic dynamometers are available in a wide range of capacities between 300 and 25,000 kW, with some portable units having capacities as low as 75 kW [26]. The largest ever built [27], absorbing up to about 75,000 kW (100,000 Hp), has been used to test propulsion systems for nuclear submarines. Maximum speeds match the operating speeds of the prime movers that they are built to test and therefore generally decrease with increasing capacity. High-speed gas turbine and aerospace engine test equipment can operate as high as 30,000 rpm [25].

In 1855, Jean B. L. Foucault (1819–1868) [22] demonstrated the conversion of mechanical work into heat by rotating a copper disk between the poles of an electromagnet. This simple means of developing drag torque, based on *eddy currents*, has, since circa 1935, been widely exploited in dynamometers. Figure 20.51 shows the essential features of this type of brake. Rotation of a toothed or spoked steel rotor through a spatially uniform magnetic field, created by direct current through coils in the stator, induces locally circulating (eddy) currents in electrically conductive (copper) portions of the stator. Electromagnetic forces between the rotor, which is magnetized by the uniform field, and the field arising from the eddy currents, create the drag torque. This torque, and hence the mechanical input power, are controlled by adjusting the *excitation* current in the stator coils. Electrical input power is less than 1% of the rated capacity. The dynamometer is effectively an internally short-circuited generator because the power associated with the resistive losses from the generated eddy currents is dissipated *within* the machine. Being heated by the flow of these currents, the stator must be cooled, sometimes (in smaller capacity machines) by air supplied by blowers [23], but more often by the continuous flow of water [25,27,28]. In *dry gap* eddy current brakes (the type shown in Figure 20.51), water flow is limited to passages within the stator. Larger machines are often of the *water in gap* type, wherein water also circulates around the rotor [28]. Water in contact with the moving rotor effectively acts as in a water brake, adding a nonelectromagnetic component to the total drag torque, thereby placing a lower limit to the controllable torque. Windage limits the minimum value of controllable torque in dry gap types. Since drag torque is developed

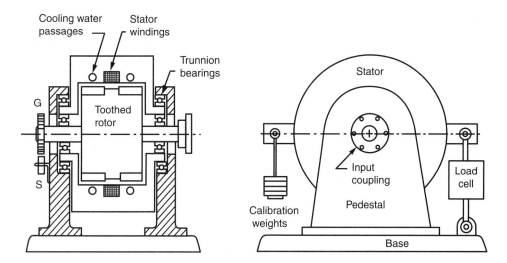

FIGURE 20.51 Cross-section (left) and front view (right) of an eddy current dynamometer. G is a gear wheel and S is a speed sensor. Hoses carrying cooling water and cable carrying electrical power to the stator are not shown.

by the motion of the rotor, it is zero at standstill for any value of excitation current. Initially rising rapidly, approximately linearly, with speed, torque eventually approaches a current limited saturation value. As in other cradled machines, the torque required to prevent rotation of the stator is measured by the reaction force acting at a fixed known distance from the rotation axis. Standard model eddy current brakes have capacities from less than 1 kW [23,27] to more than 2000 kW [27,28], with maximum speeds from 12,000 rpm in the smaller capacity units to 3600 rpm in the largest units. Special units with capacities of 3000 Hp (2238 kW) at speeds to 25,000 rpm have been built [28].

Hysteresis brakes [29] develop drag torque via magnetic attractive/repulsive forces between the magnetic poles established in a reticulated stator structure by a current through the field coil, and those created in a "drag cup" rotor by the stator field gradients. Rotation of the special steel rotor, through the spatial field pattern established by the stator, results in a cyclical reversal of the polarity of its local magnetizations. The energy associated with these reversals (proportional to the area of the hysteresis loop of the rotor material) is converted into heat within the drag cup. Temperature rise is controlled by forced air cooling from a blower or compressed air source. As with eddy current brakes, the drag torque of these devices is controlled by the excitation current. In contrast with eddy current brakes, rated drag torque is available down to zero speed. (Eddy current effects typically add only 1% to the drag torque for each 1000 rpm). As a result of their smooth surfaced rotating parts, hysteresis brakes exhibit low parasitic torques and hence cover a dynamic range as high as 200 to 1. Standard models are available having continuous power capacities up to 6 kW (12 kW with two brakes in tandem cooled by two blowers). Intermittent capacities per unit (for 5 min or less) are 7 kW. Some low-capacity units are convection cooled; the smallest has a continuous rating of just 7 W (35 W for 5 min). Maximum speeds range from 30,000 rpm for the smallest to 10,000 rpm for the largest units. Torque is measured by a strain gage bridge on a moment arm supporting the machine stator.

20.4.5.2 Driving and Universal Dynamometers

Electric generators, both AC and DC, offer another means for developing a controllable drag torque and they are readily adapted for dynamometer service by cradle mounting their stator structures. Moreover, electrical machines of these types can also operate in a motoring mode wherein they can deliver controllable *active* torque. When configured to operate selectively in either driving or absorbing modes, the machine serves as a universal dynamometer. With DC machines in the absorbing mode, the generated power is typically dissipated in a convection-cooled resistor bank. Air cooling the machine with blowers is usually adequate, since *most* of the mechanical power input is dissipated externally. Nevertheless, *all* of the mechanical input power is accounted for by the product of the reaction torque and the rotational speed. In the motoring mode, torque and speed are controlled by adjustment of both field and armature currents. Modern AC machines utilize regenerative input power converters to allow braking power to be returned to the utility power line. In the motoring mode, speed is controlled by high-power, solid-state, adjustable frequency inverters. Internal construction is that of a simple three-phase induction motor, having neither brushes, slip rings, nor commutators. The absence of rotor windings allows for higher speed operation than DC machines. Universal dynamometers are "four-quadrant" machines, a term denoting their ability to produce torque in the same or opposite direction as their rotational velocity. This unique ability allows the effective drag torque to be reduced to zero at any speed. Universal dynamometers [25,28] are available in a relatively limited range of capacities (56–450 kW), with commensurate torque (110–1900 N m) and speed (4500–13,500 rpm) ranges, reflecting their principal application in automotive engine development. Special dynamometers for testing transmissions and other vehicular drive train components insert the DUT between a diesel engine or electric motor prime mover and a hydraulic or eddy current brake [30].

20.4.5.3 Measurement Accuracy

Accuracy of power measurement (see discussion in [4]) is generally limited by the torque measurement ($\pm 0.25\%$ to $\pm 1\%$) since rotational speed can be measured with almost any desired accuracy. Torque errors can arise from the application of extraneous (i.e., not indicated) torques from hose and cable connections, from windage of external parts, and from miscalibration of the load cell. Undetected friction in the trunnion bearings of cradled dynamometers can compromise the torque measurement accuracy. Ideally, well-lubricated antifriction bearings make no significant contribution to the restraining torque. In practice, however, the unchanging contact region of the balls or other rolling elements on the bearing races makes them prone to brinelling (a form of denting) from forces arising from vibration, unsupported weight of attached devices, or even inadvertently during the alignment of connected machinery. The problem can be alleviated by periodic rotation of the (primarily outer) bearing races. In some bearing-in-bearing constructions, the central races are continuously rotated at low speeds by an electric motor while still others avoid the problem by supporting the stator on hydrostatic oil lift bearings [28].

20.4.5.4 Costs

The wide range of torque, speed, and power levels, together with the variation in sophistication of associated instrumentation, is reflected in the very wide range of dynamometer prices. Suspension systems of the type illustrated in Figure 20.49 (for which the user must supply the rotating machine) cost $4000–6000, increasing with capacity [12]. A 100-Hp (74.6 kW) *portable* water brake equipped with a strain gage load cell and a digital readout instrument for torque, speed, and power costs $4500, or $8950 with more sophisticated data acquisition equipment [26]. Stationary (and some *transportable* [23]) hydraulic dynamometers cost from $113/kW in the smaller sizes [25] down to $35/kW for the very largest [27]. Transportation, installation, and instrumentation can add significantly to these costs. Eddy current dynamometers cost from as little as $57/kW to nearly $700/kW, depending on the rated capacity, type of control system, and instrumentation [24,25,28]. Hysteresis brakes with integral speed sensors cost from $3300 to $14,000 according to capacity [29]. Compatible controllers, from manual to fully programmable for PC test control and data acquisition via an IEEE-488 interface, vary in price from $500 to $4200. The flexibility and high performance of AC universal dynamometers is reflected in their comparatively high prices of $670–2200/kW [25,28].

References

1. Pinney, C. P. and Baker, W. E., Velocity Measurement, *The Measurement, Instrumentation and Sensors Handbook*, Webster, J. G., Ed., Boca Raton, FL: CRC Press, 1999.
2. Timoshenko, S., *Strength of Materials*, 3rd ed., New York: Robert E. Kreiger, Part I, 281–290; Part II, 235–250, 1956.

3. Citron, S. J., *On-Line Engine Torque and Torque Fluctuation Measurement for Engine Control Utilizing Crankshaft Speed Fluctuations,* U.S. Patent No. 4,697,561, 1987.
4. Supplement to ASME Performance Test Codes, Measurement of Shaft Power, ANSI/ASME PTC 19.7-1980 (Reaffirmed 1988).
5. See, for example, the catalog of torque wrench products of Consolidated Devices, Inc., 19220 San Jose Ave., City of Industry, CA 91748.
6. Garshelis, I. J., Conto, C. R., and Fiegel, W. S., A single transducer for non-contact measurement of the power, torque and speed of a rotating shaft, SAE Paper No. 950536, 1995.
7. Perry, C. C., and Lissner, H. R., *The Strain Gage Primer,* 2nd ed., New York: McGraw-Hill, 1962, 9. (This book covers all phases of strain gage technology.)
8. Cullity, B. D., *Introduction to Magnetic Materials,* Reading, MA: Addison-Wesley, 1972, Section 8.5, 266–274.
9. Fleming, W. J., Magnetostrictive torque sensors—comparison of branch, cross and solenoidal designs, SAE Paper No. 900264, 1990.
10. Nonomura, Y., Sugiyama, J., Tsukada, K., Takeuchi, M., Itoh, K., and Konomi, T., Measurements of engine torque with the intra-bearing torque sensor, SAE Paper No. 870472, 1987.
11. Garshelis, I. J., *Circularly Magnetized Non-Contact Torque Sensor and Method for Measuring Torque Using Same,* U.S. Patent 5,351,555, 1994 and 5,520,059, 1996.
12. Lebow® Products, Siebe, plc., 1728 Maplelawn Road, Troy, MI 48099, Transducer Design Fundamentals/Product Listings, Load Cell and Torque Sensor Handbook No. 710, 1997, also: Torqstar™ and Torque Table®.
13. S. Himmelstein & Co., 2490 Pembroke, Hoffman Estates, IL 60195, MCRT® Non-Contact Strain Gage Torquemeters and Choosing the Right Torque Sensor.
14. Teledyne Brown Engineering, 513 Mill Street, Marion, MA 02738-0288.
15. Staiger, Mohilo & Co. GmbH, Baumwasenstrasse 5, D-7060 Schorndorf, Germany (In the U.S.: Schlenker Enterprises, Ltd., 5143 Electric Ave., Hillside, IL 60162), Torque Measurement.
16. Torquemeters Ltd., Ravensthorpe, Northampton, NN6 8EH, England (In the U.S.: Torquetronics Inc., P.O. Box 100, Allegheny, NY 14707), Power Measurement.
17. Vibrac Corporation, 16 Columbia Drive, Amherst, NH 03031, Torque Measuring Transducer.
18. GSE, Inc., 23640 Research Drive, Farmington Hills, MI 48335-2621, Torkducer®.
19. Sensor Developments, Inc., P.O. Box 290, Lake Orion, MI 48361-0290, 1996 Catalog.
20. Bently Nevada Corporation, P.O. Box 157, Minden, NV 89423, TorXimitor™.
21. Binsfield Engineering, Inc., 4571 W. MacFarlane, Maple City, MI 49664.
22. Gillispie, C. C. (Ed.), *Dictionary of Scientific Biography,* Vol. XI, New York: Charles Scribner's Sons, 1975.
23. AW Dynamometer, Inc., P.O. Box 428, Colfax, IL 61728, Traction dynamometers: portable and stationary dynamometers for motors, engines, vehicle power take-offs.
24. Roy Porter (Ed.), *The Biographical Dictionary of Scientists,* 2nd ed., New York: Oxford University Press, 1994.
25. Froude-Consine, Inc., 39201 Schoolcraft Rd., Livonia, MI 48150, F Range Hydraulic Dynamometers, AG Range Eddy Current Dynamometers, AC Range Dynamometers.
26. Go-Power Systems, 1419 Upfield Drive, Carrollton, TX 75006, Portable Dynamometer System, Go-Power Portable Dynamometers.
27. Zöllner GmbH, Postfach 6540, D-2300 Kiel 14, Germany (In the U.S. and Canada: Roland Marine, Inc., 90 Broad St., New York, NY 10004), Hydraulic Dynamometers Type P, High Dynamic Hydraulic Dynamometers.
28. Dynamatic Corporation, 3122 14th Ave., Kenosha, WI 53141-1412, Eddy Current Dynamometer—Torque Measuring Equipment, Adjustable Frequency Dynamometer.
29. Magtrol, Inc., 70 Gardenville Parkway, Buffalo, NY 14224-1322, Hysteresis Absorption Dynamometers.
30. Hicklin Engineering, 3001 NW 104th St., Des Moines, IA 50322, Transdyne™ (transmission test systems, brake and towed chassis dynamometers).

20.5 Flow Measurement

Richard Thorn

20.5.1 Introduction

Flow measurement is something that nearly everyone has experienced of. Everyday examples include the metering of household utilities such as water and gas. Similarly flowmeters are used in nearly every sector of industry, from petroleum to food manufacture and processing. It is therefore not surprising that today, the total world flowmeter market is worth over $3000 million and expected to continue growing steadily in the future.

However, what is surprising, given the undoubted importance of flow measurement to the economy, is the accuracy and technology of the most commonly used flowmeters which are poor and relatively old fashioned in comparison to instruments used to measure other measurands such as pressure and temperature. For example, the orifice plate flowmeter, which is still one of the most frequently used flowmeters in the process industry, only has a typical accuracy of ±2% of reading and was first used commercially in the late 1800s. The conservative nature of the flow measurement industry means that traditional techniques such as the orifice plate, Venturimeter, and variable area flowmeter still dominate, while ultrasonic flowmeters which were first demonstrated in the 1950s are still considered to be "new" devices by many users. This article will consider the most commonly used commercially available methods of flow measurement. For recent research developments in flow measurement see [1].

20.5.2 Terminology

The term flow measurement is a general term, and before selecting a flowmeter it is important to be sure what type of flow measurement is actually required. For a fluid flowing through a pipe, flow measurement may mean any one of six different types of measurement:

1. Point velocity measurement—the fluid's velocity at a fixed point across the pipe's cross section (m/s).
2. Mean flow velocity measurement—average fluid velocity across the cross section of the pipe (m/s).
3. Volumetric flowrate measurement—the rate of change in the volume of fluid passing through the pipe with time (m^3/s).
4. Total volume measurement—the total volume of fluid which has passed through the pipe (m^3).
5. Mass flowrate measurement—the rate of change in the mass of the fluid passing through the pipe with time (kg/s).
6. Total mass measurement—the total mass of fluid passing through the pipe (kg/s).

Although the most common type of flow measurement is that of a fluid through a closed conduit or pipe, open channel flow measurements are also regularly needed in applications such as sewage and water treatment. For further information on open channel flow measurement techniques see [2].

20.5.3 Flow Characteristics

The fluid being metered is usually a liquid or gas, and is known as single phase flow. However, there is an increasing need for the flowrate of multiphase mixtures to be measured (see Section 20.5.13.)

There are a number of important principles relating to the characteristic of flow in a pipe, which should be understood before a flowmeter can be selected and used with confidence. These are the meaning of Reynolds number, and the importance of the flow's velocity profile.

The Reynolds number Re is the ratio of the inertia forces in the flow ($\rho \bar{v} D$) to the viscous forces in the flow (η), and it can be used to determine whether a fluid flow is laminar or turbulent in nature.

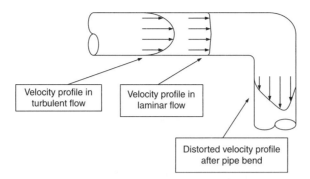

FIGURE 20.52 Flow velocity profiles in laminar and turbulent flow.

Reynolds number can be calculated using

$$\mathrm{Re} = \frac{\rho \bar{v} D}{\eta} \qquad (20.65)$$

where ρ is the density of the fluid, $\bar{v}$ is the mean velocity of the fluid, D is the pipe diameter, and η is the dynamic viscosity of the fluid. If Re is less than 2000, viscous forces in the flow dominate and the flow will be laminar. If Re is greater than 4000, inertia forces in the flow dominate and the flow will be turbulent. If Re is between 2000 and 4000 the flow is transitional and either mode can be present. The Reynolds number is mainly calculated using properties of the fluid and does not take into account factors such as pipe roughness, bends, and valves, which also affect the flow characteristic. However, the Reynolds number is a good guide to the type of flow which might be expected in most situations.

The fluid velocity across a pipe's cross section is not constant and depends on the type of flow present (Figure 20.52). In laminar flow, the velocity at the center of the pipe is twice the average velocity across the pipe cross-section and the flow profile is unaffected by the roughness of the pipe wall. In turbulent flow, pipe wall effects are less and the flow's velocity profile is flatter, with the velocity at the center being about 1.2 times the mean velocity. The exact flow profile in a turbulent flow depends on pipe wall roughness and Reynolds number. In industrial applications laminar flows are rarely encountered unless very viscous fluids are being metered. The pipe Reynolds number should always be calculated since some flowmeters are not suitable for use in both laminar and turbulent flow conditions.

A flow's velocity profile will only be symmetrical at the end of a very long pipe. Bends and obstructions such as valves will cause the profile to become distorted or asymmetric. Since the calibration of many flowmeters is sensitive to the velocity profile of the flow passing through the meter then in order to have confidence in the performance of a flowmeter, the velocity profile of the flow passing through the flowmeter should be stable and known.

20.5.4 Flowmeter Classification

Although there at least 80 different types of flowmeter commercially available, they may be all classified into nine main groups. Table 20.3 gives examples of the main types of flowmeter in each group.

Traditional flow measurement technologies are represented by the differential pressure, variable area, positive displacement, and turbine categories. Newer techniques are represented by the electromagnetic, ultrasonic, oscillatory, and mass categories. Although differential pressure flowmeters are still the most commonly used method of flow measurement, especially in the process industrial sector, in general traditional methods are being increasingly replaced by newer techniques. These techniques are now often preferred because in most cases they do not obstruct the flow, and yet match many of the traditional flowmeters in terms of accuracy and reliability.

TABLE 20.3 Main Categories of Closed Conduit Flowmeter

Type 1—differential pressure flowmeters
  Sharp edged orifice plate, chord orifice plate, eccentric orifice plate, Venturi, nozzle, Pitot tube, elbow, wedge, V-cone, Dall tube, Elliot–Nathan flow tube, Epiflo
Type 2—variable area flowmeters
  Rotameter, orifice and tapered plug, cylinder and piston, target, variable aperture
Type 3—positive displacement flowmeters
  Sliding vane, tri-rotor, bi-rotor, piston, oval gear, nutating-disc, roots, CVM, diaphragm, wet gas
Type 4—turbine flowmeters
  Axial turbine, dual-rotor axial turbine, cylindrical rotor, impeller, Pelton wheel, Hoverflo, propeller
Type 5—oscillatory flowmeters
  Vortex shedding, swirlmeter, fluidic
Type 6—electromagnetic flowmeters
  AC magnetic, pulsed DC magnetic, insertion
Type 7—ultrasonic flowmeters
  Doppler, single path transit-time, multi-path transit-time, cross-correlation, drift
Type 8—mass flowmeters
  Coriolis, thermal
Type 9—miscellaneous flowmeters
  Laser anemometer, hot-wire anemometers, tracer dilution, nuclear magnetic resonance

The following sections will consider the most popular types of flowmeter from each of the eight main categories in Table 20.3. For information on other flowmeters and those in the miscellaneous group see one of the many textbooks on flow measurement such as [3–6].

20.5.5 Differential Pressure Flowmeter

The basic principle of nearly all differential pressure flowmeters is that if a restriction is placed in a pipeline, then the pressure drop across this restriction is related to the volumetric flowrate of fluid flowing through the pipe.

The orifice plate is the simplest and cheapest type of differential pressure flowmeter. It is simply a plate with a hole of specified size and position cut in it, which can then be clamped between flanges in a pipeline (Figure 20.53). The volumetric flowrate of fluid Q in the pipeline is given by:

$$Q = \frac{C}{\sqrt{1-\beta^4}} \varepsilon \frac{\pi}{4} d^2 \sqrt{\frac{2(p_1 - p_2)}{\rho}} \qquad (20.66)$$

where p_1 and p_2 are the pressures on each side of the orifice plate, ρ is the density of the fluid upstream of the orifice plate, d is the diameter of the hole in the orifice plate, and β is the diameter ratio d/D where D is the upstream internal pipe diameter. The two empirically determined correction factors are C the discharge coefficient, and ε the expansibility factor. C is affected by changes in the diameter ratio, Reynolds number, pipe roughness, the sharpness of the leading edge of the orifice, and the points at which the differential pressure across the plate are measured. However, for a fixed geometry it has been shown that C is only dependent on the Reynolds number and so this coefficient can be determined for a particular application. ε is used to account for the compressibility of the fluid being monitored. Both C and ε can be determined from equations and tables in a number of internationally recognized

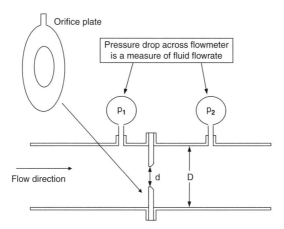

FIGURE 20.53 Flowrate measurement using an orifice plate.

documents known as standards. These standards not only specify C and ε, but also the geometry and installation conditions for the square-edged orifice plate, and two other common types of differential pressure flowmeters, the Venturi tube and nozzle. Installation recommendations are intended to ensure that fully developed turbulent flow conditions exist within the measurement section of the flowmeter. The most commonly used standard in Europe is ISO 5167-1 [7], while in the USA, API 2530 is the most popular [8]. Thus, one of the major reasons for the continued use of the orifice plate flowmeter is that measurement uncertainty (typically ±2% of reading) can be predicted without the need for calibration, as long as it is manufactured and installed in accordance with one of these international standards.

The major disadvantages of the orifice plate are its limited range and sensitivity to changes in the flow's velocity profile. The fact that fluid flow rate is proportional to the square root of the measured differential pressure limits the range of a one plate/one differential pressure transmitter combination to about 3:1. The required diameter ratio (also known as beta ratio) of the plate depends on the maximum flow rate to be measured and the range of the differential pressure transducer available.

Equation 20.66 assumes a fully developed and stable flow velocity profile, and so installation of the device is critical, particularly the need for sufficient straight pipework, upstream and downstream of the meter. Wear of the leading edge of the orifice plate can also severely alter measurement accuracy and so this device is normally only used with clean fluids.

The other two differential pressure flowmeters covered by international standards are the Venturi tube and nozzle. The Venturi tube has a lower permanent pressure less than the orifice plate, and is less sensitive to erosion and upstream disturbances. Major disadvantages are its size and cost. It is more difficult, and therefore more expensive, to manufacture than the orifice plate.

Nozzles have pressure losses similar to orifice plates but because of their smooth design they retain their calibration over a long period. However, these devices are more expensive to manufacture than the orifice plate but cheaper than the Venturi tube. The two most common nozzle designs of nozzle are covered by international Standards, with the ISA-1932 nozzle being preferred in Europe and the ASME long radius nozzle being preferred in the U.S.

There are many other types of differential pressure flowmeter, such as the segmental wedge, V-cone, elbow, and Dall tube. Each of these has advantages over the orifice plate, Venturi tube, and nozzle for specific applications. For example, the segmental wedge can be used with flows having a low Reynolds number, and a Dall tube has a lower permanent pressure loss than a Venturi tube. However, none of these instruments are yet covered by international standards and so calibration is needed to determine their accuracy.

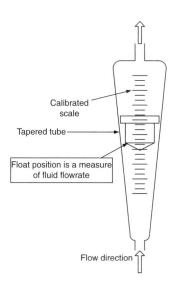

FIGURE 20.54 Tapered tube and float variable area flowmeter.

20.5.6 Variable Area Flowmeter

Variable area flowmeters are also based on using an obstruction in the flow to create a differential pressure principle, except in this case the differential pressure is constant and the area of the obstruction varies as the flowrate being measured changes. Probably the best known type of variable area flowmeter is the taper tube and float flowmeter, known almost universally as a rotameter (Figure 20.54). This type of flowmeter consists of a vertical tapered tube into which a float or bob is fitted. The fluid being metered enters the tube at the bottom and forces the float up the tube, which also increases the cross-sectional area available around the float for the fluid to pass through. Increasing the flowrate will move the float further up the tube, and so the position at which the float comes to rest is a direct function of flowrate.

Rotameters are extremely simple and reliable, and have an output which changes linearly with flowrate (unlike differential pressure flowmeters) and a typical range of 10:1 (compared to 3:1 for differential pressure flowmeters). Accuracy is typically ±2% of full scale, but will depend on range and cost of the device. In addition, the flowmeter's calibration is insensitive to changes in the velocity profile of the flow. Since the tube can be made of glass or clear plastic, a visual indication of flowrate is directly available and, of course, the flowmeter requires no external power supply in order to function. As a result such flowmeters are commonly found in many process and laboratory applications where gases or liquids need to be metered. If high temperature, high pressure, or corrosive fluids need to be metered, the rotameter's tube can be made of metal. In such cases a mechanism for detecting and displaying the position of the float is required.

A major limitation of the rotameter is that it can usually only be used vertically and so causes installation difficulties if the pipeline being metered is horizontal. Some manufacturers produce spring loaded rotameters, which can be used in any position; however, in general these have poorer accuracy than standard rotameters. Other limitations are that the calibration of the meter is dependent on the viscosity and density of the fluid being metered, and producing an electrical output signal suitable for transmission requires extra complexity. However, the use of optical or magnetic limit switches to enable the flowmeter to be used in high or low flow alarm applications is common.

20.5.7 Positive Displacement Flowmeter

Positive displacement flowmeters are based on a simple measurement principle. The flow being measured is "displaced" or moved from the inlet side of the flowmeter to the outlet side using a series of compartments of known volume. The number of compartments of fluid that have been transferred are counted to determine the total volume that has passed through the flowmeter, and if time is also measured then

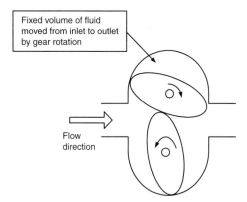

FIGURE 20.55 The oval-gear positive displacement flowmeter.

volumetric flowrate can be measured. There are many designs of positive displacement flowmeters commercially available. For liquids the most common designs are piston, sliding vane, oval-gear, bi-rotor, tri-rotor, and disc types of flowmeter while for gases roots, bellows (or diaphragm), or CVM flowmeters are popular. Despite this wide range of design all are based on the same principle and all are predominantly mechanical devices.

The advantages of positive displacement flowmeters are that they are capable of high accuracy measurement (typically ±0.5% of reading for liquids and ±1% of reading for gases) over a wide range of flowrates. They can be used to meter fluids with a wide range of viscosity and density. In addition, unlike most other flowmeters, they are insensitive to changes in flow velocity profile and so do not require long lengths of straight pipe work before and after the flowmeter.

Figure 20.55 shows the principle of the oval-gear flowmeter and illustrates the limitations of positive displacement flowmeters. They are relatively complex mechanical devices, with moving parts which of course wear with time. Their measurement accuracy depends both on the initial quality of manufacture and a regular maintenance schedule once in use. Fluids being metered should also be free of solid particles so as to reduce wear of the seals and reduce the need for excessive maintenance. Positive displacement flowmeters can also be heavy and expensive for larger pipe sizes and some designs can result in a complete blockage of the pipeline if the flowmeter seizes up.

20.5.8 Turbine Flowmeter

Like the positive displacement flowmeter, turbine (or vane) flowmeters are mechanical devices capable of achieving high measurement accuracy. The principle of operation of this type of flowmeter is that a multi-bladed rotor is placed in the flow and rotates as fluid passes through it. The rotor's speed of rotation is detected using a sensor (rf, magnetic, and mechanical types being the most common), and is proportional to the velocity of the fluid flowing through the meter. These flowmeters measure the average velocity of fluid in a pipeline, and since the pipe diameter is known, volumetric flowrate can be determined.

Despite the fact that the turbine flowmeter is a mechanical device which may appear to be old fashioned when compared to many other technologies available, it is still one of the most accurate and repeatable flowmeters available today. Measurement accuracy of better than ±0.1% of reading for liquids, and better than ±0.25% of reading for gases, is possible using this type of flowmeter. For this reason the turbine flowmeter is one of the most commonly used instruments in custody transfer applications. These flowmeters have a linear output and a range of at least 10:1, with 100:1 possible in some applications.

The main limitation of the turbine flowmeter is the fact that key mechanical components such as the rotor bearings will wear with use, and in doing so degrade the instrument's repeatability and alter its calibration. Regular maintenance and recalibration are therefore necessary with this type of flowmeter. Care should also be taken to ensure that the fluid being metered is clean, since solid particles in the flow

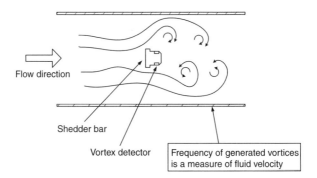

FIGURE 20.56 Principle of the vortex shedding flowmeter.

will cause more rapid bearing wear. The flowmeter's calibration is also sensitive to changes in fluid viscosity and upstream flow velocity profile.

Other types of flowmeters which use the turbine principle include the Pelton wheel and propeller meter, although they have poorer measurement accuracy than axial designs.

20.5.9 Vortex Shedding Flowmeter

The vortex shedding flowmeter, now more commonly known as the vortex flowmeter, relies on the phenomena of vortex shedding, which was first experimentally studied by Strouhal in 1878. Figure 20.56 shows the principle of the vortex flowmeter. A nonstreamlined obstruction known as a shedder bar or bluff body is placed across the middle of the flow stream. As the fluid stream in the pipe hits this obstacle it must separate to pass around it, with fluid layers nearer the surface of the bar moving slower than those further away. As a result, when these fluid layers recombine after passing the bar, vortices are generated and shed alternately from either side of the shedder bar. The frequency of generated vortices is related to the upstream velocity of the fluid and the width of the shedder bar and is defined by the K factor of the flowmeter. For a given geometry of shedder bar the K factor of a flowmeter is relatively constant over a wide range of pipe Reynolds number, and so in these circumstances the volumetric flowrate of the fluid is linearly related to the vortex shedding frequency.

The frequency of generated vortices is usually detected using sensors integrated into the sides of the shedder bar. Pressure, capacitance, thermal, and ultrasonic are the most common types of sensor used for this purpose.

The vortex flowmeter is capable of accurate measurement of liquid or gas (typically ±1% of reading) over a minimum flow range of 30:1. The flowmeter can also be used over a wide range of fluid temperatures and so is commonly used for metering process fluids at the extreme ends of the temperature range, such as liquid nitrogen and steam. The instrument's calibration is also insensitive to changes in fluid density, and so a meter's calibration holds for any fluid as long as the flowmeter is used within the Reynolds number range specified by the manufacturer. The vortex flowmeter has a simple and reliable construction and so can be used with flows containing small amounts of particles, although more extreme multiphase flows such as slurries will cause rapid wear of the shedder bar and so a change in calibration. The relatively small obstruction that the shedder bar causes results in a permanent pressure loss of about half that of an orifice plate over the same range of flowrate.

The main limitation of the vortex flowmeter is that it can only be used in turbulent flow conditions. It is, therefore, not usually suitable for use in large pipe diameters, or in applications where the flow velocity is low or the fluid viscosity high. Most manufacturers set a minimum Reynolds number of typically 10,000 at which the specified flowmeter performance can be achieved. While many flowmeters will continue operating at Reynolds numbers less than this, the generated vortex stream becomes less stable and so accuracy is reduced. At a Reynolds of less than around 3000, vortices will not be generated at all and so the flowmeter will stop operating. The vortex flowmeter is also sensitive to changes in upstream flow velocity profile and other disturbances such as swirl, and so should be installed with a sufficient straight length of pipe upstream and downstream of the measurement point. The flowmeter should not be used in applications where pipe vibration or local sources of electrical interference are high, since this will corrupt the vortex signal being detected and possibly give false readings under no-flow conditions.

20.5.10 Electromagnetic Flowmeter

The operation of the electromagnetic flowmeter is based on Faraday's law of induction, that is, when a conductor is moving perpendicular to a magnetic field, the voltage induced across the conduction is proportional its velocity. In the case of the electromagnetic flowmeter, the conductor is the fluid being metered, while the induced voltage is measured using electrodes in the pipe wall. Since in most applications the pipe wall of the flowmeter is made from a conductive material such as a stainless steel, an inner nonconducting liner is required to insulate the electrodes and prevent the generated voltage signal being dissipated into the pipe wall. Coils on the outside of the pipe are used to generate a magnetic field across the fluid, with simpler AC coil excitation methods which suffer from zero drift problems being increasingly replaced by pulsed DC excitation techniques which do not.

The electromagnetic flowmeter has a number of advantages over traditional flow measurement techniques, and some characteristics of an ideal flowmeter. The flowmeter has no moving parts, does not obstruct the pipe at all, is available in a very wide range of pipe sizes, and may be used to measure bidirectional flows. A measurement accuracy of typically ±0.5% of reading over a range of at least 10:1 is possible. The flowmeter's accuracy is also unaffected by changes in fluid viscosity and density, and may be used to meter difficult mixtures such as slurries and paper pulp.

The major limitation of the electromagnetic flowmeter is that it can only be used with fluids with a conductivity of typically greater than 5 μS/cm, although special designs are available for use with liquids with conductivities of as low as 0.1 μS/cm. The flowmeter is, therefore, not suitable for use with gases, steam, or nonconducting liquids such as oil. The flowmeter's calibration is also sensitive to changes in flow velocity profile although requiring a shorter straight length of pipe upstream of the meter than the orifice plate or turbine meter. Although electromagnetic flowmeters do not require significant maintenance, care must be taken during operation to ensure that the liner does not become damaged, and that significant deposits do not build-up on the electrodes, since these can cause changes in the calibration or in some cases cause the flowmeter to stop functioning altogether. Even if these effects are minimized, electromagnetic flowmeters will require periodic recalibration using either traditional techniques or an electronic calibrator now available as an accessory from most manufacturers.

20.5.11 Ultrasonic Flowmeter

The dream of producing a universal non-invasive flowmeter has been the catalyst for the many different ultrasonic flowmeter configurations, which have been investigated over the last 40 years [9]. However, most ultrasonic flowmeters commercially available today can be placed into one of two categories—Doppler and transit-time.

The ultrasonic Doppler flowmeter is based on the Doppler shift principle. Ultrasound at a frequency of typically 1 MHz is transmitted at an angle into the moving fluid being monitored. Some of this energy will be reflected back by acoustic discontinuities such as particles, bubbles, or turbulent eddies. The difference in frequency between the transmitted and received signals (the Doppler frequency shift) is directly proportional to the velocity of the flow.

The ultrasonic transducers, which are used to transmit and receive the ultrasound, are usually located in a single housing that can be fixed onto the outside of the pipe, and so a simple clamp-on flowmeter, which is easy to install and completely noninvasive, is possible.

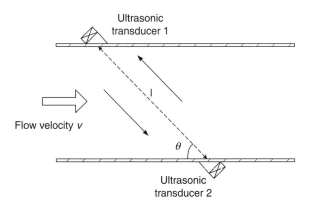

FIGURE 20.57 Principle of the transit-time ultrasonic flowmeter.

The reflected ultrasound does not consist of a single frequency, but a spread of frequencies resulting from reflections from a range of different sized discontinuities, which may also be travelling at different velocities travelling through the detection area. For liquids the frequency of the transmitted ultrasound may lie in the range from 500 kHz up to a few megahertz. At 500 kHz discontinuities must have a diameter of approximately 50 μm in order to reflect ultrasound back to the receiver. Increase in the operating frequency will allow the detection of smaller particles, but at the cost of reducing the penetration of the transmitted signal into the fluid. The flowmeter is also sensitive to changes in flow velocity profile and the spatial distribution of discontinuities in the flow. As a result the accuracy of Doppler flowmeters is poor, typically ±5% of full scale. However, this can be improved by calibrating the flowmeter on-line. Since there is a large acoustic mismatch between steel and air, clamp-on Doppler flowmeters cannot be used for metering gas flows or, of course, totally clean liquids where there are insufficient reflecting particles or bubbles to produce a reliable Doppler signal.

Figure 20.57 illustrates the basic principle of the ultrasonic transit-time flowmeter. Two ultrasonic transducers are mounted on either side of the pipe, so that ultrasound can be transmitted across the fluid flowing in the pipe. The difference in the time it takes for a pulse of ultrasound to travel between transducer 1 and 2 (with the flow) and transducer 2 and 1 (against the flow) is given by

$$\Delta T = \frac{2lv \cos \theta}{c^2 - v^2 \cos^2 \theta} \tag{20.68}$$

Since the velocity of sound in the fluid c is much greater than the velocity of the fluid v, then

$$\Delta T \approx \frac{2lv \cos \theta}{c^2} \tag{20.69}$$

Therefore, if the velocity of sound in the fluid is constant, then there is a linear relationship between ΔT and v.

Although this method is elegant and straightforward in principle, in practice there are difficulties since ΔT can be small, and the change in ΔT that occurs with changing fluid velocity is even smaller (typically fractions of microsecond per meter). In addition, as Equation 20.68 shows, if the temperature of the fluid changes then c will change. Measurement of, and correction for, changes in the fluid temperature are usually needed. Transit-time flowmeters, therefore, require the measurement complex signal conditioning and processing.

The flow velocity v, which is calculated using Equation 20.68, is the average velocity along the transmission path between the two transducers, and so the flowmeter's calibration will be very dependent on the flow velocity profile. The problem of flow regime dependency can be significantly reduced by using a configuration of several parallel ultrasonic beams and averaging the measured mean velocity along each beam to calculate the overall fluid velocity. This is analogous to numerical integration, and a wide range of multibeam configurations have been proposed, each with their own advantages.

Single path ultrasonic transit-time flowmeters have a typical accuracy of ±2% of reading over a range of at least 100:1. Although this type of flowmeter can be used with liquids or gases, clamp-on designs can only be used with liquids. Multibeam flowmeters have an improved accuracy, but are more expensive. However, they are finding increased use in high value applications like the custody transfer of natural gas. Unlike most other flowmeters, the cost of transit-time flowmeters does not increase significantly with pipe diameter. Transit-time flowmeters are intended for use with clean fluids, although most can still operate if there are a small amount of impurities present in the flow.

20.5.12 Coriolis Flowmeter

The Coriolis flowmeter can be used to measure the mass flowrate of a fluid directly. As the name suggests, the principle of operation makes use of the Coriolis effect discovered by Gustave Coriolis in 1835. The fluid being metered passes through a vibrating tube, and as a result of the Coriolis force acting on it, the fluid will accelerate as it moves towards the point of maximum vibration and decelerate as it moves away. This will result in flexure of the tube, the amount of flexure being directly proportional to the mass flowrate of the fluid.

The first commercial Coriolis flowmeter used a U-shaped tube, but now many different configurations exist, including dual loops and straight through designs. Each design has its own advantages, with factors such as accuracy, repeatability, and pressure drop varying from design to design.

Whichever design is used, the Coriolis flowmeter is a high accuracy instrument, which may be used to meter gas or liquid with an accuracy of typically ±0.25% of reading. Measurement range varies with design, but 100:1 is possible for U-tube designs and 50:1 for straight tube designs. Since the flowmeter measures mass directly, changes in density, viscosity, pressure, and temperature do not effect the calibration of the flowmeter. The flowmeter is also not affected by changes in flow velocity profile or other flow disturbances such as swirl. The flowmeter does not obstruct the flow at all, and can be used to meter flow in both directions. However, the pressure drop across U-tube designs can be a limitation with viscous fluids.

The major disadvantage of the Coriolis flowmeter is its cost, which is high in comparison to most other flowmeters. This cost may be justified in applications where the product cost is high, or where mass flowrate of the fluid is required. The cost and weight of the Coriolis flowmeter increase significantly with increasing pipe diameter, and as a result are usually limited to pipe diameters with diameters less than 100 mm.

Unlike most of the flowmeters discussed so far, the Coriolis flowmeter can meter some difficult two-phase flows. For example, reliable measurements of the mass flowrate of liquid/gas mixtures are possible if the gas component is small and well distributed, and therefore the mixture is acting like a pseudo-homogeneous fluid. The percentage of gas that can be tolerated by the flowmeter will depend on the viscosity of the liquid component. The less viscous the liquid the more likely the gas is to separate out and cause problems. Liquid/solids flows (slurries) may also be metered, although the user has to compromise between avoiding particle dropout and avoiding excessive fluid velocities which would result in accelerated wear of the flow tube.

20.5.13 Two-Phase Flow

There is a growing demand in areas such as the petroleum and food industries to be able to measure two-phase flows such as liquid with entrained gas, or liquid with solids. Yet the measurement of such flows nearly always presents difficulties to the process engineer.

For example, an ideal flowmeter would be able to directly measure the volumetric flowrate of a liquid whether it be all liquid or contain a second phase of gas. Unfortunately most of the flowmeters described above can usually only be used to meter two-phase flows when the second component is small. A review of the performance of conventional flowmeters in two-phase flows can be found in [10].

The alternative to direct flow measurement is to use an inferential method. An inferential method for liquid/gas flow would require the velocity of the gas and the liquid phases (v_g and v_l) and cross-sectional fraction of the gas phase (α) to be independently measured in order to calculate the volumetric flowrate of the mixture Q_m:

$$Q_m = v_l A(1 - \alpha) + v_g A \alpha \tag{20.69}$$

The overall uncertainty of the flowrate measurement would depend on the accuracy with which the independent measurements can be made. The velocity of the liquid and gas phases cannot be assumed to be the same, and the way the gas is distributed in the liquid (the flow regime) will change depending on factors such as gas content, individual phase velocity, and pressure. Even in a simple case such as this, it is clear that multiphase flow measurement is by no means straightforward. For this reason commercial multiphase flowmeters are generally expensive and targeted at very specific applications.

The most common two-phase flows are liquid/gas (e.g., gas in water), liquid/liquid (e.g., water in oil), gas/solid (pneumatically conveyed solids) and liquid/solids or slurries (e.g., coal in oil). Each presents its own measurement problems and it is not feasible to discuss all possible metering combinations for these types of flow in a overview article such as this. For further details of two-phase flow measurement techniques see [11].

20.5.14 Flowmeter Installation

No matter how good a flowmeter is, correct installation is essential if it is to measure with the uncertainty specified by the manufacturer. The calibration of most flowmeters is sensitive to changes in flow velocity profile and in such cases the flowmeter should be installed to ensure that a fully developed and stable velocity profile is present at the measurement point. Manufacturers' data sheets will contain recommendations for the minimum straight length of pipe required before and after a flowmeter to ensure that this is the case. Flow conditioners (or flow straighteners) can be used to correct a distorted velocity profile and remove swirl in applications where long straight lengths are not possible. However, the limitation of all flow conditioners is that they restrict the flow and so produce an unrecoverable pressure drop in the pipeline.

Installation should always ensure that the pipe is completely full at the metering point and that no unwanted second phase is present. In liquid flows entrained gas can be produced from a number of sources, including cavitation and leaking seals. While in a gas flow, an unwanted liquid phase can be produced by processes such as condensation. In most cases flowmeters will produce metering errors if a second phase is present in the flow. While it is possible to remove unwanted phases from the flow upstream of the metering point, it is better to take care with the process, pipework design, and flowmeter installation to ensure that this situation does not occur in the first place.

20.5.15 Flowmeter Selection

There is no such thing as a flowmeter which is equally good for all applications and given the large number of commercial flowmeters and the variety of data sheets available, the choice can sometimes appear bewildering. While at first sight more than one flowmeter may meet a particular application, selecting the most appropriate can be more difficult. In general the best flowmeter will be the one that can meet the performance specification at the lowest total cost (this is a combination of purchase price and cost of maintenance).

There are five factors which can be considered when trying to decide which flowmeter to use. These are the type of fluid to be metered, process conditions, installation conditions, performance requirements, and economic factors. Information required when considering the fluid to be metered are, whether it is a single phase fluid or whether it contains a second component such as gas or solids, the fluid viscosity and density, whether the fluid is corrosive, and if it is a gas whether it is dry or wet. Factors to be considered under process conditions include the pipeline temperature and pressure, and the ambient conditions outside of the pipeline. Installation conditions covers information such as the pipe diameter, the pipe Reynolds number, the orientation of pipework at the measurement point, the length of straight pipework available, whether the flow is pulsating, the need for any flow conditioning, whether an external power source is available, and if the measurement is being made in a hazardous environment. Performance requirements cover the accuracy, repeatability, range, and dynamic response required by the flowmeter. Finally, economic factors cover issues such as the initial cost of the flowmeter, installation cost, maintenance cost, and the type of training required.

Most flow measurement textbooks also contain flowmeter selection charts (e.g., [3, 4, 6]), and an international standard is now available on the selection and application of flowmeters [12].

References

1. *Flow Measurement and Instrumentation,* Oxford: Elsevier Science.
2. Grant, D. M., Open channel flow measurement, in D. W. Spitzer (Ed.), *Flow Measurement: Practical Guides for Measurement and Control,* 2nd ed., Research Triangle Park, NC: ISA, 2001.
3. Miller, R. W., *Flow Measurement Engineering Handbook,* 3rd ed., New York: McGraw Hill, 1996.
4. Baker, R. C., *Flow Measurement Handbook,* Cambridge: Cambridge University Press, 2000.
5. Webster, J. G. (Ed.), *Mechanical Variables Measurement: Solid, Fluid, and Thermal,* Boca Raton: CRC Press, 2000.
6. Spitzer, D. W. (Ed.), *Flow Measurement: Practical Guides for Measurement and Control,* 2nd ed., Research Triangle Park, NC: ISA, 2001.
7. International Organisation for Standardization, ISO5167-1, Measurement of Fluid Flow by Means of Pressure Differential Devices—Part 1: Orifice plates, nozzles and Venturi tubes inserted in circular cross-section conduits running full, Geneva, Switzerland, 1991.
8. American Petroleum Institute, API 2530, Manual of Petroleum Measurement Standards Chapter 14—Natural Gas Fluids Measurement, Section 3—Orifice Metering of Natural Gas and Other Related Hydrocarbon Fluids, Washington, 1985.
9. Lynnworth, L. C., *Ultrasonic Measurements for Process Control: Theory, Techniques, Applications,* Boston: Academic Press, 1989.
10. National Engineering Laboratory, UK, *Effects of Two-Phase Flow on Single-Phase Flowmeters,* Flow Measurement Guidance Note No. 3, 1997.
11. Rajan, V. S. V., Ridley, R. K., and Rafa, K. G., Multiphase flow measurement techniques—a review, *Journal of Energy Resource Technology,* 115, 151–161, 1993.
12. British Standards Institution, BS7405, Guide to the Selection and Application of Flowmeters for Measurement of Fluid Flow in Closed Conduits, London, 1991.

20.6 Temperature Measurements

Pamela M. Norris and Bouvard Hosticka

20.6.1 Introduction

Temperature is often cited as the most widely monitored parameter in science and industry, yet the exact definition of temperature is elusive. The simplest definition would relate temperature to the average kinetic energy of the individual molecules that comprise the system. As the temperature increases, the molecular activity also increases, and thus the average kinetic energy increases. This is an adequate

TABLE 20.4 Fixed Points Used in ITS$_{90}$*

Triple point of hydrogen	13.8033 K
Triple point of neon	24.5561 K
Triple point of oxygen	54.3584 K
Triple point of argon	83.8058 K
Triple point of mercury	234.3156 K
Triple point of water	273.16 K
Melting point of gallium	302.9146 K
Freezing point of indium	429.7485 K
Freezing point of tin	505.078 K
Freezing point of zinc	692.677 K
Freezing point of aluminum	933.573 K
Freezing point of silver	1234.93 K
Freezing point of gold	1337.33 K
Freezing point of copper	1357.77 K

*Magnum (1990) includes the full definition of these points.

definition for the discussion of temperature measuring techniques presented here. While this definition may help us understand the concept of temperature, it does not help us assign a numerical value to temperature or provide us with a convenient method for measuring temperature. The zeroth law of thermodynamics, formulated in 1931 more than half a century after the first and second laws, lays the foundation for all temperature measurement. It states that if two bodies are in thermal equilibrium with a third body, they are also in thermal equilibrium with each other. By replacing the third body with a thermometer, we can state that two bodies are in thermal equilibrium if both have the same temperature reading even if they are not in contact.

The zeroth law does not enable the assignment of a numerical value for temperature. For that we must refer to a standard scale of temperature. Two absolute temperature scales are defined such that the temperature at zero corresponds to the theoretical state of no molecular movement of the substance. This leads to the Kelvin scale for the SI system and the Rankine scale for the English system. There are other two-point scales derived by identifying two arbitrary defining points for temperature. These are usually defined as the temperature at which a pure substance undergoes a change in phase. Familiar defining points are the freezing and boiling point of water for 0°C and 100°C, respectively. A wide range of such phase changes, many of them triple points where all three phases are in equilibrium, have been accepted as the defining points of the International Practical Temperature Scale of 1990 (ITS$_{90}$) shown in Table 20.4. These can be used directly as calibration points for temperature monitors as long as the substances are pure and the other conditions, such as pressure, which are included in the defining points are met. Within the ITS$_{90}$ guidelines are standard means of interpolating temperatures between the defined points. For example, platinum resistors are used in the range from 13.8 to 1235 K. The resistance is fitted to the temperature through a higher-order polynomial that may be simplified for more limited ranges between defined temperature points. The difference between a linear interpolation of resistance between the defined points and the higher-order polynomial interpolation never exceeds 2 mK (Magnum and Furukawa, 1990).

Another complication that is encountered in any discussion of temperature measurement is the fact that temperature is an intrinsic rather than an extrinsic property. Thus, temperature can not be added, subtracted, and divided in the same way that measured extrinsic properties such as length or voltage can be manipulated.

Any property that changes predictably in response to temperature can be used in a temperature sensor. The discussion of temperature measuring devices given here subdivides the devices based on the measuring principle. Discussion will begin with a series of thermometers that rely upon the differential expansion coefficients of the materials, be they solid, liquid, or gas. Mercury thermometers, perhaps the most well known and widely used of all temperature measuring devices, belong to this category. We will then move on to devices that rely upon phase change. Next we will discuss electrical temperature sensors and transducers. Included in this category are thermocouples, RTDs, and thermistors, as well as integrated circuit temperature sensors. The final category of temperature sensors will be noncontact sensors. A separate discussion of temperature measurements on the microscale is provided at the end. Many of the techniques discussed in the microscale section will be derivatives of those introduced earlier in the discussion, but alterations, ranging from minor to quite major, must be made to enable small-scale and/or quick response temperature measurements.

20.6.2 Thermometers That Rely Upon Differential Expansion Coefficients

Thermometers that rely upon differential expansion coefficients are by far the most common and familiar direct reading temperature monitors. These thermometers can be divided into categories depending on the state of the materials used. Each of these deserves a separate discussion.

20.6.2.1 Gas versus Solid

The gas bulb thermometer, which is used to determine absolute zero from extrapolation of the change in pressure of a simple gas in a metal sphere with a change in temperature, is an example of a gas versus solid thermometer. If the metal bulb had the same expansion coefficient as the fill gas, the pressure inside would remain constant and it would not be a thermometer. Instead, the gas follows the ideal gas law, which indicates that, at a constant volume, the pressure is linearly related to the temperature and the vessel containing the gas changes linearly with the volumetric expansion coefficient of the metal making up the bulb. The thermal expansion coefficient of the metal is usually ignored unless very precise predictions of absolute zero are required.

While a large metal sphere with a pressure gage attached is not a very convenient means of measuring temperature, except as a demonstration or research tool, the bulb can be made quite small and connected via a small capillary tube to a remote pressure gage. In this miniaturized configuration the gas bulb thermometer becomes a practical means of measuring temperature. As long as the device operates in the ideal gas region, the pressure gage can be graduated to read temperature directly, since pressure is linearly related to temperature.

Some major limitations on gas bulb thermometers are that the instrument should be calibrated specifically for a particular installation since the length of the heated capillary, as well as the ambient pressure and temperature at the pressure gage, will influence the accuracy of the device. These limitations can be overcome at the expense of complication by using bimetallic elements in the pressure gage to compensate for the temperature at that point or by having a parallel capillary with no bulb follow the main capillary up to the point of measurement and have the parallel capillary equipped with a pressure gage linked to subtract its effects from the main gage. Also, any damage that changes the volume of the bulb, such as a dent, will shift the calibration. This style of instrument should not be confused with vapor pressure thermometers that can take on an identical exterior form, but instead of being filled with an ideal gas, they contain a two-phase fluid and the saturation pressure of the fluid is measured. This type of temperature sensor is discussed in further detail in another section.

20.6.2.2 Liquid versus Solid

The common mercury and glass thermometer is an example of a liquid versus solid temperature sensor. The thermal expansion of liquids, although not as great as gasses, is generally much greater than that of solids and for many applications the expansion coefficient of the glass can be ignored. However, for precision measurements, the expansion of the glass can introduce significant errors. There are two common means of dealing with the glass expansion coefficient. Thermometers intended for reading the temperature of a liquid bath might have a specified submergence depth indicated by a mark on the stem. It is assumed that the rest of the thermometer is at standard lab conditions. This is not always a good assumption, however, and a more precise way of handling the glass expansion coefficient compared to that of mercury is to use a pair of total submergence thermometers. One measures the temperature of the liquid and the other measures the temperature in the immediate vicinity of the exposed stem. A simple stem correction formula supplied by the thermometer manufacturer can then be applied to determine the temperature of the bath.

The ratio of the volume of the bulb to the bore of the capillary determines the resolution of the thermometer. The amount of liquid initially in the thermometer determines its range. The accuracy of the bore and graduation markings determines its precision. The temperature read from a liquid-glass thermometer is only valid if the liquid in the bore is continuous from the bulb to the point of reading. Separations can be eliminated by either contracting the fluid entirely into the bulb or, in some cases, by expanding it into a reservoir at the top of the thermometer. Mercury is useful from near its freezing temperature (about −40°C) to around 500°C. The boiling point of mercury is only 357°C at standard pressure, so mercury thermometers designed for very high temperatures must be pressurized with an inert cover gas when sealed. Alcohol or other liquids can be used in place of mercury, but the accuracy is generally not as great. The temperature range of alcohol extends from about −200°C to +250°C.

An alternative means of using the difference in expansion coefficients of liquids and solids to measure temperature is to use a system filled completely with a liquid and to monitor the change in the volume of the liquid by the position of a bourdon tube or bellows. If the volume-measuring element has a high spring constant, the compressibility of the liquid might have to be considered (Doebelin, 1990). This scheme has the same disadvantages as the gas bulb thermometer discussed above, as well as the same compensation means for overcoming these disadvantages.

20.6.2.3 Gas, Liquid, and Solid

Some early thermometers consisted of a gas bulb connected to a sealed U-tube containing mercury or another liquid. This, in effect, is really a gas pressure thermometer using a mercury manometer to indicate pressure. To be accurate, the various coefficients of expansion of all three phases must be considered. These instruments are rarely used where accuracy is important. The only example still widely used is a style of minimum-maximum thermometer for ambient air measurements where a small metal fiber is displaced in either leg of the manometer by the mercury. The fiber has enough friction in the tube and is not wetted by the mercury so that it remains free in the glass tube after the mercury shifts. The indicator on the gas bulb side stays at the minimum temperature while the indicator on the other leg stays at the maximum temperature as the mercury recedes. Once the minimum and maximum temperatures are observed, the fibers can be repositioned to the top of the two mercury columns by either centrifugal force (slinging the whole thermometer) or with a magnet if iron fibers are used.

20.6.2.4 Solid versus Solid

Bimetallic thermometers consist of two metals with differing temperature expansion coefficients bonded together. As the temperature varies from the temperature at which the metals were initially bonded, the metals expand by differing amounts and the composite experiences a shearing force. The most common means of monitoring the shearing is to allow the metal composite to bend in response to temperature changes. The form of the composite can take on many configurations varying in complexity from a simple leaf fixed at one end with a pointer on the other to a small helix fixed at one end and a turning shaft at the other that is linked to a pointer, possibly through a gear train. The shaft is supported on fine bearings with the pointer as much as a meter away from the bimetallic helix. Stick thermometers with a dial at one end are an example of the latter.

Since the temperature variations produce a force, there must always be some gradated restoring force applied to the bimetallic strip. The most common application is to use the bimetallic strip itself as a restoring spring. The final position of the strip is a balance between the shear imposed by the differing temperature coefficients and the spring constant of the strip. There are instances when bimetallic thermometers are required to actuate a switch. In these cases the load imposed by the switch must be overcome by the shear forces in the strip and the designer must consider it as an external load. The temperature range of bimetallic thermometers is limited by the annealing temperature or phase transformation of the metals. Bimetallics are thus mainly used well below 700°C, and they can be permanently damaged if the metals change their properties or the bonding between the different metals fails. A common pair of metals is a nickel steel, such as Invar with a very low thermal expansion coefficient, bonded to a brass alloy with a high thermal expansion coefficient.

Sensors

20.6.3 Thermometers That Rely Upon Phase Changes

Phase transitions of pure substances at specified pressures are used in the ITS$_{90}$ to define several of the temperature points. This concept of a phase change being a function of temperature, as well as pressure and the type of material, can be exploited in several forms as a means of determining the temperature of a system by observing either the phase change itself or the conditions at which the two phases are in equilibrium. Several useful applications of this are discussed below.

20.6.3.1 Liquid to Gas

A common remote thermometer consists of a bulb containing a liquid-gas two-phase fluid connected via a capillary tube to a pressure gage. As long as both phases are present, the pressure read on the gage yields the saturation pressure of the fluid. This arrangement overcomes many of the disadvantages of the gas versus solid thermometers with the same outward appearance described in section "Gas versus Solid" above. By monitoring the saturation pressure, the indicated temperature is independent of the temperature of the rest of the system and is insensitive to the actual volume of the bulb and capillary. The fluid is typically an organic solvent such as ethane selected for the particular temperature range desired. To keep the two-phase fluid entirely in the bulb, the pressure can be transmitted through the capillary using a single-phase fluid such as oil. Few fluids have a linear saturation curve. Therefore, most pressure gages have a notably nonlinear scale when graduated into units of temperature. Special compensation springs within the pressure gage can be used to allow for a nearly linear temperature scale, but the extra complication is rarely warranted.

The temperature range of liquid to gas thermometers is limited by the two phases of the fluid and they are typically useful from −40°C to 300°C, although a single instrument rarely will operate over more than about a 150°C span. If the saturation pressure of the fluid is very much greater than 100 kPa, a simple pressure gage referenced to atmospheric pressure can be used. Otherwise best accuracy is obtained using an absolute pressure gage. The volume of the bulb must be large compared to the change in volume of the capillary and bourdon tube in the pressure gage so that both phases are always present in the bulb. The size of the bulb keeps these thermometers from being used for point measurements.

20.6.3.2 Reversible Phase Change Thermostats

Fixed-point thermostats may be constructed based upon the phase change of a particular sensing element. An example is used in mechanical ice-point references where the sudden expansion of water as it freezes is sensed to cycle the cooling system to maintain a two-phase bath. The actual melting and freezing of the ice maintains the reference temperature. Waxes of various melting points can be used in a similar manner.

Another example is a magnetic switch held closed by a permanent magnet until the Curie temperature is reached at which point the magnet loses its magnetism, or more properly, changes from a ferromagnetic material to a paramagnetic material, and the switch opens. When the material cools, it regains its ferromagnetism, which closes the heater switch. The magnetic material can be selected to have the appropriate Curie temperature.

20.6.3.3 Fixed Temperature Indicators

Any substance that changes phase at a fixed temperature can be used as a temperature indicator. Numerous examples exist, the most common being a crayon made of a wax with a defined melting point. A mark is made on the object whose temperature needs to be monitored, and if the wax melts, its temperature is higher than the crayon point. These fixed temperature indicators are generally irreversible and can take on many forms in addition to crayons.

A variation that can be either reversible or irreversible is a paint containing suspended solids of the wax or similar material that melts at the desired indication temperature. As long as the particles are solid and scatter light, the paint appears opaque, but when they melt and turn into a liquid with a refractive index close to that of the base paint, it appears clear. These indicators can be made in a series of spots with varying temperature points to help monitor actual temperature attained rather than just a go, no-go indication. The spots can be made reversible for real-time indications or irreversible to indicate the maximum temperature reached over the monitoring period, although the irreversible ones are much more common than the reversible kind.

Although waxes are widely used, any material that has a distinct phase change at a defined temperature can be used in such a monitor. Imaging techniques other than visual can also be used to determine the phase change. One such application involves using gadolinium or another element that readily absorbs neutrons in a suitable form and observing the melting within the interior of an assembly by means of neutron radioscopy. The temperature range for systems based upon observing melting solids ranges from near ambient to several thousand Kelvin.

20.6.4 Electrical Temperature Sensors and Transducers

A sensor in this context is an element that varies an electrical parameter as a function of temperature. This electrical parameter is then converted to a useful electrical function, such as linear voltage to temperature, with added electronics. The sensor and added electronics make up a transducer. The variation of electrical characteristics with temperature is both a source of measurement possibilities as well as the bane of all electrical measuring systems, since the unwanted change of such things as the gain of an amplifier with temperature causes thermal errors to occur. More effort is expended on eliminating temperature induced electrical variations than is spent exploiting them for temperature measurement.

20.6.4.1 Thermocouples

There is a relationship between the temperature of a conductor and the kinetic energy of the free electrons. Thus, when a metal is subjected to a temperature gradient, the free electrons will diffuse from the high temperature region to the low temperature region where they have a lower kinetic energy. The electron concentration gradient creates a voltage gradient since the lattice atoms that constitute the positive charges are not free to move. This voltage gradient will oppose the further diffusion of electrons in the wire and a stable equilibrium will be established with no current flow.

The "thermal power" of a material relates the balance of thermal diffusion of the electrons to the electrical conductivity of the metal and is unique for every conductor and usually varies with temperature. The electrical conductivity of the material has a strong influence on the thermal power since it defines the ability of a material to support a voltage gradient. Thus, a SINGLE conductor with its ends at differing temperatures will have a voltage difference between the ends. The trick is to be able to measure the voltage at both ends of the conductor and thus determine the temperature difference between those ends. If we use the same type of wire to measure the voltage across the original wire, the second wire will develop exactly the same voltage difference when its ends are exposed to the same temperatures as the original wire. Therefore, this effect cannot be measured with a pair of similar wires. But because the voltage gradient is a function of the thermal power, which is different for each type of metal, a second conductor of a different type of wire can be used to measure the original voltage gradient. Only a conductor with either no electron mobility or infinite conductivity could be used to measure the absolute voltage gradient associated with the temperature gradient of the original conductor. This is not a practical proposition, so only the difference in the temperature-induced electron gradient between two conductors can ever be measured. This is the basis of thermocouples.

In practical terms, whenever two metals are joined together and the junction is at a different temperature than the free ends of the conductors, the free ends will have a potential difference between them, that is a function of the absolute temperature at the junction and the temperature of the free ends. The relationship between voltage difference and temperature difference will be characteristic of the chosen pair of conductors. Rather than speak of the free ends of the two wires, it is normal to refer to a second junction in the circuit. This is valid and reminds us that there is always a second junction to consider even if the two wires from the thermocouple are attached to a metering circuit. Somewhere within the meter, the circuit is completed and the second junction is formed.

From the previous explanation, all of the classic thermocouple laws can be derived. These laws can be summarized and find application as follows:

Law 1. A third metal introduced in the circuit with both ends of the third metal at an isothermal point does not affect the thermally induced voltage of the original pair.

There are two important implications associated with this law. The first means that the nature of the electrical contact between the wires at the junction is not critical, and that the thermocouple itself can be made up of two wires that are soldered, brazed, welded, or swaged together. In all these cases, a third metal is present at the junction whether it is the filler in soldered and brazed connections, the intermediate alloy produced by welding dissimilar metals, or the metal swage holding the ends of the wire together. This does not mean that there are not other concerns involved with how the junction is formed. It will obviously not do to solder wires together and then use the junction above the melting point of the solder. Most commercially prepared thermocouples are welded for that reason. This law also allows for a metallic item whose temperature is being measured to serve as the actual junction by attaching the thermocouple leads directly to it. This might be done to avoid the time needed to transfer energy between the object and an independent thermocouple assembly. The second implication of this law allows a measuring circuit made of conductors other than those used in the thermocouple to be inserted in the circuit as long as both connections between the measuring circuit and the two thermocouple wires are at the same temperature.

Law 2. The temperatures along the wires do not affect the thermally induced voltage characteristic of the temperature of the two junctions.

This means that the thermocouple leads can be conveniently routed through various temperature regions, and that only the temperatures at the junction and the monitoring location are important in determining the voltage.

Law 3. Each metal has its own voltage gradient for a given temperature gradient independent of the wire used to monitor that voltage.

This means that each type of metal can be calibrated against a standard and that the calibration is valid for each type of thermocouple that can be made from this wire.

Thus far the discussion has been limited to open circuits because this avoids the complications of energy being carried away from a hot junction by the electrons or the I^2R losses in the conductors that both produce heat and reduce the measured voltage. Using high-impedance amplified voltmeters or differential voltmeters having infinite impedance when balanced, allows the practical open circuit voltage to be measured and eliminates these sources of error. However, since the wires used in thermocouples are usually metals with reasonable heat conduction, energy may be inadvertently removed from the measured system by simple heat conduction along the wires.

To actually use a thermocouple the temperature at one junction must be known and some sort of calibration table or polynomial curve fit must be used to convert the measured voltage to temperature at the junction. There are published tables and polynomials for common pairs of metals used in thermocouples referenced to the ice point of water (Croarkin et al., 1993). These are based upon an average alloy but the alloy actually purchased cannot be precisely the same as the one represented by the table. This unavoidable variation of alloy content and application of standard tables is the major source of error in thermocouple readings. Despite the best efforts of the manufacturers, this variation can lead to as much as 2% error in the reported voltage for a given temperature measurement. Individual spools of wire or assembled thermocouples can be calibrated to minimize this source of error.

The temperature of the reference junction must be known to determine the temperature of the measuring junction from the measured voltage. If the temperature of one of the reference junctions is not the same as the reference temperature of the calibration table, the voltage associated with the known temperature of the reference junction can be algebraically added to the measured voltage to determine the voltage that would have been measured if the reference junction were, in fact, at the defined temperature of the table. This is not as difficult as it first appears. If the reference junction is kept in an ice bath, no correction is needed when using the normal calibration tables. To make this easy, there are commercial ice reference refrigerators specifically designed to be used for thermocouple measurements. There are also electronic sensors that give a voltage or current output that is proportional to its absolute temperature that can be incorporated in a measuring system. These can be used to measure the temperature of the isothermal terminal block of the instrument and then, by either analog circuitry or computer software, the thermocouple voltage associated with that temperature is added to the appropriate measured voltage. This scheme allows direct reading of the voltage associated with the unknown temperature. The requirement that the electronic sensor be at the same temperature as the terminals for the thermocouple cannot be over-emphasized. A multichannel scanner or strip chart recorder may have a terminal board extending over several hundred millimeters in length. If the temperature is electronically measured at one point along this length and a power supply or other heat source in the scanner causes a temperature gradient across the terminal board, serious errors might occur. Care must be taken to assure that what is called an isothermal block is truly isothermal. In the case of the scanner mentioned above, the terminal board and multiplexer had to be removed from its parent instrument and wrapped in insulation before accurate measurements could be obtained.

There are a wide variety of commercial thermocouples in various configurations and materials available depending upon the measuring requirements encountered. As well as thermocouples, a whole industry exists to provide readers, controllers, connectors, wires, and all else that is needed for use. Typical thermocouple readers will have an electronic reference temperature and accommodations for the nonlinear relation between voltage and temperature built in so that it is a simple matter to plug in the matching type of thermocouple and read the temperature. Wires for the standard types are color coded with the outer jacket indicating the wire pair and the color of the individual wires indicating polarity.

It is unfortunate that there is almost a perverse nonuniformity of standards across the world. For example, in the United States (ANSI/MC96-1, 1982) a yellow thermocouple wire sheath indicates a Chromel–Alumel pair with the red lead indicating the negative side when reading elevated temperatures. (In the U.S. system, red is uniformly the negative lead.) Whereas in Japan (JIS-C 1610, 1981), a yellow sheath indicates Iron–Constantan with the red lead representing the positive side.

Oxidation or contamination by unintentional alloying of the wire at high temperature can cause the calibration to shift. The calibration can also shift if the alloy changes along the length of the thermocouple after being subjected to steep temperature gradients at high temperature, which will allow the metals of the alloy to diffuse through one another. For normal uses, however, thermocouples are quite stable, easy to use, and reliable, with types suitable for use from near absolute zero to over 2000°C.

The physical size and thermal mass of the thermocouples define their spatial resolution and time constant, but they can be made with extremely fine wires or films to limit their size down to the micron range at the expense of ruggedness and actual power output. Unlike other electrical thermal sensors, thermocouples can be mounted in direct electrical contact with the measured surface, thereby further improving the time response of the measurement. This concept can be extended by having the actual junction formed by a third metal that is vapor deposited to bridge the insulation between the measuring wires. The junctions then consist of the thin film of metal, which can theoretically reduce the time constant to less than 1 μs (Deobelin, 1990). The main disadvantages of thermocouples are their nonlinear voltage to temperature response, the requirement to know the reference temperature by means other than using a thermocouple, and their relatively low accuracy unless specifically calibrated.

20.6.4.2 Resistance Temperature Devices

Most materials show a variation in electrical resistance with temperature. For metals, the resistance goes up with temperature in nearly a linear manner. Platinum is the preferred metal for practical resistance temperature measurements and indeed is the specified means of interpolating between the many defined points on the ITS_{90} scale. Metals other than platinum can be used for specific applications. For example, one way of measuring the temperature of the windings in motors or generators is to measure their resistance while operating under load. The copper windings themselves act as resistance temperature devices (RTDs) in this case. The discussion that follows is specifically for platinum RTDs, but the concepts apply to all metals.

For precise measurements over a wide temperature range, a higher order polynomial should be applied to determine temperature from resistance. For practical measurements over a narrow range of a few hundred degrees, the resistance can be treated as linear with temperature. The platinum resistors specified by ITS_{90} are not the same as commercially available in probes. ITS_{90} specifies pure, unstrained platinum wire made into sensors, typically with a resistance of 25.5 Ω at 0°C (Mangum and Furukawa, 1990). Most RTD elements use commercially pure platinum with a resistance of 100 Ω or 50 Ω at 0°C, although higher nominal values are sometimes used to minimize the effect of contact and lead resistances in the circuit and lower resistances are used at high temperatures. There are several differing standard platinum curves that reflect varying standard purity platinum alloys. While all alloys are nominally pure platinum, the first-order coefficient for temperatures from 0°C to 100°C of the European standard (DIN 43 760) is 3.85 mΩ/Ω-K, while the American standard is 3.92 mΩ/Ω-K. Besides these two common coefficients, there are several other coefficients listed for pure platinum. When purchasing RTDs and RTD reading equipment, the user should be aware of the standard that applies.

RTD elements can be fabricated as a wire wound onto insulated bobbins or films deposited on insulating substrates. The size of both the support and the wire, as well as any insulating encapsulation, will determine the response time of the element. Films are usually smaller than wire wound sensors and thus have a quicker time response. Due to the need to insulate the RTD from the measured surface, however, and also to avoid straining the element, even small thin film RTDs have time constants that are measured in tenths of seconds. In the special case of using a bare wire to measure gas temperatures, the time constant can be considerably reduced to the tens of microseconds; however, due to the self heating described below, this is more useful as a form of local anemometry rather than temperature measurement.

The measurement of temperature using a RTD is simply a matter of determining its resistance. The relation between resistance and temperature is absolute, so no reference temperature is needed, unlike with thermocouples. However, measuring the resistance of the element is not always simple. There are three conventional techniques, each of which has its own disadvantages, as will be discussed shortly.

Common to all techniques for measuring remote sensors is the complication of the lead wire resistance. As mentioned above, all metals have a variation in resistance with temperature and thus any lead wires will act as RTDs. Thus, if the temperature of the wires between the RTD and the reading mechanism varies, an error will be introduced in the temperature reading unless steps are taken to accommodate the lead temperature effects. For this reason, commercial RTDs are available with two, three, or four wires going to the actual terminals of the resistor, depending on the technique used to handle the lead effects as shown in Figure 20.58. The arrangement of Figure 20.58a is used where the lead lengths are

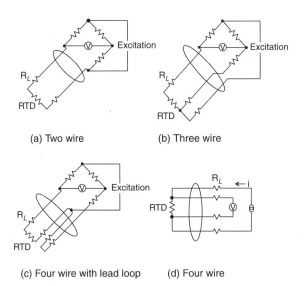

FIGURE 20.58 Various styles of commercial RTD probes and their application in reading circuits.

short, the resistance of the RTD is high compared to the lead resistance, or where high accuracy is not required. Three wire RTDs, Figure 20.58b, allow an equal length of lead wire to be included in each side of the bridge so that changes in lead resistances are felt equally on both sides and the balance of the bridge is not affected. It should be noted that four-wire RTDs of the type shown in Figure 20.58d can be used with readers designed for two-, three-, or four-wire RTDs and unless expense or size is a concern, they are recommended to allow for future upgrades of the reader. A fairly rare alternate configuration of a four wire RTD probe is shown in Figure 20.58c where two of the lead wires form a closed loop. This arrangement usually is needed only when multiple RTDs are in the bridge and should not be confused with the other type of four-wire RTD shown in Figure 20.58d, which has two wires going to each side of the resistor.

Another problem common to all reading schemes is that of self-heating of the RTD due to the measuring current through the resistor. This can be minimized by using small currents, having a high heat transfer coefficient between the sensor and the measured process, or by using low duty time pulsed measurements. Although steps can be taken to minimize it, self-heating can never be eliminated. It is such a ubiquitous characteristic of RTDs that it can be exploited as a means of determining the heat transfer coefficient of a system, such as when platinum resistors are used as anemometers.

As indicated above, the measurement of temperature using an RTD simply requires determining its resistance. The three conventional techniques are briefly discussed here.

Fully Balanced Bridges
A primitive but highly accurate means of determining the resistance of an RTD is to use it in one leg of a wheatstone bridge and have a calibrated variable resistor in the opposite leg such that when the bridge is brought to balance, as indicated by a null current on a galvanometer, the resistance of the RTD is the same as the calibrated variable resistor. There are several variations on this theme to eliminate most sources of error by splitting the error-forming device between legs of the bridge. For example, if the RTD is remote from the bridge, a three-wire lead configuration can be used to put an equal length of lead wire in both sides of the bridge with the "corner" of the bridge now defined at the remote RTD location as shown in Figure 20.58b. Ultimately the accuracy of the reading is limited by the accuracy of the resistances in the bridge, which can be made quite accurate. Since the variable resistor must be physically adjusted to null the bridge, this technique is not readily adapted to data loggers or temperature displays.

Unbalanced Bridges
If the calibrated variable resistor in the above scheme is replaced by a fixed resistor, the extent of the imbalance of the bridge can be measured with a voltmeter in place of the nulling galvanometer. This technique is fraught with errors since the imbalance of the bridge is not linear with the resistance of the RTD and the reading of the voltmeter is also proportional to the excitation current. Nevertheless, this can be a practical means of measuring temperatures over a narrow range if the values of the resistors in the far legs of the bridge are much higher than the RTD and the opposite leg resistance is close to that of the RTD. As with the other bridge techniques, three lead wires can be arranged in opposite legs of the bridge for remote sensors to compensate for lead resistance changes. If opposite legs of the bridge both consist of RTDs, then the bridge imbalance is proportional to the difference in the temperature of the two RTDs and the only way to keep equal lead lengths in both legs is to use the special four-wire type shown in Figure 20.58c.

Direct Voltage versus Current Measurements
If a constant current is passed through a resistor, the voltage across it is proportional to resistance. This is a simple concept but had to await the advent of accurate constant current sources and high impedance voltage amplifiers to become a practical replacement to the bridge techniques. By using four lead wires, as shown in Figure 20.58d, the current circuit can be made independent of the voltage sensing circuit and the lead resistances have no affect on the reading.

20.6.4.3 Thermistors

Thermistors are bulk semiconductors made from an oxide of nickel, cobalt, manganese, or other metal. The oxide is ground to a fine powder and then sintered to produce the actual thermistor material that is then incorporated into a sensor. Thermistors are resistance temperature sensing devices with several notable differences from RTDs such as their large negative temperature coefficients, and extreme nonlinear response. The resistance of a thermistor is usually so large (several thousand ohms) that lead wire resistance is rarely a concern. Thus, they are inevitably two-wire devices unless multiple thermistors or components are included in the probe. There must be some means of electrical bonding between the wire leads and the thermistor semiconductor. This bonding and the typical epoxy encapsulation places a limit on the maximum usable temperature, even though the thermistor itself is a refractory material.

There are several schemes for dealing with the nonlinearity of thermistors, ranging from applying a correction curve with a computer to having multiple thermistors with differing characteristics complete with nonthermal resistors as a bridge within a single encapsulated probe. For moderate temperatures spanning 200 K, a simple external bridge can be used to linearize the signal.

Although not normally called thermistors, germanium, silicon, and carbon are semiconductors that can also be used to monitor temperatures by measuring their resistance. Germanium is used for very precise measurements at cryogenic temperatures down to less than 1 K. The change in resistance can be very large and very nonlinear, but still very repeatable with a typical unit going from 7000 Ω at 2 K to 6 Ω at 60 K (Doebelin, 1990). Silicon can be used at room temperature and, depending on its doping, can have a very steep temperature curve. It is rarely used as a temperature sensor since other methods work better over its useful range of −200°C to 200°C. Carbon resistors out of a parts drawer found in any laboratory can be used for cryogenic measurements from 1 to 20 K, but they must be individually calibrated.

20.6.4.4 Integrated Circuit Temperature Sensors

The base-to-emitter voltage drop of a transistor operating at a constant current is a simple function of absolute temperature. Thus, any transistor can be used as a temperature sensor. In reality, this is much more of a problem with building thermally stable electronics than a convenient means of measuring temperature. Integrated circuits are available that monitor the collector current, amplify, and linearize the base-to-emitter voltage to yield an output that is proportional to absolute temperature. Common integrated circuit temperature sensors are available with outputs of 10 mV/K, or 1 μA/K. The temperature range over which they may be used is limited to −50°C to 150°C by the construction techniques of integrated circuits. This makes them very useful for referencing one junction of the thermocouple and most ambient temperature measurements. Although not intrinsically water proof, the ICs are small metal cans or plastic cases resembling signal transistors and can be potted or used in thermowells.

The IC sensors with a voltage output are commonly two terminal devices, with a possible optional lead for trimming the response. When a small current of about 1 mA is allowed to pass through it, it will have a voltage drop directly proportional to the absolute temperature (National, 2000). Even simpler IC transducers are available with separate excitation and signal leads. These are usually calibrated to 10 mV/°F or °C. These have an inherent limitation of not being able to measure below a few degrees above 0°F or 0°C unless both positive and negative power supplies are available.

Voltage output ICs are very convenient where the temperature being monitored is local to the readout and the voltage drop across the lead wires is not a concern, but for remote sensors, which require long lines, current sensors are preferred. Current sensors are also two terminal devices that behave as high impedance current sources so whatever lead resistance present may increase the voltage, but will not affect the current through the sensor (Analog, 1997). Both types can be individually adjusted by trimming resistances on the chip with a laser during manufacture to provide the rated output or they can have an external adjustment lead. Even with trimming and calibration, the accuracy over the entire span from −50°C to +150°C is rarely better than two or three degrees. Several individual ICs may be hooked up to give minimum or average temperature. Voltage ICs are placed in parallel for minimum temperature and in series for average temperature, while the current devices are connected in series for minimum and parallel for average. In addition to such simple applications of constant current or voltage sources based upon temperature, there are a wide variety of novel circuits to derive almost any function imaginable as a basis of temperature measurement.

Although there is a very small area on the silicon chip of the IC, which is temperature sensitive, it is convenient to regard the entire chip, its case, and the bonded lead wires as the sensor. This increase in thermal mass lowers the time response of the device to several seconds. Self-heating and heat transfer through the leads are also of concern and limit the applicability of these devices in critical measurements.

20.6.5 Noncontact Thermometers

All of the previously discussed temperature monitoring systems implied that the sensor of whatever type is in physical contact with the object being monitored, or in some special cases is the actual object being measured. Often times it is impractical to make this physical connection and noncontact modes of temperature measurement have been developed to overcome this objection. Almost all of these techniques require that the infrared emissions from the surface of the object be measured, but in a few special cases other surface optical properties such as reflectance can be exploited to determine the temperature remotely.

20.6.5.1 IR Emission Thermometers

Any object above absolute zero emits electromagnetic radiation whose spectrum is related to its surface temperature and surface emissivity. By characterizing the spectrum, the temperature of the object can be determined directly and absolutely. The microwave background of the universe at 3 K, and the temperature-dependent color of stars are extreme examples of this phenomena. Temperature can still be determined from the emitted surface without using a spectrometer. If two bodies are allowed to come into thermal equilibrium with each other and the temperature of one body is known, the temperature of the other is also known. This is the basis of all previously discussed temperature-measuring devices assuming conduction as the principle means of heat transfer. This can be extended to noncontact thermometers since radiation heat transfer is also a valid means of two bodies coming into thermal equilibrium. Many IR thermometers are based upon this phenomenon.

In its simplest form, an IR thermometer would consist of a temperature sensor for monitoring the temperature of an isolated object called the detector, and this detector would only be subject to radiative heat transfer with the surface whose temperature is to be measured. This would work assuming that both the surface and detector behave as black bodies, that there is no heat loss from the detector to the surroundings, and that the field of view of the detector is restricted to the object under measurement and otherwise totally unobstructed. Each one of these assumptions has to be considered when going from the ideal case to a real IR thermometer.

The concept of a black body is an idealization where all radiant energy is completely absorbed by the surface. Under this assumption, the radiant energy is a function only of the temperature of the surface. The only alternatives to being absorbed by the surface are to be reflected by the surface or transmitted through the material. The emissivity, which describes the deviation of a real surface from a black body, is then just one minus its reflectance minus its transmittance. If the emissivity is less than one but independent of wavelength, then it is a gray body. Few real materials are either black bodies or gray bodies, thus emissivity corrections must be made which will often be a function of the temperature being measured. If the surface behaves like a gray body over a limited range of wavelengths, the intensity at a few wavelengths in this range can be measured to estimate the entire spectral shape. It is fortunate that many real objects are close to gray over a narrow range of wavelengths around 750 nm and the spectral shape as a function of temperature for gray objects in the temperature range of 500–3000°C is well enough behaved at these wavelengths that the spectral shape, and thus the temperature, can be measured with just two points. The emissivity of a surface can be determined in conjunction with taking its temperature by illuminating the surface with an IR laser and measuring the amount of the known laser light that is reflected. The total IR during illumination is the sum of the reflected laser light and the thermally emitted IR. If care is taken that only short pulses or low power laser light is used, to avoid heating the measured surface, this will yield the reflectance at the wavelength of the laser and thus its emissivity which, for opaque surfaces, is one minus reflectance.

Heat transfer to and from the detector by means other than thermal radiation exchange with the monitored surface will require that the temperature of the detector be regarded as only representing the temperature of the monitored surface rather than being the same as the monitored surface. Heat transfer from the detector to the instrument can be via conduction or radiation and is accommodated by calibrating the instrument against a black body of known temperature. When this is done, the temperature of the instrument is usually monitored and circuitry may be set up to have the reference junction for the detector monitor the instrument casing temperature. Alternatively the actual photon flux can be measured by electronic means using photodiodes, photoresistive cells, or other such electronic photon sensors sensitive to IR. When this is done, narrow band filters are often used to limit the response of the detector to a particular wavelength of IR radiation to avoid counting visible photons that may be reflected from the surface or to limit the response to a particular wavelength where atmospheric interferences are minimized.

Optical components are often employed to limit the field of view of the detector so that a defined portion of the surface to be measured is brought to focus on the detector. As long as the entire detector sees the surface of interest, the distance between the detector and the surface is not important, except as it relates to IR absorption by the H_2O, CO_2, or other IR active gasses in the air.

A variation of the IR techniques discussed above is the disappearing filament pyrometer. This device superimposes the image of a tungsten filament whose temperature is a known function of current through the filament onto the view through a telescope. The walls of a furnace or other incandescent surface are observed through the telescope while adjusting the filament current until it just disappears in the background glow. The temperature of the filament then matches the temperature of the incandescent surface and can be determined from the current through the filament. A simple refinement is to put a narrow band red filter in the telescope so that the color is the same for both the target and the filament, and it becomes a single wavelength brightness comparison rather than radiation color comparison. If the emissivity of both the filament and the surface is unity, this can be very accurate. If not equal to one, when a monochromatic filter is used, only the emissivity at that one wavelength needs to be known for accurate temperature determinations.

20.6.6 Microscale Temperature Measurements

As the microelectronics industry surges forward with increasingly higher operating frequencies and increasingly smaller device dimensions, measurement techniques with high spatial and/or temporal resolution are becoming increasingly important. Few techniques are available that can actually measure temperature on a microscale, that is, sub-micron spatial resolution and/or sub-microsecond temporal resolution. However, many techniques that are being developed concentrate on observing the differential temperature on a microscale. The transient thermoreflectance technique, for example, utilizes a femtosecond pulsed laser to heat and probe the transient reflectance of the sample to enable observation of thermal transport on a sub-picosecond time scale. The technique involves relating the measured reflectivity changes to temperature changes using the material's complex index of refraction (Rosei and Lynch, 1972).

The three most common methods of observing microscale thermal phenomena include thin film thermocouples, thin film microbridges, and optical techniques. Nanometer scale thermocouples are typically used in conjunction with an atomic force microscope (AFM). This technique is nondestructive since the AFM brings the probe into contact with the sample very carefully. Thin film microbridges are patterned metallic thin films, usually thinner than 100 nanometers with a width that depends on the application. This technique relies on the fact that the electrical resistance of the microbridge is a strong function of temperature. The microbridge must be deposited onto the sample surface, therefore the technique is neither noncontact nor nondestructive. Optical techniques typically use a laser as the heating source

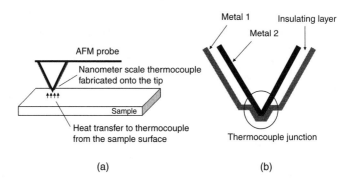

FIGURE 20.59 (a) Diagram showing the use of a scanning thermal microscope probe. (b) Schematic of a nanometer scale thermocouple maufactured onto the tip of a commercially available AFM cantilever.

and/or the thermal probe. The thermal effects can be observed optically in a number of different ways. Thermoreflectance techniques rely on the temperature dependence of reflectance (Paddock and Eesley, 1986), while photothermal techniques monitor the deflection of the probe beam by thermal expansion that results at the surface (Welsh and Ristau, 1995). "Mirage" techniques use the fact that the air just above the surface is also heated, which causes changes in the index of refraction that bends the probe beam by varying amounts depending on the change in temperature (Gonzales et al., 2000).

20.6.6.1 Scanning Thermal Microscopy

The scanning thermal microscopy (SThM) is perhaps the best example of an actual temperature measurement on sub-micron length scales. The nanometer scale thermocouple is comprised of thin metallic films deposited directly onto commercially available AFM probes. Majumdar published a comprehensive review of SThM and includes a description of several methods for manufacturing these nanometer thermocouples (Majumdar, 1999). Figure 20.59a shows a diagram of a scanning thermal microscope probe and Figure 20.59b is a schematic of a typical thermocouple junction. There are several factors that affect the spatial resolution of the measurement. These factors include the tip size of the thermocouple which can be on the order of 20 and 50 nm, the mean free path of the energy carrier of the material to be characterized, and the mechanism of heat transfer between the sample and the thermocouple, which is ultimately the limiting factor.

Operation of the AFM cantilever is identical to that of a standard AFM probe. Ideally, the thermocouple would quickly come to thermal equilibrium once in contact with the sample without affecting the temperature of the surface. Practically, a certain amount of thermal energy is transferred between the sample and the thermocouple, which affects the sample temperature, and there is also thermal resistance which delays the measurement and limits the spatial resolution. Once the sample and the thermocouple are brought into contact, there is solid–solid thermal conduction from the sample to the thermocouple. There is also thermal conduction through the gas surrounding the thermocouple tip and through a liquid layer that condenses in the small gap between the tip and the sample. Shi et al. (2000) demonstrated that conduction through this liquid layer dominates the heat transfer under normal atmospheric conditions.

20.6.6.2 Transient Thermoreflectance Technique

The transient thermoreflectance (TTR), while not capable of monitoring temperature directly, is an optical technique that enables measurement of temperature changes with sub-picosecond temporal resolution. This technique is fully noncontact and relies on the fact that reflectivity is a function of temperature. The TTR experimental setup (Paddock and Eesley, 1986; Elsayed-Ali et al., 1991; Hostetler et al., 1997) shown in Figure 20.60 can be employed to monitor the thermoreflectance response of a metallic sample after the absorption of an ultra-short laser pulse. The pulses from a femtosecond laser operating at 76 MHz are separated into two beams, an intense "pump" beam, which is used to heat the film, and a low power "probe" beam, which is used to monitor the reflectivity. The pump beam passes through an acousto-optic modulator that

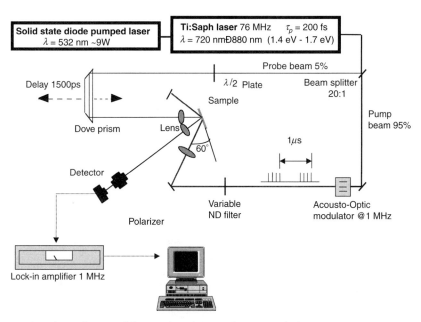

FIGURE 20.60 Experimental setup of the transient thermoreflectance technique.

effectively chops the beam on and off at a frequency of 1 MHz, resulting in thermomodulation. The probe beam passes through a dovetail prism mounted on a movable stage, which is used to increase the optical path length of the probe beam and hence the time delay between the pump and probe pulses. The reflection of the probe beam, which is centered in the heated area, is monitored by a photodiode and sent to a lock-in amplifier set to the thermomodulation frequency of 1 MHz. This yields the temporal relaxation profile of the sample.

Employing the TTR method as a temperature probe involves relating the measured reflectivity changes to temperature changes using the material's complex index of refraction. In most metals and dielectrics, the complex index of refraction depends weakly on temperature (Price, 1947). In wavelength ranges where the reflection coefficient is large, the reflectivity can be described by the linear sum of a large static contribution and a small temperature-dependent modulated contribution. The corresponding change in reflectivity is $\sim 10^{-5}$/K. The lock-in detection at 1 MHz enables resolution of the small transient signal. By comparing the transient thermal response of a surface to the appropriate heat conduction model, thermophysical properties such as the thermal diffusivity and the thermal boundary resistance can be measured (Hostetler et al., 1997; Hostetler et al., 1998; Smith et al., 2000).

20.6.7 Closing Comments

A wide variety of sensors are available for monitoring the parameter we refer to as temperature. The choice of the appropriate sensor is highly dependent upon the actual physical configuration of the measured material, as well as the required precision, accuracy, and display or processing of the temperature. While thermocouples may be an excellent choice for situations involving electrical logging of a remote process, a gas-bulb thermometer may be adequate and more appropriate for monitoring remote temperatures divorced from electricity. The physical geometry, which often limits access to the area of interest, is another important consideration. It is also important to consider the accuracy requirement, as well as the spatial and temporal resolution desired. This discussion is meant to provide a cursory overview of a wide array of temperature-sensing techniques. There are many excellent, comprehensive references and the designer is referred to these for more details. Temperature measurement often resembles an art rather than a science, with new and creative techniques for monitoring thermal responses in continuous development.

As nanotechnology progresses, many more advances in the area of sub-micron/sub-microsecond temperature measurements will become vital, since many of the traditional means of measuring temperature are not easily adapted to small local temperature measurements.

References

Analog Devices Data Sheet for AD590 Temperature Transducer, 1997.
Croarkin, M.C., et al., 1993, "Temperature-electromotive force reference functions and tables for the letter-designated thermocouple types based on the ITS-90," National Institute of Standards and Technology, Monograph 175.
Doebelin, E.O., *Measurement Systems, Application and Design*, McGraw-Hill, New York, 1990.
Elsayed-Ali, H.E., Juhasz, T., Smith, G.O., and Bron, W.E., 1991, "Femtosecond thermoreflectivity and thermotransmissivity of polycrystalline and single-crystalline gold films," *Phys. Rev. B*, Vol. 43, pp. 4488–4491.
Gonzales, E.J., Bonevich, J.E., Stafford, G.R., White, G., and Josell, D., 2000, "Thermal transport through thin films: mirage technique measurements on aluminum/titanium multilayers," *J. Mater. Res.*, Vol. 15, pp. 764–771.
Hostetler, J.L., Smith, A.N., and Norris, P.M., 1997, "Thin-film thermal conductivity and thickness measurements using picosecond ultrasonics," *Micro. Thermophys. Eng.*, Vol. 1, pp. 237–244.
Hostetler, J.L., Smith, A.N., and Norris, P.M., 1998, "Simultaneous measurement of thermophysical and mechanical properties of thin films," *Int. J. Thermophys.*, Vol. 19, pp. 569–577.
Majumdar, A., 1999, "Scanning thermal microscopy," *Ann. Rev. Mater. Sci.*, Vol. 29, pp. 505–585.
Mangum, B.W. and Furukawa, G.T., 1990, "Guidelines for realizing the international temperature scale of 1990 (ITS-90)," National Institute of Science and Technology, Technical Note 1265.
National Semiconductor Data Sheet for LM135 series Temperature Sensors, DS005698, 2000.
Paddock, C.A. and Eesley, G.L., 1986, "Transient thermoreflectance from thin metal films," *J. Appl. Phys.*, Vol. 60, pp. 285–290.
Price, D.J., 1947, "The temperature variation of the emissivity of metals in the near infrared," *Proc. Phys. Soc.* (London), Vol. 59, pp. 131.
Rosei, R. and Lynch, D.W., 1972, "Thermomodulation spectra of Al, Au, and Cu," *Phys. Rev. B*, Vol. 10, pp. 474–483.
Shi, L., Plyasunov, S., Bachtold, A., McEuen, P.L., and Majumdar, A., 2000, "Scanning thermal microscopy of carbon nanotubes using batch-fabricated probes," *Appl. Phys. Lett.*, Vol. 77, pp. 4295–4297.
Smith, A.N., Hostetler, J.L., and Norris, P.M., 2000, "Thermal boundary resistance measurements using a transient thermoreflectance technique." *Micro. Thermophys. Eng.*, Vol. 4, No. 1, pp. 51–60.
Welsh, E. and Ristau, D., 1995, "Photothermal measurements on optical thin films," *Appl. Opt.*, Vol. 34, pp. 7239–7253.

20.7 Distance Measuring and Proximity Sensors*

Jorge Fernando Figueroa and H. R. (Bart) Everett

20.7.1 Distance Measuring Sensors

20.7.1.1 Introduction

Range sensors are used to measure the distance from a reference point to an object. A number of technologies have been applied to develop these sensors, the most prominent being light/optics, computer vision, microwave, and ultrasonic. Range sensors may be of contact or noncontact types.

---

*Significant portions of this chapter were condensed from "Sensors for Mobile Robots", by H. R. Everett, with permission from A. K. Peters, Ltd., Natick, MA.

20.7.1.2 Noncontact Ranging Sensors

Sensors that measure the actual distance to a target of interest with no direct physical contact are referred to as noncontact ranging sensors. There are at least seven different types of ranging techniques employed in various implementations of such distance measuring devices (Everett et al., 1992):

- Triangulation
- Time of flight (pulsed)
- Phase-shift measurement (CW)
- Frequency modulation (CW)
- Interferometry
- Swept focus
- Return signal intensity

Noncontact ranging sensors can be broadly classified as either *active* (radiating some form of energy into the field of regard) or *passive* (relying on energy emitted by the various objects in the scene under surveillance). The commonly used terms *radar* (radio direction and ranging), *sonar* (sound navigation and ranging), and *lidar* (light direction and ranging) refer to *active* methodologies that can be based on any of several of the above ranging techniques. For example, *radar* is usually implemented using time-of-flight, phase-shift measurement, or frequency modulation. *Sonar* typically is based on time-of-flight ranging, since the speed of sound is slow enough to be easily measured with fairly inexpensive electronics. *Lidar* generally refers to laser-based schemes using time-of-flight or phase-shift measurement.

For any such active (reflective) sensors, effective detection range is dependent not only on emitted power levels, but also the following target characteristics:

- *Cross-sectional area*—determines how much of the emitted energy strikes the target.
- *Reflectivity*—determines how much of the incident energy is reflected versus absorbed or passed through.
- *Directivity*—determines how the reflected energy is redistributed (i.e., scattered versus focused).

Many noncontact sensors operate based on the physics of wave propagation. A wave is emitted at a reference point, and the range is determined by measuring either the propagation time from reference to target, or the decrease of intensity as the wave travels to the target and returns to the reference. Propagation time is measured using time-of-flight or frequency modulation methods.

Ranging by Time-of-Flight

Time-of-flight (TOF) is illustrated in Figures 20.61 and 20.62. A gated wave (a burst of a few cycles) is emitted, bounced back from the target, and detected at the receiver located near the emitter. The emitter and receiver may physically be both one sensor. The receiver may also be mounted on the target. The TOF is the time elapsed from the beginning of the burst to the beginning of the return signal. The distance is defined as $d = c \cdot TOF/2$ when emitter and receiver are at the same location, or $d = c \cdot TOF$ when the receiver is attached to the target. The accuracy is usually 1/4 of the wavelength when detecting the return signal, as its magnitude reaches a threshold limit. Gain is automatically increased with distance to maintain accuracy. Accuracy may be improved by detecting the maximum amplitude, as shown in Figure 20.63. This makes detecting the time of arrival of the wave less dependent on the amplitude of the signal. Ultrasonic, RF, or optical energy sources are typically employed; the relevant parameters

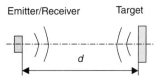

FIGURE 20.61 A wave is emitted and bounced from a target object. The distance d is determined from the speed of travel of the wave, c, and the time-of-flight, TOF as $d = (1/2) \cdot c \cdot$ TOF.

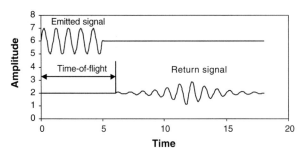

FIGURE 20.62 Definition of time-of-flight.

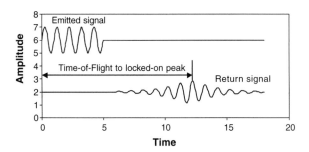

FIGURE 20.63 TOF to the maximum amplitude of the received signal for improved accuracy.

involved in range calculation, therefore, are the speed of sound in air (roughly 0.305 m/ms), and the speed of light (0.305 m/ns).

Potential error sources for TOF systems include the following:

- Variations in the speed of propagation, particularly in the case of acoustical systems
- Uncertainties in determining the exact time of arrival of the reflected pulse (Figueroa & Lamancusa, 1992)
- Inaccuracies in the timing circuitry used to measure the round-trip time of flight
- Interaction of the incident wave with the target surface

Propagation Speed—For most applications, changes in the propagation speed of electromagnetic energy are for the most part inconsequential and can basically be ignored, with the exception of satellite-based position-location systems. This is not the case, however, for acoustically based systems, where the speed of sound is markedly influenced by temperature changes, and to a lesser extent by humidity. (The speed of sound is actually proportional to the square root of temperature in degrees Rankine; an ambient temperature shift of just 30° can cause a 1-ft error at a measured distance of 35 ft.)

Detection Uncertainties—So-called *time-walk errors* are caused by the wide dynamic range in returned signal strength as a result of (1) varying reflectivity of target surfaces, and (2) signal attenuation to the fourth power of distance due to spherical divergence. These differences in returned signal intensity influence the rise time of the detected pulse, and in the case of fixed-threshold detection will cause the less reflective targets to appear further away (Lang et al., 1989). For this reason, *constant fraction timing discriminators* are typically employed to establish the detector threshold at some specified fraction of the peak value of the received pulse (Vuylsteke et al., 1990; Figueroa & Doussis, 1993).

Timing Considerations—The relatively slow speed of sound in air makes TOF ranging a strong contender for low-cost acoustically based systems. Conversely, the propagation speed of electromagnetic energy can place severe requirements on associated control and measurement circuitry in optical or RF implementations. As a result, TOF sensors based on the speed of light require sub-nanosecond timing circuitry to measure distances with a resolution of about a foot (Koenigsburg, 1982). More specifically, a desired resolution of 1 mm requires a timing accuracy of 3 ps (Vuylsteke et al., 1990). This capability is somewhat expensive to realize and may not be cost effective for certain applications, particularly at close range where high accuracies are required.

Surface Interaction—When light, sound, or radio waves strike an object, any detected echo represents only a small portion of the original signal. The remaining energy reflects in scattered directions and can be absorbed by or pass through the target, depending on surface characteristics and the angle of incidence of the beam. Instances where no return signal is received at all can occur because of specular reflection at the object surface, especially in the ultrasonic region of the energy spectrum. If the transmission source approach angle meets or exceeds a certain critical value, the reflected energy will be deflected outside the sensing envelope of the receiver. Scattered signals can reflect from secondary objects as well, returning to the detector at various times to generate false signals that can yield questionable or otherwise noisy data. To compensate, repetitive measurements are usually averaged to bring the signal-to-noise ratio within acceptable levels, but at the expense of additional time required to determine a single range value.

Ultrasonic TOF Systems

Ultrasonic TOF ranging is today the most common noncontact technique employed, primarily due to the ready availability of low-cost systems and their ease of interface. Over the past few decades, much research has been conducted in investigating applications in mobile robotics for world modeling and collision avoidance, position estimation, and motion detection. Several researchers have assessed the effectiveness of ultrasonic sensors in exterior settings (Pletta et al., 1992; Langer & Thorpe, 1992; Pin & Watanabe, 1993; Hammond, 1994). In the automotive industry, BMW now incorporates four piezoceramic transducers (sealed in a membrane for environmental protection) on both front and rear bumpers in its Park Distance Control system (Siuru, 1994).

The Polaroid ranging module is an active TOF device developed for automatic camera focusing and determines the range to target by measuring elapsed time between transmission of an ultrasonic waveform and the detected echo (Biber et al., 1980). Probably the single most significant sensor development is from the standpoint of its catalytic influence on the robotics research community and industrial applications; this system is the most widely found in the literature (Koenigsburg, 1982; Moravec & Elfes, 1985; Everett, 1985; Kim, 1986; Arkin, 1989; Borenstein & Koren, 1990). Representative of the general characteristics of a number of such ranging devices, the Polaroid unit soared in popularity as a direct consequence of its extremely low cost (Polaroid offers both the transducer and ranging module circuit board for less than $50), made possible by high-volume usage in its original application as a camera auto-focus sensor.

The most basic configuration consists of two fundamental components: (1) the ultrasonic transducer, and (2) the ranging module electronics. A choice of transducer types is now available. In the original instrument-grade electrostatic version (Figure 20.64), a very thin metalized diaphragm mounted on a machined backplate forms a capacitive transducer (Polaroid, 1981). A smaller diameter electrostatic transducer (*7000-Series*) has also been made available, developed for the Polaroid *Spectra* camera (Polaroid, 1987). A ruggedized piezoelectric (*9000-Series*) *environmental transducer* introduced for applications that may be exposed to rain, heat, cold, salt spray, and vibration is able to meet or exceed guidelines set forth in the SAE J1455 January 1988 specification for heavy-duty trucks. The range of the Polaroid system runs from about 0.3 m (1 ft) out to 10.5 m (35 ft), with a half-power (−3 dB) beam dispersion angle of approximately 12° for the original instrument-grade electrostatic transducer. A typical operating cycle is as follows.

- The control circuitry fires the transducer and waits for an indication that transmission has begun.
- The receiver is blanked for a short period of time to prevent false detection due to residual transmit signal ringing in the transducer.
- The received signals are amplified with increased gain over time to compensate for the decrease in sound intensity with distance.

FIGURE 20.64 From left to right: (1) the original instrument grade electrostatic transducer, (2) *9000-Series* environmental transducer, and (3) *7000-Series* electrostatic transducer. (Courtesy Polaroid Corp.)

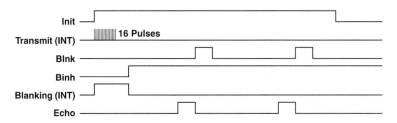

FIGURE 20.65 Timing diagrams for the 6500-*Series Sonar Ranging Module* executing a multiple-echo-mode cycle with blanking input. (Courtesy Polaroid Corp.)

- Returning echoes that exceed a fixed-threshold value are recorded and the associated distances calculated from elapsed time.

In the *single-echo* mode of operation for the *6500-series* module, the *blank* (BLNK) and *blank-inhibit* (BINH) lines are held low as the *initiate* (INIT) line goes high to trigger the outgoing pulse train. The *internal blanking* (BLANKING) signal automatically goes high for 2.38 ms to prevent transducer ringing from being misinterpreted as a returned echo. Once a valid return is received, the echo (ECHO) output will latch high until reset by a high-to-low transition on INIT. For *multiple-echo* processing, the *blank* (BLNK) input must be toggled high for at least 0.44 ms after detection of the first return signal to reset the *echo* output for the next return, as shown in Figure 20.65 (Polaroid, 1990).

Laser-Based TOF Systems

Laser-based TOF ranging systems, also known as *laser radar* or *lidar*, first appeared in work performed at the Jet Propulsion Laboratory, Pasadena, CA, in the 1970s (Lewis & Johnson, 1977). Laser energy is emitted in a rapid sequence of short bursts aimed directly at the object being ranged. The TOF of a given pulse reflecting off the object is used to calculate distance to the target based on the speed of light. Accuracies for early sensors of this type could approach a few centimeters over the range of 1–5 m (NASA, 1977; Depkovich & Wolfe, 1984).

Schwartz Electro-Optics, Inc. (SEO), Orlando, FL, produces a number of laser TOF rangefinding systems employing an innovative *time-to-amplitude-conversion* scheme to overcome the sub-nanosecond timing requirements necessitated by the speed of light. As the laser fires, a precision film capacitor begins discharging from a known set point at a constant rate, with the amount of discharge being proportional

FIGURE 20.66 The *LRF-200* series rangefinder. (Courtesy Schwartz Electro Optics, Inc.)

FIGURE 20.67 The Class 1 (eye-safe) *LD90-3 series* TOF laser rangefinder is a self-contained unit available in several versions with maximum ranges of 150–500 m under average atmospheric conditions. (Courtesy RIEGL USA.)

to the round-trip time-of-flight (Gustavson & Davis, 1992). An analog-to-digital conversion is performed on the sampled capacitor voltage; at the precise instant a return signal is detected, whereupon the resulting digital representation is converted to range and time-walk corrected using a look-up table.

The *LRF-X* series rangefinder shown in Figure 20.66 features a compact size, high-speed processing, and an ability to acquire range information from most surfaces (i.e., minimum 10% Lambertian reflectivity) out to a maximum of 100 m. The basic system uses a pulsed InGaAs laser diode in conjunction with an avalanche photodiode detector and is available with both analog and digital (RS-232) outputs.

RIEGL Laser Measurement Systems, Horn, Austria, offers a number of commercial products (i.e., laser binoculars, surveying systems, "speed guns," level sensors, profile measurement systems, and tracking laser scanners) employing short-pulse TOF laser ranging. Typical applications include lidar altimeters, vehicle speed measurement for law enforcement, collision avoidance for cranes and vehicles, and level sensing in silos.

The RIEGL *LD90-3 series* laser rangefinder (Figure 20.67) employs a near-infrared laser diode source and a photodiode detector to perform TOF ranging out to 500 m with diffuse surfaces, and to over 1000 m in the case of cooperative targets. Round-trip propagation time is precisely measured by a quartz-stabilized clock and converted to measured distance by an internal microprocessor, using one of two available algorithms. The *clutter suppression* algorithm incorporates a combination of range measurement averaging and noise rejection techniques to filter out backscatter from airborne particulates, and is, therefore, useful when operating under conditions of poor visibility (Riegel, 1994). The *standard measurement* algorithm, on the other hand, provides rapid range measurements without regard for noise suppression, and can subsequently deliver a higher update rate under more favorable environmental conditions.

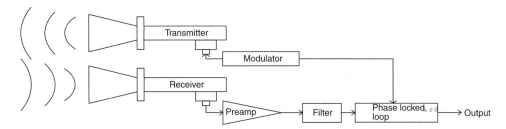

FIGURE 20.68 The microwave sensor, unlike the motion detector, requires a separate transmitter and receiver. (Adapted from Williams, 1989.)

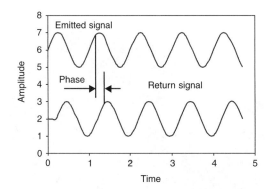

FIGURE 20.69 Range from phase measurement.

Worst-case range measurement accuracy is ±5 cm, with typical values of around ±2 cm. The pulsed near-infrared laser is Class-1 eye-safe under all operating conditions.

Microwave Range Sensors

Microwave technology may be used to measure motion, velocity, range, and direction of motion (Figure 20.68). The sensors are rugged since they have no moving parts. They can be operated safely in explosive environments, because the level of energy used is very low (no risk for sparks). Their operating temperatures range from −55°C to +125°C. They can work in environments with dust, smoke, poisonous gases, and radioactivity (assuming the components are hardened for radiation). Typically microwave sensors are used to measure ranges from 25 to 45,000 mm, but longer ranges are possible depending on power and object size. The reflected power returning to the receiver decreases as the fourth power of the distance to the object. Typical wavelength used ranges from 1 to 1000 mm.

Time-of-flight is in the order 2 ns per foot of range (reach the target and return). This translates into 10.56 ms per mile of range. Measuring short ranges may pose a problem. For 1 in. resolution, the circuit must resolve 167 ps. An alternate method more suitable to measure short distances is based on a frequency sweep of the signal generator. In this case, the return signal remains at the initial frequency (usually 10.525 GHz), and it is compared with the current frequency changed by a sweep rate. For example, to measure a range of 3 ft, one may sweep at 5 MHz/ms. After 6 ns, the frequency changes by 30 Hz (6 ns × 5 MHz/0.001 s). In this case, 0.0256 mm (0.001 in.) may be resolved easily. When using this method, a signal amplifier that increases gain with frequency is necessary. See section "Frequency Modulation" for more details on frequency modulation methods.

Phase Measurement

Time-of-flight (TOF) is defined as a phase shift between emitted and received signals when the distance is less than one wavelength (Figure 20.69). Given a phase shift f, the distance is calculated as

$d = \phi\lambda/4\pi = \phi c/4\pi f$ if the emitter and receiver are at the same location, or $d = \phi\lambda/2\pi = \phi c/2\pi f$ if the receiver is attached to the target, where c is the speed of travel, ϕ is the measured phase, and f is the modulation frequency.

The phase shift between outgoing and reflected sine waves can be measured by multiplying the two signals together in an electronic mixer, then averaging the product over many modulation cycles (Woodbury et al., 1993). This integrating process can be relatively time consuming, making it difficult to achieve extremely rapid update rates. The result can be expressed mathematically as follows (Woodbury et al., 1993):

$$\lim_{T\to\infty} \frac{1}{T}\int_0^T \sin\left(\frac{2\pi c}{\lambda}t + \frac{4\pi d}{\lambda}\right) \sin\left(\frac{2\pi c}{\lambda}\right) dt \tag{20.70}$$

which reduces to

$$A\cos\left(\frac{4\pi d}{\lambda}\right) \tag{20.71}$$

where t is the time, T is the averaging interval, and A is the amplitude factor from gain of integrating amplifier.

From the earlier expression for ϕ, it can be seen that the quantity actually measured is in fact the *cosine* of the phase shift and not the phase shift itself (Woodbury et al., 1993). This situation introduces a so-called *ambiguity interval* for scenarios where the round-trip distance exceeds the modulation wavelength λ (i.e., the phase measurement becomes ambiguous once ϕ exceeds 360°). Conrad and Sampson (1990) define this ambiguity interval as the maximum range that allows the phase difference to go through one complete cycle of 360°:

$$R_a = \frac{c}{2f} \tag{20.72}$$

where R_a is the ambiguity range interval.

Referring to Equation 20.73, it can be seen that the total round-trip distance $2d$ is equal to some integer number of wavelengths $n\lambda$ plus the fractional wavelength distance x associated with the phase shift. Since the cosine relationship is not single-valued for all of ϕ, there will be more than one distance d corresponding to any given phase-shift measurement (Woodbury et al., 1993):

$$\cos\phi = \cos\left(\frac{4\pi d}{\lambda}\right) = \cos\left(\frac{2\pi(x + n\lambda)}{\lambda}\right) \tag{20.73}$$

where

$d = (x + n\lambda)/2$ = true distance to target,
x = distance corresponding to differential phase ϕ,
n = number of complete modulation cycles.

Careful re-examination of Equation 20.73, in fact, shows that the cosine function is not single-valued even within a solitary wavelength interval of 360°. Accordingly, if only the cosine of the phase angle is measured, the ambiguity interval must be further reduced to half the modulation wavelength, or 180° (Scott, 1990). In addition, the slope of the curve is such that the rate of change of the nonlinear cosine function is not constant over the range of $0 \leq \phi \leq 180°$, and is in fact zero at either extreme. The achievable accuracy of the phase-shift measurement technique thus varies as a function of target distance, from best-case

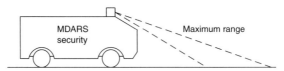

FIGURE 20.70 By limiting the maximum distance measured to be less than the range *ambiguity interval* R_a, erroneous distance measurements can be avoided.

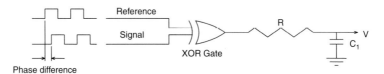

FIGURE 20.7 At low frequencies typical of ultrasonic systems, a simple phase-detection circuit based on an *exclusive-or* gate will generate an analog output voltage proportional to the phase difference seen by the inputs. (Adapted from Figueroa & Barbieri, 1991a.)

performance for a phase angle of 90° to worst case at 0 and 180°. For this reason, the useable measurement range is typically even further limited to 90% of the 180° ambiguity interval (Chen et al., 1993).

A common solution to this problem involves taking a second measurement of the same scene but with a 90° phase shift introduced into the reference waveform, the net effect being the sine of the phase angle is then measured instead of the cosine. This additional information (i.e., both sine and cosine measurements) can be used to expand the phase angle ambiguity interval to the full 360° limit previously discussed (Scott, 1990). Furthermore, an overall improvement in system accuracy is achieved, as for every region where the cosine measurement is insensitive (i.e., zero slope), the complementary sine measurement will be at peak sensitivity (Woodbury et al., 1993).

Nevertheless, the unavoidable potential for erroneous information as a result of the ambiguity interval is a detracting factor in the case of phase-detection schemes. Some applications simply avoid such problems by arranging the optical path in such a fashion as to ensure the maximum possible range is always less than the ambiguity interval (Figure 20.70). Alternatively, successive measurements of the same target using two different modulation frequencies can be performed, resulting in two equations with two unknowns, allowing both x and n (in the previous equation) to be uniquely determined. Kerr (1988) describes such an implementation using modulation frequencies of 6 and 32 MHz.

For square-wave modulation at the relatively low frequencies typical of ultrasonic systems (20–200 kHz), the phase difference between incoming and outgoing waveforms can be measured with the simple linear circuit shown in Figure 20.71 (Figueroa & Barbieri, 1991a). The output of the *exclusive-or* gate goes high whenever its inputs are at opposite logic levels, generating a voltage across capacitor C_1 that is proportional to the phase shift. For example, when the two signals are in phase (i.e., $\phi = 0$), the gate output stays low and V is zero; maximum output voltage occurs when ϕ reaches 180°. While easy to implement, this simplistic approach is limited to very low frequencies and may require frequent calibration to compensate for drifts and offsets due to component aging or changes in ambient conditions (Figueroa & Lamancusa, 1992).

Extended Range Phase Measurement Systems
Figueroa and Barbieri (1991a; 1991b) report an interesting method for extending the ambiguity interval in ultrasonic phase-detection systems through frequency division of the received and reference signals. Since the span of meaningful comparison is limited (best case) to one wavelength, λ, it stands to reason that decreasing the frequency of the phase detector inputs by some common factor will increase λ by a similar amount. The concept is illustrated in Figure 20.72. Due to the very short wavelength of ultrasonic energy (i.e., about 0.25 in. for the Polaroid system at 49.1 kHz), the total effective range is still only 4 in.

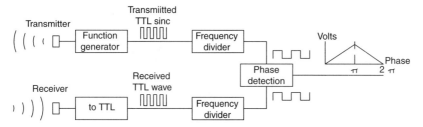

FIGURE 20.72 Dividing the input frequencies to the phase comparator by some common integer value will extend the ambiguity interval by the same factor, at the expense of resolution. (Adapted from Figueroa & Barbieri, 1991a.)

after dividing the detector inputs by a factor of 16. Due to this inherent range limitation, ultrasonic phase-detection ranging systems are not extensively applied in mobile robotic applications, although Figueroa and Lamancusa (1992) describe a hybrid approach used to improve the accuracy of TOF ranging for three-dimensional position location.

An ingenious method to measure range using phase information was developed by Young and Li (1992). The method reconstructs the total range by piecing together multiple consecutive phase chunks that reset every 2π radians of phase difference between emitted and received signals. This is another method that overcomes the limitation of phase-based systems to ranges shorter than one acoustic wavelength. The discontinuities at every 2π radians are eliminated by first taking the derivative of the phase, resulting in a smooth signal with sharp pulses (impulses) at the location of each discontinuity. Subsequently, the pulses are ignored and the result is integrated and multiplied by a constant to reconstruct the overall range. The method was tested with an experiment that employed 40 kHz transducers. Distances from 40 to 400 mm were measured with errors from ±0.1629 to ±0.4283 mm.

Laser-based continuous-wave ranging originated out of work performed at the Stanford Research Institute in the 1970s (Nitzan et al., 1977). Range accuracies approach those achievable by pulsed laser TOF methods. Only a slight advantage is gained over pulsed TOF rangefinding, however, since the difficult time-measurement problem is replaced by the need for fairly sophisticated phase-measurement electronics (Depkovich & Wolfe, 1984). In addition, problems with the phase-shift measurement approach are routinely encountered in situations where the outgoing energy is simultaneously reflected from two target surfaces at different distances from the sensor, as for example when scanning past a prominent vertical edge (Hebert & Krotkov, 1991).

The system electronics are set up to compare the phase of a single incoming wave with that of the reference signal and are not able to cope with two superimposed reflected waveforms. Adams (1993) describes a technique for recognizing the occurrence of this situation in order to discount the resulting erroneous data.

Frequency Modulation

This is a method devised to improve the accuracy in detecting the time-of-arrival of the wave to the receiver. Instead of a single frequency wave, a frequency modulated wave of the form $f = f_0 + kt$ is emitted. The difference between the emitted and received frequency at any time is $\Delta f = kt - k(t - t_f) = kt_f$ (Figure 20.73). The advantage of this method is that one does not need to know exactly when the wave arrived to the receiver. However, accurate real-time frequency measurement electronics must be used, and the transducers must respond within the frequency band sweep. Modulation other than linear is also possible in order to improve signal-to-noise ratio and hence accuracy.

The signal is reflected from a target and arrives at the receiver at time $t + T$:

$$T = \frac{2d}{c}$$

where T is the round-trip propagation time, d is the distance to target, and c is the speed of travel.

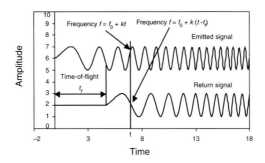

FIGURE 20.73 The frequency difference between the emitted signal and received signal is proportional to the time-of-flight at any given time.

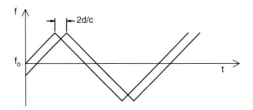

FIGURE 20.74 The received frequency curve is shifted along the time axis relative to the reference frequency.

The received signal is compared with a reference signal taken directly from the transmitter. The received frequency curve (Figure 20.74) will be displaced along the time axis relative to the reference frequency curve by an amount equal to the time required for wave propagation to the target and back. (There might also be a vertical displacement of the received waveform along the frequency axis, due to the Doppler effect.) These two frequencies when combined in the mixer produce a beat frequency F_b:

$$F_b = f(t) - f(T + t) = kT$$

where k is a constant.

This beat frequency is measured and used to calculate the distance to the object:

$$d = \frac{F_b c}{4 F_r F_d}$$

where

d = range to target,
c = speed of light,
F_b = beat frequency,
F_r = repetition (modulation) frequency,
F_d = total FM frequency deviation.

Distance measurement is therefore directly proportional to the difference or beat frequency and is as accurate as the linearity of the frequency variation over the counting interval.

Advances in wavelength control of laser diodes now permit this ultrasonic and radar ranging technique to be used with lasers. The frequency or wavelength of a laser diode can be shifted by varying its temperature. Consider an example where the wavelength of an 850-nm laser diode is shifted by 0.05 nm in 4 μs: the corresponding frequency shift is 5.17 MHz/ns. This laser beam, when reflected from a surface

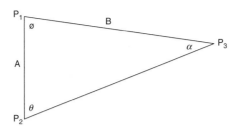

FIGURE 20.75 Triangulation ranging systems determine range B to target point P_3 by measuring angles f and q at points P_1 and P_2.

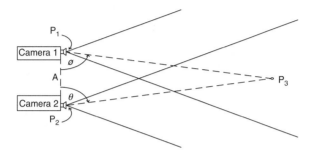

FIGURE 20.76 Passive stereoscopic ranging system configuration.

1 m away, would produce a beat frequency of 34.5 MHz. The linearity of the frequency shift controls the accuracy of the system.

The frequency-modulation approach has an advantage over the phase-shift measurement technique in which a single distance measurement is not ambiguous. (Recall that phase-shift systems must perform two or more measurements at different modulation frequencies to be unambiguous.) However, frequency modulation has several disadvantages associated with the required linearity and repeatability of the frequency ramp, as well as the coherence of the laser beam in optical systems. As a consequence, most commercially available FMCW ranging systems are radar based, while laser devices tend to favor TOF and phase-detection methods.

Triangulation Ranging

Triangulation ranging is based upon an important premise of plane trigonometry, which states that given the length of a side and two angles of a triangle, it is possible to determine the length of the other sides and the remaining angle. The basic *Law of Sines* can be rearranged as shown below to represent the length of side B as a function of side A and the angles θ and ϕ:

In ranging applications, length B would be the desired distance to the object of interest at point P_3 (Figure 20.75) for known sensor separation baseline A.

Triangulation ranging systems are classified as either *passive* (use only the ambient light of the scene) or *active* (use an energy source to illuminate the target). Passive stereoscopic ranging systems position directional detectors (video cameras, solid-state imaging arrays, or position sensitive detectors) at positions corresponding to locations P_1 and P_2 (Figure 20.76). Both imaging sensors are arranged to view the same object point, P_3, forming an imaginary triangle. The measurement of angles θ and ϕ in conjunction with the known orientation and lateral separation of the cameras allows the calculation of range to the object of interest.

Active triangulation systems, on the other hand, position a controlled light source (such as a laser) at either point P_1 or P_2, directed at the observed point P_3. A directional imaging sensor is placed at the remaining triangle vertex and is also aimed at P_3. Illumination from the source will be reflected by the target, with a portion of the returned energy falling on the detector. The lateral position of the spot as seen by the detector provides a quantitative measure of the unknown angle ϕ, permitting range determination by the *Law of Sines*.

The performance characteristics of triangulation systems are to some extent dependent on whether the system is active or passive. Passive triangulation systems using conventional video cameras require special ambient lighting conditions that must be artificially provided if the environment is too dark. Furthermore, these systems suffer from a correspondence problem resulting from the difficulty in matching points viewed by one image sensor with those viewed by the other. On the other hand, active triangulation techniques employing only a single detector do not require special ambient lighting, nor do they suffer from the correspondence problem. Active systems, however, can encounter instances of no recorded strike because of specular reflectance or surface absorption of the light.

Limiting factors common to all triangulation sensors include reduced accuracy with increasing range, angular measurement errors, and a *missing parts* (also known as *shadowing*) problem. *Missing parts* refers to the scenario where particular portions of a scene can be observed by only one viewing location (P_1 or P_2). This situation arises because of the offset distance between P_1 and P_2, causing partial occlusion of the target (i.e., a point of interest is seen in one view but otherwise occluded or not present in the other). The design of triangulation systems must include a tradeoff analysis of the offset: as this baseline measurement increases, the range accuracy increases, but problems due to directional occlusion worsen.

Stereo Disparity

The first of the triangulation schemes to be discussed, *stereo disparity* (also called *stereo vision, binocular vision*, and *stereopsis*) is a passive ranging technique modeled after the biological counterpart. When a three-dimensional object is viewed from two locations on a plane normal to the direction of vision, the image as observed from one position is shifted laterally when viewed from the other. This displacement of the image, known as *disparity*, is inversely proportional to the distance to the object. Humans subconsciously *verge* their eyes to bring objects of interest into rough registration (Burt et al., 1992). Hold up a finger a few inches away from your face while focusing on a distant object and you can simultaneously observe two displaced images in the near field. In refocusing on the finger, your eyes actually turn inward slightly to where their respective optical axes converge at the finger instead of infinity.

Most implementations use a pair of identical video cameras (or a single camera with the ability to move laterally) to generate the two disparity images required for stereoscopic ranging. The cameras are typically aimed straight ahead viewing approximately the same scene, but (in simplistic cases anyway) do not possess the capability to *verge* their center of vision on an observed point, as can human eyes. This limitation makes placement of the cameras somewhat critical because stereo ranging can take place only in the region where the fields of view overlap. In practice, analysis is performed over a selected range of disparities along the Z axis on either side of a perpendicular plane of zero disparity called the *horopter* (Figure 20.77). The selected image region in conjunction with this disparity range defines a three-dimensional volume known as the *stereo observation window* (Burt et al., 1993).

More recently there has evolved a strong interest within the research community for dynamically reconfigurable camera orientation (Figure 20.78), often termed *active vision* in the literature (Aloimonos et al., 1987; Swain & Stricker, 1991; Wavering et al., 1993). The widespread acceptance of this terminology is perhaps somewhat unfortunate in view of potential confusion with stereoscopic systems employing an active illumination source (see section 4.1.3). *Verging stereo*, another term in use, is perhaps a more appropriate choice. *Mechanical verging* is defined as the process of rotating one or both cameras about the vertical axis in order to achieve zero disparity at some selected point in the scene (Burt et al., 1992).

There are four basic steps involved in the stereo ranging process (Poggio, 1984):

- A point in the image of one camera must be identified (Figure 20.79, left).
- The same point must be located in the image of the other camera (Figure 20.79, right).
- The lateral positions of both points must be measured with respect to a common reference.
- Range Z is then calculated from the disparity in the lateral measurements.

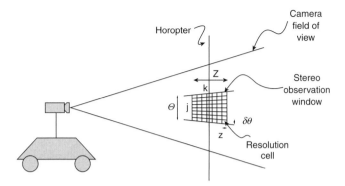

FIGURE 20.77 The *stereo observation window* is that volume of interest on either side of the plane of zero disparity known as the *horopter*. (Courtesy David Sarnoff Research Center.)

FIGURE 20.78 This stereoscopic camera mount uses a pair of lead-screw actuators to provide reconfigurable baseline separation and vergence as required. (Courtesy Robotic Systems Technology, Inc.)

On the surface this procedure appears rather straightforward, but difficulties arise in practice when attempting to locate the specified point in the second image (Figure 20.79). The usual approach is to match "interest points" characterized by large intensity discontinuities (Conrad & Sampson, 1990). Matching is complicated in regions where the intensity and/or color are uniform (Jarvis, 1983b). Additional factors include the presence of shadows in only one image (due to occlusion) and the variation in image characteristics that can arise from viewing environmental lighting effects from different angles. The effort to match the two images of the point is called *correspondence*, and methods for minimizing this

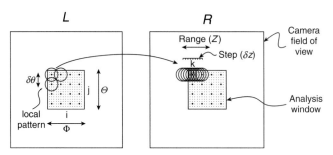

FIGURE 20.79 Range Z is derived from the measured disparity between interest points in the left and right camera images. (Courtesy David Sarnoff Research Center.)

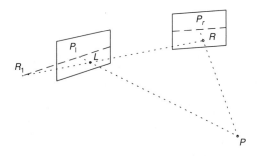

FIGURE 20.80 The *epipolar surface* is a plane defined by the lens centerpoints L and R and the object of interest at P. (Adapted from Vuylsteke et al., 1990.)

computationally expensive procedure are widely discussed in the literature (Nitzan, 1981; Jarvis, 1983a; Poggio, 1984; Loewenstein, 1984; Vuylsteke et al., 1990; Wildes, 1991).

Probably the most basic simplification employed in addressing the otherwise overwhelming *correspondence* problem is seen in the *epipolar restriction* that reduces the two-dimensional search domain to a single dimension (Vuylsteke et al., 1990). The *epipolar surface* is a plane defined by the point of interest P and the positions of the left and right camera lenses at L and R, as shown in Figure 20.80. The intersection of this plane with the left image plane defines the *left epipolar line* as shown. As can be seen from the diagram, since the point of interest P lies in the *epipolar plane*, its imaged point P_l must lie somewhere along the *left epipolar line*. The same logic dictates that the imaged point P_r must lie along a similar *right epipolar line* within the right image plane. By carefully aligning the camera image planes such that the *epipolar lines* coincide with identical scan lines in their respective video images, the correspondence search in the second image is constrained to the same horizontal scan line containing the point of interest in the first image. This effect can also be achieved with nonaligned cameras by careful calibration and rectification (resampling).

To reduce the image processing burden, most correspondence schemes monitor the overall scene at relatively low resolution and examine only selected areas in greater detail. A *foveal representation* analogous to the acuity distribution in human vision is generally employed as illustrated in Figure 20.81, allowing an extended field-of-view without loss of resolution or increased computational costs (Burt et al., 1993). The high-resolution *fovea* must be shifted from frame to frame in order to examine different regions of interest individually. Depth acuity is greatest for small disparities near the horopter and falls off rapidly with increasing disparities (Burt et al., 1992).

Active Triangulation
Rangefinding by *active triangulation* is a variation on the *stereo disparity* method of distance measurement. In place of one camera is a laser (or LED) light source aimed at the surface of the object of interest.

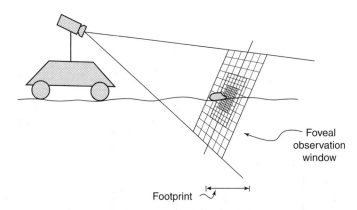

FIGURE 20.81 The *foveal* stereo representation provides high acuity near the center of the *observation window*, with decreasing resolution towards the periphery. (Courtesy David Sarnoff Research Center.)

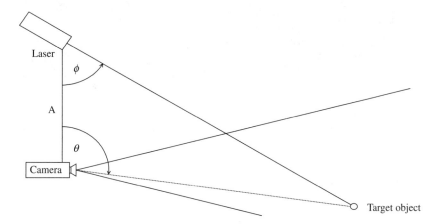

FIGURE 20.82 An active triangulation-ranging configuration employing a conventional CCD array as the detector.

The remaining camera is offset from this source by a known distance A and configured to hold the illuminated spot within its field of view (Figure 20.82).

For one- or two-dimensional array detectors such as vidicon or CCD cameras, the range can be determined from the known baseline distance A and the relative position of the laser-spot image on the image plane. For mechanically scanned single-element detectors such as photodiodes or phototransistors, the rotational angles of the detector and/or source are measured at the exact instant the detector observes the illuminated spot. The trigonometric relationships between these angles and the baseline separation are used (in theory) to compute the distance. To obtain three-dimensional information for a volumetric region of interest, laser triangulators can be scanned in both azimuth and elevation. In systems where the source and detector are self-contained components, the entire configuration can be moved mechanically. In systems with movable optics, the mirrors and lenses are generally scanned in synchronization while the laser and detector remain stationary.

Drawbacks to active triangulation include the *missing parts* situation, where points illuminated by the light source cannot be seen by the camera and vice versa (Jarvis, 1983b), as well as surface absorption or specular reflection of the irradiating energy (see Chapter 9). On the positive side, however, point-source illumination of the image effectively eliminates the correspondence problem encountered in stereo disparity rangefinders. There is also no dependence on scene contrast, and reduced influence from ambient lighting effects. (Background lighting is effectively a noise source that can limit range resolution.)

FIGURE 20.83 HERMIES IIB employed an active stereoscopic ranging system with an external laser source that could be used to designate objects of interest in the video image. (Courtesy Oak Ridge National Laboratory.)

Active Stereoscopic

Due to the computationally intensive complexities and associated resources required for establishing correspondence, passive stereoscopic methods were initially limited in practical embodiments to very simple scenes (Blais et al., 1988). One way around these problems is to employ an active source in conjunction with a pair of stereo cameras. This active illumination greatly improves system performance when viewing scenes with limited contrast. Identification of the light spot becomes a trivial matter; a video frame representing a scene illuminated by the source is subtracted from a subsequent frame of the same image with the light source deactivated. Simple thresholding of the resultant difference image quickly isolates the region of active illumination. This process is performed in rapid sequence for both cameras, and the lateral displacement of the centroid of the spot is then determined.

Alignment between the source and cameras is not critical in active stereoscopic ranging systems; in fact, the source does not even have to be located on board the robot. For example, Kilough and Hamel (1989) describe two innovative configurations using external sources for use with the robot HERMIES IIB, built at Oak Ridge National Laboratory. A pair of wide-angle black-and-white CCD cameras are mounted on a pan-and-tilt mechanism atop the robot's head, as shown in Figure 20.83. Analog video outputs from the cameras are digitized by a frame grabber into a pair of 512 by 384-pixel arrays, with offboard image processing performed by a *Hypercube* at a scaled-down resolution of 256 by 256. The initial application of the vision system was to provide control of a pair of robotic arms (from the Heathkit *HERO-1* robot) employed on HERMIES.

To accomplish this task, a near-infrared LED is attached to the end of the HERO-1 arm near the manipulator and oriented so as to be visible within the field of view of the stereo camera pair. A sequence of images is then taken by each camera, with the LED first *on* and then *off*. The *off* representations are subtracted from the *on* representations, leaving a pair of difference images, each comprised of a single bright dot representing the location of the LED. The centroids of the dots are calculated to precisely determine their respective coordinates in the difference-image arrays. A range vector to the LED can then be easily calculated, based on the lateral separation of the dots as perceived by the two cameras. This technique establishes the actual location of the manipulator in the reference frame of the robot. Experimental results indicated a 2-in. accuracy with a 0.2-in. repeatability at a distance of approximately 2 ft (Kilough and Hamel, 1989).

A near-infrared solid-state laser mounted on a remote tripod was then used by the operator to designate a target of interest within the video image of one of the cameras. The same technique described above was repeated, only this time the imaging system toggled the laser power *on* and *off*. A subsequent differencing operation enabled calculation of a range vector to the target, also in the robot's reference frame. The difference in location of the gripper and the target object could then be used to effect both platform and arm motion. The imaging processes would alternate in near-real-time for the gripper and the target, enabling the HERMIES robot to drive over and grasp a randomly designated object under continuous closed-loop control.

Structured Light

Ranging systems that employ *structured light* are a further refined case of active triangulation. A pattern of light (either a line, a series of spots, or a grid pattern) is projected onto the object surface while the camera observes the pattern from its offset vantage point. Range information manifests itself in the distortions visible in the projected pattern due to variations in the depth of the scene. The use of these special lighting effects tends to reduce the computational complexity and improve the reliability of three-dimensional object analysis (Jarvis, 1983b; Vuylsteke et al., 1990). The technique is commonly used for rapid extraction of limited quantities of visual information of moving objects (Kent, 1985), and thus lends itself well to collision avoidance applications. Besl (1988) provides a good overview of *structured-light* illumination techniques, while Vuylsteke et al. (1990) classify the various reported implementations according to the following characteristics:

- The number and type of sensors
- The type of optics (i.e., spherical or cylindrical lens, mirrors, multiple apertures)
- The dimensionality of the illumination (i.e., point or line)
- Degrees of freedom associated with scanning mechanism (i.e., zero, one, or two)
- Whether or not the scan position is specified (i.e., the instantaneous scanning parameters are not needed if a redundant sensor arrangement is incorporated)

The most common *structured-light* configuration entails projecting a line of light onto a scene, originally introduced by P. Will and K. Pennington of IBM Research Division Headquarters, Yorktown Heights, NY (Schwartz, undated). Their system created a plane of light by passing a collimated incandescent source through a slit, thus projecting a line across the scene of interest. (More recent systems create the same effect by passing a laser beam through a cylindrical lens or by rapidly scanning the beam in one dimension.) Where the line intersects an object, the camera view will show displacements in the light stripe that are proportional to the depth of the scene. In the example depicted in Figure 20.84, the lower the reflected illumination appears in the video image, the closer the target object is to the laser source. The exact relationship between stripe displacement and range is dependent on the length of the baseline

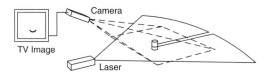

FIGURE 20.84 A common structured-light configuration used on robotic vehicles projects a horizontal line of illumination onto the scene of interest and detects any target reflections in the image of a downward-looking CCD array.

between the source and the detector. Like any triangulation system, when the baseline separation increases, the accuracy of the sensor increases, but the *missing parts* problem worsens.

Three-dimensional range information for an entire scene can be obtained in relatively simple fashion through striped lighting techniques. By assembling a series of closely spaced two-dimensional contours, a three-dimensional description of a region within the camera's field of view can be constructed. The third dimension is typically provided by scanning the laser plane across the scene. Compared to single-point triangulation, striped lighting generally requires less time to digitize a surface, with fewer moving parts because of the need to mechanically scan only in one direction. The drawback to this concept is that range extraction is time consuming and difficult due to the necessity of storing and analyzing many frames.

An alternative structured-light approach for three-dimensional applications involves projecting a rectangular grid of high-contrast light points or lines onto a surface. Variations in depth cause the grid pattern to distort, providing a means for range extraction. The extent of the distortion is ascertained by comparing the displaced grid with the original projected patterns as follows (LeMoigue & Waxman, 1984):

- Identify the intersection points of the distorted grid image.
- Label these intersections according to the coordinate system established for the projected pattern.
- Compute the disparities between the intersection points and/or lines of the two grids.
- Convert the displacements to range information.

The comparison process requires correspondence between points on the image and the original pattern, which can be troublesome. By correlating the image grid points to the projected grid points, this problem can be somewhat alleviated. A critical design parameter is the thickness of the lines that make up the grid and the spacing between these lines. Excessively thin lines will break up in busy scenes, causing discontinuities that adversely affect the intersection points labeling process. Thicker lines will produce less observed grid distortion resulting in reduced range accuracy (LeMoigue and Waxman, 1984). The sensor's intended domain of operation will determine the density of points required for adequate scene interpretation and resolution.

Magnetic Position Measurement Systems

Magnetic tracking uses a source element radiating a magnetic field (three axes) and a small sensor (three axes) that reports its position and orientation with respect to the source. Competing systems provide various multi-source, multi-sensor systems that will track a number of points at up to 100 Hz in ranges from 3 to 20 ft (Polhemus Incorporated, and Ascension Technologies). They are generally accurate to better than 0.1 in. in position and 0.1° in rotation. Magnetic systems do not rely on line-of-sight from source to object, as do optical and acoustic systems, but metallic objects in the environment will distort the magnetic field, giving erroneous readings. They require cable attachment to a central device (as do LEDs and acoustic systems). Current technology is quite robust and widely used for single or double hand-tracking, head-mounted devices, biomechanical analysis, graphics (digitization in 3D), stereotaxic localization, etc.

Magnetic field sources can be AC or DC. DC sources may emit pulses rather than continuous radiation in order to minimize interference from other magnetic sources. Using pulsed systems allows measurement of existing magnetic fields in the environment during the inactive period. Knowledge of these magnetic fields external to the system is used to improve accuracy and to overcome sensitivity to metals.

Figure 20.85 shows a typical transmitter-drive electronics (courtesy of Ascension Technologies). It provides DC current pulses to each antenna of the transmitter, one antenna at a time. The transmitter consists of a core about which the X, Y, and Z antennae are wound. While a given transmitter antenna is activated with current, readings are taken from all three antennae of the sensor. Initially the transmitter is shut off so that the sensor can measure the x, y, and z components of the earth's magnetic field. During operation, the computer sends to the digital-to-analog (D/A) converter a number that represents the amplitude of the current pulse to be sent to the selected transmitter antenna. The D/A converter converts

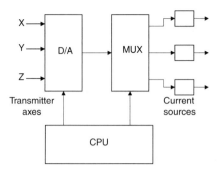

FIGURE 20.85 Magnetic positioning system: transmitting circuit. (Courtesy of Ascension Technologies.)

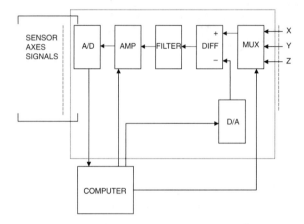

FIGURE 20.86 Magnetic positioning system: receiving circuit. (Courtesy of Ascension Technologies.)

this amplitude to an analog control voltage. This control voltage goes to the multiplexer (MUX), which connects it to the X, Y, or Z transmitter current source.

The sensor consists of three orthogonal antennae sensitive to DC magnetic fields. Many technologies can be used to implement the DC sensor. The Flock (Ascension Technologies) uses a three-axis fluxgate magnetometer. The output from the sensor goes to the signal processing electronics. As detailed in Figure 20.86, the sensor signal processing electronics consists of a multiplexer (MUX), which, on command from the computer, switches the desired X, Y, or Z sensor antenna signal, one at a time, to the differential amplifier (DIFF). The differential amplifier subtracts from this antenna signal the previously measured component of the earth's magnetic field. It outputs only that part of the received signal that is due to the transmitted field. The output from the differential amplifier is then filtered to remove noise and amplified. The analog-to-digital converter converts the DC signal to a digital format that can be read by the computer.

20.7.1.3 Other Distance Measuring Methods

The following methods are used to measure displacement, and thus can be used to infer distance travelled for certain applications.

Odometry

This is one of many methods to measure position and it is an indirect method of determining range. Range is determined by measuring the rotation of a wheel as it traverses from the reference to the target location. Wheel rotation is measured using angular encoders that may be digital or analog in nature.

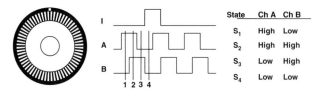

FIGURE 20.87 Optical incremental angular encoder.

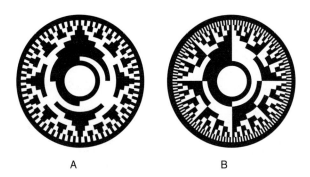

FIGURE 20.88 Absolute optical encoder.

Angular Optical Encoders

These devices encompass a light source, optics to shape and guide the light, a coded wheel with transparent and opaque sections, and a light detector array. There are two types of optical encoders: incremental and absolute.

Incremental Angular Optical Encoders. A schematic is shown in Figure 20.87. The wheel is opaque except for the slots along the circumference. Two rows of slots displaced by a 90° phase are used to determine rotation and direction. As the wheel rotates, two pulse chains 90° out of phase with each other are generated (Channels A and B). Distance is determined by counting the number of pulses (and quarter pulses for increased resolution) from the initial arbitrary zero position. The reference is lost when power is interrupted. The direction of motion is defined by determining which pulse chain leads the other.

An index pulse that appears once per revolution is also usually available (Channel Z). The resolution depends upon the number of slots around the circumference. Larger wheels can accommodate higher resolution. The total distance traveled depends upon the system used to count pulses. Since pulse counting is done outside the sensor, distances in the meters may be measured.

Decoding of the pulses (Channels A, B, and Z) to obtain angular displacement is done using specialized chips (Hewlett Packard makes a family of chips), or full-fledged integrated circuit boards. Data acquisition boards with counters may also be programmed to decode range from these sensors.

Commercially available units have a maximum operating speed of about 6000 rpm, maximum counts per revolution of about 360,000, a maximum resolution of 0.001°, and a frequency response of up to 150 kHz.

Absolute Angular Optical Encoders. These encoders have a wheel which is coded in such a way that each angular slot represents a number of bits that may be either *on* (transparent) or *off* (opaque) (Figure 20.88). Therefore, the angular position of the wheel has an absolute value given by the code of the angular slot currently aligned with the optics. Its position is known even after turning the power *off* and *on* again. When the optics crosses the line between two slots, the pattern changes to indicate an increment of one unit. However, the position is uncertain if the wheel stops with the optics right on the line. To decrease this uncertainty, the patterns are defined according to the Grey Code. In this coding scheme, only one bit of the pattern changes from one slot to the next. Thus, the uncertainty is minimized to one unit.

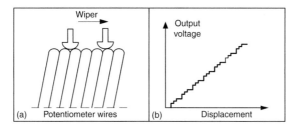

FIGURE 20.89 Potentiometer: principle of operation.

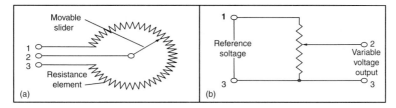

FIGURE 20.90 Potentiometer: circuit representation.

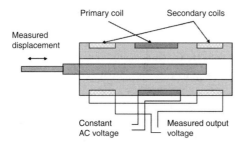

FIGURE 20.91 Linear variable differential transformer.

These sensors are suitable to measure small ranges, in the order of hundreds of millimeters. Larger wheels can accommodate more slots and more bits per slot. Commercial units of 11 bits are available, with a resolution of ±1/2 of the least significant bit, and a frequency response of 100 K 11-bit words per second.

Linear Optical Encoders. These are the same as angular encoders, except that instead of a coded wheel, they have a coded bar and a slider that carries the optical and electronic components. Distance is measured along the bar. In commercial units, the maximum measuring distance is about 2.150 m, the maximum resolution 0.08 μm, and the maximum operating speed 508 mm/s.

Potentiometers
Potentiometers are variable electrical resistance transducers. They consist of a winding and a sliding contact. As the sliding contact moves along the winding, the resistance changes in linear relationship with the distance from one end of the potentiometer (Figure 20.89). The variable resistance is wired as a voltage divider so that the output voltage is proportional to the distance traveled by the wiper (Figure 20.90). The resolution is defined by the number of turns per unit distance, and loading effects of the voltage divider circuit should be considered.

Linear Variable Differential Transformers
The linear variable differential transformer (LVDT) generates an AC signal whose magnitude is related to the displacement of a moving core (Figure 20.91). As the core changes position with respect to the coils, it changes the magnetic field, and thence the voltage amplitude in the secondary coil.

LVDT resolution depends on the instruments used to measure voltage. 25-μm resolution can be achieved. Stationary (low frequency) signals may be measured using an AC meter. High frequency signals require specialized electronics for demodulation or a data acquisition system to process the signal using a PC.

A rotary variable differential transformer (RVDT) operates under the same principle as the LVDT and is available with a range of approximately ±40°.

20.7.2 Proximity Sensors

Proximity sensors, used to determine the presence (as opposed to actual range) of nearby objects, were developed to extend the sensing range beyond that afforded by direct-contact tactile or haptic sensors. Recent advances in electronic technology have significantly improved performance and reliability, thereby increasing the number of possible applications. As a result, many industrial installations that historically have used mechanical limit switches can now choose from a variety of alternative noncontact devices for their close (between a fraction of an inch and a few inches) sensing needs. Such *proximity sensors* are classified into several types in accordance with the specific properties used to initiate a switching action:

- Magnetic
- Inductive
- Ultrasonic
- Microwave
- Optical
- Capacitive

The reliability characteristics displayed by these sensors make them well suited for operation in harsh or otherwise adverse environments, while providing high-speed response and long service lives. Instruments can be designed to withstand significant shock and vibration, with some capable of handling forces over 30,000 Gs and pressures of nearly 20,000 psi (Hall, 1984). Burreson (1989) and Peale (1992) discuss advantages and tradeoffs associated with proximity sensor selection for applications in challenging and severe environments. In addition, proximity devices are valuable when detecting objects moving at high speed, when physical contact may cause damage, or when differentiation between metallic and nonmetallic items is required. Ball (1986), Johnson (1987), and Wojcik (1994) provide general overviews of various alternative proximity sensor types with suggested guidelines for selection.

20.7.2.1 Magnetic Proximity Sensors

The simplest form of magnetic proximity sensor is the *magnetic reed switch*, schematically illustrated in Figure 20.92. A pair of low-reluctance ferromagnetic reeds are cantilevered from opposite ends of a hermetically sealed tube, arranged such that their tips overlap slightly without touching. The extreme ends of the reeds assume opposite magnetic polarities when exposed to an external magnetic flux, and the subsequent attractive force across the gap pulls the flexible reed elements together to make electrical contact (Hamlin, 1988).

Available in both *normally open* and *normally closed* configurations, these inexpensive and robust devices are commonly employed as door- and window-closure sensors in security applications. Some problems

FIGURE 20.92 The hermetically sealed *magnetic reed switch*, shown here with normally open contacts, is filled with inert gas and impervious to dust and corrosion.

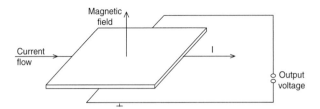

FIGURE 20.93 In 1879, E.H. Hall discovered a small transverse voltage was generated across a current-carrying conductor in the presence of a static magnetic field, a phenomenon now known as the *Hall effect*. (Adapted from Lenz, 1990.)

can be encountered with this type of sensor due to contact bounce, structural vibration, and pitting of the mating surfaces in the case of inductive or capacitive loads (Burreson, 1989), prompting most designers to opt instead for the more reliable solid-state Hall-effect magnetic sensor.

The Hall effect, as it has come to be known, was discovered by E.H. Hall in 1879. Hall noted a very small voltage was generated in the transverse direction across a conductor carrying a current in the presence of an external magnetic field (Figure 20.93), in accordance with the following equation (White, 1988):

$$V_h = \frac{R_h I B}{t}$$

where

V_h = Hall voltage,
R_h = material-dependent Hall coefficient,
I = current in amps,
B = magnetic flux density (perpendicular to I) in Gauss, and
t = element thickness in centimeters.

It was not until the advent of semiconductor technology (heralded by the invention of the transistor in 1948) that this important observation could be put to any practical use. Even so, early silicon implementations were plagued by a number of shortcomings that slowed popular acceptance, including high cost, temperature instabilities, and otherwise poor reliability (McDermott, 1969). Subsequent advances in integrated circuit technology (i.e., monolithic designs, new materials, and internal temperature compensation) have significantly improved both stability and sensitivity. With a 100-mA current flow through indium arsenide (InAs), for example, an output voltage of 60 mV can be generated with a flux density (B) of 10 kG (Hines, 1992). Large-volume applications in the automotive industry (such as distributor timing in electronic ignition systems) helped push the technology into the forefront in the late 1970s (White, 1988). Potential robotic utilization includes position and speed sensing, motor commutation (Manolis, 1993), guidepath following, and magnetic compasses.

The linear relationship of output voltage to transverse magnetic field intensity is an important feature contributing to the popularity of the modern *Hall-effect sensor*. To improve stability, *linear Hall-effect sensors* are generally packaged with an integral voltage regulator and output amplifier. The output voltage V_o fluctuates above and below a zero-field equilibrium position (usually half the power supply voltage V_{cc}), with the magnitude and direction of the offset determined by the field strength and polarity, respectively (White, 1988). (Note also that any deviation in *field direction* away from the perpendicular will also affect the magnitude of the voltage swing.) Frequency responses over 100 kHz are easily achieved (Wood, 1986).

The addition of a *Schmitt-trigger* threshold detector and an appropriate output driver transforms the linear Hall-effect sensor into a digital *Hall-effect switch*. Most commercially available devices employ transistor drivers that provide an open-circuit output in the absence of a magnetic field (Wood, 1986).

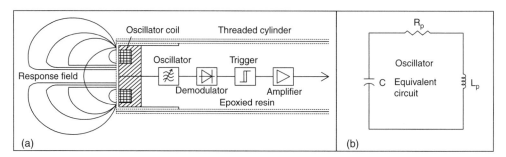

FIGURE 20.94 (a) Block diagram of a typical *ECKO type* inductive proximity sensor. (Adapted from Smith, 1985.), and (b) equivalent oscillator circuit. (Adapted from Carr, 1987.)

The detector trip point is set to some nominal value above the zero-field equilibrium voltage, and when this threshold is exceeded, the output driver toggles to the *on* state (*source* or *sink*, depending on whether PNP or NPN transistor drivers are employed). A major significance of this design approach is the resulting insensitivity of the Hall-effect switch to reverse magnetic polarity. While the mere approach of the south pole of a permanent magnet will activate the device, even direct contact by the north pole will have no effect on switching action, as the amplified output voltage actually falls further away from the *Schmitt-trigger* setpoint. Switching response times are very rapid, typically in the 400-ns range (Wood, 1986).

20.7.2.2 Inductive Proximity Sensors

Inductive proximity switches are today the most commonly employed industrial sensors (Moldoveanu, 1993) for detection of ferrous and nonferrous metal objects (i.e., steel, brass, aluminum, copper) over short distances. Cylindrical configurations as small as 4 mm in diameter have been available for over a decade (Smith, 1985). Because of the inherent ability to sense through nonmetallic materials, these sensors can be coated, potted, or otherwise sealed, permitting operation in contaminated work areas, or even submerged in fluids. Frequency responses up to 10 kHz can typically be achieved (Carr, 1987).

Inductive proximity sensors generate an oscillatory RF field (i.e., 100 kHz to 1 MHz) around a coil of wire typically wound around a ferrite core. When a metallic object enters the defined field projecting from the sensor face, eddy currents are induced in the target surface. These eddy currents produce a secondary magnetic field that interacts with field of the probe, thereby loading the probe oscillator. The effective impedance of the probe coil changes, resulting in an oscillator frequency shift (or amplitude change) that is converted into an output signal proportional to the sensed gap between probe and target.

A block diagram of a typical inductive proximity sensor is depicted in Figure 20.94(a). The oscillator comprises an active device (i.e., a transistor or IC) and the sensor probe coil itself. An equivalent circuit (Figure 20.94(b)) representing this configuration is presented by Carr (1987), wherein the probe coil is modeled as an inductor L_p with a series resistor R_p, and the connecting cable between the coil and the active element shown as a capacitance C. In the case of a typical Collpitts oscillator, the probe-cable combination is part of a resonant frequency tank circuit.

As a conductive target enters the field, the effects of the resistive component R_p dominate, and resistive losses of the tank circuit increase, loading (i.e., damping) the oscillator (Carr, 1987). As the gap becomes smaller, the amplitude of the oscillator output continues to decrease, until a point is reached where oscillation can no longer be sustained. This effect gives rise to the special nomenclature of an eddy-current-killed oscillator (ECKO) for this type of configuration. Sensing gaps smaller than this minimum threshold (typically from 0.005 to 0.020 in.) are not quantified in terms of an oscillator amplitude that correlates with range, and thus constitute a dead-band region for which no analog output is available.

Monitoring the oscillator output amplitude with an internal threshold detector creates an *inductive proximity switch* with a digital *on/off* output (Figure 20.95). As the metal target approaches the sensor face, the oscillator output voltage falls off as shown, eventually dropping below a preset *trigger level*, whereupon the threshold comparator toggles from an *off* state to an *on* state. Increasing the gap distance causes the

TABLE 20.4 Nominal Sensing Ranges for Material Other Than Mild Steel Must Be Adjusted Using the Above Attenuation Factors

Material	Attenuation Factor
Cast Iron	1.10
Mild Steel	1.00
Stainless Steel	0.70–0.90
Brass	0.45
Aluminum	0.40
Copper	0.35

Source: Smith, 1985.

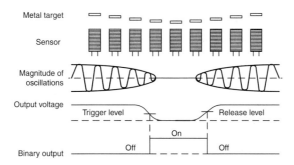

FIGURE 20.95 A small difference between the trigger and release levels (*hysteresis*) eliminates output instability as the target moves in and out of range. (Adapted from Moldoveanu, 1993.)

voltage to again rise, and the output switches *off* as the *release level* is exceeded. The intentional small difference between the trigger level and the release level, termed *hysteresis*, prevents output instabilities near the detection threshold. Typical hysteresis values (in terms of gap distance) range from 3% to 20% of the maximum effective range (Damuck & Perrotti, 1993).

Effective sensing range is approximately equal to the diameter of the sensing coil (Koenigsburg, 1982) and is influenced by target material, size, and shape. The industry standard target (for which the nominal sensing distance is specified) is a 1-mm-thick square of mild steel of the same size as the diameter of the sensor, or three times the nominal sensing distance, whichever is greater (Flueckiger, 1992). For ferrous metals, increased target thickness has a negligible effect (Damuck & Perrotti, 1993). More conductive nonferrous target materials such as copper and aluminum result in reduced detection range, as illustrated in Table 20.4. For such nonferrous metals, greater sensing distances (roughly equivalent to that of steel) can be achieved with thin-foil targets having a thickness less than their internal field attenuation distance (Smith, 1985). This phenomenon is known as the *foil effect* and results from the full RF field penetration setting up additional surface eddy currents on the reverse side of the target (Damuck & Perrotti, 1993).

There are two basic types of inductive proximity sensors: (1) *shielded* (Figure 20.96a) and (2) *unshielded* (Figure 20.96b). If an unshielded device is mounted in a metal surface, the close proximity of the surrounding metal will effectively saturate the sensor and preclude operation altogether (Swanson, 1985). To overcome this problem, the shielded configuration incorporates a coaxial metal ring surrounding the core, thus focusing the field to the front and effectively precluding lateral detection (Flueckiger, 1992). There is an associated penalty in maximum effective range, as shielded sensors can only detect out to about half the distance of an unshielded device of equivalent diameter (Swanson, 1985).

Mutual interference between inductive proximity sensors operating at the same frequency can result if the units are installed with a lateral spacing of less than twice the sensor diameter. This interference typically manifests itself in the form of an unstable pulsing of the output signal, or reduced effective range, and is most likely to occur in the situation where one sensor is undamped and the other is in the hysteresis range (Smith, 1985). Half the recommended 2*d* lateral spacing is generally sufficient for elimination of mutual

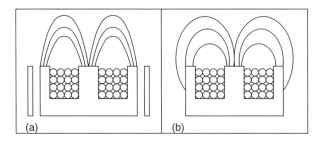

FIGURE 20.96 *Shielded* inductive sensors (a) can be embedded in metal without affecting performance, while the *unshielded* variety (b) must be mounted on nonmetallic surfaces only. (From Flueckiger, 1992.)

interaction in the case of shielded sensors (Gatzios & Ben-Ari, 1986). When mounting in an opposed facing configuration, these minimal separation distances should be doubled.

20.7.2.3 Capacitive Proximity Sensors

The *capacitive proximity sensor* is very similar to the previously discussed *inductive proximity sensor*, except that the capacitive type can reliably detect dielectric materials in addition to metals. Effective for short-range detection out to a few inches, such sensors react to the variation in electrical capacitance between a probe (or plate) and its surrounding environment. As an object draws near, the changing geometry and/or dielectric characteristics within the sensing region cause the capacitance to increase. This change in capacitance can be sensed in a number of different ways: (1) an increase in current flow through the probe (Hall, 1984), (2) initiation of oscillation in an RC circuit (McMahon, 1987), or (3) a decrease in the frequency of an ongoing oscillation (Vranish et al., 1991). Typical industrial applications include level sensing for various materials (i.e., liquids, pellets, and powders) and product detection, particularly through nonmetallic packaging.

20.7.2.4 Ultrasonic Proximity Sensors

All of the preceding proximity sensors relied on target presence to directly change some electrical characteristic or property (i.e., inductance, capacitance) associated with the sense circuitry itself. The ultrasonic proximity sensor is an example of a *reflective* sensor that responds to changes in the amount of emitted energy returned to a detector after interaction with the target of interest. Typical systems consist of two transducers (one to transmit and one to receive the returned energy), although the relatively slow speed of sound makes it possible to operate in the transceiver mode with a common transducer. The transmitter emits a longitudinal wave in the ultrasonic region of the acoustical spectrum (typically 20–200 kHz), above the normal limits of human hearing.

Ultrasonic proximity sensors are useful over distances out to several feet for detecting most objects, liquid and solid. If an object enters the acoustical field, energy is reflected back to the receiver. As is the case with any reflective sensor, maximum detection range is dependent not only on emitted power levels, but also on the target cross-sectional area, reflectivity, and directivity. Once the received signal amplitude reaches a preset threshold, the sensor output changes state, indicating detection. Due in part to the advent of low-cost microcontrollers, such devices have for most situations been replaced by more versatile ultrasonic ranging systems that provide a quantitative indicator of distance to the detected object (Section 20.7.1.2 "Ultrasonic TOF Systems").

20.7.2.5 Microwave Proximity Sensors

Microwave proximity sensors operate at distances of 5–150 ft or more (Williams, 1989) and are very similar to the ultrasonic units discussed above, except that electromagnetic energy in the microwave region of the RF energy spectrum is emitted. The FCC has allocated 10.50–10.55 and 24.075–24.175 GHz for microwave field-disturbance sensors of this type (Schultz, 1993). When the presence of a suitable target reflects sufficient energy from the transmitting antenna back to a separate receiving antenna (see Figure 20.68 in Section 20.7.1.2 "Microwave Range Sensors"), the output changes state to indicate an object is present within the field of view. An alternative configuration employing a single transmit/receive antenna monitors the Doppler shift induced by a moving target to detect relative motion as opposed to presence. These sensors are usually larger than inductive and capacitive sensors, and they are best suited to detect larger objects.

20.7.2.6 Optical Proximity Sensors

Optical (photoelectric) sensors commonly employed in industrial applications can be broken down into three basic groups: (1) *opposed*, (2) *retroreflective*, and (3) *diffuse*. (The first two of these categories are not really "proximity" sensors in the strictest sense of the terminology.) Effective ranges vary from a few inches out to several hundred feet. Common robotic applications include floor sensing, navigational referencing, and collision avoidance. Industrial applications include sensing presence at a given maximum range (for counting, or to work on a part), sensing intrusion for safety systems, alignment, etc. Modulated near-infrared energy is typically employed to reduce the effects of ambient lighting, thus achieving the required signal-to-noise ratio for reliable operation. Visible-red wavelengths are sometimes used to assist in installation alignment and system diagnostics.

Actual performance depends on several factors. Effective range is a function of the physical characteristics (i.e., size, shape, reflectivity, and material) of the object to be detected, its speed and direction of motion, the design of the sensor, and the quality and quantity of energy it radiates or receives. Repeatability in detection is based on the size of the target object, changes in ambient conditions, variations in reflectivity or other material characteristics of the target, and the stability of the electronic circuitry itself. Unique operational characteristics of each particular type can often be exploited to optimize performance in accordance with the needs of the application.

Opposed Mode

Commonly called an "electric eye" at the time, the first of these categories was introduced into a variety of applications back in the early 1950s, to include parts counters, automatic door openers, annunciators, and security systems. Separate transmitting and receiving elements are physically located on either side of the region of interest; the transmitter emits a beam of light, often supplied in more recent configurations by an LED that is focused onto a photosensitive receiver. Any object passing between the emitter and receiver breaks the beam, disrupting the circuit. Effective ranges of hundreds of feet or more are routinely possible and often employed in security applications.

Retroreflective Mode

Retroreflective sensors evolved from the *opposed* variety through the use of a mirror to reflect the emitted energy back to a detector located directly alongside the transmitter. *Corner-cube retroreflectors* (Figure 20.97) eventually replaced the mirrors to cut down on critical alignment needs. Corner-cube prisms have three mutually perpendicular reflective surfaces and a hypotenuse face; light entering through the hypotenuse face is reflected by each of the surfaces and returned back through the face to its source. A good retroreflective target will return about 3000 times as much energy to the sensor as would be reflected from a sheet of white typing paper (Banner, 1993). In most factory automation scenarios, the object of interest is detected when it breaks the beam, although some applications call for placing the retroreflector on the item itself.

FIGURE 20.97 Corner-cube retroreflectors are employed to increase effective range and simplify alignment. (Adapted from Banner, 1993.)

FIGURE 20.98 *Diffuse-mode proximity sensors* rely on energy reflected directly from the target surface.

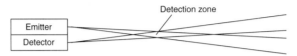

FIGURE 20.99 *Diffuse proximity sensors* configured in the *convergent mode* can be used to ascertain approximate distance to an object.

Diffuse Mode

Optical proximity sensors in the *diffuse* category operate in similar fashion to *retroreflective* types, except that energy is returned from the surface of the object of interest, instead of from a *co-operative reflector* (Figure 20.98). This feature facilitates random object detection in unstructured environments.

There are several advantages of this type of sensor over ultrasonic ranging for close-proximity object detection. There is no appreciable time lag since optical energy propagates at the speed of light, whereas up to a full second can be required to update a sequentially fired ultrasonic array of only 12 sensors. In addition, optical energy can be easily focused to eliminate adjacent sensor interaction, thereby allowing multiple sensors to be fired simultaneously. Finally, the shorter wavelengths involved greatly reduce problems due to specular reflection, resulting in more effective detection of off-normal surfaces. The disadvantage, of course, is that no direct range measurement is provided, and variations in target reflectivity can sometimes create erratic results. One method for addressing this limitation is discussed in the next section.

Convergent Mode

Diffuse proximity sensors can employ a special geometry in the configuration of the transmitter with respect to the receiver to ensure more precise positioning information. The optical axis of the transmitting LED is angled with respect to that of the detector, so the two intersect only over a narrowly defined region as illustrated in Figure 20.99. It is only at this specified distance from the device that a target can be in position to reflect energy back to the detector. Consequently, most targets beyond this range are not detected. This feature decouples the proximity sensor from dependence on the reflectivity of the target surface and is useful where targets are not well displaced from background objects.

References

Adams, M.D., "Amplitude modulated optical range data analysis in mobile robotics," *IEEE International Conference on Robotics and Automation*, Atlanta, GA, pp. 8–13, 1993.

Aloimonos, J., Weiss, I., Bandyopadhyay, A., "Active vision," *First International Conference on Computer Vision*, pp. 35–54, 1987.

Arkin, R.C., "Motor-schema-based mobile robot navigation," *International Journal of Robotics Research*, Vol. 8., No. 4, pp. 92–112, Aug., 1989.

Ascension Technologies, P.O. Box 527, Burlington, VT 05402, USA. www.ascension-tech.com.

Ball, D., "Sensor selection guide," *Sensors*, pp. 50–53, April, 1986.

Banner, *Handbook of Photoelectric Sensing*, Banner Engineering Corp., Minneapolis, MN, 1993.

Besl, P.J., "Range imaging sensors," GMR-6090, General Motors Research Laboratory, 1988.

Biber, C., Ellin, S., Shenk, E., "The polaroid ultrasonic ranging system," Audio Engineering Society, 67th Convention, New York, NY, Oct.–Nov., 1980.

Blais, F., Rioux, M., Domey, J., Beraldin, J.A., "A very compact real time 3-D range sensor for mobile robot applications," SPIE Vol. 1007, Mobile Robots III, Cambridge, MA, Nov., 1988.

Borenstein, J., Koren, Y., "Real-time obstacle avoidance for fast mobile robots in cluttered environments," *IEEE International Conference on Robotics and Automation*, Vol. CH2876-1, Cincinnati, OH, pp. 572–577, May, 1990.

Burreson, B., "Magnetic proximity switches in severe environments," *Sensors*, pp. 28–36, June, 1989.

Burt, P.J., Anadan, P., Hanna, K., van der Wal, G., "A front end vision processor for unmanned vehicles," *Advanced Image Processing Group*, David Sarnoff Research Center, Princeton, NJ, April, 1992.

Burt, P.J., Anadan, P., Hanna, K., van der Wal, G., Bassman, R., "A front end vision processor for vehicle navigation," *International Conference on Intelligent Systems*, pp. 653–662, Feb., 1993.

Carr, W.W., "Eddy current proximity sensors," *Sensors*, pp. 23–25, Nov., 1987.

Chen, Y.D., Ni, J., Wu, S.M., "Dynamic calibration and compensation of a 3-D lasar radar scanning system," *IEEE International Conference on Robotics and Automation*, Atlanta, GA, Vol. 3, pp. 652–664, May, 1993.

Damuck, N., Perrotti, J., "Getting the most out of your inductive proximity switch," *Sensors*, pp. 25–27, Aug., 1993.

Depkovich, T., Wolfe, W., "Definition of requirements and components for a robotic locating system," Final Report MCR-83-669, Martin Marietta Denver Aerospace, Denver, CO, Feb., 1984.

Everett, H.R., "A multi-element ultrasonic ranging array," *Robotics Age*, pp. 13–20, July, 1985.

Everett, H.R., DeMuth, D.E., Stitz, E.H., "Survey of collision avoidance and ranging sensors for mobile robots," *Technical Report 1194*, Naval Command Control and Ocean Surveillance Center, San Diego, CA, Dec., 1992.

Figueroa, J.F., Barbieri, E., "Increased measurement range via frequency division in ultrasonic phase detection methods," *Acustica*, Vol. 73, pp. 47–49, 1991a.

Figueroa, J.F., Barbieri, E., "An ultrasonic ranging system for structural vibration measurements," *IEEE Transactions on Instrumentation and Measurement*, Vol. 40, No. 4, pp. 764–769, Aug., 1991b.

Figueroa, J.F., Lamancusa, J.S., "A method for accurate detection of time of arrival: analysis and design of an ultrasonic ranging system," *Journal of the Acoustical Society of America*, Vol. 91, No. 1, pp. 486–494, Jan., 1992.

Fernando Figueroa and Evangelos Doussis, "A hardware-level method to improve the range and accuracy of an ultrasonic ranging system," *Acustica*, Vol. 78, No. 4, pp. 226–232, May, 1993.

Flueckiger, N., "Inductive proximity sensors: theory and applications," *Sensors*, pp. 11–13, May, 1992.

Gatzios, N.E., Ben-Ari, H., "Proximity control primer," *Sensors*, pp. 47–49, April, 1986.

Gustavson, R.L., Davis, T.E., "Diode-laser radar for low-cost weapon guidance," *SPIE*, Vol. 1633, Laser Radar VII, Los Angeles, CA, pp. 21–32, Jan., 1992.

Hall, D.J., "Robotic sensing devices," Report No. CMU-RI-TR-84-3, Carnegie-Mellon University, Pittsburgh, PA, March, 1984.

Hamlin, "The versatile magnetic proximity sensor," *Sensors*, pp. 16–22, May, 1988.

Hammond, W., "Vehicular use of ultrasonic systems," *Technical Report*, Cybermotion, Salem, VA, May, 1994.

Hebert, M., Krotkov, E., "3-D measurements from imaging laser radars: how good are they?" *International Conference on Intelligent Robots and Systems*, pp. 359–364, 1991.

Hines, R., "Hall effect sensors in Paddlewheel Flowmeters," *Sensors*, pp. 32–33, Jan., 1992.

Jarvis, R.A., "A perspective on range finding techniques for computer vision," *IEEE Transactions on Pattern Analysis and Machine Intelligence*, Vol. PAMI-1, No. 2, pp. 122–139, March, 1983a.

Jarvis, R.A., "A laser time-of-flight range scanner for robotic vision," *IEEE Transactions on Pattern Analysis and Machine Intelligence*, Vol. PAMI-5, No. 5, pp. 505–512, Sep., 1983b.

Kent, E.W., et al., "Real-time cooperative interaction between structured light and reflectance ranging for robot guidance," *Robotica*, Vol. 3, pp. 7–11, Jan.–March, 1985.

Kerr, J.R., "Real time imaging rangefinder for autonomous land vehicles," *SPIE*, Vol. 1007, Mobile Robots III, pp. 349–356, Nov., 1988.

Kilough, S.M., Hamel, W.R., "Sensor capabilities for the HERMIES experimental robot," American Nuclear Society, Third Topical Meeting on Robotics and Remote Systems, Charleston, SC, CONF-890304, Section 4-1, pp. 1–7, March, 1989.

Kim, E.J., "Design of a phased sonar array for a mobile robot," *Bachelor's Thesis*, MIT, Cambridge, MA, May, 1986.

Koenigsburg, W.D., "Noncontact distance sensor technology," GTE Laboratories, 40 Sylvan Rd., Waltham, MA, pp. 519–531, March, 1982.

Lang, S., Korba, L., Wong, A., "Characterizing and modeling a sonar ring," *SPIE Mobile Robots IV*, Philadelphia, PA, pp. 291–304, 1989.

Langer, D., Thorpe, C., "Sonar based outdoor vehicle navigation and collision avoidance," *International Conference on Intelligent Robots and Systems*, IROS'92, Raleigh, NC, July, 1992.

LeMoigue, J., Waxman, A.M., "Projected light grids for short range navigation of autonomous robots," *Proceedings, 7th IEEE Conference on Pattern Recognition*, Montreal, Canada, pp. 203–206, 30 July–2 Aug., 1984.

Lenz, J.E., "A review of magnetic sensors," *Proceedings of the IEEE*, Vol. 78, No. 6, June, 1990.

Lewis, R.A., Johnson, A.R., "A scanning laser rangefinder for a robotic vehicle," *5th International Joint Conference on Artificial Intelligence*, pp. 762–768, 1977.

Loewenstein, D., "Computer vision and ranging systems for a ping pong playing robot," *Robotics Age*, pp. 21–25, Aug., 1984.

Manolis, S., "Resolvers vs. rotary encoders for motor commutation and position feedback," *Sensors*, pp. 29–32, March, 1993.

McDermott, J., "The hall effect: success at 90," *Electronic Design 21*, pp. 38–45, 11 Oct., 1969.

McMahon, V.C., "Solutions from capacitive proximity switches," *Sensors*, pp. 31–33, May, 1987.

Moldoveanu, A., "Inductive proximity sensors: fundamentals and standards," *Sensors*, pp. 11–14, June, 1993.

Moravec, H.P., Elfes, A., "High resolution maps from wide angle sonar," *IEEE International Conference on Robotics and Automation*, St. Louis, MO, pp. 116–121, March, 1985.

NASA, "Fast accurate rangefinder," *NASA Tech Brief*, NPO-13460, Winter, 1977.

Nitzan, D., et al. "The measurement and use of registered reflectance and range data in scene analysis," *Proceedings of IEEE*, Vol. 65, No. 2, pp. 206–220, Feb., 1977.

Nitzan, D., "Assessment of robotic sensors," *Proceedings of 1st International Conference on Robotic Vision and Sensory Controls*, pp. 1–11, 1–3 April, 1981.

Peale, S., "Speed/Motion sensing in challenging environments," *Sensors*, pp. 45–46, Jan., 1992.

Pletta, J.B., Amai, W.A., Klarer, P., Frank, D., Carlson, J., Byrne, R., "The remote security station (RSS) final report," Sandia Report SAND92-1947 for DOE under Contract DE-AC04-76DP00789, Sandia National Laboratories, Albuquerque, NM, Oct., 1992.

Poggio, T., "Vision by man and machine," *Scientific America*, Vol. 250, No. 4, pp. 106–116, April, 1984.

Polaroid, "Polaroid ultrasonic ranging system user's manual," Publication No. P1834B, Polaroid Corporation, Cambridge, MA, Dec., 1981.

Polaroid, "Technical specifications for polaroid electrostatic transducer," 7000-Series Product Specification ITP-64, Polaroid Corporation, Cambridge, MA, June, 1987.

Polaroid, "6500-series sonar ranging module," Product Specifications PID 615077, Polaroid Corporation, Cambridge, MA, 11 Oct., 1990.

Polhemus Incorporated, a Rockwell Collins Company, 40 Hercules Drive, P.O. Box 560, Colchester, VT 05446 (www.polhemus.com).

Schwartz, J.T., "Structured light sensors for 3-D robot vision," Technical Report No. 65, Courant Institute of Mathematical Sciences, New York University, undated.

Scott, M.W., "Range imaging laser radar," US Patent 4,935,616, June 19, 1990.

Siuru, B., "The smart vehicles are here," *Popular Electronics*, Vol. 11, No. 1, pp. 41–45, Jan., 1994.

Smith, J.W., "Design and application of inductive proximity sensors," *Sensors*, pp. 9–14, Nov., 1985.

Swain, M.J., Stricker, M., eds., *Promising Directions in Active Vision*, Report from the National Science Foundation Active Vision Workshop, University of Chicago, IL, 1991.

Swanson, R., "Proximity switch application guide," *Sensors*, pp. 20–28, Nov., 1985.

Vranish, J.M., McConnel, R.L., Mahalingam, S., "Capaciflector collision avoidance sensors for robots," Product Description, NASA Goddard Space Flight Center, Greenbelt, MD, Feb., 1991.

Vuylsteke, P., Price, C.B., Oosterlinck, A., "Image sensors for real-time 3-D acquisition, part 1," in *Traditional and Non-Traditional Robotic Sensors*, T.C. Henderson, ed., NATO ASI Series, Vol. F63, Springer-Verlag, pp. 187–210, 1990.

Wavering, A.J., Fiala, J.C., Roberts, K.J., Lumia, R., "TRICLOPS: a high-powered trinocular active vision system," *IEEE International Conference on Robotics and Automation*, pp. 410–417, 1993.

White, D., "The hall effect sensor: basic principles of operation and application," *Sensors*, pp. 5–11, May, 1988.

Wildes, R.P., "Direct recovery of 3-D scene geometry from binocular stereo disparity," *IEEE Transactions on Pattern Analysis and Machine Intelligence*, Vol. 13, No. 8, pp. 761–774, Aug., 1991.

Williams, H., "Proximity sensing with microwave technology," *Sensors*, pp. 6–15, June, 1989.

Wojcik, S., "Noncontact presence sensors for industrial environments," *Sensors*, pp. 48–54, Feb., 1994.

Wood, T., "The hall effect sensor," *Sensors*, pp. 27–36, March, 1986.

Woodbury, N., Brubacher, M., Woodbury, J.R., "Noninvasive tank gauging with frequency-modulated laser ranging," *Sensors*, pp. 27–31, Sep., 1993.

Young, M.S., Li, Y.C., "A high precision ultrasonic system for vibration measurements," *Rev. Sci. Instrum.*, Vol. 63, No. 11, pp. 5435–5441, Nov., 1992.

20.8 Light Detection, Image, and Vision Systems

Stanley S. Ipson

20.8.1 Introduction

Light detectors span a broad spectrum of complexity. The simplest are single sensors whose output signals are easy to interpret and to interface to other components like microprocessors. In contrast, the image sensors in video and digital cameras, incorporating arrays of up to several million detectors, produce output signals which are complicated to interface and require powerful processors to interpret. Regardless of complexity, the purpose of a light detector is to measure light, and Section 20.8.2 introduces a number of radiometric terms that are employed in the characterization of light, light sources, and detectors. However, manufacturers often specify the performance of their devices using photometric units, which take into account the human visual response to light, and so it is necessary to understand both radiometric and photometric measures of light. Sources of light are briefly discussed in Section 20.8.3. There are several types of light detector in common use and the principles of operation and characteristics of the most widely used, including pyroelectric, photoresistive, photodiode, and phototransistor are summarized in Section 20.8.4. Vision systems have optical components to form an image and an image sensor to convert the light image into an electrical signal. Image formation is reviewed in Section 20.8.5, before introducing the most widely used detectors, based on charge-coupled device (CCD) technology and complementary metal oxide semiconductor (CMOS) technology, in Section 20.8.6. The elements required to complete a vision system are discussed briefly in the final section.

20.8.2 Basic Radiometry

Visible light is electromagnetic energy radiated with very short wavelengths in the range between about 400 and 700 nm. At shorter wavelengths, to about 30 nm, is invisible ultraviolet light and at longer wavelengths, up to about 0.3 mm, is invisible infrared radiation. Although electromagnetic radiation displays wave behavior including interference and diffraction, it can also behave like a stream of particles and is emitted and absorbed by matter in discrete amounts of energy called photons. The energy ε of a light

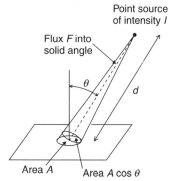

FIGURE 20.100 A point source of intensity I emits radiant power F into the solid angle subtended by the area A. The irradiance at distance d from the source is I/d^2. When the dimensions of A are small compared with d, the solid angle can be approximated by $A\cos(\theta)/d^2$.

photon with wavelength λ is given by

$$\varepsilon = \frac{hc}{\lambda} \tag{20.74}$$

where h is Planck's constant (6.6×10^{-34} J s), c is the speed of light (3×10^8 m s^{-1}) [1]. The most fundamental concept in the measurement of light is radiant power, sometimes called radiant flux (F), which is the flow of energy (photons) per unit time across a specified region in space. It is measured in watts and applies equally to visible and invisible radiation. The corresponding photometric unit is the lumen (lm), which takes into account the varying sensitivity of the eye to light of different wavelengths. One watt of radiation with a wavelength of 555 nm is defined equal to 683 lm. At other wavelengths the number of lumens is reduced (half response at 510 and 610 nm) according to the bell-shaped CIE standard eye-response curve. The remaining radiometric terms, irradiance, intensity, and radiance, are measures of the concentration of light flux. Irradiance (E) is the total radiant power falling on unit area of a surface and is measured in W m^{-2}. The corresponding photometric quantity is illuminance, measured in lm m^{-2} (lux). Radiant intensity (I) is a measure of a point source's ability to illuminate a surface, which decreases as the square of the distance d to the surface. It is measured in W sr^{-1} and its photometric equivalent is luminous intensity measured in candelas (lm sr^{-1} or cd). The irradiance from a point source of intensity I, falling on a small area A of a surface at distance d with normal inclined at an angle θ to the source as shown in Figure 20.100, is given by

$$E = \frac{F}{A} = \frac{I\cos\theta}{d^2} \tag{20.75}$$

Although few real sources would seem to be good approximations to a point source (stars in the night sky are exceptions), it is often a good approximation to calculate the irradiance of a surface (detector) by assuming the source has a specified intensity. The error caused by ignoring the spatial extent of the source is less than 1%, if the distance to the source is greater than ten times the largest dimension of the source [2]. When the distance is five (three) times the source size, the error is nearer 4% (9%). Radiant intensity is the most easily measured property of a light source and is often quoted as the performance parameter of a source. Some point sources radiate uniformly in all directions and have an intensity which is independent of direction. Other point sources emit nonuniformly. A Lambertian point source is a source whose intensity varies with direction as $I_0\cos\theta$, where θ is the angle between the measurement direction and the direction of maximum intensity I_0. Many light sources have an intensity which falls off with angle more rapidly than in the case of a Lambertian emitter, which has half intensity at an angle of 60° from the forward direction.

When a source has appreciable spatial extent, its radiance (R) in a given direction is defined as the radiant intensity of the source in that direction divided by the area of the source projected in the same direction, and is measured in W sr^{-1} m^{-2}. Conversely, the intensity of a source is the product of its area and radiance. Radiance is important in connection with optical systems. In particular, the radiance of the image produced by a lens is equal to that of the object, apart from losses due to absorption and reflection. This fact is used to calculate the image illuminance from an object of specified radiance using a specified lens. The photometric equivalent of radiance is luminance, sometimes loosely called brightness, and is measured in cd m^{-2}. Apart from color, brightness is the only property of light that we can perceive. Because it is normalized by size, the radiance of sources with different sizes can be compared. For example, the radiance of the sun is about 1.3×10^6 W sr^{-1} m^{-2}, the radiance of a 1000 W mercury arc lamp is about 10^7 W sr^{-1} m^{-2}, and the radiance of a 1-mW He–Ne laser with beam diameter of 1 mm and a beam divergence of 1 mrad is about 1.6×10^9 W sr^{-1} m^{-2}. The radiance of most sources increases as the viewing direction approaches the direction normal to the source surface. An extended Lambertian source is an exception, the intensity and projected area vary in the same way with viewing direction, so the radiance is independent of viewing direction.

It is often necessary to estimate the response of a light detector to different light sources. Exact calculations are difficult because the required information may not be available, so it is often better to make a simple estimate and then adjust the equipment to produce the required response. If the source width is small compared with its distance d from the detector then the radiation falling on the detector can be estimated by assuming that the source is a Lambertian point source of intensity I equal to the product of its brightness B and its area S. The irradiance falling on the detector is then given by [2]

$$E = \frac{I \cos \phi}{d^2} = \frac{BS \cos \theta \cos \phi}{d^2} \qquad (20.76)$$

where θ and ϕ are the angles between the normals of the source and detector, respectively, and the line connecting source and detector. If the source is circular, its solid angle S/d^2 can be approximated by $\pi \alpha^2$, providing α the angle in radians equal to the radius of the source divided by its distance from the detector is small. Light detectors have a directional response to radiation, which may be too wide or too narrow for the intended application. The extent of the angular response of a detector to light can be reduced using a collimating tube. Alternatively, if the detector is placed at the focal point of a lens the angular response can be decreased or increased using positive or negative lenses. If the source is effectively a point and all the light brought to focus by the lens falls on the detector, then there is the added advantage of the sensitivity increasing by a factor equal to the ratio of the lens-to-detector area.

20.8.3 Light Sources

The choice of a light detector should take into account the nature of the light source, which might be daylight, a tungsten filament lamp, a quartz halogen lamp, a fluorescent tube, a light emitting diode, etc. The distinct properties of light sources arise partly from their construction and partly from the physical processes which lead to the emission of light [3]. Many sources are thermal in nature; that is, their light emission is due to their high temperature. An object heated to incandescence, such as the filament in a tungsten filament lamp, emits a broad continuous spectrum of electromagnetic radiation, with an intensity that depends on the temperature and its surface emissivity. At any given temperature, no surface emits more radiation than a completely black surface, which has emissivity 1.0 at all wavelengths. It can be important when designing light detecting systems to be aware that the visible light from such lamps (and also fluorescent lamps) is often accompanied by significant amounts of invisible radiation, which detectors may be sensitive to, even if the eye is not. Many light sources operate at temperatures near room temperature and hence are not in thermal equilibrium. Luminescence is the general term used to describe the production of light at a greater rate than that due to the temperature of the body. Common examples of such sources are light-emitting diodes (LEDs), zinc-sulfide electroluminescent panels, and the electron-beam excited phosphors in computer monitor and TV screens. The major properties of sources include: total radiant or luminous power output; efficiency in converting electrical power into radiant power; spectral composition of the output; directionality of the output radiation; area of the

TABLE 20.5 The Characteristics of a Number of Different Types of Light Source

Description	Size	Electrical Input	Light Output	View Angle	Spectral Type
Ultra-bright yellow LED	10 mm dia.	20 mA, 2.1 V	14 cd	4°	Peak at 590 nm
Infrared GaAlAs LED	5 mm dia.	0.1 A, 1.9 V	16 mW sr^{-1}	80°	Peak at 880 nm
Infrared LED	5 mm dia.	0.1 A, 1.9 V	135 mW sr^{-1}	8°	Peak at 880 nm
Small filament lamp	11 mm dia.	6 V, 0.3 A	11 lm	360°	Black body
Miniature fluorescent tube	300 × 16 mm dia.	8 W	480 lm	360°	White
Standard fluorescent tube	1500 × 26 mm dia.	58 W	4800 lm	360°	White
Tungsten halogen dichroic	51 mm dia.	12 V, 20 W	3300 cd	12°	3000 K
Tungsten halogen dichroic	51 mm dia.	12 V, 20 W	460 cd	36°	3000 K

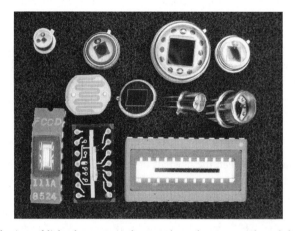

FIGURE 20.101 A collection of light detectors is shown. Along the top row from left to right are four silicon photodiodes with areas of 1, 5, 41.3, and 7.5 mm^2, the last with a photometric color correction filter. Along the middle row from left to right are a CdS photoresistor, a pyroelectric detector, a phototransistor, and a quadrant silicon photodiode containing four separate sensing elements. Along the bottom row from left to right are a 256-element linear CCD, a 64-element charge integrating CMOS array, and a 16-element linear silicon photodiode in a 24-pin d.i.l. package. The diode pitch is 1 mm.

emitting surface; lamp size and operating temperature. Table 20.5 lists characteristics of a number of common types of light sources taken from the lamp suppliers data sheets.

20.8.4 Light Detectors

A light detector converts the radiant power it absorbs into a change of a device parameter such as resistance, surface charge, current, or voltage. A number of light detectors are shown in Figure 20.101. Some signal conditioning electronics may also be needed to convert the basic output from the detector into a more useful voltage signal, for example, for digitization by an analog-to-digital converter (ADC). This may be integrated into the detector or require external components. Light detectors can be divided into two main types, thermal or photon devices. In thermal detectors, the heating effect of the absorbed radiation results in a change in a temperature dependent parameter, such as electrical resistance (in bolometers) or thermoelectric emf (in thermopiles). The output of thermal detectors is usually proportional to the radiant power absorbed in the detector, and provided the absorption efficiency is the same at all wavelengths, the output is independent of wavelength. The most widely used type of thermal detector is the pyroelectric detector, which is discussed in the next section. Photon detectors, in contrast to thermal detectors, depend on the generation of free charge by the absorption of individual photons. This photon-induced charge causes a change in device resistance, in the case of photoresistors, or an

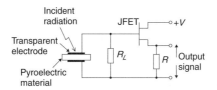

FIGURE 20.102 The basic components within a pyroelectric detector are indicated. An increase in radiation falling on the pyroelectric material causes its temperature to rise and the charge on its surface to change. A transient current flows through the resistor R_L which is of the order of 10^{11} Ω. The JFET reduces the output impedance to R.

output current or output voltage, in the case of photodiodes and transistors. All these photon detectors require a minimum photon energy to create mobile electrons and consequently have a maximum wavelength, dependent on the detector material, beyond which they do not operate. On the other hand, photon detectors generally respond faster to changes in radiation level than thermal detectors and are more sensitive.

20.8.4.1 Pyroelectric Detectors

Pyroelectric detectors employ a ferroelectric ceramic material (such as lead zirconate or lithium tantalate) which has molecules with a permanent electric dipole moment [4]. Below a critical temperature, known as the Curie temperature, the dipoles are partly aligned and give rise to a net electrical polarization for the whole crystal. As the material is heated and its temperature rises, increased thermal agitation of the molecules reduces the net polarization, which falls to zero at the Curie temperature. The basic detector, shown in Figure 20.102, consists of a thin slab of ferroelectric material fabricated so that the polarization is normal to the large area faces on which transparent electrodes are evaporated. These are connected together via a load resistor (up to 10^{11} Ω). An increase in radiation falling on the detector makes its temperature rise and causes the captive surface charge, which is proportional to the polarization, to change. This causes a change in the charge induced in the electrodes and a current to flow in the load resistance. Because of the large value of the load resistor used in pyroelectric detectors, an impedance matching circuit, such as a JFET source following circuit, is usually built into the detector as shown in Figure 20.102. Pyroelectric detectors only respond to changing irradiation and typically can detect radiation powers down to about 10^{-8} W at 1 Hz. Because they respond to the heating caused by absorption of the radiation, they have a wide spectral response. They are useful as low-cost infrared detectors, intruder alarms, and fire detectors.

20.8.4.2 Photon Detectors

The most widely used photon detectors are made from a semiconducting material. In semiconductors, the electrons fill the available energy levels in the material up to the top of the valence band (VB), which is separated from the bottom of the empty conduction band (CB) by an energy gap E_g, which is characteristic of the material. These energy bands are completely full or empty, respectively, only at a temperature of absolute zero (0 K). At a higher temperature, an equilibrium is reached between the thermal excitation of electrons across the gap (producing free electrons in the CB and positively charged free holes in the VB) and the recombination of pairs of free electrons and holes. The equilibrium number of free electrons and holes increases rapidly with temperature (T) according to the Boltzmann factor $\exp(-E_g/kT)$, where k is Boltzmann's constant (1.38×10^{-23} J/K). This equilibrium is disturbed when photons, with energy greater than E_g, are absorbed by electrons which are excited across the gap. When the radiation source is removed, the number of excess electrons and holes quickly falls back to zero over a time period governed by the recombination time of the material. While excess free charge is present there is a measurable change in the electrical conductivity and this is used in photoresistive (also called photoconductive) detectors. Alternatively, in junction detectors, the rate of generation of photocharge is converted to an output current, or voltage. All semiconductor photon detectors have a relatively narrow

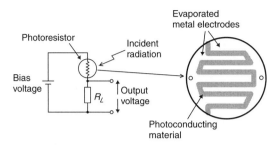

FIGURE 20.103 A simple light detector circuit employing a photoresistor is shown. An increase in light illumination causes the resistance of the photoresistor to decrease and the output voltage to increase. The comb-like pattern typically employed in photoresistors gives a relatively large active area of photoconducting material and a small electrode spacing resulting in high sensitivity.

spectral response, which peaks at a wavelength about hc/E_g. Photoresistors and junction detectors are discussed in more detail in the following sections.

Photoresistors
The electrical conductivity of a semiconductor is the sum of two terms [5], one contributed by electrons and the other by holes, as follows:

$$\sigma = ne\mu_n + pe\mu_p \qquad (20.77)$$

Each term is proportional to $n(p)$ the number of electrons (holes) per unit volume in the conduction (valence) band, the electron (hole) mobility $\mu_n(\mu_p)$, and the magnitude of the charge of the electron e. The increase in conductivity, caused by the absorption of photons increasing n and p, is the basis for the operation of the photoresistive detector. This consists of a slab of semiconductor material on the faces of which electrodes are deposited to allow the resistance to be monitored, as illustrated in Figure 20.103. The photon-induced current is proportional to the length of the electrodes and inversely proportional to their separation, hence the typical comb-like electrode geometry of photoresistors, shown in Figure 20.73. Because the resistance R_C is inversely proportional to conductivity, the variation of R_C with incident power P_D is very nonlinear and is often expressed in the form

$$\log_{10} R_C = a - b \log P_D \qquad (20.78)$$

where a and b are constants. Cadmium sulfide is commonly used as a detector of visible radiation because it is low cost and its response is similar to that of the human eye. Other photoconductive materials include lead sulfide, with a useful response from 1000 to 3400 nm, indium antimonide with a useful response out to 7000 nm, and mercury cadmium telluride with peak sensitivity in the range 5000–14,000 nm. The wavelength range 5000–14,000 nm is of importance because it covers the peak emission from bodies near and above ambient temperature and also corresponds to a region of good transmission through the atmosphere. Photoconductive devices used for the detection of long wavelength infrared radiation should be cooled because of the noise caused by fluctuations in the thermal generation of charge. As a rough rule of thumb, because of the Boltzmann factor, a detector with energy gap E_g should be cooled to a temperature less than $E_g/25k$.

Junction Detectors
In photoresistors, the rate of generation of electron–hole pairs by the absorption of radiation, combined with recombination at a rate characteristic of the device, results in an increase in free charge and therefore electrical conductivity. In junction photodetectors [6], such as photodiodes and phototransistors, newly generated electron–hole pairs separate before they can recombine so that a photon-induced electric

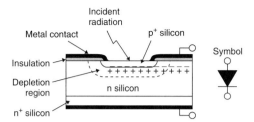

FIGURE 20.104 The basic structure of a typical silicon photodiode is illustrated. A space charge, or depletion region, is formed by the diffusion of mobile charge across the surface between the p-type and n-type silicon. It extends furthest into the n-type silicon because this is more lightly doped than the p-type silicon. Any electron hole pairs generated in this region are prevented from recombining by the presence of the electric field, which sweeps them apart, allowing them to contribute to the photon generated current. The p-type region is made thin to allow photons to penetrate into the depletion region.

current can be detected. The separation of electrons and holes takes place in the electric field associated with a P–N junction fabricated in a semiconductor material, which is usually silicon. The structure of a typical silicon photodiode is shown in Figure 20.104. The substrate material is lightly doped n-type silicon, which is pure group IV silicon into which has been added a small amount of a group V impurity element. This contributes free electrons to the conduction band of the silicon leaving the impurity atoms ionized and with a positive charge. A region of heavily doped p-type silicon is formed on the top face of the substrate by adding a group III impurity element, by diffusion or ion implantation for example. The group III atoms contribute free holes to the valence band leaving negatively charged impurity ions. The P–N junction is the boundary surface between the p-type and n-type regions on which the opposite impurity concentrations are equal. The mobile electrons and holes diffuse across the boundary from the side where they are in the majority, to the side where they are in the minority. There they recombine leaving a region containing unscreened positive impurity ions on one side of the junction and a region containing unscreened negatively charged impurity ions on the other. The charged region is called the space charge or depletion region, because it is depleted of free charge. The movement of mobile charge continues until the diffusion driving force is balanced by the opposing electric field created in the depletion region by the separation of charge. When equilibrium has been established, the voltage across the depletion region, called the built-in voltage, is about 0.6 V (for silicon). The depletion region extends much further into the n-type silicon than into the p-type silicon for the photodiode shown in Figure 20.104 because of the very different doping concentrations. The p-type region is made very thin, so that radiation can pass through it, and metallic contacts are made to the p-type and n-type materials. An ohmic contact forms between a metal and heavily doped silicon and to ensure a good ohmic contact to the lightly doped n-type material, an intermediate more heavily doped n-type region is included as shown.

Electron–hole pairs formed in the depletion region when light with wavelength less than hc/E_g is absorbed are separated by the electric field in this region and can be detected in two ways. If the photodiode is left open circuit, a voltage V_p appears across the diode, varying logarithmically with the incident irradiance P_D as follows:

$$V_p = \frac{kT}{e} \ln\left(\frac{\eta P_D A e \lambda}{hci_0}\right) \qquad (20.79)$$

where η is the probability of a photon being absorbed, A is the active area of the photodiode, and i_0 is the dark current due to thermal generation. This is the photovoltaic mode of operation. If the diode is operated with a reverse bias, a photon-generated current i_p flows given by the following expression:

$$i_p = \frac{\eta P_D A e \lambda}{hc} \qquad (20.80)$$

TABLE 20.6 The Characteristics of a Number of Different Types of Light Detector

Description	Active Region	Response	Spectral Response	Dark Current	Response Time	Acceptance Angle
Medium area silicon photodiode	41.3 mm^2	0.5 A W^{-1} peak	800 nm peak, range 350–1100 nm	4 nA	25 ns	NA
Ultra high speed silicon photodiode	0.5 mm^2	0.35 A W^{-1}	800 nm peak, range 400–1000 nm	10 nA	1 ns	NA
Filtered silicon photodiode	7.5 mm^2	7 nA lux^{-1}	560 nm peak, range 460–750 nm	2 nA	3.5 μs	100°
16 photodiode array on 1 mm pitch	Each diode 0.66 mm^2	0.6 A W^{-1}	900 nm peak, range 400–1100 nm	0.1 nA	4 ns	NA
Silicon phototransistor	0.7 mm^2	9 μA lux^{-1}	880 nm peak, range 450–1100 nm	0.3 μA	15 μs	30°
Silicon phototransistor	0.7 mm^2	2 μA lux^{-1}	880 nm peak, range 450–1100 nm	0.3 μA	15 μs	80°
CdS photoconductor	6.3 mm dia.	9 kΩ at 10 lux, 400 Ω at 1000 lux	530 nm peak	NA	100 ms	NA

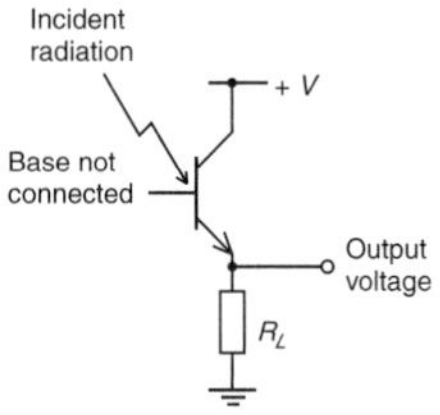

FIGURE 20.105 A simple phototransistor light detector circuit is shown. Photon-generated current flowing in the base-collector diode may be amplified several hundred times by transistor action. Although the photon-generated current is much larger than in an equivalent photodiode, response time of the phototransistor is much longer.

In this photoconductive mode, the current through the photodiode varies linearly with light irradiance. The dark current i_0 varies rapidly with temperature and limits the sensitivity of the device but the photoconductive mode generally has faster response, better stability, and wider dynamic range than the photovoltaic mode. The responsivity K_D of the detector is defined by the relation $i_p = K_D P_D$ and is less than 1 A W^{-1} for a silicon diode. In the ideal case, K_D varies linearly with wavelength, according to Equation 20.80, up to the threshold value set by the energy gap. Photodiodes are available with a wide variety of characteristics differing in sensitivity (area), speed of response, spectral response, and acceptance angle. They are available with single devices or multiple devices (quad, linear array) in a single package.

The output signals from photodiodes needs amplification for many applications. This may be provided by a separate amplifier or by providing internal gain as in the phototransistor. This is constructed so that radiation can fall on the base region of the transistor and the resulting base current is then internally amplified. Often there is no external connection to the base and the amplified photocurrent is monitored using the simple circuit shown in Figure 20.105. A typical phototransistor has a responsivity several hundred times higher than that of a photodiode but the frequency response is relatively poor. Phototransistors are often integrated with a spectrally matched LED into a single sensor package to act as a proximity sensor, as in end-of-tape sensors, coin detectors, and level sensors. For reference, the characteristics of several different types of discrete light detector are listed in Table 20.6.

By fabricating many small light detectors in a closely spaced array, it is possible to measure light intensity at an array of points over a region. This is ideal for electronic imaging applications involving video and still cameras. Image sensors designed for this purpose are discussed in the section titled "Image Sensors," but first it is useful to consider the formation of the images which the detectors sense.

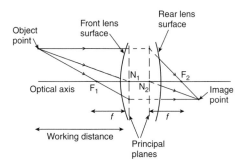

FIGURE 20.106 The cardinal points of a multi-element lens operating in a single medium (usually air) are indicated. The principal points and nodal points then coincide at N_1 and N_2 and the front and rear focal lengths are equal (f). Three rays from an object point are traced through the lens to the corresponding image point using the properties of the cardinal points. In the case shown the image magnification is 0.5, so the image is some distance behind the rear focal point F_2. For distant objects the image plane would coincide with the plane transverse to the optical axis passing through F_2. Lenses are normally corrected for aberrations assuming that the object distance will be greater than the image distance. In this case, for close-up work when the image distance is greater than the object distance, the image quality is improved by reversing the lens.

20.8.5 Image Formation

Although perfect images are formed by small pinholes, lenses are needed to form bright images and range from simple single-component lenses used to increase the amount of light falling on a single detector or in low-cost cameras to complex zoom lenses, with between 14 and 20 components, capable of producing high quality images of varying size. The two most important properties of a lens are its focal length f, which determines the imaging behavior, and its light-gathering power or speed, specified by an f-number $f_\#$. A lens has an optical axis passing through the central axis of each of its components along which a ray of light passes without deviation. A lens is characterized, regardless of its complexity, by six cardinal points [2] spaced along the optical axis as illustrated in Figure 20.106, for a positive converging lens. The position and magnification of the image of an object can be determined using these cardinal points, which include two focal points, two nodal points, and two principal points. The nodal points have the property that a ray outside the lens travelling towards one nodal point emerges from the lens in a parallel direction, appearing to come from the other nodal point. The focal point is the point which a ray of light incident on the lens parallel and close to the optical axis converges to (positive lens) or diverges from (negative lens) after passing through the lens. The point where the lines colinear with the ray on the two sides of the lens intersect defines a point on the principal plane. The point where the optical axis intersects the principal plane is called the principal point. There are two nodal points, focal points and principal planes, because light can be incident on the lens from either side. The front and back focal lengths of the lens are the distances between the front and back focal points and their corresponding principal planes. In the normal situation, when the lens is operating in a single medium, such as air, the positions of the nodal points and principal points coincide and the front and back focal lengths are equal. In general, when the lens construction is asymmetric, the front and back focal points are at different distances from the corresponding external lens surface. In the case of an ideal thin lens, the principal planes coincide with the lens center but in multi-element lenses they may be separated by +20 to −10 mm, depending on the lens design. A lens of focal length f produces an image in best focus at a distance v when the object is at distance u where

$$\frac{1}{u} + \frac{1}{v} = \frac{1}{f} \tag{20.81}$$

and the distances are measured to the corresponding principal planes. The image magnification m, defined as the ratio of image to object sizes, is equal to the ratio of the image to object distances and is related to the total distance between object and image D_T by

$$D_T = \frac{f(m+1)^2}{m} + D_N \qquad (20.82)$$

where D_N is the separation between the principal planes. Lenses generally have a focusing range around 5–10% of the focal length giving a maximum magnification of 0.05–0.1. For larger magnifications extension rings can be fitted between the lens and sensor mounting. The extension size required, with the lens focused at infinity, is simply the product of the magnification and the focal length, since image distance is $f(m+1)$. When a lens is focused to produce an optimally sharp image for a particular object distance u, there is range of closer and further object distances over which the image is still acceptably sharp. This range is called the depth-of-field or depth-of-view F_o and its size depends on the f-number of the lens, the magnification, and the acceptable blur spot size C in the image plane [7]. The blur spot size depends on the image sensor and for 35 mm film C is usually assumed to be between 0.02 and 0.033 mm, while for an image sensor array C is the separation between the individual detector elements, typically about 0.01 mm. Depth-of-field decreases with magnification and for m greater than 0.1 is calculated using the following formula:

$$F_o = 2f_\# \frac{C(m+1)}{m^2} \qquad (20.83)$$

The accuracy of alignment required of the image sensor depends on the depth-of-focus F_i, which is the longitudinal range of image positions over which the image is acceptably sharp. Sensor alignment is most critical when the lens f-number is small and the image magnification is also small. When m is small F_i reduces to $2Cf_\#\, v/f$ and equals $m^2 F_o$. When m is large, the depth-of-focus is not so critical.

It is frequently necessary to relate the lighting of a scene to the image irradiation falling on a sensor. As accurate calculations are difficult, it is usually best to make a simple estimate and then make fine adjustments to the lighting or lens aperture. When the object of interest in the scene is not a light source but is visible because it is reflecting light, then its luminance must be estimated from the radiation falling on it and the reflection coefficient R_o of its surface [3]. For example, if the object is a Lambertian surface, with illumination L_o, then its luminance is given by

$$B = L_o \frac{R_o}{\pi} \qquad (20.84)$$

When a lens is used to form an image of the object, the illuminance on the optical axis in the image plane L_S in lux is related to the luminance of the object B in cd m^{-2} by

$$L_s = \frac{TB\pi}{[2f_\#(m+1)]^2} \qquad (20.85)$$

where $f_\#$ is defined as the ratio of the focal length to diameter of the effective lens aperture and losses in the lens are characterized by a transmission coefficient T [7]. Due to a number of geometrical factors, the image illumination falls off with angle θ from the optic axis as $\cos^4\theta$. Near the axis, the illuminance varies only slowly with angle but at an angle of 30° it has fallen by 44%. Equation 20.85 is the basis for rating the speed of lenses by their f-numbers and indicates that the smaller the f-number, the greater the image illuminance. This formula is appropriate when the magnification is small, but for close-up work, when the image distance is significantly greater than the focal length, the f-number $f_\#$ should be replaced by $(m+1)f_\#$ when calculating image illuminance.

Lenses are manufactured to match standard image sizes such as the 36 mm × 24 mm 35 mm photographic format and the standard television sensor sizes 1″, 2/3″, 1/2″, 1/3″, and 1/4″. These sizes are defined to be twice the horizontal dimension of a rectangular image with 4:3 aspect ratio so that, for example, a 1″ sensor has a width of 12.7 mm, a height of 9.5 mm, and a diagonal length of 15.9 mm. Lens sizes are similarly specified to allow easy matching of lenses to sensors. The maximum angular field-of-view F_{OV} of a lens focused at infinity is given by

$$F_{OV} = 2\tan^{-1}\left(\frac{C_F}{2f}\right) \quad (20.86)$$

where C_F is the diagonal of the sensor format. For example, a 35-mm lens with a focal length of 55 mm has a field of view of 43°. Because image distortion and sharpness worsen towards the edges of the field of view it is acceptable, for example, to use a 2/3″ lens with a 1/2″ sensor, but not the converse. A 35-mm camera lens generally performs much better, at relatively low cost, than a corresponding C-mount lens supplied for a TV camera but a C-mount to Pentax, Cannon, or Nikon mount converter is then required.

20.8.6 Image Sensors

The generation of an output signal from a standard image sensor involves up to four operations: the conversion of the spatial distribution of light irradiance in the image plane into a corresponding spatial distribution of charge; the accumulation and storage of this charge near the point where it is generated; the transfer or read-out of this charge; and the conversion of the charge to an output voltage signal. Each of these operations can be achieved in many ways. Vacuum tube sensors such as vidicons, for example, use a photoconductive detector material and a scanning electron beam for read-out while CMOS sensors use a photodiode detector and a readout bus. Solid state devices are currently the most widely used types, so only CCD and CMOS sensors are considered here. In these devices the image irradiance is measured on a one- or two-dimensional array of sample regions with positions fixed during fabrication. Each sample is called a picture element or pixel and the greater the number of pixels, the higher the resolution with which the image can be recorded. Area sensors are manufactured with numbers of pixels ranging from tens of thousands to several million. Color sensors are achieved by placing color filters over the individual pixels, in a mosaic or stripe pattern, and interpolating the color values at pixels where necessary from the neighboring values. In such color devices the color resolution is lower than the luminance resolution, but this is not important for many applications because the resolution of the human eye is worse for color than for luminance. More expensive color cameras use three precisely aligned sensors, one for each primary color. Cameras incorporating such sensors generally produce either a television standard signal [8] (RS-170 monochrome and NTSC color for 525 American television or CCIR monochrome and PAL color for European television) or a digital signal such as RS-423, USB or IEE 1394 Fire Wire, which can be readily connected to a computer.

20.8.6.1 Charge-Coupled Devices

In a charge-coupled device [9] an isolated packet of charge, of between 10 and 10^6 electrons, is moved through the semiconductor, from a position in one CCD cell to a position in an adjacent cell, by applying a sequence of voltage pulses to gate electrodes. In CCD-based light sensors, photon generated charge packets accumulate in photosites, which are modified CCD cells, and are then transported through other CCD cells to another modified cell with a readout amplifier attached. A CCD is fabricated on a single crystal wafer of P-type silicon and consists of a one- or two-dimensional array of charge storage cells, on centers typically about 10 μm apart. The operation of a 3-phase CCD cell is illustrated in Figure 20.107. Each cell has three closely spaced electrodes (gates) on top, insulated from the silicon by a thin layer of silicon dioxide. A positive voltage applied to one of these gates will attract and store any free charge generated in the silicon due to light or thermal action while free holes are repelled and collected by the substrate electrode. Lower voltages on the adjacent gates isolate it from the neighboring cells,

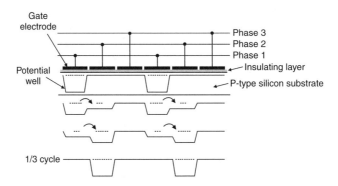

FIGURE 20.107 The movement of charge from one potential well to the next in a 3-phase CCD is illustrated. Each CCD cell has three gate electrodes. In the upper potential diagram, the well is formed under the first electrode in each CCD cell by voltage applied to the phase 1 line. As the voltage is reduced on the phase 1 line and increased on the phase 2 line, the original potential wells collapse and new ones form under the second gate in each cell, causing any charge present in the wells to move sideways as indicated. Two more cycles are required to complete the movement of charge into the first well of the next CCD cell.

creating a localized potential well within the cell. A cell of size 8 μm × 8 μm can hold about 200,000 electrons before saturating. Cells designed to be light sensitive have electrodes made of semitransparent polysilicon so that light can penetrate into the storage region, while cells intended only for charge transport are covered with a surface layer opaque to light. During operation, the voltages on the electrodes are held constant for a time (integration time) to allow packets of charge to accumulate on the photosites in proportion to the local irradiance. At the end of this time a sequence of voltage pulses are applied to the electrodes to transfer the packets of charge from one storage cell to the next until they reach a sensing amplifier, which generates a voltage which is about 0.6 μV per electron. The charge transfer efficiency (CTE) of real devices is less than 100% and between 99.95% and 99.999% of the stored charge is moved to the next cell, depending on the precise construction and clocking frequency. This allows devices to be manufactured with a line of many hundreds or thousands of storage cells feeding a single amplifier.

Although there are many variations in CCD construction, the basic characteristics of CCDs from different manufactures are similar. CCD light sensors have an inherently linear variation of output voltage with irradiance, from the minimum useful level set by noise to the maximum useful level set by saturation of the output amplifier or by the limited capacity of the charge storage cells. The dynamic range of the device is defined as the ratio of the maximum output signal to the output resulting from noise. Manufacturers sometimes quote noise figures as peak-to-peak or as root-mean-square values (typically five times smaller), but the former value is more relevant for imaging applications. Due to manufacturing limitations, the photosites do not have identical sensitivity and dark signal characteristics. For example, photoresponse nonuniformity (PRNU) is easy to measure using uniform sensor illumination and is typically 5–10%. Its effects can be removed, if necessary, by calibration. The basic spectral response of a silicon sensor extends from 200 to 1100 nm, with a maximum sensitivity of about 1 μA of generated charge per microwatt of incident radiation, but this is modified by the electrode structures formed on the surface of the silicon. Longer wavelength photons penetrate more deeply than shorter wavelength photons and the short wavelength response is typically worsened to 450 nm by absorption in the surface layers. Infrared photons may generate electrons some distance from the point of entry into the silicon, with the result that the charge may be collected by a different cell. This reduces the resolution of the device and if infrared operation is not required, but the illumination contains infrared (for example, from a tungsten lamp), an infrared reflecting filter (a hot-mirror filter) is often used. If the widest possible spectral response is required, devices have the substrate thinned and are operated with the illumination falling on the back surface, which is free of electrodes. Back illuminated devices are fragile and costly but are used in specialist low-light applications like astronomy and biology.

All CCD cells accumulate charge linearly with time due to thermally generated electrons produced within the cells and at electrode interfaces. Like the photoresponse, this dark signal varies from cell to cell and can be compensated for by calibration. These thermally generated contributions are most significant for low-light level applications and can be reduced by cooling the sensor using either a thermoelectric cooler, a Joule Thomson cooler, or a liquid nitrogen dewar. The dark signal reduces by 50% for every 7°C reduction in temperature and at −60°C, produced by a Peltier cooler, the dark signal is typically reduced to about one electron per pixel per second. Another important temperature dependent characteristic of the CCD sensor, which improves with cooling, is the noise floor of the output amplifier which is proportional to $T^{1/2}$ and typically equivalent to about 300 electrons at room temperature. A CCD device used in astronomy illustrates the performance achieved by cooling. Operated at about −110°C, this device has a readout noise of about 10 electrons, a dark current less than 0.3 electrons per minute, and a quantum efficiency for converting visible photons into electrons of between 70% and 80%. Light may be integrated for periods of hours compared with the approximately 1/8 s to 1/4 s integration period of the dark adapted eye. Compared with photographic film previously used for low-light level imaging in astronomy, cooled CCDs are from 10 to 100 times more sensitive, linear in response rather than nonlinear, and have a much greater dynamic range so that both faint and bright objects can be recorded in the same exposure.

The transfer of charge from one cell to the next takes time and the CTE worsens with increasing clocking speed and with cooler temperatures. This limits the number of cells which can be used to transport charge from a photosite to the readout amplifier. It also limits the rate at which data can be transferred out of the CCD and the resulting image transfer rate. However, there are many variations in CCD technology aiming to improve performance. For example, virtual-phase CCDs [10] have some of the electrodes replaced by ion-implanted regions resulting in improved blue response and higher sensitivity, because of the removal of some of the blocking surface gates and simpler drive circuitry due to the lower number of gates per cell. The biggest contribution to the dark signal is defects at interfaces and a manufacturing technique known as pinning can be used to passivate the interface states, producing an order of magnitude improvement in a dark signal as well as improved quantum efficiency and CTE. The readout noise performance can be improved by a signal-processing technique called correlated double sampling. This involves taking the output as the difference between two signals, one with the charge signal present and one without, so that major noise components are cancelled. A number of architectures are employed in CCD devices [11]. Several of these, including linear devices and area devices of the full-frame, frame transfer, and interline transfer types, are discussed in the following sections.

Linear Charge-Coupled Devices
A linear CCD sensor consists of a line of up to several thousand photosites and an adjacent parallel CCD shift register terminated by a sensing amplifier. Each photosite is separated from a shift register cell by a transfer gate. During operation a voltage is applied to each photosite gate to create empty storage wells, which then accumulate amounts of charge proportional to the integral of the light intensity over time. A transfer pulse at the end of the integration period causes all the accumulated charge packets to be transferred through the transfer gates to the shift register cells. The charges are clocked through the shift register to the sensing amplifier producing a sequence of voltage pulses with amplitudes proportional to the integrated light falling on the photosites. In practice it is common for shift registers to be placed on both sides of the photosites with alternate photosites connected by transfer gates to the right and left registers. These halve the time required to clock out all the data. There is a limit to the number of electrons (typically 1000–2000 times the area of the photosite in μm^2) which can be stored in a cell, before electrons start to spill over into adjacent cells. This blooming effect is a problem with images containing intense highlights. It is reduced by about a factor of 100 by adding antiblooming gates between adjacent photosites and transfer gates and channel stops between adjacent photosites. The voltage on the antiblooming gates is set at a value which allows surplus charge to drain away instead of entering the transfer gates and shift register. By clocking this voltage, variable integration times which are less than the frame pulse to frame pulse exposure time can also be attained.

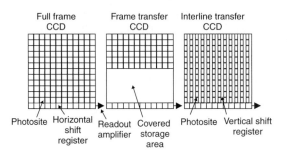

FIGURE 20.108 The three basic architectures used in area CCDs are illustrated. In the full-frame transfer CCD most of the device area is employed as photosites. Photon-generated charge is transferred down each column one cell at a time into the horizontal shift register where it must all be transferred to the readout amplifier before another vertical movement of charge can take place. The frame-transfer CCD reduces the need for a mechanical shutter to prevent charge smearing, which would otherwise occur, by providing a covered storage area into which all the photon-generated charge can be rapidly shifted vertically at the end of the integration period. The interline-transfer CCD allows all the photon-generated charge to be transferred to the covered vertical shift registers in one step, virtually eliminating this source of charge smearing.

Area Charge-Coupled Devices

Three basic architectures are used in area CCDs and are illustrated in Figure 20.108. The full-frame CCD consists of an imaging area separated from a horizontal CCD shift register by a transfer gate. In the imaging area each photosite is one stage of a vertical shift register separated from neighboring shift registers by channel stops and anti-blooming structures. During the light integration period, the vertical clocks are stopped and the photosites collect photoelectrons. At the end of this period the charge is clocked out vertically, one row at a time into the horizontal shift register. The charge in the horizontal shift register is then very rapidly shifted towards the output amplifier by the application of a horizontal clock signal. For example, the RA1001J, 1024 × 1024 pixel full-frame CCD from EG&G Reticon achieves a readout rate of 30 frames per second. To avoid image smear during the readout period, full-frame sensors must be operated with external shutters or used in low-light level applications requiring very long integration times compared with the readout time, as in astronomy.

The frame-transfer CCD greatly reduces the need for an external shutter by providing a light-shielded storage section into which the entire image charge is shifted at a rate limited primarily by CTE considerations. The charge is read from the storage region during the next integration period without any further image smearing. In some devices, such as the EG&G Reticon RA1102, the storage area is split into two on opposite sides of the imaging area. This improves performance by halving the maximum number of transfers required to reach the nearest storage region. With sensors designed for interlaced operation, as opposed to the non-interlaced progressive scan readout mode, this reduction occurs automatically. Each integration period then corresponds to one video field and only half the number of rows in the frame is required at any one time. For example, to produce an interlaced video frame containing 576 image lines (CCIR standard), a frame transfer sensor with only 288 rows of storage is required. By changing the clock signals, the odd field can be displaced vertically by half a line width relative to the even field. This ensures that the odd and even lines contain different information and reduces aliasing because the cell width is twice the separation between the lines in the frame. Many companies produce frame-transfer CCD sensors and cameras including Cohu, Dalsa, EG&G Reticon, EEV, Kodak, Philips, and Thomson-CSF.

The interline-transfer (ILT) architecture virtually eliminates image smear by providing each column of photosites with an adjacent light-shielded vertical CCD shift register into which the charge is transferred by a transfer pulse. The contents of all the vertical shift registers are then shifted simultaneously one pixel at a time into a horizontal shift register where they are rapidly shifted to an output amplifier. This approach makes it easy to achieve short integration times and true "stop-motion" exposure control with progressive scan. It also increases the "dead space" between the active pixels reducing the sensitivity of the image sensing area and increasing aliasing effects compared with frame-transfer sensors. For the latter, the fill factor, which is the percentage of the imaging area which is light sensitive, can be close to 100% whereas it is usually less than 50% for interline-transfer devices. Localized bright objects tend to produce vertical streaks in an ILT device because strong light can leak under the narrow light shield covering the vertical shift registers, causing image smearing similar to that in a full frame device. For interlaced operation, two adjacent pixels, for example, 1 and 2, 3 and 4, etc. are transferred to a single shift register cell on one field and in the next field pixels 2 and 3, 4 and 5, etc. are transferred together. This is rather similar to the interlaced operation of a frame transfer CCD. Many companies manufacture ILT CCD sensors and cameras including, Hitachi, NEC, Panasonic, Pulnix, and Sony.

20.8.6.2 CMOS Sensors

CMOS image sensors are based on a technology that is older than CCD technology [12]. However, CCD sensors originally offered better image quality than CMOS devices could match so they came to dominate the market. There is now renewed interest in the older technology because it potentially offers major advantages over CCDs. The CMOS process used in sensors is similar to that which has been highly developed in order to manufacture dynamic RAM and consequently should be able to produce cheap, small high-resolution, randomly addressed, low-power sensors. It is also possible to integrate image sensing, control, processing, and interfacing on the same chip, so that a camera on a chip is possible using CMOS technology, but not with CCD technology. As a result of recent research and development, several manufacturers are now claiming to have achieved CMOS sensors providing similar quality to that of mainstream CCDs. Manufacturers producing CMOS sensors include Fillfactory, National Semiconductor, Philips, ST Microelectronics, and Y Media.

A CMOS sensor consists of an array of photodiodes, which are connected to readout amplifiers by bus lines and MOS switches. The principle of readout is illustrated in Figure 20.109. Each pixel is connected to an output amplifier by a switch whose control line is connected to a digital shift register. Shifting a bit through the register connects the photodiodes sequentially to the output amplifier. Random readout of the photodiodes can be achieved by replacing the shift register by an address decoder connected to an address bus. Two-dimensional arrays of photodiodes are connected in a configuration similar to a cross-point switching matrix with a switch and diode at each cross point and separate vertical and horizontal shift registers. To scan the array, the vertical shift register turns on a complete row of switches and the photodiodes in that row output their signals into vertical bus lines. These are connected, in turn,

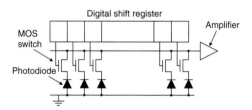

FIGURE 20.109 The principle of readout in a one-dimensional CMOS sensor is illustrated. Each pixel is connected to an output amplifier by a MOS switch whose control line is connected to the digital shift register. Shifting a bit through the register connects the photodiodes, in turn, to the output amplifier.

FIGURE 20.110 A typical CMOS three-transistor active pixel is shown. Transistor T_1 is connected to the reset line allowing the capacitance of the photodiode to be reset at the start of photo-current integration. In the continuous mode CMOS device this transistor is connected to act as high value resistor. Transistors T_2 and T_3 allow the signal to be transferred from the photodiode to the column amplifier, via the column bus line.

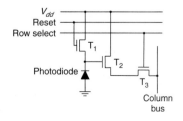

to an output amplifier by a set of switches connected to the horizontal shitch register. Using two separate vertical shift registers and separating row select and row reset functions enables a rolling curtain type of electronic shutter to be implemented with an exposure ratio equal to the number of rows.

Photodiode arrays generally have less extensive electrode structures over each sensing element compared with CCD arrays and consequently the spectral response is smoother and extends further at the blue end of the spectrum, which is an advantage for color sensors. The peak quantum efficiency is also higher ranging from 60% to 80% compared with 10% to 60% for photogates, leading to almost twice the electrical output power for a given light input power. On the other hand, photodiodes have higher noise levels than CCDs because of the reverse-bias leakage current. The photodiode can be operated in integrating mode or in continuous mode. In the former, the photodiode capacitance is reset to a reference reverse bias and then allowed to float. The charge on the photodiode capacitance is then discharged by photon-generated current and leakage currents. After a specified integration time, the remaining charge can be read and the difference from the reference value is proportional to the diode irradiance if the leakage sources are negligible. Both passive and active pixel devices have been developed. In the latter, the charge on the photodiode is read out through a MOS field effect transistor (MOSFET), which converts charge to voltage and provides gain. Figure 20.110 illustrates a typical three-transistor active pixel. Transistor T_1 resets the photodiode and after an integration period the pixel signal is read out through the source-follower transistor T_2 and the selection transistor T_3. Two disadvantages of this approach are a low fill-factor, because of the pixel area used by the amplifying transistor, and the increased pixel nonuniformity because of variations in transistor characteristics. In the ibis range of CMOS sensors designed by FillFactory, a small area photodiode is used which reduces dark current and kTC noise while also increasing the charge-to-voltage conversion factor (9 μV per electron at output in ibis1). The pixel architecture introduces a small potential barrier, which prevents photon-induced electrons from being collected by structures in the pixel other than the photodiode. This allows the photodiode to collect most of the electrons generated in the substrate beneath the pixel effectively increasing the fill-factor [13]. In integrating devices, the fixed pattern noise resulting from variations in MOSFET thresholds can be reduced by correlation double sampling. The pixel is sampled at the end of the integration period and the pixel is reset and sampled again. The difference in the two samples is the measure of the light intensity and is free of pixel offsets. Photoresponse nonuniformity is harder to control. It is caused by variations in the photodiode collection volume, junction capacitance, and gate capacitance of the MOSFET amplifier. In the ibis4 1280×1024 pixel sensor, PRNU is quoted as less than 10% peak-to-peak with half saturation in the neighborhood.

A light sensor with a continuous pixel response has the advantage that pixels can be accessed in any order without an integration time so an image processing algorithm could decide, on-the-fly, which pixel or group of pixels to read next. In the Fuga range of sensors designed by FillFactory, continuous mode operation is achieved by passing the photon-generated current through a series resistance [14]. Because the photon-induced current is very small, the series resistance must be very large and it is realized by a MOSFET operated in weak inversion. The resulting current-to-voltage conversion is logarithmic and the devices have a very wide dynamic range (six orders of magnitude or 120 dB) with a quoted dark limit of 10^{-4} W m^{-2}. The output signal of an individual pixel cannot respond instantaneously to a change in irradiance because of the RC time constant of the photodiode capacitance and the series resistance. A typical value for this time constant is a fraction of a millisecond. The fixed pattern noise of these devices due to pixel nonuniformity is very large, around 50–100%, and cannot be reduced by correlated double sampling because of the continuous response. However, it is static and can be greatly reduced using correction values stored in a look-up table implemented in PROM, so that fully corrected monochrome or color cameras with 1024×1024 pixels are available with Fuga sensors.

20.8.7 Vision Systems

Machine vision is used in a wide variety of applications including manufacturing operations, measurements in science and engineering, remote surveillance, and robotic guidance. In machine vision systems the principal imaging component is not just a sensor chip, but a complete camera, like those shown in

FIGURE 20.111 Three solid state cameras are shown. The nearest is an inexpensive single board camera with a CMOS sensor and 4-mm lens. The middle one is a miniature CCIR interline frame-transfer CCD camera, dwarfed by the 9.5- to 75-mm C-mount zoom lens. The rear camera is a high quality Cohu 4712 monochrome frame-transfer CCD camera fitted with a 16-mm C-mount lens and 20-mm extension tube.

Figure 20.111, including sensing array, associated electronics, output signal format, and lens. Depending on the application the camera could be RS-170/CCIR monochrome, NTSC/PAL color, progressive scan, variable scan, or line scan. Five major system parameters which govern the choice of camera are field of view, resolution, working distance, depth of field, and image data acquisition rate. Color may also be important to the application, but otherwise monochrome images are preferred because they require less memory and process faster. As a rule of thumb, for size measurement applications, the sensor should have a number of pixels at least equal to twice the ratio of the largest to smallest object sizes of interest. Lighting should be arranged to illuminate the objects of interest so that the best possible images can be acquired. Lighting might be ambient, high-frequency fluorescent, LED, incandescent, or quartz halogen.

A frame grabber or video capture card, usually in the form of a plug-in board which is installed in the computer, is often required to interface the camera to a host computer. Camera suppliers can recommend compatible frame grabbers. The frame grabber will store the image data from the camera in on-board, or system memory, sampling and digitizing analog data as necessary. In some cases the camera may output digital data, which is compatible with a standard computer interface like USB 2.0 or IEE-1394 Fire Wire, so a separate frame grabber may not be needed. The computer is often a PC or Macintosh and should be as fast as possible to keep the time needed to process each image as short as possible, or to allow more processing to be done in the time available. Machine vision software is needed to create the program which processes the image data. This may come in many forms, including C libraries of device drivers and functions, ActiveX controls, and point and click programming environments which allow easy assembly of image processing operations. When an image has been analyzed the system must be able to communicate the result to control the process or to pass information to a database. This requires a digital I/O interface or network connection. The human eye and brain can identify objects and interpret scenes under a wide variety of conditions. Machine vision systems are far less versatile so the creation of a successful system requires careful consideration of all elements of the system and precise identification of the goals to be accomplished, which should be kept as simple as possible.

References

1. Wilson, J. and Hawkes, J. F. B., *Optoelectronics: An Introduction,* Prentice-Hall International, London, 1983.
2. Jenkins, F. A. and White, H. E., *Fundamentals of Physical Optics,* 4th ed., McGraw-Hill, New York, 1981.

3. Hewitt, H. and Vause, A. S., *Lamps and Lighting*, Edward Arnold, London, 1966.
4. Fraden, J., Pyroelectric thermometers, in *The Measurement, Instrumentation and Sensors Handbook*, Webster, J. G., Ed., CRC Press, 1999, chap. 32.
5. Schuermeyer, F. and Pickenpaugh, T., Photoconductive sensors, in *The Measurement, Instrumentation and Sensors Handbook*, Webster, J. G., Ed., CRC Press, 1999, chap. 56.
6. Sze, S. M., *Semiconductor Devices*, John Wiley and Sons, New York, 1985.
7. Ray, S. F., *Applied Photographic Optics*, 2nd. ed., Focal Press, Oxford, 1994
8. CCIR, *Characteristics of Monochrome and Colour Television Systems*, Recommendations and Reports of the CCIR, Vol. XI, Part 1: Broadcasting Service (Television), Section IIA, 1982.
9. Amelio, G. F., Charge coupled devices, *Scientific American*, 176, 1974.
10. Sheu, L. and Kadekodi, N., Linear CCDs, Advances in linear solid-state sensors, *Electronic Imaging*, August, 72, 1984.
11. Rutherford, D. A., A new generation of cameras tackles tomorrow's challenges, *Photonics Spectra*, September, 119, 1989.
12. Asano, A., MOS sensors continue to improve their image, *Advanced Imaging*, 42-44f, 1989.
13. Dierickx, B., Meynants, G., and Scheffer, D., Near 100% fill factor in CMOS active pixels, in *Proc. IEEE Workshop on Charge-Coupled & Advanced Image Sensors*, Brugge, Belgium, P.1, 1997.
14. Ricquier, N. and Dierickx, B., Pixel structure with logarithmic response for intelligent and flexible imager architectures, *Microelectronics Engineering*, 19, 631, 1992.

20.9 Integrated Microsensors

Chang Liu

20.9.1 Introduction

The purpose of this section is to provide the general audience in the mechatronics field with information about micro-integrated sensors. It is my wish that an avid reader interested in the sensors area would be able to understand common fabrication techniques and sensing principles, and develop rudimentary background to guide the selection of commercialized sensors and development of custom sensors in the future.

Contents for this section are organized as follows. First, the general history of microsensors is discussed. This is followed by a brief discussion about major fabrication methods for microsensors. Commonly used principles for sensors are reviewed next. Sensing of a physical parameter of interest can be achieved using various structures and under different sensing principles. Examples of sensors, along with their structures and fabrication techniques, are provided to familiarize the readers with the configurations and fabrication methods for each.

The microsensors research area covers diverse disciplines such as materials, microfabrication, electronics, and mechanics. A comprehensive coverage of all aspects is beyond the scope of this book. We will focus on a few primary sensing principles and frequently used sensors examples. A reader would be able to grasp a glimpse of the sensors field from the perspectives of sensing principles and of application areas. References for further in-depth studies are provided when appropriate.

20.9.1.1 Definition of Integrated Microsensors

Integrated microsensors refer to sensors or arrays of sensors that are developed using microfabrication technology. The characteristic length scale of individual microsensors ranges between 1 μm and 1 mm. Nanosensors refer to sensors with characteristic length scale on the order of 1 nm to 1 mm. In this text, we are mainly concerned about sensors for detecting physical variables such as force, pressure, tactile contact, acceleration, rotation, temperature, and acoustic waves. Chemical sensors, used for sensing the concentration of chemicals or pH values, are beyond the scope of this book.

A brief historical overview of microsensors development is presented here. The microsensors are made possible by using integrated microfabrication technology, first developed for making integrated circuits. Since the invention of the first transistor in 1947 and the successful demonstration of the integrated circuits in 1971, technologies and equipment for building integrated and miniature circuit components on semiconductor substrates (e.g., silicon and GaAs) have improved rapidly. The integration density doubles every 12–18 months following the empirical Moore's law. The integrated circuit technology revolutionized the modern society by enabling low cost analog signal processors, digital logical units, computer memories, and CCD cameras. These achievements should serve as evidence of the power of integration and miniaturization technology.

Advanced signal processors such as analog ASIC (application specific integrated circuits) and CPU (central processing units) are merely one aspect of a highly intelligent mechtronics system. Sensors are of critical importance for mechantronics systems to interact with the physical world. In the 1970s, a few researchers experimented with using IC fabrication technology to realize mechanical transduction elements on a silicon chip. H. Nathansan [1] developed floating gate transistors where the gate is made of a suspended cantilever beam and its distance to the conducting channel can be adjusted using electrostatic forces. Work by several pioneering researchers resulted in the first commercial pressure sensors [2], accelerometers [3], integrated gas chromatometers, as well as the ink jet printer nozzle array [4,5].

There are several important advantages associated with integrated microsensors compared with conventional macroscopic sensors. Miniaturization of sensors means that such sensors offer better spatial resolution. In many cases, reduced inertia and thermal mass translate into higher mechanical resonant frequency and lower thermal time constants. Since such sensors are fabricated using photolithography methods, their costs can be low, as many identical units are made in parallel (if the demand is sufficiently large). Further, since the geometric features of sensor components are defined by precision photolithography, the uniformity and repeatability of the performance of such sensors are significantly improved over conventional sensors. The capability to monolithically integrate sensors and integrated circuits reduces the path length between sensors and circuits and increases the signal-to-noise ratio.

20.9.1.2 Fabrication Process of Integrated Microsensors

As mentioned previously, the microfabrication technology for microsensors was developed based on integrated circuits-compatible platforms. As a result, silicon has historically been a predominant material for microintegrated sensors. In other words, a majority of sensors are now made with silicon wafer as a substrate. However, in recent years, new materials such as polymers are being applied to microfabrication. Polymer materials offer lower costs than single crystal silicon wafers and, in some cases, simpler processability, compared with semiconducting silicon. There are few examples of sensors that are entirely based on polymer these days, because certain key sensing elements are not yet available in polymer format. However, with the advancement of polymer microfabrication technology and organic semiconductors, all-polymer sensors can be predicted for future use.

Microsensors and mechanical elements (notably cantilevers and diaphragms) can be made from silicon substrates in a variety of ways. The two primary categories of fabrication methods are called bulk micromachining and surface micromachining. In bulk machining, a portion of the silicon substrate is removed using chemical wet etch or plasma-assisted dry etch to render freestanding mechanical members. For silicon substrates, the following wet chemical etchants are frequently used: potassium hydroxide (KOH), ethylene-diamine pyrocatecol (EDP), or tetramethyl ammonium hydroxide (TMAH). Dry etching methods use AC-excited plasma as an energetic source to selectively and anisotropically remove the substrate materials. A review of etching solutions commonly used in the silicon microfabrication industry and their respective etch rates on various materials can be found in reference 6.

In surface micromachining methods, a freestanding structure typically resides within a thin region near the substrate surface. The fabrication process involves only layers on the surface of a substrate, hence the name surface micromachining. The fabrication process is typically referred to as a sacrificial etching method as well. First, a thin-film solid layer called the sacrificial layer is placed on the wafer surface. This is followed by the deposition of a structural layer, which constitutes the mechanical structure (e.g., cantilever or membrane). An etchant that selectively etches the sacrificial layer with a much greater rate than the structural layer is used to remove the sacrificial layer without damaging the structure layer, leaving the structural material freestanding.

In the ensuing section, the fabrication process related to specific examples of sensors will be discussed to illustrate specific uses of bulk and surface micromachining methods. Both surface and bulk micromachining techniques offer advantages and disadvantages. For example, the surface micromachining process is generally compatible with established integrated circuit foundries because no substrate etching is involved. The bulk micromachining also involves somewhat lengthy substrate removal. However, a bulk silicon micromachining process is capable of realizing single crystal silicon structures with extremely low intrinsic mechanical stress and bending. For development of custom sensors, the selection of a fabrication process must be done carefully to achieve desired device characteristics and fabrication yield and to reduce overall sensor costs.

20.9.1.3 Resources Regarding Sensors

The community that develops microfabrication methods and microintegrated sensors frequently publishes archival results in the *Journal of Microelectromechanical Systems (MEMS), Journal of Sensors and Actuators,* and the *Journal of Micromechanics and Microengineering*. Major conferences in this area include: (1) the IEEE workshop on solid-state sensors and actuators (held biannually at the Hilton Head island, South Carolina); (2) the International Conference on Solid-State Sensors and Actuators (held biannually at international venues); and (3) International Conference on Micro-Electro-Mechanical Systems (held annually at international venues).

Interested readers can find more in-depth discussions on microsensors in a number of reference books [7,8] and websites [9].

20.9.2 Examples of Micro- and Nanosensors

20.9.2.1 Basic Sensing Principles

Microsensors are based on a number of transduction principles, including electrostatic, piezoresistive, piezoelectric, and electromagnetic (including optical sensing). The fundamental principles of these sensing methods are discussed in the following.

Electrostatic Sensing

In electrostatic sensing, a physical variable of interest, such as force or vibration, is transduced into mechanical displacement of a cantilever beam or a membrane. A schematic diagram of a typical transduction structure is shown below (Figure 20.112). The moving cantilever or membrane forms a capacitor with a reference, typically immobile, electrode. For the structure shown in the diagram below, the displacement of the top plate induces changes of the capacitance. The electrostatic sensing is also commonly referred to as capacitive sensing. The changes in capacitance are used to provide information about the parameter of interests. The electrostatic sensing principle can be used for accelerometers, acoustic sensors, rotation gyros, pressure sensors, tactile sensors, and infrared sensors. Respectively in these examples, the external excitations responsible for member movements are inertia force, air mass vibration, Coriolis force, pressure, contact force, and thermal bimetallic bending due to absorbed energy and increased temperature.

The electromechanical model of a simple capacitive sensor shown in Figure 20.112a is illustrated in Figure 20.112b. The top electrode is supported by two suspension beams with a combined equivalent force constant (spring constant) of k. The capacitance value of a parallel plate capacitor is expressed as

$$C = \frac{\varepsilon_r \varepsilon_0 A}{d}$$

where A is the area of the electrode, d is the distance between two electrodes, ε_r and ε_0 are the relative permitivity of the media and the permitivity of vacuum, respectively. When the distance between the

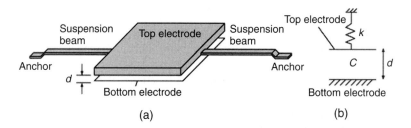

FIGURE 20.112 Schematic diagram of a parallel plate capacitor: (a) perspective view, (b) electromechanical model.

two electrodes changes by Δd, the first-order estimate of the change in capacitance is

$$\Delta C = \frac{C}{d}\Delta d$$

When a force F is applied to the top plate, the change of capacitance value is

$$\Delta C = \frac{C}{d}\frac{F}{k}$$

Piezoresistive Sensing

In a piezoresistive sensor, the magnitude of a mechanical displacement is measured by the amount of stress it induces in a mechanical member. A stress-sensitive resistor (called a piezoresistor) located strategically on the mechanical member experiences a change of resistance as a result of the applied stress. Many materials, including metals, alloys, and doped silicon, exhibit piezoresistive characteristics. The applied stress causes the lattice of a material to deform, thereby inducing changes in the resistivity as well as the dimensions of a resistor. The change of resistance (ΔR) as a function of applied strain ε is

$$\frac{\Delta R}{R_0} = G\varepsilon$$

where R_0 is the value of the resistor in the unstressed state, and G is the piezoresistive gauge factor.

Using doped silicon as a piezoresistive sensor, the overall footprint of the sensor can be made quite small and yet have a respectable value, that is, 1 kΩ. Unlike the capacitive sensor method, which requires significant plate area to achieve significant capacitance value, the piezoresistive sensor is more area efficient. As a result, the piezoresistive sensing is more likely to be used for sensors with characteristic length below 10 μm. However, the capacitive measurement method is more generally applicable, whereas the optimal piezoresistive sensors involve silicon with proper doping concentration.

Piezoelectric Sensing

A piezoelectric material is one that produces electrical polarization (internal electric field) when an external mechanical strain is applied. A piezoelectric material also exhibits reverse piezoresistivity. Namely, a mechanical strain (or displacement) will result when a voltage (or electric field) is applied on the material itself. The reverse piezoelectricity is commonly used as an actuation principle for producing mechanical movement or force.

One advantage of piezoelectric sensing over piezoresistive sensing lies in the fact that piezoelectric sensors are self-generating, that is, a potential difference will be created without external power supply. However, high quality piezoelectric films with consistent and uniform performance characteristics require dedicated machinery and calibrated processes. Such a technical barrier prevents thin film piezoelectric materials to be as widely used as piezoresistive elements. However, high quality piezoelectric films are increasingly becoming available through commercial services. Commonly used piezoelectric materials in microfabricated sensors include sputtered zinc oxide and lead zirconate titanate (PZT).

Temperature Sensing

Temperature sensing is used not only for measuring the temperature of the ambient but also for inferring heat transfer. Temperature sensors can be made of thermal resistors or thermal couples. A thermal resistor is a thin film resistive element whose resistance changes as a function of the temperature. This is explained by changes in resistivity as well as dimensions. Doped semiconductor materials (e.g., single crystal silicon or polycrystalline silicon) exhibit temperature coefficients of resistors (TCR, denoted α) on the order of $-0.1\%/°C$ to $5\%/°C$. For a thermal resistor, the normalized changes in resistance (R) are related to the change in temperature (T) by

$$\frac{\Delta R}{R} = \alpha T$$

Thermal couples are made of two different materials with different Seebeck coefficients. The voltage induced by a single thermal couple junction is proportional to the difference in temperatures at the junction and of the ambient. For more information about thermal couples, readers can refer to references.

Temperature sensing is also commonly achieved using a thermal bimetallic beam. A composite beam made of two materials with different thermal expansion coefficients will bend as the two parts expand with different speed. The amount of mechanical bending, sensed electrostatically or using piezoresistors, corresponds to the applied temperature. Such sensing principle has been used for making uncooled infrared sensors.

20.9.2.2 Pressure Sensors

Pressure sensors are important for industrial and automotive control and monitoring. Existing micropressure sensors consist of diaphragms that deform in response to pressure differences. Using micromachining technology, the diaphragms can be made very thin, hence greatly increasing the sensitivity of sensors over conventional pressure sensors with thick diaphragms. Integrated microfabrication technology also enables sensors to be made in conjunction with signal-processing circuits. Since the distance between the sensor diaphragm and the signal processors are close, the noise is generally much lower compared with conventional sensors.

The diaphragm is a critical element in a pressure sensor. It can be made by either surface micromachining or bulk micromachining techniques. Hence, we classify micropressure sensors according to the methods of forming the diaphragms. In each category of pressure sensors, the displacement of the diaphragm can be determined by using piezoresistive sensing or capacitive sensing.

Over the past two decades, many micromachined pressure sensors have been developed, some commercialized successfully for automotive and machinery applications. The intent of this section is not to present an exhaustive summary of all the work that has been accomplished, but rather discuss several representative devices with the purpose of (1) providing the readers with a general overview of the available technologies; and (2) providing leads to the existing body of literature in this area.

Bulk Micromachined Pressure Sensors

The schematic diagram of a bulk micromachined pressure sensor is illustrated in the diagram below. The diaphragm will bend when a differential pressure is applied across it. A common technique for sensing the diaphragm displacement is by using piezoresistors embedded in the diaphragms. However, it should be noted that other sensing principles are also feasible.

In the diagram below (Figure 20.113), four piezoresistors are embedded near the edge of the diaphragm. The locations of sensors are selected carefully such that under a given displacement at the center of the diaphragm, the magnitude of the stress is the greatest at the sensor locations. There are a number of choices for the diaphragms and the piezoresistors. A number of possible materials and their relative merits are summarized in the Table 20.7. Three distinct pressure sensor architectures and their respective fabrication processes are discussed in the following paragraphs.

Sensors

TABLE 20.7 A List of Possible Materials for Diaphragm and the Piezoresistive Sensors

Diaphragm Material	Piezoresistor Material	Relative Merits
Single crystal silicon	Doped single crystal silicon	Relatively difficult to control the thickness of the diaphragm
Silicon nitride thin film	Polycrystalline silicon	Easy to form thin diaphragms; involved LPCVD polysilicon

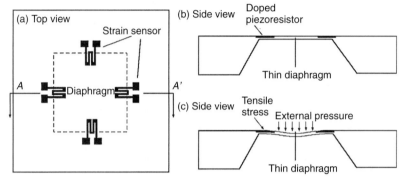

FIGURE 20.113 Schematic diagram of a bulk micromachined pressure sensor.

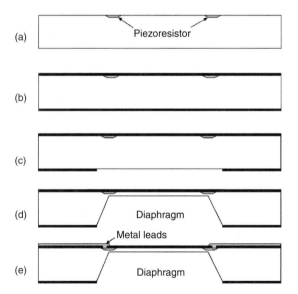

FIGURE 20.114 Schematic diagram illustrating major steps in the microfabrication process of a bulk micromachined pressure sensor.

The fabrication process for a pressure sensor using plain silicon wafer as the substrate is shown in Figure 20.114. In the first step, the wafer is selectively doped with boron or phosphorous atoms to create piezoresistors on the front side (a). The wafer is then passivated with a thermally grown silicon dioxide thin film (b). In the ensuing step, the silicon dioxide film on the backside is patterned and selectively etched to expose the silicon (c). The exposed silicon material will be etched when the wafer is immersed in an anisotropic silicon etchant (d). In order to form the silicon diaphragm with desired thickness,

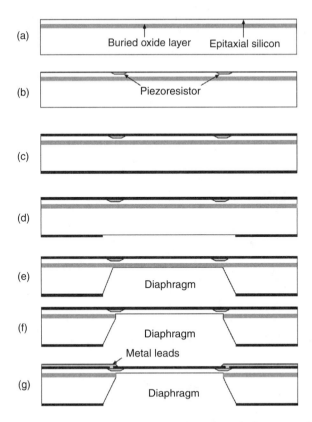

FIGURE 20.115 Schematic diagram of an alternative process for realizing bulk micromachined pressure sensors.

workers resort to timed etch with precise knowledge of calibrated wafer thickness and etch rates. However, this step must be performed with caution as the etch rate may vary with time and locations on a wafer. Typically, the resultant thickness of the diaphragm is large (30–50 μm) to ensure sufficient yield of devices. In the final step, the oxide on the front side of the wafer is patterned to provide contact vias for metal lead wires (e).

To circumvent the problem of process uncertainty of the aforementioned process, wafers with barrier layers can be used (Figure 20.115). For example, it is possible to use a silicon-on-insulator (SOI) wafer with a thin film of silicon on top of a silicon dioxide layer. The silicon and the oxide layers lie on top of the bulk silicon substrate (a). Following steps similar to the ones discussed above, one can form piezoresistors (b) and open windows in silicon oxide on the backside of the wafer (d). The anisotropic etchant of silicon has minimal etch rate on the silicon oxide, hence the through-wafer etch will automatically stop when the buried oxide layer is exposed. This allows a professional engineer to perform adequate overetch to ensure that diaphragms on all devices reach the same thickness (e). This self-limiting etching behavior reduces the complexity of process control and is conducive to reducing the process costs. The oxide layer is then selectively removed using hydrofluoric acid, which does not etch silicon. Hence a thin silicon diaphragm, with the thickness defined by the thickness of the epitaxial silicon layer specified during the SOI wafer manufacturing, can be formed efficiently. Finally, via holes are opened on the frontside and metal leads are deposited and patterned (g).

Although this process is advantageous over the one introduced earlier, it has a few shortcomings. For example, although the process discussed above is much more efficient in terms of controlling the diaphragm thickness, the SOI wafers used in the process are more expensive than ordinary silicon wafers. Even with SOI wafers, the thickness of the silicon diaphragm is typically 2–10 μm. In order to further

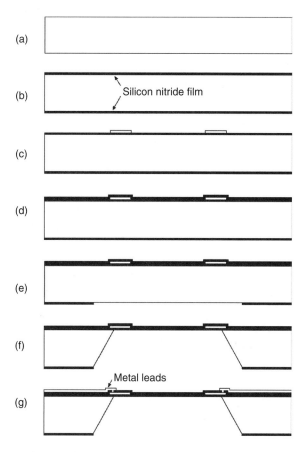

FIGURE 20.116 Schematic diagram of major process steps for realizing a bulk micromachined pressure sensor with a silicon nitride diaphragm.

increase the pressure sensitivity, it is advantageous to reduce the thickness of the diaphragm further. However, this would be difficult to achieve if silicon is the diaphragm material. In the following, a process of using silicon nitride thin film will be discussed.

An alternative sensor structure uses silicon nitride thin film as the diaphragm and deposited and doped polycrystalline silicon as the piezoresistive sensor. The process is described in Figure 20.116. Starting with a bare silicon wafer (a), a layer of silicon nitride film is deposited using LPCVD methods (b). The wafer is then coated with a layer of polysilicon with suitable doping concentration (c). The polysilicon is patterned and defined. This is followed by the deposition of yet another thin film silicon nitride to protect the polysilicon film during the ensuing silicon etching (d). The thickness of the two LPCVD silicon nitride layers is the thickness of the finished diaphragm. A window is opened on the backside of the wafer to expose the silicon material. The silicon is etched in an anisotropic etchant, which does not attack the silicon nitride film. In other words, the selectivity between silicon and silicon nitride is high. Following the formation of the diaphragm, the silicon nitride on top of the polysilicon resistors is selectively patterned and metal leads are formed (g).

Surface Micromachined Pressure Sensors

The surface micromachining process does not require the removal of silicon substrate, which is time-consuming and not fully compatible with integrated circuit processes at the present because of the silicon etchants used. For producing low-cost, high-performance integrated sensors, surface micromachining

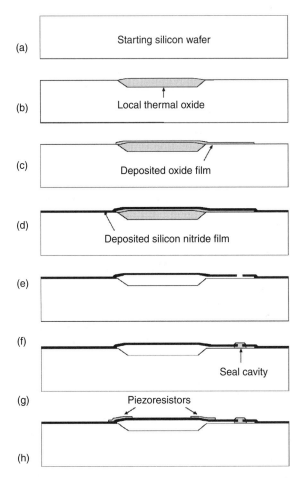

FIGURE 20.117 Schematic diagram of major steps for making a surface micromachined pressure sensor with silicon nitride diaphragm.

offers important advantages. An examplary surface micromachining process is described in the following paragraph.

The fabrication process for a surface micromachined pressure sensor is shown in Figure 20.117. It starts with a silicon substrate (a) with the front side polished. A local thermal oxidation process is performed first to form a silicon oxide well with a typical thickenss of 1.3 μm. The thermal oxide is part of the sacrificial layer that will be removed at a later stage. Using a process called low-pressure chemical vapor deposition (LPCVD), a thin layer of oxide is again deposited over the wafer surface. This oxide layer is patterned using the photolithography method (c). The entire wafer is coated with a silicon nitride thin film deposited by LPCVD technique as well (d). The silicon nitride film is patterned and etched to produce an access hole on top of the underlying oxide layer (e). Through this access hole, hydrofluoric acid removes the oxide materials inside the cavity. The etch rate of the acid on silicon nitride is negligible (f). After the cavity is emptied and dried, another layer of LPCVD silicon nitride is deposited to seal the opening in the original silicon nitride layer (g). Following this step, polycrystalline silicon with suitable doping concentration is deposited on top of the wafer and patterned to form the piezoresistors (h).

It should be noted that piezoresistive sensing, though dominant in the methods reviewed, is not the only sensing mechanism available. Capacitive sensing and piezoelectric sensing are also feasible and have been demonstrated in the past. However, discussions of these methods are beyond the scope of this text.

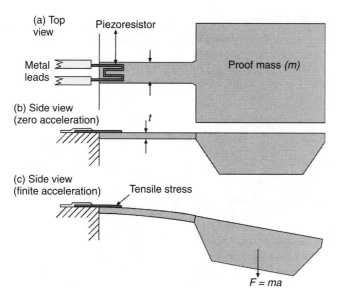

FIGURE 20.118 Schematic diagram of a bulk micromachined accelerometer.

Pressure Sensors Made of Non-Silicon Materials

For certain applications such as monitoring of internal combustion engine, pressure sensors are required to sustain high temperature of operation. In such cases, silicon is not the optimal material because high temperature causes doped silicon junctions to fail.

Work has also been done to implement polymer materials for pressure sensors. Though such devices are relatively few, they represent an important development trend for future sensors.

20.9.2.3 Accelerometers

Bulk Micromachined Accelerometers

Acceleration sensors (or so-called inertial measurement units, IMU) are important for monitoring acceleration and vibration experienced by a subject, such as an automobile, a machine, or a building. Low-cost accelerometers used in automobile airbag deployment systems can reduce the costs and enhance driver safety. Micromachined sensors can be made small and sufficiently low-cost to be used in smart projectiles, for example, concrete penetrating bombs. Small, multi-axial accelerometers can also be applied in writing instruments (smart pens) for handwriting recognition.

A representative bulk micromachined accelerometer is illustrated in Figure 20.118. A SEM micrograph of a prototype sensor is shown in Figure 20.119. A silicon proof mass is attached to the end of a cantilever beam. At the base of the cantilever beam lies a piezoresistive element. Supposing the mass of the proof mass is m, and the magnitude of the acceleration is a, one can estimate the sensor output following a few simple analysis steps. First, a concentrated force with a magnitude of $F = ma$ is applied in the center of the proof mass according to Newton's first law. Secondly, the force translates into a torque loading at the base of the cantilever with the magnitude being

$$M = F\left(l + \frac{L}{2}\right) = ma\left(l + \frac{L}{2}\right)$$

The magnitude of the strain experienced at the surface of the cantilever beam, where the piezoresistors are located, is

$$\varepsilon = \frac{Mt}{2EI}$$

FIGURE 20.119 A SEM micrograph of a prototype bulk micromachined pressure sensor (Junjun Li).

Here, the term t is the thickness of the beam, E is the modulus of elasticity of the cantilever beam material, and I is the momentum of inertia associated with the beam cross section. Supposing the cross section of the cantilever beam is a rectangle with a width w and a thickness t, the moment of inertia is

$$I = \frac{wt^3}{12}$$

Note that the moment of inertia is strongly related to the thickness of the beam. If the thickness of beam is reduced to half, the magnitude of I is reduced by eight times, and the sensitivity of the sensor increases by eight fold.

Surface Micromachined Accelerometers
Surface micromachined accelerometers offer the potential advantage of ready integration with signal processing circuits. As a result, various types of surface micromachined versions have been made in the past decade. A successful commercial product has been made by analog devices for sensing automobile acceleration to deploy airbags in the events of collision. The structure, operational principle, and fabrication process for such a sensor is briefly discussed in this section.

The sensor consists of two sets of interdigited comb-finger-shaped electrodes as shown in Figure 20.120. One set of fingers is stationary and fixed to the substrate. Another set if suspended by cantilever springs to the substrate. Capacitors are formed between each pair of comb-like fingers. When an external acceleration is applied along the horizontal axis, an inertia force is applied to the moving set of fingers and causes the moving fingers to displace. The amount of displacement is related to the magnitude of the acceleration and the force constant of the supporting springs. The relative motion of the two sets of fingers result in changes of the overall capacitance value between the two sets of fingers. The minute capacitance change is sensed and processed by a signal-processing circuit consisting of an $\Sigma - \Delta$ A/D conversion stage [10].

The fabrication process for such a sensor according to the A–A cross-section is illustrated in Figure 20.121. First, transistors for signal processing circuits are first made on a silicon substrate (a). A sacrificial silicon dioxide layer is deposited onto the wafer surface (b), followed by the deposition of a polycrystalline silicon

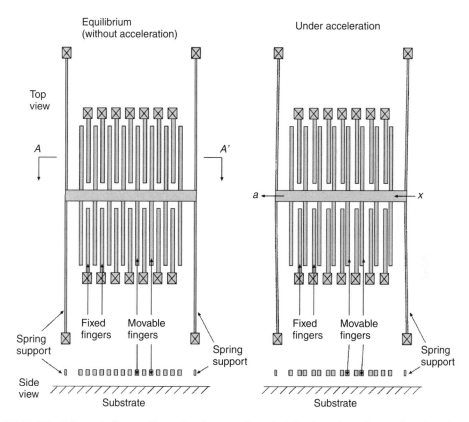

FIGURE 20.120 Schematic diagram illustrating the operation principle of a surface micromachined accelerometer.

layer (c). The polycrystalline silicon is patterned and etched to form the comb fingers (e). Subsequently, the oxide layer is removed by using a web etchant (hydrofluoric acid) that etches polycrystalline with negligible rates. In areas where the polysilicon is anchored to the substrate, a via hole is patterned and etched in the sacrificial layer before step (c).

20.9.2.4 Tactile Sensors

Tactile sensors are most widely used for robotics applications to provide tactile sensations for object handling. The sensor density on a human fingertip is on the order of $100/cm^2$. Such a high sensor density can be achieved using microfabrication technology.

An arrayed tactile sensor is illustrated in Figure 20.122. A two-dimensional array of individual sensor elements provides two-dimensional mapping of contact force and shear force. The schematic cross-sectional diagram of an array in contact with an arbitrary object is shown in Figure 20.122b. As an object contacts a sensor beam, the amount of displacement corresponds to the contact force as well as the surface topology.

The fabrication process of the tactile sensor is discussed in the following and illustrated in Figure 20.123. Starting with a silicon wafer (a), a local ion implantation is first conducted to produce piezoresistors (b). A thermal oxide film is grown to provide passivation to the entire wafer. The oxide layer on the bottom of the wafer is patterned and etched to expose silicon substrates (c). An anisotropic silicon etch is performed to remove silicon from the backside of the wafer (d). The oxide film on the front of the wafer is then patterned and etched using plasma anisotropic etch to create free-standing cantilever beams (e and f). Metal thin film is then deposited and patterned to provide lead wires (g).

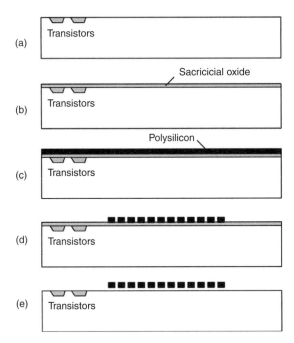

FIGURE 20.121 Schematic diagram of the fabrication process for a surface micromachined accelerometer illustrated in the previous figure.

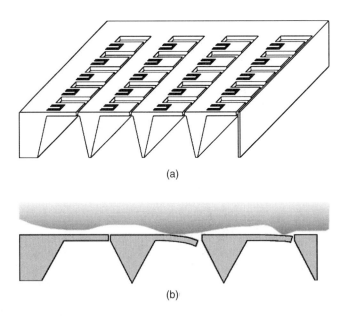

FIGURE 20.122 Schematic diagram of an array tactile sensor: (a) perspective view, (b) cross-sectional view.

20.9.2.5 Flow Sensors

Sensors for monitoring the flow rate of fluid (air or liquid) and for measuring the drag force exerted on an object moving in a fluid have important applications in robotics applications. Existing flow sensors are based on a number of principles, notably thermal and momentum transfer principles.

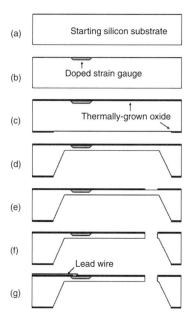

FIGURE 20.123 Schematic diagram of the microfabrication process for realizing a tactile sensor.

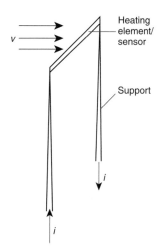

FIGURE 20.124 Perspective diagram of a thermal-transfer based flow sensor (anemometer).

Flow Sensors Based on Heat Transfer Principles

For sensors based on thermal transfer principles, a heated element is used with temperature slightly above the temperature of the ambient fluid (Figure 20.124). The heat is generally created by passing current through a resistive element. An ideal element to serve as the heating element is doped polysilicon resistor. The resistivity is generally lower than what can be achieved using metal resistors of the same dimension, hence the resistance value is greater and the heating element can be made smaller.

The movement of the fluid creates velocity-dependant forced convection of heat, thereby reducing the temperature of the heated element accordingly. The temperature of the element is therefore used to provide information about the flow rate and direction. Such sensors are commonly referred to as hot-wire anemometers. Micromachined hot-wire anemometers have been demonstrated by several groups [11,12].

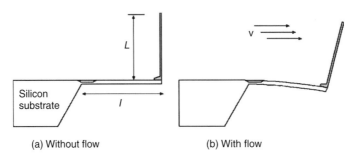

FIGURE 20.125 Schematic diagram illustrating the operation principle of a momentum-transfer based flow sensor.

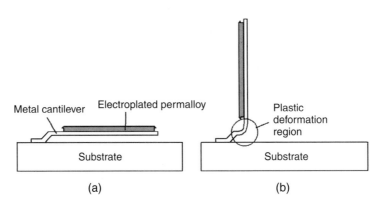

FIGURE 20.126 Schematic diagram of the plastic deformation magnetic assembly process.

Flow Sensors Based on Momentum Transfer Principles

For sensors based on momentum transfer principles, a mechanical member is bent by the momentum imparted by a moving fluid (Figure 20.125). The amount of the bending is used to decipher the strength of the fluid flow. The schematic diagram of an exemplary flow sensor is shown in the figure below. It consists of a vertical shaft attached to the end of a cantilever beam (a). When an external flow is exerted, it will apply a distributed force onto the vertical shaft, hence causing the cantilever to bend. The extent of the bending, as sensed by the embedded piezoresistor, is proportional to the average flow rate.

The fabrication process is similar to the tactile sensor except for the attachment of the integrated vertical shaft. A number of techniques for assembling three-dimensional microstructures using efficient integrated processes have been developed in the past. For example, three-dimensional structures can be realized using hinged microstructures and using solder joints or polymer joints. Recently, a process called the plastic deformation magnetic assembly (PDMA) has been developed. In the following paragraph the PDMA process is briefly discussed.

The PDMA technique is discussed using a simple surface micromachined cantilever as an example. As shown in the diagram below (Figures 20.126 and 20.127), a single-clamped cantilever made of a ductile metal (e.g., gold or aluminum) is suspended from the substrate. A piece of Permalloy, a ferromagnetic alloy made by electrodeposition, is attached to the cantilever. When a magnetic field is applied from underneath the wafer, the magnetic piece will be magnetized and will experience a magnetic torque M. The torque lifts the cantilever beam away from the substrate. If the amount of bending is significant, the ductile metal will be displaced permanently due to plastic deformation at the hinge region.

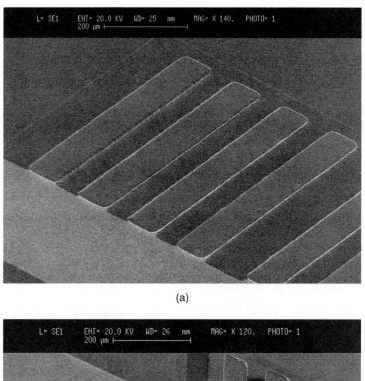

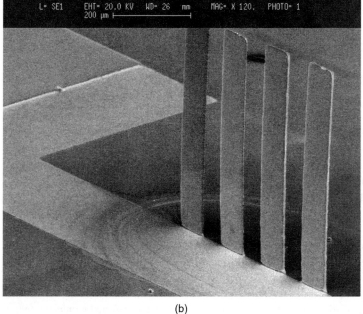

FIGURE 20.127 SEM micrographs of cantilever beams (a) while in plane, and (b) after PDMA assembly.

20.9.3 Future Development Trends

Miniaturization and integration of circuits has resulted in revolution in the society so far. It drastically reduces the costs and increased the performance of circuits. Without the integrated circuits technology, the information age would not have dawned on the human society.

It is conjectured that integrated microsensors are likely to produce as broad and deep an impact on the society as the integrated circuits. Sensors can be used for robotics sensing, smart buildings, smart toys, automotive safety and control, and industrial control. However, in order to realize the advantages of integrated sensors, a number of technical barriers must be overcome. Two important barriers are (1) high R&D costs of integrated sensors and (2) reliable and robust packaging of sensors.

The development of microintegrated sensors involves high development costs and long time-to-market. The surface micromachined accelerometer developed by analog devices costs tens of millions of dollars and took more than 5 years to produce. Why do integrated sensors cost so much to build? Sensors are developed using a group-up approach. The development cycle of a sensor begins at the level of physical principles. The cost of sensor development includes expertise for material selection, design generation, prototype process development, and characterization.

Such a development cost and speed is not tolerable in applications where only a small amount of custom sensors is required. Standard sensing modules, low-cost, flexible foundry fabrication processes, and advanced computer simulation and prototyping tools are required to advance the state-of-the-art of microintegrated sensors.

Future sensors will involve more non-silicon materials. For example, polymer materials can be used to reduce the costs while high-temperature materials maybe used for high-temperature sensing applications (e.g., monitoring of conditions in engines).

20.9.4 Conclusions

A brief historical overview of the development of microfabrication technology and microintegrated sensors is presented. Common sensing principles, including capacitive, piezoresistive, and piezoelectric sensing, are discussed. Four important case studies of sensors are undertaken. For each type of sensor applications—pressure sensors, acceleration sensors, tactile sensors, and flow sensors—the sensor architectures and fabrication processes are reviewed. Interested readers may find more in-depth information in the references provided in this section.

References

1. Nathanson, H.C., Newell, W.E., Wickstrom, R.A., and Davis J.R. Jr., "The resonant gate transistor," *IEEE Transactions on Electron Devices*, Vol. ED-14, No. 3, pp. 117–113, March 1967.
2. Petersen, K.E., "Silicon as a mechanical material," *Proceedings of the IEEE*, Vol. 70, No. 5, pp. 420–457, May 1982.
3. Angell, J.B., Terry, S.C., and Barth, P.W., "Silicon micromechanical devices," *Scientific American*, Vol., 248, pp. 44–55, April 1983.
4. Siewell, G.L., Boucher, W.R., and McClelland, P.H., "The ThinkJet orifice plate: a part with many functions," *Hewlett-Packard Journal*, May 1985, pp. 33–37.
5. Allen, R.R., Meyer, J.D., and Knight, W.R., "Thermodynamics and hydrodynamics of thermal ink jets," *Hewlett-Packard Journal*, May 1985, pp. 21–27.
6. Williams, K.R., and Muller, R.S., "Etch rates for micromachining processing," *Journal of Microelectromechanical Systems*, Vol. 5, No. 4, pp. 256–268, December 1996.
7. Kovacs, G.T.A., Micromachined transducers sourcebook, McGraw-Hill, 1998.
8. Trimmer, W.S., Micromechanics and MEMS—Classis and seminal papers to 1990, IEEE Press, 1997.
9. WWW site http://mems.isi.edu.
10 Yun, W., Howe, R.T., and Gray, P.R., "Surface micromachined, digitally force-balanced accelerometer with integrated CMOS detection circuitry," *Technical Digest, IEEE Solid-State Sensor and Actuator Workshop*, pp. 21–25, Hilton Head, SC, June 1992.
11. Jiang, F., Tai, Y.C., Ho, C.M., Karan, R., and Garstenauer, M., "Theoretical and experimental studies of micromachined hot-wire anemometers," *Technical Digest, International Electron Devices Meeting 1994*, San Francisco, CA, pp. 139–142, December 1994.
12. Ebefors, T., Kalvesten, E., and Stemme, G., "Three dimensional silicon triple-hot-wire anemometer based on polyimide joints," *Proceedings of the 11th International Workshop on MEMS*, pp. 93–98, January 1998.

21
Actuators

	21.1	Electromechanical Actuators 21-1
		Introduction • Type of Electromechanical Actuators—Operating Principles • Power Amplification and Modulation—Switching Power Electronics • References
	21.2	Electrical Machines 21-33
		DC Motor • Armature Electromotive Force (emf) • Armature Torque • Terminal Voltage • Methods of Connection • Starting DC Motors • Speed Control of DC Motors • Efficiency of DC Machines • AC Machines • Motor Selection • References
	21.3	Piezoelectric Actuators 21-51
		Piezoeffect Phenomenon • Constitutive Equations • Piezomaterials • Piezoactuating Elements • Application Areas • Piezomotors (Ultrasonic Motors) • Piezoelectric Devices with Several Degrees of Freedom • References
	21.4	Hydraulic and Pneumatic Actuation Systems 21-63
		Introduction • Fluid Actuation Systems • Hydraulic Actuation Systems • Modeling of a Hydraulic Servosystem for Position Control • Pneumatic Actuation Systems • Modeling a Pneumatic Servosystem • References
	21.5	MEMS: Microtransducers Analysis, Design, and Fabrication 21-97
		Introduction • Design and Fabrication • Analysis of Translational Microtransducers • Single-Phase Reluctance Micromotors: Microfabrication, Modeling, and Analysis • Three-Phase Synchronous Reluctance Micromotors: Modeling and Analysis • Microfabrication Aspects • Magnetization Dynamics of Thin Films • Microstructures and Microtransducers with Permanent Magnets: Micromirror Actuator • Micromachined Polycrystalline Silicon Carbide Micromotors • Axial Electromagnetic Micromotors • Conclusions • References

George T.-C. Chiu
Purdue University

Charles J. Fraser
University of Abertay Dundee

Habil Ramutis Bansevicius
Rymantas Tadas Tolocka
Kaunas University of Technology

Massimo Sorli
Stefano Pastorelli
Politecnico di Torino

Sergey Edward Lyshevski
University of Rochester

21.1 Electromechanical Actuators

George T.-C. Chiu

21.1.1 Introduction

As summarized in the previous chapter, a mechatronics system can be partitioned into function blocks illustrated in Figure 21.1. In this chapter, we will focus on the actuator portion of the system. Specifically, we will present a general discussion of the types of electromechanical actuators and their interaction

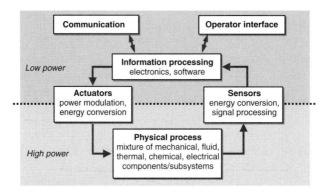

FIGURE 21.1 Mechatronic system.

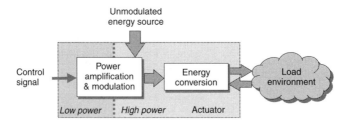

FIGURE 21.2 Actuator functional diagram.

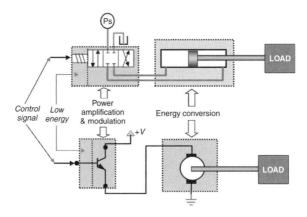

FIGURE 21.3 Electrohydraulic and electromechanical actuators.

with the load or physical environment. In addition, we will look at the electronic components that are essential for modulating the necessary electrical energy.

From an energy perspective, a mechatronic system can be separated into a relatively higher energy (power) portion that interacts with the physical world and a relatively low energy (power) portion that process the data, see Figure 21.1. Sensor and actuators are the interfacing devices that accomplish the tasks of energy modulation and energy conversion. Therefore, an actuator can be viewed as having the structure depicted in Figure 21.2. Typically, actuators are considered only as energy conversion devices. However, with the proliferation of power electronics, we will take a more inclusive view of actuators that also includes power amplification. An electrohydraulic linear actuator, see Figure 21.3, can also be similarly classified, where the spool valve is the power amplification/modulation block with spool position as the control signal and the hydraulic pressure/flow is the energy source. The hydraulic cylinder acts as the energy conversion device that converts fluidic energy to mechanical energy. For a typical electromechanical actuator, such as a DC motor (Figure 21.3), the power amplification block is the motor driver that amplifies signal level (low current/power) control signal to the higher power (large current) signal that is used to convert electrical energy to mechanical energy through the electromagnetic principle.

In this chapter, we will first present an overview of common types of electromechanical actuators. They will be classified by the respective energy conversion mechanism. The power electronic components, such as diodes, thyristors, and transistors, which are used for power amplification and modulation, will be presented followed by discussion of common power amplification building blocks. We will conclude by discussing some issues related to interfacing with electromechanical actuators.

21.1.2 Type of Electromechanical Actuators—Operating Principles

Converting electrical energy to mechanical energy is the common thread among different electromechanical actuators. Physics provided us with many different mechanisms either through direct conversion such as piezoelectric or through an intermediate medium such as a magnetic field. We will present an overview of the more common electromechanical actuators by their energy conversion mechanism: electromagnetic, electrostatic, and piezoelectric. The following discussion is intended to provide introductory information about the types of electromechanical actuation and is by no means exhaustive. Detailed discussion of each can be found in subsequent chapters, where they will be discussed in more detail.

21.1.2.1 Electromagnetics—Magnetic Field

Electromagnetic is the most widely utilized method of energy conversion for electromechanical actuators. One of the reasons for using magnetic fields instead of electric fields is the higher energy density in magnetic fields. The air gap that separates a stationary member (stator) and a moving member of an electromechanical actuator is where the electromechanical energy conversion takes place. The amount of energy per unit volume of air gap for magnetic fields can be five orders of magnitude higher than that of electric fields.

Lorentz's law of electromagnetic forces and Faraday's law of electromagnetic induction are the two fundamental principles that govern electromagnetic actuators. Before going into the detail of electromagnetics, we will first introduce the concept of magnetic field and flux.

Magnetic flux ϕ exists due the presence of a magnetic field. The magnetic field strength $\vec{H}$ (in A/m) and the magnetic flux density $\vec{B}$ (in tesla [T]) are related by the permeability of the material. In a vacuum, the magnetic flux density is directly proportional to the magnetic field strength and is expressed by

$$\vec{B} = \mu_0 \cdot \vec{H} \qquad (21.1)$$

where $\mu_0 = 4[\text{T m/A}]$ is the *permeability constant*. For other magnetic or ferromagnetic materials the relationship is given by

$$\vec{B} = \mu_r(\vec{H}) \cdot \mu_0 \cdot \vec{H} \qquad (21.2)$$

where $\mu_r(\vec{H})$ is the *relative permeability* of the material. Figure 21.4 shows typical B-H and μ-H curves.

Lorentz's Law of Electromagnetic Force

When a current carrying conductor is placed in a magnetic field, it will be subjected to an induced force given by

$$\vec{F} = \vec{i} \times \vec{B} \qquad (21.3)$$

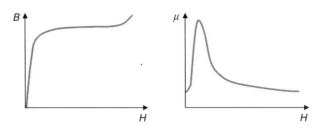

FIGURE 21.4 μ-H diagram and B-H diagram.

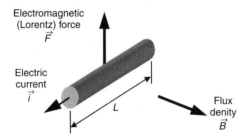

FIGURE 21.5 Lorentz's electromagnetic force.

where $\vec{F}$ is the force vector, $\vec{i}$ is the current vector, and $\vec{B}$ is the magnetic flux density. The force is called the *electromagnetic force* or the Lorentz force. If a conductor of length L carrying constant current i is placed in a constant (independent of location) field B, as shown in Figure 21.5, the magnitude of the resultant Lorentz force $\vec{F}$ exerted by the field B on the conductor is

$$F = |\vec{F}| = BLi \tag{21.4}$$

Faraday's Law of Electromagnetic Induction

The motion of a conductor in a magnetic field will produce an electromotive force (emf), or electric potential, across the conductor given by

$$\text{emf} = E = -\frac{d\phi}{dt} \tag{21.5}$$

where $\phi = \oint \vec{B} \cdot d\vec{A}$ is the magnetic flux. For a conductor of length L moving at a constant speed v in a constant (independent of location) magnetic field that is perpendicular to the area A, as shown in Figure 21.6, the magnitude of the induced electromotive force (electric potential) is

$$\text{emf} = E = BLv \tag{21.6}$$

There are two methods to generate a desired magnetic field $\vec{H}$, or equivalently, a desired magnetic flux density $\vec{B}$. One is to use a permanent magnet and the other is to utilize the Boit–Savart law.

Boit–Savart Law

A long (infinite), straight, current carrying conductor induces a magnetic field around the conductor, see Figure 21.7. The flux density at a perpendicular distance r from the conductor is

$$B = \frac{\mu_r \mu_0}{2\pi r} \cdot i \tag{21.7}$$

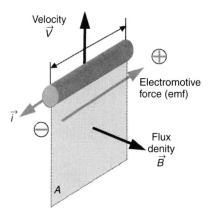

FIGURE 21.6 Motion induced electromotive force.

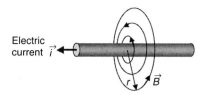

FIGURE 21.7 Magnetic field generated by current carrying conductor.

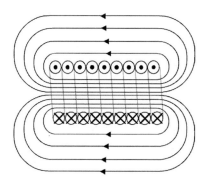

FIGURE 21.8 Coil (solenoid) induced magnetic field.

where i is the electric current. If we bend the straight current carrying conductor into a helical coil (solenoid) with N turns, it will induce a corresponding magnetic field as depicted in Figure 21.8. If the length of the coil L is much greater than its diameter D, the flux density follows the right-hand rule and the magnitude inside the coil is approximately

$$B = \mu \frac{N}{L} \cdot i \tag{21.8}$$

where $\mu = \mu_r \cdot \mu_0$ is the permeability of the material inside the coil and i is the current through the winding. This field can be intensified by inserting a ferromagnetic core into the solenoid by increasing the permeability. Coil induced magnetic fields is widely utilized in electromagnetic devices for generating controlled magnetic fields and are often referred to as *electromagnets*.

FIGURE 21.9 Assorts of solenoid actuators. (Courtesy of Shih Hsing Industrial Co., Ltd.)

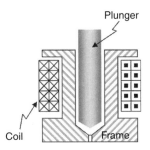

FIGURE 21.10 A typical solenoid.

Solenoid Type Devices

Solenoids, see Figure 21.9, is the simplest electromagnetic actuators that are used in linear as well as rotary actuations for valves, switches, and relays. As the name indicates, a solenoid consists of a stationary iron frame (stator), a coil (solenoid), and a ferromagnetic plunger (armature) in the center of the coil, see Figure 21.10.

As the coil is energized, a magnetic field is induced inside the coil. The movable plunger moves to increase the flux linkage by closing the air gap between the plunger and the stationary frame. The magnetic force generated is approximately proportional to the square of the applied current i and is inverse proportional to the square of the air gap δ, which is the stroke of the solenoid, i.e.,

$$F \propto \frac{i^2}{\delta^2} \tag{21.9}$$

As shown in Figure 21.11, for strokes less than 0.060 in., the flat face plunger is recommended with a pull or push force three to five times greater than 60° plungers. For longer strokes up to 0.750 in., the 60° plunger offers the greatest advantage over the flat face plunger. When the coil is de-energized, the field decreases and the plunger will return to the original location either by the load itself or through a return spring.

All linear solenoids basically pull the plunger into the coil when energized. Push-type solenoids are implemented by extending the plunger through a hole in the back-stop, see Figure 21.12. Therefore, when energized, the plunger is still pulled into the coil, but the extended producing a pushing motion from the back end of the solenoid. Return motion, upon de-energizing the coil, is provided by the load itself (i.e., the weight of the load) and/or by a return spring, which can be provided as an integral part of the solenoid assembly.

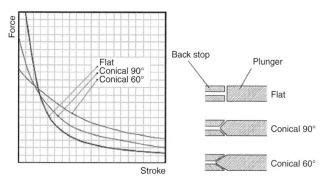

FIGURE 21.11 Typical force-stroke curve of solenoids. (Courtesy of Magnetic Sensor Systems.)

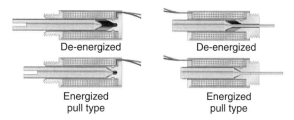

FIGURE 21.12 Push and pull type solenoids. (Courtesy of Ledex® & Dormeyer® Products.)

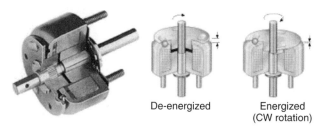

FIGURE 21.13 Rotary solenoid. (Courtesy of Ledex® & Dormeyer® Products.)

Rotary solenoids utilize ball bearings that travel down inclined raceways to convert linear motion to rotary motion. When the coil is energized, the plunger assembly is pulled towards the stator and rotated through an arc determined by the coining of the raceways, see Figure 21.13. An *electromechanical relay* (*EMR*) is a device that utilizes a solenoid to close or open a mechanical contact (switch) between high power electrical leads. A relay performs the same function as a power transistor in that relatively small electrical energy is used to switch a large amount of currents. The difference is that a relay has the capability of controlling much larger current level. Variations on this mechanism are possible: some relays have multiple contacts, some are encapsulated, some have built-in circuits that delay contact closure after actuation, and some, as in early telephone circuits, advance through a series of positions step by step, as they are energized and de-energized.

Design/Selection Considerations. Force, stroke, temperature, and duty cycle are the four major design/ selection considerations for solenoids. A linear solenoid can provide up to 30 lb of force from a unit less than 2¼ in. long. A rotary solenoid can provide well over 100 lb of torque from a unit also less than 2¼ in. long. As shown in Figure 21.11, the relationship between force and stroke can be modified by changing the design of some internal components. Higher performance, e.g., force output, can be

TABLE 21.1 Temprature Rating for Electrical Insulations

Insulation Classification		Temperature Rating	
Class A	Class 105	105°C	221°F
Class E	Class 120	120°C	248°F
Class B	Class 130	130°C	266°F
Class F	Class 155	155°C	311°F
Class H	Class 180	180°C	356°F
Class N	Class 200	200°C	392°F

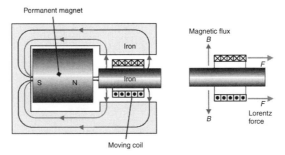

FIGURE 21.14 Voice-coil motor.

achieved by increasing the current to the coil winding. However, higher current tends to increase the winding temperature. As the winding temperature increases, the wire resistance increases. This will reduce the output force level. Solenoids are often rated as operating under continuous duty cycle or intermittent duty cycle. A solenoid rated for 100% duty cycle may be energized at its rated voltage continuously because its total coil temperature will not exceed maximum allowable ratings, while an intermittent duty cycle solenoid has an associated allowable "on" time which must not be exceeded. Intermittent duty coils provide considerably higher forces than continuous duty solenoids. The maximum operating temperature for a solenoid is determined by the rated temperature of the insulation material used in the winding (see Table 21.1).

Voice-Coil Motors (VCMs)
As the name indicates, the voice-coil motor was originally developed for loudspeakers. It is now extensively used in moving read/write heads in hard disk drives. Since the coil is in motion, VCM is also referred to as a *moving-coil* actuator. The VCM consists of a moving coil (armature) in a gap and a permanent magnet (stator) that provides the magnetic field in the gap, see Figure 21.14. When current flows through the coil, based on the Lorentz law, the coil experiences electromagnetic (Lorentz) force F

$$\vec{F} = \vec{i} \times \vec{B}$$

Since most voice-coils are designed so that the flux is perpendicular to the current direction, the resultant Lorentz force can be written as

$$F_{\text{VCM}} = \gamma BNl \cdot i = K_F \cdot i \Rightarrow F_{\text{VCM}} \propto i \qquad (21.10)$$

where l is the coil length per turn, B is the flux density, N is the number of turns in the coil, i is the current, and γ is a coil utilization factor. It is important to know that the force is proportional to the applied current amplitude and the proportional constant K_F is often called the *force constant*.

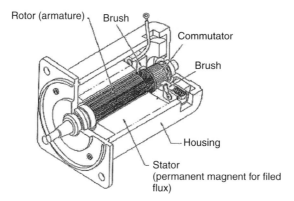

FIGURE 21.15 Permanent magnet DC motor. (From T. Keujo and S. Nagamori, *Permanent-Magnet and Brushless DC Motors,* 1985, by permission of Oxford University Press.)

The coil is usually suspended in the gap by springs and attached to the load such as the diaphragm of an audio speaker, the spool of a hydraulic valve, or the read/write head of the disk drive. The linear relationship between the output force and the applied current and the bidirectional capability makes the voice coil more attractive than solenoids. However, since the controlled output of the voice coil is force, some type of closed loop control or some type of spring suspension is needed.

Design/Selection Consideration. From Equation 21.10 we see that the force constant depends on the flux density and the amount of wires that can be packed into the gap. There are two options to increase the force constant. One is to increase the flux density, which can be achieved by using stronger magnetic material and the other is to increase either N or l, i.e., to pack more turns and/or make a larger diameter coil.

Given a fixed gap volume, using higher gauge (thinner) wires is the only way to increase the number of turns. However, higher gauge wires have larger resistance, which will increase the resistive heating of the winding and limit the allowable current. In addition, the additional insulation will also occupy more volume and tends to reduce the effect of increasing N. In summary, to improve the performance of the voice coil, a designer can either choose a better magnetic material or to make the motor bigger by either making the coil wider (increase D) or longer (increase N).

Electric Motors

Electric motors are the most widely used electromechanical actuators. They can either be classified based on functionality or electromagnetic characteristics. The differences in electric motors are mainly in the rotor design and the method of generating the magnetic field. Figure 21.15 shows the composition of a permanent magnet DC motor. Some common terminologies for electric motors are:

Stator is the stationary outer or inner housing of the motor that supports the material that generates the appropriate stator magnetic field. It can be made of permanent magnet or coil windings.

Field coil (system) is the portion of the stator that is responsible for generating the stator (field) magnetic flux.

Rotor is the rotating part of the motor. Depending on the construction, it can be a permanent magnet or a ferromagnetic core with coil windings (armature) to provide the appropriate armature field to interact with the stator field to create the torque.

Armature is the rotor winding that carries current and induces a rotor magnetic field.

Air gap is the small gap between the rotor and the stator, where the two magnetic fields interact and generate the output torque.

Brush is the part of a DC motor through which the current is supplied to the armature (rotor). For synchronous AC motors, this is done by *slip rings.*

Commutator is the part of the DC motor rotor that is in contact with the brushes and is used for controlling the armature current direction. Commutation can be interpreted as the method to control the current directions in the stator and/or the armature coils so that a desired relative stator and rotor magnetic flux direction is maintained. For AC motors, commutation is done by the AC applied current as well as the design of the winding geometry. For stepping motors and brushless DC (BLDC) motors, commutations are done in the drive electronics and/or motor commands.

Torque generation in an electric motor is either through the interaction of the armature current and the stator magnetic field (Lorentz Law) or through the interaction of the stator field and the armature field. Table 21.2 summarizes the common classification of electric motors. The next chapter will give a detailed discussion of the operation of various electric motors and the associated design considerations.

TABLE 21.2 Electric Motor Classification

Command Input	Classification — Magnetic Field		Description
DC motors	Permanent magnet		Permanent magnets are used to generate the stator magnetic field. Electrical current is supplied directly into the armature winding of the rotor through the brushes and commutators
	Electro-magnets	Shunt wound	A stator (field) winding is used as electromagnet. Stator winding is connected in parallel with the armature winding
		Series wound	A stator (field) winding is used as electromagnet. Stator winding is connected in series with the armature winding
		Compound wound	Two stator (field) windings are used as electromagnet. The stator windings are connected, one in series and one in parallel, with the armature winding
		Separate wound	A stator (field) winding is used as electromagnet. Both the stator and armature fields are individually energized
AC motors	Single-phase	Induction	Single stator winding with squirrel-cage rotor. No external connection to the rotor. Torque generation is based on the electromagnetic induction between the stator and rotor. AC current provides the commutation of the fields. Rotor speed is slightly slower than the rotating stator field (slip)
		Synchronous	Permanent magnet rotor or rotor winding with slip ring commutation. Rotating speed is synchronized with the frequency of the AC source
	Poly-phase	Induction	Similar to single-phase induction motor but with multiple stator windings. Self-starting
		Synchronous	Similar to single-phase synchronous motor but with multiple stator windings for smoother operation
	Universal		Essentially a single-phase AC induction motor with similar electrical connection as a *series wound* DC motor. Can be driven by either AC or DC source
Stepper Motors	Permanent magnet		Permanent magnet rotor with stator windings to provide matching magnetic field. By applying different sequence (polarity) of coil current, the rotor PM field will align to match induced stator field
	Variable reluctance		Teethed ferromagnetic rotor with stator windings. Rotor motion is the result of the minimization of the magnetic reluctance between the rotor and stator poles
	Hybrid		Multi-toothed rotor with stator winding. The rotor consists of two identical teethed ferromagnetic armatures sandwiching a permanent magnetic
Brushless DC motors	Poly-phase	Synchronous	Essentially a poly-phased AC synchronous motor but using electronic commutation to match rotor and stator magnetic fields. Electronic commutation enables using a DC source to drive the synchronous motor

FIGURE 21.16 MEMS comb actuator uses electrostatic actuation. (Courtesy of Sandia National Laboratories, MEMS and Novel Si Science and Technology Department, SUMMIT Technologies, www.mems.sandia.com.)

21.1.2.2 Electrostatics—Electrical Field

Since electrical fields have lower energy density than magnetic fields, typical applications of electrical field forces are limited to measurement devices and accelerating charge particles, where the required energy density is small. Recently, with the proliferation of microfabrication technology, it is possible to apply the small electrostatic forces to microelectromechanical actuators, such as comb actuators (see Figure 21.16). The advantage of electrostatic actuation is the higher switching rate and less energy loss as compared to the electromagnetic actuation. However, the limitation in force, travel, and high operating voltage still needs to be addressed. Electrostatic actuation is the main actuation for moving charged toner particles in electrophotographic (xerographic) processes, e.g., laser printers.

21.1.2.3 Piezoelectric

Piezoelectric is the property of certain crystals that produces a voltage when subjected to mechanical deformation, or undergoes mechanical deformation when subjected to a voltage. When a piezoelectric material is under mechanical stress, it produces an asymmetric displacement in the crystal structure and in the charge center of the affected crystal ions. The result is charge separation. An electric potential proportional to the mechanical strain can be measured. This is called the *direct piezoelectric effect*. Conversely, the material will have deformation without volume change when electric potential is applied. This *reciprocal piezoelectric effect* can be used to produce mechanical actuation. There are two categories of piezoelectric materials: sintered ceramics, such as lead-zirconate-titinate (PZT), and polymers, such as polyvinylidence fluoride (PVDF). Piezoceramics have a larger force output and are used more as actuators. PVDFs tend to generate larger deformation and are used more for sensor applications.

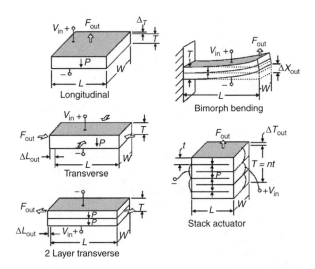

FIGURE 21.17 Common piezoelectric actuation geometries. (Courtesy of Piezo Systems, Inc.)

The coupling between the electrical and mechanical property of the material can be modeled by the following sets of linear constitutive equations:

$$\begin{cases} S = s^E \cdot T + d \cdot E \\ D = d \cdot T + \varepsilon^T \cdot E \end{cases} \Leftrightarrow \begin{cases} E = -g \cdot T + (\varepsilon^T)^{-1} \cdot D \\ S = s^D \cdot T + g \cdot D \end{cases} \quad (21.11)$$

where

- E = electric field strength [V/m]
- D = charge-density (dielectric) displacement [C/m^2]
- T = stress [N/m^2]
- S = strain
- s = compliance
- ε = permittivity [F/m]
- d, g = piezoelectric coupling coefficients

Design/Selection Considerations

Figure 21.17 shows the common orientation for piezoelectric actuation. With a typical strain of less than 0.3%, the amount of deflection or deformation is usually the limited factor for piezoelectric actuators. The most common architectures are the stacked and bending actuation. Piezoelectric actuators are most suited for high bandwidth, large force, and small stroke/deflection applications. They are widely used in noise and acoustical applications, as well as optical applications, where precision motion is critical.

Piezoelectric actuators are usually specified in terms of their free deflection and blocked force. Free deflection (X_f) refers to displacement attained at the maximum recommended voltage level when the actuator is completely free to move and is not asked to exert any force. Blocked force (F_b) refers to the force exerted at the maximum recommended voltage level when the actuator is totally blocked and not allowed to move. Figure 21.18 shows the static performance curve of a typical piezoelectric actuator (force vs. deflection). Generally, a piezo actuator must deform a specified amount and exert a specified force, which determines its operating point on the force vs. deflection line. An actuator is considered optimized for a particular

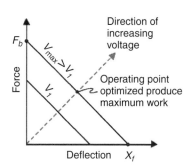

FIGURE 21.18 Static performance curve of a typical piezoelectric actuator. (Courtesy of Piezo Systems, Inc.)

application if it delivers the required force at one half its free deflection. High operating voltage, hysteresis, creep, and fatigue are the main mechanical design considerations.

21.1.2.4 Efficiency

Efficiency is one of the major considerations for any energy conversion process. In most cases, the wasted energy is converted to heat and increases the device temperature. For electromechanical actuators, heat (temperature) is one of the most prominent performance-limiting factor as well as failure mode. As device temperature increases, the underlying conversion efficiency will suffer and dump more energy into heat, which further increases the device temperature. This is often referred to as *thermal runaway*. Therefore, it is very important when designing electromechanical actuators to prevent thermal runaway and guarantee that under normal operating condition the actuator system achieves thermal equilibrium. The equilibrium temperature should be maintained below the lowest rated temperature of the components, such as the electrical insulation for the windings. The temperature rating for electrical insulations are listed in Table 21.1.

21.1.3 Power Amplification and Modulation—Switching Power Electronics

As described in the previous section and depicted in Figure 21.2, there are two main functions in an extended definition of an actuator for mechatronics systems. We have introduced a few energy conversion mechanisms and the associated actuators. In the second part of this chapter, we will focus on the power amplification and modulation portion of the actuator. This part of the actuator is traditionally called the *power amplifier* or the *driver* for the corresponding actuator. However, as miniaturization and system integration become more pervasive, power electrics are being embedded into either the controller (information processing unit) or the actuator. It is also the portion where intelligence and additional functionality/feature can be incorporated. For electromechanical actuators, the unmodulated energy source is electricity. The power amplifier acts as a buffer between the low energy part of the system, where actuation command is given in low energy electrical signals, and the high energy density electrical signal that will be converted.

Power amplification can roughly be categorized into two methods, linear and switching. The main advantage of linear power amplification is the "cleanness" of the signal as compared to the switching amplifiers. The main drawback is in efficiency, where linear amplifiers tend to run hotter than similar sized switching amplifiers. However, as with any engineering design, this is only a rule-of-thumb; the designer needs to analyze the application and select or design the appropriate driver.

Switching amplifiers are made of semiconductor components such as diodes and transistors. These semiconductor devices either function as a switching element that controls the current flow to the energy conversion element such as a winding coil, or as an amplification element that modulates the amount of current flowing into the winding coil. Another advantage of using switching type power amplifiers is that, with switching, the amplifier stage can be directly controlled by a digital signal from an information processing device (see Figure 21.1) such as a microcontroller or a microprocessor. This eliminates the need

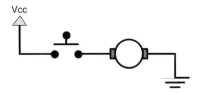

FIGURE 21.19 DC motor under switching control.

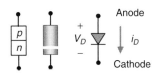

FIGURE 21.20 Diode.

for an ADC, which reduces the cost and size of the required electronics. Pulse-width modulation (PWM) is one good example using binary signal to control electromechanical actuators. Figure 21.19 shows one simple example of using a switching device to interface with a PMDC motor. As discussed in the previous section, torque is generated when current is flowing into the armature winding of a PMDC motor. To turn the motor on and off, we can connect the motor with a DC power source in series with a switch. When the switch is closed, current flows through the motor and the motor turns. If the switch is opened, current stops and the motor will eventually stop. Of course, more sophisticated circuit and switching design is needed for actual implementation.

Nevertheless, this example illustrates the fundamentals of switching power amplification. In this section, we will focus on switching amplifiers by introducing the fundamental building blocks.

21.1.3.1 Semiconductors

Semiconductors are typically materials consisting of elements from group IV of the periodic table, e.g., silicon (Si), germanium (Ge), and cadmium sulfide. Unlike conductors and insulators, semiconductors' current-carrying capability is significantly affected by the temperature and the amount of incident photons and the type and amount of impurities in the material. By introducing carefully controlled group V or III elements (called *dopants*) into the semiconductors, we can increase or decrease the number of valence electrons in the semiconductors, respectively. Depending on the type of the dopants, semiconductors can be separated into:

- *n-type semiconductors*: semiconductors doped with *donor* elements (e.g., arsenic or phosphor group V elements) that result in one additional electron freed (*free electron*) from the crystal lattice as a charge carrier that is available for conducting.
- *p-type semiconductors*: semiconductors doped with *acceptor* elements (e.g., boron or gallium group III elements) that results in a missing electron in the lattice structure, which is called a *hole*. Holes can be viewed as positive charge carriers or places that accept free electrons.

As will be discussed shortly, the interaction between the n-type and p-type semiconductors under different orientation forms the basis for all the semiconductor electronic devices. One of the more interesting aspects of modern electronics is the variety of features that can be obtained with a simple switching device that opens or closes a connection in a controlled manner. We will discuss a few electronic elements that are widely used, mainly as a controlled switching element, in power electronics for constructing power amplifiers/drivers for electromechanical actuators.

21.1.3.2 Diodes

A diode is a two-terminal electronic device that is constructed by joining a p-type and an n-type semiconductor together to form a *pn junction*. Figure 21.20 shows the schematic symbol of a generic diode. The terminal associated with the p-type material is called the *anode* and the terminal associated with the n-type material is called the *cathode*. If the anode has higher electrical potential (>0.7 V) than

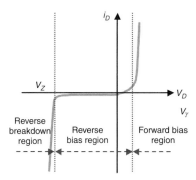

FIGURE 21.21 Diode characteristics.

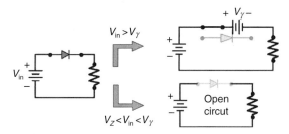

FIGURE 21.22 Approximating a diode in a circuit.

the cathode, the diode is said to be *forward biased*, i.e., $V_D > 0.7$ V. Conversely, if $V_D < 0.7$ V, the diode is *reverse biased*.

As shown in Figure 21.21, depending on the applied voltage, a diode can operate in three different regions:

- *Forward biased region*: $V_D > V_\gamma$, where V_γ is called the *forward bias voltage* and is typically around 0.7 V for silicon and 0.3 V for germanium. The diode acts as a closed switch, and the anode and cathode become short-circuited with a slight reverse potential (forward voltage drop) that is equal to V_γ, see Figure 21.22.
- *Reverse biased region*: $V_Z < V_D < V_\gamma$, where V_Z is the *reverse breakdown voltage* of the diode. The diode acts as an opened switch and the circuit is open, see Figure 21.22.
- *Breakdown region*: $V_D < V_Z$. The diode again acts as a closed switch and a large current flows through the diode. This is called the *avalanche* effect. If the magnitude of the reverse current i_D is larger than the critical reverse bias current, the device will fail.

A diode is analogous to a fluid check valve, which allows fluid (current) to flow in only one direction if the forward pressure is sufficient to overcome the spring force, see Figure 21.23. Table 21.3 summarizes properties of some typical diodes.

Maximum allowable current through a diode and the reverse breakdown voltage are the two major design considerations for diodes. The voltage across a diode times the current it carries is the power loss across the diode that is completely converted into heat. The temperature of a diode can rise rapidly due to its small size and mass. For safe operation, the temperature of the diode junction should not exceed 200°C. To improve heat transfer, diodes are commonly mounted on metallic *heat sinks*. Signal diodes are rated between $1/2$ and 1 W. Power diodes can be rated as high as several hundred kilowatts.

TABLE 21.3 Typical Diode Properties

Relative Power	Maximum Average Forward Current [A]	Voltage Drop at Maximum Average Forward Current [V]	Maximum Peak Forward Current [A]	Reverse Breakdown Voltage [V]	Maximum Junction Temperature [°C]
Low	1	0.8	30	1000	175
Medium	12	0.6	240	1000	200
High	100	0.6	1600	1000	200
Very High	1000	1.1	10,000	2000	200

Source: T. Wildi, *Electrial Machines, Drivers and Power Systems,* Prentice Hall, 2000.

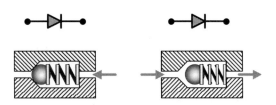

FIGURE 21.23 Analogy between diodes and check valves.

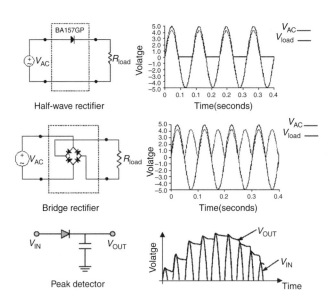

FIGURE 21.24 Common diode applications.

For switching applications, the *reverse-recovery time* is another important design parameter. The reverse-recovery time imposes an upper bound on the frequency at which the diode can be switched on and off. Attempts to operate a diode above this frequency will result in a decrease in switching efficiency and may cause severe overheating.

Diodes are widely used in electronic power circuits. They are most widely used for rectification and peak detection. Figure 21.24 illustrates some of the common diode applications. If multiple diodes are to be used, diode array can be used. In general, the term diode array implies four or more diodes in a single package. The most efficient packaging scheme is typically eight diodes or more in a dual inline package (DIP). Other packages are the single inline package (SIP), the flat pack, and even a surface mount diode array. Although multiple diode arrays can incorporate different type diodes, the most popular arrays incorporate a fast, small signal diode such as the 1N4148, and the core driver arrays, which employ a fast switching, higher current, 100-mA diode.

Zener Diode

Recall the current–voltage curve of a diode shown in Figure 21.21. If a diode is reverse biased to the breakdown region, a large reverse current will flow through the diode. For most diodes, this voltage is usually larger than 50 V and may exceed kilovolts. *Zener (Avalanche)* diodes are a class of diodes that exhibit a steep breakdown curve with a well-defined reverse breakdown voltage V_Z. This unique breakdown characteristic makes Zener diodes good candidates for building *voltage regulators*, since they can maintain a stable source voltage under variable supply as well as varying load impedance. Figure 21.25 shows the special symbol that represents a Zener diode.

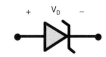

FIGURE 21.25 Zener (avalanche) diode.

To use the Zener diode as a voltage regulator, it should be reverse biased with a supply voltage higher than the rated reverse breakdown voltage V_Z, see Figure 21.26. For an ideal Zener diode, in Figure 21.26, $V_S > V_Z$, the voltage across the load will equal to V_Z; hence the load current i_{load} and the Zener diode current can be written as

$$i_{load} = \frac{V_Z}{R_{load}} \quad \text{and} \quad i_Z = \frac{V_S - V_Z}{R_S}$$

Zener diodes are often rated by their power dissipation, which is

$$P_{Zmax} = i_{Zmax} \cdot V_Z$$

Therefore, when selecting Zener diode for voltage regulation applications, it is important to ensure that i_{Zmax} does not exceed the allowable limit. The most common range of the reverse breakdown voltage for Zener diodes is from 3.3 to 75 V. However, voltages out of this range are available. Some typical power ratings for Zener diodes are 1/4, 1/2, 1, 5, 10, and 50 W.

21.1.3.3 Thyristors

A thyristor, or a *silicon-controlled rectifier* (SCR), is a 4-layer semiconductor switch, similar to a diode, but with an additional terminal to control the instant of conduction. A thyristor has three terminals: an anode, a cathode, and a gate, see Figure 21.27. One can think of a thyristor as a controllable diode that the gate terminal provides as a mean of precise control of the instance when the thyristor is to be turned on, i.e., it is a controlled switch.

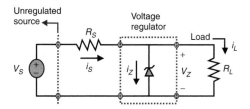

FIGURE 21.26 Use Zener diode as simple voltage regulator.

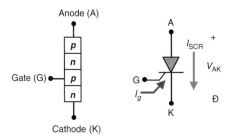

FIGURE 21.27 Thyristor and its schematic symbol.

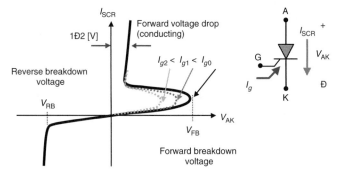

FIGURE 21.28 Thyristor (SCR) characteristics.

Figure 21.28 shows the current–voltage characteristics of a thyristor. To turn on (short or conduct) a thyristor, two conditions have to be satisfied:

1. The anode (A) and cathode (K) terminal has to be forward biased, i.e., the anode voltage needs to be higher than the cathode voltage.
2. A gate current I_G has to flow into the gate for a sufficient amount of time, typically, a few microseconds. The gate current can be generated by a short positive voltage pulse applied across the gate (G) and cathode (K) terminal. The minimum amount of gate current that is required to turn on a thyristor is called the *latching current*.

When the thyristor is turned on, the amount of current flowing through the device I_{SCR} is limited by the rest of the circuit impedance. Once the thyristor is turned on, the gate terminal loses control of the device, i.e., we cannot use the gate to turn off the device. The thyristor will only turn off if the anode current I_{SCR} goes to zero, after which the gate terminal can assert control to turn on the device again. Obviously, the thyristor can also be switched on by exceeding the forward breakdown voltage V_{FB}. However, this is usually considered a design limitation and switching is normally controlled with a gate voltage. If the gate (G) terminal is shorted with the cathode (K), the thyristor cannot be turned on, even if V_{AK} is forward biased. One can think of the thyristor as a normally opened switch with a detent. Once the switch is closed, no additional control is needed. Figure 21.29 shows the operation of a thyristor driving a simple resistive load under a sinusoidal bipolar source voltage V_S. In Figure 21.31, the gate voltage V_G will be the command or control input.

When the thyristor is reverse biased, the gate (G) to cathode (K) terminals should not be forward biased to prevent reverse breakdown of the first pn junction of the thyristor, see Figure 21.27. The reverse breakdown voltage V_{RB}, the latching current, the current and power rating, and the rate of rise of voltage are the more important design parameters for selecting a thyristor. When the voltage across the thyristor is suddenly applied or increased rapidly, the thyristor may turn on even if the gate current (voltage) is

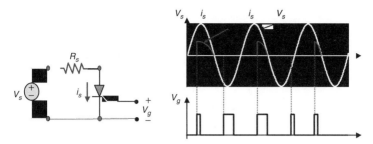

FIGURE 21.29 Thyristor driving a resistive load.

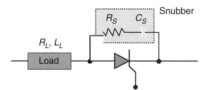

FIGURE 21.30 Snubber circuit.

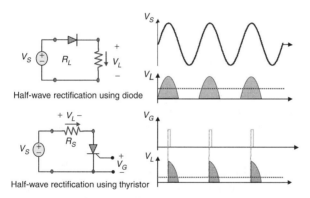

FIGURE 21.31 Controlled (thyristor) and uncontrolled (diode) rectifications.

zero. A typical rate of voltage change that will induce thyristor turn-on is about 50 V/μs. To prevent undesired conduction due to a large rate (high frequency) of voltage variation, a *snubber circuit*, see Figure 21.30, is often connected in parallel with the thyristor to filter out the high frequency voltage variations. The snubber circuit is essentially a passive RC low-pass filter. The selection of the snubber resistance R_S and capacitance C_S can use the following formula:

$$C_S = \frac{V_{Amax}^2}{L_L(dV/dt)_{max}^2} \quad \text{and} \quad R_S = 2\sqrt{\frac{L_L}{C_S}} - R_L \quad (21.12)$$

where R_L and L_L are the load inductance and load resistance, respectively. V_{Amax} is the maximum anode voltage and $(dV/dt)_{max}$ is the maximum expected rate of raise of voltage across the anode and cathode.

Unlike diodes used in rectifier circuits that can only rectify the input AC voltage to an average DC voltage, thyristors can be used to build controlled rectifiers that can rectify AC sources and modulate the average output DC voltage by modulating the firing timing of the gate voltage/current, see Figure 21.31.

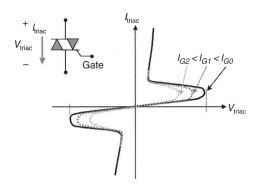

FIGURE 21.32 Triac characteristic.

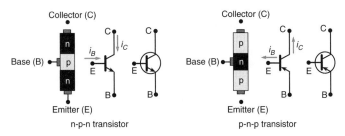

FIGURE 21.33 Biploar junction transistors (BJTs).

Triac

The thyristor can only be turned on in the forward biased direction. A *triac* is a controlled switch that is equivalent to a pair of thyristors that are connected in an anti-parallel configuration, see Figure 21.32. As depicted in Figure 21.32, a triac can be turned on in both the reverse and forward directions.

21.1.3.4 Transistors

A *transistor* is a semiconductor device that has three or more terminals and can provide power amplification and switching. As we have seen in the previous discussions, electronic switching can be accomplished through either diodes or thyristors. Diode switching does not provide any control freedom. A thyristor is a three-terminal device and the third (gate) terminal can be used to control and switching instant. However, one drawback of thyristor switching is that the switching control is only in one direction, i.e., the gate terminal can only be used to turn on the device. The thyristor switch can only be turned off by dropping the anode current to zero.

A transistor is a special semiconductor device that can be used for *power amplification* by modulating a relatively large current between or voltage across two terminals using a small control current or voltage, and *switching* by effectively opening and closing the connection between two terminals using a controlled signal on the third terminal.

Transistors form the basis of modern electronics and are the fundamental building blocks for digital electronics, operational amplifiers, and power electronics. There are three common types of transistors, bipolar junction transistors (BJTs), metal-oxide field effect transistors (MOSFETs), and insulated gate bipolar transistors (IGBTs).

Bipolar Junction Transistors (BJTs)

A bipolar junction transistor is a three-layer device that is made of the p-n-p or the n-p-n combinations of semiconductors, see Figure 21.33. BJTs have three terminals connected to each of the three layers called

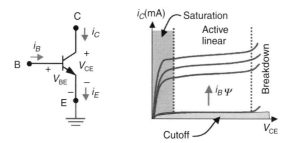

FIGURE 21.34 Characteristics of a common emitter n–p–n BJT.

collector (C), emitter (E), and base (B). Figure 21.34 shows the operation of an n–p–n type BJT under a common emitter type connection. BJTs can operate in three regions:

1. *Cutoff*—When the base-emitter voltage is less than the turn-on voltage V_γ, the base current i_B will be negligible. The transistor is in the *cutoff* region and no current will flow through the collector and emitter terminal, i.e.,

$$\begin{cases} V_{BE} < V_\gamma \\ i_B = 0 \end{cases} \Rightarrow \begin{cases} i_C \approx 0 \\ V_{CE} \geq 0 \end{cases}$$

Typically, $V_\gamma = 0.6 - 0.7$ V. In this mode, the transistor from C to E can be viewed as an open connection. This is analogous to the closed flow control valve.

2. *Active Linear*—When $V_{BE} = V_\gamma$, the transistor is in the *active linear* region, where

$$V_{BE} = V_\gamma \quad \text{and} \quad \begin{cases} i_C = \beta \cdot i_B \\ V_{CE} > V_\gamma \end{cases}$$

In this mode, the transistor can be viewed as a current-controlled current amplifier, where the collector current i_C is proportional to the base current i_B. The proportionality constant (current amplification factor or current gain) $\beta = 20 \sim 200$, is often denoted as h, hf, or h_{FE} in the data sheets. In this mode, the connection between the terminals C and E can be viewed as closed. This is analogous to a partially opened flow control valve, where the amount of the fluid (current) flow is proportional to the size of the valve opening (base current magnitude). The power dissipation across the transistor P_{BJT} is

$$P_{BJT} = i_C \cdot V_{CE}$$

3. *Saturation*—When the base current i_B is larger than the maximum available collector current i_C, the transistor is in the *saturation* region, where

$$\begin{cases} i_B > i_C/\beta \\ V_{BE} = V_\gamma \end{cases} \quad \text{and} \quad V_{CE} = V_{SAT} \approx 0.2 \text{ V} \tag{21.13}$$

In this mode, the transistor can be viewed as a closed switch between the terminals C and E. The collector current i_C is controlled (determined) by the collector circuit. This is analogous to a completely opened flow control valve, where the flow is determined by the source and the load. Note also that when the transistor is in saturation, the collector-emitter voltage drop is maintained at a small value called the saturation voltage V_{SAT}.

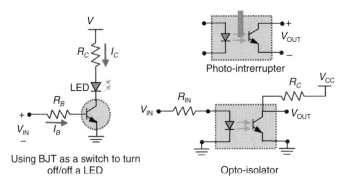

FIGURE 21.35 Some examples of using BJT.

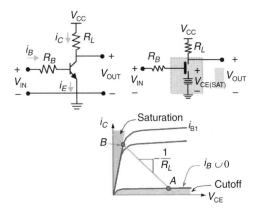

FIGURE 21.36 BJT as a current controlled switch.

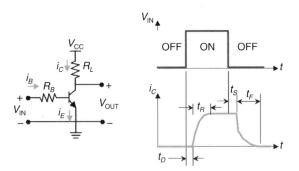

FIGURE 21.37 BJT switching characteristics.

In summary, when the transistor is saturated, it acts as a closed switch. When a transistor is in the cutoff region, it acts as an open switch. When it is in the active region, it acts as a current (i_B) controlled current (i_C) amplifier. Although this is a very simplistic approximation, it is very useful for designing and understanding power electronics and interfacing electromechanical systems. Figure 21.35 illustrates some examples of BJT devices and applications.

When carefully controlling the base-emitter voltage V_{BE} and base current i_B, the transistor can be made to operate between the cutoff and the saturation region, which act as a switch, see Figure 21.36. Realistically, transistor switching is not instantaneous (see Figure 21.37). The turn-on time t_{ON} of the transistor is the

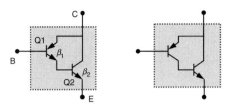

FIGURE 21.38 Two type of Darlington transistor pairs.

sum of the delay time t_D and the rise time t_R. Similarly, the turn-off time t_{OFF} is the sum of the storage time t_S and the fall time t_F. The turn-on and turn-off time of a transistor limits the maximum switching frequency. Typical switching frequency for a power BJT is between 2 and 20 kHz. Generally speaking, BJTs can switch at a higher frequency than thyristors but can handle less power. Power BJTs can handle currents up to several hundred amperes and V_{CE} up to about 1 kV.

Power dissipation is a key design constraint for BJTs. Recall that if the BJT is used in the active linear region (linear amplifier), the power dissipation is $P_{BJT} = i_C \cdot V_{CE}$ with $V_{CE} > V_\gamma$. With a large collector current and considering the small volume and thermal mass of the device, the transistor is not very efficient when operating in the active linear region. On the other hand, when the BJT is switching between saturation and cutoff, the collector current will be small (during cutoff) and V_{CE} will be small (during saturation). The switching power dissipation is much smaller compared with the active linear mode of operation. This makes switching much more efficient.

One design consideration working with BJT is to supply adequate base current, especially when the transistor is to operate in the saturated region, see Equation 21.13. This may require large input power and may overload the input stage. As will be discussed later, this is also the main reason that BJTs are less used in switching power electronics and are being replaced by devices such as MOSFET and IGBT, which require much less control current. One solution to this constraint is to increase the current gain β. A simple and elegant implementation to increase the effective current gain of a BJT is the Darlington pair configuration.

Darlington Transistor Pairs

A *Darlington transistor pair* connects two BJT transistors to form an effective three terminal device that has increased current gain, see Figure 21.38. In Figure 21.38, let β_1 and β_2 be the current gains of the two transistors, then the relationship between the base current of transistor Q1 and the collector current of transistor Q2 is

$$i_{C2} = \beta_2 \cdot i_{B2} = \beta_2 \cdot (\beta_1 \cdot i_{B1}) = (\beta_2 \cdot \beta_1) \cdot i_{B1} = \beta_D \cdot i_{B1}$$

Therefore, the effective current gain for the Darlington transistor pair is the product of the two individual current gains, i.e., $\beta_D = \beta_1 \cdot \beta_2$. For a typical Darlington pair, this can be in the range of 500–10,000. The trade-off for using Darlington pair configuration is the additional space (real estate) needed for two transistors instead of one.

Metal-Oxide-Semiconductor Field Effect Transistor (MOSFET)

MOSFET is a type of field effect transistor (FET). FETs are voltage controlled three terminal devices respectively called drain (D), source (S), and gate (G). The terms come from the analogy of overhead tank system that uses a gate valve to control the water flow from source to drain. MOSFET uses a metal plate as the gate terminal and it is insulated from the p- or n-type silicon substrate by a thin layer of oxide (see Figure 21.39). When a gate voltage V_G is applied to the gate plate, an electrostatic field induces reverse charges at the gate and the substrate. The charges at the substrate initiate transistor type characteristics by forming either an n-type channel or a p-type channel. Hence, the n- or p-type MOSFET classifications (see Figure 21.39).

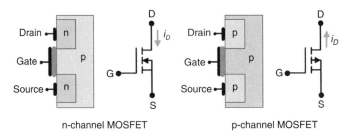

FIGURE 21.39 Metal-oxide-semiconductor (MOS) field effect transistor (FET).

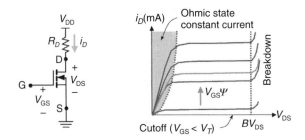

FIGURE 21.40 Enhancement mode MOSFET characteristic.

For a majority of the power amplification and modulation applications, MOSFETs are designed to operate in the *enhancement* mode. Figure 21.40 illustrates the *enhancement* mode characteristic of an n-channel MOSFET:

1. *Cutoff*—When the potential across the gate and the substrate (source) V_{GS} is less that the turn-on (threshold) voltage V_T, the MOSFET is in the *cutoff* region and there is negligible current flow through the drain (D) terminal, i.e.,

$$\begin{cases} V_{GS} < V_T \\ i_G = 0 \end{cases} \Rightarrow \begin{cases} i_D \approx 0 \\ V_{DS} \approx V_{DD} \end{cases}$$

Typically, $V_T \approx 1\text{–}2$ V. In this mode, the transistor from D to S can be viewed as an open connection.

2. *Active Region*—When the $V_{GS} > V_T$, the MOSFET is in the *active* region, where

$$V_{GS} > V_T \quad \text{and} \quad \begin{cases} i_D \propto (V_{GS} - V_T)^2 \\ V_{DS} > V_{GS} - V_T \end{cases}$$

In this mode, the transistor can be viewed as a voltage-controlled current amplifier, where the drain current i_C is proportional to square of the difference between the gate-source voltage and the threshold voltage. The drain current is controlled by the gate-source voltage V_{GS}. The power dissipation across the transistor P_{FET} is

$$P_{FET} = i_D \cdot V_{DS}$$

3. *Ohmic State*—When V_{GS} is large enough so that the drain current is determined by the drain source circuit, the MOSFET is in *saturation* and

$$V_{GS} \gg V_T \quad \text{and} \quad \begin{cases} i_D = V_{DD}/R_D \\ V_{DS} \approx i_D \cdot R_{ON}(V_{DS}) < V_{GS} - V_T \end{cases} \tag{21.14}$$

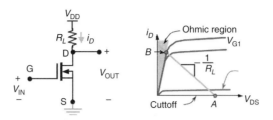

FIGURE 21.41 MOSFET as a voltage controlled switch.

In this mode, the transistor can be viewed as a closed switch between the terminals D and S with a voltage controlled resistance R_{ON}. The drain current i_D is controlled (determined) by the drain circuit. At rate current, the V_{DS} drop during saturation ranges from 2 to 5 V.

When operating in the enhancement mode, a MOSFET behaves very similar to a BJT. Instead of base current, the MOSFET behavior is determined by the gate voltage. When carefully controlling the gate voltage of a MOSFET, the transistor can be made to operate as a voltage controlled switch (Figure 21.41) that operates between the cutoff (point A) and the Ohmic (point B) region.

One advantage of a MOSFET device is that the MOSFET has significantly larger input impedance as compared to BJT. This simplifies the circuit that is needed to drive the MOSFET since the magnitude of the gate current is not a factor. This also implies that a MOSFET is much more efficient than BJTs as well as it can be switching at a much higher frequency. Typical MOSFET switching frequency is between 20 and 200 kHz, which is an order of magnitude higher than BJTs. Power MOSFETs can carry drain currents up to several hundreds of amperes and V_{DS} up to around 500 V.

Field effect is one of the key reasons why MOSFET has better switching performance than BJT. However, static field is also one of its main failure modes. MOSFETs are very sensitive to static voltage. Since the oxide insulating the gate and the substrate is only a thin film (in the order of a fraction to a few micrometer), high static voltage can easily break down the oxide insulation. A typical gate breakdown voltage is about 50 V. Therefore, static electricity control or insulation is very important when handling MOSFET devices.

Comparing BJT with MOSFET, we can conclude the following:

- Both can be used as current amplifiers.
  - BJT is a current-controlled amplifier where the collector current i_C is proportional to the base current i_B.
  - MOSFET is a voltage-controlled amplifier where the drain current i_D is proportional to the square of the gate voltage V_G.
- Both can be used as three terminal switches or voltage inverters.
  - BJT: switching circuit give rise to TTL logics.
  - MOSFET: switching circuit give rise to CMOS logics.
- BJT usually has larger current capacity than similar sized MOSFET.
- MOSFET has much higher input impedance than BJT and is normally off, which translates to less operating power.
- MOSFETs are more easily fabricated into integrated circuit.
- MOSFETs are less prone to go into thermal runaway.
- MOSFETs are susceptible to static voltage (exceed gate breakdown voltage ~50 V).
- BJT has been replaced by MOSFET in low-voltage (<500 V) applications and is being replaced by IGBT in applications at voltages above 500 V.

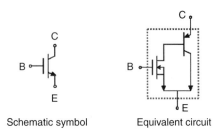

FIGURE 21.42 Insulated gate bipolar transistor (IGBT).

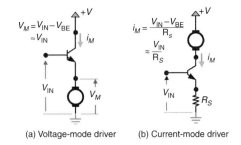

FIGURE 21.43 Two basic linear motor amplifiers.

Insulated Gate Bipolar Transistor (IGBT)

IGBT is a voltage-controlled transistor that has the terminals identified in the same way as BJTs. IGBT is a four-layer device that has the similar construction of a MOSFET with an additional p layer. Figure 21.42 shows the schematic symbol and equivalent circuit for an IGBT. IGBT has the combined characteristics of the BJT and MOSFET. Similar to MOSFET, it has high input impedance and high switching frequency. It also has high power handling capacity like the BJT.

21.1.3.5 Typical Power Amplifiers for Electromechanical Actuators

Power amplification and modulation for electromechanical actuators are classified into two basic categories, based on the methods the respective power electronics are driven. Linear amplifiers drive the BJTs in their active linear region. Switching amplifiers drive the transistors in on-off switch mode. Depending on the control objective, the command signal (Figure 21.2) to the amplifier can be either a voltage or current command that intends to modulate the electric energy delivered to the energy conversion device. Since most of the electromechanical actuator involves driving an inductance load such as a coil winding of an electromagnet or the rotor of a DC motor, in the following discussion, we will use the DC motor as an example of an inductance load for the power amplifier.

Linear Amplifiers

Figure 21.43 shows the basic drive circuit for linear voltage and current control amplifiers. Both schemes have the following commonalities:

1. The input command voltage $V_i(t)$ is applied to the base of the transistor.
2. The electric power needed to driver the load is provided by a DC supply.
3. The transistors are driven in the active linear region.

Voltage Control (Mode) Amplifier

In Figure 21.43a, the motor is driven as a load of an emitter circuit. If the base-emitter voltage is ignored, the voltage across the motor V_M is directly controlled by the input voltage V_{IN} and the current is supplied by the power supply. The amount of current, on the other hand, depends on the applied voltage, speed, and the motor parameter.

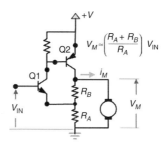

FIGURE 21.44 Variable gain voltage-mode amplifier.

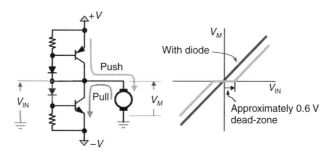

FIGURE 21.45 Bipolar voltage-mode amplifier.

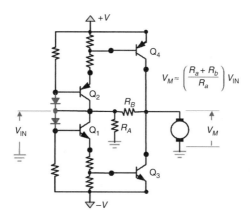

FIGURE 21.46 Bipolar variable gain voltage-mode amplifier.

To provide adjustable voltage gain, Figure 21.44 shows a variable gain voltage-mode amplifier, where the apparent amplifier gain from the command input V_{IN} to the applied motor (winding) voltage V_M can be approximated by $(R_A + R_B)/R_A$, which can be adjusted by proper selection of the resistors R_A and R_B. If a large motor current is required, transistor Q2 can be replaced by a Darlington transistor pair.

The amplifiers shown in Figures 21.43 and 21.44 can only drive the current through the motor (load) in one direction. Hence, they are also called *unipolar* amplifiers. To provide bidirectional current flow, two transistors can be connected with the motor in a *push-pull* type configuration, as shown in Figure 21.45. The two diodes in the circuit are used to eliminate the dead-zone created by the base-emitter voltage drop for the transistors. Notice that a bipolar voltage source is needed for this configuration. Figure 21.46 shows a bipolar voltage-mode driver with variable gain $(R_A + R_B)/R_A$. Similarly, if larger motor current is required, transistors Q3 and Q4 can be replaced by Darlington transistor pairs.

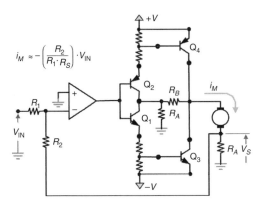

FIGURE 21.47 Bipolar variable gain current-mode amplifier.

Current Control (Mode) Amplifier

As previously discussed, in many electromagnetic actuators, the output force or torque of the device has strong correlation with the winding current, e.g., for a permanent magnet DC motor and a voice coil actuator, the output torque and force are proportional to the input current. Therefore, in many motion control applications, it is more desirable to have a voltage-to-current conversion (*current-mode amplifier*) at the power stage, where the input voltage command is proportional to the current flowing into/out of the motor (winding). Figure 21.43b shows a basic circuit for a current-mode amplifier. The relationship between the emitter (motor) current i_M and the input voltage command V_{IN} is

$$i_M = \frac{V_{IN} - V_{BE}}{R_S}$$

If the base-emitter voltage is ignored, the voltage across the motor current i_M is proportional to the input voltage V_{IN}, i.e., $i_M \approx (1/R_S) \cdot V_{IN}$.

Figure 21.47 shows a basic bipolar current-mode amplifier. An Op-Amp is used to close the current loop. The resistor R_S, often called the *sensing resistor*, is used to sense the motor current for feedback to the Op-Amp. Depending on the desired current magnitude, the sensing resistor needs to have adequate power rating to dissipate the heat ($i_M^2 \cdot R_S$) generated by flowing current through the resistor. For a zeroth order approximation, at steady state, the Op-Amp will try to equalize the potential at the positive and the negative terminals, i.e., it will try to make

$$V_S \approx -\left(\frac{R_2}{R_1}\right) \cdot V_{IN},$$

which implies

$$i_M \approx -\left(\frac{R_2}{R_S \cdot R_1}\right) \cdot V_{IN}.$$

Although a current amplifier tends to have a linear relationship between the command input and the winding current, there is practical limitation due to the limited source voltage. In Figure 21.47, the supply voltage is ±V. Assuming that the motor winding has resistance R_M, the maximum current i_{MAX} the voltage source can supply is upper bounded by

$$i_{MAX} < \frac{V}{R_M + R_S}$$

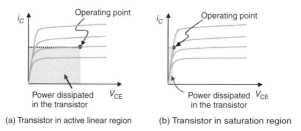

FIGURE 21.48 Power dissipation in transistors.

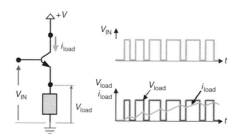

FIGURE 21.49 Simple switching amplifier with switching input.

The above bound has not considered the effect of the back-emf that will be induced in the winding if the winding is moving. Hence, the amount of current available for a current-mode amplifier is limited and needs to be considered when working with a current-mode amplifier.

Switching Amplifiers

Linear amplifiers are simple and do not generate electrical noises. However, since the final stage transistors are operating in the active linear region, significant power is dissipated into heat; this reduces the efficiency of the device as well as requires large heat sinks to protect the components. However, as shown in Figure 21.48, when operating in the saturation region, if the collector-emitter voltage drop is in the order of 1 V or less, the power loss across the transistor is significantly less, given the same amount of current flow. The trade-off is that additional circuits are needed to provide the modulation for current or voltage control.

Figure 21.49 shows a simple switching amplifier that is simply a transistor connecting a load. It is essentially the same as the basic linear amplifier shown in Figure 21.43a. The difference is in the way the transistor is controlled. For a switching amplifier, the input (base) voltage only takes on two values (states), high and low. When the base (input) voltage is high, the transistor is turned on in the saturation mode and current will flow through the load. If we neglect the collector-emitter voltage drop, the voltage across the load is approximately the supply voltage. When the base voltage is low, the transistor is turned off in the cutoff state and no voltage is applied to the load. If the load has a low pass characteristic, the average current/voltage across the load will be proportional to the turn-on time. Therefore, if the switching frequency is sufficiently high (relative to the load impedance), the effective voltage/current across the load can be modulated by the percent high input voltage, e.g., if the V_{IN} is high 80% of the time, the average voltage across the load will be close to 80% of the supplied voltage V. This is the so-called *pulse-width modulation* (PWM). Another benefit of using switching amplifiers is that V_{IN} can be directly interfaced with a digital device without the need for a DAC.

Push-Pull (Class B) Power Amplifier

The switching amplifier shown in Figure 21.49 is unipolar, i.e., it can only drive current through the load in one direction. Figure 21.52 shows a simple *push-pull* (*Class B*) type power stage to supply bi-directional current to the load. The circuit in Figure 21.50 is very similar to the bipolar voltage-mode amplifier shown

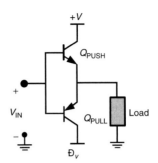

FIGURE 21.50 Switching push-pull amplifier.

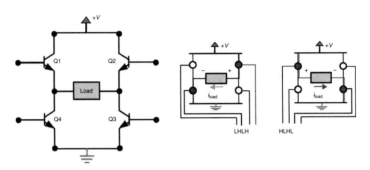

FIGURE 21.51 H-bridge driver.

in Figure 21.45. The difference is also in the way the transistors are controlled. When the base voltage V_{IN} is sufficiently positive (+V), the push transistor Q_{push} will be turned on and the pull transistor Q_{pull} will be turned off. This results in a load current flowing from positive supply to ground. If the base voltage is sufficiently negative (−V), Q_{push} will be turned off and Q_{pull} will be turned on, which results in a current flow from ground to the negative supply. To modulate the load voltage/current, PWM can also be used. This configuration is also called a *half H-bridge* driver or half-bridge driver for short. From an implementation perspective, this device requires both a positive supply and a negative supply, which tends to increase the complexity and cost of the circuit.

H-Bridge Driver
H-bridge configuration is a neat solution to achieve bipolar operation with unipolar supply. Figure 21.51 shows a simple H-bridge circuit driving a load. An H-bridge consists of four transistors that are connected in a Wheatstone bridge configuration. By turning on/off different pairs of transistors (Q1-Q3) or (Q2-Q4), bipolar voltage across the load can be achieved using a unipolar supply, see Figure 21.51. In many applications, the transistors pairs in the H-bridge can be directly driven by the output of a digital device (TTL or CMOS). The n-p-n or n-channel transistors can be turn on to saturation by a high output from the digital port and turned off by a low output. If large amount of current is required for the load, Darlington pairs can be used in place of the individual transistors. Since MOSFETs have larger input impedance and faster switching characteristics, they are replacing BJTs in almost all switching applications.

Pulse-Width Modulation (PWM)
PWM is one of the more common ways of encoding analog information using digital signal. A PWM signal is a wave of fixed frequency and varying duty cycle (pulse width). The duty cycle in PWM context refers to the percentage of time that the signal is in the active state—usually this means a state of logic 1, see Figure 21.52. In essence, PWM encodes (modulates) the information in the time domain rather than the voltage domain as with analog signals.

PWM actuation has several advantages over the use of D/A converters and linear components. One is the efficiency where the switching amplifiers are more efficient than their linear counter parts. Another

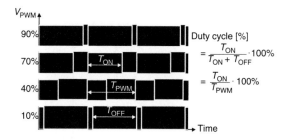

FIGURE 21.52 Pulse-width modulation signals.

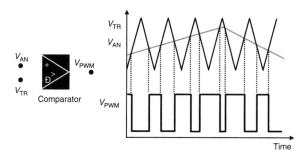

FIGURE 21.53 Generating PWM signal from analog signal.

advantage is there is no need for D/A conversion. Digital signal is maintained from the microprocessor/microcontroller to the power amplifier. In additional to having better noise rejection capability, this also reduces the need for a DAC and tends to make the circuit simpler and more cost effective. One drawback for using PWM and switching amplifiers as a whole is that the high frequency switching induces radio frequency interference (RFI) and electromagnetic interference (EMI).

The fixed PWM carrier frequency is one main design consideration. Ideally, the PWM frequency should be high enough to avoid generating audible switching noise, which mean that it should be greater than 20 kHz. However, there are a few factors that put an upper bound on the carrier frequency. Switching losses of switching devices tends to increase as the switching frequency increases. This reduces the efficiency of switching components and amplifiers. Higher PWM carrier frequency requires faster switching components that cost more. The amount of current going through the device also limits the switching rate. In general, sub-horsepower devices and office/desktop equipments usually use PWM at 20–40 kHz. For larger scale industrial applications, the PWM frequency tends to be less than 500 Hz. Another commonly specified design parameter is the PWM resolution. This is required for generating PWM from a digital source. The PWM resolution is equivalent to the quantization resolution for ADC. An 8-bit PWM means that there are $2^8 = 256$ different pulse widths per PWM carrier signal period.

PWMs are widely adopted in the field and almost all microcontrollers and microprocessors have at least one PWM output port. PWM signal can be easily generated from analog signal by comparing the analog signal with a periodic triangular signal through a comparator, see Figure 21.53.

21.1.3.6 Interfacing Considerations

We will conclude this section by discussing some issues relating to interfacing between the electromechanical actuator and the power amplification device.

Driving Inductive Load
A majority of the electromechanical actuators use coils (windings) to convert electrical energy to magnetic energy. From a power driver viewpoint, windings are resistive and inductive loads. Inductors are energy

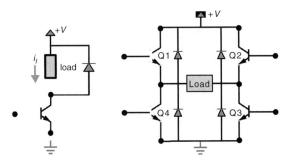

FIGURE 21.54 Using diodes to reduce swithcing voltage when driving inductive loads.

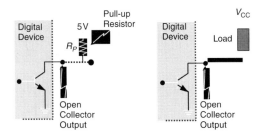

FIGURE 21.55 Open-collector output.

storage elements, where the energy is stored in the induced magnetic field. The voltage across an ideal inductor $V_l(t)$ is

$$V_l(t) = L \cdot \frac{d}{dt} i_l(t) \qquad (21.15)$$

where $i_l(t)$ is the current going through the inductor and L is the inductance. When the current to the inductor is suddenly switched off, e.g., by switching off a driving transistor, Equation 21.15 indicates that there will be a large transient voltage build-up across the inductor. If not properly suppressed this transient voltage can shorten or even damage the driving transistor. This is sometimes called *inductor kickback*.

A simple method of reducing the instantaneous switching voltage surge is to create a loop for the excess energy to flow. This can be done by placing diodes in parallel with the load, see Figure 21.54. Figure 21.54 illustrates two methods of using *flyback* or *free-wheeling* diodes to suppress switching voltage surge when driving inductive loads.

Open-Collector Output

For some digital devices, the output stage (pin) is simply the collector of a transistor. This is called an *open-collector* output, see Figure 21.55. Since the output of the device is only the collector of a transistor, it has no output drive capacity. The output value can be measured through a pull-up resistor, see Figure 21.55. Open-collector output is convenient for driving electromechanical devices if the output transistor can sink adequate current, see Figure 21.57.

Isolation

Recall that the power amplification/modulation part of an electromechanical actuator contains both low- and high-energy signals, see Figure 21.2. For safety and reliability reasons, it is desired to prevent transients or noise spikes in the high power side of the system from the signal processing (low power) side of the circuit. Mechanical relay is one option. *Optoisolators* or *optocouplers* use light to couple the high and low energy side of the device. Typically, an LED source is combined with either a phototransistor or photo thyristor, see Figure 21.35. In addition to signal isolation, optoisolators also help to reduce ground loop issues between the logic and power side of the circuit.

Grounding

It is important to provide common ground among the different devices. For electromechanical actuators, the high energy side is often switching at high frequency; if the ground point of the high energy side of the circuit is directly connected to the ground of the low energy side of the circuit, switching noise may propagate through the ground wire and negatively affect the operation of the low energy side of the system. It is recommended that separate common grounds are established for the high and low energy side and the two grounds are then connected at the power supply. In addition, an adequate-sized ground plane needs to be provided to minimize the possibility of differences among grounding points.

21.2 Electrical Machines

Charles J. Fraser

The utilization of electric motors as the power source in a mechatronic application is substantial. Electric motors, therefore, often feature as the prime mover in a variety of driven systems. It is usually the mechanical features of the application that determines the type of electric motor to be employed. The torque–speed characteristics of the motor and the driven system are therefore very important. It is perhaps then a paradox that while the torque–speed characteristics of the motor are readily available from the supplier, the torque–speed characteristics of the driven system are often quite obscure.

21.2.1 DC Motor

All conventional electric motors consist of a stationary element and a rotating element, which are separated by an air gap. In DC motors, the stationary element consists of salient "poles," which are constructed of laminated assemblies with coils wound round them to produce a magnetic field. The function of the laminations is to reduce the losses incurred by eddy currents. The rotating element is traditionally called the "armature" and this consists of a series of coils located between slots around the periphery of the armature. The armature is also fabricated in laminations, which are usually keyed onto a location shaft. A very simple form of DC motor is illustrated in Figure 21.56.

The single coil is located between the opposite poles of a simple magnet. When the coil is aligned in the vertical plane, the conventional flow of electrons is from the positive terminal to the negative terminal. The supply is through the brushes, which make contact with the commutator segments. From Faraday's laws of electromagnetic induction, the "left-hand rule," the upper part of the coil will experience a force acting from right to left. The lower section will be subject to a force in the opposite direction. Since the

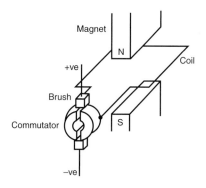

FIGURE 21.56 Single-coil, 2-pole DC motor.

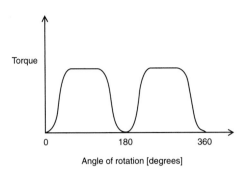

FIGURE 21.57 Torque variation through one revolution.

coil is constrained to rotate, these forces will generate a torque, which will tend to make the coil turn in the anti-clockwise direction. The function of the commutator is to ensure that the flow of electrons is always in the correct direction as each side of the coil passes the respective poles of the magnet. The commutator incorporates brass segments, separated by insulating mica strips. The carbon brushes make sliding contact with the commutator.

When the coil lies in the horizontal direction, there is maximum magnetic flux linking the coil but a minimum rate of change of flux linkages. On the other hand, when the coil is in the vertical plane, there is zero flux linking the coil but the rate of change of flux linkages is a maximum. The resultant change in torque acting on the coil through one revolution is as shown in Figure 21.57.

If two coils physically displaced by 90° are used in conjunction with two separate magnets, also displaced by 90°, then the output torque is virtually constant. With the introduction of a second coil, the commutator needs to have four separate segments. In a typical DC machine there may be as many as 36 coils, which would require a 72-segment commutator.

The simple DC motor of Figure 21.56 can be improved in perhaps three obvious ways. Firstly, the number of coils can be increased, the number of turns in each coil can be increased, and finally the number of magnetic poles can be increased. A typical DC machine would therefore normally incorporate four poles, wired in such a way that each consecutive pole has the opposite magnetic polarity to each of its immediate neighboring poles. If the torque generated in the armature coils are to assist one another then while one side of the coil is passing under a north pole, the other side must be passing under a south pole. With a two-pole machine the armature coils are wound with one side of the coil diametrically opposite the other. In a four-pole machine the coils are wound such that one side of the coil is displaced 90° from the other. The size of the machine will generally determine how many coils and the number of turns on each coil which can be accommodated.

21.2.2 Armature Electromotive Force (emf)

If a conductor cuts a magnetic flux, a voltage of 1 V will be induced in the conductor if the flux is cut at the rate of 1 Wb/s. Denoting the flux per pole as Φ and the speed (in rev/s), as N, for a single turn coil and two-pole machine, the emf induced in the coil is given as

$$E_{coil} = \frac{\text{flux per pole}}{\text{time for half rev}} = \frac{\Phi}{1/2N} = 2N\Phi \qquad (21.15)$$

For a machine having Z_s armature conductors connected in series, i.e., $Z_s/2$ turns, and $2p$ magnetic poles, the total induced emf is

$$E = \frac{2N\Phi Z_s 2p}{2} = 2N\Phi Z_s p \qquad (21.16)$$

The induced emf or back emf will oppose the applied voltage. Since the emf is directly proportional to the motor speed then on startup, there will be no back emf generated. This will have consequences on the current, which will be drawn by the coils, and some measures will have to be taken to counteract this effect. This topic will be considered later.

21.2.3 Armature Torque

The force on a current carrying conductor is given as

$$F = BLI \tag{21.17}$$

where B is the magnetic flux density under a pole, I is the current flowing in the conductor, and L is the axial length of the conductor.

The torque on one armature conductor is, therefore,

$$T = Fr = B_{av} LI_a r \tag{21.18}$$

where r is the radius of the armature conductor about the center of rotation, I_a is the current flowing in the armature conductor, L is the axial length of the conductor, and B_{av} is the average flux density under a pole.

Given that $B_{av} = \Phi/[(2\pi rL)/2p]$, the resultant torque per conductor is

$$T = \frac{\Phi 2pLI_a r}{2\pi rL} = \frac{\Phi pI_a}{\pi} \tag{21.19}$$

For Z_s armature conductors connected in series, the total torque (in Nm) on the armature is given by

$$T = \frac{\Phi pI_a Z_s}{\pi} \tag{21.20}$$

21.2.4 Terminal Voltage

Denoting the terminal voltage by V, in normal running conditions we have a balanced electrical system where:

$$V = E + I_a R_a \tag{21.21}$$

Since the number of poles and number of armature conductors are fixed, then from Equation 21.16 we have a proportionality relationship between the speed, the induced emf, and the magnetic flux, i.e.,

$$N \propto \frac{E}{\Phi} \tag{21.22}$$

Using Equation 21.21

$$N \propto \frac{(V - I_a R_a)}{\Phi} \tag{21.23}$$

Since the value of $I_a R_a$ is normally less than about 5% of the terminal voltage then to a reasonable approximation

$$N \propto \frac{V}{\Phi} \tag{21.24}$$

Similarly Equation 21.19 provides a proportionality relationship between the torque, the armature current, and the magnetic flux, i.e.,

$$T \propto I_a \Phi \tag{21.25}$$

Equation 21.24 shows that the speed of the motor is directly proportional to the applied voltage and inversely proportional to the magnetic flux. All methods of speed control for DC motors are based on this proportionality relationship.

Equation 21.25 indicates that the torque of a given DC motor is directly proportional to the product of the armature current and the flux per pole. It is obvious therefore that speed control methods which are based on altering the magnetic flux will also have an effect on the output torque.

21.2.5 Methods of Connection

21.2.5.1 The Shunt-Wound Motor

The shunt-wound motor (Figure 21.58) is wired such that the armature and field coils are connected in parallel with the supply.

Under normal operating conditions, the field current will be constant. As the armature current increases, the armature reaction effect will weaken the field and the speed will tend to increase. However, the induced voltage will decrease due to the increasing armature voltage drop and this will tend to decrease the speed. The two effects are not self cancelling and overall the motor speed will fall slightly as the armature current increases.

The motor torque increases approximately linearly with the armature current until the armature reaction starts to weaken the field. These general characteristics are shown in Figure 21.59 where it can also be seen that no torque is developed until the armature current is large enough to overcome the constant losses in the machine. Figure 21.60 shows the derived torque-speed characteristic.

Since the torque increases dramatically for a slight decrease in speed, the shunt-wound motor is particularly suitable for driving equipment like pumps, compressors, and machine tool elements where the speed must remain "constant" over a wide range of load conditions.

21.2.5.2 The Series-Wound Motor

The series-wound motor is shown in Figure 21.61. As the load current increases, the induced voltage, E, will decrease due to the armature and field resistance drops. Because the field winding is connected in series with the armature, the flux is directly proportional to the armature current. Equation 21.24 therefore

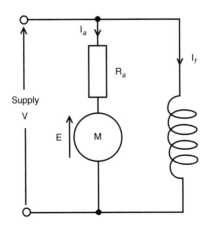

FIGURE 21.58 The shunt-wound motor.

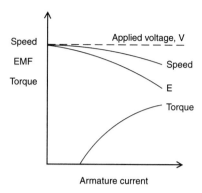

FIGURE 21.59 The shunt-wound motor load characteristics.

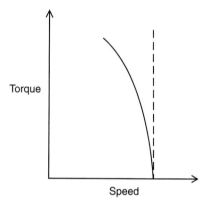

FIGURE 21.60 The shunt-wound torque–speed characteristics.

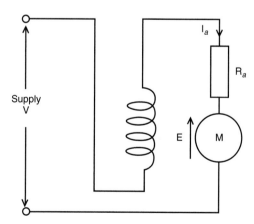

FIGURE 21.61 The series-wound motor.

suggests that the speed–armature current characteristic will take the form of a hyperbola. Similarly, Equation 21.25 indicates that the torque–armature current characteristic will be approximately parabolic. These general characteristics are illustrated in Figure 21.62, along with the derived torque–speed characteristic in Figure 21.63. The general characteristics indicate that if the load falls to a particularly low value

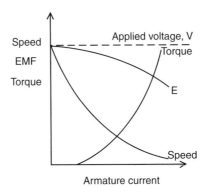

FIGURE 21.62 The series-wound motor load characteristics.

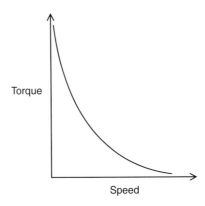

FIGURE 21.63 The series-wound motor torque–speed characteristics.

then the speed may become dangerously high. A series-wound motor should never be used, therefore, in situations where the load is likely to be suddenly relaxed.

The main advantage of the series-wound motor is that it provides a large torque at low speeds. Series-wound motors are eminently suitable, therefore, for applications where a large starting torque is required. This includes, for example, lifts, hoists, cranes, and electric trains.

21.2.5.3 The Compound-Wound Motor

Compound-wound motors are produced by including both series and shunt fields. The resulting characteristics of the compound-wound motor fall somewhere in between those of the series-wound and the shunt-wound machines.

21.2.6 Starting DC Motors

With the armature stationary, the induced emf is zero. If while at rest, the full voltage is applied across the armature winding then the current drawn would be massive. A typical 40-kW motor might have an armature resistance of about 0.06 Ω. If the applied voltage is 240 V, the current drawn is 4000 A. This current would undoubtedly blow the fuses and thereby cut off the supply to the machine. To limit the starting current a variable external resistance is connected in series with the armature. On start-up the full resistance is connected in series. As the machine builds up speed and increases the back emf, the external resistance can be reduced until at rated speed the series resistance is disconnected. Alternatively, a series resistance can be momentarily activated in conjunction with the starter switch.

Actuators

21.2.7 Speed Control of DC Motors

Equation 21.24 shows that the speed of a DC motor is influenced both by the applied voltage and the magnetic flux. A change in either one of these parameters will therefore effect a change in the motor speed.

21.2.7.1 Field Regulator

For shunt-wound and compound-wound motors a variable resistor, called a "field regulator," can be incorporated in series with the field winding to reduce the flux. For the series-wound motor the variable resistor is connected in parallel with the field winding and is called a "diverter." Figures 21.64–21.66 show the various methods of weakening the field flux for shunt-, compound-, and series-wound motors.

In all of the above methods, the flux can only be reduced and from Equation 21.24 this implies that the speed can only be increased above the rated speed. The speed may in fact be increased to about three or four times the rated speed. The increased speed, however, is at the expense of reduced torque since the torque is directly proportional to the flux which is being reduced.

21.2.7.2 Variable Armature Voltage

Alternatively, the speed can be increased from standstill to rated speed by varying the armature voltage from zero to rated value. Figure 21.67 illustrates one method of achieving this.

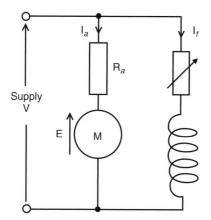

FIGURE 21.64 Speed control by flux reduction: shunt-wound motor.

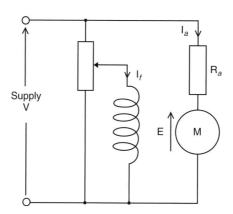

FIGURE 21.65 Speed control by flux reduction: compound-wound motor.

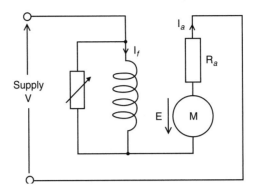

FIGURE 21.66 Speed control by flux reduction: series-wound motor.

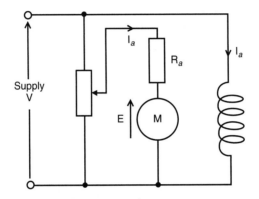

FIGURE 21.67 Speed control by varying the armature voltage.

The potential divider carries the same current as the motor, and this limits this method of speed control to small machines. Additionally much of the input energy is dissipated in the variable resistance, which consequently renders the system inefficient.

21.2.7.3 Ward Leonard Drive

In this case the variable DC voltage for the speed controlled motor is obtained from a separate DC generator, which is in itself driven by an induction motor.

The field coil for the DC generator is supplied from a center-zero potential divider. When the wiper arm is in the center position, the speed controlled motor is at a standstill. By moving the wiper arm away from the center position the speed of the motor is increased in either clockwise or anti-clockwise direction. The Ward Leonard drive is smooth and accurate in either direction and also provides for very responsive braking. The complexity, however, makes it a very expensive system, and it is only used in high quality applications.

21.2.7.4 Chopper Control

Figure 21.68 shows a thyristor circuit connected in series with the armature of a DC motor. The thyristor circuit is triggered such that it operates essentially as a high speed ON/OFF switch. The output waveform across the armature terminals is depicted in Figure 21.69. The ratio of time on to time off, i.e., the "mark/space ratio," can be varied with the result that the average voltage supplied to the armature is effectively varied between zero and fully on. The frequency of the signal may be up to about 3 kHz and the timing circuit is quite complex. Speed control of DC motors using thyristors, however, is effective and relatively inexpensive.

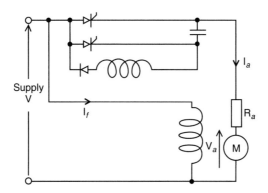

FIGURE 21.68 Speed control using thyristors.

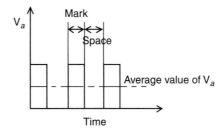

FIGURE 21.69 Voltage across armature terminals.

21.2.8 Efficiency of DC Machines

The losses in DC machines can be generally classified as

1. **Armature losses:** This is the I^2R loss in the armature winding, often referred to as the "copper loss."
2. **Iron loss:** This loss is attributable to magnetic hysteresis and eddy currents in the armature and field cores.
3. **Commutator losses:** This loss is related to the contact resistance between the commutator brushes and segments. The total commutator loss is due to both mechanical friction and a voltage loss across the brushes.
4. **Excitation loss:** In shunt-wound machines, this power loss is due to the product of the shunt current and the terminal voltage.
5. **Bearing friction and windage:** Bearing friction is approximately proportional to the speed, but windage loss varies with the cube of the speed. Both of these losses are fairly minor unless the machine is fitted with a cooling fan, in which case the windage loss can be quite significant.

Despite the variety and nature of the losses associated with DC machines, they have nonetheless a very good performance with overall efficiencies, often in excess of 90%.

21.2.9 AC Machines

21.2.9.1 Synchronous Motors

Synchronous motors are so called because they operate at only one speed, i.e., the speed of a rotating magnetic field. The production of the rotating magnetic field may be actioned using three, 120° displaced,

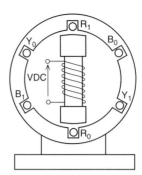

FIGURE 21.70 Simple synchronous motor.

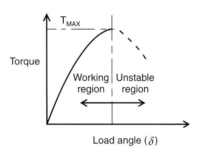

FIGURE 21.71 Torque characteristic for a synchronous motor.

stator coils supplied with a three-phase current. The rotational speed of the field is related to the frequency of the currents.

$$N_s = \frac{60f}{\text{number of pole pairs}} \quad (21.26)$$

where N_s is the speed of the field in revolutions per minute and f is the frequency of the supply currents.

The mechanical construction is shown in Figure 21.70. The rotor field is supplied from a DC source and the stator coils are supplied with a three-phase current. The rotating magnetic field is induced by the stator coils and the rotor, which may be likened to a permanent bar magnet, aligns itself to the rotating flux produced in the stator. When a mechanical load is driven by the shaft, the field produced by the rotor is pulled out of alignment with that produced by the stator. The angle of misalignment is called the "load angle." The characteristics of synchronous motors are normally presented in terms of torque against load angle, as shown in Figure 21.71.

The torque characteristic is basically sinusoidal with

$$T = T_{max} \sin \delta \quad (21.27)$$

where T_{max} is the maximum rated torque and δ is the load angle.

It is evident from Equation 21.27 that synchronous motors have no starting torque and the rotor must be run up to synchronous speed by some alternative means. One method utilizes a series of short-circuited copper bars inserted through the outer extremities of the salient poles. The rotating magnetic flux induces currents in these "grids" and the machine accelerates as if it were a cage-type induction motor, see following section. A second method uses a wound rotor similar to a slip-ring induction motor. The machine is run up to speed as an induction motor and is then pulled into synchronism to operate as a synchronous motor.

The advantages of the synchronous motor are the ease with which the power factor can be controlled and the constant rotational speed of the machine, irrespective of the applied load. Synchronous motors,

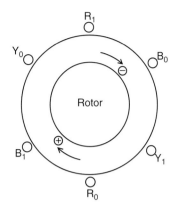

FIGURE 21.72 Schematic representation of an induction motor.

however, are generally more expensive and a DC supply is a necessary feature of the rotor excitation. These disadvantages coupled with the requirement for an independent starting mode make synchronous motors much less common than induction motors.

21.2.9.2 Induction Motors

The stator of an induction motor is much like that of an alternator and in the case of a machine supplied with three-phase currents, a rotating magnetic flux is produced. The rotor may be either of two basic configurations, which are the "squirrel cage" or the slip-ring type. In the squirrel cage motor the rotor core is laminated and the conductors consist of uninsulated copper, or aluminium, bars driven through the rotor slots. The bars are brazed or welded at each end to rings or plates to produce a completely short-circuited set of conductors. The slip-ring machine has a laminated core and a conventional three-phase winding, similar to the stator, and connected to three slip-rings on the locating shaft. Figure 21.72 shows a schematic representation of an induction motor having three stator coils displaced by 120°.

If the stator coils are supplied with three-phase currents, a rotating magnetic field is produced in the stator. Consider the single rotor coil shown in the figure. At standstill the rotating field will induce a voltage in the rotor coil since there is a rate of change of flux linking the coil. If the coil forms a closed circuit, the induced emf will circulate a current in the coil. The resultant force on the current carrying conductor is a consequence of Equation 21.17 and this will produce a torque, which will accelerate the rotor. The rotor speed will increase until the electromagnetic torque is balanced by the mechanical load torque. The induction motor will never attain synchronous speed because if it did there would be no relative motion between the rotor coils and the rotating field. Under these circumstances there would be no emf induced in the rotor coils and subsequently no electromagnetic torque. Induction motors, therefore, always run at something less than synchronous speed. The ratio of the difference between the synchronous speed and the rotor speed to the synchronous speed is called the "slip"; i.e.,

$$s = \frac{N_s - N}{N_s} \tag{21.28}$$

The torque–slip characteristic is shown in Figure 21.73. With the rotor speed equal to the synchronous speed, i.e., $s = 0$, the torque is zero. As the rotor falls below the synchronous speed the torque increases almost linearly to a maximum value dictated by the total of the load torque and that required to overcome the rotor losses. The value of slip at full load varies between 0.02 and 0.06. The induction motor may be regarded as a constant speed machine. The difficulties, in fact, of varying the speed constitute one of the induction motor's main disadvantages.

On start-up, the slip is equal to unity and the starting torque is sufficiently large enough to accelerate the rotor. As the rotor runs up to its full load speed the torque increases in essentially inverse proportion to the slip. The start-up and running curves merge at the full load position.

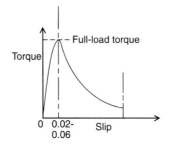

FIGURE 21.73 Torque–slip characteristic for an induction motor.

21.2.9.3 Starting Induction Motors

As with DC motors, the current drawn during starting of AC motors is very large, up to about five times full load current. A number of devices are therefore employed to limit the starting current but they all involve the use of auxiliary equipment, which is usually quite expensive.

Star-Delta Starter

With the machine at standstill and the starter in the "start" position, the stator coils are connected in the star pattern. As the machine accelerates up to running speed, the switch is quickly moved over to the "run" position, which reconnects the stator windings in the delta pattern. By this simple expedient, the starting supply current is reduced to about one third of what it would have been had the stator windings been connected up in the delta pattern on start-up.

Autotransformer Starter

The autotransformer represents an alternative method of reducing the starting current drawn by an induction motor. The autotransformer incorporates a star connection, which is supplied from a mid-point tapping on each phase. The voltage supplied to the stator is, therefore, one half of the supply voltage. With such an arrangement the supply current and the starting torque are both only one quarter of the values, which would be applied to the motor when the full voltage is supplied. After the motor has accelerated, the starter device is moved to the "run" position thereby connecting the motor directly across the supply and opening the star-connection of the autotransformer. Unfortunately, the starting torque is also reduced and the device is generally expensive since it has to have the same rating as the motor.

Rotor Resistance

With slip-ring induction motors, it is possible to include additional resistance in series with the rotor circuit. The inclusion of extra resistance in the rotor provides for reduced starting current and improved starting torque.

21.2.9.4 Braking Induction Motors

Induction motors may be brought to a standstill by either "plugging" or by "dynamic braking."

1. **Plugging:** This is a technique where the direction of the rotating magnetic field is reversed. This is brought about by reversing any two of the supply leads to the stator. The current drawn during plugging is very large, and machines which are regularly plugged must be specially rated.
2. **Dynamic braking:** In this braking method the stator is disconnected from the AC supply and reconnected to a DC source. The direct current in the stator produces a stationary unidirectional field and as the rotor will always tend to align itself with the field, it will therefore come to a standstill.

21.2.9.5 Speed Control of Induction Motors

Under normal circumstances, the running speed of an induction motor will be about 94–98% of the synchronous speed, depending on the load. With the synchronous speed given by Equation 21.26, it is clear that the speed may be varied either by changing the frequency of the supply current, or by changing the number of poles.

Change of Supply Current Frequency

Solid state variable-frequency drives first began to appear in 1968. They were originally applied to the control of synchronous AC motors in the synthetic fiber industry and rapidly gained acceptance in that particular market. In more recent times they have been used in applications to pumping, synchronized press lines, conveyor lines, and to a lesser extent in the machine-tool industry as spindle drives. Modern AC variable-frequency motors are available in power ratings ranging from 1 to 750 kW and with speed ranges from 10/1 to 100/1.

The synchronous and squirrel cage induction motors are the types most commonly used in conjunction with solid-state, adjustable frequency inverter systems. In operation the motor runs at, or near, the synchronous speed determined by the input current frequency. The torque available at low speed, however, is decreased and the motor may have to be somewhat oversized to ensure adequate performance at the lower speeds. The most advanced systems incorporate a digital tachogenerator to supply a corrective feedback signal which is compared against a reference frequency. This gives a speed regulation of about 3%. Consequently, the AC variable-frequency drive is generally used only for moderate to high power velocity control applications, where a wide range of speed is not required. The comparative simplicity of the AC induction motor is usually sacrificed to the complexity and cost of the control electronics.

Change of Number of Poles

By bringing out the ends of the stator coils to a specially designed switch it becomes possible to change an induction motor from one pole configuration to another. To obtain three different pole numbers, and hence three different speeds, a fairly complex switching device would be required.

Changing the number of poles gives a discrete change in motor speed with little variation in speed over the switched range. For many applications, however, two discrete speeds are all that is required and changing the number of poles is a simple and effective method af achieving this.

Changing the Rotor Resistance

For slip-ring induction motors additional resistance can be coupled in series with the rotor circuit. It has already been stated that this is a common enough method used to limit the starting current of such machines. It can also be used as a method of marginal speed control. Figure 21.74 shows the torque characteristics of a slip-ring induction motor for a range of different resistances connected in series with the rotor windings.

As the external resistance is increased from R_1 to R_3, a corresponding reduction in speed is achieved at any particular torque. The range of speeds is increased at the higher torques.

The method is simple and therefore inexpensive, but the reduction in speed is accompanied with a reduction in overall efficiency. Additionally, with a large resistance in the rotor circuit, i.e., R_3, the speed changes considerably with variations in torque.

Reduced Stator Voltage

By reducing the applied stator voltage a family of torque–speed characteristics are obtained, as shown in Figure 21.75. It is evident that as the stator voltage is reduced from V_1 to V_3, a change in speed is effected

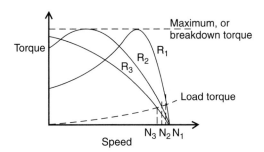

FIGURE 21.74 Torque–speed characteristics for various rotor resistances.

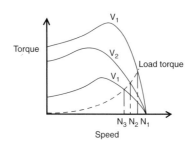

FIGURE 21.75 Torque–speed characteristics for various stator voltages.

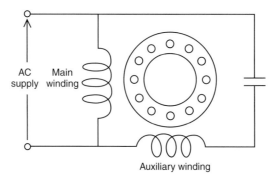

FIGURE 21.76 Capacitor motor.

at any particular value of torque. This is provided, of course, that the torque does not exceed the maximum load torque available at the reduced stator voltage. This latter point is obviously a limiting factor, which places a constraint on this method of speed control. Generally only very small speed ranges can be obtained using variable stator supply voltage.

21.2.9.6 Single-Phase Induction Motors

The operation of an induction motor depends upon the creation of a rotating magnetic field. A single stator coil cannot achieve this and all of the so-called single-phase induction motors use some or other external means of generating an approximation to a two-phase stator supply. Two stator coils are, therefore, used and these are displaced by 90°. Ideally the currents which supply each coil should have a phase difference of 90°. This then gives the two-phase equivalent of the three-phase induction motor.

The Shaded Pole Motor

The stator of the shaded pole motor consists of a salient pole single-phase winding and the rotor is of the squirrel cage type. One half of the stator features a copper "shading ring." When the exciting coil is supplied with alternating current, the flux produced induces a current in the shading ring. The phase difference between the currents in the exciting coil and the shading ring is relatively small and the rotating field produced is far from ideal. In consequence the shaded pole motor has a poor performance and an equally poor efficiency due to the continuous losses in the shading rings. Shaded pole motors have a low starting torque and are used only in light duty applications such as small fans and blowers or other easily started equipment. Their advantage lies in their simplicity and low cost of manufacture.

The Capacitor Motor

The stator has two windings physically displaced by 90°. A capacitor is connected in series with the auxiliary winding such that the currents in the two windings have a large phase displacement (see Figure 21.76). The current phase displacement can be made to approach the ideal 90°, and the performance of the capacitor motor closely resembles that of the three-phase induction motor.

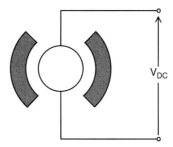

FIGURE 21.77 A DC permanent magnet motor.

The Universal Motor

These are small DC series-wound motors that operate at about the same speed and power on direct current, or on single-phase current with approximately the same root mean square voltage. If alternating current is supplied, the stator and rotor field strengths vary sinusoidally in magnitude but with the same phase relationship. As the applied voltage changes polarity, so do the armature and field currents. Equation 21.25 suggests that under these conditions the applied torque will not reverse polarity and will remain at all times positive. The universal, or plain-series motor, is used mainly in small domestic appliances such as hair dryers, electric drills, vacuum cleaners, hedge trimmers, etc.

21.2.9.7 The DC Permanent Magnet (PM) Motor

The DC permanent magnet (PM) motor is a continuous rotation electromagnetic actuator that can be directly coupled to its load. Figure 21.77 shows the schematic representation of a PM motor. The PM motor consists of an annular brush ring assembly, a permanent magnet stator ring, and a laminated wound rotor. They are particularly suitable for servo systems where size, weight, power, and response times must be minimized and where high position and rate accuracies are required.

The response times for PM motors are very fast and the torque increases directly with the input current, independently of the speed or the angular position. Multiple pole machines maximize the output torque per watt of rotor power. Commercial PM motors are available in many sizes from 35 mN m at about 25 mm diameter to 13.5 N m at about 3 m diameter.

Direct drive rate and position systems using PM motors utilize DC tachogenerators and position sensors in various forms of closed-loop feedback paths for control purposes.

21.2.9.8 The Stepper Motor

A stepper motor is a device that converts a DC voltage pulse train into a proportional mechanical rotation of its shaft. In essence, stepper motors are a discrete version of the synchronous motor. The discrete motion of the stepper motor makes it ideally suited for use with a digitally based control system such as a microcontroller. The speed of a stepper motor may be varied by altering the rate of the pulse train input. Thus, if a stepper motor requires 48 pulses to rotate through one complete revolution, then an input signal of 96 pulses per second will cause the motor to rotate at 120 rev/min. The rotation is actually carried out in finite increments of time; however, this is visually indiscernible at all but the lowest speeds.

Stepper motors are capable of driving a 2.2-kW load with stepping rates from 1000 to 20,000 per second in angular increments from 180° down to 0.75°.

There are three basic types of stepper motor, viz.

1. **Variable reluctance:** This type of stepper motor has a soft iron multi-toothed rotor with a wound stator. The number of teeth on the rotor and stator, together with the winding configuration and excitation determines the step angle. This type of stepper motor provides small to medium sized step angles and is capable of operation at high stepping rates.
2. **Permanent magnet:** The rotor used in the PM type stepper motor consists of a circular permanent magnet mounted onto the shaft. PM stepper motors give a large step angle ranging from 45° to 120°.

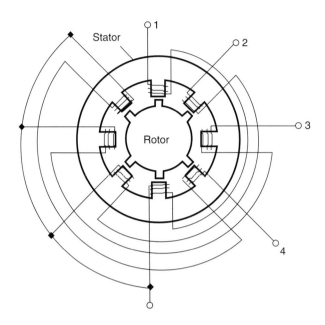

FIGURE 21.78 Variable reluctance stepper motor.

3. **Hybrid:** The hybrid stepper motor is a combination of the previous two types. Typically the stator has eight salient poles, which are energized by a two-phase winding. The rotor consists of a cylindrical magnet, which is axially magnetized. The step angle depends on the method of construction and is generally in the range 0.9°–5°. The most popular step angle is 1.8°.

The principle of operation of a stepper motor can be illustrated with reference to a variable reluctance, four-phase machine. This motor usually has eight stator teeth and six rotor teeth, see Figure 21.78.

If phase 1 of the stator is activated alone, two diametrically opposite rotor teeth align themselves with the phase 1 teeth of the stator. The next adjacent set of rotor teeth in the clockwise direction are then 15° out of step with those of the stator. Activation of the phase 2 winding on its own, would cause the rotor to rotate a further 15° in the anti-clockwise direction to align the adjacent pair of diametrically opposite rotor teeth. If the stator windings are excited in the sequence 1, 2, 3, 4, then the rotor will move in consecutive 15° steps in the anti-clockwise direction. Reversing the excitation sequence will cause a clockwise rotation of the rotor.

21.2.9.9 Stepper Motor Terminology

Pull-out torque: The maximum torque that can be applied to a motor, running at a given stepping rate, without losing synchronism.
Pull-in torque: The maximum torque against which a motor will start, at a given pulse rate, and reach synchronism without losing a step.
Dynamic torque: The torque developed by the motor at very slow stepping speeds.
Holding torque: The maximum torque that can be applied to an energized stationary motor without causing spindle rotation.
Pull-out rate: The maximum switching rate at which a motor will remain in synchronism while the switching rate is gradually increased.
Pull-in rate: The maximum switching rate at which a loaded motor can start without losing steps.
Slew range: The range of switching rates between pull-in and pull-out in which a motor will run in synchronism but cannot start or reverse.

The general characteristics of a typical stepper motor are given in Figure 21.79.

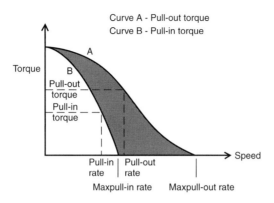

FIGURE 21.79 Stepper motor characteristics.

During the application of each sequential pulse, the rotor of a stepper motor accelerates rapidly towards the new step position. However, on reaching the new position there will be some overshoot and oscillation unless sufficient retarding torque is provided to prevent this happening. These oscillations can cause rotor resonance at certain pulse frequencies resulting in loss of torque, or perhaps even pull-out conditions. As variable reluctance motors have very little inherent damping, they are more susceptible to resonances than either of the permanent magnet, or the hybrid types. Mechanical and electronic dampers are available, which can be used to minimize the adverse effects of rotor resonance. If at all possible, the motor should be selected such that its resonant frequencies are not critical to the application under consideration.

Owing to their unique characteristics, stepper motors are widely used in applications involving positioning, speed control, timing, and synchronized actuation. They are prevalent in X-Y plotters, floppy disc head drives, printer carriage drives, numerically controlled machine tool slide drives, automatic teller machines, and camera iris control mechanisms.

By far the most severe limitation on the purely electric stepper motor is its power handling capability. Currently this is restricted to about 2.25 kW.

21.2.9.10 Brushless DC Motors

These motors have position feedback of some kind so that the input waveforms can be kept in the proper timing with respect to the rotor position. Solid-state switching devices are used to control the input signals and the brushless DC motor can be operated at much higher speeds with full torque available at those speeds. The brushless motor can normally be rapidly accelerated from zero to operating speed as a PM motor. On reaching operating speed, the motor can then be switched over to synchronous operation.

The brushless motor system consists of a wound stator, a permanent magnet rotor, a rotor position sensor, and a solid state switching assembly. The wound stator can be made with two or more input phases. Figure 21.80 gives the schematic representation of a two-phase brushless motor.

The torque output of phase A is

$$T = I_A(Z\Phi/2\pi)\sin(p\theta/2) = I_A K_T \sin(p\theta/2) \qquad (21.29)$$

where

I_A = current in phase A,
$K_T = (Z\Phi/2\pi)$ = torque constant of the motor,
p = number of poles, and
θ = angular position of the rotor.

In the expression for the torque constant, Z is the total number of conductors, and Φ is the magnetic flux.

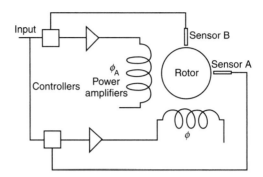

FIGURE 21.80 Two-phase brushless motor.

In a similar manner, the torque output of phase B is

$$T = I_B K_T \sin(p\theta/2) \qquad (21.30)$$

If the motor currents are arranged to be supplied in the following relationships

$$I_A = I\sin(p\theta/2) \quad \text{and} \quad I_B = I\cos(p\theta/2)$$

then the total torque for a two-pole motor becomes

$$T = T_A + T_B = IK_T[\sin^2(p\theta/2) + \cos^2(p\theta/2)] = IK_T \qquad (21.31)$$

Equation 21.31 shows that if all of the above conditions are satisfied then the brushless DC motor operates in a similar manner to the conventional DC motor, i.e., the torque is directly proportional to the armature current. Note that the armature current in this context refers to the stator windings. Excitation of the phases may be implemented with sinusoidal, or square wave inputs. The sine wave drive is the most efficient but the output transistors in the drive electronics must be capable of dissipating more power than that dissipated in square wave operation. Square wave drive offers the added advantage that the drive electronics can be digitally based. The brushless DC motor will duplicate the performance characteristics of a conventional DC motor only if it is properly commutated. Proper commutation involves exciting the stator windings in a sequence that keeps the magnetic field produced by the stator approximately 90 electrical degrees ahead of the rotor field. The brushless DC motor therefore relies heavily on the position feedback system for effective commutation. It might also be apparent that the brushless motor as described is not strictly a DC machine, but a form of AC machine with position feedback.

21.2.10 Motor Selection

For the mechatronics engineer the main concerns regarding electric motors will be those of selection for purpose. At the very least the motor must be capable of matching the power requirements of the driven load. In all cases, therefore, the motor power available should be enough to cope with the anticipated demands of the load. Other requirements are the need for the motor to have enough torque available on start-up to overcome the static friction, accelerate the load up to the working speed, and be able to handle the maximum overload. Too much excess motor torque on start-up might result in a violent initial acceleration. Some systems therefore require a "soft start" whereby the motor torque is gradually increased to allow the load to accelerate gently.

The operating speed of the motor will be fixed by the point at which the torque supplied by the motor is just balanced by the torque requirements of the load. At any other condition, the motor and load will be either accelerating or decelerating. Correct matching of a motor to a driven machine can only be confidently accomplished if both the motor and the load torque–speed characteristics are known. The motor torque–speed characteristics are usually provided by the supplier. The driven machine torque–speed characteristics can be something of an enigma.

Friction devices like industrial sanders, buffers, and polishing machines have a torque–speed characteristic that is initially very high, but drops sharply once motion is established. Continued acceleration usually sees the torque requirement of the load decrease further but at a slower rate than that at start-up. The difference between the static and dynamic friction accounts for this behavior.

Fans and blowers have a torque–speed characteristic that increases parabolically from zero as the speed increases. Such machines do not, therefore, need much motor torque to enable them to start.

High inertia devices like machine tool drives, rolling mills, and electric lifts require a large torque on start-up to overcome the inertia. Once motion is established the torque requirements tend to decrease with increasing speed. The series-wound DC motors are ideal for these types of loads.

This brief discussion of rotating electrical machines is in no way comprehensive. A fuller discourse on AC and DC machines is given both by Gray [1] and Sen [2]. Orthwein [3] presents an interesting practical discussion on the mechanical applications of AC and DC motors and Kenjo and Nagamori [4] provide a detailed in-depth study of permanent-magnet DC motors.

References

1. Gray, C. B. (1989), *Electrical Machines and Drive Systems,* Longmans Scientific and Technical, Harlow.
2. Sen, P. C. (1989), *Principles of Electric Machines and Power Electronics,* Wiley, Chichester.
3. Orthwein, W. (1990), *Machine Component Design,* West Publishing, St Paul, Minnesota.
4. Kenjo, T. & Nagamori, S. (1985), *Permanent Magnet and Brushless dc Motors,* Monographs in Electrical & Electronic Engineering, Clarendon Press, Oxford.

21.3 Piezoelectric Actuators

Habil Ramutis Bansevicius and Rymantas Tadas Tolocka

21.3.1 Piezoeffect Phenomenon

The piezoelectric effect was discovered by the Curie brothers in 1880 [1]. The direct piezoelectric effect contains the ability of certain materials, which are called piezomaterials, to generate electric charge in proportion to externally applied force. The effect is reversible and then is called an inverse piezoelectric effect. The effects have been used for actuating/sensing functions in engineering applications.

21.3.2 Constitutive Equations

Coupled electric and mechanical constitutive equations of piezoelectric materials for one-dimension medium are:

$$S = s^E T + d^t E, \tag{21.31}$$

$$D = \varepsilon^T E + dT, \tag{21.32}$$

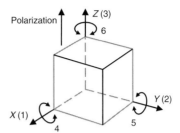

FIGURE 21.81 Directions for indexing piezoelectric constants.

where S is strain, T is stress (N/m^2), E is electric field strength (V/m), D is electric displacement (C/m^2), s^E is the compliance of the material (m^2/N) when the electric field strength is constant and ε^T is the permittivity (F/m) under constant stress, and d is piezoelectric constant (m/V or C/N). Superscript t stands for matrix transpose.

The first members on the right side of the equations refer to the mechanical properties of an elastic body (Equation 21.31) and to the electric properties of a medium (Equation 21.32), while the second ones describe the piezoeffect.

Piezoelectric properties of piezoelectric materials depend on the directions of electrical and mechanical inputs/outputs. Thus, these properties are described by constants with two subscripts, first of which is related to electrical and second to mechanical action or response directions in the orthogonal axes X, Y, and Z system. Indexes 1, 2, and 3 are prescribed to these axes respectively. Piezoelectric material polarization direction coincides with that of the poling electric field and is referred by convention to the axis Z with index 3. Subscripts 4, 5, and 6 are used additionally for describing shear distortions in respect to the directions X, Y, and Z (Figure 21.81).

Indexes show possible piezomaterial operation mode thickness expansion, transverse expansion, thickness shear, and face shear. The mode of motion depends on the location of electrodes relative to the polarization axis and the displacement that is used. The poling electric field direction causes elongation in this direction and contraction in the perpendicular ones, which are three times smaller than the first one. The reverse field causes contraction along the electric field direction; the elongation in perpendicular directions and displacement proportions are the same. The piezomaterial properties are identical along axes 1 and 2. The main constants characterizing the piezoeffect are:

- d_{ij} (piezoelectric constants) strain or charge coefficients expressed in C/N or m/V (according to direct or reverse piezoeffect). They relate to the strain developed by the electric field in the absence of mechanical stress (Equation 21.31), and to the electric charge per unit area by the applied stress under zero electric field (Equation 21.32). Example: symbol d31 means that electrodes are perpendicular to axis 3 (electric field along it) and stress or strain is along axis 1.

- g_{ij} (voltage constants) relate to open circuit electric field developed per applied mechanical stress or strain developed per applied charge density and is expressed in Vm/N or m^2/C. The relation between d_{ij} and g_{ij} is as follows:

$$g_{ij} = \frac{d_{ij}}{\varepsilon^T} \quad (21.33)$$

- k_{ij} (electromechanical coupling factors)—energy ratios describing conversion from mechanical to electrical energy and vice versa. Factor k^2 is the ratio of converted energy to input energy at operating frequencies far from resonant.

$$k^2 = \frac{d^2}{s^E \varepsilon^T} \quad (21.34)$$

- Factor k_p refers to the plane mode operation (strain or stress is equal in all directions perpendicular to axis 3).

21.3.3 Piezomaterials

Materials that exhibit a significant and useful piezoelectric effect fall into three main groups: natural (quartz, tourmaline, topaz, Rochelle salt) and synthetic crystals (lithium sulfate, ammonium dihydrogen phosphate), polarized ferroelectric ceramics, and certain polymer films. The main piezomaterials for engineering applications are ferroelectric ceramics and PZT (lead zirconate titanate). The latter is characterized by high coupling factors, and piezoelectric and dielectric constants over extended temperature and stress ranges. Barium titanate, lead magnesium niobate, modified lead titanate, and bismuth titanate compounds are used, as well as other special compositions. Because of their natural asymmetric structure, crystal materials exhibit the effect without further processing. After sintering ferroelectric ceramics, domains are of statistically distributed orientation and must be artificially polarized by a strong electric field, while the material is heated above its Curie point (the temperature at which a piezomaterial becomes completely depolarized) and then slowly cooled with the field applied. The remnant polarization being retained, the material exhibits the piezoeffect. Piezopolymer—polyvinylidene fluoride (PVDF) is a special class of polymer that exhibits a high degree of piezoelectric activity. They are used for manufacturing piezofilms of low thickness (less than 30 mm), which may be laminated on the structural materials. The values of piezo constants for PVDF piezofilms are lower than those for piezoceramics.

New piezoelectric materials such as single crystals and relaxors are now under research. Single crystals of natural or man-made materials exhibit the piezoelectric properties that might be offered by polycrystalline element (ferroelectric ceramics) with perfectly aligned domains, which is impossible; thus, single crystals demonstrate improved performance. For example, single crystal lead magnesium niobate lead titanate exhibits ten times the strain of comparable polycrystalline PZT element. A large variety of materials, such as lithium niobate, lead zirconate, niobate lead titanate, lithium tetraborate, and so forth, are used to fabricate single crystal piezoelectric elements. Relaxor materials such as lead nickel niobate are less sensitive to temperature than the conventional ones, because the transition between piezoelectric behavior and the loss of such a capability cover some temperature range (Curie range) and do not occur at a specific Curie point. They exhibit higher electromechanical coupling factors than conventional ceramics (greater than 0.9), which makes them attractive for actuators, transducers, and so forth.

Polypropylene foils with foam structure, caused by the enclosed micrometer scale vapor locks, exhibit piezoelectric behavior after polarization, and are being introduced owing to higher piezoelectric constants values than polyvinylidene fluoride.

21.3.4 Piezoactuating Elements

Piezoactuating elements are produced in the wide range of sizes as squares, rectangles, rings, disks, spheres, hemispheres, bars, cylinders, and special elements. The typical thickness is from 0.2–20 mm and up to 100 mm length or external diameter. Due to the unique properties of the material, they can be used directly as actuators. The main disadvantages are the dependency of characteristic values on the age and temperature and the necessity of high- voltage power supply.

The piezoeffect is linearly dependent on the applied electric field. Possible strength depends on piezomaterial short circuit resistance and is in the range 1–2 kV/mm. Maximum possible relative change in length is of the order of 0.1%. The shortest time of expansion to reach the nominal displacement is 1/3 of the period at resonant frequency oscillations of the mechanical system containing a piezoelement. The piezoelement resonant properties are described by frequency constant Ni, which is the resonance frequency fr multiplied by the linear dimension of the piezoelement related to the specific resonance.

Voltage being changed (it remains constant after the change), the piezomaterial continues to expand/contract in the same direction. This drift, known as creep, can be estimated by

$$\Delta L(t) = \Delta L \left(1 + \gamma \lg\left(\frac{t}{0.1}\right)\right), \tag{21.35}$$

where ΔL is 0.1 s expansion after the positioning process and γ is the drift factor. It depends on the design and mechanical load and lies between 0.01 and 0.02. Hysteresis is common in the piezomaterials as well. PZT hysteresis is a fairly constant fraction of the stroke, and the width of the hysteresis curve for it can be as large as 20% of the stroke. Due to compensation strategies, hysteresis errors decrease down to 3%. A piezoelement with electrodes laminated onto it is an electrical capacitor. Because of extremely high piezomaterial internal resistance (more than 100 MOhm), only small discharge current flows if piezomaterial remains static in the expanded state. Thus, the piezoelement is separated from the source of high voltage, and its expansion decreases slowly. This in turn causes a change in the charge, which results in a current.

$$i = \frac{dQ}{dt} = C\frac{dV}{dt}, \qquad (21.36)$$

where Q is the charge, C is the capacity, and V is the voltage.

The same equation is valid to describe the flow of electric energy when the material is experiencing a change in strain, and to define the time required to build up voltage in the piezoelement.

Table 21.4 shows different piezoelement sensing/actuating possibilities for some shape cases. Note: F is the force and ε_3^T is dielectric permittivity of the material at constant stress in direction 3, K_i^T is relative dielectric constant ($K_i^T = \varepsilon_i^T/\varepsilon_0$), and ε_0 is dielectric permittivity in vacuum.

Piezoactuators are able to generate a force of several thousand Newtons in a stroke range of hundreds micrometers (without amplifiers) with sub-nanometer resolution and response time of millisecond range. Resonant frequencies of industrially applied actuators are in the range from several to hundreds kiloHertz. The basic types of piezoactuators used are stacked and are of laminar design. Stacked actuators consist of some thin layers of piezoactive material between metallic electrodes in parallel connection (Figure 21.82a).

This way of connection allows greater stroke at lower voltage. Stacked actuators may be produced by cut-and-bond and tape-casting methods. The first one is realized by stacking a large number of thin piezoelectric disks with copper shims in between each disk and positive electrodes to opposite sides of the actuator stack. Usually, these disks are 0.3–1 mm thick. Displacement in this case is in proportion to the number n of disks, if no external load is applied and operating mode is d_{33}.

$$\Delta L = Vnd_{33}. \qquad (21.37)$$

Relationship between the piezoactuator's generated displacement and external load F is a linear function

$$\Delta L = \frac{F_0 - F}{c}, \qquad (21.38)$$

where F_0 and c are blocked force and piezoactuator stiffness.

Blocked force is the force that is generated at maximum voltage when the actuator is not allowed to move. The displacement attained at maximum voltage when the actuator is not loaded is called free displacement.

If the piezoactuator is loaded by spring, the following relationship is valid:

$$\Delta L = \Delta L_{max} \frac{c}{c + c_s}, \qquad (21.39)$$

where ΔL_{max} is free displacement and c_s is the spring stiffness.

Such stacked actuators allow us to achieve the stroke of 100-μm range. The principal disadvantage of such technology is high voltage, necessary to get sufficient electric field strength, because the minimum

TABLE 21.4 Piezoelement Sensing/Actuators Possibilities

Action Mode (L, length; W, width; T, thickness; D, diameter)	Generated Voltage, V	Displacement, ΔL ($\Delta T, \Delta D, \Delta X$)	Capacitance, C
Transverse length mode: $L > 3W > 3T$	$V = \dfrac{g_{31}}{W} F$	$\Delta l = \dfrac{d_{31} L}{T} V$	$C = \dfrac{\varepsilon_3^T LW}{T}$
Thickness extension mode: $D > 5T$	$V = \dfrac{4 T g_{33}}{\pi D^2} F$	$\Delta T = d_{33} V$	$C = \dfrac{\pi \varepsilon_3^T D^2}{4T}$
Radial mode: $D > 5T$	Not applied	$\Delta D = \dfrac{d_{31} D}{T} V$	$C = \dfrac{\pi}{4} K_3^T \varepsilon_0 \dfrac{D^2}{T}$
Longitudinal mode: $L > 3D$	$V = \dfrac{4L}{\pi D^2} g_{33} F$	$\Delta l = d_{33} V$	$C = \dfrac{\pi D^2}{4L} K_3^T \varepsilon_0$
Thickness shear mode: $W > 5T, L > 5T$	$V = \dfrac{g_{15}}{W} F$	$\Delta x = d_{15} V$	$C = \dfrac{LW}{T} K_1^T \varepsilon_0$
Radial mode of a thin wall cylinder: $D_o > 8T$	$V = \dfrac{1}{2} g_{31} D_0 p$	$\Delta D_m = \dfrac{d_{31} D_m}{T} V$	$C = \dfrac{2 \pi K_3^T \varepsilon_0 L}{\ln(D_0 / D_i)}$
Longitudinal mode of a thin wall cylinder: $L > 5T$	$V = \dfrac{g_{31}}{\pi D_m} F$	$\Delta l = \dfrac{d_{31} L}{T} V$	Note: D_o, D_i and D_m are outer, inner and mean diameters of the cylinder; p is the pressure.

Note: F is the force and ε_3^T is dielectric permittivity of the material at constant stress in direction 3, K_i^T is relative dielectric constant ($K_i^T = \varepsilon_i^T / \varepsilon_0$), and ε_0 is dielectric permittivity in vacuum.

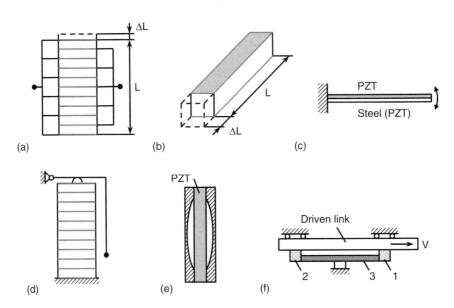

FIGURE 21.82 Piezoactuators: (a) stacked; (b) laminated; (c) unimorph; (d) amplified; (e) mooney; and (f) inchworm motor.

layer thickness is limited. The high-voltage actuators, consisting of 0.5 mm thickness discs disks, require about 1000 V and the low-voltage ones, consisting of 0.1-mm thickness disks, require about 200 V electric field.

Tape-casting technologies make it possible to harness the full potential of the piezoelectric effect at low voltages, enabling to reach the maximum permissible field strength already at 100 V and less. This technology uses a number of tape-cast ceramic layers with thickness of several tens of micrometers laminated together with screen-printed internal electrodes for producing multi-layer piezoactuating elements. External connections are provided to the internal electrodes by screen printing conductive materials onto the sides of the ceramic laminate. It allows to get much thinner ceramic layers, a customized internal structure, and flexibility of the component design. All this opens new possibilities for high volume and low-cost production. Piezoelectric multilayer elements can be manufactured from a number of different materials, each optimized to match application-specific functions and performance. Ceramic multilayer actuators may be made very small (e.g., 1 mm × 1 mm × 0.2 mm), give up to 5 μm stroke, and control up to 20 kN forces with resonant frequency 0.5–10 MHz. The produced multi-layer ceramic may be used as a finished transducer and in stacked design.

Laminar design actuators are based on transversal effect (mode d_{31}) and consist of piezoelectric layers with films bonded on to the structure (Figure 21.82b). They are of lower stiffness and displacement than the stacked ones. The actuator displacement and stiffness depend on the length/thickness ratio, and its influence is opposite. Therefore piles consisting of several strips are often used in making a design similar to the stacked design.

If a long stroke is required, bending piezoelements based on transversal effect can be used. All kind of bending elements feature greater deflection, lower stiffness, lower resonant frequency, response speed, and life time in comparison with stacked design. A unimorph and a bimorph are devices of such kind, which allow the displacement to reach up to 1000 μm. A unimorph is a composite cantilever of two layers (Figure 21.82c). One of them, an elastic shim, is of structural material and the other is of piezo-material. The shim is of constant length and the element bends in order to compensate the different behavior when the ceramics expand or contract. Two ceramic strips with the shim in between are used in bimorph, and one of them expands while the second one contracts. Disk or ring multilayer-bending piezoactuators consist of circular elements with a diameter of a few centimeters. When voltage is applied, the component deforms, taking partly a spherical shape, and reaches displacement up to 200 μm.

Displacement amplification up to 2000 μm can be achieved by various constructive means. The stiffness of such designs decreases with the square of the displacement amplification ratio and is much smaller than in the stack design. Along with levers (Figure 21.82d), monolithic hinge lever mechanisms made of structural material by cutting constricted hinges are used to function as lever mechanisms. The second mechanisms are free of backlash. Moonie piezoactuator consists of a piezoelectric disk sandwiched between two metal plates (Figure 21.82e) with cavities bonded together. The small radial displacement of the disk is transformed into a much longer axial displacement, normal to the surface of the metal plates. Forces and displacements achieved are in between conventional multi-layer and bimorph actuators.

If absolute positioning of high accuracy is necessary, the controlled operation of piezoactuators is used. Sensors detect the actual displacement, and the input voltage is controlled by their signals. This leads to positioning with high accuracy, avoiding hysteresis effect and changes in position because of fluctuating forces.

An inchworm motor is used for positioning in hundreds of mm range with resolution up to 2 nm and up to 2 mm/s speed. This motor is composed of three piezoceramics elements (Figure 21.82f). Two elements (1 and 2) are used to clamp and release the driven link, and element 3 is fixed on the frame and used to shift the driven link when expanding. By applying drive voltages with a certain phase difference, this device can move simulating the worm motion, and the working cycle of stepping is as follows: 1e; 3e/ driven link moves by step; 2e; 1c; 3c/ driven link moves by step; 2c. Here *e* stands for expansion and *c* for contraction.

21.3.5 Application Areas

Owing to the inherent properties of piezomaterials, actuators with a lot of engineering advantages can be developed. Some of them are: compactness and light weight, rapid response, practically unlimited resolution, no magnetic field, large force generation, broad operating frequency range, high stability, solid state, low power consumption, proportional to the applied voltage displacement, and 50% and more energy conversion efficiency.

They are used in micromanipulation, noise and vibration suppression systems, valves, lasers and optics, ultrasonic motors, positioning devices, relays, pumps, in automotive industry, industrial automation systems, telecommunications, computers, and so forth. Some of the applications are shown in Figure 21.83.

Suppression of oscillations: Piezoactive material-based dampers convert mechanical oscillations into electrical energy. The generated energy is then shunted to dissipate the energy as heat, that is, oscillation energy is dispersed. The principle scheme is given in [2].

Microrobot: The microrobot platform legs are piezoactuators. By applying voltage to the electrodes, the piezo–legs are lengthened, shortened, or bent in any direction in a fine movement.

Micropump: The diaphragm is actuated by a piezoactuator; input and output check valves are subsequently opened for liquid or gas pumping. The advantages are fast switching and high compression rate.

Microgripper: The piezoactuator works on contraction for gripping motion based on the compliant mechanism. The gripper is of very small size and almost of any required geometrical shape.

Micromanipulator: Due to the practically unlimited resolution, piezoactuators are used in numerous positioning applications.

Microdosage device: Piezoactuators allow high precision dosage of wide variety of liquids in the range of nanoliters for various applications.

Multilayer ceramics and single crystals are expanding their application. Multi-layer ceramics as finished transducers are used for stabilization of hard drives, tuning the lasers, adjusting mirrors in optical switches, and other instrumentation applications. Their stacked design is applied for fuel injection, optical switches, and medical instrumentation where nanorange is required. Single crystals have found applications in

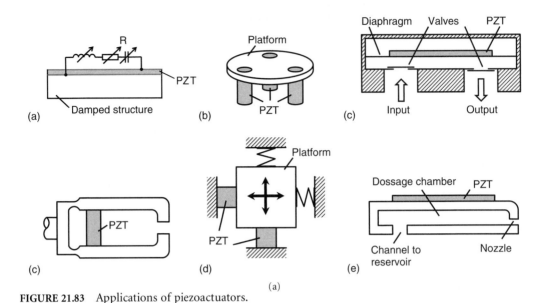

FIGURE 21.83 Applications of piezoactuators.

acoustical, optical, and wireless communication as well as actuators. Some of them exhibit useful combinations of electro-optic and piezoelectric properties and are exploited for surface acoustic wave and electro-optic devices.

21.3.6 Piezomotors (Ultrasonic Motors)

Ultrasonic motors, from the first attempt to design them by Williams and Brown in 1942 [3] and first developments in Eastern Europe in the early 1970's [4–6], up to commercializing them in Japan in the last decade of the XX twentieth century [7–9], find wider and wider applications as converters of high-frequency mechanical oscillations (dozens of kHz) into continuous motion. If piezomaterials are used as oscillators, such motors may be called piezomotors. Various types of piezomotors may be developed, and the basic ideas are given in Table 21.5. Piezomotors are mainly based on producing elliptical motion in the contact area between input and output links. The oscillations in the contact point of the oscillating stator and the driven link are excited in them in such a way that the trajectory of this point is elliptical, and thus the friction is used to involve this link into the motion. The shape of trajectory and frequency predetermines the velocity of the driven link. For this purpose, oblique impact on the output link or the traveling wave is made use of.

In piezomotors, making use of oblique impact, friction force transmits motion and energy between the input and output links. This may be realized by two oscillatory motions (normal and tangential components) u_y and u_x in the contact area with phase difference φ, which is used also to change the direction of the output link motion. Both motions can be realized by one or two active links oscillating resonantly. Various oscillations (longitudinal, transversal, shear, and torsional) offer possibilities to develop different kinds of piezomotors.

Traveling wave motion piezomotors are based on frictional interaction between the traveling wave motion in the elastic body and the output link, that is, its principle of operation is similar to the harmonic traction transmission. Wave propagating along the surface (Rayleigh wave) of the input link forms the elliptical motion in the contact area. Rayleigh wave is a coupled wave of longitudinal and shear waves, thus each surface point in the elastic medium moves along an elliptical locus. Flexural, shear, torsional, and longitudinal waves are used. Traveling wave in the piezoceramic transducer is excited by the electrical field. It can be generated by two standing waves whose phases differ by 90° to each other, both in time and in space:

$$u = u_0 \cos(kx - \omega t) = u_0 \cos kx \cos \omega t + u_0 \cos(kx - \pi/2)\cos(\omega t - \pi/2). \quad (21.40)$$

In practice, this is accomplished by dividing the electrodes of the converters into n equal parts and connecting them to the n-phase generator of electrical oscillations, where $n \geq 3$; phases are shifted between adjacent electrodes being $2\pi/n$, or by using discrete converters. Application of piezoceramics sectors with alternating polarization allows us to simplify design. Generation of more than two standing waves increases output power.

Piezomotors can be developed by using only one standing wave, excited in the stator that moves its vibratory tips, generating flat-elliptical movement when they bend because of the restriction by the rotor. A one-phase driving voltage signal is needed and electric circuits are less complicated than in the traveling wave case, but alternating the direction of the output link rotation can sometimes be complicated.

Piezomotors with frictional anisotropy of contact are based on oscillatory motion variations in the normal direction of active link contact during the oscillation cycle. This is achieved by superposing additional periodic actions in the contact. The distinguishing feature is τ_c/T, which is the ratio of the reduced duration of the contact to the oscillations period. The contact anisotropy can be achieved in two ways: (a) by locking the active link in a specified segment of the trajectory (Table 21.5, case C, a), and (b) by superimposing oscillations of higher frequencies (Table 21.5, case C, b), in the direction of basic oscillations, or in the perpendicular direction of basic oscillations (Table 21.5, case C, c), normal or tangential plane.

Piezomotors with asymmetrical oscillations are based on the asymmetry of inertia forces in non-harmonic high frequency oscillations, multiple frequency oscillations (Table 21.5, case D, a), or forces of dry friction with nonlinear relationship between viscosity of some liquids and velocity (Table 21.5, case D, b). Asymmetric cycles of oscillations are generated by summing the harmonics of multiple frequencies. The amplitude of each harmonic is chosen by using the electrode shape variations and the area of divided electrodes or changing the amplitude of the voltage supplied. Shift in voltage supply phases is used. The piezomotor efficiency in this case is lower, but the designs of devices are characterized by higher up to 0.002 μm resolution in the translational drive. Besides, this leads to designing piezomotors of limited dimensions in both coordinates, which in turn is very important in a number of applications.

Piezomotors are easily miniaturized, thus, micromotors are successfully developed. Surface acoustic wave (SAW) motors are a good example. Piezomaterial is used as elastic substrate in SAW motor development and, in this case, Rayleigh wave is generated in it by an interdigital transducer that comprises two sets of interpenetrating metallic electrodes, fabricated photolithographically on the surface of the piezoelectric substrate. Their geometry depends on the generated wave length. The driven link - slider is arranged and preloaded on the elastic substrate so that friction force is sufficient to drive it when the surface wave propagates. The slider has many projections of 10–50 μm in diameter in order to control contact conditions. The SAW linear motors operate at 10 and more MHz, giving output force of several Newtons, and the slider speed is more than 1 m/s with the positioning resolution in the order of 10 nm and stroke in the order of 1 cm. Higher frequency of the generated wave guides to miniaturization, and keeping the slider in contact is the main problem in it.

Advantages of piezomotors are large torque in the low speed range, wide speed range, standstill force without excitation, high stiffness, high resolution, excellent controllability, small time constant, simple structure and compactness, flexibility in the shape of the motor, high power/weight ratio, silent operation since no speed reduction gears are required, and no electromagnetic induction. They give up to 1 Nm torque and are operating at the velocity of some hundreds rpm, with resolution and time constant of μm and ms range. Disadvantages are the requirement for wear-resistant materials and less durability due

TABLE 21.5 Piezomotors Operating Principles

Basic Idea	Schematic of Realization	Remarks
A. Elliptic motion in the contact: two motion components with phase difference	1. One active link 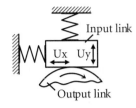 2. Two active links 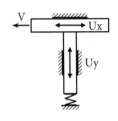	$u_y = u_{y0} \sin(\omega t + \varphi)$ $u_x = u_{x0} \sin \omega t$ $u_y = u_{y0} \sin(\omega t + \varphi)$ $u_x = u_{x0} \sin \omega t$ where u_{y0}, u_{x0}, ω and φ are amplitudes, angular frequency and phase of oscillatory motions of piezoelements
B. Elliptic motion in the contact area: travelling wave 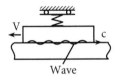	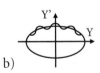	$u = u_0 \cos 2\pi/\lambda(u^* - ct)$ here u_0, λ, and c are amplitude, length and velocity of wave
C. Frictional anisotropy of contact	a)  b) c)	Usually $\dfrac{\tau_c}{T} \geq 0.05$ here τ_c and T are the duration of contact and oscillation period

Actuators

TABLE 21.5 Piezomotors Operating Principles (continued)

Basic Idea	Schematic of Realization	Remarks
D. Asymmetrical oscillations cycles	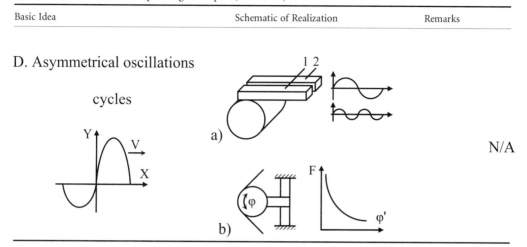	N/A

to frictional drive, suitability only for low and medium power (less than 100 W) applications, and need of high frequency power supply.

The application of piezomotors is expanding in those markets where their specific advantages are particularly useful. Good examples of implementation are the *Canon* camera's auto focus system, and *Seiko* wristwatches. *New Scale Technologies'* attempts to create tiny piezomotors to achieve auto focus and zoom optics moving in a mobile phone camera show that they meet new challenges. Piezomotors are expected to use applications in such areas as automotive accessories, opto-mechanical systems, automatic control and adjustment systems, robotics, military, medical, printing equipment, wearable instruments, precision tool moving mechanisms, and so forth.

20.3.7 Piezoelectric Devices with Several Degrees of Freedom

Piezoelectric motors with several degrees of freedom have led to a new class of mechanisms, changing their kinematic structure under control. They are based on active kinematic pair exploitation. If one or both links of the kinematic pair are made from piezoactive material, it is possible to generate static displacement of its elements and quasi-static or resonant oscillations, resulting in generating forces or torque in the contact area of links. The required motion of one link relative to the other is obtained. Because of controllable degrees of freedom, such kinematic pairs can be defined as active. An active kinematic pair is characterized by:

- Control of number of degrees of freedom. The simplest one is to control friction in the pair, usually when the elements of the pair are closed by force. Here, either the friction coefficient or the magnitude of the force executing the closure can be varied. This is achieved by excitation of high frequency tangential or normal oscillations in the contact area of the pair.
- Generation of forces or torque in the contact area between links. The direction of generated forces or torque is controlled by a spacial phase shift of oscillations, for example, by activating particular electrodes among the sectioned ones in the piezoelectric transducer.
- Possibilities to realize additional features: self- diagnostics, multi-functionality, self-repair, and self-adaptation.

An example of a positioning system on a plane is given in Figure 21.84. The hollow piezoceramic cylinder with electrodes on its internal and external surfaces is situated on the plain plate. The electrodes

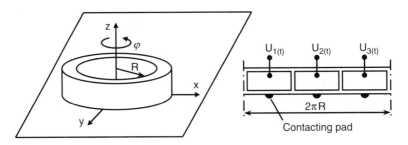

FIGURE 21.84 Positioning system.

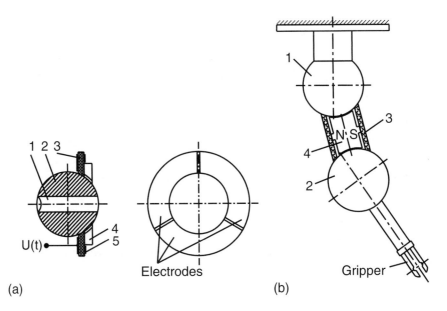

FIGURE 21.85 Peizoelectric devices with several degrees of freedom: (a) robot's eye and (b) Peizoelectric robot.

placed on the inner wall of the cylinder are sectioned as shown in its development. Activation of any of the electrodes with alternating voltage of resonant frequency $U_i(t)$, $i = 1,2,3$ results in excitation of oscillations in the piezoceramics, which generate the driving forces, acting in the corresponding direction in the contact zone of the plate and contacting pads of the cylinder. Generation of axial traveling wave oscillations of the cylinder by activation of all electrodes with a phase shift results in the rotation of the cylinder around its axis.

A practical example is Robot's eye (Figure 21.85a) in which miniature CCD camera 1 is fixed in the passive sphere 2, contacting with piezoelectric ring 3 with sectioned electrodes (in this case into 3 symmetric parts). Constant pressure in the contact zone is developed by permanent magnet 4. The system is a kinematic pair possessing a variable number of degrees of freedom. Activation of any of the electrodes results in the rotation of the sphere around the axis, position of which can be controlled by changing the activated electrode. Generation of travelling wave oscillations of the ring results in the rotation of the sphere around the axis of the ring. Approximately 2 angular second resolutions in every direction can be achieved higher than the requirements for robot vision systems.

Another practical example is a piezoelectric robot (Figure 21.85b), consisting of two spheres 1 and 2, made from passive material (e.g., steel), with a piezoelectric transducer 3 between them. The contact forces are ensured with the help of permanent magnet 4. Consisting of two spherical joints (two kinematic pairs) the robot has up to 6 degree of freedom. The structure control and motions (rotations) in all directions are achieved in the same manner as in the previous examples. The same active link is being used here in two kinematic pairs to increase redundancy in the system. Using direct piezoeffect it is possible to extract additional information (with the help of additional electrodes on the active link) on the forces and torque, acting on the gripper and on the state of contacting surfaces. This information is used to reduce positioning errors and to correct the motion trajectory.

References

Cady, W.G., *Piezoelectricity*, Dover publications, New York, 1964.
Volkov, V., Some theoretical problems in modern techniques of diagnostics in mechanical systems, in *Proc. Int. AMSE Conf. Systems Analysis, Control and Design*, Lyon, France, 1994, pp. 205–212.
Williams, W. and Brown, W.J., *Piezoelectric Motor*, US Patent 2,439,499, 1942.
Ragulskis, K. and Bansevitchyus, R., Vibromotors – high frequency vibration motion converters, in *Proc. Int. Symp. Exploitation of Vibration*, East Kilbridge, Glasgow, 1974, pp. 1–16.
Ragulskis, K. and Bansevicius, R., *Vibromotors*, Mokslas, Vilnius, 1981, (in Russian).
Ragulskis, K., Bansevicius, R. Barauskas, and G. Kulvietis, *Vibromotors for Precision Microrobots*, Hemisphere Publishing Corporation, 1988.
Sashida, T. and Kenjo, T., *An Introduction to Ultrasonic Motors*, Oxford Science Publications, 1993, Oxford University Press, New York.
Ueha, S. and Tomikawa, Y., *Ultrasonic Motors, Theory and Application*, Oxford Science Publications, Oxford Press, Oxford, 1993.
Uchino, K., *Piezoelectric Actuators and Ultrasonic Motors*, Kluwer Academic Publishers, MA, 1997.

21.4 Hydraulic and Pneumatic Actuation Systems

Massimo Sorli and Stefano Pastorelli

21.4.1 Introduction

The primary function of an actuation system is to influence the controlled system so as to obtain the desired movement or action. This objective is made possible by the actuation system, which converts the primary energy with which the actuator operates into the final mechanical energy.

There are three main types of power with which actuation systems work: electric power, hydraulic power, and pneumatic power. The first envisages the use of electric actuators such as motors, solenoids, and electromagnets. The remaining two envisage the use of cylinders (linear motors) and rotary motors, substantially similar in form and dimensions, the motion of which is respectively governed by a fluid considered uncompressible in an initial approximation (a hydraulic liquid, mineral oil generally, or a liquid with lower viscosity) and by a compressible fluid (compressed air or a generic gas).

Other types of energy are available but are fairly unusual in automatic systems. Chemical energy and thermal energy, which cause a change of phase in a material or the thermodynamic expansion of the systems into a mechanical movement, can be considered in this category.

The characteristics of fluid servosystems are examined below, with particular reference to systems which permit continuous control of one of the two physical magnitudes which express the fluid power: pressure and flow rate. In general, pressure control is carried out in cases in which it is necessary to create a determined force or torque law, while flow rate control is used to carry out controls on kinematic magnitudes such as position, speed, and acceleration.

Continuous control of a force or of a speed can be effectively realized with a fluid actuation device, with evident advantages compared with electric actuation, such as the possibility of maintaining the system under load without any limitation and with the aid of adequate control devices, the possibility of carrying out linear movements directly at high speeds, without devices for transforming rotary motion to linear, and the possibility of having high bandwidths, in particular in hydraulic systems, as these have limited dimensions and therefore low inertia.

21.4.2 Fluid Actuation Systems

An actuation system, which is part of an automatic machine, consists of a power part and a control part as illustrated in Figure 21.86. The power part comprises all the devices for effecting the movements or actions. The control part provides for the processing of the information and generates the automated cycle and the laws of variation of the reference signals, in accordance with the governing procedures implemented and with the enabling and feedback signals arriving from the sensors deployed on the operative part. The order signals coming from the control part are sent to the operative part by means of the interface devices which convert and amplify the signals, where necessary, so that they can be used directly by the actuators. These interfaces can be the speed drives or the contactors of the electric motors, the distributor valves in hydraulic and pneumatic actuators.

Figure 21.87 illustrates a fluid actuation system. The power part consists of the actuator—a double-acting cylinder in the case in the figure—the front and rear chambers of which are fed by a 4/2 distributor valve, which constitutes the fluid power adjustment interface.

The valve switching command is the order from the control part. This order is sent in accordance with the movement strategy, determined by the desired operating cycle of the cylinder in the control part, on the basis of the feedback signals from the sensors in the cylinder, represented in the figure by the limit switches.

Then there are discontinuous actuation systems and continuous actuation systems, depending on the type of automation realized, while retaining the control part and the actuation part. The first are effective when used in discontinuous automation, typical of assembly lines and lines for the alternating handling of machine parts or components; on the other hand, continuous actuation systems are found in continuous process plants and as continuous or analog control devices for the desired magnitudes, and constitute fluid servosystems.

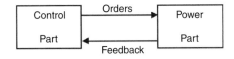

FIGURE 21.86 Actuation system.

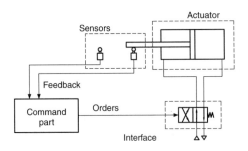

FIGURE 21.87 Fluid power actuation system.

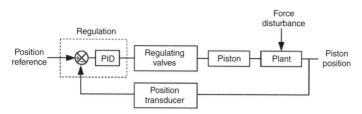

FIGURE 21.88 Scheme of a fluid power servosystem.

Fluid actuators, whether they are linear (cylinders) or rotary (motors) are continuous systems as they can determine the positioning of the mobile component (of the rod with respect to the cylinder liner; of the shaft with respect to the motor casing) at any point in the stroke. Performance of the usual cylinders and motors is currently highly influenced by the action of friction (static and dynamic) developed by contacts between mobile parts. This action, in pneumatic systems in particular, gives rise to the well-known phenomenon of stick-slip, or intermittent motion at very low movement speeds, due to the alternation of conditions of friction and adherence in the motion of the mobile element in the actuator. Given the nature of the friction itself, the presence of devices suitable for sustaining the mobile components of the actuator and maintaining the correct pressure conditions, such as supports and gaskets, gives rise to nonlinear conditions in the equilibrium of the actuator, increasing the level of difficulty in obtaining high precision in positioning the system. To overcome these problems in specific applications it is necessary to use actuators without seals, for example, with fluid static and/or fluid dynamic bearings.

The interface element, indicated as a distributor in the figure, takes on a crucial role in the definition of the operating mode of the actuator. Indeed, in the case in which it is only necessary to create reciprocating movements, with positioning of the actuator at the end of its stroke, it is only necessary to use a two- or three-position distributor valve, with digital operation. This is the solution shown in Figure 21.87.

If, on the other hand, it is necessary to have continuous control of the position and force transmitted, it is necessary to use devices which are not digital now, but which are continuous, such as proportional valves and servovalves, or it is necessary to use digital devices operating with control signal modulation, for example those of the PWM (Pulse Width Modulation) type.

The actuation system therefore becomes a fluid servosystem, such as the one outlined in Figure 21.88, for example. A practical construction of a hydraulic linear servoactuator having the same working scheme of Figure 21.88 is shown in Figure 21.89. It consists of a cylinder, a valve, and a position transducer integrated in a single device.

A controlled, fluid-actuated system is a classical mechatronic system, as it combines mechanical and fluid components, and control and sensing devices, and normally requires a simulation period for defining the size and characteristics of the various elements so as to comply with the desired specifications.

The standardized symbols for the different components of hydraulic and pneumatic fluid systems, and the definitions of the associated circuits, are defined in the standard, ISO 1219 "Fluid power systems and components—Graphic symbols and circuit diagrams; Part 1: Graphics symbols, Part 2: Circuit diagrams."

21.4.2.1 Fluid Servosystems

Fluid servosystems are devices for controlling a generically mechanical output power, either by controlling a kinematic magnitude (servosystems for controlling position or speed) or by controlling an action (servosystems for controlling the force, torque, or pressure).

The output magnitude control action is obtained by controlling the fluid power, that is, by the power of the fluid passing through the components of the servosystem.

Two large classes of fluid servosystems are usually present in current applications: hydraulic servosystems, in which the operating fluid is a liquid, and pneumatic servosystems, in which the fluid used is compressed air. The working pressure in hydraulic servosystems is typically comprised between 150 and 300 bar, while in the case of pneumatic systems, the pressure values are generally below 10 bar.

FIGURE 21.89 Hydraulic servocylinder (Hanchen).

The first group obviously includes hydraulic oils, that is, fluid with high viscosity, now traditionally used in servosystems in which a high controlled pressure is requested, but also combustible fluids, such as automotive or aeronautical petrols (JPA, JPB,...), used in all the applications found in the fuel circuits of combustion engines. Other servosystems include those which use both industrial and seawater as the working fluid. The latter solution has unquestionable advantages in all naval and off-shore applications.

Pneumatic servosystems include all the industrial applications for automation of production and process automation, and also the vehicular applications on means of air, sea, road, and rail transport. The compressed air in these applications is generated by compressors using air drawn in from the environment. Further applications include those in which the working fluid is not compressed air but a particular gas. In this regard, there are servosystems with refrigerant fluids in the gaseous stage, in both vehicular and industrial cryogenic systems, with fuel gases (LPG, methane, propane) in domestic applications, and with nitrogen in high-pressure applications.

It can be seen from this preliminary analysis that fluid servosystems are present both in the realization of a product, being integral parts of the automated production process, along with the electric servomechanisms, and as controlled actuation devices integrated in the product itself; in this regard we can mention generic servoactuators installed on aeroplanes and increasingly in road vehicles today.

21.4.3 Hydraulic Actuation Systems

The components of a hydraulic actuation system are:

- The pump, that is, the hydraulic power generation system
- The actuator, that is, the element which converts hydraulic power into mechanical power
- The valve, that is, the hydraulic power regulator
- The pipes for connecting the various components of the actuation system
- The filters, accumulators, and reservoirs
- The fluid, which transfers the power between the various circuit elements
- The sensors and transducers
- The system display, measurement, and control devices

Actuators

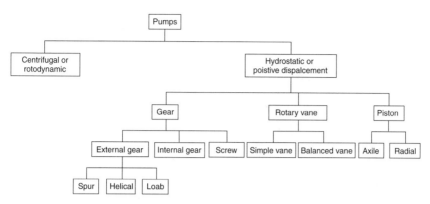

FIGURE 21.90 Pumps classification.

21.4.3.1 Pumps

Pumps transform electrical or mechanical energy into hydraulic energy. They constitute the fluid flow generator of the hydraulic system, as the pressure is determined by the fluid resistance downstream from the generator. The main types of pumps are shown in Figure 21.90.

Centrifugal pumps permit high deliveries with low pressures. They do not have internal valves but have a large clearance between the rotary part and stator part and guarantee a sufficiently stationary flow. Vice versa, hydrostatic or positive displacement pumps, which are those most commonly used, guarantee high pressures with limited deliveries. They have elements such as valves and caps, which permit separation of the delivery zone from the intake zone, and they may introduce pulses in the flow in the delivery line and generally require the use of a fluid with sufficient lubricating properties and load capacity, so as to reduce the friction between the sliding parts of the pump. There are constant displacement and variable displacement pumps.

The main positive displacement pumps belong to the gear, rotary vane, and piston types.

Gear Pumps

Gear pumps are subdivided into pumps with external gears, pumps with internal gears, and screw pumps. In all cases, the pump is made up of two toothed wheels inserted into a casing with little slack so as to minimize leakage.

Figure 21.91 is a photograph of a pump with external gears. The opposed rotation of the wheels causes the transfer of the oil trapped in the space between the teeth and walls of the gear from the intake to the outlet. Depending on the form of the teeth, there are external gear pumps of the spur gear, helical gear, and lobe gear types.

Pumps with internal gears are functionally similar to the above, but in this case the gears rotate in the same direction. Figure 21.92 is a section plane of a two-stage pump. In screw pumps, which may have one or more rotors, the elements have helical toothing similar to a threaded worm screw. Transfer of the fluid takes place in an axial direction following rotation of the screw. These types of pump guarantee very smooth transfer of the flow, with reduced pulsation and low noise levels.

The usual rotation speeds are between 1000 and 3000 rpm, with powers between 1 and 100 kW. Delivery pressures can reach 250 bar, with higher values in the case of the pumps with external gears. The flow transferred is a function of the pump displacement and the angular input speed, with values comprised between 0.1 and 1000 cm^2/rev. Double pumps can be used to increase these values. Gear pumps have high performance levels, with values around 90%.

Rotary Vane Pumps

Vane pumps (Figure 21.93) generally consist of a stator and a rotor, which can rotate eccentrically with respect to one another. Vanes can move in special slits placed radially in the stator or in the rotor and

FIGURE 21.91 External spur gear pump (Casappa).

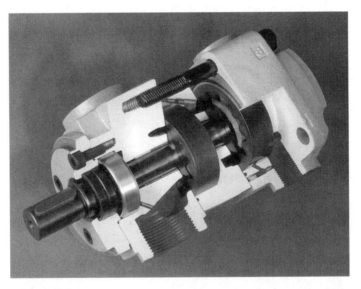

FIGURE 21.92 Internal gear pump (Truninger).

delimit appropriate variable volumes. In Figure 21.93, as in most constructions, the vanes are borne by the rotor which can rotate inside the stator. Rotation leads to the displacement of volumes of fluid enclosed between two consecutive vanes from the intake environment to input into the delivery environment. This type of pump permits a range of working pressures up to 100 bar and, compared with gear pumps, guarantees lower pulsing of the delivery flow and greater silence.

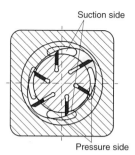

FIGURE 21.93 Rotary vane pump.

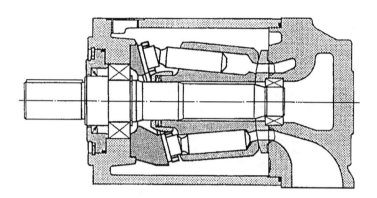

FIGURE 21.94 Axial piston swash plate pump (Bosch Rexroth).

Piston Pumps

Volumetric piston pumps can have one or more cylinders; that is, there may be one or more cylinders with a piston sliding in each of them. Transfer of the volume of fluid from intake to delivery is determined by the displacement of the piston inside the cylinder, which is provided with input and output valves or shutters. Depending on the geometrical arrangement of the cylinders with respect to the rotating motor shaft, piston pumps are subdivided into axial pumps (bent axis type and swash plate type) and radial pumps. Figure 21.94 shows the plan of a fixed-displacement axial piston pump, of the swash plate type. The working pressure range available with the aid of piston pumps is greater than in the previous cases, being able to reach pressures in the order of 400–500 bar but with the disadvantage of more uneven flow.

21.4.3.2 Motion Actuators

Motion actuators convert the hydraulic energy of the liquid under pressure into mechanical energy. These actuators are therefore volumetric hydraulic motors and are distinguished, on the basis of the type of movement generated, similar to what has been said about pumps, into rotary motors, semi-rotary motors or oscillating ones, which produce limited rotation by the output shaft, and into linear reciprocating motors, that is hydraulic cylinders.

Rotary and Semi-Rotary Motors

In construction terms, rotary motors are identical to rotary pumps. Therefore gear, vane, and piston motors, radial or axial, are available. Obviously, the operating principle is the opposite of what has been said for pumps. The symbols of hydraulic rotary motors are shown in Figure 21.95. Semi-rotary motors generate the oscillating motion either directly, by means of the rotation of a vane connected to the output shaft, or indirectly, by coupling with a rack, driven by a piston, with a toothed wheel connected to the output shaft, as in the example in Figure 21.96. The semi-rotary vane motors produce high instantaneous torsional torque on the output shaft; for this reason they are also called hydraulic torque-motors.

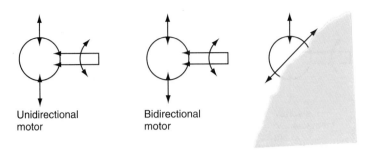

FIGURE 21.95 Symbols of hydraulic rotary motors.

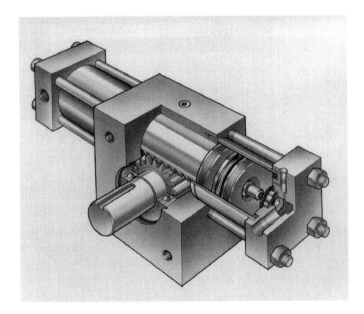

FIGURE 21.96 Hydraulic rotary actuator (Parker Hannifin).

Linear Actuators
Linear hydraulic motors constitute the most commonly used type of actuator. They provide a rectilinear movement realized by the stroke of a rod connected to a piston sliding inside the cylinder. A distinction is made between single acting and double acting cylinders. The former only permit a single work stroke and therefore the pressure of the fluid is exerted on the surface of the piston in one single direction; the retract stroke is made by means of the force applied externally to the cylinder rod, or with the aid of a helical spring incorporated with the actuator inside a chamber. The latter permit both strokes, so that the fluid acts alternately on both faces of the piston, generating both the advance and retract strokes. Double acting cylinders may have a single rod or a double through rod. These are composed of a tube closed at the ends by two heads, and a mobile piston inside the barrel bearing one or two rods connected externally to the load to move. As it is fitted with sealing gaskets, the piston divides the cylinder into two chambers. By sending the oil under pressure into one of the chambers through special pipes in the heads, a pressure difference is generated between the two surfaces of the piston and a thrust transmitted to the outside by the rod. Figure 21.97 shows the constructional solution of a hydraulic double acting cylinder with a single rod. Single rod actuators are also known as asymmetrical cylinders because the working area on the rod side is smaller than the area of the piston, as it is reduced by the section of the rod itself.

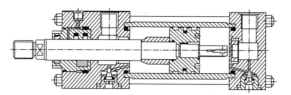

FIGURE 21.97 Single rod double-acting piston actuator (Atos).

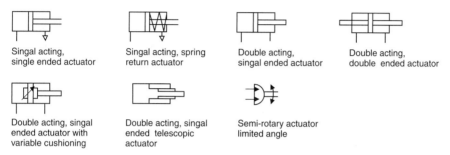

FIGURE 21.98 Actuators symbols.

This involves actuating forces and feed speeds which are different in the two directions, with the same feed pressure in the two thrust chambers.

Hydraulic actuators are able to support external overloads, as, if the load exceeds the available thrust force, the rod stops or reverses, but generally does not suffer any damage. Cylinders may get damaged however, or at least suffer a drop in performance, when they have to support loads which are not applied along the axis of the rod, that is, with components in the radial direction, as reactions are generated on the rod supports and piston bearings, which leads to fast wear of the same and reduces the tightness with oil leakage as a result.

The main features of a linear actuator are its bore, its stroke, its maximum working pressure, the type of working fluid, and the way its connections are fitted.

The symbols of the different types of actuators can be seen in Figure 21.98.

21.4.3.3 Valves

Valves are the components in hydraulic circuits that carry out the task of regulating the hydraulic power sent to the actuator. Their role is to turn the oil flow on or off or to divert it according to needs, thereby permitting adjustment of the two fundamental physical magnitudes of fluid transmission: pressure and flow rate. They are subdivided as follows on the basis of the operations they carry out:

- Directional valves
- On-off valves
- Pressure regulator valves
- Flow-rate regulator valves

In servomechanism applications valves with the continuous positioning of the moving components in them, said flow proportional valves or servovalves, and pressure proportional valves are used.

Directional Valves

Directional valves determine the passage and the flow direction of the oil current by means of the movement of appropriate moving parts contained in them, actuated from outside. Directional valves,

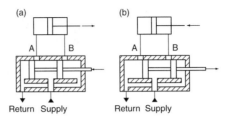

FIGURE 21.99 Scheme of four-way two-position valve.

also known as distributors, are distinguished according to the type of mobile element and therefore of their internal structure, by the number of possible connections with external pipes and by the number of switching positions.

The mobile element can be a poppet type or a spool type. Poppet valves are indifferent to fluid type and are not affected by impurities in the fluid, but require high actuating forces as it is not possible to compensate for the hydraulic forces of the oil pressure. Spool valves permit simultaneous connection to several ways and different switching schemes and therefore are more common because of their variability. The number of possible connections is defined by the number of hydraulic connections or ways present on the external body of the valve. The number of switching positions corresponds to the number of connection schemes which a valve makes it possible to obtain by means of appropriate movements of the mobile element.

Figure 21.99 shows the operating scheme of a four-way, two-position spool valve (indicated as 4/2) connected to a double acting linear actuator. In the first position (Figure 21.99a) the supply is in communication through output A with the rear chamber of the cylinder, while the front chamber discharges through port B. In this configuration, the piston effects an advance stroke with the rod coming out. In the second position, (Figure 21.99b), the result of the movement of the slide valve is that the feed and discharge conditions of the two chambers are inverted, and therefore, a retract stroke is effected.

A directional valve with several positions is represented symbolically by means of quadrants side by side depicting the connections made by each position. Figure 21.100, for example, shows some directional valve symbols in accordance with ISO standards. The central configuration of the three-position valves, which is normally the rest position, is linked with the geometry of the valve spool and of the associated seats.

Directional valves can be controlled in various ways (Figure 21.100): manually, by applying muscle power; mechanically, by means of devices such as cams, levers, etc.; hydraulically and pneumatically, by means of fluids under pressure; and electromagnetically, directly or piloted, depending on whether the positioning force is generated directly by the electromagnet placed in line with the slide valve, or by means of a hydraulic fluid, the direction of which is managed by a pilot valve which is smaller than the main controlled valve.

On–Off Valves

On–off valves are unidirectional valves, which permit the fluid to flow in one direction only. Because they impede flow in the opposite direction they are also called nonreturn or check valves. On-off valves are normally placed in the hydraulic circuit between the pump and the actuator so that, when the generator stops, the fluid contained in the system is not discharged into the reservoir but remains in the piping. This prevents a waste of energy for subsequent refilling and guarantees positioning of the actuator under load.

Constructively, check valves consist of an actuator, with ball or piston, which in the impeded flow configuration is maintained in contact against its seat by the thrust of a spring (nonreturn valve), or by the pressure difference between inlet and outlet (unidirectional valve).

Pressure Regulator Valves

There are essentially two types of pressure regulator valves: pressure limiter valves or relief valves, and pressure reduction valves.

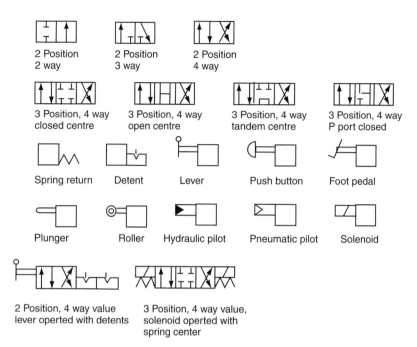

FIGURE 21.100 Valves symbols.

Relief valves guarantee correct operation of the system, preventing the pressure from exceeding danger levels in the system itself. There is always one maximum pressure valve in a hydraulic circuit to discharge any excess flow not used by the system back towards the reservoir. This is because the generator, or positive-displacement pump, provides a continuous flow of fluid which, if not absorbed by the user and in the absence of a relief or maximum pressure valve, would let the pressure in the system increase to unacceptable values. Pressure limiter valves can be direct-acting or piloted. The first provides the force of a spring with a fixed preload as the force contrasting the pressure of an obturator or an adjustable one, which guarantees the maximum opening pressure. The latter replaces the action of the spring with that of the hydraulic control fluid managed by a pilot valve.

The function of the pressure regulator valves is to maintain a constant pressure valve downstream from them, independently from variations in the upstream pressure. The regulated pressure value can be set manually, by means of a pilot signal, or by an electrical analog command. In the latter case, pressure regulator valves may operate in closed electrical loops, as they have an internal transducer to measure the controlled pressure.

Flow-rate Regulator Valves
A flow-rate regulator valve makes it possible to control the intensity of the flow of fluid passing through it. Functionally it operates as a simple restriction, similar to an orifice, with a variable area. The flow passing through a restriction is a function of the area of passage and of the difference in the pressures upstream and downstream from the component. The simple restriction is therefore sensitive to the load, as the flow rate also depends on the pressure drop at its ends, which is established by the other components in the circuit.

In the case of a pressure-compensated flow regulator valve, the flow rate is found to be maintained sufficiently constant above a minimum pressure stage (typically 10 bar) as an exclusive function of the external manual or electrical set-point. In this case, the valve has two restrictions in series, one of which is fixed and the other automatically variable, so as to maintain the pressure drop constant on the fixed restriction and guarantee the constancy of the flow rate.

The symbols for flow regulator valves in accordance with ISO standards are given in Figure 21.101.

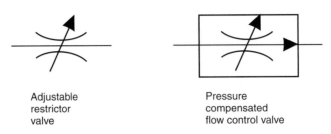

FIGURE 21.101 Symbols of flow control valves.

Proportional Valves and Servovalves

Servovalves began to appear at the end of the 1930s and were mainly used in the military and aeronautical fields. The first commercial versions appeared in the mid-50s. Servovalves and proportional valves are widely used today in the civil field, in the aeronautical, aerospace, automotive, and industrial sectors. In general, they are used for the continuous control of the displacement, speed, and force of a hydraulic actuator from which high performance is requested in terms of positioning precision, or accuracy in up and running conditions, and of working frequency bandwidth amplitude, both in open and closed loop control configurations.

A servovalve or proportional valve is a fluid component capable of producing a controlled output as a function of an input of electrical type. The device converting the electric signal into an action of the spool or poppet of the valve is electromagnetic, of the torque motor or proportional solenoid type. The torque motor converts a small DC current into torque acting on the rotor plate, in bipolar mode. Proportional solenoids produce a unidirectional force on the mobile armature function of the current circulating in the winding, with the characteristic of maintaining this force approximately constant within the cursor work displacement range. The torque motor, with lower current and inductance values, has shorter response times than the servosolenoid, which operates with notably higher currents, but generates lower mechanical power outputs. The torque motor, therefore, constitutes the pilot stage usually found in servovalves, while the servosolenoid used in proportional valves acts directly on the valve spools.

The magnitude directly controlled by the servovalve or proportional valve can be a flow rate or a pressure difference, depending on the type.

Servovalves and proportional valves are usually distinguished on the basis of the following characteristics:

- Input signals
- Precision
- Hysteresis
- Linearity between input and output
- Dead band
- Bandwidth

Input signals are characterized by the type of signal and range of variation. Current signals (±10 mA or 4–20 mA) or voltage ones (0–10 V) are typical. Precision is intended as the difference between the desired value and the value effectively achieved. It is provided as a percentage of the full scale value. The hysteresis derives from the different behavior shown by the component with ascending settings and corresponding points descending. Its value expresses the percentage ratio between the maximum deviation and the full scale value. Linearity by nature is a characteristic that can be assessed over the entire working range. It can be expressed in an absolute manner as the maximum percentage deviation of the input/output relation of its linear regression. In general, better linearity is requested in position control compared with the cases of speed, pressure, or force controls. The dead band determines the minimum input value at which an output variation is obtained. Unlike the above, bandwidth is a

Actuators

TABLE 21.6 Main Typical Differences between Servovalves and Proportional Hydraulic Valves

	Servovalve	Proportional Valve
Electromechanical converter	Bidirectional torque motor (0.1 ÷ 0.2 W) with nozzle-flapper or jet pipe	Unidirectional servosolenoid (10 ÷ 20 W)
Input current	100 ÷ 200 mA	<3 A
Flow rate	2 ÷ 200 l/min (two stage type) with valve pressure drop = 70 bar	10 ÷ 500 l/min (single stage type) with valve pressure drop = 10 bar
Hysteresis	<3% (<1% with dither)	<6% (<2% with electric feedback)
Bandwidth	>100 Hz depending on the amplitude of the input and of the supply pressure	<100 Hz depending on the amplitude of the input
Radial clearance of the spool	1 µm (aerospace) 4 µm (industrial)	2 ÷ 6 µm
Dead band of the spool	<5% of the stroke	Overlap 10–20% of the stroke, less if compensated

characteristic of the dynamic type. This is because it refers to a frequency diagram of the component and defines the frequency at which the response drops by 3 dB below the low frequency value. Normally, a bandwidth two to five times greater is required for continuous control valves compared with that required by the system.

The main differences between servovalves and proportional valves are shown in Table 21.6. Except for the traditional difference in the electromechanical conversion device, there are overlaps in the static and dynamic characteristics in many components available on the market.

On the basis of the generically superior static and dynamic characteristics, servovalves are commonly used in closed loop controls while proportional valves are used in open loop systems.

Servovalves

Two-stage models are very common in the context of servovalves, where a first pilot stage converts a low power electric signal into a pressure difference capable of acting on the slide valve of the second stage, usually four-way and symmetrical. Flow rate control servovalves are divided into two categories on the basis of how they make the electric–hydraulic conversion:

- Nozzle-flapper
- Jet-pipe

An example of nozzle flapper servovalves is shown in Figure 21.102. It comprises two stages: the former consists of a torque motor, the flapper, and a system of nozzles and chokes, while the latter consists mainly of the spool valve and output ports. The torque motor, constituted by the motor coil, the magnet, the armature, and the polepiece, is capable of transmitting a torque to the flapper which undergoes an angular displacement, thereby obstructing one of more calibrated nozzles to a greater degree. This operation causes a pressure difference at the ends of the spool, thereby causing the latter to move until the feedback wire, which connects the spool and the flapper, returns the flapper to the central position. Through the flow metering slot carried out in the bushing, the spool thereby permits communication between the various ports. The feedback wire is an elastic flexional element that provides the feedback between the main power stage (spool valve) and the first stage (torque motor).

This type of valve usually requires a greater degree of oil filtration, as nozzle-flappers are more sensitive to contaminants compared with the jet-pipe system.

Figure 21.103 shows a schematic section of an example of a servovalve of the jet-pipe type. It is connected to orifice P (supply) of the servovalve by means of a filter and flexible hose. It should be noted that, unlike the nozzle-flapper valve, it is not necessary to filter the entire incoming oil flow, but only what is called the control flow (the one going through the jet-pipe); this is certainly an advantage in terms of economic running and sensitivity to solid contamination. Starting with a standardized input voltage, the amplifier produces a voltage increase in one torque motor coil and an identical reduction in

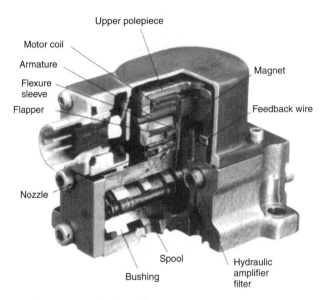

FIGURE 21.102 Nozzle-flapper servovalve (Moog).

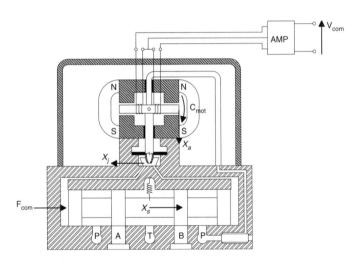

FIGURE 21.103 Jet-pipe servovalve scheme.

the other. This provokes an imbalance of the forces at the ends of the motor armature, generating a torque which tends to make the armature itself rotate and the jet-pipe with it. The displacement of the nozzle involves a different distribution of the control flow between the two pipes below it and, as a result, a pressure difference is created at the ends of the spool with a consequent net force on it, which causes its displacement. A spring element connects the spool and the jet-pipe, creating position feedback. The displacement of the spool and jet-pipe deforms the feedback spring, giving rise to a force proportional to it, and this makes it possible to balance the torque applied to the jet-pipe and the force generated at the ends of the spool. In this way, the system finds an equilibrium position, proportional to the input voltage. The feedback spring also produces centering of the slide valve at rest, making the presence of centering springs superfluous. Figure 21.104 shows the block diagram of the jet-pipe servovalve components with annotations of the physical magnitudes present in Figure 21.103.

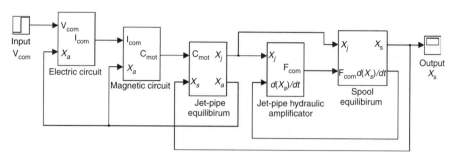

FIGURE 21.104 Block diagram of a jet-pipe servovalve.

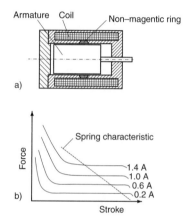

FIGURE 21.105 Proportional servosolenoid: (a) solenoid section scheme, (b) solenoid characteristics.

Proportional Valves

Proportional valves can be subdivided into proportional in flow and proportional in pressure (relief and pressure reduction) valves. In the first case, the action of the servosolenoid armature (Figure 21.105a) displaces the main spool of the valve, which is checked by a spring on the valve body.

The characteristics of the force generated by the servosolenoid, along the entire possible stroke of the spool, is a function of the input current only, as indicated in Figure 21.105b. For all possible values of the input current, the equilibrium of the magnetic force supplied by the solenoid and of the feedback forces of the spring is determined by reaching a certain position value.

In cases in which precise positioning of the slide valve is requested, position feedback is introduced. The photograph of a proportional valve of this type is shown in Figure 21.106. The input signal to the servosolenoid, sent by the feedback module, is the error compensated by a PID network between the reference signal and the feedback signal from the position transducer LVDT. The valve's accuracy and repeatability is improved by using the position feedback, as the hysteresis errors and those due to friction between the moving parts are partially compensated.

Flow proportional valves can have two stages. In this case, the outputs of the pilot proportional valve feed the end chambers of a spool valve of greater size, permitting greater controlled flows to be obtained while reducing dynamic performance at the same time. An example of a two-stage flow proportional valve is shown in Figure 21.107.

In pressure regulator proportional valves, the action of the servosolenoid acts on a conical needle in such a way so as to regulate the pressure in the chamber upstream from the needle itself. Figure 21.108 shows the plan of a pilot operated proportional relief valve.

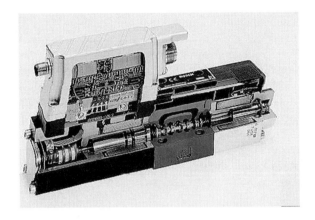

FIGURE 21.106 Flow proportional valve (Bosch Rexroth).

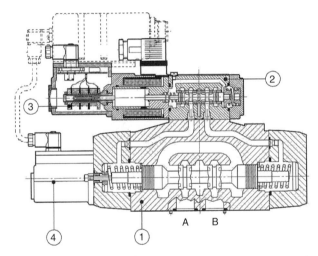

FIGURE 21.107 Double stage flow proportional valve: (1) main spool valve stage, (2) pilot stage, (3) LVDT of the pilot stage, (4) LVDT of the main stage (Atos).

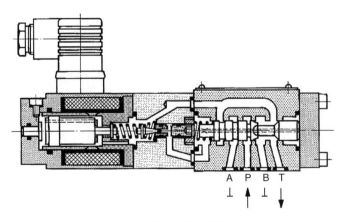

FIGURE 21.108 Proportional pressure relief valve (Bosch Rexroth).

21.4.4 Modeling of a Hydraulic Servosystem for Position Control

Figure 21.109 shows the scheme of a hydraulic servo system for position control constituted by a double acting, double ended actuator controlled in closed loop by a four-way servovalve. The x position of the piston is determined by the equilibrium of the forces acting on it: external force, thrust due to the pressures P_1 and P_2 acting in the chambers of the ram, friction force, and force of inertia. The pressures P_1 and P_2 are determined by the oil flows Q_{C1} and Q_{C2} entering and leaving the chambers. Flow rate Q_{FI} represents the oil leakage flow between the piston and the barrel. The flow proportional valve controls the oil flow on the basis of the reference signal *ref* from a compensator G_C. The input to the compensator is the error e_V between the signal V_{SET}, corresponding to the desired rod position x_{SET}, and the feedback signal $V_{F/B}$, corresponding to the effective rod position x measured by a position transducer LVDT.

The actuator is modeled by considering the equations of flow continuity in the chambers and the dynamic equilibrium equation for the rod. The continuity equation is expressed in a general form:

$$\rho\left(\sum Q_{IN} - \sum Q_{OUT}\right) = \frac{d(\rho V)}{dt} = \rho\frac{dV}{dt} + V\frac{d\rho}{dt} \qquad (21.39)$$

where

$\sum Q_{IN}$ = sum of the flows in volume entering
$\sum Q_{OUT}$ = sum of the flows in volume leaving
ρ = density
V = volume
t = time

From the definition of the compressibility modulus of the oil β, P being the pressure in the chamber considered:

$$\frac{dV}{V} = -\frac{d\rho}{\rho} = -\frac{dP}{\beta} \qquad (21.40)$$

We obtain

$$\sum Q_{IN} - \sum Q_{OUT} = \frac{dV}{dt} + \frac{V}{\beta}\frac{dP}{dt} \qquad (21.41)$$

The continuity equation for chamber 1 is

$$Q_{C1} - Q_{FI} = A_c \dot{x} + \frac{V_0 + A_c x + V_{sm}}{\beta}\frac{dP_1}{dt} \qquad (21.42)$$

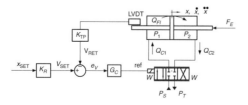

FIGURE 21.109 Scheme of a hydraulic servosystem with position control.

The continuity equation for chamber 2 is

$$Q_{FI} - Q_{C2} = A_c \dot{x} + \frac{V_0 - A_c x + V_{sm}}{\beta} \frac{dP_2}{dt} \qquad (21.43)$$

where

Q_{C1} = flow entering chamber 1
Q_{C2} = flow leaving chamber 2
Q_{FI} = leakage flow between piston and barrel
A_c = thrust section = $(D_{al}^2 - D_{st}^2)\pi/4$
D_{al} = bore diameter
D_{st} = rod diameter
V_0 = volume of the chambers with piston centered = $A_c L/2$
L = stroke
x = piston displacement ($x = 0$ in centered position)
V_{sm} = dead band volume

The dynamic equilibrium equation of the piston is

$$P_1 A_c - P_2 A_c - F_e - M\ddot{x} - F_A = 0 \qquad (21.44)$$

where

M = translating mass
F_e = external force
F_A = force of friction = $\gamma \dot{x} + F_{ATT} \text{sign}(\dot{x})$
γ = coefficient of viscous friction
F_A = force of coulomb friction

Leaks can be modeled as resistances in laminar and steady-state conditions of the following type:

$$R = \frac{\Delta P}{Q} \qquad (21.45)$$

where

R = resistance
Q = flow rate
ΔP = pressure difference

In the case of an annular pipe, we get

$$Q = \frac{\pi D h^3 (1 + 1.5\varepsilon^2)}{12 \mu l} \Delta P \qquad (21.46)$$

where

D = seat diameter
h = meatus thickness = $(D - d)/2$
d = spool diameter
ε = eccentricity = $2e/(D - h)$
e = distance between seat axis and spool axis
μ = dynamic viscosity
l = meatus length

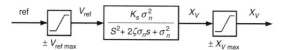

FIGURE 21.110 Block diagram of the valve.

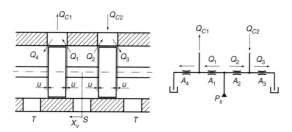

FIGURE 21.111 Reference scheme of the flow rates through the proportional valve.

The dynamic behavior of the electromechanical converter and of the spool valve can be identified with a linear model of the second order between the reference *ref* and the slide valve position x_V, as indicated in Figure 21.110, where K_S (m/V) is the static gain of the valve, σ_n (rad/s) the natural frequency of the valve, and ζ the damping factor of the valve.

The flows regulated by the proportional valves are indicated in detail in the plan in Figure 21.111. Having defined the areas A_1, A_2, A_3, A_4, functions of the spool displacement x_V, on the basis of the geometry, the flows transiting through the valve as a function of the pressures are given by

$$Q_1 = A_1 C_d \operatorname{sign}(P_S - P_1) \sqrt{\frac{2}{\rho}|P_S - P_1|} \tag{21.47}$$

$$Q_4 = A_4 C_d \operatorname{sign}(P_1 - P_T) \sqrt{\frac{2}{\rho}|P_1 - P_T|} \tag{21.48}$$

$$Q_2 = A_2 C_d \operatorname{sign}(P_S - P_2) \sqrt{\frac{2}{\rho}|P_S - P_2|} \tag{21.49}$$

$$Q_3 = A_3 C_d \operatorname{sign}(P_2 - P_T) \sqrt{\frac{2}{\rho}|P_2 - P_T|} \tag{21.50}$$

where P_S is the supply pressure, P_T is the pressure of the discharge reservoir, and C_d is the flow coefficient of the metering ports.

Therefore we get

$$Q_{C1} = Q_1 - Q_4 \tag{21.51}$$

$$Q_{C2} = Q_3 - Q_2 \tag{21.52}$$

The system of equations given above can be linearized in a working neighborhood, defined by the passage area of the valve A_{V0}, load pressure drop $P_{L0} = P_1 - P_2$, and piston position x_0. In the hypothesis

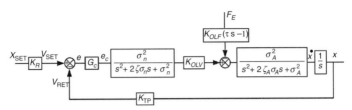

FIGURE 21.112 Block diagram of the linearized model of an hydraulic servosystem with position control.

that the leaks are negligible, the diagram of the linearized system is shown in Figure 21.112, which indicates:

G_C — compensator transfer function [V/V]

$K_{OLV} = \dfrac{K'_S K_Q A_C}{A_C^2 - K_{PQ}\gamma}$ — static speed gain $\left[\dfrac{m/s}{V}\right]$

$K_{TP} = K_R$ — static gain of the position transducer [V/m]

$K_{OLF} = \dfrac{K_{PQ}}{A_C^2 - K_{PQ}\gamma}$ — force constant $\left[\dfrac{m}{sN}\right]$

σ_n — natural frequency of the proportional valve [rad/s]
ζ — proportional valve damping factor

$\sigma_A = \sqrt{\dfrac{C_0}{M}[1-(K_{PQ}\gamma/A_C^2)]}$ — hydraulic resonance frequency [rad/s]

$\zeta_A = \dfrac{\gamma - (MK_{PQ}C_0/A_C^2)}{2\sqrt{C_0 M[1-(K_{PQ}\gamma/A_C^2)]}}$ — hydraulic damping factor

$\tau = \dfrac{A_C^2}{C_0 K_{PQ}}$ — force disturbance time constant [s]

where, in the above expressions:

$K'_S = \dfrac{A_{V\max}}{V_{\max}}$ — proportional valve static gain $\left[\dfrac{m^2}{V}\right]$

$K_Q = \left.\dfrac{\partial Q}{\partial A_V}\right|_{A_{V0},P_{L0}}$ — proportional valve flow gain $\left[\dfrac{m^3}{s\,m^2}\right]$

A_C — piston area [m^2]

$K_{PQ} = \left.\dfrac{\partial Q}{\partial P_L}\right|_{A_{V0},P_{L0}}$ — proportional valve flow-pressure gain $\left[\dfrac{m^3}{s\,Pa}\right]$

γ — coefficient of viscous friction $\left[\dfrac{N}{s/m}\right]$

$C_0 = \dfrac{4\beta A_C^2}{V_T}$ — hydraulic stiffness in centred position [N/m]

V_T — total volume of the two chambers [m^3]
M — mass of the moving parts [kg]

The open loop transfer function is

$$G_{OL} = \frac{V_{RET}}{e} = G_C \cdot K_{OLV} \cdot K_{TP} \frac{\sigma_n^2}{s^2 + 2\zeta\sigma_n s + \sigma_n^2} \cdot \frac{\sigma_A^2}{s^2 + 2\zeta_A \sigma_A s + \sigma_A^2} \cdot \frac{1}{s} \quad (21.53)$$

Supposing G_C to be constant, the static gain in open loop is

$$K_0 = G_C \cdot K_{OLV} \cdot K_{TP} = \frac{G_C K'_S K_Q A_C K_{TP}}{A_C^2 - K_{PQ}\gamma} \quad (21.54)$$

The closed loop transfer function is

$$G_{CL} = \frac{x}{X_{set}} = \frac{K_R G_C K_{OLV} \sigma_n^2 \sigma_A^2}{(s^2 + 2\zeta\sigma_n s + \sigma_n^2)(s^2 + 2\zeta_A \sigma_A s + \sigma_A^2)s + G_C K_{OLV} \sigma_n^2 \sigma_A^2 K_{TP}}$$

$$= \frac{1}{a_5 s^5 + a_4 s^4 + a_3 s^3 + a_2 s^2 + a_1 s + 1} \quad (21.55)$$

where

$$a_5 = \frac{1}{\sigma_n^2 \sigma_A^2 K_0}, \quad a_4 = \frac{2}{\sigma_n \sigma_A K_0}\left(\frac{\zeta_A}{\sigma_n} + \frac{\zeta}{\sigma_A}\right)$$

$$a_3 = \frac{1}{K_0}\left(\frac{1}{\sigma_n^2} + \frac{4\zeta_A \zeta}{\sigma_n \sigma_A} + \frac{1}{\sigma_A^2}\right), \quad a_2 = \frac{2}{K_0}\left(\frac{\zeta}{\sigma_n} + \frac{\zeta_A}{\sigma_A}\right), \quad a_1 = \frac{1}{K_0}$$

The transfer function between output and disturbance, said dynamic compliance, is

$$G_{FCL} = \frac{x}{F_e} = -\frac{K_{OLF}(\tau s - 1)(s^2 + 2\zeta\sigma_n s + \sigma_n^2)\sigma_A^2}{(s^2 + 2\zeta\sigma_n s + \sigma_n^2)(s^2 + 2\zeta_A \sigma_A s + \sigma_A^2)s + G_C K_{OLV} \sigma_n^2 \sigma_A^2 K_{TP}} \quad (21.56)$$

Static compliance is

$$\left.\frac{x}{F_e}\right|_{s=0} = \frac{K_{OLF}}{G_C K_{OLV} K_{TP}} = \frac{K_{PQ}}{G_C K'_S K_Q A_C K_{TP}} = \frac{1}{G_C K'_S K_P A_C K_{TP}} \quad (21.57)$$

having put

$$K_{PQ} = \frac{K_Q}{K_P}, \quad K_P = \left.\frac{\partial P_L}{\partial A_V}\right|_{A_{V0}, P_{L0}} \quad \text{pressure gain } \left[\frac{Pa}{m^2}\right]$$

Finally, a static stiffness is defined equal to

$$\left.\frac{F_e}{x}\right|_{s=0} = G_C K'_S K_P A_C K_{TP} \tag{21.58}$$

The predominant time constant, which is obtainable from the closed loop transfer function, is the coefficient $a_1 = 1/K_0$.

In conclusion, the following general considerations can be drawn:

- The speed gain K_{OLV}, and therefore the open loop static gain K_0, depend to a considerable degree on the flow gain K_Q and increase with increases in K_Q. K_Q increases as P_S increases, decreases as ΔP_{L0} increases, and does not vary with A_{V0}. In the hypothesis of γ below 1000 Ns/m, the effect of K_{PQ} is modest, practically negligible.
- The force constant K_{OLF} depends on the flow-pressure gain K_{PQ} and increases with it. $|K_{PQ}|$ increases with A_{V0} and with ΔP_{L0}, while it decreases as P_S increases; therefore $|K_{OLF}|$ decreases as P_S increases. Leaks lead to an increase in K_{OLF}.
- Static stiffness depends considerably on the pressure gain of the valve and increases with it. Given that K_P decreases with the leaks, these lead to a reduction in static stiffness. Furthermore, given that $|K_{PQ}|$ increases with A_{V0} while K_Q does not vary with A_{V0}, the pressure gain decreases with the increase in A_{V0}, and therefore static stiffness decreases if the valve is working in greater opening conditions.
  Furthermore, given that $|K_{PQ}|$ decreases as P_S increases, while K_Q increases as P_S increases, the pressure gain increases as P_S increases and therefore, the static stiffness increases with P_S.
- The hydraulic resonance frequency increases with the increase in hydraulic stiffness C_0 and decreases with the increase of the mass M. It is practically uninfluenced by the flow-pressure gain K_{PQ} if $\gamma < 1000$ Ns/m.
- The predominant time constant is inversely proportional to the speed gain, decreases as K_Q increases, that is it decreases as P_S increases, increases as ΔP_{L0} increases, and is indifferent to variations in A_{V0}.

21.4.5 Pneumatic Actuation Systems

Just as described for the hydraulic system, the components of a pneumatic actuation system are:

- The compressed air generation system, consisting of the compressor, the cooler, possibly a dryer, the storage tank, and the intake and output filters
- The compressed air treatment unit, usually consisting of the FRL assembly (filter, pressure regulator, and possibly a lubrifier), which permits filtration and local regulation of the supply pressure to the actuator valve
- The valve, that is, the regulator of the pneumatic power
- The actuator, which converts the pneumatic power into mechanical power
- The piping
- The sensors and transducers
- The system display, physical magnitude measurement, and control devices

Some of the components of the pneumatic actuation system such as the compressors, treatment units, and some valves used in pneumatic servosystems are described below. The actuators are similar in function and construction to hydraulic ones, though they are built slightly lighter because of the lower working pressure.

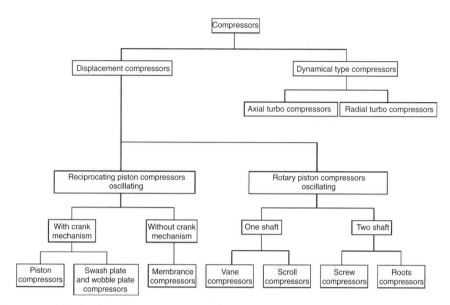

FIGURE 21.113 Classification of pneumatic compressors.

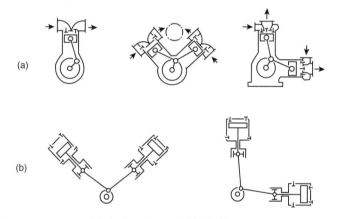

FIGURE 21.114 Piston compressors: (a) single action and (b) double action.

21.4.5.1 Compressors

The types of compressors used to produce compressed air are summarized in Figure 21.113. In volumetric compressors, the air or gas is sucked in by means of a valve in the compression chamber where its volume is reduced to cause compression of the gas. Opening of the delivery valve, when a predetermined pressure has been reached, results in the distribution of the air mass to the user.

Vice versa, in dynamic compressors or turbocompressors, the kinetic energy is converted into pressure energy transferred to the gas as a result of the rotary motion of the impeller.

Alternating piston compressors determine the compression of the gas as an effect of the motion of the piston, moved by a connecting rod and crank mechanism, inside a gas-tight cylinder. They can be single and double acting, with one or more pistons and one or more stages (Figure 21.114). They make it possible to obtain pressures of hundreds bar, where there are several stages, and flow rates of thousands of cubic meters per hour, in the case of several cylinders. Vane compressors (Figure 21.115) have a rotor, fitted eccentrically with respect to the axis of the cylinder in which it rotates, which leads to a certain

FIGURE 21.115 Rotary vane compressor (Pneumofore).

number of vanes which can move radially with respect to its axis. In the continuous rotation motion of the rotor, the vanes are centrifuged in contact with the seat of the stator, isolating chambers whose volume varies progressively with the angular stroke, guaranteeing input suction on the one hand, and a compressed gas output on the other. The compression pressures are below 15 bar, with maximum flow rates of 500 m^3/h. Compared with reciprocating piston compressors, they have less flow pulsation, fewer vibrations, and are more compact.

Screw compressors have two rotors rotating in opposite directions inside a stator, one with convex lobes and the other with concave lobes. The coupling of the profiles of the two rotors leads to a reduction of the volume during the angular stroke and consequent compression of the gas. With pressures typically below 15 bar, they provide a sufficiently continuous flow, up to values of about 3000 m^3/h.

In the same way, Roots compressors, also known as superchargers, are made up of two figure-of-eight-shaped rotors counter-rotating inside a stator in such a way as to transport volumes of gas from suction to delivery. Their efficiency is low because of the leakage between the rotors themselves and between the lobes and the casing, and they are, therefore, used for low compression pressures, below 2 bar. However, they do permit operation without lubrication, like screw compressors, so that oil-free air can be obtained.

Both axial and radial dynamic compressors are used to obtain high compressed air flow rates from a few thousand to 100,000 m^3/h.

21.4.5.2 Compressed Air Treatment Units

Pneumatic supply to a servosystem is generally provided by a local gas treatment unit, consisting of a filter, connected to a compressed gas distribution and generation network, a pressure regulator, and in case a lubricator L. Figure 21.116 shows an example of an integrated filter device and pressure regulator. The air first passes through the filter and is filtered by the deflector while the solid and liquid impurities in contact with the walls are deposited on the bottom of the cup, also as an effect of the conical bottom screen, located below the porous cylindrical element in sintered bronze or fabric. The filtered air then flows into the inlet of the pressure regulator, made up of an obturator in equilibrium between pressure

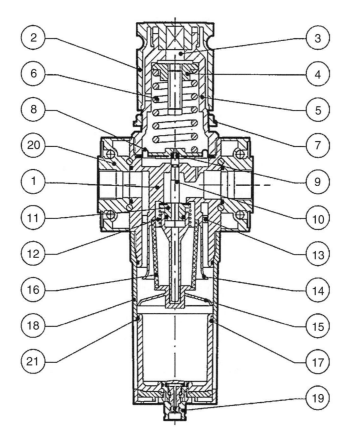

FIGURE 21.116 Pneumatic filter/pressure reducer (Metal Work).

forces. Control of the downstream pressure is determined by the position of the main obturator, which regulates the flow towards the outlet. The passage aperture is closed when the force due to the downstream pressure, acting on a diaphragm and on a translating piston, is in equilibrium with the force of the top spring, the preload of which is set by the rotation of the control knob. Vice versa, if the pressure force is below the desired value, the flow sent to the user tends to compensate for the pressure error with consequent closing of the obturator again when the set point has been reached. The opposite situation occurs if the regulated pressure is above the requested value, so that an aperture passage opens between the user and discharge.

21.4.5.3 Pneumatic Valves

Pneumatic valves are functionally similar to those used in hydraulic systems, so that reference should be made to the general considerations described above. In particular, this is also valid for the directional valves of the digital and proportional types. Even in pneumatic systems, there are digital spool or poppet two-, three-, or four-way distributors, with two or three working positions, and actuated manually, mechanically, pneumatically, and electrically.

Flow proportional valves are substantially similar to hydraulic ones and are available both with the torque motor electromechanical converter (servovalve), and with servosolenoid acting directly on the spool.

As well as these components for controlling the gas flow, digital electrically controlled two- or three-way valves are also used, and their control signals are modulated using PWM (pulse width modulation), PFM (pulse frequency modulation), PCM (pulse code modulation), PNM (pulse number modulation), or a combination of these.

As far as pressure regulation valves are concerned, three-way pressure proportional valves are available for pneumatic actuation which convert an electrical reference signal with standardized input into a controlled output pressure with good dynamics and high precision.

PWM (Pulse Width Modulation) Valves

The structure of PWM valves is similar to the corresponding electrically controlled unistable digital valves, but uses a technique for modulating the width of the pulses sent to the solenoid for supplying proportional control of the flow rate. This technique envisages that the input voltage reference analog signal V_{REF} (for example 0–10 V) is converted by a special driver into a digital V_{PWM} (ON/OFF) signal with pulse duration proportional to the input signal. Alternatively, the modulated signal can be generated directly by a digital controller, such as a PLC.

Figure 21.117 shows the PWM operating principle. The digital voltage signal sent to the valve solenoid is made up of a pulse train, with a constant amplitude, with a constant period T, but with the duration t of every pulse being a linear function of the analog value of the reference voltage. The average valve opening value, and therefore an initial approximation of the generated flow, is a function of the duration t of the pulse, in particular of the duty cycle t/T, and increases as the latter increases.

PWM valves generally do not have any feedback, so that the value of the downstream pressure, and therefore of the flow rate, depends on the type of pneumatic circuit present.

Figure 21.118 shows two plans which depict operation of the two-way, two-position valves, with PWM, used as a flow regulator (Figure 21.118a) and as a pressure regulator (Figure 21.118b). In plan a, the valve proportionally controls the flow which transits between the two points at pressure P_S (feed pressure) and at pressure P_V (downstream pressure) maintained constant. In this case, this flow is only a function of the aperture of the valve and therefore the proportionality is of the linear type. In plan b, the cross-fitted valves control the pressure P_R, for example, within a fluid capacity of volume V, respectively regulating the mass flow rate G_1 and discharge flow G_2. The time gradient for the controlled pressure P_R corresponds to the resulting flow G entering the reservoir.

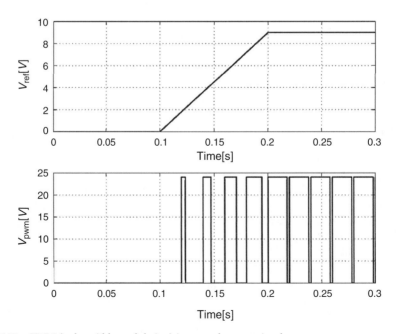

FIGURE 21.117 PWM (pulse width modulation) input and output signals.

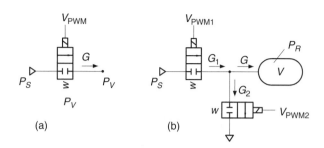

FIGURE 21.118 PWM (pulse width modulation) digital valves: (a) flow regulator and (b) pressure regulator.

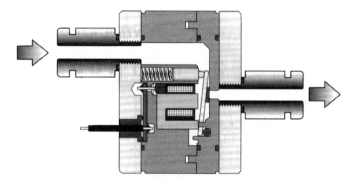

FIGURE 21.119 Two-way digital poppet valve (Matrix).

The parameters affecting the performance of a regulation made by a PWM on/off valve are as follows:

- Valve opening and closing times
- Dependence on the opening/closing times of the upstream and downstream pressures
- Valve size
- Period T or modulation carrier frequency $f = 1/T$
- Working life of the valve

While small opening/closing times and high flow capacity are always antithetical characteristics in an on/off valve, when designing the system it is always necessary to find a compromise between the need for good control resolution and linearity and a high response dynamic.

In pneumatic servosystem applications, typical carrier frequency values $f = 1/T$ range between 20 and 100 Hz, so that valves with opening/closing times of 1 ÷ 5 ms are used.

An example of a normally closed 2/2 valve that guarantees minimum opening times (below 1 ms with speed-up command) can be seen in Figure 21.119. These characteristics are obtained by reducing the mass of the moving parts, with the use of a poppet connected to a small oscillating bar, while practically eliminating the friction between the parts in relative motion.

Proportional Pressure Regulator Valves

These valves are normally three-way, with double poppets or with spool. Poppet valves operate in a similar way to pressure regulator valves. In the same way as with pressure regulators, the poppet which separates the high pressure environment from the regulated pressure one is in equilibrium between the force due to the regulated pressure and that exerted by the action of the control block. The latter can directly be the force of the servosolenoid armature, or that due to a pressure controlled by the control block which acts on a piston or on a diaphragm linked with the poppet.

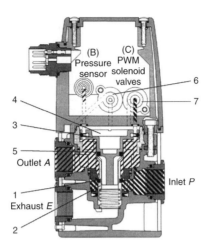

FIGURE 21.120 Pressure proportional valve (Parker).

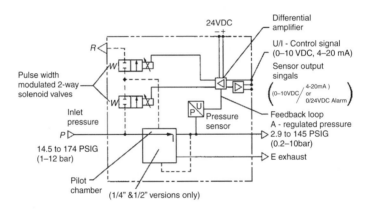

FIGURE 21.121 Pneumatic control scheme of the pressure proportional valve.

Figure 21.120 shows an example of a pressure proportional valve with double poppets. The ports at supply pressure, controlled pressure, and discharge are respectively indicated by P, A, and E. In the position indicated in the figure, the supply poppet 2 is at the top end of its stroke, as the seal 1 is against the fixed seat. In the same way, the regulating poppet 3 is in contact with the poppet 2 by means of the seal 5. The opening of the feed aperture, between the port P and the port A, is determined by the equilibrium of the forces acting on the piston of poppet 3, in particular the force F_R of the regulation pressure P_R in the servochamber 4 directed downwards, and the force F_C due to the action of the regulated pressure P_c on the outlet, directed upwards. If $F_R = F_C$, the moving bodies in the valve are in the positions shown in the figure, so that the chamber at controlled pressure P_C is isolated both from supply and discharge. If $F_R > F_C$, then the two poppets move downwards and the feed aperture is opened so as to convey the air mass towards the output and rebalance the pressure P_C at the desired value. In the opposite case, if $F_R < F_C$, the regulating poppet moves upwards, but while remaining at the top end of its stroke, the seal 5 opens and permits the passage of the masses from port A to the exhaust E.

In Figure 21.120, 6 and 7 indicate the PWM on/off valves, which regulate the pressure P_R of the servochamber 4.

The pneumatic control plan of the valve is shown in Figure 21.121. The two 2-way PWM valves receive the modulated control signal from the regulation block. These are fitted in such a way that one controls

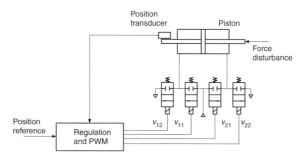

FIGURE 21.122 Scheme of a pneumatic servosystem with two-ways digital valves.

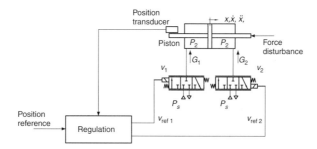

FIGURE 21.123 Scheme of a pneumatic servosystem with three-ways proportional valves.

a flow entering the servochamber 4 (see Figure 21.120) while the other controls the flow exiting towards discharge. By means of appropriate action, the control signal is converted into a pressure proportional signal.

21.4.6 Modeling a Pneumatic Servosystem

The circuitry plan of a pneumatic servosystem capable of controlling the position, speed, or force can be similar to that shown in Figure 21.109 for hydraulic actuation. The signal of the transducer of the desired magnitude must be specially fed back in a closed loop on the regulator depending on the controlled magnitudes.

In the plan in Figure 21.122, the axial position of the piston, fed back by means of the position transducer, is determined by controlling the pressure in thrust chambers 1 and 2 by means of the flow proportional interfaces. The position reference is compared with the feedback signal and the error is compensated in a control regulator. On the basis of the valve opening strategy used, the signal is sent to the regulating valves which feed the chambers of the piston, hypothesized to be symmetrical. The pressure forces acting on the thrust surfaces of the piston oppose the external force disturbance. The circuitry plan hypothesizes the use of four digital valves each with two unistable ways, electrically controlled. This solution makes it possible to use small-sized valves, with resulting high bandwidth, which must be compatible with the overall bandwidth requested by the pneumatic servosystem. In this solution, the proportionality of the opening of the valves is obtained by pulse width modulation of the digital signal. Each pair of valves V_{11}, V_{12} and V_{21}, V_{22} constitutes a three-way valve the output of which is connected to a piston chamber, so that the scheme in Figure 21.122 can be equivalent to that in Figure 21.123 with the three-way analogically controlled valves V_1 and V_2.

The cylinder model envisages a system with three-differential equations, two of continuity of the air mass in the chambers and one of dynamic translation equilibrium.

The following magnitudes are differentiated (the subscripts 1 and 2 refer, respectively, to the rear 1 and front 2 chambers of the pistons):

A	piston thrust section
F_e	disturbance of force acting on the piston rod
G	mass flow rate of air entering the chamber
M	mass of the translating parts of the piston
n	air polytropic coefficient
P	cylinder chamber pressure
P_i	initial cylinder chamber pressure
P_{amb}	ambient pressure
R	air constant
T_i	initial cylinder chamber air temperature
x	rod position measured starting from x_0
x_0	piston half stroke
x_m	dead band
γ	coefficient of viscous friction

The continuity and equilibrium equations are given by:

$$\frac{dP_1}{dt} = \frac{G_1 n R T_{1i}}{A_1(x_0 + x_{m1} + x)(P_1/P_{1i})^{(1-n)/n}} - \frac{P_1 n}{(x_0 + x_{m1} + x)} \frac{dx}{dt} \tag{21.58}$$

$$\frac{dP_2}{dt} = \frac{G_2 n R T_{2i}}{A_2(x_0 + x_{m2} - x)(P_2/P_{2i})^{(1-n)/n}} + \frac{P_2 n}{(x_0 + x_{m2} - x)} \frac{dx}{dt} \tag{21.59}$$

$$\frac{d^2 x}{dt^2} = \frac{(P_1 - P_{amb})A_1 - (P_2 - P_{amb})A_2 - F_e - \gamma dx/dt}{M} \tag{21.60}$$

The flow proportional valve V_1 is modeled as a variable section pneumatic resistance. The equations used for calculating mass flow rate G through a pneumatic resistance, characterized by a conductance C and by a critical ration b, in accordance with ISO 6358, which connects two environments A and B, with respective pressures of P_A and P_B, taken to be positive in the $A \to B$ direction, are

$$\text{sonic flow:} \quad G = \rho_0 P_A C \quad \text{for} \quad 0 < \frac{P_B}{P_A} \leq b \tag{21.61}$$

$$\text{subsonic flow:} \quad G = \rho_0 P_A C \sqrt{1 - \left(\frac{P_B/P_A - b}{1 - b}\right)^2} \quad \text{for} \quad b < \frac{P_B}{P_A} \leq 1 \tag{21.62}$$

$$\text{sonic flow:} \quad G = -\rho_0 P_B C \quad \text{for} \quad 0 < \frac{P_A}{P_B} \leq b \tag{21.63}$$

$$\text{subsonic flow:} \quad G = -\rho_0 P_B C \sqrt{1 - \left(\frac{P_A/P_B - b}{1 - b}\right)^2} \quad \text{for} \quad b < \frac{P_A}{P_B} \leq 1 \tag{21.64}$$

Hypothesizing a bipolar reference signal, it can be assumed that the range $V_{ref} > 0$ corresponds to the supply–user connection, while the field $V_{ref} < 0$ corresponds to the user–discharge connection. The appropriate equation of flow above must be rewritten in the same way.

Calculation of the conductance of the flow proportional valve is made considering the static and dynamic link between the reference voltage V_{ref} and the opening of the passage aperture A_V in accordance with modeling of the second order of the type:

$$\frac{d^2 A_V}{dt^2} + 2\zeta\sigma_n \frac{dA_V}{dt} + \sigma_n^2 A_V = K_s \sigma_n^2 V_{ref} \qquad (21.65)$$

where ζ is the damping factor, σ_n is the valve's natural frequency, and K_s is its area static gain.

Assuming a static relation of the linear type between the opening A_V and the conductance C, as an initial approximation, we get

$$C = K_c A_V = K_c K_s V_{ref} = K_V V_{ref} \qquad (21.66)$$

where K_V is the flow static gain of the valve, function of the maximum conductance C_{max}, and of the maximum value of the reference voltage $V_{ref\,max}$:

$$K_V = \frac{C_{max}}{V_{ref\,max}} \qquad (21.67)$$

The complete dynamic relation between reference voltage and conductance is, therefore,

$$\frac{d^2 C}{dt^2} + 2\zeta\sigma_n \frac{dC}{dt} + \sigma_n^2 C = K_c \sigma_n^2 V_{ref} \qquad (21.68)$$

The nonlinear model of the pneumatic servosystem with the position reference x_{set} and the force disturbance F_e as inputs, is made up of a nonlinear system of nine equations, of order eight overall, of the type:

a) $C_1 = C_1(V_{ref\,1}, t)$ order 2   conductance of valve V_1 (see (21.68))
b) $C_2 = C_2(V_{ref\,2}, t)$ order 2   flow rate of valve V_1 (see (21.68))
c) $G_1 = G_1(C_1, P_1)$ order 0   flow rate of valve V_1 (see (21.61)–(21.64))
d) $G_2 = G_2(C_2, P_2)$ order 0   flow rate of valve V_2 (see (21.61)–(21.64))
e) $G_1 = G_1(P_1, \dot{P}_1, x, \dot{x})$ order 1   continuity chamber 1 (see (21.58))
f) $G_2 = G_2(P_2, \dot{P}_2, x, \dot{x})$ order 1   continuity chamber 2 (see (21.59))
g) $x = x(F_e, P_1, P_2, \ddot{x})$ order 2   piston equilibrium (see (21.60))
h) $V_{ref\,1} = V_{ref\,1}(x_{set}, x_{ret})$ order 0 V_1 valve control
i) $V_{ref\,2} = V_{ref\,2}(x_{set}, x_{ret})$ order 0 V_2 valve control

If we want to carry out a linear analysis, it can be assumed that the equations a), b), g), h), i) are already written in linear form.

As far as the flow rates of valves c) and d) are concerned, it is hypothesized that the flow rate for each of them is subsonic in feed, with $V_{ref} > 0$, and sonic in discharge, with $V_{ref} < 0$. This means that for valve V_1, for example, the pressure P_1 must be within the range $bP_s < P_1 \leq P_s$ in feed and in the range $P_1 \geq P_{amb}/b$

in discharge. This hypothesis is physically acceptable; hypothesizing $b = 0.3$, $P_s = 10$ bar, $P_{amb} = 1$ bar, we get that P_1 can vary between 3.33 bar and 10 bar. It is the same for P_2.

Linearizing the subsonic feed flow rate curve with a secant passing through the points $P_1 = P_s$, $G_1 = 0$ and $P_1 = b_1^* P_s$, $G_1 = G_{1\,sonic}$, of angular coefficient K_{L1}, we get

$$G_1 = K_{L1}(P_s - P_1) \tag{21.69}$$

where

$$K_{L1} = \frac{G_{1\,sonic}}{P_s - b_1^* P_s} = \frac{\rho_n C_1 P_s}{P_s(1 - b_1^*)} = \frac{\rho_n C_1}{1 - b_1^*} \tag{21.70}$$

or

$$G_1 = \frac{\rho_n(P_s - P_1)}{(1 - b_1^*)} C_1 \tag{21.71}$$

In the neighborhood $P_1 = P_{1r}$ and $C_1 = C_{1r} = 0$ (the subscript r indicates *of reference*), we get

$$G_1 = \frac{\rho_n(P_s - P_{1r})}{(1 - b_1^*)} C_1 = K_{11} C_1 \tag{21.72}$$

Equation 21.72 is valid for $V_{refl} > 0$, to which $C_1 > 0$ corresponds analytically.
The flow rate discharge is expressed by

$$G_1 = \rho_n C_1 P_1 \tag{21.73}$$

which in the neighborhood $P_1 = P_{1r}$ becomes

$$G_1 = \rho_n P_{1r} C_1 = K_{12} C_1 \tag{21.74}$$

Equation 21.74 is valid for $V_{refl} < 0$, to which $C_1 < 0$ corresponds analytically.
Calculating a mean of slopes K_{11} and K_{12} we get a mean slope K_{mean} given by

$$K_{mean} = \frac{K_{11} + K_{12}}{2} = \frac{\rho_n(P_s - b_1^* P_{1r})}{2(1 - b_1^*)} \tag{21.75}$$

The linearized flow rate as a function of C_1 therefore becomes

$$G_1 = \frac{\rho_n(P_s - b_1^* P_{1r})}{2(1 - b_1^*)} C_1 = K_1 C_1 \tag{21.76}$$

Actuators

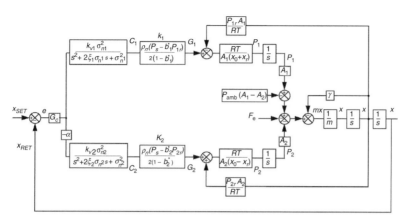

FIGURE 21.124 Block diagram of the linearized model of a pneumatic servosystem with position control.

In the same way for the valve V_2 we get

$$G_2 = \frac{\rho_n(P_s - b_2^* P_{2r})}{2(1 - b_2^*)} C_2 = K_2 C_2 \qquad (21.77)$$

The continuity equations of the mass in the piston chambers e) and f), linearized in the reference neighborhood defined by

$$x = x_r, \quad P_1 = P_{1r}, \quad P_2 = P_{2r},$$

$$\dot{P}_1 = \dot{P}_{1r} = 0, \quad \dot{P}_2 = \dot{P}_{2r} = 0, \quad \dot{x} = \dot{x}_r = 0$$

$$x_{m1} = x_{m2} = 0, \quad n = 1$$

become

$$G_1 = \frac{P_{1r} A_1}{RT} \dot{x} + A_1 \frac{x_0 + x_r}{RT} \dot{P}_1 \qquad (21.78)$$

$$G_2 = -\frac{P_{2r} A_2}{RT} \dot{x} + A_2 \frac{x_0 - x_r}{RT} \dot{P}_2 \qquad (21.79)$$

The block diagram of the linearized model is shown in Figure 21.124.

By applying the Laplace transforms of the system of linearized equations, assuming identical valves, we get:

$$\bar{x} = \frac{\sigma_n^2}{(s^2 + 2\zeta\sigma_n s + \sigma_n^2)} \frac{\sigma_A^2}{(s^2 + 2\zeta_A \sigma_A s + \sigma_A^2)} G_c K_{OLV} \bar{e} - \frac{\sigma_A^2}{(s^2 + 2\zeta_A \sigma_A s + \sigma_A^2)} K_{OLF} s \bar{F}_e + \text{C.I.} \qquad (21.80)$$

where C.I. indicates the initial conditions, K_{OLV} is the static gain in speed, K_{OLF} is the gain of the force disturbance, σ_A and ζ_A are, respectively, the actuator's natural frequency and the damping factor, and G_c is the compensator law.

This result is shown in the block diagram in Figure 21.125. Figure 21.126, on the other hand, shows the closed loop block diagram with position feedback. Obvious similarities can be seen when this plan is compared with that for a hydraulic servosystem.

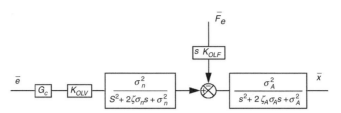

FIGURE 21.125 Block diagram of the open loop model.

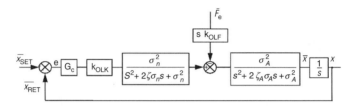

FIGURE 21.126 Block diagram of the closed loop model.

In particular, hypothesizing that:

$P_{1r} = P_{2r} = \delta P_s$ with $\delta \in 0.6 - 0.9$
$A_1 = A_2 = A$ double ended actuator
$n = 1$ isothermal transformation

we can express the significant parameters of the pneumatic servosystem:

$$K_{OLV} = \frac{RT\rho_n}{2} K_V \frac{1}{(1-b^*)} \frac{1}{A} \frac{(1/\delta - b^*)}{(1-b^*)} \tag{21.81}$$

$$K_{OLF} = \frac{x_0[1-(x_r/x_0)^2]}{2\delta P_s A} \tag{21.82}$$

$$\sigma_A = \sqrt{\frac{2\delta P_s A}{x_0 m[1-(x_r/x_0)^2]}} \tag{21.83}$$

$$\zeta_A = \gamma \sqrt{\frac{x_0[1-(x_r/x_0)^2]}{8\delta P_s A m}} \tag{21.84}$$

On the basis of the design specifications, it is possible to choose the size and characteristics of the servosystem components, operating first on the linearized model, and then checking the complete effectiveness of the choice made with a complete nonlinear system model.

References

Andersen, B. W., *The Analysis and Design of Pneumatic Systems*, Wiley, New York, 1967.
Bouteille, D., Belforte, G., *Automazione Flessibile, Elettropneumatica e Pneumatica*, Tecniche Nuove, Milano, 1987.

Belforte, G., D'Alfio, N., *Applicazioni e Prove Dell'automazione a Fluido*, Levrotto & Bella, Torino, 1997.
Belforte, G., Manuello Bertetto, A., Mazza, L., *Pneumatica: Corso Completo*, Tecniche Nuove, Milano, 1998.
Blackburn, J. F., Reethof, G., Shearer, J. L., *Fluid Power Control*, MIT Press, Cambridge, 1960.
Dransfield, P., *Hydraulic Control Systems—design and Analysis of their Dynamics*, Springer, Berlin, 1981.
Esposito, A., *Fluid Power with Applications*, 5th ed., Prentice-Hall, Upper Saddle River, NJ, 2000.
Gotz, W., *Hydraulics. Theory and Applications*, Robert Bosch Automation Technology Division Training, Ditzingen, 1998.
Hehn, A. H., *Fluid Power Troubleshooting*, 2nd ed., Dekker, New York, 1995.
Introduction to Hydraulic Circuits and Components, The University of Bath, Bath, 2000.
Introduction to Control for Electrohydraulic Systems, The University of Bath, Bath, 1999.
Jacazio, G., Piombo, B., *Meccanica Applicata Alle Macchine 3: Regolazione e servomeccanismi*, Levrotto & Bella, Torino, 1994.
Johnson, J. E., *Electrohydraulic Servo Systems*, 2nd ed., Penton IPC, Cleveland, 1977.
Johnson, J. L., *Design of Electrohydraulic Systems for Industrial Motion Control*, Penton IPC, Cleveland, 1991.
Johnson, J. L., *Basic Electronics for Hydraulic Motion Control*, Penton IPC, Cleveland, 1992.
Lewis, E. E., Stern, H., *Design of Hydraulic Control Systems*, McGraw-Hill, New York, 1962.
Mang, T., Dresel, W., *Lubricants and Lubrications*, Wiley-VCH, Weinheim, 2001.
McCloy, D., Martin, H. R., *The Control of Fluid Power*, Longman, London, 1973.
Merritt, H. E., *Hydraulic Control Systems*, Wiley, New York, 1967.
Moog, *Technical Bulletins*, 101–152, Moog, New York.
Muller, R., *Pneumatics. Theory and Applications*, Robert Bosch Automation Technology Division Training, Ditzingen, 1998.
Nervegna, N., *Oleodinamica e Pneumatica*, Politeko, Torino, 1999.
Parr, A., *Hydraulics and Pneumatics: A Technician's and Engineer's Guide*, 2nd ed., Butterworth Heinemann, Oxford, 1998.
Shetty, D., Kolk, R. A., *Mechatronics System Design*, PWS publishing company, Boston, 1997.
Tonyan, M. J., *Electronically Controlled Proportional Valves: Selection and Application*, Dekker, New York, 1985.
Viersma, T. J., *Analysis, Synthesis and Design of Hydraulic Servosystems and Pipelines*, Elsevier, Amsterdam, 1980.
Yeaple, F., *Fluid Power Design Handbook*, 3rd ed., Dekker, New York, 1996.

21.5 MEMS: Microtransducers Analysis, Design, and Fabrication*

Sergey Edward Lyshevski

21.5.1 Introduction

In many applications (from medicine and biotechnology to aerospace and security), the use of nano- and microscale structures, devices, and systems is very important [1–4]. This chapter discusses the analysis, modeling, design, and fabrication of electromagnetic-based microscale structures and devices (microtransducers controlled by ICs). It is obvious that to attain our objectives and goals, the synergy of multidisciplinary engineering, science, and technology must be utilized. In particular, electromagnetic

---

*This section is a part of the book: S. E. Lyshevski, *MEMS and NEMS: Systems, Devices, and Structures*, CRC Press, Boca Raton, FL, 2001.

theory and mechanics comprise the fundamentals for analysis, modeling, simulation, design, and optimization, while fabrication is based on the micromachining and high-aspect-ratio techniques and processes, which are the extension of the CMOS technologies developed to fabricate ICs. For many years, the developments in microelectromechanical systems (MEMS) have been concentrated on the fabrication of microstructures adopting, modifying, and redesigning silicon-based processes and technologies commonly used in integrated microelectronics. The reason for refining of conventional processes and technologies as well as application of new materials is simple: in general, microstructures are three-dimensional with high aspect ratios and large structural heights in contrast to two-dimensional planar microelectronic devices. Silicon structures can be formed from bulk silicon micromachining using wet or dry processes, or through surface micromachining. Metallic micromolding techniques, based upon photolithographic processes, are also widely used to fabricate microstructures. Molds are created in polymer films (usually photoresist) on planar surfaces, and then filled by electrodepositing metal (electrodeposition plays a key role in the fabrication of the microstructures and microdevices, which are the components of MEMS). High-aspect ratio technologies use optical, e-beam, and x-ray lithography to create trenches up to 1 mm deep in polymethylmethacrylate resist on the electroplating base (called seed layer). Electrodeposition of magnetic materials and conductors, electroplating, electroetching, and lift-off are extremely important processes to fabricate microscale structures and devices. Though it is recognized that the ability to use and refine existing microelectronics fabrication technologies and materials is very important, and the development of novel processes to fabricate MEMS is a key factor in the rapid growth of affordable MEMS, other emerging areas arise. In particular, devising, design, modeling, analysis, and optimization of novel MEMS are extremely important. Therefore, recently, the MEMS theory and microengineering fundamentals have been expanded to thoroughly study other critical problems such as the system-level synthesis and integration, synergetic classification and analysis, modeling and design, as well as optimization. This chapter studies the fabrication, analysis, and design problems for electromagnetic microstructures and microdevices (microtransducers with ICs). The descriptions of the fabrication processes are given, modeling and analysis issues are emphasized, and the design is performed.

21.5.2 Design and Fabrication

In MEMS, the fabrication of thin film magnetic components and microstructures requires deposition of conductors, insulators, and magnetic materials. Some available bulk material constants (conductivity σ, resistivity ρ at 20°C, relative permeability μ_r, thermal expansion t_e, and dielectric constant—relative permittivity ε_r) in SI units are given in Table 21.7.

TABLE 21.7 Material Constants

Material	σ	ρ	μ_r	$t_e \times 10^{-6}$	ε_r
Silver	6.17×10^7	0.162×10^{-7}	0.9999998	NA	
Copper	5.8×10^7	0.172×10^{-7}	0.99999	16.7	
Gold	4.1×10^7	0.244×10^{-7}	0.99999	14	
Aluminum	3.82×10^7	0.26×10^{-7}	1.00000065	25	
Tungsten	1.82×10^7	0.55×10^{-7}	NA	NA	
Zinc	1.67×10^7	0.6×10^{-7}	NA	NA	
Cobalt	NA	NA	250	NA	
Nickel	1.45×10^7	0.69×10^{-7}	600 nonlinear	NA	
Iron	1.03×10^7	1×10^{-7}	4000 nonlinear	NA	
Si				2.65	11.8
SiO$_2$				0.51	3.8
Si$_3$N$_4$				2.7	7.6
SiC				3.0	6.5
GaAs				6.9	13
Ge				2.2	16.1

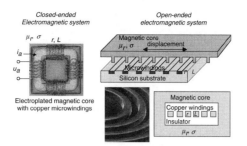

FIGURE 21.127 Closed-ended and open-ended electromagnetic systems in microtransducers (toroidal microstructures with the insulated copper circular conductors wound around the magnetic material and linear micromotor) with magnetic cores (stator and rotor electroplated thin films).

Although MEMS topologies and configurations vary (see the MEMS classification concept [2]), in general, electromagnetic microtransducers have been designed as the closed-ended, open-ended, and integrated electromagnetic systems. As an example, Figure 21.127 illustrates the microtoroid and the linear micromotor with the closed-ended and open-ended electromagnetic systems, respectively. The copper windings and magnetic core (microstructures) can be made through electroplating, and Figure 21.129 depicts the electroplated circular copper conductors which form the windings (10 μm wide and thick with 10 μm spacing) deposited on the insulated layer of the magnetic core.

The comprehensive electromagnetic analysis must be performed for microscale structures and devices. For example, the torque (force) developed and the voltage induced by microtransducers depend upon the inductance, and the microdevice's efficiency is a function of the winding resistance, resistivity of the coils deposited, eddy currents, hysteresis, etc. Studying the microtoroid, consider a circular path of radius R in a plane normal to the axis. The magnetic flux intensity is calculated using the following formula:

$$\oint_s \mathbf{H} \cdot d\mathbf{s} = 2\pi R H = Ni$$

where N is the number of turns. Thus, one has

$$H = \frac{Ni}{2\pi R}$$

The value of H is a function of R, and therefore, the field is not uniform.

Microwindings must guarantee the adequate inductance in the limited footprint area with the minimal resistance. For example, in the microtransducers and low power converters, 0.5 μH (or higher) inductance is required at high frequency (1–10 MHz). Compared with the conventional minidevices, the thin film electromagnetic microtransducers have lower efficiency due to higher resistivity of thin films, eddy currents, hysteresis, fringing effect, and other undesirable phenomena, which usually have the secondary (negligible) effect in the miniscale and conventional electromechanical devices. The inductance can be increased by ensuring a large number of turns, using core magnetic materials with high relative permeability, increasing the cross-sectional core area, and decreasing the path length. In fact, at low frequency, the formula for inductance is

$$L = \frac{\mu_0 \mu_r N^2 A}{l}$$

where μ_r is the relative permeability of the core material, A is the cross-sectional area of the magnetic core, and l is the magnetic path length.

Using the reluctance $\mathcal{R} = l/(\mu_0 \mu_r A)$, one has $L = N^2/\mathcal{R}$. For the electromagnetic microtransducers, the flux is a very important variable, and using the *net* current, one has $\Phi = Ni/\mathcal{R}$.

It is important to recall that the inductance is related to the energy stored in the magnetic field, and

$$L = \frac{2W_m}{i^2} = \frac{1}{i^2}\int_v \mathbf{B} \cdot \mathbf{H} \, dv.$$

Thus, one has

$$L = \frac{1}{i^2}\int_v \mathbf{B} \cdot \mathbf{H} \, dv = \frac{1}{i^2}\int_v \mathbf{H} \cdot (\nabla \times \mathbf{A}) \, dv = \frac{1}{i^2}\int_v \mathbf{A} \cdot \mathbf{J} \, dv = \frac{1}{i}\oint_l \mathbf{A} \cdot d\mathbf{l} = \frac{1}{i}\oint_s \mathbf{B} \cdot d\mathbf{s} = \frac{\Phi}{i}$$

or

$$L = \frac{N\Phi}{i}$$

We found that the inductance is the function of the number of turns, flux, and current.

Making use of the equation

$$L = \frac{\mu_0 \mu_r N^2 A}{l}$$

one concludes that the inductance increases as a function of the squared number of turns. However, a large number of turns requires the high turn density (small track width and spacing so that many turns can be fitted in a given footprint area). However, reducing the track width leads to an increase in the conductor resistance, decreasing the efficiency. Therefore, the design trade-off between inductance and winding resistance must be studied. To achieve low resistance, one must deposit thick conductors with the thickness in the order of tens of micrometers. In fact, the dc resistance is found as $R = \rho_c l_c/A_c$, where ρ_c is the conductor resistivity, l_c is the conductor length, A_c is the conductor cross-sectional area. Therefore, the most feasible process for deposition of conductors is electroplating. High-aspect-ratio processes ensure thick conductors and small track widths and spaces (high-aspect-ratio conductors have a high thickness to width ratio). However, the footprint area is limited not allowing to achieve a large conductor cross-sectional area. High inductance value can also be achieved by increasing the magnetic core cross-sectional area using thick magnetic cores with large A. However, most thin film magnetic materials are thin film metal alloys, which generally have characteristics not as good as the bulk ferromagnetic materials. This results in the eddy current and undesirable hysteresis effects, which increase the core losses and decrease the inductance. It should be emphasized that eddy currents must be minimized.

As illustrated, magnetic cores and microwindings are key components of microstructures, and different magnetic and conductor materials and processes to fabricate microtransducers are employed. Commonly, the *permalloy* (nickel$_{80\%}$-iron$_{20\%}$ alloy) thin films are used. It should be emphasized that *permalloy* as well as other materials (e.g., amorphous cobalt-phosphorous) are soft magnetic materials that can be made through electrodeposition. In general, the deposits have nonuniform thickness and composition due to the electric current nonuniformity over the electrodeposition area. Furthermore, hydrodynamic effects in the electrolyte also usually increase nonuniformity (these nonuniformities are reduced by choosing a particular electrochemicals). The inductance and losses remain constant up to a certain frequency (which is a function of the layer thickness, materials used, fabrication processes, etc.), and in the high frequency operating regimes, the inductance rapidly decreases and the losses increase due to the eddy current and hysteresis effects. For example, for the *permalloy* ($Ni_{80\%}Fe_{20\%}$) thin film magnetic core and copper winding, the inductance decreases rapidly above 1, 3, and 6 MHz for the 10, 8, and 5 μm thick layers, respectively. It should be emphasized that the skin depth of the magnetic core thin film as a function of the magnetic properties and the frequency f is found as $\delta = \sqrt{1/(\pi f \mu \sigma)}$, where μ and σ are the permeability and conductivity of the magnetic core material, respectively. The total power losses are found using the Pointing vector $\Xi = \mathbf{E} \times \mathbf{H}$, and the total power loss can be approximately derived using the expression for the power crossing the conductor surface within the area, e.g.,

$$P_{average} = \int_s \Xi_{average} ds = \frac{1}{4}\int_s \sigma \delta E_o^2 e^{-2/\delta \sqrt{\pi f \mu \sigma}} ds$$

It is important to emphasize that the skin depth (depth of penetration) is available, and for the bulk copper $\delta_{cu} = 0.066/\sqrt{f}$.

In general, the inductance begins to decrease when the ratio of the lamination thickness to skin depth is greater than one. Thus, the lamination thickness must be less than skin depth at the operating frequency f to attain the high inductance value. In order to illustrate the need to comprehensively study microinductors, we analyze the toroidal microinductor (1 mm by 1 mm, 3 μm core thickness, 2000 permeability). The inductance and winding resistances are analyzed as the functions of the operating frequency. Modeling results indicate that the inductance remains constant up to 100 kHz and decreases for the higher frequency. The resistance increases significantly at frequencies higher than 150 kHz (the copper microconductor thickness is 2 μm, and the dc winding resistance is 10 Ω). The decreased inductance and increased resistance at high frequency are due to hysteresis and eddy current effects.

The skin depth in the magnetic core material depends on the permeability and the conductivity. The $Ni_{x\%}Fe_{100-x\%}$ thin films have a relative permeability in the range from 600 to 2000, and the resistivity is in the order of 20 $\mu\Omega$ cm. It should be emphasized that the materials with high resistivity have low eddy current losses and allow one to deposit thicker layers as the skin depth is high. Therefore, high resistivity magnetic materials are under consideration, and the electroplated FeCo thin films have 100–130 $\mu\Omega$ cm resistivity. Other high resistivity materials, which can be deposited by sputtering, are FeZrO and CoHfTaPd (sputtering has advantages for the deposition of laminated layers of magnetic and insulating materials because magnetic and insulating materials can be deposited in the same process step). Electroplating, as a technique for deposition of laminated multilayer structures, in general, requires different processes to deposit magnetic and insulating materials (layers).

The major processes involved in the electromagnetic microtransducer fabrication are etching and electroplating magnetic vias and through-holes, and then fabricating the inductor-type microstructures on top of the through-hole wafer using multilayer thick photoresist processes [5–7]. For example, let us use the silicon substrate (100-oriented n-type double-sided polished silicon wafers) with a thin layer of thermally grown silicon dioxide (SiO_2). Through-holes are patterned on the topside of the Si–SiO_2 wafer (photolithography process) and then etched in the KOH system (different etch rate can be attained based upon the concentration and temperature). Then, the wafer is removed from the KOH solution with 20–30 μm of silicon remaining to be etched. A seed layer of Ti–Cu or Cr–Cu (20–40 nm and 400–500 nm thickness, respectively) are deposited on the backside of the wafer using electron beam evaporation. The copper acts as the electroplating seed layer, while a titanium (or chromium) layer is used to increase adhesion of the copper layer to the silicon wafer. On the copper seed layer, a protective NiFe thin film layer is electroplated directly above the through-holes to attain protection and stability. The through-holes are fully etched again (in the KOH system), and then the remaining SiO_2 is stripped (using the BHF solution) to reveal the backside metal layers. Then, the titanium adhesion layer is etched in the HF solution (chromium, if the Cr–Cu seed layer is used, can be removed using the $K_3Fe(CN)_6$–NaOH solution). This allows the electroplating of through-holes from the exposed copper seed layer. The empty through-holes are electroplated with NiFe thin film. This forms the magnetic vias. Because the KOH-based etching process is crystallographically dependent, the sidewalls of the electroplating mold are the 111-oriented crystal planes (54.7° angular orientation to the surface). As a result of these 54.7°-angularly oriented sidewalls, the electroplating can be nonuniform. To overcome this problem, the through-holes can be over-plated and polished to the surface level [5–7]. After the through-hole plating and polishing, the seed layer is removed, and 10–20 μm coat (e.g., polyimide PI2611) is spun on the backside and cured at 300°C to cover the protective NiFe layer. Now, the microinductor can be fabricated on the topside of the wafer. In particular, the microcoils are fabricated on top of the through-hole wafer with the specified magnetic core geometry (e.g., plate- or horseshoe-shaped) parallel to the surface of the wafer. The microcoils must be wounded around the magnetic core to form the electromagnetic system. Therefore, the additional structural layers are needed (for example, the first level is the conductors that are the bottom segments of each microcoil turn, the second level includes the magnetic core and vertical conductors which connect the top and bottom of each microcoil turn segment, and the third level consists of the top conductors that are connected to the electrical vias, and thus form microcoil turns wounded around the magnetic core). It is obvious that the insulation (dielectric) layers are required to insulate the magnetic core and microcoils. The fabrication can be performed through the electron beam evaporation of the Ti-Cu seed layer, and then, 25–35 μm electroplating molds are formed (AZ-4000 photoresist can be used). The copper microcoils are electroplated on the top of the mold through electroplating. After electroplating is completed, the photoresist is removed with acetone. Then, the seed layer is removed (copper is etched in the H_2SO_4 solution, while the titanium adhesion layer is etched by the HF solution). A new layer of the AZ-4000 photoresist is spun on the wafer to insulate the bottom conductors from the magnetic core. The vias' openings are patterned at the ends of the conductors, and the photoresist is cured forming the insulation layer. In addition to insulation, the hard curing leads to reflow of the photoresist serving the planarization purpose needed to pattern additional layers. Another seed layer is deposited from which electrical vias and magnetic core are patterned and electroplated. This leads to two lithography sequential steps, and the electrical vias (electroplated Cu) and magnetic core (NiFe thin film) are electroplated using the same seed layer. After the vias and magnetic core are completed, the photoresist and seed layers are removed. Then, the hard curing is performed. The top microconductors are patterned and deposited from another seed layer using the same process as explained above for the bottom microconductors. The detailed description of the processes described and the fabricated microtransducers are available in [5–7]. We have outlined the fabrication of microinductors because these techniques can be adopted and used to fabricate microtransducers. It also must be emphasized that the analysis and design can be performed using the equations given.

21.5.3 Analysis of Translational Microtransducers

Figure 21.128 illustrates a microelectromechanical device (translational microtransducer) with a stationary member (magnetic core with windings) and movable member (microplunger), which can be fabricated using the micromachining technology. Our goal is to perform the analysis and modeling of the microtransducer developing the lamped-parameter mathematical model. That is, the goal is to derive the differential equations which model the microtransducer steady-state and dynamic behavior.

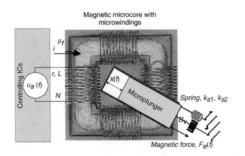

FIGURE 21.128 Schematic of the translational microtransducer with controlling ICs.

Applying Newton's second law for translational motion, we have

$$F(t) = m\frac{d^2x}{dt^2} + B_v\frac{dx}{dt} + (k_{s1}x + k_{s2}x^2) + F_e(t)$$

where x denotes the microplunger displacement, m is the mass of a movable member (microplunger), B_v is the viscous friction coefficient, k_{s1} and k_{s2} are the spring constants, and $F_e(t)$ is the magnetic force

$$F_e(i,x) = \frac{\partial W_c(i,x)}{\partial x}$$

It should be emphasized that the restoring/stretching force exerted by the spring is given by $(k_{s1}x + k_{s2}x^2)$.
Assuming that the magnetic system is linear, the coenergy is found to be $W_c(i,x) = \frac{1}{2}L(x)i^2$ and the electromagnetic force developed is given by

$$F_e(i,x) = \frac{1}{2}i^2\frac{dL(x)}{dx}$$

In this formula, the analytic expression for the term $dL(x)/dx$ must be found. The inductance is

$$L(x) = \frac{N^2}{\mathcal{R}_f + \mathcal{R}_g} = \frac{N^2 \mu_f \mu_0 A_f A_g}{A_g l_f + 2A_f \mu_f (x + 2d)}$$

where $\mathcal{R}_f$ and $\mathcal{R}_g$ are the reluctances of the magnetic material and air gap; A_f and A_g are the cross-sectional areas; l_f and $(x + 2d)$ are the lengths of the magnetic material and the air gap.
Thus

$$\frac{dL}{dx} = -\frac{2N^2 \mu_f^2 \mu_0 A_f^2 A_g}{[A_g l_f + 2A_f \mu_f (x + 2d)]^2}$$

Using Kirchhoff's law, the voltage equation for the electric circuit is

$$u_a = ri + \frac{d\psi}{dt}$$

where the flux linkage ψ is $\psi = L(x)i$.
Thus, one obtains

$$u_a = ri + L(x)\frac{di}{dt} + i\frac{dL(x)}{dx}\frac{dx}{dt}$$

Therefore, the following nonlinear differential equation results:

$$\frac{di}{dt} = -\frac{r}{L(x)}i + \frac{2N^2 \mu_f^2 \mu_0 A_f^2 A_g}{L(x)[A_g l_f + 2A_f \mu_f (x + 2d)]^2}iv + \frac{1}{L(x)}u_a$$

Augmenting this equation with the differential equation and the *torsional-mechanical* dynamics

$$F(t) = m\frac{d^2x}{dt^2} + B_v\frac{dx}{dt} + (k_{s1}x + k_{s2}x^2) + F_e(t)$$

three nonlinear differential equations for the considered translational microtransducer are found to be

$$\frac{di}{dt} = -\frac{r[A_g l_f + 2A_f \mu_f(x+2d)]}{N^2 \mu_f \mu_0 A_f A_g} i + \frac{2\mu_f A_f}{A_g l_f + 2A_f \mu_f(x+2d)} iv + \frac{A_g l_f + 2A_f \mu_f(x+2d)}{N^2 \mu_f \mu_0 A_f A_g} u_a$$

$$\frac{dv}{dt} = \frac{N^2 \mu_f^2 \mu_0 A_f^2 A_g}{m[A_g l_f + 2A_f \mu_f(x+2d)]^2} i^2 - \frac{1}{m}(k_{s1}x + k_{s2}x^2) - \frac{B_v}{m}v$$

$$\frac{dx}{dt} = v$$

The derived differential equations represent the lumped-parameter mathematical model of the microtransducer. Although, in general, the high-fidelity modeling must be performed integrating nonlinearities (for example, nonlinear magnetic characteristics and hysteresis) and secondary effects, the lumped-parameter mathematical models as given in the form of nonlinear differential equations have been validated for microtransducers. It is found that the major phenomena and effects are modeled for the current, velocity, and displacement (secondary effects such as Coulomb friction, hysteresis and eddy currents, fringing effect and other phenomena have not been modeled and analyzed). However, the lumped-parameter modeling provides one with the capabilities to attain reliable preliminary steady-state and dynamic analysis using primary circuitry and mechanical variables. It is also important to emphasize that the voltage, applied to the microwinding, is regulated by ICs. The majority of ICs to control microtransducers are designed using the pulse-width-modulation topologies. The switching frequency of ICs is usually 1 MHz or higher. Therefore, as was shown, it is very important to study the microtransducer performance at the high operating frequency. This can be performed using Maxwell's equations, which will lead to the high-fidelity mathematical models [2].

21.5.4 Single-Phase Reluctance Micromotors: Microfabrication, Modeling, and Analysis

Consider the single-phase reluctance micromachined motors as illustrated in Figure 21.129.

The emphases are concentrated on the analysis, modeling, and control of reluctance micromotors in the rotational microtransducer applications. Therefore, mathematical models must be found. The lumped-parameter modeling paradigm is based upon the use of the circuitry (voltage and current) and mechanical (velocity and displacement) variables to derive the differential equations using Newton's and Kirchhoff's laws. In these differential equations, the micromotor parameters are used. In particular, for the studied micromotor, the parameters are the stator resistance r_s, the magnetizing inductances in the *quadrature* and *direct* axes L_{mq} and L_{md}, the average magnetizing inductance $\bar{L}_m$, the leakage inductance L_{ls}, the moment of inertia J, and the viscous friction coefficient B_m.

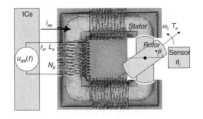

FIGURE 21.129 Single-phase reluctance micromotor with ICs and rotor displacement (position) sensor.

The expression for the electromagnetic torque was derived in [8]. In particular,

$$T_e = L_{\Delta m} i_{as}^2 \sin 2\theta_r$$

where $L_{\Delta m}$ is the half-magnitude of the sinusoidal magnetizing inductance L_m variations,

$$L_m(\theta_r) = \bar{L}_m - L_{\Delta m} \cos 2\theta_r$$

Thus, to develop the electromagnetic torque, the current i_{as} must be fed as a function of the rotor angular displacement θ_r. For example, if $i_{as} = i_M \mathrm{Re}(\sqrt{\sin 2\theta_r})$, then

$$T_{eaverage} = \frac{1}{\pi} \int_0^\pi L_{\Delta m} i_{as}^2 \sin 2\theta_r \, d\theta_r = \frac{1}{4} L_{\Delta m} i_M^2$$

The micromotor under our consideration is the synchronous micromachine, and the obtained expression for the phase current is very important to control the microtransducer. In particular, the Hall-effect position sensor should be used to measure the rotor displacement, and the ICs must feed the phase current as a nonlinear function of θ_r. Furthermore, the electromagnetic torque is controlled by changing the current magnitude i_M.

The mathematical model of the single-phase reluctance micromotor is found using Kirchhoff's and Newton's second laws. In particular, we have

$$u_{as} = r_s i_{as} + \frac{d\psi_{as}}{dt} \quad \text{(circuitry equation—Kirchhoff's law)}$$

$$T_e - B_m \omega_r - T_L = J \frac{d^2\theta_r}{dt^2} \quad \text{(torsional–mechanical equation—Newton's law)}$$

Here, the electrical angular velocity ω_r and displacement θ_r are used as the mechanical system variables. From $u_{as} = r_s i_{as} + \frac{d\psi_{as}}{dt}$ and the flux linkage equation $\psi_{as} = (L_{ls} + \bar{L}_m - L_{\Delta m} \cos 2\theta_r) i_{as}$, using the *torsional-mechanical* dynamics, one obtains a set of three first-order nonlinear differential equations which models single-phase reluctance micromotors. In particular, we have

$$\frac{di_{as}}{dt} = -\frac{r_s}{L_{ls} + \bar{L}_m - L_{\Delta m} \cos 2\theta_r} i_{as} - \frac{2 L_{\Delta m}}{L_{ls} + \bar{L}_m - L_{\Delta m} \cos 2\theta_r} i_{as} \omega_r \sin 2\theta_r + \frac{1}{L_{ls} + \bar{L}_m - L_{\Delta m} \cos 2\theta_r} u_{as}$$

$$\frac{d\omega_r}{dt} = \frac{1}{J}(L_{\Delta m} i_{as}^2 \sin 2\theta_r - B_m \omega_r - T_L)$$

$$\frac{d\theta_r}{dt} = \omega_r$$

As the mathematical model is found and the micromotor parameters are measured, nonlinear simulation and analysis can be straightforwardly performed to study the dynamic responses and analyze the micromotor efficiency. In particular, the resistance, inductances, moment of inertia, viscous friction coefficient, and other parameters can be directly measured or identified based upon micromotor testing. The steady-state and dynamic analysis based upon the lumped-parameter mathematical model is straightforward. However, the lumped-parameter mathematical models simplify the analysis, and thus, these models must be compared with the experimental data to validate the results.

The disadvantage of single-phase reluctance micromotors are high torque ripple, vibration, noise, low reliability, etc. Therefore, let us study three-phase synchronous reluctance micromotors.

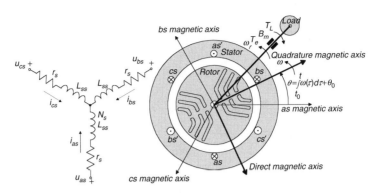

FIGURE 21.130 Three-phase synchronous reluctance micromotor.

21.5.5 Three-Phase Synchronous Reluctance Micromotors: Modeling and Analysis

Our goal is to address and solve a spectrum of problems in analysis, modeling, and control of synchronous reluctance micromachines. The electromagnetic features must be thoroughly analyzed before attempt to control micromotors. In fact, electromagnetic features significantly restrict the control algorithms to be applied. Depending upon the conceptual methods employed to analyze synchronous reluctance micromachines, different control laws can be designed and implemented using ICs. Analysis and control of synchronous reluctance micromotors can be performed using different modeling, analysis, and optimization concepts. Complete lumped-parameter mathematical models of synchronous reluctance micromotors in the *machine* (*abc*) and in the *quadrature, direct*, and *zero* (*qd*0) variables should be developed in the form of nonlinear differential equations. In particular, the circuitry lumped-parameters mathematical model is found using the Kirchhoff's voltage law. We have, see Figure 21.130,

$$\mathbf{u}_{abcs} = \mathbf{r}_s \mathbf{i}_{abcs} + \frac{d\psi_{abcs}}{dt}$$

where u_{as}, u_{bs}, and u_{cs} are the phase voltages; i_{as}, i_{bs}, and i_{cs} are the phase currents; ψ_{as}, ψ_{bs}, and ψ_{cs} are the flux linkages,

$$\psi_{abcs} = \mathbf{L}_s \mathbf{i}_{abcs}, \qquad \mathbf{r}_s = \begin{bmatrix} r_s & 0 & 0 \\ 0 & r_s & 0 \\ 0 & 0 & r_s \end{bmatrix}$$

$$\mathbf{L}_s = \begin{bmatrix} L_{ls} + \bar{L}_m - L_{\Delta m}\cos(2\theta_r) & -\frac{1}{2}\bar{L}_m - L_{\Delta m}\cos 2\left(\theta_r - \frac{1}{3}\pi\right) & -\frac{1}{2}\bar{L}_m - L_{\Delta m}\cos 2\left(\theta_r + \frac{1}{3}\pi\right) \\ -\frac{1}{2}\bar{L}_m - L_{\Delta m}\cos 2\left(\theta_r - \frac{1}{3}\pi\right) & L_{ls} + \bar{L}_m - L_{\Delta m}\cos 2\left(\theta_r - \frac{2}{3}\pi\right) & -\frac{1}{2}\bar{L}_m - L_{\Delta m}\cos 2(\theta_r + \pi) \\ -\frac{1}{2}\bar{L}_m - L_{\Delta m}\cos 2\left(\theta_r + \frac{1}{3}\pi\right) & -\frac{1}{2}\bar{L}_m - L_{\Delta m}\cos 2(\theta_r + \pi) & L_{ls} + \bar{L}_m - L_{\Delta m}\cos 2\left(\theta_r + \frac{2}{3}\pi\right) \end{bmatrix}$$

$$\bar{L}_m = \frac{1}{3}(L_{mq} + L_{md}) \quad \text{and} \quad L_{\Delta m} = \frac{1}{3}(L_{md} - L_{mq})$$

The micromachine parameters are the stator resistance r_s, the magnetizing inductances in the *quadrature* and *direct* axes L_{mq} and L_{md}, the average magnetizing inductance $\bar{L}_m$, the leakage inductance L_{ls}, the moment of inertia J, and the viscous friction coefficient B_m.

The expressions for inductances are nonlinear functions of the electrical angular displacement θ_r. Hence, the *torsional–mechanical* dynamics must be used. Taking note of the Newton's second law of rotational motion, and using ω_r and θ_r (electrical angular velocity and displacement) as the state variables (mechanical variables), one obtains

$$T_e - B_m \frac{2}{P}\omega_r - T_L = J\frac{2}{P}\frac{d\omega_r}{dt}, \quad \frac{d\theta_r}{dt} = \omega_r$$

where T_e and T_L are the electromagnetic and load torques.

Torque Production Analysis—Using the coenergy, the electromagnetic torque, which is a nonlinear function of the micromotor variables (phase currents and electrical angular position) and micromotor parameters (number of poles P and inductance $L_{\Delta m}$), is found to be [8],

$$T_e = \frac{P}{2}L_{\Delta m}\left[i_{as}^2 \sin 2\theta_r + 2i_{as}i_{bs}\sin 2\left(\theta_r - \frac{1}{3}\pi\right) + 2i_{as}i_{cs}\sin 2\left(\theta_r + \frac{1}{3}\pi\right) \right.$$
$$\left. + i_{bs}^2 \sin 2\left(\theta_r - \frac{2}{3}\pi\right) + 2i_{bs}i_{cs}\sin 2\theta_r + i_{cs}^2 \sin 2\left(\theta_r + \frac{2}{3}\pi\right) \right]$$

To control the angular velocity, the electromagnetic torque must be regulated. To maximize the electromagnetic torque, ICs must feed the following phase currents as functions of the angular displacement measuring or observing (sensorless control) θ_r

$$i_{as} = \sqrt{2}i_M \sin\left[\theta_r + \frac{1}{3}\varphi_i \pi\right]$$

$$i_{bs} = \sqrt{2}i_M \sin\left[\theta_r - \frac{1}{3}(2-\varphi_i)\pi\right]$$

$$i_{cs} = \sqrt{2}i_M \sin\left[\theta_r + \frac{1}{3}(2+\varphi_i)\pi\right]$$

Thus, for $\varphi_i = 0.3245$, one obtains

$$T_e = \sqrt{2}PL_{\Delta m}i_M^2$$

That is, T_e is maximized and controlled by changing the magnitude of the phase currents i_M. Furthermore, it is no torque ripple (in practice, based upon the experimental results, and performing the high-fidelity modeling integrating nonlinear electromagnetics using Maxwell's equations, one finds that there exists the torque ripple which is due to the cogging torque, eccentricity, bearing, pulse-width-modulation, and other phenomena).

The majority of ICs are designed to control the phase voltages u_{as}, u_{bs}, and u_{cs}. Therefore, the three-phase balance voltage set is important. We have

$$u_{as} = \sqrt{2}u_M \sin\left[\theta_r + \frac{1}{3}\varphi_i \pi\right]$$

$$u_{bs} = \sqrt{2}u_M \sin\left[\theta_r - \frac{1}{3}(2-\varphi_i)\pi\right]$$

$$u_{cs} = \sqrt{2}u_M \sin\left[\theta_r + \frac{1}{3}(2+\varphi_i)\pi\right]$$

where u_M is the magnitude of the supplied voltages.

The mathematical model of synchronous reluctance micromotors in the *abc* variables is found to be

$$\frac{di_{as}}{dt} = \frac{1}{L_D}\Big\{(r_s i_{as} - u_{as})(4L_{ls}^2 + 3\bar{L}_m^2 - 3L_{\Delta m}^2 + 8\bar{L}_m L_{ls} - 4L_{ls}L_{\Delta m}\cos 2\theta_r)$$

$$+ (r_s i_{bs} - u_{bs})\Big[3\bar{L}_m^2 - 3L_{\Delta m}^2 + 2\bar{L}_m L_{ls} + 4L_{ls}L_{\Delta m}\cos 2\Big(\theta_r - \frac{1}{3}\pi\Big)\Big]$$

$$+ (r_s i_{cs} - u_{cs})\Big[3\bar{L}_m^2 - 3L_{\Delta m}^2 + 2\bar{L}_m L_{ls} + 4L_{ls}L_{\Delta m}\cos 2\Big(\theta_r + \frac{1}{3}\pi\Big)\Big]$$

$$+ 6\sqrt{3}L_{\Delta m}^2 L_{ls}\omega_r(i_{cs} - i_{bs}) + (8L_{\Delta m}L_{ls}^2\omega_r + 12L_{\Delta m}\bar{L}_m L_{ls}\omega_r)$$

$$\times\Big(\sin 2\theta_r i_{as} + \sin 2\Big(\theta_r - \frac{1}{3}\pi\Big)i_{bs} + \sin 2\Big(\theta_r + \frac{1}{3}\pi\Big)i_{cs}\Big)\Big\}$$

$$\frac{di_{bs}}{dt} = \frac{1}{L_D}\Big\{(r_s i_{as} - u_{as})\Big[3\bar{L}_m^2 - 3L_{\Delta m}^2 + 2\bar{L}_m L_{ls} + 4L_{ls}L_{\Delta m}\cos 2\Big(\theta_r - \frac{1}{3}\pi\Big)\Big]$$

$$+ (r_s i_{bs} - u_{bs})\Big[4L_{ls}^2 + 3\bar{L}_m^2 - 3L_{\Delta m}^2 + 8\bar{L}_m L_{ls} - 4L_{ls}L_{\Delta m}\cos 2\Big(\theta_r + \frac{1}{3}\pi\Big)\Big]$$

$$+ (r_s i_{cs} - u_{cs})[3\bar{L}_m^2 - 3L_{\Delta m}^2 + 2\bar{L}_m L_{ls} + 4L_{ls}L_{\Delta m}\cos 2\theta_r]$$

$$+ 6\sqrt{3}L_{\Delta m}^2 L_{ls}\omega_r(i_{as} - i_{cs}) + (8L_{\Delta m}L_{ls}^2\omega_r + 12L_{\Delta m}\bar{L}_m L_{ls}\omega_r)$$

$$\times\Big(\sin 2\Big(\theta_r - \frac{1}{3}\pi\Big)i_{as} + \sin 2\Big(\theta_r + \frac{1}{3}\pi\Big)i_{bs} + \sin 2\theta_r i_{cs}\Big)\Big\}$$

$$\frac{di_{cs}}{dt} = \frac{1}{L_D}\Big\{(r_s i_{as} - u_{as})\Big[3\bar{L}_m^2 - 3L_{\Delta m}^2 + 2\bar{L}_m L_{ls} + 4L_{ls}L_{\Delta m}\cos 2\Big(\theta_r + \frac{1}{3}\pi\Big)\Big]$$

$$+ (r_s i_{bs} - u_{bs})(3\bar{L}_m^2 - 3L_{\Delta m}^2 + 2\bar{L}_m L_{ls} + 4L_{ls}L_{\Delta m}\cos 2\theta_r)$$

$$+ (r_s i_{cs} - u_{cs})\Big[4L_{ls}^2 + 3\bar{L}_m^2 - 3L_{\Delta m}^2 + 8\bar{L}_m L_{ls} - 4L_{ls}L_{\Delta m}\cos 2\Big(\theta_r - \frac{1}{3}\pi\Big)\Big]$$

$$+ 6\sqrt{3}L_{\Delta m}^2 L_{ls}\omega_r(i_{bs} - i_{as}) + (8L_{\Delta m}L_{ls}^2\omega_r + 12L_{\Delta m}\bar{L}_m L_{ls}\omega_r)$$

$$\times\Big[\sin 2\Big(\theta_r + \frac{1}{3}\pi\Big)i_{as} + \sin 2\theta_r i_{bs} + \sin 2\Big(\theta_r - \frac{1}{3}\pi\Big)i_{cs}\Big]\Big\}$$

$$\frac{d\omega_r}{dt} = \frac{P^2}{4J}L_{\Delta m}\Big[i_{as}^2\sin 2\theta_r + 2i_{as}i_{bs}\sin 2\Big(\theta_r - \frac{1}{3}\pi\Big) + 2i_{as}i_{cs}\sin 2\Big(\theta_r + \frac{1}{3}\pi\Big)$$

$$+ i_{bs}^2\sin 2\Big(\theta_r - \frac{2}{3}\pi\Big) + 2i_{bs}i_{cs}\sin 2\theta_r + i_{cs}^2\sin 2\Big(\theta_r + \frac{2}{3}\pi\Big)\Big] - \frac{B_m}{J}\omega_r - \frac{P}{2J}T_L$$

$$\frac{d\theta_r}{dt} = \omega_r$$

Here

$$\bar{L}_m = \frac{1}{3}(L_{mq} + L_{md}), \quad L_{\Delta m} = \frac{1}{3}(L_{md} - L_{mq}) \quad \text{and} \quad L_D = L_{ls}(9L_{\Delta m}^2 - 4L_{ls}^2 - 12\bar{L}_m L_{ls} - 9\bar{L}_m^2)$$

The mathematical model can be simplified. In particular, in the rotor reference frame, we apply the Park transformation [8]

$$\mathbf{u}_{qd0s}^r = \mathbf{K}_s^r \mathbf{u}_{abcs}, \quad \mathbf{i}_{qd0s}^r = \mathbf{K}_s^r \mathbf{i}_{abcs}, \quad \mathbf{\psi}_{qd0s}^r = \mathbf{K}_s^r \mathbf{\psi}_{abcs}$$

$$\mathbf{K}_s^r = \frac{2}{3} \begin{bmatrix} \cos\theta_r & \cos\left(\theta_r - \frac{2}{3}\pi\right) & \cos\left(\theta_r + \frac{2}{3}\pi\right) \\ \sin\theta_r & \sin\left(\theta_r - \frac{2}{3}\pi\right) & \sin\left(\theta_r + \frac{2}{3}\pi\right) \\ \frac{1}{2} & \frac{1}{2} & \frac{1}{2} \end{bmatrix}$$

where u_{qs}, u_{ds}, u_{0s}, i_{qs}, i_{ds}, i_{0s}, and ψ_{qs}, ψ_{ds}, ψ_{0s} are the $qd0$ voltages, currents, and flux linkages.

Using the circuitry and *torsional–mechanical* dynamics, one finds the following nonlinear differential equations to model synchronous reluctance micromotors in the rotor reference frame:

$$\frac{di_{qs}^r}{dt} = -\frac{r_s}{L_{ls} + L_{mq}} i_{qs}^r - \frac{L_{ls} + L_{md}}{L_{ls} + L_{mq}} i_{ds}^r \omega_r + \frac{1}{L_{ls} + L_{mq}} u_{qs}^r$$

$$\frac{di_{ds}^r}{dt} = -\frac{r_s}{L_{ls} + L_{md}} i_{ds}^r + \frac{L_{ls} + L_{mq}}{L_{ls} + L_{md}} i_{qs}^r \omega_r + \frac{1}{L_{ls} + L_{md}} u_{ds}^r$$

$$\frac{di_{0s}^r}{dt} = -\frac{r_s}{L_{ls}} i_{0s}^r + \frac{1}{L_{ls}} u_{0s}^r$$

$$\frac{d\omega_r}{dt} = \frac{3P^2}{8J}(L_{md} - L_{mq}) i_{qs}^r i_{ds}^r - \frac{B_m}{J}\omega_r - \frac{P}{2J} T_L$$

$$\frac{d\theta_r}{dt} = \omega_r$$

One can easily observe that this model is much simpler compared with the lumped-parameter mathematical model derived using the *abc* variables.

To attain the balanced operation, the *quadrature* and *direct* currents and voltages must be derived using the *direct* Park transformation $\mathbf{i}_{qd0s}^r = \mathbf{K}_s^r \mathbf{i}_{abcs}$, $\mathbf{u}_{qd0s}^r = \mathbf{K}_s^r \mathbf{u}_{abcs}$. Hence, the $qd0$ voltages u_{qs}^r, u_{ds}^r, and u_{0s}^r are found using the three-phase balance voltage set. In particular, we have

$$u_{qs}^r = \sqrt{2} u_M, \quad u_{ds}^r = 0, \quad u_{0s}^r = 0$$

We derived the mathematical models of three-phase synchronous reluctance micromotors. Based upon the differential equations obtained, nonlinear analysis can be performed, and the phase currents and voltages needed to guarantee the balance operating conditions can be found. The results reported can be straightforwardly used in nonlinear simulation.

21.5.6 Microfabrication Aspects

The fabrication of electromechanical microstructures and microtransducers can be made through deposition of the conductors (coils and windings), magnetic core, insulating layers, as well as other microstructures (movable and stationary members and their components). The order of the processes, materials, and sequential steps are different depending on the MEMS which must be devised, designed, analyzed, and optimized first.

21.5.6.1 Conductor Thin Films Electrodeposition

The conductors (microcoils to make windings) in microstructures and microtransducers can be fabricated by electrodepositing the copper and other low resistivity metals. Electrodeposition of metals is made by immersing a conductive surface in a solution containing ions of the metal to be deposited. The surface is electrically connected to an external power supply, and current is fed through the surface into the solution. In general, the reaction of the metal ions (Metal^{x+}) with x electrons (xe^-) to form metal (Metal) is Metal^{x+} + xe^- = Metal.

To electrodeposit copper on the silicon wafer, the wafer is typically coated with a thin conductive layer of copper (seed layer) and immersed in a solution containing cupric ions. Electrical contact is made to the seed layer, and current is flowed (passed) such that the reaction $Cu^{2+} + 2e^- \rightarrow Cu$ occurs at the wafer surface. The wafer, which is electrically interacted such that the metal ions are changed to metal atoms, is the cathode. Another electrically active surface (anode) is the conductive solution to make the electrical path. At the anode, the oxidation reaction occurs that balances the current flow at the cathode, thus maintaining the electric neutrality. In the case of copper electroplating, all cupric ions removed from solution at the wafer cathode are replaced by dissolution from the copper anode. According to the Faraday law of electrolysis, in the absence of secondary reactions, the current delivered to a conductive surface during electroplating is proportional to the quantity of the metal deposited. Thus, the metal deposited can be controlled varying the electroplating current (current density) and the electrodeposition time.

The hydrated Cu ions reaction is

$$Cu^{++} \rightarrow Cu(H_2O)_6^{++}$$

and the cathode reactions are

$$Cu^{++} + 2e^- \rightarrow Cu, \quad Cu^{++} + e^- \rightarrow Cu^+, \quad Cu^+ + e^- \rightarrow Cu, \quad 2Cu^+ \rightarrow Cu^{++} + Cu, \quad H^+ + e^- \rightarrow \frac{1}{2}H_2$$

The copper electroplating solution commonly used is $CuSO_4$–$5H_2O$ (250 g/l) and H_2SO_4 (25 ml/l).

The basic processes are shown in Figure 21.131, and the brief description of the sequential steps and equipment that can be used are given.

It must be emphasized that commonly used magnetic materials and conductors do not adhere well to silicon. Therefore, as was described, the adhesion layers (e.g., titanium Ti or chromium Cr) are deposited on the silicon surface prior to the magnetic material electroplating.

The electrodeposition rate is proportional to the current density and, therefore, the uniform current density at the substrate seed layer is needed to attain the uniform thickness of the electrodeposit. To achieve the selective electrodeposition, portions of the seed layer are covered with the resist (the current density at the mask edges nonuniform degrading electroplating). In addition to the current density, the deposition rate is also a nonlinear function of temperature, solution (chemicals), pH, direct/reverse current or voltage waveforms magnitude, waveform pulses, duty ratio, plating area, etc. In the simplest form, the thickness and electrodeposition time for the specified materials are calculated as

$$\text{Thickness}_{material} = \frac{\text{Time}_{electroplating} \times \text{Current}_{density} \times \text{Weight}_{molecular}}{\text{Faraday}_{constant} \times \text{Density}_{material} \times \text{Electron}_{number}}$$

$$\text{Time}_{electroplating} = \frac{\text{Thickness}_{material} \times \text{Faraday}_{constant} \times \text{Density}_{material} \times \text{Electron}_{number}}{\text{Current}_{density} \times \text{Weight}_{molecular}}$$

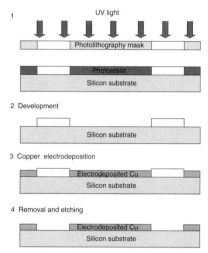

FIGURE 21.131 Electrodeposition of copper and basic processes: silicon, kapton, and other substrates, can be used. After clearing, the silicon substrate is covered with a 5–10 nm chromium or titanium and 100–200 nm copper seed layer by sputtering. The copper microcoils (microstructures) are patterned using the UV photolitography. The AZ-4562 photoresist can be spincoated and prebaked on a ramped hot plate at 90–100°C (ramp 30–40% with initial temperature 20–25°C) for one hour. Then, the photoresist is exposed in the Karl Suss Contact Masker with the energy 1200–1800 mJ cm^2. The development is released in 1:4 diluted alkaline solution (AZ-400) for 4–6 min. This gives the photoresist thickness 15–25 μm. Copper is electroplated with a three-electrode system with a copper anode and a saturated calomel reference electrode (the current power supply is the Perkin Elmer Current Source EG&G 263). The Shipley sulfate bath with the 5–10 ml/l brightener to smooth the deposit can be used. The electrodeposition is performed at 20–25°C with magnetic stirring and the dc current density 40–60 mA/cm^2 (this current density leads to smooth copper thin films with the 5–10 nm rms roughness for the 10 μm thickness of the deposited copper thin film). The resistivity of the electrodeposited copper thin film (microcoils) is 1.6–1.8 $\mu\Omega$ cm (close to the bulk copper resistivity). After the deposition, the photoresist is removed.

It was emphasized that electroplating is used to deposit thin-film conductors and magnetic materials. However, microtransducers need the insulation layers, otherwise the magnetic core and coils as well as multilayer microcoils themself will be short-circuited. Furthermore, the seed layers are embedded in microfabrication processes. As the magnetic core is fabricated on top of the microcoils (or microcoils are made on the magnetic core), the seed layer is difficult to remove because it is at the bottom or at the center of the microstructure. The mesh seed layer can serve as the electroplating seed layer for the lower conductors, and as the microstructure is made, the edges of the mesh seed layer can be exposed and removed through plasma etching [6]. Thus, the microcoils are insulated. It should be emphasized that relatively high aspect ratio techniques must be used to fabricate the magnetic core and microcoils, and patterning as well as surface planarization issues must be addressed.

21.5.6.2 NiFe Thin Films Electrodeposition

Magnetic cores in microstructures and microtransducers must be made. For example, the electroplated $Ni_{x\%}Fe_{100-x\%}$ thin films, such as *permalloy* $Ni_{80\%}Fe_{20\%}$, can be deposited to form the magnetic core of microtransducers (actuators and sensors), inductors, transformers, switches, etc. The basic processes and sequential steps used are similar to the processes for the copper electrodeposition and the electroplating is done in the electroplating bath. The windings (microcoils) must be insulated from magnetic cores, and therefore, the insulation layers must be deposited. The insulating materials used to insulate the windings from the magnetic core are benzocyclobutene, polyimide (PI-2611), etc. For example, the cyclotene 7200-35 is photosensitive and can be patterned through photolithography. The benzocyclobutene,

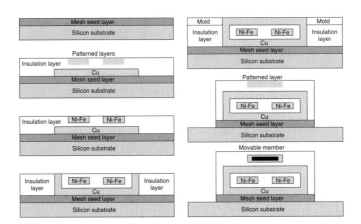

FIGURE 21.132 Basic fabrication sequential steps for the microtransducer fabrication.

used as the photoresist, offers good planarization and pattern properties, stability at low temperatures, and exhibits negligible hydrophilic properties.

The sketched fabrication process with sequential steps to make the electromagnetic microtransducer with a movable member is illustrated in Figure 21.132. On the silicon substrate, the chromium–copper–chromium (Cr–Cu–Cr) mesh seed layer is deposited (through electron-beam evaporation) forming a seed layer for electroplating. The insulation layer (polyimide Dupont PI-2611) is spun on the top of the mesh seed layer to form the electroplating molds. Several coats can be done to obtain the desired thickness of the polyimide molds (one coat results in 8–12 μm insulation layer thickness). After coating, the polyimide is cured (at 280–310°C) in nitrogen for 1 h. A thin aluminum layer is deposited on top of the cured polyimide to form a hard mask for dry etching. Molds for the lower conductors are patterned and plasma etched until the seed layer is exposed. After etching the aluminum (hard mask) and chromium (top chromium–copper–chromium seed layer), the molds are filled with the electroplated copper, applying the described copper electroplating process. One coat of polyimide insulates the lower conductors and the magnetic core (thus, the insulation is achieved). The seed layer is deposited, mesh-patterned, coated with polyimide, and hard-cured. The aluminum thin layers (hard mask for dry etching) are deposited, and the mold for the magnetic cores is patterned and etched until the seed layer is exposed. After etching the aluminum (hard mask) and the chromium (top chromium–copper–chromium seed layer), the mold is filled with the electroplated $Ni_{x\%}Fe_{100-x\%}$ thin films (electroplating process). One coat of the insulating layer (polyimide) is spin-cast and cured to insulate the magnetic core and upper conductors. The via holes are patterned in the sputtered aluminum layer (hard mask) and etched through the polyimide layer using oxygen plasma. The vias are filled with the electroplated copper (electroplating process). A copper–chromium seed layer is deposited and the molds for the upper conductors are formed using thick photoresist. The molds are filled with the electroplated copper and removed. Then, the gap for the movable member is made using the conventional processes. After removing the seed layer, the passivation layer (polyimide) is coated and cured to protect the top conductors. The polyimide is masked and etched to the silicon substrate. The bottom mesh seed layer is wet etched and the microtransducer (with the ICs to control it) is diced and sealed.

Electroplated aluminum is the needed material to fabricate microstructures. In particular, aluminum can be used as the conductor to fabricate microcoils as well as mechanical microstructures (gears, bearing, pins, reflecting surfaces, etc.). Advanced techniques and processes for the electrodeposition of aluminum are documented in [9].

As was reported, the magnetic core of microstructures and microtransducers must be fabricated. Two major challenges in fabrication of high-performance microstructures are to make electroplated magnetic thin films with good magnetic properties as well as planarize microstructures (stationary and movable

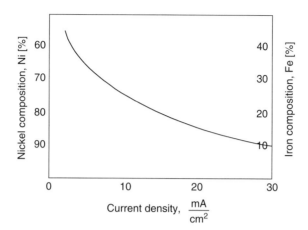

FIGURE 21.133 Nickel and iron compositions in $Ni_{x\%}Fe_{100-x\%}$ thin films as the functions of the current density.

members) [10]. Electroplating and micromolding techniques and processes are used to deposit NiFe alloys ($Ni_{x\%}Fe_{100-x\%}$ thin films), and $Ni_{80\%}Fe_{20\%}$ is called *permalloy*, while $Ni_{50\%}Fe_{50\%}$ is called *orthonol*.

Let us document the deposition process. To deposit $Ni_{x\%}Fe_{100-x\%}$ thin films, the silicon wafer is covered with a seed layer (for example, 15–25 nm chromium, 100–200 nm copper, and 25–50 nm chromium) deposited using electron beam evaporation. The photoresist layer (e.g., 10–20 μm Shipley STR-1110) is deposited on the seed layer and patterned. Then, the electrodeposition of the $Ni_{x\%}Fe_{100-x\%}$ is performed at 20–30°C using a two-electrode system, and the current density is in the range from 1 to 30 mA/cm^2. The temperature and pH should be maintained within the recommended values. High pH causes highly stressed NiFe thin films, and the low pH reduces leveling and cause chemical dissolving of the iron anodes resulting in disruption of the bath equilibrium and nonuniformity. High temperature leads to hazy deposits, and low temperature causes high current density burning. For deposition, the pulse-width-modulation (with varied waveforms, different forward and reverse magnitudes, and controlled duty cycle) can be used applying commercial or in-house made pulsed power supplies. Denoting the duty cycle length as T, the forward and reverse pulses lengths are denoted as T_f and T_r. The pulse length T can be 5–20 μs, and the duty cycle (ratio T_f/T_r) can be varied from 1 to 0.1. The ratio T_f/T_r influences the percentage of Ni in the $Ni_{x\%}Fe_{100-x\%}$ thin films, e.g., the composition of $Ni_{x\%}Fe_{100-x\%}$ can be regulated based upon the desired properties, which will be discussed later. However, varying the ratio T_f/T_r, the changes of the Ni are relatively modest (from 85% to 79%), and therefore, other parameters must vary to attain the desired composition.

It must be emphasized that the nickel (and iron) composition is a function of the current density, and Figure 21.133 illustrates the nickel (iron) composition in the $Ni_{x\%}Fe_{100-x\%}$ thin films.

The $Ni_{80\%}Fe_{20\%}$ thin films of different thickness (which is a function of the electrodeposition time) are usually made at the current density 14–16 mA/cm^2. This range of the current density can be used to fabricate a various thickness of *permalloy* thin films (from 500 nm to 50 μm). The rms value of the thin film roughness is 4–7 nm for the 25 μm thickness. It should be emphasized that to guarantee good surface quality, the current density should be kept at the specified range, and usually to change the composition of the $Ni_{x\%}Fe_{100-x\%}$ thin films, the reverse current is controlled.

To attain a good deposit of the *permalloy*, the electroplating bath contains $NiSO_4$ (0.7 mol/l), $FeSO_4$ (0.03 mol/l), $NiCl_2$ (0.02 mol/l), saccharine (0.016 mol/l) as leveler (to reduce the residual stress allowing the fabrication of thicker films), and boric acid (0.4 mol/l).

The air agitation and saccharin were added to reduce internal stress and to keep the Fe composition stable. The deposition rate varies linearly as a function of the current density (the Faraday law is obeyed), and the electrodeposition slope is 100–150 nm cm^2/min mA. The *permalloy* thin films' density is 9 g/cm^3 (as for the bulk *permalloy*).

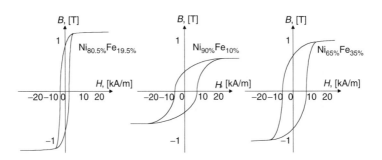

FIGURE 21.134 B–H curves for different $Ni_{x\%}Fe_{100-x\%}$ permalloy thin films.

The magnetic properties of the $Ni_{80\%}Fe_{20\%}$ (permalloy) thin films are studied, and the field coercivity (H_c) is a function of the thickness. For example, H_c = 650 A/m for 150 nm thickness and H_c = 30 A/m for 600 nm films.

Other $Ni_{80\%}Fe_{20\%}$ (deposited at 25°C) and $Ni_{50\%}Fe_{50\%}$ (deposited at 55°C) electroplating solutions are:

- $Ni_{80\%}Fe_{20\%}$: $NiSO_4$–$6H_2O$ (200 g/l/), $FeSO_4$–$7H_2O$ (9 g/l), $NiCl_2$–$6H_2O$ (5 g/l), H_3BO_3 (27 g/l), saccharine (3 g/l), and pH (2.5–3.5);
- $Ni_{50\%}Fe_{50\%}$: $NiSO_4$–$6H_2O$ (170 g/l/), $FeSO_4$–$7H_2O$ (80 g/l), $NiCl_2$–$6H_2O$ (138 g/l), H_3BO_3 (50 g/l), saccharine (3 g/l), and pH (3.5–4.5).

To electroplate $Ni_{x\%}Fe_{100-x\%}$ thin films, various additives and components (available from M&T Chemicals and other suppliers) can be used to control the internal stress and ductility of the deposit, keep the iron content solublized, obtain bright film and leveling of the process, attain the desired surface roughness, and most importantly to guarantee the desired magnetic properties.

In general, the *permalloy* thin films have optimal magnetic properties at the following composition: 80.5% of Ni and 19.5% of Fe. For $Ni_{80.5\%}Fe_{19.5\%}$ thin films, the material magnetostriction has zero crossing. Films with minimal magnetostriction usually have optimal coercivity and permeability properties, and, in general, the coercivity (depending on the films thickness) is 20 A/m (and higher as the thickness decreases), and permeability is from 600 to 2000. Varying the composition of Fe and Ni, the characteristics of the $Ni_{x\%}Fe_{100-x\%}$ thin films can be changed. The composition of the $Ni_{x\%}Fe_{100-x\%}$ thin films is controlled by changing the current density, T_f/T_r ratio (duty cycle), bath temperature (varying the temperature, the composition of Ni can be varied from 75% to 92%), reverse current (varying the reverse current in the range 0–1 A, the composition of Ni can be changed from 72% to 90%), air agitation of the solution, paddle frequency (0.1–1 Hz), forward and reverse pulses waveforms, etc. The B–H curves for three different $Ni_{x\%}Fe_{100-x\%}$ thin films are illustrated in Figure 21.134. The $Ni_{80.5\%}Fe_{19.5\%}$ thin films have the saturation flux density 1.2 T, remanence B_r = 0.26 T-A/m, and the relative permeability 600–2000.

It must be emphasized that other electroplated permanent magnets (NiFeMo, NiCo, CoNiMnP, and other) and micromachined polymer magnets exhibit good magnetic properties and can be used as the alternative solution to the $Ni_{x\%}Fe_{100-x\%}$ thin films widely used.

21.5.6.3 NiFeMo and NiCo Thin Films Electrodeposition

To attain the desired magnetic properties (flux density, coercivity, permeability, etc.) and thickness, different thin film alloys can be used based upon the microstructure's and microtransducer's design, applications, and operating envelopes (temperature, shocks, radiation, humidity, etc.). As was discussed, the $Ni_{x\%}Fe_{100-x\%}$ thin films can be effectively used, and the desired magnetic properties can be readily achieved varying the composition of Ni and Fe. For sensors, the designer usually maximizes the flux density and permeability and minimizes the coercivity. The $Ni_{x\%}Fe_{100-x\%}$ thin films have the flux density up to 1.2 T, coercivity 20 (*permalloy*) to 500 A/m, and permeability 600–2000 (it was emphasized that the magnetic properties also depend upon the thickness). Having emphasized the magnetic properties of the $Ni_{x\%}Fe_{100-x\%}$ thin films, let us perform the comparison. It was reported in the literature that [11]:

$Ni_{79\%}Fe_{17\%}Mo_{4\%}$ thin films have the flux density 0.7 T, coercivity 5 A/m, and permeability 3400, $Ni_{85\%}Fe_{14\%}Mo_{1\%}$ thin films have the flux density 1–1.1 T, coercivity 8–300 A/m, and permeability 3000–20000, $Ni_{50\%}Co_{50\%}$ thin films have the flux density 0.95–1.1 T, coercivity 1200–1500 A/m, and permeability 100–150 ($Ni_{79\%}Co_{21\%}$ thin films have the permeability 20).

In general, high flux density, low coercivity, and high permeability lead to high-performance MEMS. However, other issues (affordability, compliance, integrity, operating envelope, fabrication, etc.) must be also addressed while making the final choice. It must be emphasized that the magnetic characteristics, in addition to the film thickness, are significantly influenced by the fabrication processes and chemicals (materials) used.

The magnetic core in microstructures and microtransducers must be made. Two major challenges in fabrication of high-performance microstructures and microtransducers are to make electroplated magnetic thin films with good magnetic properties as well as planarize the stationary and movable members. Electroplating and micromolding techniques and processes are used to deposit NiFe, NiFeMo, and NiCo thin films. In particular, the $Ni_{80\%}Fe_{20\%}$, NiFeMo, and NiCo (deposited at 25°C) electroplating solutions are:

- $Ni_{80\%}Fe_{20\%}$: $NiSO_4$–$6H_2O$ (200 g/l), $FeSO_4$–$7H_2O$ (9 g/l), $NiCl_2$–$6H_2O$ (5 g/l), H_3BO_3 (27 g/l), and saccharine (3 g/l). The current density is 10–25 mA/cm^2 (nickel foil is used as the anode);
- NiFeMo: $NiSO_4$–$6H_2O$ (60 g/l/), $FeSO_4$–$7H_2O$ (4 g/l), Na_2MoO_4–$2H_2O$ (2 g/l), NaCl (10 g/l), citric acid (66 g/l), and saccharine (3 g/l). The current density is 10–30 mA/cm^2 (nickel foil is used as the anode);
- $Ni_{50\%}Co_{50\%}$: $NiSO_4$–$6H_2O$ (300 g/l/), $NiCl_2$–$6H_2O$ (50 g/l), $CoSO_4$–$7H_2O$ (30 g/l), H_3BO_3 (30 g/l), sodium lauryl sulfate (0.1 g/l), and saccharine (1.5 g/l). The current density is 10–25 mA/cm^2 (nickel or cobal can be used as the anode).

The most important feature is that the $Ni_{x\%}Fe_{100-x\%}$–NiFeMo–NiCo thin films (multiplayer nanocomposites) can be fabricated shaping the magnetic properties of the resulting materials to attain the desired performance characteristics through design and fabrication processes.

21.5.6.4 Micromachined Polymer Permanent Magnets

Electromagnetic microactuators can be deviced and fabricated using micromachined permanent magnet thin films including polymer magnets (magnetically hard ceramic ferrite powder imbedded in epoxy resin). Different forms and geometry of polymer magnets are available. Thin-film disks and plates are uniquely suitable for microactuator applications. For example, to actuate the mirrors in optical devices and optical MEMS, permanent magnets are used in rotational and translational (linear) microtransducers, microsensors, microswitches, etc. These polymer magnets have thickness ranging from hundreds of micrometers to several millimeters. Excellent magnetic properties can be achieved. For example, the micromachined polymer permanent-magnet disk with 80% strontium ferrite concentration (4 mm diameter and 90 μm thickness), magnetized normal to the thin-film plane (in the thickness direction), has the intrinsic coercivity H_{ci} = 320,000 A/m and a residual induction B_r = 0.06 T [12]. Permanent-magnet polymer magnets with thickness up to several millimeters can be fabricated by the low-temperature processes. To make the permanent magnets, the Hoosier Magnetics Co. strontium ferrite powder (1.1–1.5 μm grain size) and Shell epoxy resin (cured at 80°C for 2 h) can be used [12]. The polymer matrix contain a bisphenol-A-based epoxy resin diluted with cresylglycidyl ether and the aliphatic amidoamine is used as for curing. To prepare the polymer magnet composites, the strontium ferrite powder is mixed with the epoxy resin in the ball-mill rotating system (0.5 rad/s for many hours). After the aliphatic amidoamine is added, the epoxy is deposited and patterned using screen-printing. Then, the magnet is cured at 80°C for 2 h and magnetized in the desired direction.

It must be emphasized that magnets must be magnetized. That is, in addition to fabrication processes, one should study other issues, for example, the magnetization dynamics. The magnetic field in thin films are modeled, analyzed, and simulated solving differential equations, and the analytic and numerical results will be covered.

21.5.7 Magnetization Dynamics of Thin Films

The magnetic field, including the magnetization distribution, in thin films are modeled, analyzed, and simulated solving differential equations. The dynamic variables are the magnetic field density and intensity, magnetization, magnetization direction, wall position domain, etc. The thin films must be magnetized. Therefore, let us study the magnetization dynamics for thin films. To attain high-fidelity modeling, the magnetization dynamics in the angular coordinates is described by the Landay–Lifschitz–Gilbert equations [13]:

$$\frac{d\psi}{dt} = -\frac{\gamma}{M_s(1+\alpha^2)}\left(\sin^{-1}\psi\frac{\partial E(\theta,\psi)}{\partial \theta} + \alpha\frac{\partial E(\theta,\psi)}{\partial \psi}\right)$$

$$\frac{d\theta}{dt} = -\frac{\gamma\sin^{-1}\psi}{M_s(1+\alpha^2)}\left(\alpha\sin^{-1}\psi\frac{\partial E(\theta,\psi)}{\partial \theta} - \frac{\partial E(\theta,\psi)}{\partial \psi}\right)$$

where M_s is the saturation magnetization; $E(\theta, \psi)$ is the total Gibb's thin film free energy density; γ and α are the gyromagnetic and phenomenological constants.

The total energy consists the magnetocrystalline anisotropy energy, the exchange energy, and the magnetostatic self-energy (stray field energy) [14]:

$$E = \int_v \left(\frac{k_{exh}}{J_s^2}\sum_{j=1}^{3}(\nabla J_j)^2 - \frac{k_J}{J_s^2}(\mathbf{a}_J \mathbf{J})^2 - \frac{1}{2}\mathbf{J}\cdot\mathbf{H}_D - \mathbf{J}\cdot\mathbf{H}_{ex}\right)dv \quad \text{(Zeeman energy)}$$

$$\frac{\partial \mathbf{J}}{\partial t} = -|\gamma|\mathbf{J}\times\mathbf{H}_{eff} + \frac{\alpha}{J_s}\mathbf{J}\times\frac{\partial \mathbf{J}}{\partial t} \quad \text{(Gilbert equation)}$$

where $\mathbf{J}$ is the magnetic polarization vector, $\mathbf{H}_D$ and $\mathbf{H}_{ex}$ are the demagnetizing and external magnetic fields, $\mathbf{H}_{eff}$ is the effective magnetic field (sum of the applied, demagnetization, and anisotropy fields), k_{exh} and k_J are the exchange and magnetocrystalline anisotropy constants, $\mathbf{a}_J$ is the unit vector parallel to the uniaxial easy axis.

Using the vector notations, we have

$$\frac{d\mathbf{M}}{dt} = -\frac{\gamma}{1+\alpha^2}\left(\mathbf{M}\times\mathbf{H}_{eff} + \frac{\alpha}{M_s}\mathbf{M}\times(\mathbf{M}\times\mathbf{H}_{eff})\right)$$

Thus, using the nonlinear differential equations given, the high-fidelity modeling and analysis of nanostructured nanocomposite permanent magnets can be performed using field and material quantities, parameters, constants, etc.

21.5.8 Microstructures and Microtransducers with Permanent Magnets: Micromirror Actuator

The electromagnetic microactuator (permanent magnet on the cantilever flexible beam and spiral planar windings controlled by ICs fabricated using CMOS-MEMS technology) is illustrated in Figure 21.135.

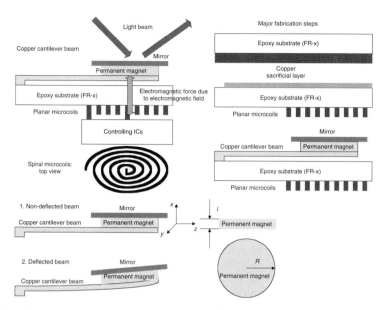

FIGURE 21.135 Electromagnetic microactuator with controlling ICs.

The electromagnetic microactuators can be made using conventional surface micromachining and CMOS fabrication technologies through electroplating, screen printing, lamination processes, sacrificial layer techniques, photolithography, etching, etc. In particular, the electromagnetic microactuator studied can be made on the commercially available epoxy substrates (e.g., FR series), which have the one-sided laminated copper layer (the copper layer thickness, which can be from 10 μm and higher, is defined by the admissible current density and the current value needed to establish the desired magnetic field to attain the specified mirror deflection, deflection rate, settling time, and other steady-state and dynamic characteristics). The spiral planar microcoils can be made on the one-sides laminated copper layer using photolithography and wet etching in the ferric chloride solution. The resulting x-μm thick N-turn microwinding will establish the magnetic field (the number of turns is a function of the footprint area available, thickness, spacing, outer-inner radii, geometry, fabrication techniques and processes used, etc.). After fabrication of the planar microcoils, the cantilever beam with the permanent magnet and mirror is fabricated on other side of the substrate. First, a photoresist sacrificial layer is spin-coated and patterned on the substrate. Then, a Ti–Cu–Cr seed layer is deposited to perform the copper electroplating (if the copper is used to fabricate the flexible cantilever structure). The second photoresist layer is spun and patterned to serve as a mold for the electroplating of the copper-based cantilever beam. The copper cantilever beam is electroplated in the copper-sulfate-based plating bath. After the electroplating, the photoresist plating mold and the seed layer are removed releasing the cantilever beam structure. It must be emphasized that depending upon the permanent magnet used, the corresponding fabricated processes must be done before or after releasing the beam. The permanent-magnet disk is positioned on the cantilever beam free end (for example, the polymer magnet can be screen-printed, and after curing the epoxy magnet, the magnet is magnetized by the external magnetic field). Then, the cantilever beam with the fabricated mirror is released by removing the sacrificial photoresist layer using acetone. It must be emphasized that the studied electromagnetic microactuator is fabricated using low-cost (affordable), high-yield micromachining—CMOS technology, processes, and materials. The most attractive feature is the application of the planar microcoils, which can be easily made. The use of the polymer permanent magnets (which have good magnetic properties) allows one to design high-performance electromagnetic microactuators. It must be emphasized that the polysilicon can be used to fabricate the cantilever beam and other permanent magnets can be applied.

In the article [12], the vertical electromagnetic force F_{ze}, acting on the permanent-magnet, is given by

$$F_{ze} = M_z \int_v \frac{dH_z}{dz} \, dv,$$

where M_z is the magenetization; H_z is the vertical component of the magnetic field intensity produced by the planar microwindings (H_z is a nonlinear function of the current fed or voltage applied to the microwindings, number of turns, microcoils, geometry, etc.; therefore, the thickness of the microcoils must be derived based on the maximum value of the current needed and the admissible current density).

The magnetically actuated cantilever microstructures were studied also in articles [15,16], and the expressions for the electromagnetic torque are found as the functions of the magnetic field using assumptions and simplifications which, in general, limit the applicability of the results. The differential equations which model the electromagnetic and *torsional-mechanical* dynamics can be derived. In particular, the equations for the electromagnetic field are found using electromagnetic theory, and the electromagnetic filed intensity H_z is controlled changing the current applied to the planar microwindings. The steady-state analysis, performed using the small-deflection theory [17], is also valuable. The static deflection of the cantilever beam x can be straightforwardly found using the force and beam quantities. In particular, $x = (l^3/3EJ)F_n$, where, l is the effective length of the beam; E is the Young's (elasticity) modulus; J is the equivalent moment of inertia of the beam with permanent magnet and mirror, and for the stand-alone cantilever beam with the rectangular cross section $J = \frac{1}{12} wh^3$; w and h are the width and thickness of the beam; F_n is the net force, which is normal to the cantilever beam.

In general, assuming that the magnetic flux is constant through the magnetic plane (loop), the torque on a planar current loop of any size and shape in the uniform magnetic field is

$$\mathbf{T} = i\mathbf{s} \times \mathbf{B} = \mathbf{m} \times \mathbf{B}$$

where i is the current and $\mathbf{m}$ is the magnetic dipole moment [Am^2].

Thus, the torque on the current loop always tends to turn the loop to align the magnetic field produced by the loop with the permanent-magnet magnetic field causing the resulting electromagnetic torque.

For example, for the current loop shown in Figure 21.136, the torque (in Nm) is found to be

$$\mathbf{T} = i\mathbf{s} \times \mathbf{B} = \mathbf{m} \times \mathbf{B} = 1 \times 10^{-3} \left[(1 \times 10^{-3})(2 \times 10^{-3}) \mathbf{a}_z \right] \times (-0.5\mathbf{a}_y + \mathbf{a}_z) = 1 \times 10^{-9} \mathbf{a}_x$$

The electromagnetic force is found as

$$\mathbf{F} = \oint_l i d\mathbf{l} \times \mathbf{B}$$

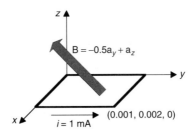

FIGURE 21.136 Rectangular planar loop in a uniform magnetic field with flux density $\mathbf{B} = -0.5\mathbf{a}_y + \mathbf{a}_z$.

Actuators

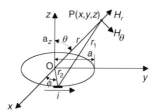

FIGURE 21.137 Planar current loop.

In general, the magnetic field quantities are derived using

$$\mathbf{B} = \frac{\mu_0}{4\pi} i \oint_l \frac{d\mathbf{l} \times \mathbf{r}_0}{r^2} \quad \text{or} \quad \mathbf{H} = \frac{1}{4\pi} i \oint_l \frac{d\mathbf{l} \times \mathbf{r}_0}{r^2}$$

and the Ampere circuital law gives

$$\oint_l \mathbf{H} \cdot d\mathbf{l} = i_{\text{total}} \quad \text{or} \quad \oint_l \mathbf{H} \cdot d\mathbf{l} = Ni$$

Making use of these expressions and taking note of the variables defined in Figure 21.137, we have

$$\mathbf{H} = \frac{1}{4\pi} i \oint_l \frac{d\mathbf{l} \times \mathbf{r}_1}{r_1^3} \quad \text{and} \quad \mathbf{B} = \frac{\mu_0}{4\pi} i \oint_l \frac{d\mathbf{l} \times \mathbf{r}_1}{r_1^3}$$

where $d\mathbf{l} = \mathbf{a}_\phi a\, d\phi = (-\mathbf{a}_x \sin\phi + \mathbf{a}_y \cos\phi) a\, d\phi$ and $\mathbf{r}_1 = \mathbf{a}_x (x - a\cos\phi) + \mathbf{a}_y (y - a\sin\phi) + \mathbf{a}_z z$. Hence,

$$d\mathbf{l} \times \mathbf{r}_1 = [\mathbf{a}_x z\cos\phi + \mathbf{a}_y z\sin\phi - \mathbf{a}_z(y\sin\phi + x\cos\phi - a)] a\, d\phi.$$

Then, neglecting the small quantities ($a^2 \ll r^2$), we have

$$r_1^3 = (x^2 + y^2 + z^2 + a^2 - 2ax\cos\phi - 2ay\sin\phi)^{3/2} \approx r^3 \left(1 - \frac{2ax}{r^2}\cos\phi - \frac{2ay}{r^2}\sin\phi\right)^{3/2}$$

Therefore, one obtains

$$\frac{1}{r_1^3} = \frac{1}{r^3}\left(1 + \frac{3ax}{r^2}\cos\phi + \frac{3ay}{r^2}\sin\phi\right)$$

Thus,

$$\mathbf{B} = \frac{\mu_0 a}{4\pi} i \int_0^{2\pi} [\mathbf{a}_x z\cos\phi + \mathbf{a}_y z\sin\phi - \mathbf{a}_z(y\sin\phi + x\cos\phi - a)] a \frac{1}{r^3}\left(1 + \frac{3ax}{r^2}\cos\phi + \frac{3ay}{r^2}\sin\phi\right) d\phi$$

$$= \frac{\mu_0 a^2}{4\pi r^3} i \left[\mathbf{a}_x \frac{3xz}{r^2} + \mathbf{a}_y \frac{3yz}{r^2} - \mathbf{a}_z\left(\frac{3x^2}{r^2} + \frac{3y^2}{r^2} - 2\right)\right]$$

Furthermore, using the coordinate transformation equations, in the spherical coordinate system one has

$$\mathbf{B} = \frac{\mu_0 a^2}{4\pi r^3} i(2\mathbf{a}_r \cos\theta + \mathbf{a}_\theta \sin\theta)$$

We have the expressions for the far-field components

$$B_r = \frac{\mu_0 a^2 \cos\theta}{2\pi r^3} i, \quad B_\theta = \frac{\mu_0 a^2 \sin\theta}{4\pi r^3} i, \quad B_\phi = 0$$

(due to the symmetry about the z axis, the magnetic flux density does not have the B_ϕ component).

Using the documented technique, one can easily find the magnetic vector potential. In particular, in general

$$\mathbf{A} = \frac{\mu_0}{4\pi} i \oint_l \frac{d\mathbf{l}}{r_1}$$

Assuming that $a^2 \ll r^2$, gives the following expression:

$$\frac{1}{r_1} = \frac{1}{r}\left(1 + \frac{ax}{r^2}\cos\phi + \frac{ay}{r^2}\sin\phi\right)$$

Therefore,

$$\mathbf{A} = \frac{\mu_0 a}{4\pi} i \int_0^{2\pi} (-\mathbf{a}_x \sin\phi + \mathbf{a}_y \cos\phi)\frac{1}{r}\left(1 + \frac{ax}{r^2}\cos\phi + \frac{ay}{r^2}\sin\phi\right) d\phi$$

$$= \frac{\mu_0 a}{4\pi r^3} i(-\mathbf{a}_x y + \mathbf{a}_y x)$$

Hence, in the spherical coordinate system, we obtain

$$\mathbf{A} = (\mathbf{A} \cdot \mathbf{a}_r)\mathbf{a}_r + (\mathbf{A} \cdot \mathbf{a}_\phi)\mathbf{a}_\phi + (\mathbf{A} \cdot \mathbf{a}_\theta)\mathbf{a}_\theta$$

$$= \frac{\mu_0 a}{4\pi r^2} i \mathbf{a}_\phi \sin\theta = A_\phi \mathbf{a}_\phi$$

It should be emphasized that the equations derived can be expressed using the magnetic dipole.

However, in the microtransducer studied, high-fidelity analysis should be performed. Hence, let us perform the comprehensive analysis.

The vector potential is found to be

$$A_\phi(r,\theta) = \frac{\mu_0 a i}{4\pi} \int_0^{2\pi} \frac{\cos\phi \, d\phi}{\sqrt{a^2 + r^2 - 2ar\sin\theta\cos\phi}}$$

and

$$B_r = \frac{1}{r\sin\theta} \frac{\partial(\sin\theta A_\phi)}{\partial\theta}, \quad B_\theta = -\frac{1}{r}\frac{\partial(rA_\phi)}{\partial r}, \quad B_\phi = 0$$

Making use of the following approximation:

$$A_\phi(r,\theta) = \frac{\mu_0 a i}{4\pi} \int_0^{2\pi} \frac{\cos\phi \, d\phi}{\sqrt{a^2 + r^2 - 2ar\sin\theta\cos\phi}} \approx \frac{\mu_0 a^2 r \sin\theta i}{4(a^2 + r^2)^{3/2}} \left(1 + \frac{15 a^2 r^2 \sin^2\theta}{8(a^2 + r^2)^2} + \cdots\right)$$

one finds

$$B_r(r,\theta) = \frac{\mu_0 a^2 \cos\theta i}{2(a^2 + r^2)^{3/2}} \left(1 + \frac{15 a^2 r^2 \sin^2\theta}{4(a^2 + r^2)^2} + \cdots\right)$$

$$B_\theta(r,\theta) = -\frac{\mu_0 a^2 \sin\theta i}{4(a^2 + r^2)^{5/2}} \left(2a^2 - r^2 + \frac{15 a^2 r^2 \sin^2\theta(4a^2 - 3r^2)}{8(a^2 + r^2)^2} + \cdots\right)$$

$$B_\phi = 0$$

One can specify three regions:

- near the axis $\theta \ll 1$,
- at the center $r \ll a$,
- in far-field $r \gg a$.

The electromagnetic torque and field depend upon the current in the microwindings and are nonlinear functions of the displacement.

The expression for the electromagnetic forces and torques must be derived to model and analyze the *torsional-mechanical* dynamics. Newton's laws of motion can be applied to study the mechanical dynamics in the Cartesian or other coordinate systems (e.g., previously for the translational motion in the *x*-axis, we used

$$\frac{dv}{dt} = \frac{1}{m}(F_e - F_L) \quad \text{and} \quad \frac{dx}{dt} = v$$

to model the translational *torsional-mechanical* dynamics of the electromagnetic microactuators using the electromagnetic force F_e and the load force F_L).

For the studied microactuator, the rotational motion can be studied, and the electromagnetic torque can be approximated as

$$T_e = 4R^2 t_{tf} M H_p \cos\theta$$

where R and t_{tf} are the radius and thickness of the permanent-magnet thin-film disk; M is the permanent-magnet thin film magnetization; H_p is the field produced by the planar windings; θ is the displacement angle.

Then, the microactuator rotational dynamics is given by

$$\frac{d\omega}{dt} = \frac{1}{J}(T_e - T_L) \quad \text{and} \quad \frac{d\theta}{dt} = \omega$$

where T_L is the load torque, which integrates the friction and disturbances torques.

It should be emphasized that more complex and comprehensive mathematical models can be developed and used integrating the nonlinear electromagnetic and six-degree-of-freedom rotational–translational motions (*torsional-mechanical* dynamics) of the cantilever beam. As an illustration we consider the high-fidelity modeling of the electromagnetic system.

21.5.8.1 Electromagnetic System Modeling in Microactuators with Permanent Magnets: High-Fidelity Modeling and Analysis

In this section we focus our efforts to derive the expanded equations for the electromagnetic torque and force on cylindrical permanent-magnet thin films, see Figure 21.135. The permanent-magnet thin film is assumed to be uniformly magnetized and the equations are developed for two orientations of the magnetization vector (the orientation is parallel to the axis of symmetry, and the orientation is perpendicular to this axis). Electromagnetic fields and gradients produced by the planar windings should be found at a point in inertial space, which coincides with the origin of the permanent-magnet axis system in its initial alignment. Our ultimate goal is to control microactuators, and thus, high-fidelity mathematical models (which will result in viable analysis, control, and optimization) must be derived. To attain our objective, the complete equations for the electromagnetic torque and force on a cylindrical permanent-magnet thin films are found.

The following notations are used: A, R, and l are the area, radius, and length of the cylindrical permanent magnet; **B** is the magnetic flux density vector; $\mathbf{B}_e$ is the expanded magnetic flux density vector; $[\partial \mathbf{B}]$ is the matrix of field gradients [T/m]; $[\partial \mathbf{B}_e]$ is the matrix of expanded field gradients [T/m]; **F** and **T** are the total force and torque vectors on the permanent-magnet thin film; i is the current in the planar microwinding; **m** is the magnetic moment vector [A m^2]; **M** is the magnetization vector [A/m]; **r** is the position vector (x, y, z are the coordinates in the Cartesian system),

$$\mathbf{r} = \begin{bmatrix} x \\ y \\ z \end{bmatrix}$$

T_r is the inertial coordinate vector-transformation matrix; W and Π are the work and potential energy; θ is the Euler orientation for the 3-2-1 rotation sequence; ∇ is the gradient operator; subscript ij represents partial derivative of i component in j-direction; subscript $(ij)k$ represents partial derivative of ij partial derivative in k-direction; ¯(bar over a variable) indicates that it is referenced to the microactuator coordinates.

21.5.8.2 Electromagnetic Torques and Forces: Preliminaries

The equations for the electromagnetic torque and force on a cylindrical permanent-magnet thin film are found by integrating the equations for torques and forces on an incremental volume of the permanent-magnet thin film with magnetic moment **M**dv over the volume. Figure 21.135 illustrates the microactuator with the cylindrical permanent-magnet thin film in the coordinate system, which consists of a set of orthogonal body-fixed axes that are initially aligned with a set of orthogonal x-, y-, z-axes fixed in the inertial space.

The equations for the electromagnetic torque and force on an infinitesimal current can be derived using the fundamental relationship for the force on a current-carrying-conductor element in a uniform magnetic field. In particular, for a planar current loop (planar microwinding) with constant current i in the uniform magnetic field **B** (vector **B** gives the magnitude and direction of the flux density of the external field), the force on an element $d\mathbf{l}$ of the conductor is found using the Lorentz force law

$$\mathbf{F} = \oint_l i\, d\mathbf{l} \times \mathbf{B}$$

Assuming that the magnetic flux is constant through the magnetic loop, the torque on a planar current loop of any size and shape in the uniform magnetic field is

$$\mathbf{T} = i\oint_l \mathbf{r} \times (d\mathbf{l} \times \mathbf{B}) = i\oint_l \left((\mathbf{r} \cdot \mathbf{B}) d\mathbf{l} - \mathbf{B} \oint_l \mathbf{r} \cdot d\mathbf{l} \right)$$

Using Stokes's theorem, one has

$$\mathbf{T} = i\left(\oint_s d\mathbf{A} \times \nabla(\mathbf{r} \cdot \mathbf{B}) - \mathbf{B}\oint_s (\nabla \times \mathbf{r}) \cdot d\mathbf{A}\right) = i\oint_s d\mathbf{A} \times \mathbf{B}$$

or

$$\mathbf{T} = i\mathbf{A} \times \mathbf{B} = \mathbf{m} \times \mathbf{B}$$

The electromagnetic torque **T** acts on the infinitesimal current loop in a direction to align the magnetic moment **m** with the external field **B**, and if **m** and **B** are misaligned by the angle θ, we have

$$T = mB \sin \theta$$

The incremental potential energy and work are found as

$$dW = d\Pi = T\, d\theta = mB\sin\theta\, d\theta \quad \text{and} \quad W = \Pi = -mB\cos\theta = -\mathbf{m} \cdot \mathbf{B}$$

Using the electromagnetic force, we have

$$dW = -d\Pi = \mathbf{F} \cdot d\mathbf{r} = -\nabla\Pi \cdot d\mathbf{r}$$

and

$$\mathbf{F} = -\nabla\Pi = \nabla(\mathbf{m} \cdot \mathbf{B}) = (\mathbf{m} \cdot \nabla)\mathbf{B}$$

21.5.8.3 Coordinate Systems and Electromagnetic Field

The transformation from the inertial coordinates to the permanent-magnet coordinates is

$$\bar{\mathbf{r}} = T_r \mathbf{r} = \begin{bmatrix} \cos\theta_y\cos\theta_z & \cos\theta_y\sin\theta_z & -\sin\theta_y \\ \sin\theta_x\sin\theta_y\cos\theta_z - \cos\theta_x\sin\theta_z & \sin\theta_x\sin\theta_y\sin\theta_z + \cos\theta_x\cos\theta_z & \sin\theta_x\cos\theta_y \\ \cos\theta_x\sin\theta_y\cos\theta_z + \sin\theta_x\sin\theta_z & \cos\theta_x\sin\theta_y\sin\theta_z - \sin\theta_x\cos\theta_z & \cos\theta_x\cos\theta_y \end{bmatrix} \begin{bmatrix} x \\ y \\ z \end{bmatrix}$$

$$\mathbf{r} = \begin{bmatrix} x \\ y \\ z \end{bmatrix}, \quad \bar{\mathbf{r}} = \begin{bmatrix} \bar{x} \\ \bar{y} \\ \bar{z} \end{bmatrix}$$

We use the transformation matrix

$$T_r = \begin{bmatrix} \cos\theta_y\cos\theta_z & \cos\theta_y\sin\theta_z & -\sin\theta_y \\ \sin\theta_x\sin\theta_y\cos\theta_z - \cos\theta_x\sin\theta_z & \sin\theta_x\sin\theta_y\sin\theta_z + \cos\theta_x\cos\theta_z & \sin\theta_x\cos\theta_y \\ \cos\theta_x\sin\theta_y\cos\theta_z + \sin\theta_x\sin\theta_z & \cos\theta_x\sin\theta_y\sin\theta_z - \sin\theta_x\cos\theta_z & \cos\theta_x\cos\theta_y \end{bmatrix}$$

If the deflections are small, we have

$$T_{rs} = \begin{bmatrix} 1 & \theta_z & -\theta_y \\ -\theta_z & 1 & \theta_x \\ \theta_y & -\theta_x & 1 \end{bmatrix}$$

It should be emphasized that we use the 3-2-1 orthogonal transformation matrix for the *z-y-x* Euler rotation sequence, and θ_x, θ_y, θ_z are the rotation Euler angle about the *x*, *y*, and *z* axes.

The field **B** and gradients of **B** produced by the microcoils fixed in the inertial frame and expressed assuming that the electromagnetic fields can be described by the second-order Taylor series. Expanding **B** about the origin of the *x*, *y*, *z* system as a Taylor series, we have [18]

$$\mathbf{B}_e = \mathbf{B} + (\mathbf{r} \cdot \nabla)\mathbf{B} + \frac{1}{2}(\mathbf{r} \cdot \nabla)^2 \mathbf{B}$$

or

$$B_{ei} = B_i + \frac{\partial B_i}{\partial \mathbf{r}}\mathbf{r} + \frac{1}{2}\mathbf{r}^T \frac{\partial^2 B_i}{\partial \mathbf{r}^2}\mathbf{r}$$

where

$$\frac{\partial B_i}{\partial \mathbf{r}} = \begin{bmatrix} \frac{\partial B_i}{\partial x} & \frac{\partial B_i}{\partial y} & \frac{\partial B_i}{\partial z} \end{bmatrix} \quad \text{and} \quad \frac{\partial^2 B_i}{\partial \mathbf{r}^2} = \begin{bmatrix} \frac{\partial \frac{\partial B_i}{\partial x}}{\partial x} & \frac{\partial \frac{\partial B_i}{\partial x}}{\partial y} & \frac{\partial \frac{\partial B_i}{\partial x}}{\partial z} \\ \frac{\partial \frac{\partial B_i}{\partial y}}{\partial x} & \frac{\partial \frac{\partial B_i}{\partial y}}{\partial y} & \frac{\partial \frac{\partial B_i}{\partial y}}{\partial z} \\ \frac{\partial \frac{\partial B_i}{\partial z}}{\partial x} & \frac{\partial \frac{\partial B_i}{\partial z}}{\partial y} & \frac{\partial \frac{\partial B_i}{\partial z}}{\partial z} \end{bmatrix}$$

We denote

$$B_{ij} = \frac{\partial B_i}{\partial j} \quad \text{and} \quad B_{(ij)k} = \frac{\partial \frac{\partial B_i}{\partial j}}{\partial k}$$

Then,

$$\frac{\partial B_i}{\partial \mathbf{r}} = [B_{ix} \quad B_{iy} \quad B_{iz}] \quad \text{and} \quad \frac{\partial^2 B_i}{\partial \mathbf{r}^2} = \begin{bmatrix} B_{(ix)x} & B_{(ix)y} & B_{(ix)z} \\ B_{(iy)x} & B_{(iy)y} & B_{(iy)z} \\ B_{(iz)x} & B_{(iz)y} & B_{(iz)z} \end{bmatrix}$$

Hence, the first-order gradients are given as

$$B_{eij} = B_{ij} + \frac{\partial \frac{\partial B_i}{\partial j}}{\partial \mathbf{r}}\mathbf{r} = B_{ij} + [B_{(ij)x} \quad B_{(ij)y} \quad B_{(ij)z}]\mathbf{r}$$

The expanded field is expressed in the permanent-magnet coordinates as

$$\bar{\mathbf{B}}_e = \bar{\mathbf{B}} + (\bar{\mathbf{r}} \cdot \bar{\nabla})\bar{\mathbf{B}} + \frac{1}{2}(\bar{\mathbf{r}} \cdot \bar{\nabla})^2 \bar{\mathbf{B}}$$

where $\bar{\mathbf{B}} = T_r \mathbf{B}$ and $\bar{\nabla} = T_r \nabla$.

Using $\mathbf{r} = T_r^T \bar{\mathbf{r}}$, one has

$$B_{ei} = B_i + \frac{\partial B_i}{\partial \mathbf{r}} T_r^T \bar{\mathbf{r}} + \frac{1}{2} \bar{\mathbf{r}}^T T_r \frac{\partial^2 B_i}{\partial \mathbf{r}^2} T_r^T \bar{\mathbf{r}}$$

and

$$\bar{\mathbf{B}}_e = T_r \begin{bmatrix} B_x + \dfrac{\partial B_x}{\partial \mathbf{r}} T_r^T \bar{\mathbf{r}} + \dfrac{1}{2} \bar{\mathbf{r}}^T T_r \dfrac{\partial^2 B_x}{\partial \mathbf{r}^2} T_r^T \bar{\mathbf{r}} \\ B_y + \dfrac{\partial B_y}{\partial \mathbf{r}} T_r^T \bar{\mathbf{r}} + \dfrac{1}{2} \bar{\mathbf{r}}^T T_r \dfrac{\partial^2 B_y}{\partial \mathbf{r}^2} T_r^T \bar{\mathbf{r}} \\ B_z + \dfrac{\partial B_z}{\partial \mathbf{r}} T_r^T \bar{\mathbf{r}} + \dfrac{1}{2} \bar{\mathbf{r}}^T T_r \dfrac{\partial^2 B_z}{\partial \mathbf{r}^2} T_r^T \bar{\mathbf{r}} \end{bmatrix}$$

21.5.8.4 Electromagnetic Torques and Forces

Now let us derive the fields and gradients at any point in the permanent magnet using the second-order Taylor series approximation. To eliminate the transformations between the inertial and permanent magnet coordinate systems and simplify the second-order negligible small components, we assume that the relative motion between the magnet and the reference inertial coordinate is zero and the T_{rs} transformation matrix is used (otherwise, the second-order gradient terms will lead to cumbersome results).

The magnetization (the magnetic moment per unit volume) is constant over the volume of the permanent-magnet thin films, and $\mathbf{m} = \mathbf{M}v$.

Assuming that the magnetic flux is constant, the total electromagnetic torque and force on a planar current loop (microwinding) in the uniform magnetic field is

$$\bar{\mathbf{T}} = \int_v (\bar{\mathbf{M}} \times \mathbf{B}_e + \bar{\mathbf{r}} \times (\bar{\mathbf{M}} \cdot \nabla)\mathbf{B}_e) dv$$

$$\bar{\mathbf{F}} = \int_v (\bar{\mathbf{M}} \cdot \nabla)\mathbf{B}_e \, dv$$

where

$$(\bar{\mathbf{M}} \cdot \nabla)\mathbf{B}_e = [\partial \mathbf{B}_e]\bar{\mathbf{M}} = \begin{bmatrix} B_{exx} & B_{exy} & B_{exz} \\ B_{eyx} & B_{eyy} & B_{eyz} \\ B_{ezx} & B_{ezy} & B_{ezz} \end{bmatrix} \begin{bmatrix} M_{\bar{x}} \\ M_{\bar{y}} \\ M_{\bar{z}} \end{bmatrix}$$

Case 1: Magnetization Along the Axis of Symmetry
For orientation of the magnetization vector along the axis of symmetry (*x*-axis) of the permanent-magnet thin films, we have

$$(\bar{\mathbf{M}} \cdot \nabla)\mathbf{B}_e = [\partial \mathbf{B}_e]\bar{\mathbf{M}} = M_{\bar{x}} \begin{bmatrix} B_{exx} \\ B_{exy} \\ B_{exz} \end{bmatrix}$$

Thus, in the expression $\bar{\mathbf{T}} = \int_v (\bar{\mathbf{M}} \times \mathbf{B}_e + \bar{\mathbf{r}} \times (\bar{\mathbf{M}} \cdot \nabla)\mathbf{B}_e) dv$, the terms are

$$\bar{r} \times (\overline{M} \cdot \nabla) B_e = M_{\bar{x}} \begin{bmatrix} -B_{exy}\bar{z} + B_{exz}\bar{y} \\ B_{exx}\bar{z} - B_{exz}\bar{x} \\ -B_{exx}\bar{y} + B_{exy}\bar{x} \end{bmatrix} \quad \text{and} \quad \overline{M} \times B_e = M_{\bar{x}} \begin{bmatrix} 0 \\ -B_{ez} \\ B_{ey} \end{bmatrix}$$

Therefore,

$$T_{\bar{x}} = M_{\bar{x}} \int_v (B_{exz}\bar{y} - B_{exy}\bar{z}) dv$$

$$T_{\bar{y}} = -M_{\bar{x}} \int_v B_{ez} dv + M_{\bar{x}} \int_v (B_{exx}\bar{z} - B_{exz}\bar{x}) dv$$

and

$$T_{\bar{z}} = M_{\bar{x}} \int_v B_{ey} dv + M_{\bar{x}} \int_v (B_{exy}\bar{x} - B_{exx}\bar{y}) dv$$

The terms in the derived equations must be evaluated.

Let us find the analytic expression for the electromagnetic torque $T_{\bar{x}}$. In particular, we have

$$\int_v B_{exz}\bar{y} \, dv = B_{xz} \int_v \bar{y} \, dv + B_{(xx)z} \int_v \bar{x}\bar{y} \, dv + B_{(xy)z} \int_v \bar{y}^2 dv + B_{(xz)z} \int_v (\bar{z}\bar{y}) \, dv$$

where

$$\int_v \bar{y} \, dv = 0, \quad \int_v \bar{x}\bar{y} \, dv = 0, \quad \int_v \bar{z}\bar{y} \, dv = 0$$

and

$$\int_v \bar{y}^2 dv = \int_{-\frac{1}{2}l}^{\frac{1}{2}l} \int_{-R}^{R} \int_{-\sqrt{R^2-z^2}}^{\sqrt{R^2-z^2}} \bar{y}^2 \, d\bar{y} d\bar{z} d\bar{x} = \frac{1}{4}\pi l R^4 = \frac{1}{4} v R^4$$

Therefore,

$$M_{\bar{x}} \int_v B_{exz}\bar{y} \, dv = M_{\bar{x}} \frac{1}{4} B_{(xy)z} v R^4$$

Furthermore,

$$M_{\bar{x}} \int_v B_{exy}\bar{z} \, dv = M_{\bar{x}} \frac{1}{4} B_{(xy)z} v R^4$$

Thus, for $T_{\bar{x}}$, one has

$$T_{\bar{x}} = M_{\bar{x}} \int_v (B_{exz}\bar{y} - B_{exy}\bar{z}) dv = M_{\bar{x}} \left(\frac{1}{4} B_{(xy)z} v R^4 - \frac{1}{4} B_{(xy)z} v R^4 \right) = 0$$

Then, for $T_{\tilde{y}}$, we obtain

$$T_{\tilde{y}} = -M_{\tilde{x}}\int_v B_{ez}\,dv + M_{\tilde{x}}\int_v (B_{exx}\tilde{z} - B_{exz}\tilde{x})\,dv$$

$$= M_{\tilde{x}}\left[-\left(B_z + B_{(zx)x}\frac{1}{24}l^2 + B_{(zy)y}\frac{1}{8}R^2 + B_{(zz)z}\frac{1}{8}R^2\right)v + B_{(xx)z}\left(\frac{1}{4}R^2 - \frac{1}{12}l^2\right)v\right]$$

$$= -vM_{\tilde{x}}\left[B_z + B_{(xx)z}\left(\frac{1}{4}R^2 - \frac{1}{8}l^2\right) + B_{(yy)z}\frac{1}{8}R^2 + B_{(zz)z}\frac{1}{4}R^2\right]$$

Finally, we obtain the expression for $T_{\tilde{z}}$ as

$$T_{\tilde{z}} = M_{\tilde{x}}\int_v B_{ey}\,dv + M_{\tilde{x}}\int_v (B_{exy}\tilde{x} - B_{exx}\tilde{y})\,dv$$

$$= vM_{\tilde{x}}\left[B_y + B_{(xx)y}\left(\frac{1}{8}l^2 - \frac{1}{4}R^2\right) - B_{(yy)y}\frac{1}{8}R^2 - B_{(yz)z}\frac{1}{8}R^2\right]$$

Thus, the following electromagnetic torque equations result:

$$T_{\tilde{x}} = 0$$

$$T_{\tilde{y}} = -vM_{\tilde{x}}\left[B_z + B_{(xx)z}\left(\frac{1}{4}R^2 - \frac{1}{8}l^2\right) + B_{(yy)z}\frac{1}{8}R^2 + B_{(zz)z}\frac{1}{4}R^2\right]$$

$$T_{\tilde{z}} = vM_{\tilde{x}}\left[B_y + B_{(xx)y}\left(\frac{1}{8}l^2 - \frac{1}{4}R^2\right) - B_{(yy)y}\frac{1}{8}R^2 - B_{(yz)z}\frac{1}{8}R^2\right]$$

The electromagnetic forces are found as well. In particular, from

$$F_{\tilde{x}} = M_{\tilde{x}}\int_v B_{exx}\,dv$$

$$F_{\tilde{y}} = M_{\tilde{x}}\int_v B_{exy}\,dv$$

and

$$F_{\tilde{z}} = M_{\tilde{x}}\int_v B_{exz}\,dv$$

using the expressions for the expanded magnetic fluxes, e.g.,

$$\int_v B_{exx}\,dv = \int_v (B_{xx} + B_{(xx)x}\tilde{x} + B_{(xx)y}\tilde{y} + B_{(xx)z}\tilde{z})\,dv$$

and performing the integration, one has the following expressions for the electromagnetic forces as the function of the magnetic field:

$$F_{\tilde{x}} = vM_{\tilde{x}}B_{xx}, \quad F_{\tilde{y}} = vM_{\tilde{x}}B_{xy}, \quad F_{\tilde{z}} = vM_{\tilde{x}}B_{xz}$$

Case 2: Magnetization Perpendicular to the Axis of Symmetry

For orientation of the magnetization vector perpendicular to the axis of symmetry, the following equation is used to find the electromagnetic torque:

$$\bar{T} = \int_v (\bar{M} \times B_e + \bar{r} \times (\bar{M} \cdot \nabla) B_e) \, dv$$

where

$$(\bar{M} \cdot \nabla) B_e = [\partial B_e] \bar{M} = M_{\bar{z}} \begin{bmatrix} B_{exz} \\ B_{eyz} \\ B_{ezz} \end{bmatrix} \qquad \bar{r} \times (\bar{M} \cdot \nabla) B_e = M_{\bar{z}} \begin{bmatrix} -B_{eyz}\bar{z} + B_{ezz}\bar{y} \\ B_{exz}\bar{z} - B_{ezz}\bar{x} \\ B_{exz}\bar{y} + B_{eyz}\bar{x} \end{bmatrix}$$

and

$$\bar{M} \times B_e = M_{\bar{z}} \begin{bmatrix} -B_{ey} \\ B_{ex} \\ 0 \end{bmatrix}$$

Thus,

$$T_{\bar{x}} = -M_{\bar{z}} \int_v B_{ey} \, dv + M_{\bar{z}} \int_v (B_{exz}\bar{y} - B_{eyz}\bar{z}) \, dv$$

$$T_{\bar{y}} = M_{\bar{z}} \int_v B_{ex} \, dv + M_{\bar{z}} \int_v (B_{ezz}\bar{z} - B_{ezz}\bar{x}) \, dv$$

$$T_{\bar{z}} = M_{\bar{z}} \int_v (B_{eyz}\bar{x} - B_{exz}\bar{y}) \, dv$$

Expressing the fluxes and performing the integration, we have the following expressions for the torque components as the function of the magnetic field:

$$T_{\bar{x}} = -v M_{\bar{z}} \left(B_y + B_{(xx)y} \frac{1}{24} l^2 + B_{(yy)y} \frac{1}{8} R^2 + B_{(yz)z} \frac{1}{8} R^2 \right)$$

$$T_{\bar{y}} = v M_{\bar{z}} \left[B_x + B_{(xz)y} \left(\frac{3}{8} R^2 - \frac{1}{12} l^2 \right) + B_{(xx)x} \frac{1}{24} l^2 + B_{(xy)y} \frac{1}{8} R^2 \right]$$

$$T_{\bar{z}} = v M_{\bar{z}} B_{(xy)y} \left(\frac{1}{12} l^2 - \frac{1}{4} R^2 \right)$$

The electromagnetic forces are found to be

$$F_{\bar{x}} = M_{\bar{z}} \int_v B_{exz} \, dv = v M_{\bar{z}} B_{xz}$$

$$F_{\bar{y}} = M_{\bar{z}} \int_v B_{eyz} \, dv = v M_{\bar{z}} B_{yz}$$

$$F_{\bar{z}} = M_{\bar{z}} \int_v B_{ezz} \, dv = v M_{\bar{z}} B_{zz}$$

Thus, the expressions for the electromagnetic force and torque components are derived. These equations provide one with a clear perspective on how to model, analyze, and control the electromagnetic forces and torques changing the applied magnetic field because the terms

$$B_{ij} = \frac{\partial B_i}{\partial j} \quad \text{and} \quad B_{(ij)k} = \frac{\partial \frac{\partial B_i}{\partial j}}{\partial k}$$

can be viewed as the control variables. It must be emphasized that the electromagnetic field (B_{ij} and $B_{(ij)k}$) is controlled by regulating the current in the planar microwindings and designing the microwindings (or other radiating energy microdevices). As was discussed, the derived forces and torques must be used in the torsional-mechanical equations of motion for the microactuator, and, in general, the six-degree-of-freedom microactuator mechanical dynamics results. These mechanical equations of motion are easily integrated with the derived electromagnetic equations, and closed-loop systems can be designed to attain the desired microactuator performance. These equations guide us to the importance of electromagnetic features in the modeling, analysis, and design of microactuators.

21.5.8.5 Some Other Aspects of Microactuator Design and Optimization

In addition to the electromagnetic-mechanical (electromechanical) analysis and design, other design and optimization problems are involved. As an example, let us focus our attention on the planar windings. The ideal planar microwindings must produce the maximum electromagnetic field, minimizing the footprint area, taking into consideration the material characteristics, operating conditions, applications, power requirements, and many other factors. Many planar winding parameters and characteristics can be optimized, for example, the dc resistance must be minimized to improve the efficiency, increase the flux, decrease the losses, etc. To attain good performance, in general, microwindings have the concentric circular current path and no interconnect resistances. For N-turn winding, the total dc resistance r_t is found to be

$$r_t = \frac{2\pi\rho}{t_w} \sum_{k=1}^{N} \frac{1}{\ln(r_{Ok}/r_{Ik})}$$

where ρ is the winding material resistivity, t_w is the winding thickness, r_{Ok} and r_{Ik} are the outer and inner radii of the k-turn winding, respectively.

To achieve the lowest resistance, the planar winding radii can be optimized by minimizing the resistance, and the minimum resistance is denoted as r_{tmin}. In particular, making use of first- and second-order necessary conditions for minimization, one has

$$\frac{dr_t}{dr_w} = 0 \quad \text{and} \quad \frac{d^2 r_t}{dr_w^2} > 0$$

where r_w is the inner or outer radius of an arbitrary turn of the optimized planar windings from the standpoint of minimizing the resistance.

Then, the minimum value of the microcoil resistance is given by

$$r_{tmin} = \frac{2\pi\rho}{t_w} \frac{N}{(r_{OR}/r_{IR})}$$

where r_{OR} and r_{IR} are the outer and inner radii of the windings (i.e., $r_{O\ Nth\ microcoil}$ and $r_{I\ 1st\ microcoil}$), respectively.

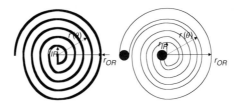

FIGURE 21.138 Planar spiral microwinding.

Thus, using the number of turns and turn-to-turn spacing, the outer and inner radii of the *k*-turn winding are found as

$$\frac{r_{Ok}}{r_{Ik}} = \left(\frac{r_{OR}}{r_{IR}}\right)^{1/N}$$

For spiral windings, the *averaging* (*equivalency*) concept should be used because the outer and inner radii are the functions of the planar angle, see Figure 21.138. Finally, it should be emphasized that the width of the *N*th microcoil is specified by the rated voltage current density versus maximum current density needed, fabrication technologies used, material characteristics, etc.

21.5.9 Micromachined Polycrystalline Silicon Carbide Micromotors

Articles [19,20] report the silicon-based fabrication of reluctance micromotors. This section is focused on a new enabling technology to fabricate microtransducers. Multilayer fabrication processes at low temperature and micromolding techniques were developed to fabricate SiC microstructures and salient-pole micromotors, which can be used at a very high temperature (400°C and higher) [21,22]. This was done through the SiC surface micromachining. Advantages of the SiC micromachining and SiC technologies (high temperature and ruggedness) should be weighted against fabrication drawbacks because new processes must be designed and optimized. Reactive ion etching is used to pattern SiC thin films; however, many problems, such as masking, low etch rates, and poor etch selectivity, must be addressed and resolved. Articles [21,22] report two single-layer reactive ion etching-based polycrystalline SiC surface micromachining processes using polysilicon or SiO_2 as the sacrificial layer. In addition, the micromolding process, used to fabricated polysilicon molds in conjunction with polycrystalline SiC film deposition and mechanical polishing to pattern polycrystaline SiC films, are introduced. The micromolding process can be used for single- and multilayer SiC surface micromachining.

The micromotor fabrication processes are illustrated in Figure 21.139. A 5–10 μm thick sacrificial molding polysilicon is deposited through the LPCVD on a 3–5 μm sacrificial thermal oxide. The rotor-stator mold formation can be made on the polished (chemical-mechanical polishing) polysilicon surface, enabling the 2 μm fabrication features using standard lithography and reactive ion etching. After the mold formation and delineation, the SiC is deposited on the wafer using atmospheric pressure chemical vapor deposition reactor. In particular, the phosphorus-doped (*n*-type) polycrystalline SiC films are deposited on the SiO_2 sacrificial layers at 1050°C with 0.5–1 μm/h rate (deposition is not selective, and SiC will be deposited on the surfaces of the polysilicon molds as well). Mechanical polishing of SiC is needed to expose the polysilicon and planarize the wafer surface (in [21,22], the polishing was done with 3 μm diameter diamond suspension, 360 N normal force, and 15 rad/sec pad rotation—the removal rate of SiC was reported to be 100 nm/min). The wafers are polished until the top surface of the polysilicon mold is exposed (polishing must be stopped at once due to the fast polishing rate). The flange mold is fabricated through the polysilicon and the sacrificial oxide etching (using the KOH and BHF, respectively). The 0.5 μm bearing clearance low-temperature oxide is deposited and annealed at 1000°C. Then, the 1 μm polycrystalline SiC film is deposited and patterned by reactive ion etching to make the bearing.

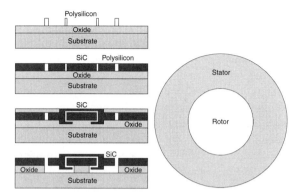

FIGURE 21.139 Fabrication of the SiC micromotors: cross-sectional schematics.

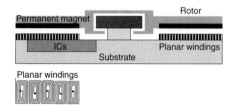

FIGURE 21.140 Slotless axial electromagnetic micromotor (cross-sectional schematics) with controlling ICs.

The release begins with the etching (BHF solution) to strip the left-over bearing clearance oxide. The sacrificial mold is removed by etching (KOH system) the polysilicon. It should be emphasized that the SiC and SiO are not etched during the mold removal step. Then, the moving parts of the micromotor were released. The micromotor is rinsed in water and methanol, and dried with the air jet.

Using this fabrication process, the micromotor with the 100–150 μm rotor diameter, 2 μm airgap, and 21 μm bearing radius, was fabricated and tested in [21, 22]. The rated voltage was 100 V and the maximum angular velocity was 30 rad/s. For silicon and polysilicon micromotors, two of the most critical problems are the bearing and ruggedness. The application of SiC reduces the friction and improves the ruggedness. These contribute to the reliability of the SiC-based fabricated micromachines.

21.5.10 Axial Electromagnetic Micromotors

The major problem is to devise novel microtransducers in order to eliminate fabrication difficulties and guarantee affordability, efficiency, reliability, and controllability of MEMS. In fact, the electrostatic and planar micromotor fabricated and tested to date are found to be inadequate for a wide range of applications due to difficulties associated and the cost. Therefore, this section is devoted to devising novel affordable rotational micromotors.

Figure 21.140 illustrates the devised axial topology micromotor, which has the *closed-ended* electromagnetic system. The stator is made on the substrate with deposited microwindings (printed copper coils can be made using the fabrication processes described as well as using a double-sided substrate with one-sided deposited copper thin films through conventional photolithography processes). The bearing post is fabricated on the stator substrate and the bearing hold is a part of the rotor microstructure. The rotor with permanent-magnet thin films rotates due to the electromagnetic torque developed. It is important to emphasize that the stator and rotor are made using conventional well-developed processes and materials.

It is evident that conventional silicon and SiC technologies can be used. The documented micromotor has a great number of advantages. The most critical benefit is the fabrication simplicity. In fact, axial micromotors can be straightforwardly fabricated and this will enable their wide applications as microactuators and microsensors. However, the axial micromotors must be designed and optimized to attain good performance. The optimization is based upon electromagnetic, mechanical, and thermal design. The micromotor optimization can be carried out using the steady-state concept (finite element analysis) and dynamic paradigms (lumped-parameters models or complete electromagnetic-mechanical-thermal high-fidelity mathematical models derived as a set of partial differential equations using Maxwell's, *torsional-mechanical*, and heat equations). In general, the nonlinear optimization problems are needed to be addressed, formulated, and solved to guarantee the superior microtransducer performance. In addition to the microtransducer design, one must concentrate the attention on the ICs and controller design. In particular, the circuitry is designed based upon the converter and inverter topologies (e.g., hard- and soft-switching, one-, two-, or four-quadrant, etc.), filters and sensors used, rated voltage and current, etc. From the control prespective, the electromagnetic features must be thoroughly examined. For example, the electromagnetic micromotor studied is the synchronous micromachine. Therefore, to develop the electromagnetic torque, the voltages applied to the stator windings must be supplied as the functions of the rotor angular displacement. Therefore, the Hall-effect sensors must be used, or the so-called sensorless controllers (the rotor position is observed or estimated using the directly measured variables) must be designed and implemented using ICs. This brief discussion illustrates a wide spectrum of fundamental problems involved in the design of integrated microtransducers with controlling and signal processing ICs.

21.5.11 Conclusions

The critical focus themes in MEMS development and implementation are rapid synthesis, design, and prototyping through synergetic multi-disciplinary system-level research in electromechanics. In particular, MEMS devising, modeling, simulation, analysis, design and optimization, which is relevant to cognitive study, classification, and synthesis must be performed. As microtransducers and MEMS are devised, the fabrication techniques and processes are developed and carried out. Devising microtransducers is the closed evolutionary process to study possible system-level evolutions based upon synergetic integration of microscale structures and devices in the unified functional core. The ability to devise and optimize microtransducers to a large extent depends on the validity and integrity of mathematical models. Therefore, mathematical models for different microtransducers were derived and analyzed. It is documented that microtransducer modeling, analysis, simulation, and design must be based on reliable mathematical models which integrate nonlinear electromagnetic features. It is important to emphasize that the secondary phenomena and effects, usually neglected in conventional miniscale electromechanical motion devices (modeled using lamped-parameter models and analyzed using finite element analysis techniques) cannot be ignored. The fabrication processes were described to make high-performance microtransducers.

References

1. Campbell, S. A., *The Science and Engineering of Microelectronic Fabrication*, Oxford University Press, New York, 2001.
2. Lyshevski, S. E., *Nano- and Micro-Electromechanical Systems: Fundamental of Micro- and Nano-Engineering*, CRC Press, Boca Raton, FL, 2000.
3. Lyshevski, S. E., *MEMS and NEMS: Systems, Devices, and Structures*, CRC Press, Boca Raton, FL, 2001.
4. Madou, M., *Fundamentals of Microfabrication*, CRC Press, Boca Raton, FL, 1997.
5. Kim, Y.-J. and Allen, M. G., "Surface micromachined solenoid inductors for high frequency applications," *IEEE Trans. Components, Packaging, and Manufacturing Technology*, part C, vol. 21, no. 1, pp. 26–33, 1998.
6. Park, J. Y. and Allen, M. G., "Integrated electroplated micromachined magnetic devices using low temperature fabrication processes," *IEEE Trans. Electronics Packaging Manufacturing*, vol. 23, no. 1, pp. 48–55, 2000.

7. Sadler, D. J., Liakopoulos, T. M., and Ahn, C. H., "A universal electromagnetic microactuator using magnetic interconnection concepts," *Journal Microelectromechanical Systems*, vol. 9, no. 4, pp. 460–468, 2000.
8. Lyshevski, S. E., *Electromechanical Systems, Electric Machines, and Applied Mechatronics*, CRC Press, Boca Raton, FL, 1999.
9. Frazier, A. B. and Allen, M. G., "Uses of electroplated aluminum for the development of microstructures and micromachining processes," *Journal Microelectromechanical Systems*, vol. 6, no. 2, pp. 91–98, 1997.
10. Guckel, H., Christenson, T. R., Skrobis, K. J., Klein, J., and Karnowsky, M., "Design and testing of planar magnetic micromotors fabricated by deep x-ray lithography and electroplating," *Technical Digest of International Conference on Solid-State Sensors and Actuators, Transducers 93*, Yokohama, Japan, pp. 60–64, 1993.
11. Taylor, W. P., Schneider, M., Baltes, H., and Allen, M. G., "Electroplated soft magnetic materials for microsensors and microactuators," *Proc. Conf. Solid-State Sensors and Actuators, Transducers 97*, Chicago, IL, pp. 1445–1448, 1997.
12. Lagorce, L. K., Brand, O., and Allen, M. G., "Magnetic microactuators based on polymer magnets," *Journal Microelectromechanical Systems*, vol. 8, no. 1, pp. 2–9, 1999.
13. Smith, D. O., "Static and dynamic behavior in thin permalloy films," *Journal of Applied Physics*, vol. 29, no. 2, pp. 264–273, 1958.
14. Suss, D., Schreft, T., and Fidler, J., "Micromagnetics simulation of high energy density permanent magnets," *IEEE Trans. Magnetics*, vol. 36, no. 5, pp. 3282–3284, 2000.
15. Judy, J. W. and Muller, R. S., "Magnetically actuated, addressable microstructures," *Journal Microelectromechanical Systems*, vol. 6, no. 3, pp. 249–256, 1997.
16. Yi, Y. W. and Liu, C., "Magnetic actuation of hinged microstructures," *Journal Microelectromechanical Systems*, vol. 8, no. 1, pp. 10–17, 1999.
17. Gere, J. M. and Timoshenko, S. P., *Mechanics of Materials*, PWS Press, 1997.
18. Groom, N. J. and Britcher, C. P., "A description of a laboratory model magnetic suspension test fixture with large angular capability," *Proc. Conf. Control Applications, NASA Technical Paper – 1997*, vol. 1, pp. 454–459, 1992.
19. Ahn, C. H., Kim, Y. J., and Allen, M. G., "A planar variable reluctance magnetic micromotor with fully integrated stator and coils," *Journal Microelectromechanical Systems*, vol. 2, no. 4, pp. 165–173, 1993.
20. O'Sullivan, E. J., Cooper, E. I., Romankiw, L. T., Kwietniak, K. T., Trouilloud, P. L., Horkans, J., Jahnes, C. V., Babich, I. V., Krongelb, S., Hegde, S. G., Tornello, J. A., LaBianca, N. C., Cotte, J. M., and Chainer, T. J., "Integrated, variable-reluctance magnetic minimotor," *IBM Journal Research and Development*, vol. 42, no. 5, 1998.
21. Yasseen, A. A., Wu, C. H., Zorman, C. A., and Mehregany, M., "Fabrication and testing of surface micromachined polycrystalline SiC micromotors," *IEEE Trans. Electron Device Letters*, vol. 21, no. 4, pp. 164–166, 2000.
22. Yasseen, A. A., Zorman, C. A., and Mehregany, M., "Surface micromachining of polycrystalline silicon carbide films microfabricated molds of SiO and polysilicon," *Journal Microelectromechanical Systems*, vol. 8, no. 1, pp. 237–242, 1999.

Index

A

Absolute angular optical encoders, 20-108–20-109
Absolute encoder, 20-6
Acceleration measurements techniques, 20-12
Acceleration sensors, 17-4–17-5
  accelerometers, *see* accelerometers
  dynamics and characteristics, 20-13–20-15
  in error measurements, 20-17–20-19
  in signal conditioning and biasing, 20-31–20-33
  types, 20-13
  in vibration measurements, 20-15–20-17
Accelerometers
  electromechanical, 20-23–20-24
  electrostatic, 20-27–20-29
  inertial, 20-19–20-23
  micro/nano, 20-30–20-31
  piezoelectric, 20-24–20-25
  piezoresistive, 20-25–20-26
  silicon microfabricated, 5-8
  strain-gauge, 20-26–20-27
AC inductive sensor, 20-9
AC machines
  induction motors, 21-43–21-50
  synchronous motors, 21-41–21-43
AC network analysis, 11-21–11-27
ACSL, 2-10
Active linear region, 21-21
Actuation force, of the comb-drive configuration, 5-3
Actuators
  characteristics
    backlash, 19-6
    deadband, 19-7–19-8
    eccentricity, 19-6
    errors, 19-2
    first-order system response, 19-8–19-9
    frequency response, 19-12–19-14
    impedance, 19-4–19-5
    linearity and accuracy, 19-3–19-4
    nonlinearities, 19-5
    range of, 19-1
    repeatability, 19-3
    resolution of, 19-2
    saturation, 19-7

    second-order system response, 19-9–19-11
    sensitivity of, 19-2
    static and Coulomb friction, 19-5
    system response, 19-8
  classification, 17-11–17-13
  principle of operation
    alternate current motors, 17-14
    electrical actuators, 17-13
    electromagnetic actuators, 17-14–17-15
    electromechanical actuators, 17-13–17-14
    hydraulic and pneumatic actuators, 17-15
    micro and nanoactuators, 17-18
    smart material actuators, 17-15–17-18
    stepper motors, 17-14
    ultrasonic actuators, 17-19
  selection criteria, 17-19–17-20
ADAMS, 7-6
Adaptive friction compensation, 2-4
Air-standard analysis, 12-27, 12-29
ALAMBETA device, 4-3
Alternate current motors, 17-14
Ammeter, 11-14
Ampere circuital law, 21-119
Amperes law, 7-13
Analog Devices' accelerometer, 14-9
Analog Hall sensors, 20-10–20-11
Analogies, in physical system models
  force-current analogy
    beyond one-dimensional mechanical systems, 16-3
    drawbacks, 16-2
    intuitions in processes, 16-3–16-4
    measurement as a basis, 16-3
    graphical representations, 16-8–16-9
    history, 16-2
  Maxwell's force-voltage analogy
    dependence on reference frames, 16-5
    intuitions in processes, 16-4–16-5
    systems of particles, 16-4
  thermodynamic basis
    equilibrium and steady state, 16-6–16-7

    extensive and intensive variables, 16-6
    of nocidity, 16-8
    use of inertial reference frame, 16-7–16-8
Analog-to-digital converter (ADC), 3-2
Angle representation, of rotations, 9-34–9-37
Angular acceleration, 9-31, 20-14
Angular optical encoders, 20-108
Angular velocity, 9-31
Anisotropic etching, 15-6
Anode, 21-14
ANSYS software, 14-12
Antilock braking system (ABS), 1-8–1-9, 3-3
APLAC, 14-12
Application specific integrated circuits (ASIC), 4-2
Auto CAD, 2-9
Automated paperboard container-manufacturing machine, 1-6
Axial electromagnetic micromotors, 21-132

B

Band-pass filters, 3-7
Band-stop filters, 3-7
Batch-fabricated microscale systems, 15-2
Battery charging systems, 1-7
Beam-type load cells, 20-38–20-39
Bidirectional transmission, 4-7
Bimetallic thermometers, 20-76
BIOSPICE simulator, 14-8
Bipolar junction transistor (BJTs), 21-20–21-23
Bitmap (BMP), 20-155
8-bit timer/counter, 4-5
Black, H- S-, 1-3
Blocked force, 21-12
Bode, 1-3
Boit–Savart Law, 21-4–21-5
Bond graph modeling, 2-11, 9-6–9-10
Boolean algebra statements, 4-2
Boolean functions, 14-2
Boolean methods, 2-9
Bouncing circuit, 20-2
Brayton cycle, 12-25–12-26
Breakdown region, 21-15

I-1

Buckling analysis, **8**-10–**8**-11
Bulk micromachined pressure sensor, **20**-140–**20**-143
Bulk micromachining, **15**-6
Bulk silicon micromachining process, **20**-138

C

CAD/CAE tools, **2**-9
Calibrations, in mechatronic system, **3**-9
Capacitance, defined, **11**-22
Capacitive displacement transducer, **11**-25
Capacitive force transducer, **20**-43–**20**-44
Capacitive proximity sensor, **20**-114
Capacitive sensors, **20**-8
Capacitors, **11**-22–**11**-24
Carbon nanotubes (CNTs), **5**-10
Cart flywheel, **9**-46
Cathode, **21**-14
Causality assignment, in system models, **9**-7–**9**-8, **9**-21–**9**-23
Central processing unit (CPU), **4**-2
Cesium oscillators, **18**-12–**18**-13
CFD software, **14**-12
Charge-coupled devices, **20**-129–**20**-133
Checkerboard test, **3**-13
Chemical sensors, **17**-8
Circuit elements and their i–v characteristics, **11**-5
Closed loop transfer function, **21**-83
CMOS image sensors, **20**-133–**20**-134
Coenergy, **15**-15
Color image, **20**-154
Comb-drive electrostatic actuator, **5**-3–**5**-4
Complementary metal oxide semiconductor (CMOS) technologies, **15**-2
Complex image, **20**-154
Compressibility charts, **12**-15
Compression ignition engine, **12**-28
Compression stroke, **12**-28
Computer integrated manufacturing (CIM), **3**-10
Computer operating properly (COP) errors, **3**-13
Consolidated Micromechanical Element Library (CaMEL), **14**-14
Continuity equations, for chambers, **21**-79–**21**-80
Contrast, **20**-157
Control engineering, **1**-3
Controller area network (CAN), **3**-10
Control prototyping, **2**-14–**2**-15
Control theory, **1**-3
Coordinated Universal Time (UTC), **18**-2

Copper–chromium seed layer, **21**-112
Coriolis acceleration, **5**-9
Coriolis flowmeters, **20**-71
Coulomb's law, **5**-2
Coulomb model, of mechanical systems, **9**-11–**9**-12
Coupling mechanisms, of mechanical systems, **9**-15–**9**-17
Current control (mode) amplifier, **21**-28–**21**-29
Curriculum, for mechatronics, **6**-5–**6**-9
Curvilinear acceleration, **20**-14
Cutoff region, **21**-21

D

DADS, **7**-6
D'Alembert principle, **9**-48
Damping constants, **9**-10
Darlington transistor pair, **21**-23
Dashpot resistive model, **9**-9
Data acquisition board (DAQ board), **3**-7
Data retrieval, in thermodynamics, **12**-14–**12**-15
DC magnetic sensors, **20**-9–**20**-11
DC motors, **21**-33–**21**-34
efficiency of, **21**-41
speed control of, **21**-39–**21**-40
starting of, **21**-39
Debouncing circuit, **20**-2
Debugger, **3**-13
Decoding methods, standard, **20**-5
Deformation gradient, **8**-4
Degrees of freedom, **9**-5
Depth of field, of a lens, **20**-157
Detectors
light
basic radiometry, **20**-119–**20**-121
image formation, **20**-127–**20**-129
image sensors, **20**-129–**20**-134
light sources, **20**-121–**20**-122
photon detectors, **20**-123–**20**-127
pyroelectric detectors, **20**-123
positive sensitive (PSD), **20**-4
Schmitt-trigger threshold, **20**-111
Diesel cycle, **12**-29–**12**-31
Differential equations, for electrical circuits, **11**-30–**11**-32
Differential global positioning systems (DGPS), **1**-10
Differential output encoder, **20**-6
Differential pressure flowmeters, **20**-64–**20**-66
Diffuse proximity sensors, **20**-116
Digital communications, **4**-6–**4**-7
Digital electronics, **4**-1
Digital integrated circuits, **4**-1
Digital logic, **4**-2
Digital signal processors (DSPs), **4**-3
Digital-to-analog converter (DAC), **3**-2

Digitized image, properties, **20**-153–**20**-154
Directional valves, **21**-71–**21**-72
Direct memory access (DMA), **3**-7
Direct piezoelectric effect, **21**-11
Dissipative effects, in mechanical systems, **9**-9–**9**-12
Distance measuring sensors, *see* Noncontact ranging sensors
Distortions, **20**-157
DPST (double pole single throw) switch, **20**-2–**20**-3
Dual cycle, **12**-29–**12**-31
DYMOLA, **2**-9
Dynamic compliance, **21**-83

E

Earnshaw's theorem, **7**-18–**7**-19
Edge
characteristics, **20**-160–161
definition, **20**-160
detection, **20**-158–**20**-162
Effective number of bits (ENOB), **4**-5
Effort source, defined, **9**-8
Elastic buckling/ divergence, **7**-18
Elastic system modeling, **7**-8–**7**-10
Electrical actuators, **17**-13
Electrical engineering
AC network analysis, **11**-21–**11**-27
circuit elements and their i–v characteristics, **11**-5
circuits containing dynamic elements, **11**-30–**11**-32
electrical power and sign convention, **11**-4–**11**-5
fundamentals of electric circuits, **11**-1–**11**-4
measuring devices
ammeter, **11**-14
voltmeter, **11**-14–**11**-15
nonlinear circuit elements, **11**-20–**11**-21
phasors and impedance, **11**-32–**11**-36
practical voltage and current sources, **11**-12–**11**-13
resistance and ohm's law
common resistor values, **11**-7
open and short circuits, **11**-8–**11**-10
parallel resistors and current divider rule, **11**-11
resistance strain gauge, **11**-8
series resistors and the voltage divider rule, **11**-10–**11**-11
Wheatstone bridge, **11**-11–**11**-12
resistive network analysis
mesh current method, **11**-16–**11**-17
node voltage method, **11**-15–**11**-16
Thévenin and Norton equivalent circuits, **11**-17–**11**-20

Index

time-dependent signal sources, 11-28–11-30
Electrical machines
  AC machines, 21-41
  armature electromotive force (amf), 21-34–21-35
  armature torque, 21-35
  DC motors, 21-33–21-34
  methods of connection, 21-36–21-38
  motor selection, 21-50–21-51
  terminal voltage, 21-35–21-36
Electrical power and sign convention, 11-4–11-5
Electrical temperature sensors and transducers, 20-78–20-84
Electric and magnetic circuits, dynamic principles, 7-14–7-17
Electric eye, 20-115
Electric motors, 21-9–21-10
Electroactive polymer-metal composites (EAPs), 8-13–8-16
Electrohydraulic control systems, 10-2
Electrolytic tilt sensor, 20-8
Electromagnetic actuation, 5-5–5-6
Electromagnetic actuators, 17-14–17-15
Electromagnetic flowmeters, 20-69
Electromagnetic forces, 7-10–7-14
Electromagnetic microactuators, 21-116–21-117
Electromagnetic system–geometric set, 15-5
Electromagnetic torque and force, on a cylindrical permanent-magnet thin film, 21-122–21-123
Electromechanical accelerometers, 20-23–20-24
Electromechanical actuators, 17-13–17-14
  types
    power amplifier or the driver, 21-13–21-20
    using energy conversion mechanism, 21-3–21-13
Electromechanical stability, 7-18–7-19
Electromechanical systems, models for, 7-2
Electromechanical transduction, 5-2
Electronic damping, 2-4
Electronic systems, 2-1
Electroplated aluminum, 21-112
Electrostatic accelerometers, 20-27–20-29
Electrostatic actuation, 5-3–5-5
Electrostriction, 8-5
Embedded microcomputers, 4-3
Energy flows, in the machine, 2-2
Energy methods, for modeling mechanical system
  determination and checking of constitutive relations, 9-29–9-30
  and energetically-correct constitutive relations, 9-28–9-29
  multiport models, 9-28
Energy storage elements, 9-12–9-13
Enthalpy-entropy (Mollier) diagram, for water, 12-19
Equilibrium, of physical system models, 16-6–16-7
Erasable programmable ROM (EPROM), 3-8
Euler parameters and quaternions, 9-37–9-39
Euler's theorem, 9-31
Exhaust stroke, 12-28

F

Farad, 11-22
Faraday, Michael, 1-3
Faraday–Henry law of flux change, 7-15
Faraday's law, 5-2, 15-13, 21-3–21-4
Fault detection, classical way for, 2-8
Feedback control system, 1-3
Field of view, 20-156
Field programmable gate array (FPGA), 4-5
Filtering, 3-7
Finite element analysis (FEA), 14-10
Finite rotations, 9-31
Fixed-point mathematics, 3-8–3-9
Flow measurement
  flow characteristics, 20-62–20-63
  flowmeter classification, 20-63–20-64
  installation of flowmeters, 20-72
  selection of flowmeters, 20-72–20-73
  terminology, 20-62
  two-phase flows, 20-71–20-72
  types of flowmeters
    coriolis, 20-71
    differential pressure, 20-64–20-66
    electromagnetic, 20-69
    positive displacement, 20-66–20-67
    turbine (or vane), 20-67–20-68
    ultrasonic, 20-69–20-71
    variable area, 20-66
    vortex shedding, 20-68–20-69
Flow-rate regulator valves, 21-73
Flow sensors, 17-5–17-6, 20-148–20-151
Flow source, defined, 9-9
Fluid actuation system, 21-64–21-66
Fluid power system, control
  E/H system feedforward-plus-PID control, 10-10
  E/H system generic fuzzy control, 10-11–10-12
  steady-state characteristics, 10-8–10-9
  system dynamic characteristics, 10-9–10-10
Foil effect, 20-113
Force constant, 21-8
Force-current analogy
  beyond one-dimensional mechanical systems, 16-3
  drawbacks, 16-2
  intuitions in processes, 16-3–16-4
  measurement as a basis, 16-3
Force feedback accelerometers, 20-33
Forced mass–spring system, 20-15
Force measurement
  general considerations, 20-34
  Hooke's law, 20-34–20-36
Forces, produced by field distributions around electric charge, 7-11
Force sensing resistors (FSRs), 20-44
Force sensors, 5-8, 20-36–20-47
Force/torque sensors, 17-5
Ford Motor Company, 1-3
Forward biased region, 21-15
Four-stroke internal combustion engine, 12-28
Frequency-domain techniques, 1-5
Friction values, for different surfaces, 9-11
Function block diagram (FBD), 4-6

G

Gas bulb thermometer, 20-75
Gauging applications, 20-158–20-159
Gear pumps, 21-67
Geometric matching, 20-164–20-166
Geometric scaling factor, 5-2
Gibb's thin film free energy density, 21-116
Global positioning system (GPS), 18-16
GPS-based continuous traffic model, 1-9–1-10
GRAFCET, 4-6
Gray code, 20-6
Grayscale image, 20-154
Gyrator, 9-15
Gyroscopes, 5-8–5-9

H

Hall-effect sensor, 1-8, 20-111
Hall effect switches, 20-9–20-10, 20-111
Handshaking, 3-10
Harmonic drive type electrostatic motor, 5-4
Harvard architecture, 4-3
H-bridge configuration, 21-30
HC12 microcontroller input–output subsystems, 3-10

Heating/cooling system for homes and offices, 3-2–3-3
Heat sinks, 21-15
HF radio signals, 18-15
High-pass filters, 3-7
High-volume precision manufacturing, 1-6
Horizontal representation, 6-4
Human computer interface (HCI), 3-3
Hydraulic and pneumatic actuation systems
   fluid actuation system, 21-64–21-66
   hydraulic actuation system, 21-66–21-78
   scheme of a hydraulic servo system for position control, 21-79–21-84
Hydraulic and pneumatic actuators, 17-15
Hydraulic control valves, 10-4–10-5
   principle of valve control, 10-3–10-4
Hydraulic cylinder, 10-7
Hydraulic fluids, properties
   bulk modulus, 10-3
   density, 10-2
   viscosity, 10-2–10-3
Hydraulic pumps
   principle of pump operation, 10-5–10-6
   pump controls and systems, 10-6

I

IC decoder chips, 20-5
Ideal capacitor, 11-21–11-24
Ideal gas model, 12-19–12-22
Ideal inductor, 11-25–11-27
Identity, of mechatronics, 6-2–6-3
Ignition system, electronic, 1-8
Image file, 20-154–20-155
Image formation, 20-127–20-129
Image sensors, 20-129–20-134
Impedance, 11-32–11-36
   functions in mechanical system, 9-17–9-19
Inclinometer, 20-8
Incremental angular optical encoders, 20-108
Incremental encoder, 20-5
Index signal, 20-5
Induction motor
   braking method, 21-44
   brushless DC motor, 21-49–21-50
   DC permanent magnet (PM) motor, 21-47
   single-phase, 21-46–21-47
   speed control, 21-44–21-46
   starting of, 21-44
   stepper motor, 21-47–21-49
   torque–slip characteristic, 21-43
Inductive method, 20-41–20-42
Inductive proximity switches, 20-112
Inductor kickback, 21-32

Inertial accelerometers, 20-19–20-23
Inference mechanism, 2-9
Infinitestimal rotations, 9-31
Information flow, 2-3
Information processing systems
   intelligent control systems, 2-8–2-9
   model-based and adaptive control systems, 2-7
   multilevel control architecture, 2-6–2-7
   special signal processing, 2-7
   supervision and fault detection, 2-8
Information technology, 2-1
Information technology and mechatronics, 1-2
Infrared photodetectors, 20-3
Infrared sensitive device, 20-3–20-6
Infrared type sensors, 17-6
Input signals, to mechatronic systems
   analog-to-digital converters, 3-5
   transducer/sensor input, 3-3
Insulated Gate Bipolar Transistor (IGBT), 21-26
Intake stroke, 12-28
Integrated circuit temperature sensors, 20-83–20-84
Integrated microsensors
   definition, 20-136–20-137
   fabrication process, 20-137–20-138
   principles in, 20-138–20-151
Intel Corporation microprocessor, 1-7
Intelligent control systems, 2-8–2-9
Intelligent mechatronic systems, 2-9
Intelligent safety systems, 1-7
International Practical Temperature Scale of 1990 (ITS_{90}), 20-74
Internet, 4-3
Interrupt enable (IE), 3-9
Interrupt request (IRQ), 3-9
Inventions, in mechatronics, 1-3
Isentropic efficiency, 12-27
ISO Open Systems Interconnection (OSI) model, 3-10
Isotropic etching, 15-6

J

Japan Society for the Promotion of Machine Industry (JSPMI), 1-7
Joint photographic experts group format (JPEG), 20-155
1-junction, 9-6
Junction photodetectors, 20-124–20-127

K

Kinematic pairs, 21-61
Kinetic energy storage, 9-14–9-15
Kirchhoff's voltage law, 9-6, 9-20, 15-15, 21-106
Knowledge base, 2-9

L

LabVIEW Simulation Module, 13-5–13-6
LabVIEWState Diagram Toolkit, 13-5–13-6
Lagrange's equations, 7-6
   case of nonholonomic constraints, 9-49–9-50
   classical approach, 9-48–9-49
   for electromechanical systems, 7-10–7-12
   in formulation of subsystem models, 9-51–9-53
   models of mechanical systems, 9-50–9-51
   of motion for electromechanical systems, 7-15–7-17
   nonconservative effects, 9-49
Landay–Lifschitz–Gilbert equations, 21-116
Laplace operator, 9-17
Laser interferometers, 20-12
Latching current, 21-18
Lead-zirconate-titinate (PZT), 21-11
Leaks, in hydraulic system, 21-80
Legitimacy, of mechatronics, 6-3–6-4
Lens focal length, 20-156–20-157
Lenz's law, 15-13
LF radio signals, 18-15–18-16
LIGA and LIGA-like technologies, 15-7–15-8
Light detectors
   basic radiometry, 20-119–20-121
   image formation, 20-127–20-129
   image sensors, 20-129–20-134
   light sources, 20-121–20-122
   photon detectors, 20-123–20-127
   pyroelectric detectors, 20-123
Light polarizing filters, 20-3
Light sensors, 17-6–17-7
Limit switch, 20-2
Linear and rotational position sensors, 17-2–17-4
Linear and rotational sensors
   AC inductive sensor, 20-9
   capacitive sensors, 20-8
   DC magnetic sensors, 20-9–20-11
   infrared sensitive device, 20-3–20-6
   laser interferometers, 20-12
   magnetostrictive wire transducers (MTS), 20-11–20-12
   resistors, 20-7–20-8
   switches, 20-2
   tilt sensor, 20-8
   ultrasonic (US) sensors, 20-11
Linear hydraulic motors, 21-70–21-71
Linearized model, of an hydraulic servosystem with position control, 21-82
Linear optical encoders, 20-109

Index

Linear variable differential transformer (LVDT), **20**-9, **20**-109–**20**-110
Liquid-glass thermometer, **20**-76
Local area network (LAN), **4**-3
Logical values, **4**-1
Logic analyzer, **3**-13
Lorentz equation, **5**-2, **5**-5
Lorentz's law, of electromagnetic forces, **21**-3–**21**-4
Lorentz-type nonrotary actuators, **5**-6
Low-pass filters, **3**-7
Lumped parameter processes, theoretical modeling of, 2-11

M

MAGIC, **14**-5, **14**-8
Magnetic stress tensor, **7**-13
Magnetization dynamics, of thin films, **21**-116
Magnetoelastic force transducers, **20**-45
Magnetoresistive force sensors, **20**-44
Magnetostrictive sensors, **17**-7
Magnetostrictive wire transducers (MTS), **20**-11–**20**-12
Manufacturing automation protocol (MAP), **3**-10
Mass conservation law, **8**-3
Master-Slave pooling method, **4**-7
Mathematical model
  MEMS
    elementary synchronous reluctance micromotor, **15**-16–**15**-18
    translational microtransducer, **15**-14–**15**-16
    two-phase permanent-magnet stepper micromotors, **15**-18–**15**-20
    two-phase permanent-magnet synchronous micromotors, **15**-20–**15**-22
  lumped-parameter mechanical systems
    bond graph approach, **9**-24–**9**-26
    classical approach, **9**-23–**9**-24
Matlab simulations, **14**-12
MATLAB/SIMULINK, **2**-10
MATRIX-X, **2**-10
Maxwell, J- C-, **1**-3
Maxwell–Faraday equations, of electromagnetics, **7**-2
Maxwell's force-voltage analogy
  dependence on reference frames, **16**-5
  intuitions in processes, **16**-4–**16**-5
  systems of particles, **16**-4
Maxwell's equations, **8**-5, **15**-9–**15**-11
Maxwell stress tensor method, **15**-10
Mean effective pressure (mep), **12**-28–**12**-29
Mechanical control systems, **1**-7–**1**-8

Mechanical systems, 2-1
Mechanics, laws of
  electric phenomena, **8**-5–**8**-6
  equations of motion of deformable bodies, **8**-2–**8**-4
  statics and dynamics of mechatronic systems, **8**-1–**8**-2
Mechatronics
  classification of products, **1**-7
  communication of, **6**-5
  curriculum development of, **6**-5–**6**-7
  definitions, **1**-1–**1**-2, **2**-1–**2**-3
  educational programmes, **1**-2
  evolution of, **1**-7–**1**-10
  evolution of the subject of, **6**-7–**6**-9
  historical perspective, **1**-3–**1**-7
  identity of, **6**-2–**6**-3
  inventions, **1**-3
  journals, **6**-2
  key elements of, **1**-2–**1**-4
  legitimacy of, **6**-3–**6**-4
  in modern times, **1**-10–**1**-11
  selection of, **6**-4–**6**-5
Mechatronic systems
  antilock braking system (ABS), **3**-3
  common structures in, **8**-6–**8**-8
  computer aided development of, **2**-9–**2**-10
  control prototyping, **2**-14–**2**-15
  definitions, **2**-1–**2**-3
  design steps, 2-9
  functions of, **2**-3–**2**-5
  heating/cooling system for homes and offices, **3**-2–**3**-3
  historical development of, **2**-1–**2**-3
  input signals of, **3**-3–**3**-5
  integration of, **2**-5–**2**-6
  mechanical system modeling in, **9**-2–**9**-8
    difficulties in the mathematical model development, **9**-27–**9**-28
    energy methods, **9**-28–**9**-30
    Lagrange's equations, **9**-50–**9**-51
    mechanical components in mechatronic systems, **9**-8–**9**-19
    physical laws in model formulation, **9**-19–**9**-27
  microprocessor control, **3**-8
    input–output, **3**-9–**3**-11
    numerical, **3**-8–**3**-9
  modeling procedure, **2**-10–**2**-12
  output signals of, **3**-5–**3**-6
  real-time simulation, **2**-12–**2**-14
  rigid body multidimensional dynamics
    coordinate transformations, **9**-46–**9**-47
    dynamic properties, **9**-39–**9**-43
    equations of motion, **9**-43–**9**-45
    graph formulation, **9**-45–**9**-46

  signal conditioning, **3**-6–**3**-7
  kinematics, **9**-31–**9**-39
  software control, **3**-11–**3**-12
  statics and dynamics of, **8**-1–**8**-2
  testing and instrumentation, **3**-12–**3**-13
  vs conventional design systems, **2**-5
MEMCAD, **14**-13
MEMS accelerometers, **17**-9–**17**-10
MemsPro, **14**-13
Mercury switch, **20**-8
Mesh current method, **11**-16–**11**-17
Metal-Oxide-Semiconductor Field Effect Transistor (MOSFET), **21**-23–**21**-26
Microaccelerometers
  signal-conditioning circuitry, **20**-33
Microactuator technology
  electromagnetic actuation, **5**-5–**5**-6
  electrostatic actuation, **5**-3–**5**-5
Micro and nanoactuators, **17**-18
Micro and nanosensors, **17**-8
Microcomputers, **2**-4
Microcontroller firmware, **4**-5
Microcontroller network systems, **3**-10
Microcontrollers, **4**-4–**4**-5
Microelectromechanical systems (MEMS), **1**-9, **5**-1
  analysis and modeling of the microtransducer, **21**-102–**21**-104
  control of
    constrained control of nonlinear MEMS, **15**-26–**15**-29
    constrained control of nonlinear uncertain MEMS, **15**-29–**15**-34
    by the proportional-integral-derivative (PID) controllers, **15**-23
    soft-switching sliding mode control, **15**-25–**15**-26
    time-optimal controller, **15**-25
    tracking control, **15**-23–**15**-25
  design and fabrication, **21**-98–**21**-102
  design of motion microdevices, **15**-3–**15**-6
  electromagnetic fundamentals and modeling, **15**-8–**15**-11
  fabrication of
    bulk micromachining, **15**-6
    surface micromachining, **15**-6–**15**-7
    use of LIGA and LIGA-like technologies, **15**-7–**15**-8
  mathematical models, **15**-11–**15**-22
  modeling and simulation of analog and mixed-signal circuit development, **14**-7–**14**-8

Microelectromechanical systems
    (MEMS) (*Continued*)
  digital circuit development,
      14-2–14-7
  digital vs analog circuits,
      14-13–14-15
  guidelines for successful, 14-15
  resources, 14-10–14-13
  techniques and tools,
      14-8–14-13
Microelectronics, 4-1
Microfabrication, of electromechanical
    microstructures and
    microtransducers
  electrodeposition of metals,
      21-110–21-111
  NiFeMo and NiCo thin films
      electrodeposition,
      21-114–21-115
  NiFe thin films electrodeposition,
      21-111–21-114
  using micromachined permanent
      magnet thin films,
      21-115–21-116
Micromachined polycrystalline silicon
    carbide micromotors,
    21-130–21-132
Micromanipulation tool, 1-11
Micromirror actuator,
    21-116–21-130
Micro/nanoaccelerometers,
    20-30–20-31
Microprocessor control, of
    mechatronic systems,
    3-8
  input–output, 3-9–3-11
  numerical, 3-8–3-9
Microprocessor control system, 3-2
Microprocessors, 4-4–4-5
Microsensor technology, 5-2
  force sensors, 5-8
  gyroscopes, 5-8–5-9
  pressure sensors, 5-7
  silicon microfabricated
      accelerometer, 5-8
  strain measurements, 5-7
Microswitch, 20-2
Microwave proximity sensors,
    20-114–20-115
Millimeter-wave radar technology, 1-9
MOBILE, 2-9
Model-based fault detection, 2-8
Modeling, of a physical system
  bond graph, 2-11
  difficulties in the mathematical
      model development,
      9-27–9-28
  elastic system, 7-8–7-10
  energy methods, 9-28–9-30
  of hydraulic servosystem for
      position control,
      21-79–21-84
  Lagrange's equations, 9-50–9-51
  lumped parameter processes, 2-11

mechatronic system
  concept of causality, 9-7–9-8
  interconnection of components,
      9-6–9-7
  mechanical components,
      9-8–9-19
  physical variables, 9-3–9-5
  power variables, 9-3–9-5
  of MEMS, 15-11–15-22
  analog and mixed-signal circuit
      development, 14-7–14-8
  digital circuit development,
      14-2–14-7
  digital vs analog circuits,
      14-13–14-15
  guidelines for successful, 14-15
  microtransducer, 21-102–21-104
  resources, 14-10–14-13
  techniques and tools,
      14-8–14-13
  objected-oriented, 2-11
  physical laws in model
      formulation, 9-19–9-27
  pneumatic actuation system,
      21-91–21-96
MOS field effect transistor (MOSFET),
    14-3–14-4, 14-7, 20-134
Motion actuators, 21-69–21-71
Motion and play, equations of, 8-2
Motions of points, in the body, 9-31
  relative to coordinate systems,
      9-32–9-33
Motion transducers, 11-24
Multicomponent dynamometers,
    20-42
Multilayer standard protocols, 4-6
Multi-turn pot, 20-7
Multi-walled carbon nanotubes
    (MWNTs), 5-10

N

Nanomachines, 5-9–5-11
Nanotechnology, 5-1
National instruments internal image
    file format (AIPD),
    20-155
Navier–Stokes equations, of fluid
    mechanics, 7-2
NBC sensors, 17-7
NEWEUL, 7-6
Newton–Euler equation, 7-4–7-6
Newton's laws of motion, 15-10–15-11,
    15-21, 20-145, 21-121
NiFeMo and NiCo thin films
    electrodeposition,
    21-114–21-115
NiFe thin films electrodeposition,
    21-111–21-114
NODAS v 1-4, 14-13–14-15
Node voltage analysis, 11-15–11-16
Noncontact ranging sensors
  frequency modulation techniques,
      20-97–20-99

  magnetic position measurement
      systems, 20-106–20-107
  other distance measuring methods,
      20-107–20-110
  phase measurements, 20-94–20-97
  structured light methods,
      20-105–20-106
  triangulation ranging method,
      20-99–20-105
  using microwave technology, 20-94
  using time-of-flight or frequency
      modulation methods,
      20-89–20-94
Noncontact thermometers,
    20-84–20-85
Nonholonomic constraints,
    9-49–9-50
Nonlinear circuit elements,
    11-20–11-21
Nonstationary random vibrations,
    20-17
Norton equivalent circuits
  computation of the Norton
      equivalent current, 11-19
  determination of resistance, 11-18
  experimental determination
      Norton equivalents,
      11-19–11-20
Norton theorem, 11-18
N-tuple, 15-5
N-type semiconductors, 21-14
Nuclear sensors, 17-8
Numerically controlled (NC)
    machines, 1-3
Numerical simulation, of dynamic
    systems
  common simulation blocks
    continuous linear system blocks, 13-2
    discrete linear system blocks, 13-2
    nonlinear system blocks, 13-2
    and signal generation, 13-3
    table lookup blocks, 13-2
  hybrid control approach, 13-6
  ordinary differential equation
      (ODE) solvers, 13-4
  textual equations within block
      diagrams, 13-3–13-4
  timing options, 13-4
  visualization, 13-5–13-6
Nyquist, H-, 1-3
Nyquist theorem, 3-6

O

Object-oriented modeling, 2-11
Online expert system, 2-8
On–off valves, 21-72
Open-collector output, 21-32
Open-ended electromagnetic
    system, 15-3
Open loop transfer function, 21-83
Optical encoders, 20-5–20-6
Optical proximity sensors,
    20-115–20-116

Index

Optocouplers, 21-32–21-33
Optoisolators, 21-32–21-33
Ordinary differential equations (ODEs), 7-2, 13-4
Orthonol, 21-113
OTP (One Time Programmable) EPROM/ROM, 4-5
Otto cycle, 12-29–12-31
Output signals, to mechatronic systems
    actuator output, 3-6
    digital-to-analog converters, 3-5

P

Parallel mode traffic, 3-10
Partial differential equations (PDEs), 7-2
Pascal, 1-3
Pattern-matching, 20-162–20-164
Paynter's reticulated equation, of energy continuity, 9-3
Periodic signals, 11-28
Periodic vibrations, 20-15–20-17
Permalloy thin films, 21-113–21-114
Permanent-magnet DC (PMDC) motor, 9-4
Permanent-magnet polymer magnets, 21-115
Permeability constant, 21-3
Permittivity, of air, 11-24
Perspective errors, 20-157
Phasors, 11-32–11-36
Phasors and impedance, 11-32–11-36
Photocell, 20-3
Photodetectors
    infrared, 20-3
    junction, 20-124–20-127
Photodiode, 20-3
Photodiode capacitance, 20-134
Photoemitter, 20-3
Photointerrupter, 20-3–20-4
Photon detectors, 20-123–20-127
Photoreflector, 20-3–20-4
Photoresistor, 20-3, 20-124
Phototransistor, 20-3
PID control, 3-8
Piezoelastic beam, 7-9–7-10
Piezoelectric accelerometers, 20-24–20-25, 20-31–20-33
Piezoelectric actuators
    application areas, 21-57–21-58
    constitutive equations of piezoelectric materials, 21-51–21-52
    motors with several degrees of freedom, 21-61–21-63
    piezoactuating elements, 21-53–21-57
    piezoelectric effect, 21-51
    piezoelectric materials, 21-53
    ultrasonic motors, 21-58–21-61
Piezoelectric effect, 20-40, 20-42
Piezoelectric elements, 20-40
Piezoelectric sensing, 20-139–20-140
Piezoresistive accelerometers, 20-25–20-26
Piezoresistive sensor, 20-139
Piezoresistive strain gages, 5-7
Piezoresistive transducers, 20-33
Piezotransistors, 20-42
Piston, dynamic equilibrium equation of, 21-80
Planar microwindings, 21-129–21-130
Pneumatica, 1-3
Pneumatic actuation system
    modeling of a, 21-91–21-96
    types of compressors, 21-85–21-87
    valves in, 21-87–21-91
Pneumatic control elements, 1-3
Pn junction, 21-14
Poles and throws, of switch, 20-2
Polytropic process, 12-24–12-25
Polyvinylidence fluoride (PVDF), 21-11
Polzunov, 1-3
Portable network graphics (PNG), 20-155
Position sensitive detectors (PSDs), 20-4
Positive displacement flowmeters, 20-66–20-67
Possibilistic methods, 2-9
Potentiometers, 20-7, 20-109
Power amplification, 21-13
Power and energy variables, for mechanical systems, 9-3
Power bond, 9-4
Power dissipation, 21-23
Pressure regulator valves, 21-72–21-73
Pressure sensors, 5-7
Problem-based learning (PBL), 6-4
Programmable electrohydraulic valves, 10-12–10-13
Programmable logic controller (PLC), 3-8, 4-5–4-6
Proportional valves, 21-77
Proximity sensors, 17-6, 20-110–20-116
P-type semiconductors, 21-14
Pull-down voltage, 14-9
"Pull-in" voltage, 14-9
Pulse-width modulation (PWM), 3-10, 21-30–21-31
Push-pull (class B) power amplifier, 21-29–21-30
Pyroelectric detectors, 20-123

Q

Quartz crystal oscillators, 18-10–18-12

R

Radial flux microdevices, 15-5
Rankine cycle, 12-25–12-26
Rankine cycle steam turbine, 12-27
Rapid prototyping, 14-6
Real-time counter (RTC), 4-5
Real-time simulation, in design of mechatronic systems, 2-12–2-14
Reciprocal piezoelectric effect, 21-11
Rectilinear acceleration, 20-14
Registers, 3-8
Relative permeability, 21-3
Resistance and ohm's law
    common resistor values, 11-7
    open and short circuits, 11-8–11-10
    parallel resistors and current divider rule, 11-11
    resistance strain gauge, 11-8
    series resistors and the voltage divider rule, 11-10–11-11
    Wheatstone bridge, 11-11–11-12
Resistance temperature devices, 20-80–20-82
Resistive method, 20-41
Resistors, 20-7–20-8
Resolvers, 20-9
Retroreflective sensors, 20-115
Reverse biased region, 21-15
Reverse-recovery time, 21-16
Rheostat, 20-7
Ribbens, William, 3-7
Rigid body models
    constraints and generalized coordinates, 7-2–7-4
    equations of dynamics of, 7-4–7-6
    kinematics of, 7-2
    kinematic vs dynamics problems, 7-4
Rigid body multidimensional dynamics
    coordinate transformations, 9-46–9-47
    dynamic properties, 9-39–9-43
        angular momentum, 9-41–9-42
        inertia, 9-39–9-41
        kinetic energy, 9-41–9-43
    equations of motion, 9-43–9-45
    graph formulation, 9-45–9-46
    kinematics, 9-31–9-39
Ring-type load cells, 20-39–20-40
RISC (Reduced Instruction Set), 4-4
Robot's eye, 21-62
Robust control theory, 1-7
Rotary electrostatic actuators, 5-4
Rotary vane pumps, 21-67–21-68
RS-232C serial line, 4-5
Rubidium oscillators, 18-12

S

Saturation region, 21-21
Saturation voltage, 21-21
Scaling, physics of, 5-1–5-2
Scanning probe microscope (SPM) assemblers, 5-11, 8-6
Scanning thermal microscopy (SThM), 20-86

Schmitt-trigger threshold detector, **20**-111
Self-inductances, of the stator windings, **15**-21
Semiconductor gages, *see* piezoresistive strain gages
Semiconductors, **21**-14
Sensitivity, of motion transducer, **11**-24
Sensors
  calibration, **17**-11
  characteristics
    backlash, **19**-6
    deadband, **19** 7 **19** 8
    eccentricity, **19**-6
    errors, **19**-2
    first-order system response, **19**-8–**19**-9
    frequency response, **19**-12–**19**-14
    impedance, **19**-4–**19**-5
    linearity and accuracy, **19**-3–**19**-4
    nonlinearities, **19**-5
    range of, **19**-1
    repeatability, **19**-3
    resolution of, **19**-2
    saturation, **19**-7
    second-order system response, **19**-9–**19**-11
    sensitivity of, **19**-2
    static and Coulomb friction, **19**-5
    system response, **19**-8
  classification, **17**-2
  defined, **17**-1
  principle of operation
    acceleration sensors, **17**-4–**17**-5
    chemical sensors, **17**-8
    flow sensors, **17**-5–**17**-6
    force/torque sensors, **17**-5
    infrared type sensors, **17**-6
    light sensors, **17**-6–**17**-7
    linear and rotational position sensors, **17**-2–**17**-4
    magnetostrictive sensors, **17**-7
    MEMS accelerometers, **17**-9–**17**-10
    micro and nanosensors, **17**-8
    NBC sensors, **17**-7
    nuclear sensors, **17**-8
    proximity sensors, **17**-6
    smart material sensors, **17**-7
    temperature sensors, **17**-6
    thermocouples, **17**-6
    ultrasonic flow meters, **17**-6
    vision, **17**-8
  selection criteria, **17**-10
  signal conditioning, **17**-11
Separation of variables, **7**-2
Serial communications interface (SCI), **3**-10
Serial data transmission, **3**-10
Serial in-circuit debugger (SDI), **3**-10

Serial peripheral interface (SPI), **3**-10
Servovalves, **21**-74–**21**-76
Shocks, **20**-17
SiC surface micromachining, **21**-130–**21**-131
Signal bonds, **9**-4
Signal conditioning, in mechatronic system, **3**-6–**3**-7
Silicon-controlled rectifier (SCR), **21**-17
SIMPACK, **2**-10
Simple dynamic models
  compound pendulum, **7**-6–**7**-7
  gyroscopic motions, **7**-7–**7**-8
Simulation blocks, common
  continuous linear system blocks, **13**-2
  discrete linear system blocks, **13**-2
  nonlinear system blocks, **13**-2
  and signal generation, **13**-3
  table lookup blocks, **13**-2
Single-chip microcontroller, **4**-1
Single-ended output encoder, **20**-6
Single-phase reluctance micromachined motors, **21**-104–**21**-106
Single-turn pot, **20**-7
Single-walled carbon nanotubes (SWNTs), **5**-10
Sketched fabrication process, **21**-112
Smart material actuators, **17**-15–**17**-18
Smart material sensors, **17**-7
Snubber circuit, **21**-19
SoC devices, **1**-11
SoftPLC, **4**-6
Software control, of mechatronic systems, **3**-11–**3**-12
Solenoids, **21**-6–**21**-8
Spark-ignition engine, **12**-28
Spatial calibration, **20**-158
SPDT (single pole double throw) switch, **20**-2–**20**-3
Spherical-conical geometry, **15**-5
SPICE (Simulation Program with Integrated Circuit Emphasis) simulator, **14**-5, **14**-7, **14**-13
S-plane methods, **1**-5
Standard dry friction model, of mechanical systems, **9**-12
Start and stop bits, **3**-10
Static compliance, **21**-83
Static stiffness, **21**-84
Stationary random vibrations, **20**-17
Steady state, of physical system models, **16**-6–**16**-7
Stepper motors, **17**-14
Stick-slip system, **9**-12
Stiffness matrix, **8**-4
Stokes's theorem, **21**-123
Strain gage load cell, **20**-37–**20**-40
Strain-gauge accelerometers, **20**-26–**20**-27

Stress tensor, **8**-3
String pots, **20**-7
SUGAR, **14**-12
Surface micromachined accelerometers, **20**-146–**20**-147
Surface micromachined beams, **8**-6–**8**-7
Surface micromachining, **15**-6–**15**-7
Surface micromachining pressure sensor, **20**-143–**20**-145
Switches, **20**-2
Switching amplifiers, **21**-29
Symplectic gyrator, **9**-50
Synchronous detection, **20**-9
System boundary, **9**-8

T

Tactile array sensors, **20**-46
Tactile sensors, **20**-45–**20**-46, **20**-147–**20**-148
Tagged image file format (TIFF), **20**-155
Tait-Bryan or Cardan angles, **9**-37
Tape-based sensors, **20**-11
Taylor series, **21**-124–**21**-125
Telephone system, **1**-3
Temperature-entropy diagram, for water, **12**-18
Temperature measurements
  absolute temperature scales, **20**-74
  electrical temperature sensors and transducers, **20**-78–**20**-84
  International Practical Temperature Scale of 1990 (ITS$_{90}$), **20**-74
  microscale temperature measurements, **20**-85–**20**-87
  noncontact thermometers, **20**-84–**20**-85
  thermometers, **20**-75–**20**-78
  zeroth law, **20**-74
Temperature sensing, **20**-140
Temperature sensors, **17**-6
Tesla, Nikola, **1**-3
Thermal actuators, **8**-13
Thermal efficiency, of a power cycle, **12**-26
Thermal runaway, **21**-13
Thermistors, **20**-83
Thermocouples, **17**-6, **20**-78–**20**-80
Thermodynamics
  concepts and definitions
    condition and properties, **12**-2
    equilibrium, **12**-2
    irreversibilities, **12**-3
    phase and pure substance, **12**-2
    process and cycle, **12**-2
    system, **12**-1–**12**-2
    temperature, **12**-2–**12**-3
  extensive property balances
    control volume at steady state, **12**-6–**12**-8

energy balance, **12**-5
entropy balance, **12**-5–**12**-6
exergy balance, **12**-9–**12**-12
mass balance, **12**-4–**12**-5
laws of, **12**-4
of physical systems models
  equilibrium and steady state,
    16-6–**16**-7
  extensive and intensive variables,
    16-6
  of nocidity, **16**-8
  use of inertial reference frame,
    16-7–**16**-8
property relations and data
  analytical equations of state,
    12-15
  compressibility charts, **12**-15
  enthalpy–entropy (Mollier)
    diagram for water, **12**-19
  ideal gas model, **12**-19–**12**-22
  phase diagram, **12**-12–**12**-14
  sample stream data, **12**-16–**12**-17
  temperature–entropy diagram
    for water, **12**-18
  thermodynamic data retrieval,
    12-14–**12**-15
vapor and gas power cycles,
  12-23–**12**-31
Thermometers
  based on differential expansion
    coefficients, **20**-75–**20**-76
  based on phase changes,
    20-77–**20**-78
  noncontact, **20**-84–**20**-85
Thévenin equivalent circuits
  computation of Thévenin voltage,
    11-19
  determination of resistance, **11**-18
  experimental determination of
    Thévenin equivalents,
    11-19–**11**-20
  Thévenin theorem, **11**-18
Thin plate theory, **8**-8
Three-phase synchronous reluctance
  micromotors,
  21-106–**21**-109
Thyristors, **21**-17–**21**-20
Tilt sensors, **20**-8
Time and frequency, fundamentals
  Coordinated Universal Time
    (UTC), **18**-2
  definitions, **18**-1
  measurements
    accuracy, **18**-3–**18**-6
    stability, **18**-6–**18**-9
  radio time and frequency transfer
    signals
    global positioning system (GPS),
      18-16
    HF radio signals, **18**-15
    LF radio signals, **18**-15–**18**-16
  standards
    cesium oscillators, **18**-12–**18**-13

quartz crystal oscillators,
  18-10–**18**-12
rubidium oscillators, **18**-12
time interval, **18**-1
transfer techniques, **18**-13–**18**-14
Time-dependent signal sources,
  11-28–**11**-30
Time-domain methods, **1**-5–**1**-6
Token Passing method, **4**-7
Torque and power measurements
  absorption dynamometers,
    20-57–**20**-59
  apparatus for power measurement,
    20-56–**20**-57
  arrangements of apparatus,
    20-51–**20**-52
  costs, **20**-60
  driving and universal
    dynamometers, **20**-59–**20**-60
  fundamental concepts, **20**-49–**20**-51
  measurement accuracy, **20**-60
  torque transducer construction,
    operation and application,
    20-54–**20**-56
  torque transducer technologies,
    20-52–**20**-54
Torquewhirl dynamics, **9**-46–**9**-47
Torsional balances, **20**-45
Torsional-mechanical dynamics,
  21-118, **21**-121
Torsional springs, **8**-7
Torsional stiffness, of rectangular
  cross-section beams, **8**-7
Traction Control System (TCS), **1**-8
Transducers
  electromagnetic, **8**-12–**8**-13
  electrostatic, **8**-11–**8**-12
Transformer modulus, defined, **9**-15
Transient vibrations, **20**-17
Transient thermoreflectance (TTR),
  20-86–**20**-87
Transistors, **21**-20–**21**-33
Transport theorem, **8**-3
Trial-and-error methods, **1**-3
Turbine (or vane) flowmeters,
  20-67–**20**-68
Two-stroke cycles, **12**-28

U

UART (Universal Asynchronous
  Receiver/Transmitter), **4**-5
Ultrasonic actuators, **17**-19
Ultrasonic flowmeters, **17**-6,
  20-69–**20**-71
Ultrasonic motors, **21**-58–**21**-61
Ultrasonic proximity sensors, **20**-114
Ultrasonic (US) sensors, **20**-11
Understanding Automotive Electronics,
  3-7

V

Valves, **21**-71–**21**-78

Vapor and gas power cycles,
  12-23–**12**-31
Variable area flowmeters, **20**-66
Variable capacitance type electrostatic
  motor, **5**-4
VASE (VHDLAMS Synthesis
  Environment), **14**-8
Vector time derivatives, in coordinate
  systems, **9**-31–**9**-32
Vehicle Dynamics Control (VDC)
  system, **1**-8
Vertical exemplification, **6**-4
VHDL (Very Large Scale Integrated
  Circuit Hardware
  Description Language),
  14-5
Vibrating-beam accelerometers, **20**-31
Vibration and modal analysis,
  8-9–**8**-10
Villari effect, **20**-41
Viscous model, of mechanical systems,
  9-11
Vision systems, **20**-134–**20**-135
  digital images, **20**-153–**20**-155
  machine vision, **20**-158–**20**-167
  system setup and calibration,
    20-155–**20**-158
Voice-coil motors (VCMs), **21**-8–**21**-9
Voltage control (mode) amplifier,
  21-26–**21**-28
Voltmeter, **11**-14–**11**-15
Volumetric piston pumps, **21**-69
Vortex shedding flowmeters,
  20-68–**20**-69

W

Water-level float regulator, **1**-3, **1**-5
Watt's flyball governor, **1**-5
Wet etching, **15**-7
Wheatstone bridge, **11**-11–**11**-12,
  11-25, **20**-38
Wiedemann effect, **20**-45
Wikipedia definition, of mechatronics,
  1-2
Word bond graph model, **9**-4
Work and heat transfer, in internally
  reversible processes,
  12-23–**12**-31
Working Model code, **7**-6
World Wide Web (WWW), **4**-3

Y

Yasakawa Electric Company, **1**-1, **1**-7

Z

Zener diodes, **21**-17
Zero junctions, **9**-6–**9**-7
Zeroth law, of temperature, **20**-74